Key Papers in

The Development of Coding Theory

Key Papers in

The Development of Coding Theory

Edited by
Elwyn R. Berlekamp

Professor of Mathematics and of
Electrical Engineering and Computer Science
University of California, Berkeley

A volume in the IEEE PRESS Selected Reprint Series,
prepared under the sponsorship of
the IEEE Information Theory Group

IEEE
PRESS

The Institute of Electrical and Electronics Engineers, Inc. New York

IEEE PRESS SELECTED REPRINT SERIES

Clearing the Air: The Impact of the Clean Air Act on Technology
Edited by John C. Redmond, John C. Cook, and A. A. J. Hoffman
Active Inductorless Filters
Edited by Sanjit K. Mitra
A Practical Guide to Minicomputer Applications
Edited by Fred F. Coury
Power Semiconductor Applications, Volume I: General Considerations
Edited by John D. Harnden, Jr. and Forest B. Golden
Power Semiconductor Applications, Volume II: Equipment and Systems
Edited by John D. Harnden, Jr. and Forest B. Golden
Semiconductor Memories
Edited by David A. Hodges
Minicomputers: Hardware, Software, and Applications
Edited by James D. Schoeffler and Ronald H. Temple
Digital Signal Processing
Edited by Lawrence R. Rabiner and Charles M. Rader
Laser Theory
Edited by Frank S. Barnes
Integrated Optics
Edited by Dietrich Marcuse
Literature in Digital Signal Processing: Terminology and Permuted Title Index
Edited by Howard D. Helms and Lawrence R. Rabiner
Laser Devices and Applications
Edited by Ivan P. Kaminow and A. E. Siegman
Computer-Aided Filter Design
Edited by George Szentirmai
Key Papers in the Development of Information Theory
Edited by David Slepian

International Standard Book Numbers:
Clothbound: 0-87942-031-6
Paperbound: 0-87942-032-4

Library of Congress Catalog Card Number 73-87652

PRINTED IN THE UNITED STATES OF AMERICA

Preface

This volume is a collection of reprints of original papers in coding theory, with a few short notes and comments inserted by the Editor. This volume and a complementary volume on the Shannon theory, edited by D. Slepian, were both compiled at the invitation of the Administrative Committee of the IEEE Group on Information Theory under the leadership of its 1972 President, T. M. Cover. These two volumes will help to commemorate the 25th anniversary of Shannon's classic paper. The responsibility for selecting which papers to include in this volume is entirely my own. It is a reflection of my own tastes and judgments. I am sure that no other Editor would have selected precisely the same set of papers, although there would certainly be some overlap. The reader who wishes to make such comparisons should also examine I. Blake's *Selected Papers on Algebraic Coding Theory* (Dowden, Hutchinson and Ross, 1973).

If one's primary objective is to learn the subject matter, then a volume such as this one is not a particularly good way of going about it. Hindsight often enables an author to present a clearer and more complete picture of a subject than can ever be obtained by reading only the original papers. If one wants to learn how to construct algebraic codes and formulate algorithms to decode them, and if he wishes to acquire an understanding of these topics as quickly as possible, then I strongly urge him to begin by studying my book, *Algebraic Coding Theory* (McGraw-Hill, 1968). However, if one is more interested in a general impression about where coding theory has come from and how it has arrived in its present state, then this book is a good place to begin.

Contents

Part III: Decoding Algorithms for Block Codes

Part IV: Convolutional Codes and Sequential Decoding

Part V: Distance Bounds, Perfect Codes, and Weight Structure

Introduction

The coding theorists of the 1950's considered the major problems of their field to be the following:

1) How good are the best codes?
2) How can we design good codes?
3) How can we decode such codes?

It was commonly believed that the solution of problems 2) and 3) might be followed by a major technological revolution in the telecommunications industry. Although these two theoretical problems were solved, the predicted revolution did not occur, primarily because concurrent progress in various areas of applied physics reduced the price of wide-band long distance communication lines so much that the major costs of large telecommunication systems are now associated with switching rather than with transmission. Even though the engineering applications of coding theory have fallen far short of the hopes of the early enthusiasts, the theory has nevertheless found significant applications in deep space communication systems [1], military communication systems, data communication systems, information retrieval systems, and in large secondary memories for computer systems [2], [3] (where the coding is used to overcome imperfections in the manufacture of the disks rather than errors in transmission). Algebraic decoding algorithms have even been used in a sophisticated simulator of governmental and legal systems! [4]. Whether or not these applications would be sufficient to justify coding theory to a cost accountant would depend very much on his perspective. It is quite clear that the total commercial profit attributable to coding adds up to only a miniscule fraction of the annual budget of ATT or IBM, but it is also quite clear that this profit is considerably larger than the combined lifetime salaries of most of the theory's major contributors.

While the problems of coding theory have originated from real engineering situations, coding theory itself has always had a distinctly academic flavor. Within the IEEE, the Communications Society has maintained an interest in the practical problems of using coding in communication systems, and this has allowed the Group on Information Theory to concentrate more on the subject's theoretical aspects and its potential applications to novel areas. The IEEE TRANSACTIONS ON INFORMATION THEORY (IT) is now one of the three journals which publish a sizable number of papers on coding theory. The other two are *Information and Control* (I&C) and the Soviet *Problemy Peredachi Informatsii* (PPI). The translated table of contents of PPI is now republished in IT. Two other journals have devoted special issues entirely to coding theory: the October 1971 issue of the IEEE TRANSACTIONS ON COMMUNICATION TECHNOLOGY, vol. COM-19, no. 5, and the September 1972 issue of *Discrete Mathematics*, vol. 3, nos. 1, 2, 3 (North-Holland Publishing Company, Amsterdam). Each of these issues is a veritable gold mine of results, as is the special volume, *Error-Correcting Codes* (Wiley), which is the proceedings of a conference held at the University of Wisconsin in May 1968, edited by H. B. Mann.

There are at least five textbooks now available on coding theory: Peterson, Berlekamp, Lin, Peterson–Weldon, and van Lint. The relative strengths and weaknesses of these books are summarized quite accurately in their IT reviews [5]. Additional books now being written by MacWilliams–Sloane and by Massey are expected to be published in the not too distant future. There have also been a large number of surveys, of which those by Berlekamp [6], Goethals [7], and Sloane [8] are among the most recent, along with the invited series in [9]–[12].

If one evaluates theoretical subjects in terms of their interactions with other theoretical subjects, then coding theory must receive a high mark. In the early years of coding theory, nearly all of the interaction with other academic disciplines consisted of discoveries that various minor problems in coding theory could be solved by the application of elementary techniques from other fields. However, in the past decade, coding theory has matured. Not only has coding theory found applications for a number of deep results in pure mathematics (including the Davenport–Hasse theorem [13], Weyl's proof of the Reimann hypothesis for function fields [14], and Baker's recent Field-medal-winning results on Diophantine analysis [15]), but coding theory has also been able to make significant contributions to other areas [16].

Like all academic subjects, information theory has received its share of criticism. The tradition of skepticism dates all the way back to J. L. Doob's review of Shannon's 1948 paper [17]. There was even an entire conference, of sorts, devoted to skeptics [18]. Anyone who attempts to select a few papers for a volume such as this cannot avoid being impressed by the large amount of very competent work in coding theory which has been done continually over the last 25 years. The problem of editing this volume was not one of scraping up enough excellent papers to fill in between the covers, but rather one of deciding which excellent papers had to be omitted for other reasons. Among the standards which I considered in the final selection were the following: *quality, relevance, originality, completeness, historical significance, brevity, and readibility.* The manner in which I have attempted to balance these conflicting objectives is best illustrated by a few specific examples.

My notion of *relevance* has nothing at all to do with the popular political definition of that term. (After all, coding has played its most prominent role within the communications industry in the technology of deep space expolation, a program whose overall mission is now popularly considered "irrelevant" to the greater needs of society.) Perhaps my notion of the relevance of a paper is best defined as some estimate of the inverse distance between the subject matters of the paper and the subject matter of this volume. Since the reader may determine the subject matter of this volume by reading it, I will attempt to clarify its boundaries only by mentioning a few of the papers which I marginally rejected on the grounds of insufficient relevancy.

In 1969, Assmus-Mattson and Pless [19] presented some new properties of certain codes, but the major significance of their results lies not in coding theory itself, but in another branch of combinatorial mathematics (design theory). The paper by Leech and Sloane [20] presents major contributions to combinatorial geometry. While this paper has significant applications to the problem of signal design, its subject matter is still too far from the narrow-sense coding theory with which this volume is concerned. Levenshtein [21] and others have written a number of very clever papers on synchronization codes, but that subject is also considered outside the scope of this volume. My own work on the factorization of polynomials, as presented in Chapter 6 of *Algebraic Coding Theory* and in the subsequent paper [22], has some relevance to algebraic decoding, but its major applications now appear to be in the area of symbolic manipulation of algebraic expressions by computer [23]. My work on erasure-burst-correcting convolutional codes [24] arose from the problem of coding for fading channels. While this work has attracted the interest of some number theorists [25], it has had no detectable influence on the development of coding theory. The recent work of Forney [26] on the structure of convolutional codes must be classified as automata theory and system theory as well as coding theory. The work of Massey-Sain [27] has revealed a number of interesting relationships between coding theory and control theory, but I consider only one of their papers to be within the narrow scope of this volume. The Soviets have recently obtained a large number of interesting results on the asymptotic complexity of the best encoders and decoders [28]. While these results are obviously highly relevant to coding theory, I think they are better classified as a part of the modern area of computer science which is now known as "complexity theory." And finally, of course, there are a number of outstanding papers on such topics as source coding, channel capacity, error bounds, etc. All of these topics are also tangentially relevant to coding theory, but they have been assigned to the companion volume which is edited by D. Slepian. Of course, the boundary line between these two volumes was not easy to draw. Some of the papers in Slepian's volume, such as those by Huffman [29] and Berlekamp [30], are written from a constructive combinatorial point of view, but they deal with source coding and feedback systems, respectively.

The *originality* of a contribution is often difficult to evaluate. When a subject reaches a certain point in its development, certain problems become popular or "ripe" for solution, and then several authors may all find similar solutions independently. For example, the weight enumerator of Reed-Solomon codes was discovered in 1965–1966 by Assmus-Mattson-Turyn [31], Forney [32], and Kasami-Lin-Peterson [33]. The MacWilliams identities for nonlinear codes were independently formulated in the fall of 1971 in Belgium, at Murray Hill, N.J. [34], and in Southern California [35]. In the latter case, all three sets of people were responding to the same stimulus, which was the paper of Kerdock [36], preprints of which had been widely circulated in the summer of 1971. Only later did it become evident that the MacWilliams identities for nonlinear codes had been known to Zierler several years earlier!

The problem of trying to decide how to distribute credit among several independent authors is complicated by another scenario which has also occurred several times. In this story, Author A gets some exciting new results which he begins advertising in one way or another. Perhaps he gives some lectures, or circulates a preliminary version of his manuscript, or perhaps he even publishes a paper and sends out reprints. Author B then talks to someone who heard A's talk, or he glances at A's preprint or reprint, *fails to understand it*, but gets interested in the problem and begins working on it himself. Being a clever fellow, he solves the problem by a slightly different method, gets excited about it, and writes his own paper on the same problem. Unaware of A's real contribution, B makes only misleading reference to A's work, or maybe no reference at all. Eventually A reads B's paper. If he is flattered by the imitation, the whole story might yet have a happy ending, but if he is angered by the alleged plagiarism, a bitter controversy may ensue. At this point, I only want to observe that in situations such as this, the independence or lack of it is an unsolvable historical problem. After all, B might have heard only misleading reports of A's lecture, or he might not have read beyond the misleading abstract of A's preprint, or he might not have even received the reprint which A thought his secretary sent out. It is also possible that A's first lecture on the subject contained only rudimentary partial results, and that he deceived himself into thinking that he had solved the problem earlier than he actually did.

Many scholars maintain that credit for research should be given only to the first author to publish, and that an independent solution should receive credit iff its "manuscript received" date predates the publication of the first paper to appear. Although sound in principle, this simplistic rule may be difficult to apply in practice. For example, consider the Hamming codes. The first published reference to this subject was the 1948 paper of Shannon, which included the Hamming (7, 4) code as an example, complete with reference to Hamming. After reading Shannon's paper, Golay began work on error-correcting codes and obtained some impressive new results, such as the (23, 12) Golay code, as well as the rather straightforward generalization of Hamming's (7, 4) code to all of the other Hamming codes. Both were published prior to Hamming's 1950 paper. Because of this, certain scholastic purists have decided that only the (7, 4) code belongs to Hamming. They have referred to the other single-error-correcting perfect linear binary codes as "H-Golay" codes.

I see no merit to the "H-Golay" position. There is considerable unpublished evidence that Hamming was aware of all of the perfect single-error-correcting linear binary codes when he discussed the subject with Shannon in 1947 or 1948. Furthermore, it is quite clear from the published literature that the subject of error-correcting codes was founded by Hamming (not Golay).

The history of the Elias bound presents another case where the scholarly rules encounter difficulty. This result was first discovered by Elias in 1959 or 1960. By 1961, it was known to nearly everyone at MIT who was interested in coding theory. News of this bound then spread by word of mouth. For several years, the only written reference to it was an unpublished Master's thesis by Gramenopoulis, supervised by Elias. Gramenopoulis obtained a weak generalization of the bound from the binary case to the q-ary case. Several people who read this thesis, including Elias, obtained strengthened versions of the q-ary bound. But Elias never published his bounds. At first, he thought he would soon be able to improve them. Later, the publication was postponed because he was too busy with administrative duties as head of the MIT Electrical Engineering Department. Finally, he decided that the bound was so old and so well known that it was no longer new enough to merit publication. Shannon, Gallager, and I finally published it as a lemma in our 1967 I&C papers, and at that time we thought that this might be its first appearance in print. We later learned that Bassilygo had published this bound in PPI in 1964. It was already well-known in the West before that date, but in view of the lack of East–West communication in those days, and the omission of the Elias bound from those Western publications which were available in Russia (such as Peterson's book), there is every reason to believe that Bassilygo's work was independent. Since his paper was the first to appear in print with this important result, I have included it in this volume.

In two cases (Hoquenghem, Bose–Chaudhuri and Zigangirov, Jelinek), I have judged the quality of the work to be sufficiently outstanding to merit the inclusion of a pair of papers with a large overlap in contents, particularly in view of the fact that in each case the original work was not in English. However, in all other cases I have selected only one from each set of papers with similar results.

The criterion of *completeness* is used to justify the inclusion of those papers which present a final solution to a problem. Because of this criterion, I have omitted all of the papers in certain important specialized areas which are now undergoing rapid advances. One such area is concerned with the problem of determining the weight distribution of various codes, on which I include only work published before 1968. Another such area is the problem of determining the maximum number of codewords in codes of given length and distance. This problem has always played a central role in coding theory, and the best currently available reference is the paper by Sloane [37]. Sloane's paper contains so many results that I finally decided to include it, even though the subject is still undergoing very rapid development. Another area which I have omitted because of the lack of any truly comprehensive reference is the construction of codes based on finite geometries. The recent book by Peterson and Weldon [38] presents the most thorough coverage of that area that will be available for some time, although there have already been a few significant subsequent results in that area [39].

Taken together, the criteria of originality and completeness tend to overemphasize beginnings and endings, and to underplay the significance of the intermediate work. On perfect codes, I originally planned to include only Hamming, who introduced the problem in 1950, and van Lint and Tietavainen [40], who settled the last outstanding cases in 1971, but I was forced to omit all of the intermediate work except that of Lloyd [41], upon which the final proofs are heavily dependent. This pattern follows a natural tendency of most historians to focus attention on transitions at the expense of whatever steady progress occurred between them. Unfortunately, the paper by van Lint was still in press when this manuscript went to the printer.

The criterion of *historical significance* is used to justify the inclusion of two types of papers: those which inspire lots of subsequent work by others, and those which refute popular myths. The first type of historical significance is time-varying, as may be evidenced by the fact that Gallager's (1961) paper on low-density parity check codes would not have been included in this volume five years ago. While that paper was then recognized as a competent piece of work, it was viewed as only one approach among several, and probably not the most promising. The last few years, however, have seen an increasing interest in the theoretical questions related to the complexity of decoding, particularly in the USSR. Among the preliminary conclusions of these studies is the fact that, in a certain well-defined sense, low-density parity check codes are asymptotically the best possible. This result has led to a resurgence of interest in Gallager's original work, and for that reason it must be included in this volume.

The evaluation of the historical significance of a paper which disproves a myth requires some appreciation of the extent to which the myth was believed. After such a paper has appeared, the disproof may be recognized as trivial, or both the myth and its disproof may become irrelevant. Since many authors refrain from publishing their misconceptions and false conjectures, the extent of former belief in a myth cannot be determined from the literature alone. The history of science is dominated by men such as Copernicus and Columbus whose major contribution was the slaying of a popular myth. Within coding theory, the most obvious examples are Pinsker (who refuted the myth that decoding algorithms must be extremely complex at rates above R_{comp}) and Justesen (who refuted the myth that asymptotically good codes could not be described "explicitly"). Like Columbus, Pinsker was not the first to reach his conclusion. (Ziv's work was earlier.) But his work was the first to be generally understood and appreciated, and for that reason, his paper is included in this volume.

Justesen's paper refuted the myth that explicit long codes cannot be good. The literature reveals that some disbelief of this myth existed prior to Justesen. Those who had attempted to define "explicit" in any precise sense [42] had accepted the notion which is common throughout the computer science community: "explicit" means specifiable by an algorithm whose work grows algebraically rather than exponentially [43]. In this precise sense, the problem of specifying good codes explicitly was solved by Forney in 1964. There were, however, many coding theorists who clung to the more intuitive notion of "explicit" which is still commonly accepted by many mathematicians: "explicit" means specifiable by a formula of an acceptable type. (For example, those who adhere to this viewpoint might reject some iterative method for solving a differential equation as "inexplicit" no matter how quickly the method converges, while a solution which gives a formula for the answer as a summation of Bessel functions might be considered explicit.) Justesen's beautiful paper specified a class of good long codes as explicitly as anyone could desire. While Justesen's codes are no better than Forney's codes in any practical engineering sense, they have a greater aesthetic appeal. One might also hope that a comparison of the codes of Justesen and Forney might serve to raise the algorithm-consciousness of some pure mathematicians.

The criterion of *brevity* is essential if one wishes to keep the price of a volume such as this within the range of the potential buyers. If a 40-page paper is only slightly better than each of two 20-page papers, then it is not unreasonable to assume that the two 20-page papers together will be more interesting to the reader than the single 40-page paper. Among the outstanding but lengthy works in coding theory are the book by Peterson [44] and the monography of Wozencraft–Reiffen [45], Massey [46], and Forney [47]. Fortunately, the bulk of Peterson's new results were also published in his shorter IT paper, which is included in this volume in lieu of his book. I have included only excerpts of certain key sections from the monographs of Wozencraft–Reiffen and Forney, even though the unexcerpted sections of Forney's monograph contain many important results. Massey's monograph contains a wealth of significant original material (even though threshold decoding itself dates back to Reed), but each independent section of it appears to exceed this volume's threshold on length.

Papers omitted at least partly because of insufficient *readability* include my result on the distance of BCH codes [48] and Jelinek's result on the distribution of sequential decoding computation [49]. Each of these papers presented the first complete solution to a problem which had attracted a good deal of previous attention. On grounds of quality, completeness, or historical significance, such papers would merit inclusion. In each case, however, the papers are highly technical and very difficult for the nonspecialist to read. While this unreadability may or may not be partly due to deficiencies of the author's writing ability, it is mostly due to the nature of the proofs themselves. When methods are highly technical and extremely specialized, the paper cannot be expected to attract a large audience. For that reason, I have omitted the less readable papers, even when they have a high archival value. Of course, some papers of high readibility are omitted for other reasons. For example, the book by Wozencraft and Jacobs [50] presents a far clearer exposition of sequential decoding than is found in the original monograph by Wozencraft and Reiffen. However, considerations of originality and historical significance force me to include the latter and omit the former.

The list of criteria on which these papers were selected did not include *authorship*. Some readers of any volume such as this will assume that the table of contents establishes some sort of pecking order within the field. The Editor of any such volume therefore feels a definite pressure to appease this segment of his readers by including at least one paper by each of the most prominent researchers. I have resisted this temptation. Indeed, if one were to tabulate an honor roll consisting of the names of the 25 people who have contributed the most to coding theory, it would certainly not be identical to the list of authors in this table of contents. Several additional names would certainly be included, and some authors listed in the table of contents would be omitted. While I will not attempt to compile such an honor roll, I will reveal my nominations for certain work in coding theory which deserves this modest recognition as we commemorate our 25th anniversary.

Best papers: Bose–Chaudhuri and Hocquenghem (close second: Reed–Solomon).

Most influential book: Peterson, 1961.

Most influential conference: MIT, 1954 [51].

Best single published page: Golay, 1949.

Best talk to nonspecialists: Robinson (see the introduction to Section II of this volume).

Most entertaining conference paper: E. C. Posner, Madison, May 1968 [38].

Best open problem: Resolve the asymptotic discrepancy between the Elias bound and the Gilbert bound.

Our understanding of codes is now much greater than it was 25 years ago when the subject was founded by Shannon and Hamming. Most of the early problems have been solved. However, the astute reader will notice that my current nomination for the best open problem is but a sophisticated restatement of the first question stated at the beginning of this Introduction.

References

[1] E. C. Posner, "Combinatorial structures in planetary reconnaissance," in *Error-Correcting Codes,* H. B. Mann, Ed. New York: Wiley, 1969, pp. 15–46; D. R. Lum, "Test and preliminary flite results on the sequential decoding of convolutional encoded data from pioneer," in *IEEE Int. Conf. Commun.,* 1969, pp. 39.1–39.8.

[2] I. B. Oldham, R. T. Chien, and D. T. Tang, "Error detection and correction in a photo-digital storage system," *IBM J. Res. Develop.*, vol. 12, no. 6, pp. 422–430, 1968.

[3] R. T. Chien, "Memory error control: Beyond parity," *IEEE Spectrum,* vol. 10, pp. 18–23, July 1973.

[4] L. G. Foschio, J. M. Daschbach, *et al., Systems Study in Court Delay; LEADICS: Law-Engineering Analysis of Delay in Court Systems,* Law School and College Eng., Univ. Notre Dame, Notre Dame, Ind., Jan. 1972; vol. I, *Executive Summary*; vol. II, *Legal Analysis and Recommendations*; vol. III, *Engineering Sec*; vol. IV, *Appendix.* See also E. W. Henry, J. J. Uhran, Jr., and M. K. Sain, "Interactive computer simulation of court system delays," in *Proc. 3rd Pittsburgh Conf. Modeling and Simulation*, Apr. 1972, pp. 89–98; and J. J. Uhran, Jr., M. K. Sain, E. W. Henry, and D. Sharpe, "Computer model of the felony delay problem," in *IEEE Int. Conv. Dig.,* Mar. 1972, pp. 310–311.

[5] R. T. Chien, "Review of 'Algebraic coding theory,'" by E. R. Berlekamp, *IEEE Trans. Inform. Theory* (Book Reviews), vol. IT-15, pp. 509–510, July 1969.
J. L. Massey, "Review of 'An introduction to error-correcting codes,'" by S. Lin, *IEEE Trans. Inform. Theory* (Book Reviews), vol. IT-17, pp. 768–769, Sept. 1971.
E. R. Berlekamp, "Review of 'Coding theory,'" by J. H. van Lint, *IEEE Trans. Inform. Theory* (Book Reviews), vol. IT-19, p. 138, Jan. 1973.

J. L. Massey, "Review of 'Error-correcting codes,'" by W. W. Peterson and E. J. Weldon, Jr., *IEEE Trans. Inform. Theory* (Book Reviews), vol. IT-19, pp. 373-374, May 1973.
J. K. Wolf, "Review of 'Error-correcting codes,'" by H. B. Mann, Ed., *IEEE Trans. Inform. Theory* (Book Reviews), vol. IT-16, pp. 242-243, Mar. 1970.

[6] E. R. Berlekamp, "Survey of coding theory," *J. Royal Stat. Soc. (A)*, vol. 135, pp. 44-73, 1972.

[7] J. M. Goethals, "Some combinatorial aspects of coding theory," in *A Survey of Combinatorial Theory,* J. N. Srivastava, Ed. Amsterdam: North-Holland/American Elsevier, 1973, pp. 189-208.

[8] N. J. A. Sloane, "A survey of constructive coding theory, and a table of binary codes of highest known rate," *Discrete Math.*, vol. 3, pp. 265-294, 1972.

[9] J. R. Pierce, "The early days of information theory," *IEEE Trans. Inform. Theory,* vol. IT-19, pp. 3-8, Jan. 1973.

[10] D. Slepian, "Information theory in the fifties," *IEEE Trans. Inform. Theory,* vol. IT-19, pp. 145-148, Mar. 1973.

[11] A. J. Viterbi, "Information theory in the sixties," *IEEE Trans. Inform. Theory,* vol. IT-19, pp. 257-262, May 1973.

[12] J. K. Wolf, "A survey of coding theory: 1967-1972," *IEEE Trans. Inform. Theory,* vol. IT-19, pp. 381-389, July 1973.

[13] R. J. McEliece and H. Rumsey, Jr., "Euler products, cyclotomy, and coding," *J. Number Theory,* vol. 4, pp. 302-311, 1972.

[14] The application is via the paper of L. Carlitz and S. Uchiyama, "Bounds for exponential sums," *Duke Math. J.*, vol. 24, pp. 37-41, 1957. The relevance of this paper to coding theory was first noticed by D. R. Anderson, "A new class of cyclic codes," *SIAM J. Appl. Math.*, vol. 16, pp. 181-197, 1968, then extended by E. R. Berlekamp, "Weight enumeration theorems," in *Proc. 6th Annu. Allerton Conf. Circuit and Syst. Theory,* 1968, pp. 161-170. These results were first used to determine the weight enumerators of low-rate BCH codes by E. R. Berlekamp, "The weight enumerators for certain subiodes of the second order binary Reed-Muller code," *Inform. Contr.*, vol. 17, pp. 485-500, 1970.

[15] J. H. van Lint first used this result to establish the nonexistence of perfect codes under certain conditions, but his paper is now obsolete because he and Tietavainen later obtained stronger results by other methods. See [40].

[16] Details follow three paragraphs henceforth.

[17] *Math. Rev.*, vol. 10, p. 133, 1949. See also Doob's editorial in *IEEE Trans. Inform. Theory,* vol. IT-5, p. 3, Jan. 1959.

[18] A workshop on "New Directions," which was held near St. Petersburg, Fla., in April 1971, was actually devoted to confessions about the allegedly sinful ways of past directions in coding theory. No new directions were suggested, except "out." When the critics subsequently realized that the science and engineering recession had affected other areas even more than coding theory, they became less vocal.

[19] E. F. Assmus, Jr. and H. F. Mattson, Jr., "New 5-designs," *J. Combinatorial Theory,* vol. 6, pp. 122-151, 1969. V. Pless "On a new family of symmetry codes and related new five-designs," *Bull. Amer. Math. Soc.*, vol. 75, no. 6, pp. 1339-1342, 1969.

[20] J. Leech and N. J. A. Sloane, "Sphere packings and error-correcting codes," *Can. J. Math.*, vol. 23, pp. 718-745, 1971.

[21] Detailed references are given in surveys cited below [18.1], [18.2].

[22] E. R. Berlekamp, "Factoring polynomials over large finite fields," *Math. Comput.*, vol. 24, pp. 713-735, 1970.

[23] An excellent survey of this area is G. D. Collins, "Computer algebra of polynomials and rational functions," *Amer. Math. Monthly,* 1973.

[24] E. R. Berlekamp, "A class of convolution codes," *Inform. Contr.*, vol. 6, pp. 1-13, 1963; *Math. Rev.*, vol. 28, p. 2935.

[25] L. Carlitz, D. P. Roselle, and R. A. Scoville, "Some remarks on ballot-type sequences of positive integers," *J. Combinatorial Theory,* vol. 11, pp. 258-271, 1971.

[26] G. P. Forney, Jr., "Convolutional codes I: Algebraic structure," *IEEE Trans. Inform. Theory,* vol. IT-16, pp. 720-738, Nov. 1970; "Convolutional codes II: Max-likelihood decoding," and "Convolutional codes III: Sequential decoding," Center for Syst. Res., Stanford Univ., Stanford, Calif., Tech. Rep. 7004-1, June 1972.

[27] J. L. Massey and M. K. Sain, "Codes, automata, and continuous systems: Explicit interconnections," *IEEE Trans. Automat. Contr.*, vol. AC-12, pp. 644-650, Dec. 1967.

[28] Summaries and detailed references to Soviet work may be found in the following two places: R. L. Dobrushin, "Survey of Soviet research in information theory," *IEEE Trans. Inform. Theory,* vol. IT-18, pp. 703-724, Nov. 1972, and W. H. Kautz and K. N. Levitt, "A survey of progress in coding theory in the Soviet Union," *IEEE Trans. Inform. Theory,* vol. IT-15, (1, part II), pp. 197-245, Jan. 1969.

[29] D. A. Huffman, "A method for the construction of minimum redundancy codes," *Proc. IRE* (Corresp.), vol. 40, p. 1098, Sept. 1952.

[30] E. R. Berlekamp, "Block coding for the binary symmetric channel with noiseless, delayless feedback, in *Proc. Symp. Error-Correcting Codes* (Univ. Wisconsin, May 6-8, 1968). New York: Wiley, 1968, pp. 61-88.

[31] E. F. Assmus, Jr., H. F. Mattson, and R. Turyn, "Cyclic codes," *Air Force Cambridge Res. Lab. Sum. Rep. 4,* 1969.

[32] G. D. Forney, Jr., *Concatenated Codes.* Cambridge, Mass.: MIT Press, Monograph 37, 1966.

[33] T. Kasami, S. Lin, and W. W. Peterson, "Some results on cyclic codes which are invariant under the affine group," AFCRL Rep., 1966, unpublished.

[34] The paper was published jointly: F. J. MacWilliams, N. J. A. Sloane, and J. M. Goethals, "The MacWilliams identities for nonlinear codes," *Bell Syst. Tech. J.,* vol. 51, pp. 803-819, 1972.

[35] This work was primarily due to R. J. McEliece and L. Welch, with assistance from others at the Jet Propulsion Lab. They have several papers in preparation, but none that have yet appeared.

[36] A. M. Kerdock, "A class of low-rate nonlinear binary codes," *Inform. Contr.*, vol. 20, pp. 182-187, 1973.

[37] N. J. A. Sloane, "A survey of constructive coding theory, and a table of binary codes of highest known rate," *Discrete Math.*, vol. 3, pp. 265-294, 1972.

[38] *Error-Correcting Codes,* 2nd ed. Cambridge, Mass.: MIT Press, 1972.

[39] J. Lin, "On the number of information symbols in polynomial codes," *IEEE Trans. Inform. Theory,* vol. IT-18, pp. 785-794, Nov. 1972.

[40] J. H. van Lint, "Survey of results on the nonexistence of perfect codes," *Rocky Mountain J. Math.,* 1973; and A. Tietavainen, "There are no unknown perfect codes over fields," *SIAM J. Appl. Math.,* 1973, Another proof is given by V. A. Zinoviev and V. K. Leontiev, "Theorem on the nonexistence of perfect codes over Galois fields," *Probl. Peredach. Inform.*, vol. 8, no. 4, 1972. Most of the basic techniques and novel methods used to solve this (and other!) problems can be found in J. H. van Lint, *Coding Theory.* Berlin: Springer, 1971; reviewed in *IEEE Trans. Inform. Theory,* vol. IT-19, p. 138, Jan. 1973.

[41] S. P. Lloyd, "Binary block coding," *Bell Syst. Tech. J.*, vol. 36, pp. 517-535, 1957.

[42] G. D. Forney, *Concatenated Codes.* Cambridge, Mass.: MIT Press, Monograph 37, 1966.
V. V. Ziablov, "A bound on the complexity of the construction of binary linear concatenated codes," *Probl. Peredach. Inform.*, vol. 7, no. 1, pp. 5-13, 1971.

[43] The first statement in print of the widely accepted notion that algorithms which run in polynomial-bounded time are "good," but that algorithms which take longer are "bad" occurs in the introduction to J. Edmonds, "Paths, trees, and flowers," *Can. J. Math.*, pp. 449-467, 1965. See Sec. 2, p. 50. There has recently been a great deal of work on this topic by computer scientists. For example, see R. Karp, "Reducibility among combinatorial problems," in *Proc. Symp. Complexity of Combinatorial Comput.*, R. E. Miller, J. W. Thatcher, and J. Bohlinger, Ed. New York: Plenum, 1972.

[44] W. W. Peterson, *Error-Correcting Codes.* Cambridge, Mass.: MIT Press, 1961.

[45] J. M. Wozencraft and B. Reiffen, *Sequential Decoding.* Cambridge, Mass.: MIT Press, Monograph 10, 1961; *Math. Rev.*, vol. 19, p. 824.

[46] J. L. Massey, *Threshold Decoding.* Cambridge, Mass.: MIT Press, Monograph 20, 1963.

[47] G. D. Forney, Jr., *Concatenated Codes.* Cambridge, Mass.: MIT Press, Monograph 37, 1964.

[48] E. R. Berlekamp, "Long primitive binary BCH codes have distance $d \sim 2n\ 1n\ R^{-1}/\log n \cdots$," *IEEE Trans. Inform. Theory,* vol. IT-18, pp. 415–426, May 1972.

[49] F. Jelinek, "An upper bound on moments of sequential decoding effort," *IEEE Trans. Inform. Theory,* vol. IT-15, pp. 140–149, Jan. 1969.

[50] J. M. Wozencraft and I. M. Jacobs, *Principle of Communication Engineering.* New York: Wiley, 1965. See pp. 425–440 for excellent exposition of sequential decoding.

[51] The entire proceedings of this conference were published in a special issue of the *IRE Trans. Inform. Theory,* vol. IT-4, Mar. 1958.

Part I: The Early Work

Hamming was the first coding theorist whose work attracted widespread interest. He was a colleague of Shannon at Bell Telephone Laboratories in the late forties, and some of his early work appeared as an example in Shannon's classic 1948 paper. Apparently delayed because of patent considerations, Hamming's own paper appeared in 1950. Even though both were concerned with the fundamental problem of communicating over noisy channels, there was a clear difference between the combinatorial, constructive viewpoint of Hamming and the statistical, existential viewpoint of Shannon. The dichotomy between coding theory and Shannon theory has increased in subsequent years. This volume contains only papers on the former topic; some of the historically significant papers dealing primarily with the Shannon theory appear in the companion volume edited by D. Slepian [1].

The major coding theory papers of the early 1950's introduced a number of important concepts which laid the basis for the various subspecialities which have since grown up: code constructions, block decoding, weight structure, bounds on distance, and sequential decoding. While most of the post-1956 papers can be readily classified within one of these topics, the subspecialities did not yet exist so clearly in the early years. Hamming [2] was concerned both with code constructions and with bounds; Muller [3] and Reed [4] not only constructed an important class of codes and invented the notion of threshold decoding, they also gave some preliminary indications about how a large body of knowledge about finite mathematical structures (rings, fields, algebras) might be brought to bear on the coding problem. Elias [5], [6] not only introduced a remarkable asymptotic construction for block codes, he also invented convolutional codes and proved that randomly chosen codes of various types are asymptotically good. Slepian [7] exposed the mathematical foundations of the subject of linear codes.

The other sections of this book amplify the importance of the work of these early pioneers. The asymptotic bound of Gilbert [8] has not been improved in over 20 years, and it is now known that there are no other particular codes as beautiful as those discovered by Golay [9].

References

[1] D. Slepian, *Key Papers in the Development of Information Theory*. New York: IEEE Press, 1974.

[2] R. W. Hamming, "Error detecting and error correcting codes," *Bell Syst. Tech. J.*, vol. 29, pp. 147-160, 1950; *Math Rev.*, vol. 12, p. 35.

[3] D. E. Muller, "Application of Boolean algebra to switching circuit design and to error detection," *IRE Trans. Electron. Comput.*, vol. EC-3, pp. 6-12, Sept. 1954; *Math. Rev.*, vol. 16, p. 99.

[4] I. S. Reed, "A class of multiple-error-correcting codes and the decoding scheme," *IRE Trans. Inform. Theory*, vol. IT-4, pp. 38-49, Sept. 1954; *Math. Rev.*, vol. 19, p. 721.

[5] P. Elias, "Error-free coding," *IRE Trans. Inform. Theory*, vol. IT-4, pp. 29-37, 1954; *Math. Rev.*, vol. 19, p. 721.

[6] —, "Coding for noisy channels," in *IRE Conv. Rec.*, pt. 4, 1955, pp. 37-46.

[7] D. Slepian, "A class of binary signaling alphabets," *Bell Syst. Tech. J.*, vol. 35, pp. 203-234, 1956; *Math. Rev.*, vol. 17, p. 1100. —, "Some further theory of group codes," *Bell Syst. Tech. J.*, pp. 1219-1252, 1960.

[8] E. N. Gilbert, "A comparison of signalling alphabets," *Bell Syst. Tech. J.*, vol. 31, pp. 504-522, 1952.

[9] M. J. E. Golay, "Notes on digital coding," *Proc. IRE* (Corresp.), vol. 37, p. 657, June 1949.

Error Detecting and Error Correcting Codes

By R. W. HAMMING

1. Introduction

THE author was led to the study given in this paper from a consideration of large scale computing machines in which a large number of operations must be performed without a single error in the end result. This problem of "doing things right" on a large scale is not essentially new; in a telephone central office, for example, a very large number of operations are performed while the errors leading to wrong numbers are kept well under control, though they have not been completely eliminated. This has been achieved, in part, through the use of self-checking circuits. The occasional failure that escapes routine checking is still detected by the customer and will, if it persists, result in customer complaint, while if it is transient it will produce only occasional wrong numbers. At the same time the rest of the central office functions satisfactorily. In a digital computer, on the other hand, a single failure usually means the complete failure, in the sense that if it is detected no more computing can be done until the failure is located and corrected, while if it escapes detection then it invalidates all subsequent operations of the machine. Put in other words, in a telephone central office there are a number of parallel paths which are more or less independent of each other; in a digital machine there is usually a single long path which passes through the same piece of equipment many, many times before the answer is obtained.

In transmitting information from one place to another digital machines use codes which are simply sets of symbols to which meanings or values are attached. Examples of codes which were designed to detect isolated errors are numerous; among them are the highly developed 2 out of 5 codes used extensively in common control switching systems and in the Bell Relay Computers,[1] the 3 out of 7 code used for radio telegraphy,[2] and the word count sent at the end of telegrams.

In some situations self checking is not enough. For example, in the Model 5 Relay Computers built by Bell Telephone Laboratories for the Aberdeen Proving Grounds,[1] observations in the early period indicated about two or three relay failures per day in the 8900 relays of the two computers, representing about one failure per two to three million relay operations. The self-checking feature meant that these failures did not introduce undetected errors. Since the machines were run on an unattended basis over nights and week-ends, however, the errors meant that frequently the computations came to a halt although often the machines took up new problems. The present trend is toward electronic speeds in digital computers where the basic elements are somewhat more reliable per operation than relays. However, the incidence of isolated failures, even when detected, may seriously interfere with the normal use of such machines. Thus it appears desirable to examine the next step beyond error detection, namely error correction.

We shall assume that the transmitting equipment handles information in the binary form of a sequence of 0's and 1's. This assumption is made both for mathematical convenience and because the binary system is the natural form for representing the open and closed relays, flip-flop circuits, dots and dashes, and perforated tapes that are used in many forms of communication. Thus each code symbol will be represented by a sequence of 0's and 1's.

The codes used in this paper are called *systematic* codes. Systematic codes may be defined[3] as codes in which each code symbol has exactly n binary digits, where m digits are associated with the information while the other $k = n - m$ digits are used for error detection and correction. This produces a *redundancy* R defined as the ratio of the number of binary digits used to the minimum number necessary to convey the same information, that is,

$$R = n/m.$$

This serves to measure the efficiency of the code as far as the transmission of information is concerned, and is the only aspect of the problem discussed in any detail here. The redundancy may be said to lower the effective channel capacity for sending information.

The need for error correction having assumed importance only recently, very little is known about the economics of the matter. It is clear that in using such codes there will be extra equipment for encoding and correcting errors as well as the lowered effective channel capacity referred to above. Because of these considerations applications of these codes may be expected to occur first only under extreme conditions. Some typical situations seem to be:

a. unattended operation over long periods of time with the minimum of standby equipment.
b. extremely large and tightly interrelated systems where a single failure incapacitates the entire installation.
c. signaling in the presence of noise where it is either impossible or uneconomical to reduce the effect of the noise on the signal.

These situations are occurring more and more often. The first two are particularly true of large scale digital computing machines, while the third occurs, among other places, in "jamming" situations.

The principles for designing error detecting and correcting codes in the cases most likely to be applied first are given in this paper. Circuits for implementing these principles may be designed by the application of well-known techniques, but the problem is not discussed here. Part I of the paper shows how to construct special minimum redundancy codes in the following cases:

a. single error detecting codes
b. single error correcting codes
c. single error correcting plus double error detecting codes.

Part II discusses the general theory of such codes and proves that under the assumptions made the codes of Part I are the "best" possible.

PART I

SPECIAL CODES

2. Single Error Detecting Codes

We may construct a single error detecting code having n binary digits in the following manner: In the first $n - 1$ positions we put $n - 1$ digits of information. In the n-th position we place either 0 or 1, so that the entire n positions have an even number of 1's. This is clearly a single error detecting code since any single error in transmission would leave an odd number of 1's in a code symbol.

The redundancy of these codes is, since $m = n - 1$,

$$R = \frac{n}{n-1} = 1 + \frac{1}{n-1}.$$

It might appear that to gain a low redundancy we should let n become very large. However, by increasing n, the probability of at least one error in a symbol increases; and the risk of a double error, which would pass undetected, also increases. For example, if $p \ll 1$ is the probability of any error, then for n so large as $1/p$, the probability of a correct symbol is approximately $1/e = 0.3679\ldots$, while a double error has probability $1/2e = 0.1839\ldots$.

The type of check used above to determine whether or not the symbol has any single error will be used throughout the paper and will be called a *parity check*. The above was an *even* parity check; had we used an odd number of 1's to determine the setting of the check position it would have been an odd parity check. Furthermore, a parity check need not always involve all the positions of the symbol but may be a check over selected positions only.

3. Single Error Correcting Codes

To construct a single error correcting code we first assign m of the n available positions as information positions. We shall regard the m as fixed, but the specific positions are left to a later determination. We next assign the k remaining positions as check positions. The values in these k positions are to be determined in the encoding process by even parity checks over selected information positions.

[1] Franz Alt, "A Bell Telephone Laboratories' Computing Machine"—I, II. Mathematical Tables and Other Aids to Computation, Vol. 3, pp. 1–13 and 60–84, Jan. and Apr. 1948.
[2] S. Sparks, and R. G. Kreer, "Tape Relay System for Radio Telegraph Operation," *R.C.A. Review*, Vol. 8, pp. 393–426, (especially p. 417), 1947.
[3] In Section 7 this is shown to be equivalent to a much weaker appearing definition.

Let us imagine for the moment that we have received a code symbol, with or without an error. Let us apply the k parity checks, in order, and for each time the parity check assigns the value observed in its check position we write a 0, while for each time the assigned and observed values disagree we write a 1. When written from right to left in a line this sequence of k 0's and 1's (to be distinguished from the values assigned by the parity checks) may be regarded as a binary number and will be called the *checking number*. We shall require that this checking number give the position of any single error, with the zero value meaning no error in the symbol. Thus the check number must describe $m + k + 1$ different things, so that

$$2^k \geq m + k + 1$$

is a condition on k. Writing $n = m + k$ we find

$$2^m \leq \frac{2^n}{n+1}.$$

Using this inequality we may calculate Table I, which gives the maximum m for a given n, or, what is the same thing, the minimum n for a given m.

We now determine the positions over which each of the various parity checks is to be applied. The checking number is obtained digit by digit, from right to left, by applying the parity checks in order and writing down the corresponding 0 or 1 as the case may be. Since the checking number is

TABLE I

n	m	Corresponding k
1	0	1
2	0	2
3	1	2
4	1	3
5	2	3
6	3	3
7	4	3
8	4	4
9	5	4
10	6	4
11	7	4
12	8	4
13	9	4
14	10	4
15	11	4
16	11	5
	Etc.	

to give the position of any error in a code symbol, any position which has a 1 on the right of its binary representation must cause the first check to fail. Examining the binary form of the various integers we find

$$1 = 1$$
$$3 = 11$$
$$5 = 101$$
$$7 = 111$$
$$9 = 1001$$
Etc.

have a 1 on the extreme right. Thus the first parity check must use position

$$1, 3, 5, 7, 9, \cdots.$$

In an exactly similar fashion we find that the second parity check must use those positions which have 1's for the second digit from the right of their binary representation,

$$2 = 10$$
$$3 = 11$$
$$6 = 110$$
$$7 = 111$$
$$10 = 1010$$
$$11 = 1011$$
Etc.,

the third parity check

$$4 = 100$$
$$5 = 101$$
$$6 = 110$$
$$7 = 111$$
$$12 = 1100$$
$$13 = 1101$$
$$14 = 1110$$
$$15 = 1111$$
$$20 = 10100$$
Etc.

It remains to decide for each parity check which positions are to contain information and which the check. The choice of the positions 1, 2, 4, 8, $\cdots$ for check positions, as given in the following table, has the advantage of making the setting of the check positions independent of each other. All other positions are information positions. Thus we obtain Table II.

TABLE II

Check Number	Check Positions	Positions Checked
1	1	1, 3, 5, 7, 9, 11, 13, 15, 17,...
2	2	2, 3, 6, 7, 10, 11, 14, 15, 18,...
3	4	4, 5, 6, 7, 12, 13, 14, 15, 20,...
4	8	8, 9, 10, 11, 12, 13, 14, 15, 24,...
.	.	.
.	.	.
.	.	.

As an illustration of the above theory we apply it to the case of a seven-position code. From Table I we find for $n = 7$, $m = 4$ and $k = 3$. From Table II we find that the first parity check involves positions 1, 3, 5, 7 and is used to determine the value in the first position; the second parity check, positions 2, 3, 6, 7, and determines the value in the second position; and the third parity check, positions 4, 5, 6, 7, and determines the value in position four. This leaves positions 3, 5, 6, 7 as information positions. The results of writing down all possible binary numbers using positions 3, 5, 6, 7, and then calculating the values in the check positions 1, 2, 4, are shown in Table III.

Thus a seven-position single error correcting code admits of 16 code symbols. There are, of course, $2^7 - 16 = 112$ meaningless symbols. In some applications it may be desirable to drop the first symbol from the code to avoid the all zero combination as either a code symbol or a code symbol plus a single error, since this might be confused with no message. This would still leave 15 useful code symbols.

TABLE III

Position 1	2	3	4	5	6	7	Decimal Value of Symbol
0	0	0	0	0	0	0	0
1	1	0	1	0	0	1	1
0	1	0	1	0	1	0	2
1	0	0	0	0	1	1	3
1	0	0	1	1	0	0	4
0	1	0	0	1	0	1	5
1	1	0	0	1	1	0	6
0	0	0	1	1	1	1	7
1	1	1	0	0	0	0	8
0	0	1	1	0	0	1	9
1	0	1	1	0	1	0	10
0	1	1	0	0	1	1	11
0	1	1	1	1	0	0	12
1	0	1	0	1	0	1	13
0	0	1	0	1	1	0	14
1	1	1	1	1	1	1	15

As an illustration of how this code "works" let us take the symbol 0 1 1 1 1 0 0 corresponding to the decimal value 12 and change the 1 in the fifth position to a 0. We now examine the new symbol

0 1 1 1 0 0 0

by the methods of this section to see how the error is located. From Table II the first parity check is over positions 1, 3, 5, 7 and predicts a 1 for the first position while we find a 0 there; hence we write a

1 .

The second parity check is over positions 2, 3, 6, 7, and predicts the second position correctly; hence we write a 0 to the left of the 1, obtaining

0 1 .

The third parity check is over positions 4, 5, 6, 7 and predicts wrongly; hence we write a 1 to the left of the 0 1, obtaining

1 0 1 .

This sequence of 0's and 1's regarded as a binary number is the number 5; hence the error is in the fifth position. The correct symbol is therefore obtained by changing the 0 in the fifth position to a 1.

4. Single Error Correcting Plus Double Error Detecting Codes

To construct a single error correcting plus double error detecting code we begin with a single error correcting code. To this code we add one more position for checking all the previous positions, using an even parity check. To see the operation of this code we have to examine a number of cases:

1. No errors. All parity checks, including the last, are satisfied.
2. Single error. The last parity check fails in all such situations whether the error be in the information, the original check positions, or the last check position. The original checking number gives the position of the error, where now the zero value means the last check position.
3. Two errors. In all such situations the last parity check is satisfied, and the checking number indicates some kind of error.

As an illustration let us construct an eight-position code from the previous seven-position code. To do this we add an eighth position which is chosen so that there are an even number of 1's in the eight positions. Thus we add an eighth column to Table III which has:

Table IV

0
0
1
1
1
1
0
0
1
1
0
0
0
0
1
1

PART II

GENERAL THEORY

5. A Geometrical Model

When examining various problems connected with error detecting and correcting codes it is often convenient to introduce a geometric model. The model used here consists in identifying the various sequences of 0's and 1's which are the symbols of a code with vertices of a unit n-dimensional cube. The code points, labelled $x, y, z, \cdots$, form a subset of the set of all vertices of the cube.

Into this space of 2^n points we introduce a *distance*, or, as it is usually called, a *metric*, $D(x, y)$. The definition of the metric is based on the observation that a single error in a code point changes one coordinate, two errors, two coordinates, and in general d errors produce a difference in d coordinates. Thus we define the distance $D(x, y)$ between two points x and y as the number of coordinates for which x and y are different. This is the same as the least number of edges which must be traversed in going from x to y. This distance function satisfies the usual three conditions for a metric, namely,

$$D(x, y) = 0 \quad \text{if and only if } x = y$$

$$D(x, y) = D(y, x) > 0 \quad \text{if } x \neq y$$

$$D(z, y) + D(y, z) \geq D(x, z) \text{ (triangle inequality).}$$

As an example we note that each of the following code points in the three-dimensional cube is two units away from the others,

0 0 1
0 1 0
1 0 0
1 1 1 .

To continue the geometric language, a sphere of radius r about a point x is defined as all points which are at a distance r from the point x. Thus, in the above example, the first three code points are on a sphere of radius 2 about the point (1, 1, 1). In fact, in this example any one code point may be chosen as the center and the other three will lie on the surface of a sphere of radius 2.

If all the code points are at a distance of at least 2 from each other, then it follows that any single error will carry a code point over to a point that is *not* a code point, and hence is a meaningless symbol. This in turn means that any single error is detectable. If the minimum distance between code points is at least three units then any single error will leave the point nearer to the correct code point than to any other code point, and this means that any single error will be correctable. This type of information is summarized in the following table:

Table V

Minimum Distance	Meaning
1	uniqueness
2	single error detection
3	single error correction
4	single error correction plus double error detection
5	double error correction
	Etc.

Conversely, it is evident that, if we are to effect the detection and correction listed, then all the distances between code points must equal or exceed the minimum distance listed. Thus the problem of finding suitable codes is the same as that of finding subsets of points in the space which maintain at least the minimum distance condition. The special codes in sections 2, 3, and 4 were merely descriptions of how to choose a particular subset of points for minimum distances 2, 3, and 4 respectively.

It should perhaps be noted that, at a given minimum distance, some of the correctability may be exchanged for more detectability. For example, a subset with minimum distance 5 may be used for:

a. double error correction, (with, of course, double error detection).
b. single error correction plus triple error detection.
c. quadruple error detection.

Returning for the moment to the particular codes constructed in Part I we note that any interchanges of positions in a code do not change the code in any essential way. Neither does interchanging the 0's and 1's in any position, a process usually called complementing. This idea is made more precise in the following definition:

Definition. Two codes are said to be *equivalent* to each other if, by a finite number of the following operations, one can be transformed into the other:

1. The interchange of any two positions in the code symbols.
2. The complementing of the values in any position in the code symbols.

This is a formal equivalence relation ($\sim$) since $A \sim A$; $A \sim B$ implies $B \sim A$; and $A \sim B$, $B \sim C$ implies $A \sim C$. Thus we can reduce the study of a class of codes to the study of typical members of each equivalence class.

In terms of the geometric model, equivalence transformations amount to rotations and reflections of the unit cube.

6. Single Error Detecting Codes

The problem studied in this section is that of packing the maximum number of points in a unit n-dimensional cube such that no two points are closer than 2 units from each other. We shall show that, as in section 2, 2^{n-1} points can be so packed, and, further, that any such optimal packing is equivalent to that used in section 2.

To prove these statements we first observe that the vertices of the n-dimensional cube are composed of those of two $(n - 1)$-dimensional cubes. Let A be the maximum number of points packed in the original cube. Then one of the two $(n - 1)$-dimensional cubes has at least $A/2$ points. This cube being again decomposed into two lower dimensional cubes, we find that one of them has at least $A/2^2$ points. Continuing in this way we come to a two-dimensional cube having $A/2^{n-2}$ points. We now observe that a square can have at most two points separated by at least two units; hence the original n-dimensional cube had at most 2^{n-1} points not less than two units apart.

To prove the equivalence of any two optimal packings we note that, if the packing is optimal, then each of the two sub-cubes has half the points. Calling this the first coordinate we see that half the points have a 0 and half have a 1. The next subdivision will again divide these into two equal groups having 0's and 1's respectively. After $(n - 1)$ such stages we have, upon reordering the assigned values if there be any, exactly the first $n - 1$ positions of the code devised in section 2. To each sequence of the first $n - 1$ coordinates there exist $n - 1$ other sequences which differ from it by one coordinate. Once we fix the n-th coordinate of some one point, say the origin which has all 0's, then to maintain the known minimum distance of two units between code points the n-th coordinate is uniquely determined for all other code points. Thus the last coordinate is determined within a complementation so that any optimal code is equivalent to that given in section 2.

It is interesting to note that in these two proofs we have used only the assumption that the code symbols are all of length n.

7. Single Error Correcting Codes

It has probably been noted by the reader that, in the particular codes of Part I, a distinction was made between information and check positions, while, in the geometric model, there is no real distinction between the various coordinates. To bring the two treatments more in line with each other we redefine a *systematic* code as a code whose symbol lengths are all equal and

1. The positions checked are independent of the information contained in the symbol.
2. The checks are independent of each other.
3. We use parity checks.

This is equivalent to the earlier definition. To show this we form a matrix whose i-th row has 1's in the positions of the i-th parity check and 0's elsewhere. By assumption 1 the matrix is fixed and does not change from code symbol to code symbol. From 2 the rank of the matrix is k. This in turn means that the system can be solved for k of the positions expressed in terms of the other $n - k$ positions. Assumption 3 indicates that in this solving we use the arithmetic in which $1 + 1 = 0$.

There exist non-systematic codes, but so far none have been found which for a given n and minimum distance d have more code symbols than a systematic code. Section 9 gives an example of a non-systematic code.

Turning to the main problem of this section we find from Table V that a single error correcting code has code points at least three units from each other. Thus each point may be surrounded by a sphere of radius 1 with no two spheres having a point in common. Each sphere has a center point and n points on its surface, a total of $n + 1$ points. Thus the space of 2^n points can have at most:

$$\frac{2^n}{n+1}$$

spheres. This is exactly the bound we found before in section 3.

While we have shown that the special single error correcting code constructed in section 3 is of minimum redundancy, we cannot show that all optimal codes are equivalent, since the following trivial example shows that this is not so. For $n = 4$ we find from Table I that $m = 1$ and $k = 3$. Thus there are at most two code symbols in a four-position code. The following two optimal codes are clearly not equivalent:

$$\begin{matrix}0\,0\,0\,0\\1\,1\,1\,1\end{matrix} \quad \text{and} \quad \begin{matrix}0\,0\,0\,0\\0\,1\,1\,1\end{matrix}\,.$$

8. Single Error Correcting Plus Double Error Detecting Codes

In this section we shall prove that the codes constructed in section 4 are of minimum redundancy. We have already shown in section 4 how, for a minimum redundancy code of $n - 1$ dimensions with a minimum distance of 3, we can construct an n dimensional code having the same number of code symbols but with a minimum distance of 4. If this were not of minimum redundancy there would exist a code having more code symbols but with the same n and the same minimum distance 4 between them. Taking this code we remove the last coordinate. This reduces the dimension from n to $n - 1$ and the minimum distance between code symbols by, at most, one unit, while leaving the number of code symbols the same. This contradicts the assumption that the code we began our construction with was of minimum reduncancy. Thus the codes of section 4 are of minimum redundancy.

This is a special case of the following general theorem: To any minimum redundancy code of N points in $n - 1$ dimensions and having a minimum distance of $2k - 1$ there corresponds a minimum redundancy code of N points in n dimensions having a minimum distance of $2k$, and conversely. To construct the n dimensional code from the $n - 1$ dimensional code we simply add a single n-th coordinate which is fixed by an even parity check over the n positions. This also increases the minimum distance by 1 for the following reason: Any two points which, in the $n - 1$ dimensional code, were at a distance $2k - 1$ from each other had an odd number of differences between their coordinates. Thus the parity check was set oppositely for the two points, increasing the distance between them to $2k$. The additional coordinate could not decrease any distances, so that all points in the code are now at a minimum distance of $2k$. To go in the reverse direction we simply drop one coordinate from the n dimensional code. This reduces the minimum distance of $2k$ to $2k - 1$ while leaving N the same. It is clear that if one code is of minimum redundancy then the other is, too.

9. Miscellaneous Observations

For the next case, minimum distance of five units, one can surround each code point by a sphere of radius 2. Each sphere will contain

$$1 + C(n, 1) + C(n, 2)$$

points, where $C(n, k)$ is the binomial coefficient, so that an upper bound on the number of code points in a systematic code is

$$\frac{2^n}{1 + C(n, 1) + C(n, 2)} = \frac{2^{n+1}}{n^2 + n + 2} \geq 2^m.$$

This bound is too high. For example, in the case of $n = 7$, we find that $m = 2$ so that there should be a code with four code points. The maximum possible, as can be easily found by trial and error, is two.

In a similar fashion a bound on the number of code points may be found whenever the minimum distance between code points is an odd number. A bound on the even cases can then be found by use of the general theorem of the preceding section. These bounds are, in general, too high, as the above example shows.

If we write the bound on the number of code points in a unit cube of dimension n and with minimum distance d between them as $B(n, d)$, then the information of this type in the present paper may be summarized as follows:

$$B(n, 1) = 2^n$$

$$B(n, 2) = 2^{n-1}$$

$$B(n, 3) = 2^m \leq \frac{2^n}{n+1}$$

$$B(n, 4) = 2^m \leq \frac{2^{n-1}}{n}$$

$$B(n - 1, 2k - 1) = B(n, 2k)$$

$$B(n, 2k - 1) = 2^m \leq \frac{2^n}{1 + C(n, 1) + \cdots + C(n, k - 1)}\,.$$

While these bounds have been attained for certain cases, no general methods have yet been found for contructing optimal codes when the minimum distance between code points exceeds four units, nor is it known whether the bound is or is not attainable by systematic codes.

We have dealt mainly with systematic codes. The existence of non-systematic codes is proved by the following example of a single error correcting code with $n = 6$.

0 0 0 0 0 0
0 1 0 1 0 1
1 0 0 1 1 0
1 1 1 0 0 0
0 0 1 0 1 1
1 1 1 1 1 1 .

The all 0 symbol indicates that any parity check must be an even one. The all 1 symbol indicates that each parity check must involve an even number of positions. A direct comparison indicates that since no two columns are the same the even parity checks must involve four or six positions. An examination of the second symbol, which has three 1's in it, indicates that no six-position parity check can exist. Trying now the four-position parity checks we find that

1 2   5 6
2 3 4 5

are two independent parity checks and that no third one is independent of these two. Two parity checks can at most locate four positions, and, since there are six positions in the code, these two parity checks are not enough to locate any single error. The code is, however, single error correcting since it satisfies the minimum distance condition of three units.

The only previous work in the field of error correction that has appeared in print, so far as the author is aware, is that of M. J. E. Golay.[4]

[4] M. J. E. Golay, Correspondence, Notes on Digital Coding, *Proceedings of the I.R.E.*, Vol. 37, p. 657, June 1949.

Notes on Digital Coding*

The consideration of message coding as a means for approaching the theoretical capacity of a communication channel, while reducing the probability of errors, has suggested the interesting number theoretical problem of devising lossless binary (or other) coding schemes serving to insure the reception of a correct, but reduced, message when an upper limit to the number of transmission errors is postulated.

An example of lossless binary coding is treated by Shannon[1] who considers the case of blocks of seven symbols, one or none of which can be in error. The solution of this case can be extended to blocks of 2^n-1-binary symbols, and, more generally, when coding schemes based on the prime number p are employed, to blocks of $p^n-1/p-1$ symbols which are transmitted, and received with complete equivocation of one or no symbol, each block comprising n redundant symbols designed to remove the equivocation. When encoding the message, the n redundant symbols x_m are determined in terms of the message symbols Y_k from the congruent relations

$$E_m \equiv X_m + \sum_{k=1}^{k=(p^n-1)/\,p-1)-n} a_{mk}\, Y_k \equiv 0 \pmod{p}.$$

In the decoding process, the E's are recalculated with the received symbols, and their ensemble forms a number on the base p which determines univocally the mistransmitted symbol and its correction.

In passing from n to $n+1$, the matrix with n rows and $p^n-1/p-1$ columns formed with the coefficients of the X's and Y's in the expression above is repeated p times horizontally, while an $(n+1)$ st row added, consisting of $p^n-1/p-1$ zeroes, followed by as many one's etc. up to $p-1$; an added column of n zeroes with a one for the lowest term completes the new matrix for $n+1$.

If we except the trivial case of blocks of $2S+1$ binary symbols, of which any group comprising up to S symbols can be received in error which equal probability, it does not appear that a search for lossless coding schemes, in which the number of errors is limited but larger than one, can be systematized so as to yield a family of solutions. A necessary but not sufficient condition for the existence of such a lossless coding scheme in the binary system is the existence of three or more first numbers of a line of Pascal's triangle which add up to an exact power of 2. A limited search has revealed two such cases; namely, that of the first three numbers of the 90th line, which add up to 2^{12} and that of the first four numbers of the 23rd line, which add up to 2^{11}. The first case does not correspond to a lossless coding scheme, for, were such a scheme to exist, we could designate by r the number of E_m ensembles corresponding to one error and having an odd number of 1's and by $90-r$ the remaining (even) ensembles. The odd ensembles corresponding to two transmission errors could be formed by re-entering term by term all the conbinations of one even and one odd ensemble corresponding each to one error, and would number $r(90-r)$. We should have $r+r(90-r)=2^{11}$, which is impossible for integral values of r.

On the other side, the second case can be coded so as to yield 12 sure symbols, and the a_{mk} matrix of this case is given in Table I. A second matrix is also given, which is that of the only other lossless coding scheme encountered (in addition to the general class mentioned above) in which blocks of eleven ternary symbols are transmitted with no more than 2 errors, and out of which six sure symbols can be obtained.

It must be mentioned that the use of the ternary coding scheme just mentioned will always result in a power loss, whereas the coding scheme for 23 binary symbols and a maximum of three transmission errors yields a power saving of $1\frac{1}{2}$ db for vanishing probabilities of errors. The saving realized with the coding scheme for blocks of 2^n-1 binary symbols approaches 3 db for increasing n's and decreasing probabilities of error, but a loss is always encountered when $n=3$.

MARCEL J. E. GOLAY
Signal Corps Engineering Laboratories
Fort Monmouth, N. J

TABLE I

	Y_1	Y_2	Y_3	Y_4	Y_5	Y_6	Y_7	Y_8	Y_9	Y_{10}	Y_{11}	Y_{12}
X_1	1	0	0	1	1	1	0	0	0	1	1	1
X_2	1	0	1	0	1	1	0	1	1	0	0	1
X_3	1	0	1	1	0	1	1	0	1	0	1	0
X_4	1	0	1	1	1	0	1	1	0	1	0	0
X_5	1	1	0	0	1	1	1	0	1	1	0	0
X_6	1	1	0	1	0	1	1	1	0	0	0	1
X_7	1	1	0	1	1	0	0	1	1	0	1	0
X_8	1	1	1	0	0	1	0	1	0	1	1	0
X_9	1	1	1	0	1	0	1	0	0	0	1	1
X_{10}	1	1	1	1	0	0	0	0	1	1	0	1
X_{11}	0	1	1	1	1	1	1	1	1	1	1	1

	Y_1	Y_2	Y_3	Y_4	Y_5	Y_6
X_1	1	1	1	2	2	0
X_2	1	1	2	1	0	2
X_3	1	2	1	0	1	2
X_4	1	2	0	1	2	1
X_5	1	0	2	2	1	1

* Received by the Institute, February 23, 1949.
[1] C. E. Shannon, "A mathematical theory of communication," *Bell Sys. Tech. Jour.*, vol. 27, p. 418; July, 1948.

Reprinted from *Proc. IRE,* vol. 37, p. 657, June 1949.

A Comparison of Signalling Alphabets

By E. N. GILBERT

(Manuscript received March 24, 1952)

Two channels are considered; a discrete channel which can transmit sequences of binary digits, and a continuous channel which can transmit band limited signals. The performance of a large number of simple signalling alphabets is computed and it is concluded that one cannot signal at rates near the channel capacity without using very complicated alphabets.

INTRODUCTION

C. E. Shannon's encoding theorems[1] associate with the channel of a communications system a capacity C. These theorems show that the output of a message source can be encoded for transmission over the channel in such a way that the rate at which errors are made at the receiving end of the system is arbitrarily small provided only that the message source produces information at a rate less than C bits per second. C is the largest rate with this property.

Although these theorems cover a wide class of channels there are two channels which can serve as models for most of the channels one meets in practice. These are:

1. *The binary channel*

This channel can transmit only sequences of binary digits 0 and 1 (which might represent hole and no hole in a punched tape; open-line and closed line; pulse and no pulse; etc.) at some definite rate, say one digit per second. There is a probability p (because of noise, or occasional equipment failure) that a transmitted 0 is received as 1 or that a transmitted 1 is received as 0. The noise is supposed to affect different digits independently. The cpacity of this channel is

$$C = 1 + p \log p + (1 - p) \log (1 - p) \tag{1}$$

bits per digit. The log appearing in Equation (1) is log to the base 2; this convention will be used throughout the rest of this paper.

2. *The low-pass filter*

The second channel is an ideal low-pass filter which attenuates completely all frequencies above a cutoff frequency W cycles per second and which passes frequencies below W without attenuation. The channel is supposed capable of handling only signals with average power P or less. Before the signal emerges from the channel, the channel adds to it a noise signal with average power N. The noise is supposed to be white Gaussian noise limited to the frequency band $|\nu| < W$. The capacity of this channel is

$$C = W \log \left(1 + \frac{P}{N}\right) \tag{2}$$

bits per second.

Shannon's theorems prove that encoding schemes exist for signalling at rates near C with arbitrarily small rates of errors without actually giving a constructive method for performing the encoding. It is of some interest to compare encoding systems which can easily be devised with these ideal systems. In Part I of this paper some schemes for signalling over the binary channel will be compared with ideal systems. In Part II the same will be done for the low-pass filter channel.

[1] C. E. Shannon, "A Mathematical Theory of Communication," *Bell System Tech. J.*, **27**, p. 379–423 and pp. 623–656, 1948, theorems 9, 11, and 16 in particular.

PART I

THE BINARY CHANNEL

1. *Error-Correcting Alphabets*

Imagine the message source to produce messages which are sequences of letters drawn from an alphabet containing K letters. We suppose that the letters are equally likely and that the letters which the source produces at different times are independent of one another. (If the source given is a finite state source which does not fit this simple description, it can be converted into one which approximately does by a preliminary encoding of the type described in Shannon's Theorem 9.) To transmit the message over the binary channel we construct a new alphabet of K letters in which the letters are different sequences of binary digits of some fixed length, say D digits. Then the new alphabet is used as an encoding of the old one suitable for transmission over the channel. For example, if the source produced sequences of letters from an alphabet of 3 letters, a typical encoding with $D = 5$ might convert the message into a binary sequence composed of repetitions of the three letters.

00000
11100
and 00111

If $K = 2^D$, the alphabet consists of all binary sequences of length D and hence if any of the digits of a letter is altered by noise the letter will be misinterpreted at the receiving end of the channel. If K is somewhat smaller than 2^D it is possible to choose the letters so that certain kinds of errors introduced by the noise do not cause a misinterpretation at the receiver. For example, in the three letter alphabet given above, if only one of the five digits is incorrect there will be just one letter (the correct one) which agrees with the received sequence in all but one place. More generally if the letters of the alphabet are selected so that each letter differs from every other in at least $2k + 1$ out of the D places, then when k or fewer errors are made the correct interpretation of the received sequence will be the (unique) letter of the alphabet which differs from the received sequence in no more than k places. An alphabet with this property will be called a *k error correcting alphabet*[2].

Error correcting alphabets have the advantage over the random alphabets which Shannon used to prove his encoding theorems that they are uniformly reliable whereas Shannon's alphabets are reliable only in an average sense. That is, Shannon proved that the probability that a letter *chosen at random* shall be received incorrectly can be made arbitrarily small. However, a certain small fraction of the letters of Shannon's alphabets are allowed a much higher probability of error than the average. This kind of alphabet would be undesirable in applications such as the signalling of telephone numbers; one would not want to give a few subscribers telephone numbers which are received incorrectly more often than most of the others. It is only conjectured that the rate C can be approached using error correcting alphabets. The alphabets which are to be considered here are all error correcting alphabets.

A geometric picture of an alphabet is obtained by regarding the D digits of a sequence as coordinates of a point in Euclidean D dimensional space. The possible received sequences are represented by vertices of the unit cube. A k error correcting alphabet is represented by a set of vertices, such that each pair of vertices is separated by a distance at least $\sqrt{2k + 1}$

Let $K_0(D, k)$ be the largest number of letters which a D dimensional

[2] R. W. Hamming, "Error Detecting and Error Correcting Codes," *Bell System Tech. J.*, **29**, pp. 147–160, 1950.

k error correcting alphabet can contain. Except when $k = 1$, there is no general method for constructing an alphabet with $K_0(D, k)$ letters, nor is $K_0(D, k)$ known as a function of D and k. Crude upper and lower bounds for $K_0(D, k)$ are given by the following theorem.

Theorem 1. The largest number of letters $K_0(D, k)$ satisfies

$$\frac{2^D}{N(D, 2k)} \leq K_0(D, k) \leq \frac{2^D}{N(D, k)} \tag{3}$$

where

$$N(D, k) = \sum_{r=0}^{k} C_{D,r}$$

is the number of sequences of D digits which differ from a given sequence in 0, 1, $\cdots$, or k places.

Proof

The upper bound is due to R. W. Hamming and is proved by noting that for each letter S of a k error correcting alphabet there are $N(D, k)$ possible received sequences which will be interpreted as meaning S. Hence $N(D, k)\, K_0(D, k) \leq 2^D$, the total number of sequences.

The lower bound is proved by a random construction method. Pick any sequence S_1 for the first letter. There remain $2^D - N(D, 2k)$ sequences which differ from S_1 in $2k + 1$ or more places. Pick any one of these S_2 for the second letter. There remain at least $2^D - 2N(D, 2k)$ sequences which differ from both S_1 and S_2 in $2k + 1$ or more places. As the process is continued, there remain at least $2^D - rN(D, 2k)$ sequences, which differ in $2k + 1$ or more places from $S_1, \cdots, S_r$, from which S_{r+1} is chosen. If there are no choices available after choosing S_K, then $2^D - KN(D, 2k) \leq 0$ so the alphabet $(S_1, \cdots, S_K)$ has at least as many letters as the lower bound (3).

For all the simple cases (D and k not very large) investigated so far the upper bound is a better estimate of $K_0(D, k)$ than the lower bound. The upper and lower bounds differ greatly, as may be seen from a quick inspection of Table I. For example, in the case of a ten dimensional two error correcting alphabet, the bounds are 2.7 and 18.3.

2. Efficiency Graph

The first step in constructing an efficiency graph for comparing alphabets is to decide on what constitutes reliable transmission. The criterion used here is that on the average no more than one letter in 10^4 shall be misinterpreted.

TABLE I

TABLE OF $2^D/N(D, k)$

k =	1	2	3	4	5	6	7
D = 3	2						
4	3.2						
5	5.3	2					
6	9.1	2.9					
7	16	4.4	2.9				
8	28.4	6.9	2.8				
9	51.2	11.1	3.9	2			
10	93.1	18.3	5.8	2.7			
11	170.7	30.6	8.8	3.6	2		
12	315.8	51.8	13.7	5.2	2.6		
13	585.2	89.0	21.6	7.5	3.4	2	
14	1092.3	154.4	34.9	11.1	4.7	2.5	
15	2048	270.8	56.8	16.8	6.6	3.3	2

Missing entries are numbers between 1 and 2.

This sort of criterion might be appropriate for a channel transmitting English text. For other messages it is not always appropriate. For example, if the messages are telephone numbers, one would naturally require that the probability of mistaking a telephone number be small, say less than 10^{-4}. If the telephone numbers are L decimal digits long, and if the alphabet has K different letters in it (so that it takes about $L \log 10/\log K$ letters to make up a telephone number) the probability of making a mistake in a single letter should be required to be less than about

$$\frac{10^{-4} \log K}{L \log 10}$$

which gives alphabets with large K an advantage over alphabets with small K.

Since the probability that exactly r binary digits out of D shall be received incorrectly is $C_{D,r}p^r(1 - p)^{D-r}$, we achieve the required reliability with a D-dimensional k-error correcting alphabet provided p satisfies

$$\sum_{r=k+1}^{D} C_{D,r}p^r(1 - p)^{D-r} \leq 10^{-4}. \tag{4}$$

The value of p which makes the inequality hold with the equals sign determines the noisiest channel over which the alphabet can be used safely.

Let K be the number of different letters in the alphabet. Then the rate in bits per digit at which information is being recieved is

$$R = \frac{\log K}{D}. \tag{5}$$

In Equation (5) we have neglected a term which takes account of the information lost due to channel noise. This is legitimate because all but 10^{-4} of the letters are received correctly.

The worst tolerable probability p of (4) and the rate R of Equation (5) determine the noise combating ability of an alphabet. To compare different alphabets one may represent them as points on an efficiency graph of R versus p. Fig. 1 is an efficiency graph on which the values (p, R) for a number of simple error correcting alphabets have been plotted. Each point on the graph is labelled with the two numbers k, D in that order. The alphabets represented were not found by any systematic process and are not all proved to be best possible (i.e., to have the largest K) for the stated values of k and D. Fortunately, R depends on K only logarithmically so that it is not likely the points representing the best possible alphabets lie far away from the plotted points.

The solid line represents the curve

$$R = C = 1 + p \log p + (1 - p) \log (1 - p).$$

According to Shannon's theorems, all alphabets are represented by points lying below this line.

The efficiency graph only partially orders the alphabets according to

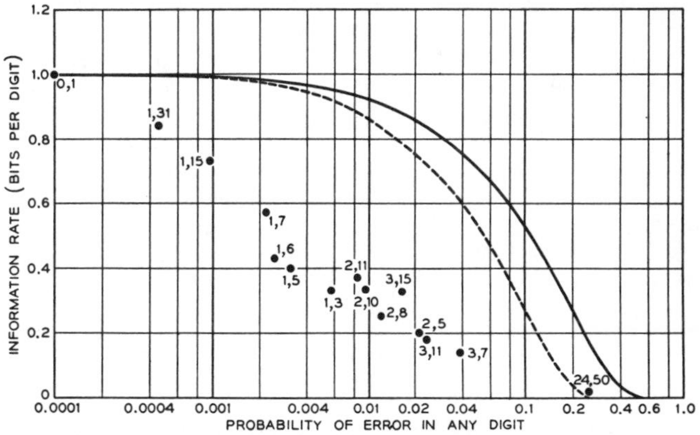

Fig. 1—Probability of error in a letter is 10^{-4}.

their invulnerability to noise. For example, it is clear that the alphabet 3, 15 is better than 2, 8. However, without further information about the channel, such as knowledge of p, there is no reasonable way of choosing between 3, 15 and 3, 7.

3. Large Alphabets

We have been unable to prove that there are error correcting alphabets which signal at rates arbitrarily close to C while maintaining an arbitrarily small probability of error for any letter. A result in this direction is the following theorem.

Theorem 2. Let any positive ϵ and δ be given. Given a channel with $p < \frac{1}{4}$ there exists an error correcting alphabet which can signal over the channel at a rate exceeding $R_0 - \epsilon$ where

$$R_0 = 1 + 2p \log 2p + (1 - 2p) \log (1 - 2p)$$

bits per digit and for which the probability of error in any letter is less than δ.

Proof

The probability of error in any letter is the sum on the left of (4). This is a sum of terms from a binomial distribution which, as is well known, tends to a Gaussian distribution with mean Dp and variance $Dp(1 - p)$ for large D. Hence there is a constant $A(\delta)$ such that all k error correcting alphabets with sufficiently large D have a letter error probability less than δ provided

$$k \geq Dp + A(\delta)\,(Dp(1 - p))^{1/2} \tag{6}$$

Let $k(D)$ be the smallest integer which satisfies (6) and consider an alphabet which corrects $k(D)$ errors and contains $K_0(D, k(D))$ letters. By Equation (5) and the lower bound of Theorem 1, this alphabet signals at a rate $R(D)$ satisfying

$$1 - \frac{1}{D} \log N(D, 2k(D)) \leq R(D).$$

Since $p < \frac{1}{4}$, $2k(D) < D/2$ for large D and hence

$$N(D, 2k(D)) < (2k(D) + 1)C_{D,2k(D)}.$$

Then an application of Stirling's approximation for factorials shows that as $D \to \infty$

$$1 - \frac{1}{D} \log N(D, 2k(D)) \to R_0 .$$

Hence by taking D large enough one obtains an alphabet with rate exceeding $R_0 - \epsilon$ and letter error probability less than δ.

The rate R_0 appears on the efficiency graph as a dotted line.

It has not been shown that no error-correcting alphabet has a rate exceeding R_0. In fact, one alphabet which exceeds R_0 in rate is easy to construct. If the noise probability p is greater than $\frac{1}{4}$, then $R_0 = 0$. The alphabet with just two letters

$$0\,0\,0\,0 \ldots 0$$

and

$$1\,1\,1\,1 \ldots 1$$

will certainly transmit information at a (small) positive rate, and with a 10^{-4} probability of errors if D is large enough, as long as $p < \frac{1}{2}$.

Using a more refined lower bound for $K_0(D, k)$ it might be shown that there are error-correcting alphabets which signal with rates near C. If one repeats the calculation that led to R_0 using the upper bound (3) (which seems to be a better estimate of the true $K_0(D, k)$) instead of the lower bound (3), one is led to the rate C instead of R_0.

The condition (4) is more conservative than necessary. The structure of the alphabet may be such that a particular sequence of more than k errors may occur without causing any error in the final letter. This is illustrated by the following simple example due to Shannon: the alphabet with just two letters

$$0\,0\,0\,0\,0\,0$$
$$1\,1\,1\,0\,0\,0$$

corrects any single error but also corrects certain more serious errors such as receiving 0 0 1 1 1 1 for 0 0 0 0 0 0. An alphabet designed for practical use would make efficient enough use of the available sequences so that any sequence of much more than k errors causes an error in the final letter; the random alphabets constructed above probably do not. If this kind of error were properly accounted for, the rate R_0 could be improved, perhaps to C.

4. Other Discrete Channels

If instead of transmitting just 0's and 1's the channel can carry more digits

$$0, 1, 2, \cdots, n$$

a similar theory can be worked out. The simplest kind of noise in this channel changes a digit into any one of the n other possible numbers with probability p/n. Then the capacity of the channel is

$$C = \log(n + 1) + p \log \frac{p}{n} + (1 - p) \log (1 - p).$$

Error-correcting alphabets for this channel can also be constructed and the criterion (4) for good transmission remains unchanged. The proof of theorem 1 can be repeated with little change using

$$N(D, k) = \sum_{r=0}^{k} C_{D,r} n^r$$

as the number of sequences which can be reached after k or fewer errors [the terms 2^D in (1) and (3) are replaced by $(n + 1)^D$]. Once more, using the lower bound, one finds an expression for R_0 which is the same as the one for C but with p replaced by $2p$.

PART II

THE LOW PASS FILTER

1. Encoding and Detection

If $f(t)$ is a signal emerging from a low pass filter (so that its spectrum is confined to the frequency band $|\nu| < W$ cycles per second) then $f(t)$ has a special analytic form given by the sampling theorem[3]

$$f(t) = \sum_{m=-\infty}^{\infty} f\left(\frac{m}{2W}\right) \frac{\sin \pi\,(2Wt - m)}{\pi(2Wt - m)} \tag{7}$$

Thus the signal is completely determined by the sequence of sample values $f(m/2W)$. The average power of the signal $f(t)$ is measured by

$$P = \lim_{T\to\infty} \frac{1}{2T} \int_{-T}^{T} f^2(t)\, dt$$

which can be expressed in terms of the sample values as follows

$$P = \lim_{M\to\infty} \frac{1}{2M} \sum_{m=-M}^{M} f^2\left(\frac{m}{2W}\right). \tag{8}$$

As in Part I, consider a message source producing a sequence of letters from an alphabet of K equally likely letters. To transmit this information over the low pass filter we must encode the sequence into a function $f(t)$ of the form (7), or in other words into a sequence of sample values $f(m/2W)$. To do this, we construct a new alphabet containing K letters which are different sequences of real numbers of some fixed length, say D places. When we let the letters of the new alphabet correspond to letters of the old one the message is translated into a sequence of real numbers which we use for the sequence $f(m/2W)$.

If the K letters of the sequence alphabet are

$$\begin{array}{l} S_1\colon\ a_{11}, \cdots, a_{1D} \\ S_2\colon\ a_{21}, \cdots, a_{2D} \\ \quad\vdots \\ S_K\colon\ a_{K1}, \cdots, a_{KD}, \end{array}$$

the expression (8) for the average power of the function $f(t)$ becomes

$$P = \frac{1}{DK}\,(d_1^2 + d_2^2 + \cdots + d_K^2) \tag{9}$$

where

$$d_i^2 = \sum_{j=1}^{D} a_{ij}^2 .$$

If the D numbers in the sequence S_i are regarded as coordinates of a point in Euclidean D dimensional space, d_i^2 represents the square of the distance from the point representing S_i to the origin.

When $f(t)$ is transmitted, the received signal will be $f(t) + n(t)$ where $n(t)$ is some (unknown) white Gaussian noise signal. The noise signals $n(t)$ are characterized by the fact that their sample values $n(m/2W)$ are independently distributed according to Gaussian laws. That is,

$$\text{Prob}\left(n\left(\frac{m}{2W}\right) \leq X\right) = \frac{1}{\sqrt{2\pi}\sigma} \int_{-\infty}^{X} e^{-y^2/2\sigma^2}\, dy. \tag{10}$$

The variance σ^2 of the distribution of noise samples is, by an application of (8), the power of this ensemble of noise signals.

[3] C. E. Shannon, "*Communication in the Presence of Noise*," *Proc. I. R. E.*, **37**, pp. 10–21, Jan. 1949.

At the receiving end of the channel, there is a detector which observes each block of D sample values $f(m/2W) + n(m/2W)$ and tries to decide which one of the K letters $S_1, \cdots, S_K$ was sent. In terms of the geometric picture, the detector divides all of D dimensional space into K non-overlapping regions $U_1, \cdots, U_K$ with the property that, if the D received sample values are represented by a point in U_i, the detector decides that S_i was sent. By Equation (10), the probability that the detector picks the wrong letter when S_i is sent is

$$p_i = \frac{1}{(2\pi)^{D/2}\sigma^D} \int \int_{\bar{U}_i} \cdots \int e^{-r_i^2/2\sigma^2}\, dy_1 \cdots dy_D \tag{11}$$

where $\bar{U}_i$ is the set of all points not in U_i and r_i is the distance from $(y_1, \cdots, y_D)$ to the point representing S_i.

For any given alphabet the best possible detector (in the sense that it minimizes the average probability of making an error in guesssing a letter) is called a *maximum likelihood detector*. The region U_i for a maximum likelihood detector consists of all points $(y_1, \cdots, y_D)$ which are closer to the point S_i than to any other letter point $S_j(r_i < r_j$ for all $j \neq i)$. To prove that this choice of U_i is best possible consider any other detector such that U_i contains a set V of points in which $r_i > r_j$. A direct calculation shows that the detector obtained by removing V from U_i and making V part of U_j has a smaller probability of error per letter. The set of points equidistant from two given points is a hyperplane. The region U_i of a maximum likelihood detector is a convex region bounded by segments of the hyperplanes

$$r_i = r_1, \qquad r_i = r_2, \cdots.$$

To compare signalling alphabets under the most favorable possible circumstances, we always compute letter error probabilities assuming that the detector is a maximum likelihood detector.

2. *Computation of error probabilities*

Exact evaluation of the letter error probability integral (11) is impossible except in a few special cases. Fortunately we are only interested in (11) when σ is small enough in comparison to the size of U_i to make the integral small. Then fairly accurate approximate formulas can be derived.

Theorem 3. Let R_{ij} be the distance between letter points S_i and S_j. Then

$$1 - \prod_{j \neq i} (1 - Q_{ij}) \leq p_i \leq \sum_{j \neq i} Q_{ij} \tag{12}$$

where

$$Q_{ij} = \frac{1}{\sqrt{2\pi}} \int_{R_{ij}/2\sigma}^{\infty} e^{-x^2/2}\, dx.$$

The proof of Theorem 3 follows from the fact that Q_{ij} is the probability that, when S_i is transmitted, the received sequence will be closer to S_j than to S_i.

In the cases to be computed Q_{ij} is a rapidly decreasing function of R_{ij} and the only terms worth keeping in (12) are the ones for which R_{ij} is the smallest of the numbers $R_{i1}, \cdots, R_{iK}$. Moreover since the Q_{ij} are all small enough so that the upper and lower bounds differ only by a few per cent, the upper bound is a good approximation to p_i. Then a simple approximate formula for the average letter error probability $p = (p_1 + \cdots + p_K)/K$ is

$$p = \frac{N}{\sqrt{2\pi}} \int_{r_0/\sigma}^{\infty} e^{-x^2/2}\, dx \tag{13}$$

where $2r_0$ is the smallest of the $K(K-1)/2$ distances R_{ij} and N is the average over all letters in the alphabet of the number of letter points which are a distance $2r_0$ away.

3. *Efficiency graph*

The efficiency graph to be described was constructed originally to compare alphabets for signalling telephone numbers of length equal to ten decimal digits. It was desired that on the average only one telephone number in 10^4 should be received incorrectly. As described in Part I section 2, if the telephone numbers are encoded into sequences of letters from an alphabet of K letters, we must require that the average probability of error in any letter be

$$p = 10^{-5} \log_{10} K \tag{14}$$

or smaller.

Given an alphabet, one can compute with the help of (13) and (14) and a table of the error integral the largest value of the noise power σ^2 which can be tolerated. The average power of the transmitted signal is P given by Equation (9). Hence we can compute the smallest signal to noise ratio

$$Y = P/\sigma^2 \tag{15}$$

which will be satisfactory.

A letter containing $\log K$ bits of information is transmitted during an interval of $D/2W$ seconds. Hence the rate at which information is received is

$$R = \frac{2W \log K}{D} \tag{16}$$

bits per second. Again Equation (16) ignores a term representing information lost due to channel noise which is negligible because the error probability is low.

The efficiency graph, Fig. 2, is a chart on which the signal to noise ratio Y in db [computed from Equation (15)] is plotted against the signalling rate per unit bandwidth $R/W = (2 \log K)/D$ for different alphabets. An alphabet is considered poor if its point on the efficiency graph lies far above the ideal curve $R/W = C/W = \log(1 + Y)$.

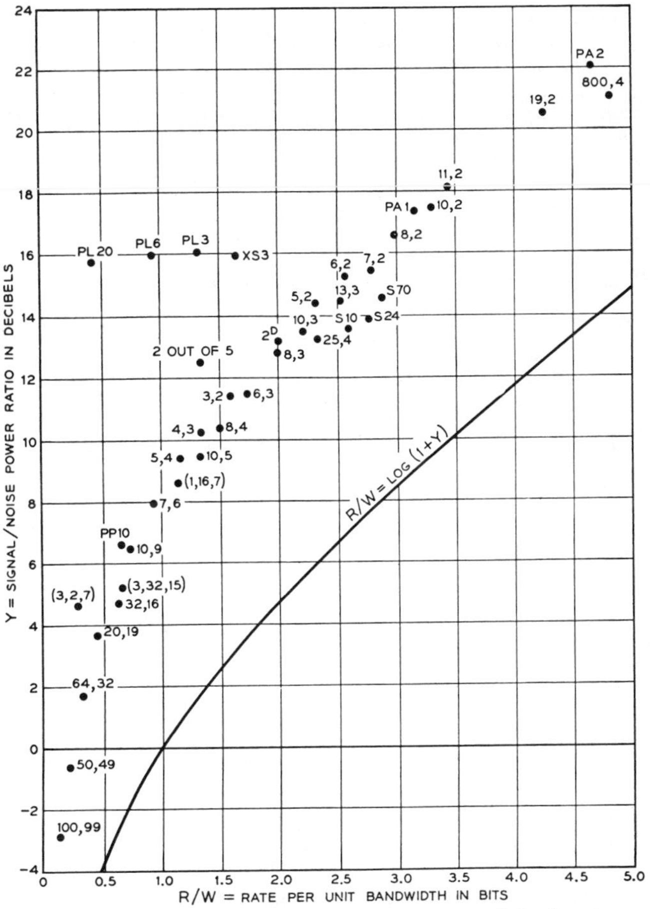

Fig. 2—Probability is 10^{-4} that an error is made in a 10 digit decimal number.

4. *The alphabets*

The alphabets which appear on the efficiency graph are the following:

excess three (*XS3*): the ten sequences of 4 binary digits which represent 3, 4, $\cdots$, and 12 in binary notation;

two out of five: the ten sequences of five binary digits which contain exactly two ones;

pulse position (*PP10*): the ten sequences of ten binary digits which contain exactly one one;

2^D *binary*: all of 2^D sequences of D binary digits.

pulse amplitude (*PAn*): the $2n + 1$ sequences of length 1 consisting of $-n, -n + 1, \cdots, n$. This alphabet gives rise to a sort of quantized amplitude modulation.

pulse length (*PLn*): the $n + 1$ sequences of n binary digits of the form $11 \cdots 10 \cdots 0$, i.e., a run of ones followed by a run of zeros.

Minimizing alphabets (K, D): The above alphabets are taken from actual practice. They are convenient because, aside from PAn, they require a signal generator with only two amplitude levels. If we ignore ease of generating the signals as a factor, a great many geometric arrangements of points suggest themselves as possible good alphabets. The principle by which one arrives at good alphabets may be described as follows. When a D and K have been determined which give the desired information rate R [by Equation (16)] try to arrange the K letter points in D dimensional space in such a way that the distances between pairs of points are all greater than some fixed distance and that the average of the K squared distances to the origin is minimized. By Equations (9) and (13) it is seen that, apart from the small influence of the factor N, this process must minimize the signal to noise ratio Y required.

Ordinarily it is difficult to prove that a configuration is a minimizing one. Even to recognize a configuration which leads to a relative minimum (*i.e.* a minimum over all nearby configurations) is not always easy. The eight vertices of a cube, for example, do not give a relative minimum. Consequently, most of the alphabets to be described are only conjectured to be "best possible." Each of them satisfies one necessary requirement of minimizing alphabets that the centroid of the point configuration (assuming a unit mass at each letter point) lies at the origin. That this condition is necessary follows from the easily derived identity

$$r_2^2 = r_1^2 - R_0^2$$

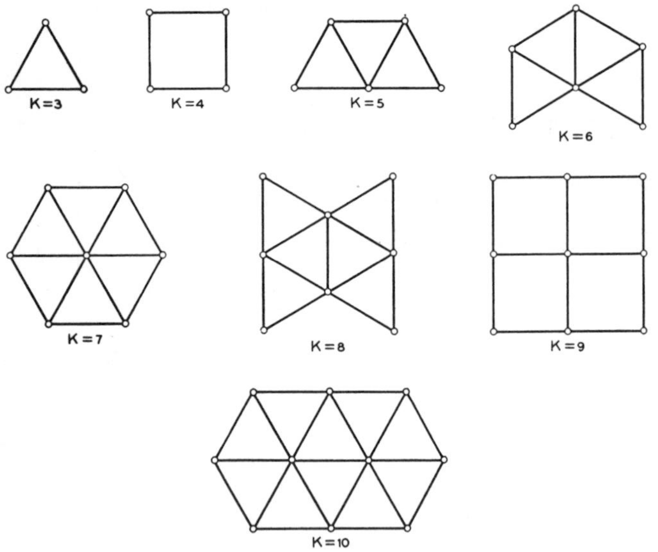

Fig. 3—Two dimensional alphabets.

where r_1 is the rms distance from the origin to the points of a configuration A, R_0 is the distance from the origin to the centroid of A, and r_2 is the rms distance from the points of A to the centroid of A.

In plotting points on the efficiency graph the notation K, D is used for the best K-letter D-dimensional alphabet which has been found. The arrangement of points for various K, 2 and K, 3 alphabets is given in Figs. 3 and 4. In these figures two points are joined by a straight line if the distance between them is 1 (which is the value we have adopted for the minimum allowed separation $2r_0$). Although not shown, the origin is always at the centroid of the figure. To aid interpretation of these diagrams we have included Fig. 5 which demonstrates how all the signals of a typical alphabet can be generated. The functions of time shown in Fig. 5 are not the code signals themselves but impulse functions which are to be passed through a low pass filter with cutoff at W c.p.s. to form the code signals.

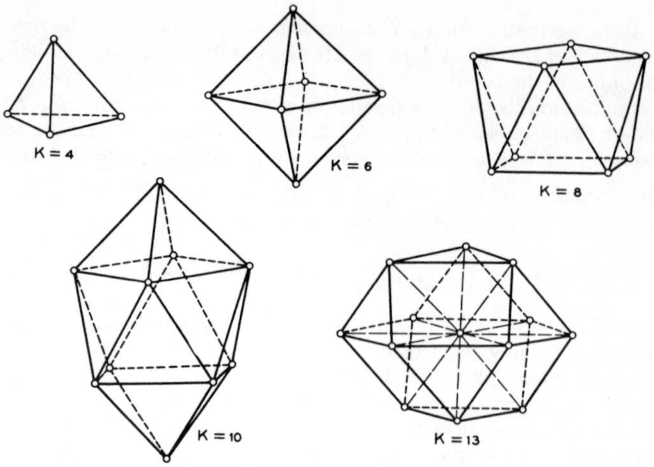

Fig. 4—Three dimensional alphabets.

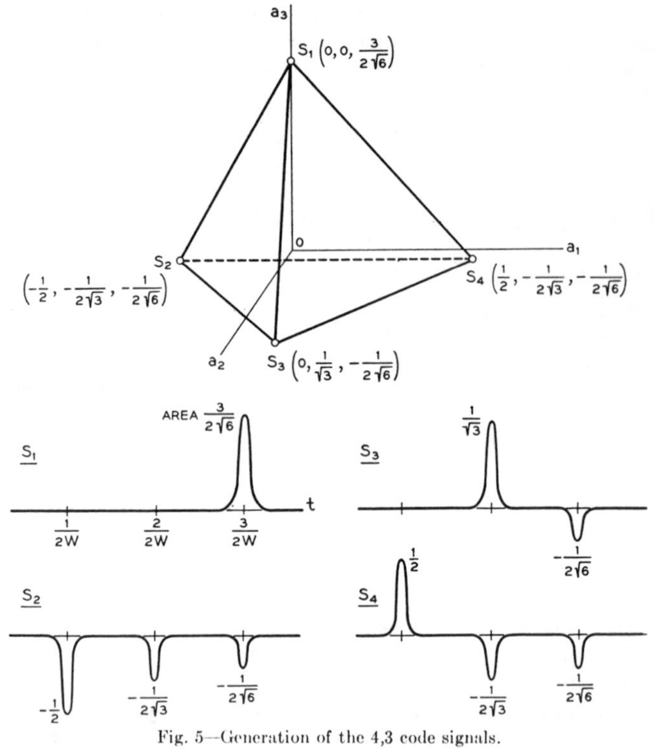

Fig. 5—Generation of the 4,3 code signals.

The best possible higher dimensional alphabets can be described more easily verbally than pictorially. In four dimensions we have found four alphabets.

The *25,4* alphabet consists of the origin and all 24 points in 4 dimensional space having two coordinates equal to zero and the remaining two equal to $1/\sqrt{2}$ or $-1/\sqrt{2}$. Each of the 24 points lies a unit distance away from the origin and its 10 other nearest neighbors; they are, in fact, the vertices of a regular solid. This alphabet has an advantage beyond its high efficiency. The code signals are composed entirely of positive and negative pulses of fixed energy and so should be easier to generate than most of the other codes which appear in this paper.

The *800, 4* alphabet is constructed in the following way: Consider a lattice of points throughout the entire 4-dimensional space formed by taking all the linear combinations with integer coefficients of a basic set of four vectors. That is, the lattice points are of the form $C_1v_1 + C_2v_2 + C_3v_3 + C_4v_4$ where $C_1, \cdots, C_4$ are integers and the v_i are the four given vectors. In connection with our problem it is of interest to know what lattice, (i.e. what choice of v_1, v_2, v_3, v_4) has all lattice points separated at least unit distance from one another and at the same time packs as many points as possible into the space per unit volume. When a solution to this "packing problem" is known, it is clear that a good alphabet can be obtained just by using all the lattice points which are contained inside a hypersphere about the origin as the letter points. Many of the two dimensional alphabets illustrated in the sketches are related in this way to the corresponding two dimensional packing problem (which is solved by letting v_1 and v_2 be a pair of unit vectors 60° apart). A solution to the four dimensional packing problem is afforded by

$$v_1 = \frac{1}{\sqrt{2}}, \frac{1}{\sqrt{2}}, 0, 0$$
$$v_2 = \frac{1}{\sqrt{2}}, 0, \frac{1}{\sqrt{2}}, 0$$
$$v_3 = \frac{1}{\sqrt{2}}, 0, 0, \frac{1}{\sqrt{2}}$$
$$v_4 = 0, \frac{1}{\sqrt{2}}, \frac{1}{\sqrt{2}}, 0.$$

This lattice contains two points per unit volume (twice as dense as the cubic lattice in which $v_1, \cdots, v_4$ are orthogonal to one another) and each point has 18 nearest neighbors. A hypersphere of radius 3 about the origin has a volume $(\pi^2/2)3^4$, about 400. Thus it contains about 800 lattice points. Take these as the code points of the 800, 4 code. Their average squared distances from the origin can be estimated as

$$\frac{\int_0^3 r^5\,dr}{\int_0^3 r^3\,dr} = \frac{2}{3}(3)^2 = 6.$$

N in Equation (13) may be estimated at 18; this is conservative because some lattice points outside the sphere are being counted.

The two remaining four dimensional alphabets belong to two families of D-dimensional alphabets.

The 4, 3; 5, 4; $\cdots$; $D + 1$, $D \cdots$ alphabets are the vertices of the simplest regular solid in D-dimensional space. For example, 4, 3 is a tetrahedron. Such a solid can be constructed from $D + 1$ vertices whose coordinates are the first $D + 1$ rows of the scheme

0	0	0	0	0	$\cdots$
1	0	0	0	0	$\cdots$
$\frac{1}{2}$	$\frac{3}{2\sqrt{3}}$	0	0	0	$\cdots$
$\frac{1}{2}$	$\frac{1}{2\sqrt{3}}$	$\frac{4}{2\sqrt{6}}$	0	0	$\cdots$
$\frac{1}{2}$	$\frac{1}{2\sqrt{3}}$	$\frac{1}{2\sqrt{6}}$	$\frac{5}{2\sqrt{10}}$	0	$\cdots$
$\frac{1}{2}$	$\frac{1}{2\sqrt{3}}$	$\frac{1}{2\sqrt{6}}$	$\frac{1}{2\sqrt{10}}$	$\frac{6}{2\sqrt{15}}$	$\cdots$
.	.	.	.	.	$\cdots$
.	.	.	.	.	$\cdots$
.	.	.	.	.	$\cdots$

The vertices all lie a distance $\sqrt{D/2(D + 1)}$ from the centroid of the figure.

6, 3; 8, 4; $\cdots$; $2D$, D, $\cdots$ are obtained by placing a point wherever any positive or negative coordinate axis intersects the sphere of radius $1/\sqrt{2}$ about the origin. Thus it follows that 6, 3 consists of the vertices of an octohedron.

Error correcting alphabets ((k, K, D)): The error correcting alphabets discussed in Part I can be converted into good alphabets for this channel by replacing all digits which equalled 0 by -1. Three error correcting alphabets appear on the chart; each is labelled by three numbers signifying (k, K, D).

Slepian alphabets (SD): Using group theoretic methods, D. Slepian has attempted to construct families of alphabets which signal at rates approaching C. Although this goal has not yet been reached, families of alphabets depending on the parameter D have been found which approach the ideal curve to within 6.2 db and then get worse as $D \to \infty$. In the simplest of these families of alphabets, $D = 2m$ is even and the letters consist of all the $2^m C_{2m,\,m}$ sequences containing m zeros, the remaining places being filled by ± 1. The best alphabet in this family is the one with $D = 24$. It lies 6.23 db away from the ideal curve and contains 1.1 x 10^{10} letters. The alphabets of this family for $D = 10$, 24, and 70 appear on the efficiency graph labelled $S10$, $S24$, and $S70$.

The conclusion to which one is forced as a result of this investigation is that one cannot signal over a channel with signal to noise level much less than 7 db above the ideal level of Equation (2) without using an unbelievably complicated alphabet. No ten digit alphabet tolerates less than 7.7 db more than the ideal signal to noise ratio.

It would be interesting to know more about good higher dimensional alphabets. They are very much more difficult to obtain. The regular solids, which provided some good alphabets in 3 and 4 dimensions, provide nothing new in 5 or more dimensions; there are only three of them and they correspond to our $D + 1$, D; $2D$, D, and 2^D binary alphabets. Worse still, the packing problem also becomes unmanageable after dimension 5.

ACKNOWLEDGMENT

The author wishes to thank R. W. Hamming, L. A. MacColl, B. McMillan, C. E. Shannon, and D. Slepian for many helpful suggestions during the investigation summarized by this paper.

Application of Boolean Algebra to Switching Circuit Design and to Error Detection

D. E. MULLER*

Summary—A solution is sought to the general problem of simplifying switching circuits that have more than one output. The mathematical treatment of the problem applies only to circuits that may be represented by "polynomials" in Boolean algebra. It is shown that certain parts of the multiple output problem for such circuits may be reduced to a single output problem whose inputs are equal in number to the sum of the numbers of inputs and outputs in the original problem. A particularly simple reduction may be effected in the case of two outputs.

Various techniques are described for simplifying Boolean expressions, called "+ polynomials," in which the operation "exclusive or" appears between terms. The methods described are particularly suitable for use with an automatic computer, and have been tested on the Illiac.

An unexpected metric relationship is shown to exist between the members of certain classes of "+ polynomials" called "nets." This relationship may be used for constructing error-detecting codes, provided the number of bits in the code is a power of two.

FOLLOWING the work of Shannon,[1] design of switching circuits has leaned heavily upon logical algebra, and systematic methods have been developed by Burkhart, Kalin, Aiken, Quine,[2,3] and others for reducing polynomial expressions in logical algebra. Much of the effectiveness of the application of these techniques has depended on the skill of the designer and upon the amount of time he is willing to spend in the manipulation of algebraic expressions which are obtained after having applied systematic reduction procedures. This has been especially true in the frequently encountered case in which more than one output is required from a particular circuit. Here, systematic methods for treating the single output circuit will be extended to the multiple output case.

Multiple Output Circuits

A switching circuit will be defined as a circuit in which voltage (or current) at any point in the circuit may take either of two possible values. These values may be arbitrarily described by the symbols 0 and 1. Such a circuit will be assumed to have p points $X^1, X^2, X^3, \cdots,$ X^p at which input voltages will be applied and q other points $Z^1, Z^2, Z^3, \cdots, Z^q$ from which outputs may be taken. It will be further assumed that all voltages in the circuit will be uniquely determined by the combined effect of the p inputs. If each of the q outputs is specified for each admissible combination of values at the p inputs, then the logical specifications for the circuit have been completely given and each output may be expressed as a logical function of the inputs

$$\begin{aligned} Z^1 &= Z^1(X^1, X^2, \cdots, X^p) \\ Z^2 &= Z^2(X^1, X^2, \cdots, X^p) \\ &\cdots\cdots\cdots\cdots \\ Z^q &= Z^q(X^1, X^2, \cdots, X^p). \end{aligned} \qquad (1)$$

In general, certain combinations of values at the inputs will never occur, and for this reason the inputs will not be entirely independent. Such a relation will be expressed by the subsidiary condition

$$g(X^1, X^2, \cdots, X^p) = 0. \qquad (2)$$

Those combinations of input values which never occur are just those for which $g=1$. Hence condition (2) completely specifies those combinations.

Algebraic manipulations may now be carried out to simplify the functional expressions (1) while making use of the subsidiary condition (2). These manipulations should tend to simplify the switching circuit corresponding to (1) according to prescribed criteria of simplicity, while maintaining the logical specifications for the circuits. Such manipulations, if carried out empirically, may be quite difficult and tedious. Often it is necessary to expand the functions Z^i so as to make them more complex before they can be simplified later. Systematic methods have therefore been developed to relieve the designer of some of the tedious work involved in reducing the functions Z^i.

A function Z^i of the inputs $X^1, X^2, \cdots, X^p$ may be expressed in canonical form

$$\begin{aligned} Z^i = {} & Z_0{}^i X^p X^{p-1} \cdots X^2 X^1 \vee Z_1{}^i X^p X^{p-1} \cdots X^2 \overline{X}^1 \\ & \vee \cdots \vee Z^i_{2^p-1} \overline{X}^p \overline{X}^{p-1} \cdots \overline{X}^2 \overline{X}^1 \end{aligned} \qquad (3)$$

where $\overline{X}^i$ represents the complement (or negation) of X^i and the symbol "$\vee$" represents the logical operation "or." In a particular term the inputs and their complements are connected by the logical operation "and." The coefficients $Z_j{}^i$ of the $j+1$ term is a constant having either the value 0 or 1, and serves to define the value of Z^i when the input values are such that the other factors in the $j+1$ term are all 1.

Expansion (3) is a special case of what may be called a polynomial in Boolean algebra. In a general polynomial, however, it will not be necessary for a term to depend on all inputs but it may be represented by a product of less than p of the inputs and complements of inputs. Thus $X^1 \vee \overline{X}^2 X^3$ and $\overline{X}^1 \vee X^4$ would also be re-

* Digital Computer Lab., University of Illinois, Urbana, Illinois.

[1] C. E. Shannon, "A symbolic analysis of relay and switching circuits" *Trans. A.I.E.E.*, vol. 57, pp. 713–723; 1938.

[2] "The Synthesis of Electronic Computing and Control Circuits," vol. XXVII. Annals of the Computation Laboratory of Harvard University, Harvard University Press, Cambridge, Mass.; 1951.

[3] W. V. Quine, "The problem of symplifying truth functions," *Amer. Math. Monthly*, vol. 59, p. 521; October, 1952.

Reprinted from *IRE Trans. Electron. Comput.*, vol. EC-3, pp. 6-12, Sept. 1954.

garded as polynomials. A different type of polynomial may be formed if the operation "exclusive or" (designated by "+") is used between terms. Such a polynomial will be referred to as a + polynomial while the previous type will be referred to as a $\vee$ polynomial. Expansion (3) may also be written as a + polynomial since all terms in (3) are disjoint (i.e., never more than one term may equal 1) and "+" may be used to replace "$\vee$" wherever it appears, giving

$$Z^i = Z_0^i X^p X^{p-1} \cdots X^2 X^1 + Z_1^i X^p X^{p-1} \cdots X^2 \overline{X}^1 + \cdots + Z^i_{2^p-1} \overline{X}^p \overline{X}^{p-1} \cdots \overline{X}^2 \overline{X}^1. \tag{4}$$

In discussing multiple output functions, results will be valid for both + polynomials and $\vee$ polynomials and the symbol "+" will be used to refer to both operations. Furthermore, the term polynomial will mean either type of polynomial. Functions in Boolean algebra such as Z^i and g will also be expressed in polynomial form.

If a suitable reduction of the functions Z^i has been achieved in polynomial form, a set of polynomials must be specified, each one of which will be used when manufacturing certain z functions. These polynomials will be written $Mj_1, j_2, \cdots, j_q$ where j_i will have the value 0 if $Mj_1, j_2, \cdots, j_q$ is used in the construction of Z^i and the value 1 if it is not. There are 2^q-1 of these polynomials, in general, since it is not possible for j_i to be 1 for all i. In reduced form the functions Z^i will be written.

$$Z^i = \sum_{j_i=0} Mj_1, j_2, \cdots, j_q \tag{5}$$

where the sum is taken over all $Mj_1, j_2, \cdots, j_q$ for which $j_i=0$. In this sum the operation between the polynomials is either "$\vee$" or "+" depending upon whether $\vee$ polynomials or + polynomials are being used. In forming the switching circuit having outputs Z^i each of the polynomials $Mj_1, j_2, \cdots, j_q$ will be manufactured first and then combined according to (5) giving the Z^i.

The problem of reducing the switching circuit producing the outputs Z^i may now be considered in two parts:

(a) The problem of simultaneously minimizing the set of polynomials $Mj_1, j_2, \cdots, j_q$.

(b) The problem of minimizing the number of connectives between such polynomials in (5). When none of the polynomials $Mj_1, j_2, \cdots, j_q$ are zero, part (b) may be ignored, since the structure of the connectives is unalterable. On the other hand, if the number of outputs is large and the number of inputs small, part (b) tends to assume importance comparable to part (a).

Theorem 1: The problem of simultaneously minimizing the polynomials used in constructing a circuit having p inputs and q outputs may be replaced by the problem of finding a minimal polynomial to represent a certain single output circuit having $p+q$ inputs.

Proof: It is not necessary to define precisely the meaning of minimization for the purposes of this theorem since the two processes are merely to be shown to be equivalent.

The imaginary single output circuit described in the above theorem is assumed to use q inputs $y^1, y^2, \cdots, y^q$ in addition to inputs $X^1, X^2, \cdots, X^p$ which are used in the multiple output circuit. The single output F of the imaginary circuit is defined as

$$F = \sum_{i=1}^{q} \bar{y}^i Z^i(X^1, X^2, \cdots, X^p). \tag{6}$$

Just as the inputs $X^1, X^2, \cdots, X^p$ are assumed to be restricted by the condition $g(X^1, X^2, \cdots, X^p)=0$ the artificial inputs y^i will be restricted by the conditions:

$$\begin{aligned} \bar{y}^i \bar{y}^j &= 0 \quad \text{when} \quad i \neq j \\ y^1 y^2 \cdots y^q &= 0. \end{aligned} \tag{7}$$

In order to express all of these conditions as a single condition they may be added, giving

$$g(X^1, X^2, \cdots, X^p) \vee y^1 y^2 \cdots y^q \vee \sum_{i \neq j} \bar{y}^i \bar{y}^j = 0. \tag{8}$$

The sum used in this last expression is understood to use the "$\vee$" operation while all other sums in this proof represent either "$\vee$" or "+" depending upon which type of polynomial is being considered.

If F as defined by (6) is minimized subject to condition (8) it will be represented as a polynomial P. From (7) it may be shown that

$$\bar{y}^i = y^1 y^2 \cdots y^{i-1} y^{i+1} \cdots y^q \tag{9}$$

and each $\bar{y}^i$ appearing in P may be replaced accordingly. After this has been done the resulting polynomial may be written.

$$F = P \equiv \sum (y^1)^{j_1} (y^2)^{j_2} \cdots (y^q)^{j_q} Mj_1, j_2, \cdots, j_q \tag{10}$$

where $Mj_1, j_2, \cdots, j_q$ is a polynomial involving X^1, $X^2, \cdots, X^p$ and the notation $(y^i)^{j_i}$ is defined by

$$\begin{aligned} (y^i)^{j_i} &= 1 \quad \text{if} \quad j_i = 0 \\ (y^i)^{j_i} &= y^i \quad \text{if} \quad j_i = 1. \end{aligned}$$

The sum in expression (10) is taken over the 2^q-1 combinations of values of the j_i in which not all of them are 1.

It now remains to identify the $Mj_1, j_2, \cdots, j_q$ in (10) with the $Mj_1, j_2, \cdots, j_q$ in (5). From (6) and (7) it may be seen that if $\bar{y}^i$ assumes the value 1 relation (6) will become $F=Z^i$. Eq. (10) then turns into (5) since terms in (10) containing y^i vanish. If (5) have not been minimized by this process then a more reduced set of $Mj_1, j_2, \cdots, j_q$ exists satisfying (5). These equations may be substituted in (6) and using (7) may be manipulated into form (10) thus giving a more reduced version of (10). Since (10) was assumed to be minimal this contradicts the hypothesis and the theorem is proved.

Eqs. (5) now represent the multiple output circuit made up of minimal polynomials $Mj_1, j_2, \cdots, j_q$.

Theorem 1 specializes in a convenient fashion when $q=2$, and Theorem 2 expresses this special case. Eqs. (7) then become

$$\bar{y}^1 \bar{y}^2 = 0, \qquad y^1 y^2 = 0.$$

Both of these equations will be automatically satisfied if the single condition $\bar{y}^2 = y^1$ is used. y^2 therefore may be eliminated by this equation and no subsidiary conditions are required.

Theorem 2: The problem of minimizing a two output circuit having p inputs, which is to be expressed in polynomial form, may be replaced by the problem of minimizing a single output circuit having $p+1$ inputs.

Equations may be specialized as follows:

$$Z^1 = M_{0,1} + M_{0,0}$$

$$Z^2 = M_{1,0} + M_{0,0}, \quad (5')$$

$$F = \bar{y}^1 Z^1 + y^1 Z^2, \quad (6')$$

$$g(X^1, X^2, \cdots, X^p) = 0, \quad (8')$$

$$F = P \equiv \bar{y}^1 M_{0,1} + y^1 M_{1,0} + M_{0,0}. \quad (10')$$

In (6′) y^1 was substituted for $\bar{y}^2$ of (6), and in (10′) $\bar{y}^1$ was substituted for y^2. In this way (7) becomes unnecessary.

By way of example the set of equations

$$Z^1 = X^3X^2X^1 + X^3\bar{X}^2\bar{X}^1 + \bar{X}^3X^2\bar{X}^1 + \bar{X}^3\bar{X}^2X^1$$

$$Z^2 = X^3X^2X^1 + X^3X^2\bar{X}^1 + X^3\bar{X}^2X^1 + \bar{X}^3X^2X^1$$

may be used with no subsidiary conditions. These equations represent a single stage binary adder where Z^1 is the output and Z^2 is the carry. Eq. (6′) becomes

$$F = \bar{y}^1X^3X^2X^1 + \bar{y}^1X^3\bar{X}^2\bar{X}^1 + \bar{y}^1\bar{X}^3X^2\bar{X}^1 + \bar{y}^1\bar{X}^3\bar{X}^2X^1 + y^1X^3X^2X^1 + y^1X^3X^2\bar{X}^1 + y^1X^3\bar{X}^2X^1 + y^1\bar{X}^3X^2X^1.$$

Again no subsidiary conditions are to be used. Using + polynomial reduction techniques, to be described in the next section, this polynomial may be reduced to

$$F = \bar{y}^1\bar{X}^3\bar{X}^2 + \bar{y}^1\bar{X}^1 + y^1X^3X^1 + y^1X^2X^1 + X^3X^2.$$

This gives

$$M_{0,1} \equiv \bar{X}^3\bar{X}^2 + \bar{X}^1$$

$$M_{1,0} \equiv X^3X^1 + X^2X^1$$

$$M_{0,0} \equiv X^3X^2$$

to complete the construction.

The multiple output circuit is represented by

$$Z^1 = \bar{X}^3\bar{X}^2 + \bar{X}^1 + X^3X^2$$

$$Z^2 = X^3X^1 + X^2X^1 + X^3X^2.$$

These expressions are not necessarily the simplest forms for Z^1 and Z^2. Further reduction, by replacing "+" with "∨" and by factoring, are outside the realm of the present discussion since the resulting expressions would then no longer be polynomials.

Reduction of + Polynomials

Reduction of ∨ polynomials has been completely analyzed by Quine[3] and by Burkhart, Kalin, and Aiken.[2] Extension of these methods to include the possibility of subsidiary conditions has been carried out by I. S. Reed.[4] Applying these methods to theorem 1 permits multiple output circuits to be treated also.

Circuit reduction by use of methods involving + polynomials presents an alternative process which usually yields considerably different results from those involving ∨ polynomials. By way of review, the important properties of the operation "+" are:

$$\begin{array}{ll} \text{i)} & a + b = a\bar{b} \vee \bar{a}b \\ \text{ii)} & a + b = b + a \\ \text{iii)} & a + (b + c) = (a + b) + c \\ \text{iv)} & a(b + c) = ab + ac \\ \text{v)} & a + a = 0 \\ \text{vi)} & a + \bar{a} = 1. \end{array} \quad (11)$$

If property (i) is taken as a definition the other properties may be directly deduced. Because of rule (v), it is evident that one need never retain duplicate terms in a polynomial. For this reason it will be assumed that duplicate terms are always to be combined in any polynomial representation. Operations may be performed upon polynomials which leave them equal to the same Boolean function but change their form. Two polynomials, P_1 and P_2 will be regarded as equivalent only if they are termwise equivalent. Such a relation will be written $P_1 \equiv P_2$ while $P_1 = P_2$ will be taken to mean that the two polynomials equal the same Boolean function, but are not necessarily equivalent. The symbol $P_1 + P_2$ will represent a polynomial containing the terms of both P_1 and P_2 with the exception that duplicate terms are combined according to (v). P_1P_2 will represent the expanded product of the two polynomials and $P_1 \cdot P_2$ will represent a polynomial having only those terms which are common to P_1 and P_2.

A general operator R_j which may be used to alter the form of a + polynomial without changing it functionally is defined by the relation

$$R_jP \equiv P + X^jM + \bar{X}^jM + M$$

where M is a + polynomial which depends on R_j and may or may not depend on P. If M is independent of P the operator R_j is its own inverse since $R_jR_jP \equiv P$. Special operators of this type may be formed in various ways. Four operators of type R_j may be defined by writing:

$$P \equiv X^jM_0 + \bar{X}^jM_1 + M_2$$

where M_0, M_1 and M_2 are polynomials which are independent of X^j. They are:

$$A_jP \equiv P + X^jM_2 + \bar{X}^jM_2 + M_2$$

$$B_jP \equiv P + X^jM_1 + \bar{X}^jM_1 + M_1$$

$$C_jP \equiv P + X^jM_0 + \bar{X}^jM_0 + M_0$$

$$D_jP \equiv P + X^j(M_0 + M_1 + M_2) + \bar{X}^j(M_0 + M_1 + M_2) + (M_0 + M_1 + M_2).$$

The symbol Q_j will be used to denote any one of these four operators, and can be shown to possess the algebraic properties

[4] Technical Memo No. 23, Lincoln Lab., M.I.T.

1) $$Q_j(P_1 + P_2) \equiv Q_jP_1 + Q_jP_2$$

2) $$Q_jQ_kP \equiv Q_kQ_jP$$

3) $$Q_jR_jP \equiv Q_jP \quad (\text{any } R_j).$$

Theorem 3: The operator $A_pA_{p-1} \cdots A_1$ reduces a polynomial P to its canonical form (4).

Proof: A_jP is a polynomial in which each term contains either X^j or $\overline{X}^j$. To see this one may write

$$A_jP \equiv P + X^jM_2 + \overline{X}^jM_2 + M_2$$
$$\equiv X^jM_0 + \overline{X}^jM_1 + M_2 + X^jM_2 + \overline{X}^jM_2 + M_2$$
$$\equiv X^j(M_0 + M_2) + \overline{X}^j(M_1 + M_2).$$

If all terms in the polynomial P contain either X^k or $\overline{X}^k$, then A_jP also possesses this property since no terms containing neither X^k nor $\overline{X}^k$ are introduced. Hence every term of $A_pA_{p-1} \cdots A_1P$ contains either each input or its complement. Therefore $A_pA_{p-1} \cdots A_1P$ has the form of (4) which is a unique canonical form for each function. Henceforth the expression $A_pA_{p-1} \cdots A_1$ will be abbreviated A.

Theorem 4: If $P_1 = P_2$ then it is possible to transform P_1 into P_2 by p operations of the general type R_j.

Proof: If P_3 represents the canonical form of P_1 and P_2, the relations $AP_1 \equiv P_3$ and $AP_2 \equiv P_3$ are satisfied. The polynomials M_2 in the expression $A_jP \equiv P + X^jM_2 + \overline{X}^jM_2 + M_2$ may now be regarded as constants which do not depend on P since they are defined by the relations $AP_1 \equiv P_3$ and $AP_2 \equiv P_3$. The resulting operators may no longer be regarded as of the type A_j since the M's involved are constants. These operators will be written $S_pS_{p-1} \cdots S_1P_1 \equiv P_3$ and $T_pT_{p-1} \cdots T_1P_2 \equiv P_3$.

The operator S_j has the same effect as A_j when used in this equation but when applied to a different polynomial it would not have the same effect since the polynomial M_2 would be altered in the case of A_j and not in the case of S_j.

Thus if

$$P \equiv X^jM_0 + \overline{X}^jM_1 + M_2$$

and

$$P' \equiv X^jM_0' + \overline{X}^jM_1' + M_2'$$

then

$$A_jP \equiv P + X^jM_2 + \overline{X}^jM_2 + M_2$$

and

$$S_jP \equiv P + X^jM_2 + \overline{X}^jM_2 + M_2$$

but

$$A_jP' \equiv P' + X^jM_2' + \overline{X}^jM_2' + M_2'$$

while

$$S_jP' \equiv P' + X^jM_2 + \overline{X}^jM_2 + M_2.$$

Since operators using constant M commute and are their own inverses it is evident that

$$P_3 \equiv S_1S_2 \cdots S_pP_1$$

and

$$P_2 \equiv T_pT_{p-1} \cdots T_1S_1S_2 \cdots S_pP_1$$
$$\equiv T_pS_pT_{p-1}S_{p-1} \cdots T_1S_1P_1.$$

The operator T_jS_j may, however, be regarded as a single operator since the M's involved may be added, and the theorem is proved.

From theorem 4 it may be seen that operations of the type R_j are sufficient to reduce any arbitrary polynomial P_1 to its minimal form P_2. Such a reduction is not in general possible simply because the required operations cannot usually be found without a knowledge of P_2.

If operators of type Q_j are combined, a variety of characteristic forms are obtained. Theorem 5 proves the existence of these forms.

Theorem 5: If $P_1 = P_2$ then $Q_pQ_{p-1} \cdots Q_1P_1 \equiv Q_pQ_{p-1} \cdots Q_1P_2$ where Q_j may represent different ones of the four operators A_j, B_j, C_j or D_j for each j, but must have the same meaning on the two sides of the equation.

Proof: Let $AP_1 \equiv AP_2 \equiv P_3$.
Then $Q_pQ_{p-1} \cdots Q_1P_3 \equiv Q_pQ_{p-1} \cdots Q_1A_pA_{p-1} \cdots A_1P_1 \equiv Q_pQ_{p-1} \cdots Q_1P_1$ by properties 1 and 3. Similarly, $Q_pQ_{p-1} \cdots Q_1P_3 \equiv Q_pQ_{p-1} \cdots Q_1P_2$ and hence $Q_pQ_{p-1} \cdots Q_1P_1 \equiv Q_pQ_{p-1} \cdots Q_1P_2$.

From this property the operator $Q_pQ_{p-1} \cdots Q_1$ may be said to yield a "characteristic" polynomial. Since four possible choices are available for each operator Q_j (it may be either A_j, B_j, C_j, or D_j) the number of such expansions is 4^p. A particularly symmetrical expansion of this type is the one produced by the operator $D_pD_{p-1} \cdots D_1$. Other expansions such as that produced by $B_pB_{p-1} \cdots B_1$ have singular metric properties which will be described later.

Simplification of polynomials is carried out with the help of operators of the type Q_j but principally one must rely on the mathematically less interesting operator H_j defined by

$$H_jP \equiv P + X^j(M_0 \cdot M_1 + M_1 \cdot M_2 + M_2 \cdot M_0)$$
$$+ \overline{X}^j(M_0 \cdot M_1 + M_1 \cdot M_2 + M_2 \cdot M_0)$$
$$+ (M_0 \cdot M_1 + M_1 \cdot M_2 + M_2 \cdot M_0)$$

where M_0, M_1 and M_2 do not involve X_j or $\overline{X}_j$ and are defined by the relation $P \equiv X^jM_0 + \overline{X}^jM_1 + M_2$. Polynomials represented by $M_0 \cdot M_1$ etc., are defined, as before, to be those containing terms common to M_0 and M_1 etc. H_j has none of the convenient algebraic properties of the Q_j's but it tends to reduce the number of terms in the polynomial to which it is applied. If "a" represents one term in a polynomial, H_j effects the following types of simplifications.

$$\begin{aligned} H_j(X^ja + \overline{X}^ja) &\equiv a \\ H_j(X^ja + a) &\equiv \overline{X}^ja \\ H_j(\overline{X}^ja + a) &\equiv X^ja \\ H_j(X^ja + \overline{X}^ja + a) &\equiv 0. \end{aligned} \tag{12}$$

One of the most elementary types of simplifications which can be applied to a polynomial is therefore $H_pH_{p-1} \cdots H_1P$ which will be denoted by HP. Although the operator H tends to simplify the polynomial to which it is applied, it will usually yield a result which is far more complex than that attained by more refined methods. In order to attain greater simplification than is possible merely by use of the H operator, one may expand each term by reversing one of the first three operations (12) whenever subsequent application of the H operator effects a still greater simplification. Such a process which will be called Method I may be explained, stepwise, as follows:

1) One starts with a polynomial P_1 to be simplified. It is first reduced to canonical form.

$$P_2 \equiv AP_1.$$

2) This result is simplified initially by use of H.

$$P_3 \equiv HP_2.$$

3) A simplification operator C_j' is constructed according to the definition

$$C_j'P \equiv H[\overline{X}^j\{H(M_0 + M_1)\} + \{M_0 + M_2\}].$$

Successive application of C_j' yields

$$P_4 \equiv C_p'C_{p-1}' \cdots C_1'P_3.$$

4) An operator B_j' is constructed according to the definition

$$B_j'P \equiv H[X^j(M_0 + M_1) + H(M_2 + M_1)]$$

and the final result P_5 is given by

$$P_5 \equiv B_p'B_{p-1}' \cdots B_1'P_4.$$

In this process the operators B_j' and C_j' have the effect of expanding the polynomial whenever it may be simplified later by application of the H operator.

An operator H' which is more effective than the H operator may be formed by use of a gate polynomial KP. If $P_1 \nabla P_2$ represents a polynomial having terms which are in either or both of the polynomials P_1 and P_2, then the polynomial K_jP may be defined as

$$K_jP \equiv (X^j \nabla \overline{X}^j \nabla 1)(M_0 \nabla M_1 \nabla M_2)$$

and

$$KP \equiv K_pK_{p-1} \cdots K_1P.$$

For purposes of notation let

$$KP \equiv X^jN_0 + \overline{X}^jN_1 + N_2$$

and let

$$J_j^0P \equiv P + X^j(M_0 \cdot N_1 \cdot N_2) + \overline{X}^j(M_0 \cdot N_1 \cdot N_2) + (M_0 \cdot N_1 \cdot N_2)$$

$$J_j^1P \equiv P + X^j(N_0 \cdot M_1 \cdot N_2) + \overline{X}^j(N_0 \cdot M_1 \cdot N_2) + (N_0 \cdot M_1 \cdot N_2)$$

$$J_j^2P \equiv P + X^j(N_0 \cdot N_1 \cdot M_2) + \overline{X}^j(N_0 \cdot N_1 \cdot M_2) + (N_0 \cdot N_1 \cdot M_2)$$

and let

$$G_j^0P \equiv H_{j-1}H_{j-2} \cdots H_1H_p \cdots H_{j+1}J_j^0P$$

$$G_j^1P \equiv H_{j-1}H_{j-2} \cdots H_1H_p \cdots H_{j+1}J_j^1P$$

$$G_j^2P \equiv H_{j-1}H_{j-2} \cdots H_1H_p \cdots H_{j+1}J_j^2P.$$

Then

$$G^0P \equiv G_p^0G_{p-1}^0 \cdots G_1^0P$$

$$G^1P \equiv G_p^1G_{p-1}^1 \cdots G_1^1P$$

$$G^2P \equiv G_p^2G_{p-1}^2 \cdots G_1^2P$$

and finally

$$H'P \equiv G^2G^1G^0HP.$$

Method II may now be described as Method I with H' substituted for H wherever it appears. Method II has the advantage of forming as many as two expansions provided later contractions makes this advantageous, and thus yields a more effective, but more time-consuming process.

Justification for the choice of these processes rather than others which involve expansion and later simplification is based mainly upon their efficiency in reducing randomly chosen polynomials. A set of twenty randomly chosen functions of 5 inputs was used for comparison of different processes. Taking the number of terms in the final polynomial as a convenient measure of the effectiveness of the processes, the results obtained from Method I, Method II and the simple HA operator are compared in Table I.

TABLE I

Twenty Random Functions of Five Inputs Were Used to Test Three Simplification Processes. The Number of Terms in the Simplified Expansion is Listed in Each Case

Using Only HA Operator	Method I	Method II
11	7	7
9	7	6
6	5	5
9	6	6
8	7	6
10	7	6
8	7	7
9	7	7
8	6	6
7	7	6
7	7	6
9	7	7
9	7	7
10	8	7
10	8	7
8	7	7
7	5	5
8	7	7
9	6	6
8	5	5

Following polynomial-type simplifications such as Methods I and II, nonsystematic manipulation may be used to further simplify the result. Two types are especially useful:

a) Between pairs of terms of the form X^ia and $\overline{X}^ib$ the operation "+" may be replaced by "$\vee$" and between triples of the form $ab+bc+ca$ one may make the same substitution. By use of skill one should attempt to make the combination of substitutions which leave the fewest "+" operations to be performed.

b) Using skill, factor the result so as to reduce the resulting expression as much as possible.

Systematic polynomial reduction processes may conveniently be carried out by the use of high speed computers. Programs for the ILLIAC have been prepared to reduce $\vee$ polynomials using the Harvard method and to reduce $+$ polynomials using Methods I and II described here. As yet none of these processes have been modified to permit the inclusion of subsidiary conditions. These programs make use of an interpretive subroutine which makes it possible to manipulate polynomials conveniently in the machine. In the memory of the machine a polynomial is represented as a set of 3^p binary digits. The position of each binary digit specifies the term. If the digit is 1 it is regarded as being present in the polynomial, and if the digit is 0 it is regarded as absent. No distinction need be made between $\vee$ polynomials and $+$ polynomials. To each input is allotted a digit of a number written in the ternary system. This digit is 0, 1, or 2 according to whether the input is present, complemented, or absent. The ternary number so obtained represents the relative position of the binary digit corresponding to a term in a polynomial. Thus to the term $\overline{X}^3X^2$ corresponds the number 102 written in the ternary system. Since 0 represents $X^3X^2X^1$ it would be placed in the first relative position and $\overline{X}^3X^2$ would be represented by a 1 in the twelfth relative position.

Using the interpretive routine it is possible to extract just those digits of the polynomial $P \equiv X^iM_0 + \overline{X}^iM_1 + M_2$ corresponding to one of the M's, say X^iM_0. By shifting these digits to a new relative position it is possible to form from these extracted digits either X^iM_0, $\overline{X}^iM_0$ or M_0. Assume that $\overline{X}^iM_0$ is formed. This result may then be combined algebraically with some other polynomial P' to form $P' + \overline{X}^iM_0$, $P' \cdot \overline{X}^iM_0$ or $P' \nabla X^iM_0$. The polynomial containing all terms not in $\overline{X}^iM_0$ may also be used when performing these combinations. Such a sequence of operations as the one described will be produced by one order in the interpretive routine. By a series of such orders the operators of the simplification processes may be formed. Special control transfer orders allow repeating a process using logical input indices 1, 2, $\cdots$, $p-1$, p and other orders permit algebraic operations to be performed without extractions. Various "red tape" orders are also provided.

Error Detection

In the theory of error detecting codes one deals with sequences of n binary digits. Such a set "a" may be written as a vector $a = (a_0, a_1, \cdots, a_{n-1})$ where a_i may take on values 0 or 1. A metric $L(a, b)$ has been defined with respect to two such vectors "a" and "b" as the number of components in which "a" and "b" differ.[5] This metric may be shown to possess all the usual metric properties. The problem of finding an error detecting code consists of finding a set of r vectors $r^0, r^1, \cdots, r^j, \cdots, r^{t-1}$ such that $L(r^j, r^k) \geqq d$ for $j \neq k$ when one is given a number d called the order of the code.

The theory of $+$ polynomials in Boolean algebra may be applied directly, when n and d are powers of two. n and d were not assumed to be restricted in this fashion in the definition of the general problem. If $n = 2^p$ and $d = 2^m$ the solutions so obtained give $t = 2^{C_p^p + C_{p-1}^p + \cdots + C_m^p}$ where C_q^p is a binomial coefficient. Components "a_i" of the vector "a" may be identified with the coefficients of the terms in expansion (4), the canonical expansion of a corresponding function "a" of p inputs in Boolean algebra. Such a canonical expansion may be regarded as a polynomial P_1.

$$a = P_1 \equiv a_0X^pX^{p-1} \cdots X^1 + a_1X^pX^{p-1} \cdots \overline{X}^1 + \cdots + a_{2^p-1}\overline{X}^p\overline{X}^{p-1} \cdots \overline{X}^1. \quad (13)$$

A characteristic polynomial P_2 may be formed by successive application of the operators Bj described previously.

$$P_2 \equiv B_pB_{p-1} \cdots B_1P_1 \equiv BP_1. \quad (14)$$

By an argument similar to that given in the proof of theorem 3, it may be seen that none of the inputs appearing in the terms of P_2 are complemented. Each input therefore is either present or absent and P_2 may be written

$$P_2 \equiv g_0X^pX^{p-1} \cdots X^1 + g_1X^pX^{p-1} \cdots X^2 + \cdots + g_{2^p-1} \quad (15)$$

The coefficients $g_0, g_1, \cdots, g_{2^p-1}$ are each either 0 or 1, and depend uniquely on $a_0, a_1, \cdots, a_{2^p-1}$. It is interesting to note that the transformation of the coefficients $a_0, a_1, \cdots, a_{2^p-1}$ to $g_0, g_1, \cdots, g_{2^p-1}$ is its own inverse. This may be seen by interchanging the role of X^i and the absence of X^i or $\overline{X}^i$ in the terms.

In expression (15) it is possible to group those terms containing a given number of inputs. There may exist C_k^p terms having k inputs but certain terms in (15) may vanish because their coefficients are zero.

Definition: A net of logical functions of order d is defined as all those functions whose expansions (as given in (15)) contain no terms having more than $p-m$ inputs, where m is defined by the relation $2^m = d$.

Theorem 6: If $r^1, r^2, \cdots, r^t$ are members of a net of order d then $L(r^i, r^j) \geqq d$ for all pairs r^i, r^j with $i \neq j$.

Proof: The theorem is proved by induction. It is true in case $p = \log_2 d$, since then the functions 0 and 1 are the only net members. Assume it is true when $p = k$ for all allowable d. It will be shown to be true when $p = k+1$.

From expansion (15) it may be noted that the members of a net are closed under the operation "$+$" since no terms containing more $p-m$ inputs can be generated by adding expansions having no such terms. Thus there is an r^l in the net such that $r^i + r^j = r^l$ for every pair r^i, r^j in the net. From the definition of the operation "$+$"

$$L(r^i, r^j) = L(r^i + r^j, 0) = L(r^l, 0).$$

Thus it is only necessary to prove that $L(r^l, 0) \geqq d$ for each non-zero member of the net r^l. If r^l, a member of the net of order d, is a function of $k+1$ inputs and is expressed in the form of (15), the $(k+1)$st input may be factored out giving

$$r^l = f^1 + X^{k+1}f^2$$

f^1 and f^2 are functions of k inputs, which are not both zero. f^1 is a member of the net of order $d/2$ and f^2 is a

[5] R. W. Hamming, "Error detecting and error correcting codes," *Bell Sys. Tech. Jour.*, vol. 29, pp. 147–160; April, 1950.

member of the net of order d. Four cases must be considered:

a) If the function f^2 is zero, then $r^l = f^1$. Regarding f^1 as a function of $k+1$ inputs the function r^l may be written as a sum of disjoint parts: $r^l = \overline{X}^{k+1}f^1 + X^{k+1}f^1$. If f^1 is written in the form of expansion 13 then it may seen that the separate parts $\overline{X}^{k+1}f^1$ and $X^{k+1}f^1$ of r^l contribute independently to $L(r^l, 0)$. Since $L(\overline{X}^{k+1}f^1, 0) \geqq d/2$ and $L(X^{k+1}f^1, 0) \geqq d/2$ the result

$$L(r^l, 0) = L(\overline{X}^{k+1}f^1, 0) + L(X^{k+1}f^1, 0) \geqq d$$

follows.

b) If the function f^1 is zero, then $r^l = X^{k+1}f^2$, and

$$L(X^{k+1}f^2, 0) \geqq d.$$

c) If f^1 and f^2 are not zero, but $f^1 = f^2$ then

$$r^l = f^2 + X^{k+1}f^2 = \overline{X}^{k+1}f^2$$

and

$$L(\overline{X}^{k+1}f^2, 0) \geqq d.$$

d) If f^1 and f^2 are not zero and not equal then

$$r^l = f^1 + X^{k+1}f^2 = \overline{X}^{k+1}f^1 + X^{k+1}f^3$$

where $f^3 = f^1 + f^2$ is not zero since $f^1 \neq f^2$ and is a member of the net of order $d/2$ by closure. Hence $L(\overline{X}^{k+1}f^1, 0) \geqq d/2$ and $L(X^{k+1}f^3, 0) \geqq d/2$ giving $L(r^l, 0) \geqq d$ since as before the expansions $\overline{X}^{k+1}f^1$ and $X^{k+1}f^3$ are disjoint.

Error detecting codes of order d may be formed therefore by use of vectors whose components are the coefficients of terms in the expansion (13) of net members. The information carried in such a code depends upon the number of coefficients in expansion (15) which are not forced to be zero by the net requirement. This number is $C_p{}^p + C_{p-1}{}^p + \cdots + C_m{}^p$ so that the number of such vectors available is $2^{C_p{}^p + C_{p-1}{}^p + \cdots + C_m{}^p}$.

For convenience expansion (15) may be used for interpreting information, and expansion (13) for transmission.

It has been shown that the members[6] of the net of order d do not always give the most numerous set of functions satisfying the relation $L(r^i, r^j) \geqq d$ and including the net members. When $d = 2^p, 2^{p-1}, 4, 2, 1$ the net does always give the most numerous set. When d takes on any other allowable value it can be shown that a more numerous set always exists which includes the net members, if sufficiently large p is used.

The case $d = 8$, $p = 5$ was investigated by use of the ILLIAC. In this case it was shown that no larger set of functions than the net members exists which satisfies $L(r^i, r^j) \geqq d$, and contains all net members. Hence the first case of a more numerous set must have $p \geqq 6$.

[6] D. E. Muller, "Metric Properties of Boolean Algebra and Their Application to Switching Circuits." Internal Report No. 46, Univ. of Illinois Graduate College, Digital Computer Laboratory.

A CLASS OF MULTIPLE-ERROR-CORRECTING CODES AND THE DECODING SCHEME

Irving S. Reed

Lincoln Laboratory - Massachusetts Institute of Technology
Cambridge, Massachusetts

I. Introduction

A procedure for constructing one-error-correcting and two-error-detecting systematic codes was introduced in a recent study by R. W. Hamming.[1] It is the purpose of this paper to exhibit some examples of n-error-correcting and (n + 1) error-detecting systematic codes for the cases where both the code length and (n + 1) are powers of two. The class of codes to be considered was developed by D. E. Muller in his recent work.[2]

The decoding scheme presented in this paper differs from Hamming's scheme in that the encoded message will be extracted directly from the possibly corrupted received code by a majority testing of the redundant relations within the code. Hamming's scheme for n = 1 was dependent first on the location of a possible digit error in the code; secondly, on the correction of that digit; and lastly, on the extraction of the message from the corrected code. By circumventing Hamming's step of error location and correction, which is quite a severe problem when n is not equal to one, we have arrived at a decoding scheme that makes a natural use of the redundancy within the code as well as being conceptually simple.

In this paper, some of the mathematical proofs of the methods discussed will be avoided for the sake of brevity of exposition. A more detailed mathematical analysis will appear elsewhere.

II. Some Mathematical Preliminaries

A code having n binary digits may be considered the element of a space, consisting of 2^n elements of the form

$$f = (f_0, \ldots f_{n-1})$$

where

$$(f_j = 0, 1) \quad \text{for} \quad (j = 0, 1, 2, \ldots n-1) .$$

This space is technically an Abelian group if the sum of any two elements f and g in the space is defined as follows:

$$f \oplus g = (f_0, f_1, \ldots f_{n-1}) \oplus (g_0, g_1, \ldots g_{n-1}) = (f_0 \oplus g_0, f_1 \oplus g_1, \ldots f_{n-1} \oplus g_{n-1}) ,$$

where $f_j \oplus g_j$ is the sum modulo two of the binary digits f_j and g_j for $(j = 0,1,2, \ldots n-1)$. If multiplication by the binary scalar α is allowed as

$$\alpha f = \alpha(f_0, f_1, \ldots f_{n-1}) = (\alpha f_0, \alpha f_1, \ldots \alpha f_{n-1}) ,$$

the Abelian group may be termed a generalized vector space of n-dimensions or a module. Finally, if the product operation

$$f \cdot g = (f_0, f_1, \ldots f_{n-1}) \cdot (g_0, g_1, \ldots g_{n-1}) = (f_0 g_0, f_1 g_1, \ldots f_{n-1} g_{n-1})$$

for f and g in the module is introduced, the space is a Boolean ring. The prime operation is defined to be

$$f' = f \oplus I$$

for f in the ring, and where I is the identity vector (1, 1, 1, ... 1).

Into this space one may further introduce a norm or length of a vector as follows:

$$\|f\| = \sum_{i=0}^{n-1} f_i$$

Reprinted from *IRE Trans. Inform. Theory*, vol. PGIT-4, pp. 38-49, Sept. 1954.

where Σ refers to ordinary addition. It is not difficult to see that the norm of the sum of two elements f and g in the ring or $\|f \oplus g\|$ is precisely the Hamming distance D(f,g) as defined in Ref. 1.

Now let n the dimension of the vector space be a power of two or $n = 2^m$. Let a vector of this space be of the form

$$f = (f_0, f_1, \dots f_{2^m-1}) ,$$

where f_j is a binary digit for $(j = 0, 1, \dots 2^m-1)$. Now the vector f may be clearly expressed as

$$f = f_0 I_0 \oplus f_1 I_1 \oplus \dots f_{2^m-1} I_{2^m-1} , \tag{1}$$

where I_j is a unit vector with the digit one in j-th coordinate of the vector and zeros elsewhere for $(j = 0, 1, \dots 2^m-1)$. Further, each unit vector I_j can be determined as a product of m vectors from the set of 2m vectors $x_1, x_2, x_3, \dots x_m, x_1', x_2', x_3', \dots x_m'$, where x_1 is a vector consisting of alternating zeros and ones, beginning with zero; x_2 is a vector consisting of alternating zero pairs and one pairs, beginning with a zero pair, and so forth, as follows:

$$x_1 = (0\,1\,0\,1\,0\,1\,0\,1 \dots 0\,1) ,$$

$$x_2 = (0\,0\,1\,1\,0\,0\,1\,1 \dots 1\,1) ,$$

$$x_3 = (0\,0\,0\,0\,1\,1\,1\,1 \dots 1\,1) ,$$

$$\vdots$$

$$x_m = (0\,0\,0\,0\,0\,0\,0\,0 \dots 1\,1) . \tag{2}$$

If $x_k^{i_k}$ is defined to be x_k' for $i_k = 0$ and x_k for $i_k = 1$, then by the rules of Boolean algebra,

$$I_j = x_1^{i_1} x_2^{i_2} \dots x_m^{i_m} , \tag{3}$$

where

$$j = \sum_{k=1}^{m} i_k 2^{k-1} \text{ with } (i_k = 0,1) \text{ for } (j = 0, 1, \dots m - 1) .$$

Combining Eqs. (1) and (3), we have

$$f = \sum_{j=0}^{2^m-1} f_j x_1^{i_1} x_2^{i_2} \dots x_m^{i_m} , \tag{4}$$

where $i_1, i_2, \dots i_m$ are the digits of the binary representation of j, and where the summation sign Σ is with respect to the sum operation $\oplus$. Equation (4) is the canonical expansion of any vector f in the Boolean algebra of 2^m dimensional vectors, consisting of binary digits.

If the identity $x_j' = I \oplus x_j$ and the distributive law of algebra is used, Eq.(4) may be expanded to obtain the following polynomial in the x_j's:

$$f = g_0 \oplus g_1 x_1 \oplus \dots \oplus g_m x_m \oplus g_{12} x_1 x_2 \oplus \dots g_{m-1,m} x_{m-1} x_m \oplus \dots$$

$$\dots \oplus g_{12 \dots m} x_1 x_2 \dots x_m . \tag{5}$$

Equation (5) can be written more explicitly as

$$f = f(0,\dots 0) \oplus \Delta_1 f(0,\dots 0)x_1 \oplus \dots \oplus \Delta_m f(0,\dots 0)x_m \oplus \Delta^2_{12} f(0,\dots 0)\, x_1x_2$$

$$\oplus \dots \oplus \Delta^m_{12\dots m} f(0,\dots 0)x_1x_2\dots x_m \quad , \tag{6}$$

where

$$f(i_1,\dots i_m) = f_j \quad \text{when} \quad j = \sum_{k=1}^{m} i_k\, 2^{k-1} \quad \text{for} \quad i_k = 0,1 \ ,$$

and the Δ's are multiple partial differences, for example,

$$\Delta_1 f(0) = f(1,0,0,\dots) \oplus f(0,0,0,\dots 0) \ ,$$

$$\Delta^2_{12} f(0) = [f(1,1,0,\dots) \oplus f(0,1,0,\dots)] \oplus [f(1,0,0,\dots) \oplus f(0,0,0,\dots)] \ ,$$

and so forth. The polynomial representation in Eq. (6) of the vector f supplies the relations between the coefficients of Eq. (5) and the scalars f_j of Eq. (4) for ($j = 0,1,2,\dots 2^m-1$). This definition of the Δ's will be expanded in another section of this paper.

III. The Generation of the Multiple Error Allowing Codes

Suppose that the dimension of the space considered in the previous section is 2^m. Consider the set Φ^m_r of all polynomials of the form (5) of degree less than or equal to r where $r \leq m$. Each such polynomial must have the form

$$g_0 \oplus g_1x_1 \oplus \dots \oplus g_mx_m \oplus \dots \oplus g_{12\dots r}\, x_1\dots x_r \oplus \dots \oplus g_{m-r+1,\dots m}$$

$$x_{m-r+1\dots}x_m \ , \tag{7}$$

and the sum of any two such polynomials is a member of the same set. This implies that Φ^m_r the set of all polynomials of type (7) or of degree less than or equal to r forms an Abelian group or submodule of the Boolean ring of 2^m dimensional vectors. Since Φ^m_r is a module, the Hamming distance between any two elements of Φ^m_r is the norm of a third element of Φ^m_r. This fact was exploited by D. E. Muller[2] in proving his Theorem 25. Muller's Theorem 25, in our terminology, may be expressed as follows:

<u>Theorem A</u>:- The norms of all non-zero vectors f of Φ^m_r satisfy

$$\| f \| \geq 2^{m-r} \quad \text{for } (m = 0,1,2,\dots) \text{ and } r \leq m \ .$$

We shall not prove this theorem here. It suffices to say that Muller proved the theorem by an induction on m and r and the properties of the Hamming distance.

By the above theorem there is at least a distance 2^{m-r} between two elements of Φ^m_r and, as a consequence, there is an open Hamming sphere of radius 2^{m-r-1} about each element of Φ^m_r in Φ^m_m (the whole vector space) which does not intersect any other such sphere. This means that it is possible to associate each element of such a sphere with the element defining the sphere or what is the same to associate an element of Φ^m_r which is less than a distance 2^{m-r-1} from an element f of Φ^m_r with f.

In order to illustrate how a message may be coded into an error-detecting code of the type described above, consider the following example: Let m = 4 and r = 1, by (7) the vectors of Φ^4_1 are of the form

$$g_0 \oplus g_1x_1 \oplus g_2x_2 \oplus g_3x_3 \oplus g_4x_4 \ . \tag{8}$$

Let the message consist of the five binary digits $(g_0, g_1, g_2, g_3, g_4)$. The code space Φ_1^4 may be regarded as generated by the four vectors x_1, x_2, x_3, x_4 and the identity vector I which may be written explicitly as follows:

$$
\begin{aligned}
x_1 &= (0\,1\,0\,1\,0\,1\,0\,1\,0\,1\,0\,1\,0\,1\,0\,1) \quad , \\
x_2 &= (0\,0\,1\,1\,0\,0\,1\,1\,0\,0\,1\,1\,0\,0\,1\,1) \quad , \\
x_3 &= (0\,0\,0\,0\,1\,1\,1\,1\,0\,0\,0\,0\,1\,1\,1\,1) \quad , \\
x_4 &= (0\,0\,0\,0\,0\,0\,0\,0\,1\,1\,1\,1\,1\,1\,1\,1) \quad , \\
I &= (1\,1\,1\,1\,1\,1\,1\,1\,1\,1\,1\,1\,1\,1\,1\,1) \quad .
\end{aligned} \tag{9}
$$

The 32 vector codes of Φ_1^4 can be obtained by scalar multiplication of the vectors of (9) by the message digits g_0, g_1, g_2, g_3, g_4 in accordance with (8). For example, the message (0 1 1 0 0) has the code vector $g_1x_1 \oplus g_2x_2$ or

$$(0\,1\,1\,0\,0\,1\,1\,0\,0\,1\,1\,0\,0\,1\,1\,0) \quad .$$

Each of the 32 codes will be a distance of at least eight from each other.

In order to practically generate the above code, one should note that the vector x_1 is the sequence of digits generated by the least significant binary stage B_1 of a binary counter of scale sixteen; x_2 is obtained from the second stage B_2; x_3 from the third stage B_3; and x_4 from the final stage B_4, as the counter goes through one period of its operation. If the message $(g_0, g_1, g_2, g_3, g_4)$ is stored in a binary register with stages A_0, A_1, A_2, A_3, A_4, then the switching function

$$C = A_0 \oplus A_1B_1 \oplus A_2B_2 \oplus A_3B_3 \oplus A_4B_4$$

will generate the code sequentially during one period of operation of the binary counter.

If one of the above codes of Φ_1^4 is corrupted during transmission so that no more than three errors are made, it is evidently possible by the previous discussion of this section to somehow extract the original message from the corrupted received code. The method by which this extraction may be accomplished will be shown by example in the next section and in general in the last section. It should be clear from the above example how the vectors of Φ_r^m may be generated for arbitray r and m where $r \leq m$.

IV. Decoding Corrupted Codes of Φ_r^m by a Majority Testing of Redundancy Relations

Let us first consider the coding space Φ_1^3. By (7), the vector of this space has the form

$$g_0 I \oplus g_1x_1 \oplus g_2x_2 \oplus g_3x_3 \quad . \tag{10}$$

The message will consist of the four binary digits (g_0, g_1, g_2, g_3), and the generating vectors of the space are

$$
\begin{aligned}
x_1 &= (0\,1\,0\,1\,0\,1\,0\,1) \quad , \\
x_2 &= (0\,0\,1\,1\,0\,0\,1\,1) \quad , \\
x_3 &= (0\,0\,0\,0\,1\,1\,1\,1) \quad , \\
I &= (1\,1\,1\,1\,1\,1\,1\,1) \quad .
\end{aligned} \tag{11}
$$

By (6) we have the following set of relations for the message digits g_j in terms of f_k, the code digits.

$$g_0 = f(0, \ldots 0) = f_0 \,, \qquad \underset{12}{\Delta} f(0\ldots) = f_0 \oplus f_1 \oplus f_2 \oplus f_3 = 0 \,,$$

$$g_1 = \underset{1}{\Delta} f(0\ldots) = f_0 \oplus f_1 \,, \qquad \underset{13}{\Delta} f(0\ldots) = f_0 \oplus f_1 \oplus f_4 \oplus f_5 = 0 \,,$$

$$g_2 = \underset{2}{\Delta} f(0\ldots) = f_0 \oplus f_2 \,, \qquad \underset{23}{\Delta} f(0\ldots) = f_0 \oplus f_2 \oplus f_4 \oplus f_6 = 0 \,,$$

$$g_3 = \underset{3}{\Delta} f(0\ldots) = f_0 \oplus f_4 \,, \qquad \underset{123}{\Delta} f(0\ldots) = \sum_{i=0}^{7} f_i = 0 \,. \qquad (12)$$

By (12) there are four relations which g_1 satisfies,

$$g_1 = f_0 \oplus f_1 = f_2 \oplus f_3 = f_4 \oplus f_5 = f_2 \oplus f_3 \oplus f_4 \oplus f_5 \oplus f_6 \oplus f_7 \,.$$

By substituting the second and third relations into the fourth relation, we have

$$g_1 = g_1 \oplus g_1 \oplus f_6 \oplus f_7 = 0 \oplus f_6 \oplus f_7 = f_6 \oplus f_7 \,.$$

Thus we obtain the four independent and disjoint relations for g_1,

$$g_1 = f_0 \oplus f_1 = f_2 \oplus f_3 = f_4 \oplus f_5 = f_6 \oplus f_7 \,.$$

These four relations are disjoint in the sense that no two of the relations have variables in common. In a similar manner, we may obtain four independent and disjoint relations for both g_2 and g_3 so that g_1, g_2, g_3 may be expressed as

$$g_1 = f_0 \oplus f_1 = f_2 \oplus f_3 = f_4 \oplus f_5 = f_6 \oplus f_7 \,,$$

$$g_2 = f_0 \oplus f_2 = f_1 \oplus f_3 = f_4 \oplus f_6 = f_5 \oplus f_7 \,,$$

$$g_3 = f_0 \oplus f_4 = f_1 \oplus f_5 = f_2 \oplus f_6 = f_3 \oplus f_7 \,.$$

Let us now suppose that the received code is the vector $(f_0, f_1, \ldots f_7)$. If there were no error in transmission of the code, all of the above relations would hold. If there were one error, three out of four of the relations would hold. If there were two errors, at least two of the g_j's would have two out of four incorrect relations. Then g_1, g_2, g_3 may be determined uniquely if one or no error occurred during transmission, and two errors may always be detected by making a majority test on the arithmetic sum of the values of the four relations for each $g_j (j = 1,2,3)$. In order to state this criterion more explicitly, let the values of the four relations for g_j be denoted by $r_{j1}, r_{j2}, r_{j3}, r_{j4}$ for $(j = 1,2,3)$, and let S_j be the arithmetic sum of $r_{j1}, r_{j2}, r_{j3}, r_{j4}$ or

$$S_j = \sum_{i=1}^{4} r_{ji} \,.$$

Then the majority decision test for g_j is

$$\begin{aligned} g_j &= 0 && \text{if } 0 \le S_j < 2 \,, \\ g_j &\text{ is indeterminate} && \text{if } S_j = 2 \,, \\ g_j &= 1 && \text{if } 2 < S_j \le 4 \text{ for } (j = 1,2,3) \,. \end{aligned}$$

With the assumption that the received code is no more than two digits in error, the majority test (13) will determine g_1, g_2, g_3 uniquely for only one or no errors, and reject the code as meaningless in the case of two errors. In the case of one error or less, g_1, g_2, g_3 may be assumed now to be determined; it remains to determine g_0. In order to find g_0, note that if, as g_1, g_2, g_3 are found, the vectors $g_1 x_1, g_2 x_2, g_3 x_3$ are added successively to the received vector, by (10) we will end with either the vector $g_0 I$ in the case of no error or with a vector of distance one from $g_0 I$. Thus to detect g_0 the following majority decision test will suffice:

$$g_0 = 0 \text{ if } \sum_{i=0}^{7} m_i < 4 \ ,$$

$$= 1 \text{ if } \sum_{i=0}^{7} m_i > 4 \ , \tag{14}$$

where m_i are the digits of the code after extraction of digits g_1, g_2, g_3 in accordance with the above procedure.

The above method of decoding may be illustrated by the following example: Suppose that the message sent was (1 0 1 1), and that during transmission an error was made in the fifth digit of the original code (1 1 0 0 0 0 1 1) so that the received code had the form (1 1 0 0 1 0 1 1). We first test for g_1, g_2, g_3 by (12) and find $g_1 = 0$, $g_2 = 1$ and $g_3 = 1$. Using (11), we add $g_1 x_1 \oplus g_2 x_2 \oplus g_3 x_3$ to the code, obtaining

$$0(0\ 1\ 0\ 1\ 0\ 1\ 0\ 1) \oplus (0\ 0\ 1\ 1\ 0\ 0\ 1\ 1) \oplus (0\ 0\ 0\ 0\ 1\ 1\ 1\ 1) \oplus (1\ 1\ 0\ 0\ 1\ 0\ 1\ 1)$$

$$= (1\ 1\ 1\ 1\ 0\ 1\ 1\ 1) = (m_0, m_1, m_2, \ldots m_3) \ .$$

Finally, by (14)

$$g_0 = 1 \ , \quad \text{since } \sum_{i=0}^{7} m_i = 7 > 4 \ .$$

Although Φ_1^3 is none other than an example of a set of one-error-correcting and two-error-detecting codes of the type described by Hamming in Ref. 1, the method of decoding considered above is different. Our procedure of decoding is advantageous in that it may be generalized in a natural way to include any of the coding spaces Φ_1^m of the second section of this paper. Before we consider the generalization by further examples, let us note a tabular way of representing the redundancy relations.

If the digits or variables of each relation are connected by lines for each of the vectors x_1, x_2, x_3 as

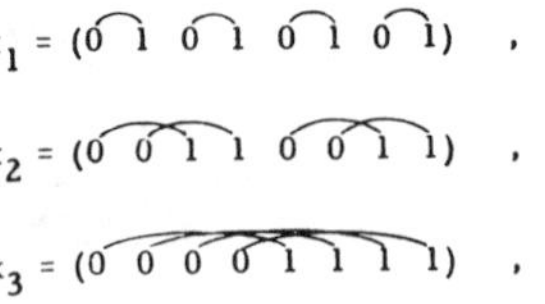

$$x_1 = (0\ 1\ 0\ 1\ 0\ 1\ 0\ 1) \ ,$$

$$x_2 = (0\ 0\ 1\ 1\ 0\ 0\ 1\ 1) \ ,$$

$$x_3 = (0\ 0\ 0\ 0\ 1\ 1\ 1\ 1) \ , \tag{15}$$

the relations of (12) become almost self-evident by their simplicity with respect to order and symmetry. This simplicity makes it possible to discover the redundancy relations for more general spaces Φ_r^m without resorting to the algebraic approach used above.

As a second example of our decoding procedure, consider the coding space Φ_1^4 introduced in the latter part of the preceding section. Each vector of this space has the form of (8), where the generating vectors are x_1, x_2, x_3, x_4 and I of (9). The first-degree redundancy relations may be determined in a manner similar to the above example and represented in a tabular manner similar to (15) as follows:

$$x_1 = (0\ 1\ 0\ 1\ 0\ 1\ 0\ 1\ 0\ 1\ 0\ 1\ 0\ 1\ 0\ 1) \ ,$$
$$x_2 = (0\ 0\ 1\ 1\ 0\ 0\ 1\ 1\ 0\ 0\ 1\ 1\ 0\ 0\ 1\ 1) \ ,$$
$$x_3 = (0\ 0\ 0\ 0\ 1\ 1\ 1\ 1\ 0\ 0\ 0\ 0\ 1\ 1\ 1\ 1) \ ,$$
$$x_4 = (0\ 0\ 0\ 0\ 0\ 0\ 0\ 0\ 1\ 1\ 1\ 1\ 1\ 1\ 1\ 1) \ . \qquad (16)$$

For instance, the eight independent and disjoint relations for g_1 are

$$g_1 = f_{2i} \oplus f_{2i+1} \quad \text{for } (i = 0,1, \ldots 7) \ .$$

If the eight values of the redundancy relations for g_j are labeled $r_{j1}, r_{j2}, \ldots r_{j8}$ for $(j = 1,2,3,4)$, and S_j is defined by

$$S_j = \sum_{i=1}^{8} r_{ji} \ ,$$

then, by an argument similar to that used in the previous example, the majority decision test for g_j is as follows:

$$\begin{aligned} g_j &= 0 && \text{if } 0 \le S_j < 4 \ , \\ g_j &\text{ is indeterminate} && \text{if } S_j = 4 \ , \\ g_j &= 1 && \text{if } 4 < S_j \le 8 \quad \text{for } (j = 1,2,3,4). \end{aligned} \qquad (17)$$

In order to determine g_0, we first add the determined vectors $g_j\mathbf{x}_j$ to the received message, assuming, of course, that no g_j is indeterminate, and we are left with the zero-degree polynomial Φ_0^4, possibly corrupted by errors. If there had been no errors, there would be sixteen zero-degree relations which g_0 satisfies, or

$$g_0 = m_j \quad \text{for } (j = 0,1,2, \ldots 15) \ ,$$

where, as in (14), m_j are the digits of the code after extraction of g_1, g_2, g_3 and g_4. Thus g_0 is determined by the majority decision test

$$\begin{aligned} g_0 &= 0 \ \text{if} \sum_{i=0}^{15} m_i < 8 \ , \\ &= 1 \ \text{if} \sum_{i=0}^{15} m_i > 8 \ . \end{aligned} \qquad (18)$$

For the above example three errors may be made in the code and the correct message obtains. If four errors are made, some of the message digits are indeterminate. It is of some interest to note that, for some cases of five errors in the code, the message may be extracted correctly. For example, suppose that the message was (0 0 0 0 0) and that the received code was (1 1 0 0 1 0 1 0 1 0 0 0 0 0 0 0). Clearly, the correct message will be extracted from this code by the above procedure.

As a final example of coding and decoding scheme, consider Φ_2^4. This space is generated by x_1, x_2, x_3, x_4 of (16) and I, as well as the quadratic variables $x_1x_2, x_1x_3, x_1x_4, x_2x_3, x_2x_4, x_3x_4$. The latter six vectors may be presented in the following tabular manner.

$$x_1x_2 = (0\ 0\ 0\ 1\ 0\ 0\ 0\ 1\ 0\ 0\ 0\ 1\ 0\ 0\ 0\ 1) \quad ,$$

$$x_1x_3 = (0\ 0\ 0\ 0\ 0\ 1\ 0\ 1\ 0\ 0\ 0\ 0\ 0\ 1\ 0\ 1) \quad ,$$

$$x_1x_4 = (0\ 0\ 0\ 0\ 0\ 0\ 0\ 0\ 0\ 1\ 0\ 1\ 0\ 1\ 0\ 1) \quad ,$$

$$x_2x_3 = (0\ 0\ 0\ 0\ 0\ 0\ 1\ 1\ 0\ 0\ 0\ 0\ 0\ 0\ 1\ 1) \quad ,$$

$$x_2x_4 = (0\ 0\ 0\ 0\ 0\ 0\ 0\ 0\ 0\ 0\ 1\ 1\ 0\ 0\ 1\ 1) \quad ,$$

$$x_3x_4 = (0\ 0\ 0\ 0\ 0\ 0\ 0\ 0\ 0\ 0\ 0\ 0\ 1\ 1\ 1\ 1) \quad . \qquad (19)$$

The messages for this example will be 11 binary digit numbers of the form $(g_0, g_1, g_2, g_3, g_4, g_{12}, g_{13}, g_{14}, g_{23}, g_{24}, g_{34})$. Each code will be sent as a vector of the form

$$g_0 \oplus g_1x_1 \oplus g_2x_2 \oplus g_3x_3 \oplus g_4x_4 \oplus g_{12}x_1x_2 \oplus g_{13}x_1x_3 \oplus g_{14}x_1x_4$$

$$\oplus\ g_{23}x_2x_3 \oplus g_{24}x_2x_4 \oplus g_{34}x_3x_4 \quad .$$

The second-degree coefficients g_{ij} of the received message are extracted first with a majority decision based on the redundancy relations illustrated in (19). Next, assuming that no indeterminacy occurred in the second-degree coefficients, the vectors $g_{ij}x_ix_j$ are added to the received code, after which we are left with a residual code from which the first-degree coefficient g_0 may be determined by test (18) after adding the vectors $g_1x_1, g_2x_2, g_3x_3, g_4x_4$ to the residual code.

This example illustrates the general principle of decoding the particular class of codes under consideration. The highest degree coefficients of a received code are extracted first; then these terms of the polynomial are subtracted out of the code, thereby leaving a residual code of the next lower degree than the original code in the special case of no errors. The operation is repeated over and over on the successive residual codes until either an indeterminacy occurs or until g_0 is extracted.

The relations of (19) illustrate the fact that there are four redundancy relations each of four variables for the second-degree coefficients g_{ij}. For example, the redundancy relations for g_{12} are

$$g_{12} = f_{4i} \oplus f_{4i+1} \oplus f_{4i+2} \oplus f_{4i+3} \quad \text{for } (i = 0,\ 1,\ 2,\ 3) \quad . \qquad (20)$$

In general, these relations will allow only one error; two errors will lead to indeterminacy. This is another example of Hamming's one-error-correction and two-error-detection codes.

It should be noted that the majority decision tests used in the above examples were, in general, overdeterminate. For instance, in the first example, if one error had been made, no more than one error would remain in the residual code after determining g_1, g_2, g_3. On the other hand, if two errors had occurred, the process of extraction would have ended before g_0 could be determined. Thus a test of only the following type would be necessary:

$$g_0 = 0 \quad \text{if } m_{i1} + m_{i2} + m_{i3} \leq 1 \quad ,$$

$$g_1 = 1 \quad \text{if } m_{i1} + m_{i2} + m_{i3} \geq 2 \quad ,$$

where i_1, i_2, i_3 are any three distinct numbers between zero and seven, inclusive. Refinements such as this, however, do not destroy the validity of the previous tests.

V. THE GENERAL DECODING PRINCIPLE

To study the general decoding scheme, illustrated by example in Section IV, it will be necessary to consider the general multinomial expansion formula (6) more carefully. Let us first define the multiple differences, used in (6) in more detail.

As in (6), $f(i_1, \dots i_m)$ is defined as

$$f(i_1,\dots i_m) = f_j \quad \text{when } j = \sum_{k=1}^{m} i_k \, 2^{k-1} \quad \text{for } (i_k = 0, 1) \quad . \tag{21}$$

The general multiple partial difference

$$\overset{p}{\underset{k_1,k_2,\dots k_p}{\Delta}} f(i_1,i_2,\dots i_m)$$

is defined inductively as

$$\underset{k}{\Delta} f(i_1,\dots i_m) = f(i_1,\dots i_{k-1}, i_k \oplus 1, i_{k+1},\dots i_m) \oplus f(i_1,\dots i_k,\dots i_m)$$

$$\overset{p}{\underset{k_1 k_2,\dots k_p}{\Delta}} f(i_1,\dots i_m) = \overset{p-1}{\underset{k_1,\dots k_{p-1}}{\Delta}} f(i_1,\dots i_{k_{p-1}}, i_{k_p} \oplus 1, i_{k_{p+1}},\dots i_m)$$

$$\oplus \overset{p-1}{\underset{k_1,\dots k_{p-1}}{\Delta}} f(i_1,\dots i_m) \tag{22}$$

With these definitions it is possible to prove by induction the validity and uniqueness of expansion (6) for any Boolean algebra of m variables, and in particular, for the Boolean algebra of 2^m dimensional vectors as described in Section II.

One evident consequence of (21) is the identity

$$f(i_1,\dots i_{k-1} i_k \oplus 1, i_{k+1},\dots i_m) = f_{i+(-1)^{i_k} 2^{k-1}} \quad . \tag{23}$$

By the use of (23) it is possible to write (22) explicitly in terms of the f_i as

$$\underset{k}{\Delta} f(i_1,\dots i_m) = f_i \oplus f_{i+(-1)^{i_k} 2^{k-1}}$$

and

$$\overset{p}{\underset{k_1,k_2,\dots k_p}{\Delta}} f(i_1,\dots i_m) = \sum_{i=1}^{2^{p-1}} f_{j_i} \oplus \sum_{i=1}^{2^{p-1}} f_{j_i + (-1)^{i_{k_p}} 2^{k_p - 1}}$$

where

$$\overset{p-1}{\underset{k_1,k_2,\dots k_{p-1}}{\Delta}} f(i_1,\dots i_m) = \sum_{i=1}^{2^{p-1}} f_{j_i} \quad \text{and} \quad j_i \neq j_s + (-1)^{i_{k_p}} 2^{k_p - 1}$$

$$\text{for } (i,s = 1,\dots 2^{p-1}) \quad . \tag{24}$$

We are now in a position to prove the following fundamental theorem on which the general decoding principle of the class of codes under consideration rests.

Theorem B:- Each highest or r-th degree coefficient of any vector or polynomial f of Φ_r^m satisfies exactly 2^{m-r} disjoint relations where each relation has precisely the form

$$\sum_{k=1}^{2^r} f_{i_k} \quad ,$$

where i_k are distinct numbers from the set $(0,1,2,\ldots 2^m - 1)$ for $(k = 1,2,\ldots 2^r)$. Disjointness of relations means that no two relations have variables f_i in common.

Proof:- Choose m and r. By (6),(7) and (24), the highest degree coefficients for an f of Φ_r^m are

$$g_{k_1\ldots k_r} = \overset{r}{\underset{k_1k_2\ldots k_r}{\Delta}} f(0,\ldots 0) = \sum_{i=1}^{2^r} f_{j_i} \quad , \tag{25}$$

where k_j are distinct integers from the set $(1,2,\ldots m)$ for $(j = 1,\ldots r)$, and j_i are distinct integers from the set $(0,1,\ldots 2^m - 1)$ for $(i = 1,2,\ldots 2^r)$. Moreover,

$$\underset{k_1\ldots k_r n_1 n_2\ldots n_t}{\Delta} f(0,\ldots 0) = 0 \tag{26}$$

for $t \geq 1$, and k_j and n_1 are distinct integers from the set $(1,2,\ldots m)$ for $(j = 1,\ldots t)$.

Let $k_1,k_2 \ldots k_r$ be a distinct set of integers from the set $(1,2,\ldots m)$. Then by (26) and (22),

$$\overset{r+1}{\underset{k_1\ldots k_r n_1}{\Delta}} f(0,\ldots 0) = \overset{r}{\underset{k_1\ldots k_r}{\Delta}} f(0,\ldots 0) \oplus \overset{r}{\underset{k_1\ldots k_r}{\Delta}} f(0,\ldots 1,\ldots 0) = 0 \tag{27}$$

where n_1 is any one of the m-r integers from the set $(1,2,\ldots m)$ which is distinct from the integers $(k_1,k_2,\ldots k_r)$. Thus, by (24) and (25), we have exhibited m-r new relations of the form required by the theorem. Each of these new relations is distinguished by the fact that the digit one appears only in the n_1-th position of the function $f(i_1,\ldots i_m)$ operated on by

$$\overset{r}{\underset{k_1\ldots k_r}{\Delta}}$$

Now define $f[n_1,n_2,\ldots n_t]$ to be $f(i_1,i_2,\ldots i_m)$ with $i_k = 1$ for $k = n_1,n_2,\ldots n_t$ and $i_k = 0$ otherwise. The theorem will be proved by induction on the subscript of n. Assume therefore that

$$\overset{r+s-1}{\underset{k_1k_2\ldots k_r n_1 n_2\ldots n_{s-1}}{\Delta}} f(0,0,\ldots 0) = \overset{r}{\underset{k_1\ldots k_r}{\Delta}} f(0,0,\ldots 0)$$

$$\oplus \overset{r}{\underset{k_1\ldots k_r}{\Delta}} f[n_1,n_2\ldots n_{s-1}] \quad . \tag{28}$$

Now, by (22) and (26) and the induction hypothesis (28),

$$\mathop{\Delta}_{k_1\ldots k_r n_1\ldots n_s}^{r+s} f(0,0,\ldots 0) = \mathop{\Delta}_{n_s}\left(\mathop{\Delta}_{k_1\ldots k_r n_1\ldots n_{s-1}}^{r+s-1} f(0,0,\ldots 0)\right)$$

$$= \mathop{\Delta}_{n_s}\left(\mathop{\Delta}_{k_1\ldots k_r}^{r} f(0,\ldots 0) \oplus \mathop{\Delta}_{k_1\ldots k_r}^{r} f\left[n_1,\ldots n_{s-1}\right]\right)$$

$$= \mathop{\Delta}_{k_1,\ldots k_r}^{r} f(0,0,\ldots 0) \oplus \mathop{\Delta}_{k_1\ldots k_r}^{r} f\left[n_1,\ldots n_{s-1}\right]$$

$$\oplus \mathop{\Delta}_{k_1\ldots k_r}^{r} f\left[n_s\right] \oplus \mathop{\Delta}_{k_1\ldots k_r}^{r} f\left[n_1,\ldots n_s\right] = 0 \quad .$$

Now, by (27) and (28), the two middle terms are equal to

$$\mathop{\Delta}_{k_1\ldots k_r}^{r} f(0,0,\ldots 0) \quad ,$$

and therefore their sum modulo 2 is zero. Hence

$$\mathop{\Delta}_{k_1\ldots n_s}^{r+s} f(0,\ldots 0) = \mathop{\Delta}_{k_1\ldots k_r}^{r} f(0,\ldots 0) \oplus \mathop{\Delta}_{k_1\ldots k_r}^{r} f\left[n_1,\ldots n_s\right] = 0 \quad ,$$

and the induction is complete. The theorem is proved when we observe that the relation

$$\mathop{\Delta}_{k_1\ldots n_s}^{r+s} f(0,\ldots 0) = 0 \text{ contributes } \binom{m-r}{s} \text{ distinct relations,}$$

$$\mathop{\Delta}_{k_1\ldots k_r} f(0,\ldots 0) = \mathop{\Delta}_{k_1\ldots k_r} f\left[n_1,n_2\ldots n_s\right] \quad ,$$

since there are $\binom{m-r}{s}$ ways of choosing s integers from m-r integers. Using all the relations (26) for the particular set $k_1\ldots k_r$ and t = 1 to t = m-r and the relation (25), we get

$$1 + \sum_{t=1}^{m-r} \binom{m-r}{t} = 2^{m-r}$$

distinct relations for $g_{k_1,k_2,\ldots k_r}$. Since these relations exhaust all variables f_{i_k}, the theorem is proved.

The above theorem shows that the generalization of the decoding principle, discussed in the last section obtains. The majority decision test for the general case can clearly be used to extract the r-th degree coefficients of Φ_r^m, where the relations used for the test are the 2^{m-r} relations of Theorem B. The (r-1)-th degree coefficients are then extracted the same way after the determined r-th order terms have been subtracted or added into the received code. This process is continued for the r-2,r-3,... degree coefficients until the message is extracted or an indeterminacy is reached.

VI. Concluding Remarks

Since there are $\binom{m}{j}$ j-th degree coefficients $g_{i_1 i_2 \ldots i_j}$ in expansion (5), there must be

$$N = \sum_{i=0}^{r} \binom{m}{i}$$

coefficients in each polynomial (7) of the coding space Φ_r^m. The coefficients of (7) constitute the message sent, thus each code of Φ_r^m contains N bits of message information. Since each element of Φ_r^m is a vector of dimension 2^m, there are 2^m-N bits of the code used to supply redundancy.

In order to illustrate the relationship of the number of message bits to number of errors corrected, consider the coding space Φ_4^7. By (29) each code of (29) has 99 bits of message information for a code of 128 bits. By Section III at least

$$2^{m-r-1} - 1 = 2^{7-4-1} - 1 = 3$$

bits of error in the code can be corrected. By Section IV and Section V four bits of error will lead undoubtedly to an indeterminacy in the message and it is likely that in some cases of five errors the correct message will be extracted by the majority decision process. Further examples of the numerical relationship of message bits to number of errors corrected may be constructed in a similar manner.

Attempts have been made with little success to investigate the structure of the complete convex set S of points, containing an element σ of Φ_r^m, whose points correspond to the element σ under the majority decision test procedure of Section V. As the second example of Section IV shows, there are in general more points in S than in a Hamming sphere of radius 2^{m-r-1} containing σ. These attempts were motivated by a desire to show that the coding system discussed here would satisfy Shannon's fundamental theorem for a discrete channel with noise (Theorem 11 in Ref. 3). So far, this fact has not been shown.

There are two generalizations of the codes discussed in this paper. In Ref. 2 Muller discusses generalizations of the binary codes, discussed here, for lengths other than 2^m. Another generalization is possible where the polynomials considered here are considered over a field of characteristic other than two; i.e., ternary codes, etc. It will not be the purpose of this paper to investigate these generalizations.

ACKNOWLEDGMENTS

The author expresses his appreciation to E. B. Rawson for his assistance in the construction of the second example of Section 4; to G. P. Dinneen for his help in the simplification of Theorem B; and to T. A. Kalin, W. B. Davenport, D. E. Muller, and O. G. Selfridge for several useful discussions.

REFERENCES

1. R. W. Hamming, Bell System Tech. J. 26, No. 2, 147 (April 1950).

2. D. E. Muller, "Metric Properties of Boolean Algebra and Their Application to Switching Circuits," Report No. 46, Digital Computer Laboratory, Univ. of Illinois (April 1953).

3. D. E. Shannon, "A Mathematical Theory of Communication," Bell System Tech. J. 27, (July,October 1948).

ERROR-FREE CODING*

Peter Elias

Department of Electrical Engineering and Research Laboratory of Electronics
Massachusetts Institute of Technology
Cambridge, Massachusetts

Introduction

This paper describes constructive procedures for encoding messages to be sent over noisy channels so that they may be decoded with an arbitrarily low error rate. The procedures are a kind of iteration of simple error-correcting codes such as those of Hamming[1] and Golay[2]; any additional systematic codes which may be discovered, such as those discussed by Reed[3] and Muller[4], may be iterated in the same way.

The procedures are not ideal; that is, the capacity of a noisy channel for the transmission of error-free information using such coding is smaller than information theory says it should be. However, the procedures do permit the transmission of error-free information at a positive rate. They also have these two properties.

(1) The codes are "systematic" in Hamming's sense: they are what Golay calls "digit codes" rather than "message codes." That is, the transmitted symbols are divided into so-called "information digits" and "check digits." The customer who has a message to send supplies the information digits which are transmitted unchanged. Periodically the coder at the transmitter computes some check digits, which are functions of past information digits, and transmits them. The customer with a short message does not have to wait for a long block of symbols to accumulate before coding can proceed, as in the case of codebook coding, nor does the coder need a codebook memory containing all possible symbol sequences. The coder needs only a memory of the past information digits it has transmitted and a quite simple computer.

(2) The error probability of the received messages is as low as the receiver cares to make it. If the coding process has been properly selected for a given noisy channel, the customer at the receiver can set the probability of error per decoded symbol (or the probability of error for the entire sequence of decoded symbols transmitted up to the present, or the equivocation of part or all of the decoded symbol sequence) at as low a value as he chooses. It will cost him more delay to get a more reliable message, but it will not be necessary to alter the coding and decoding procedure when he raises his standards, nor will it be necessary for less particular and more impatient customers using the same channel to put up with the additional delay. This is again unlike codebook processes, in which the codebook must be rewritten for all customers if any one of them raises his standards.

Perhaps the simplest way to indicate the basic behavior of such codes is to describe how one would work in a commercial telegraph system. A customer entering the telegraph office presents a sequence of symbols which are sent out immediately over a noisy channel to another office, which immediately reproduces the sequence, adds a note "the probability of error per symbol is 10^{-1}, but wait till tomorrow," and sends it off to the recipient. Next day the recipient receives a note saying "For 'sex' read 'six'. The probability of error per symbol is now 10^{-2}, but wait till next week." A week later the recipient gets another note: "For 'lather' read 'gather'. The probability of error per symbol is now 10^{-4}, but wait till next April." This flow of notes continues, the error probability dropping rapidly from note to note, until the recipient gets tired of the whole business and tells the telegraph company to stop bothering him.

Since these coding procedures are derived by an iteration of simple error-correcting and detecting codes, their performance depends on what kind of code is iterated. For a binary channel with a small and symmetric error probability, the best choice among the available procedures is the Hamming-Golay single-error-correction double-error-detection code developed by Hamming[1] for the binary case and extended by Golay[2] to the case of symbols selected from an alphabet of M different symbols, where M is any prime number. The analysis of the binary case will be presented in some detail and will be followed by some notes on diverse modifications and generalizations.

---

*This work was supported in part by the Signal Corps; the Office of Scientific Research, Air Research and Development Command; and the Office of Naval Research.

Reprinted from *IRE Trans. Inform. Theory,* vol. PGIT-4, pp. 29–37, Sept. 1954.

Iterated Hamming Codes

First-Order Check

Consider a noisy binary channel, which transmits each second either a zero or a one, with a probability $(1 - p_o)$ that the symbol will be received as transmitted, and a probability p_o that it will be received in error. Error probabilities for successive symbols are assumed to be statistically independent.

Let the receiver divide the received symbol sequence into consecutive blocks, each block consisting of N_1 consecutive symbols. Because of the assumed independence of successive transmission errors, the error distribution in the blocks will be binomial: there will be a probability

$$P(o) = (1 - p_o)^{N_1}$$

that no errors have occurred in a block, and a probability P(i)

$$P(i) = \frac{N_1!}{i!(N_1 - i)!} p_o^i (1 - p_o)^{N_1 - i} \tag{1}$$

that exactly i errors have occurred.

If the expected number of errors per received block, $N_1 p_o$, is small, then the use of a Hamming error-correction code will produce an average number of errors per block, $N_1 p_1$, after error correction, which is smaller still. Thus p_1, the average probability of error per position after error correction, will be less than p_o. An exact computation of the extent of this reduction is complicated, but some inequalities are easily obtained.

The single-error-correction check digits of the Hamming code give the location of any single error within the block of N_1 digits, permitting it to be corrected. If more errors have occurred, they give a location which is usually not that of an incorrect digit, so that altering the digit in that location will usually cause one new error, and cannot cause more than one. The double-error-detection check digit tells the receiver whether an even or an odd number of errors has occurred. If an even number has occurred and an error location is indicated, the receiver does not make the indicated correction, and thus avoids what is very probably the addition of a new error.

The single-correction double-detection code, therefore, will leave error-free blocks alone, will correct single errors, will not alter the number of errors when it is even, and may increase the number by at most one when it is odd and greater than one. This gives for the expected number of errors per block after checking

$$N_1 p_1 \leqslant \sum_{\text{even } i \geqslant 2}^{\leqslant N_1} i\,P(i) + \sum_{\text{odd } i \geqslant 3}^{\leqslant N_1} (i+1)\,P(i)$$

$$\leqslant P(2) + \sum_{i=3}^{N_1} (i+1)\,P(i)$$

$$\leqslant \sum_{i=0}^{N} (i+1)\,P(i) - P(0) - 2P(1) - P(2)$$

$$\leqslant 1 + N_1 p_o - P(0) - 2P(1) - P(2). \tag{2}$$

Substituting the binomial error probabilities from (1), expanding and collecting terms, gives, for $N_1 p_o \leqslant 3$,

$$N_1 p_1 \leqslant N_1(N_1 - 1)\,p_o^2,$$

$$p_1 \leqslant (N_1 - 1)\,p_o^2 < N_1 p_o^2. \tag{3}$$

The error probability per position can therefore be reduced by making N_1 sufficiently small. The shortest code of this type requires $N_1 = 4$, and the inequality (3) suggests that a reduction will therefore not be possible if $p_o \geq 1/3$. The fault is in the equation, however, and not the code: for $N_1 = 4$ it is a simple majority-rule code which will always produce an improvement for any $p_o < 1/2$.

A Hamming single-correction double-detection code uses C of the N positions in a block for checking purposes and the remaining N - C positions for the customer's symbols, where

$$C = \left[\log_2(N-1) + 2\right]. \tag{4}$$

(Here and later, square brackets around a number denote the largest integer which is less than or equal to the number enclosed. Logarithms will be taken to the base 2 unless otherwise specified.)

Higher Order Checks

After completing the first-order check, the receiver discards the C_1 check digits, leaving only the $N_1 - C_1$ checked information digits, with the reduced error probability p_1 per position. (It can be shown that the error probability after checking is the same for all N_1 positions in the block, so that discarding the check digits does not alter the error probability per position for the information digits.) Now some of these checked digits are made use of for further checking, again with a Hamming code. The receiver divides the checked digits into blocks of N_2; the C_2 checked check digits in each block enable it, again, to correct any single error in the block, although multiple errors may be increased by one in number. In order for the checking to reduce the expected number of errors per second-order block, however, it is necessary to select the locations of the N_2 symbols in the block with some care.

The simplest choice would be to take several consecutive first-order blocks of $N_1 - C_1$ adjacent checked information digits as a second-order block, but this is guaranteed not to work. For if there are any errors at all left in this group of digits after the first-order checking, there are certainly two or more, and the second-order check cannot correct them. In order for the error probability per place after the second-order check to satisfy the analog of (3), namely,

$$p_j \leq (N_j - 1)\, p_{j-1}^2 < N_j\, p_{j-1}^2, \tag{5}$$

it is necessary for the N_2 positions included in the second-order check to have statistically independent errors after the first check has been completed. This will be true if, and only if, each position was in a <u>different</u> block of N_1 adjacent symbols for the first-order check.

The simplest way to guarantee this independence is to put each group of $N_1 \times N_2$ successive symbols in a rectangular array, checking each row of N_1 symbols by means of C_1 check digits, and then checking each column of already checked symbols by means of C_2 check digits. The procedure is illustrated in Fig. 1. The transmitter sends the $N_1 - C_1$ information digits in the first row, computes the C_1 check digits and sends them, and proceeds to the next row. This process continues down through row $N_2 - C_2$. Then the transmitter computes the C_2 check digits for each column and writes them down in the last C_2 rows. It transmits one row at a time, using the first $N_1 - C_1$ of the positions in that row for the second-order check, and the last C_1 digits in the row for a first-order check of the second-order check digits.

After the second-order check, then, the inequality (5) applies as before, and we have for p_2, the probability of error per position,

$$p_2 < N_2\, p_1^2 < N_2\, N_1^2\, p_o^4. \tag{6}$$

The N_3 digits to be checked by the third-order check may be taken from corresponding positions in each of N_3 different $N_1 \times N_2$ rectangles, the N_4 digits in a fourth-order block from corresponding positions in N_4 such collections of $N_1 \times N_2 \times N_3$ symbols each, and so on ad infinitum. At the kth stage this gives

$$p_k < N_k^{2^0} \cdot N_{k-1}^{2^1} \cdots N_{k-j}^{2^j} \cdots N_1^{2^{k-1}} \cdot p_o^{2^k}. \tag{7}$$

It is now necessary to show that not all of the channel is occupied, in the limit, with checking digits of one order or another so that some information can also get through. The fraction of symbols used for information at the first stage is $\left[1 - (C_1/N_1)\right]$. At the kth stage, it is

$$F_k = \prod_1^k \left(1 - \frac{C_j}{N_j}\right). \tag{8}$$

It is now necessary to find a sequence of N_j for which p_k approaches zero and F_k does not, as k increases without bound. A convenient sequence is

$$N_1 = 2^n$$

$$N_j = 2^{j-1} N_1 = 2^{j+n-1}. \tag{9}$$

This gives for p_k, from (7),

$$p_k < \left(N_1 \cdot 2^{k-1}\right)^{2^0} \ldots \left(N_1 \cdot 2^{k-j}\right)^{2^{j-1}} \ldots \left(N_1 \cdot 2^0\right)^{2^{k-1}} p_o^{2^k}$$

$$< \frac{1}{N_1} (2N_1 p_o)^{2^k} \cdot 2^{-(k+1)}. \tag{10}$$

The right side of (10) approaches zero as k increases, for any $N_1 p_o \leqslant 1/2$. Thus the error probability can be made to vanish in the limit. Note that the inequality gives a much weaker kind of approach to zero for the threshold value $N_1 p_o = 1/2$ than for any smaller value of errors per first-order block.

For the same sequence of N_j, a lower bound on F_∞ can be computed. From equations (8) and (4) we have

$$F_\infty = \prod_1^\infty \left(1 - \frac{C_j}{N_j}\right) = \prod_1^\infty \left(1 - \frac{\log_2 N_j + 1}{N_j}\right)$$

$$= \prod_1^\infty \left(1 - \frac{j+n}{2^{j+n-1}}\right). \tag{11}$$

Let

$$\sigma_j = \frac{C_j}{N_j}$$

$$\sigma = \sum_1^\infty \sigma_j. \tag{12}$$

Then σ_j is monotonic decreasing in j and is less than 1 for all constructable Hamming codes, that is, for $N_1 = 2^n \geqslant 4$. This makes it possible to write the following inequalities:

$$e^{-\sigma} > F_\infty > (1 - \sigma_1)^{\sigma/\sigma_1} > 1 - \sigma. \tag{13}$$

Here the last term on the right is one of the Weierstrasse inequalities for an infinite product; the other terms are useful when $\sigma > 1$, and show that for $\sigma_1 < 1$ and $\sigma < \infty$, F_∞ is strictly positive.

Evaluating σ in the present case gives

$$\sigma = \sum_{j=1}^\infty \frac{j+n}{2^{j+n-1}} = \frac{n+2}{2^{n-1}} = \frac{2 \log 4N_1}{N_1}. \tag{14}$$

At threshold, that is, at $N_1 p_o = 1/2$, this gives

$$\sigma = 4 p_o \log \frac{2}{p_o} < 4 \left\{ p_o \log \frac{1}{p_o} + (1 - p_o) \log \frac{1}{1 - p_o} \right\} = 4E \tag{15}$$

where E is the equivocation of the noisy channel. Thus for p_o small, from (13) we have

$$F_\infty > 1 - 4E. \tag{16}$$

That is, under the specified conditions ($N_1 p_o = 1/2$, $N_1 = 2^n \geqslant 4$) the number of check digits required is never more than four times the number that would be required for an ideal code, provided that an ideal code of the check-digit type exists, which is not obvious. When E is $> 1/4$, the interior inequality shows that F_∞ is still positive.

Equivocation

Feinstein[3] has shown that it is possible to find ideal codes for which not only the probability of error, but the total equivocation, vanishes in the limit as longer and longer symbol sequences are used. This property is also true for the coding processes described here. This is a very important result in the case of codebook codes, where the message becomes infinite in the limit. For the codes under discussion here, it is a less important property, since any finite message can be received without an infinite lag, and its equivocation vanishes with the error probability per position.

The total number of binary digits checked by the kth checking stage is

$$M_k = \prod_1^k N_j. \tag{17}$$

Of these $F_k M_k$ are information digits and the remainder are checks. Using the values (9) for the N_j, we have

$$M_k = N_1^k \, 2^{\frac{k(k-1)}{2}}. \tag{18}$$

The bound (10) limits the probability of error per position. Multiplying this by M_k gives a bound on the mean number of errors per M_k digits, which is also a bound on the fraction of sequences of M_k digits which are in error after checking – a gross bound, since actually any such sequence which is in error must have many errors, and not just one. Thus for Q_k, the probability that a checked group of M_k digits is in error, we have

$$Q_k < p_k M_k \leqslant \frac{1}{4} \left(\frac{N_1}{2} \right)^{k-1} (2N_1 p_o)^{2^k} \, 2^{\frac{k(k-1)}{2}}. \tag{19}$$

At threshold ($N_1 p_o = 1/2$) this inequality does not guarantee convergence, but for $N_1 p_o < 1/2$, Q_k certainly approaches zero as k increases.

The equivocation E_k per sequence of M_k terms is bounded by the value it would have if any error in a block made all possible symbol sequences equally likely at the receiver, that is,

$$E_k < Q_k \log \frac{1}{Q_k} + (1 - Q_k) \log \frac{1}{1 - Q_k} + Q_k M_k. \tag{20}$$

Again at threshold convergence is not guaranteed, but for $N_1 p_o < 1/2$, E_k, the absolute equivocation of the block, will also vanish as k increases.

Distance Properties

At the kth stage of this coding process, a sequence of M_k binary digits has been selected as a message. Because the check digit values are determined by the information digit values, there are only $2^{F_k M_k}$ possible message sequences, rather than 2^{M_k}. Any two of these possible messages will have a "distance" from one another, defined as the number of positions in which they have different binary symbols, and the smallest such distance will be 4^k for the iterated single-correction, double-detection code. This means that by using this set of codes with a codebook, any set of errors less than one-half of the minimum distance in number can be corrected by choosing as the transmitted message the message point nearest to the received sequence.

It is easy to see that for the coding procedure just described this error-correction capability will not be realized. Any set of 2^k errors which are at the corners of a k-dimensional cube in the k-dimensional rectangle of symbol positions will not be corrected by this process, since each check will merely indicate a double error which it cannot correct. By inspecting any two of the sets of check digits at once, these errors could be located, but they will not have been corrected by the process as described above. The effective minimum of the maximum number of errors which will be corrected is therefore $2^k - 1$, rather than $2^{2k-1} - 1$.

This shows a loss of error-correction capability because of the strictly sequential use of the checking information. Without going to the extreme memory requirement of codebook techniques, a portion of this loss may be recouped by not throwing the low-order check digits away but using them to recheck after higher order checking has been done. This does not increase the maximum number of errors for which correction is always guaranteed, but it does reduce the average error probability at each stage; the exact amount of this reduction is, unfortunately, difficult to compute. This behavior, however, points up a significant feature of the coding process. If the maximum number of errors for which correction is always guaranteed were the maximum number of errors for which correction was ever guaranteed, the procedure could not transmit information at a nonzero rate; that is, the minimum distance properties of the code are inadequate for the job. It is average error-correction capability that makes transmission at a nonzero rate possible.

The Poisson Limit

Much of the above analysis has assumed that $N_j = 2^{j-1} N_1$, and part of it has further assumed that $N_1 = 2^n$. However, any series of N_j which increases rapidly enough so that σ is finite will lead to a coding process that is error-free for sufficiently small values of $N_1 p_o$. In particular, any other approximately geometric series may be used, for which

$$N_j \approx b^{j-1} N_1, \quad b > 1. \tag{21}$$

The approximation is necessary if b is not an integer. The expression for p_k analogous to (10) is then

$$p_k < \frac{1}{N_1} (bN_1 p_o)^{2^k} b^{-(k+1)}, \tag{22}$$

with a threshold at $N_1 p_o = 1/b$. The value of σ can also be bounded for this series. At threshold, the bound corresponding to (15) is

$$\sigma \leqslant \frac{b^2 p_o}{b-1} \log\left(\frac{4}{p_o\, b^{\frac{b-2}{b-1}}}\right). \tag{23}$$

Again, for $N_1 p_o$ below threshold, Q_k and E_k approach zero as k increases.

For very small p_o, the value of b that minimizes σ is b = 2. This leads to the maximum value of F_∞ given by (16). However, for very small p_o, N_1 may be made very large. The distribution of errors in the blocks then approaches the Poisson distribution, for which the probability that just i errors have occurred in a block is

$$P(i) = e^{-N_1 p_o} \cdot \frac{(N_1 p_o)^i}{i!}. \tag{24}$$

This equation may be used to derive an iterative inequality on the mean number of errors per block after single-detection, double-correction coding.

$$\begin{aligned} N_j p_j &\leqslant 1 + N_j p_{j-1} - e^{-N_j p_{j-1}} \left\{1 + 2N_j p_{j-1} + \frac{(N_j p_{j-1})^2}{2!} + \frac{(N_j p_{j-1})^4}{4!} + \dots\right\} \\ &\leqslant 1 - N_j p_{j-1}\left(2e^{-N_j p_{j-1}} - 1\right) - \frac{1}{2}\left(1 - e^{-2N_j p_{j-1}}\right). \end{aligned} \tag{25}$$

Keeping $N_j p_j$ constant gives the geometric series (21) for N_j. A joint selection of $N_1 p_o$ and b for the minimization of the bound on σ gives $N_1 p_o \approx 0.75$, $b \approx 1.75$, and an effective channel capacity

$$F_\infty \sim 1 - 3.11E, \tag{26}$$

where E is the equivocation of the binary channel. This is an improvement over (16).

Iteration of Other Codes

The analysis in the preceding sections has dealt only with iteration of the Hamming single-error-correction, double-error-detection code. Other kinds of codes may also be iterated; nor is it necessary to use the same type of code at each stage in the iterative process. The only requirement is that each code be of the check-digit, or systematic, type, so that its check digits may be computed on the basis of the preceding information digits and added on to the message.

First, the final parity check digit of a Hamming code may be omitted, destroying the double-detection feature of the code. This leads to the inequality

$$p_j \leqslant \frac{3}{2}(N_j - 1)\, p_{j-1}^2 < \frac{3}{2} N_j\, p_{j-1}^2, \tag{27}$$

in place of (5). Iterating this code alone gives a bound on σ that is only slightly smaller than (15), but the threshold becomes $N_1 p_o = 1/3$ rather than $N_1 p_o = 1/2$, and the effective channel capacity for small p_o is bounded by

$$F_\infty > 1 - 6E, \tag{28}$$

where E is the equivocation of the binary channel.

Second, the Golay[2] analogs to both kinds of Hamming code may be constructed, for M-ary channels, where M is a prime number. If there is a probability $(1 - p_o)$ that any symbol will be received correctly, and if the consecutive errors are statistically independent, the results of the binary case carry over quite directly. The inequalities (5) and (27) still hold for the two kinds of codes, since the errors as a whole are still binomially distributed in blocks. At threshold, inequalities (16) and (28) still hold for the effective channel capacity, where E is now the equivocation of a symmetrical M-ary channel; that is, of a channel in which the probability of an error taking any given symbol into any other different symbol is $p_o/(M-1)$. The result (26) for the Poisson limit also applies, with the same interpretation of E.

Third, the Reed[4]-Muller[5] codes may be treated as check digit codes, and may be iterated to give an error-proof system. For these codes, the average error-reduction capability is not known; only the minimum distance is known. Certain of the codes, such as the triple-correction quadrupole-detection code for blocks of 32 binary symbols, might provide a good starting point for an iteration which proceeds by iteration of Hamming codes. The Golay triple-correction quadruple-detection code for blocks of 24 symbols might be used in the same way. It will take considerable computation to evaluate such mixed iteration schemes.

It is not, at present, profitable to use the Reed-Muller codes for later stages in the iteration. The reason is that an efficient triple-correction quadruple-detection code should require about $C = 2 \log N$ check digits for a block of length N. The Reed-Muller codes require about $C = 1 + \log N + 1/2 \log N (\log N - 1)$ check digits for this purpose. For large N, therefore, the effective channel capacity is reduced by the large number of check digits required. There is a similar inefficiency in the Reed-Muller codes with greater error-correction capabilities, which might be removed if the average error-correction capabilities of these codes were known.

Nonrectangular Iteration

The problem of assuring statistical independence among the N_k digits checked by a kth order check, so that the inequality (5) derived on the basis of statistical independence can be used as an iterative inequality, was solved above by what might be called rectangular iteration. Each of the N_k digit positions in a check group are selected from a different sequence of M_{k-1} consecutive symbols. Thus until the kth order checking has been carried out, no two of them have been associated by lower order checking procedures in any way. This iteration solves the problem, but it makes M_k a function that grows very rapidly with k. When N_j is the geometric series (21), then

$$M_k \approx N_1^k\, b^{\frac{1}{2}k(k-1)} \tag{29}$$

This means that p_k, Q_k, and E_k decrease quite rapidly as functions of k, but much more slowly as functions of the length of the message M_k, or its information content $F_k M_k$.

Roughly speaking, if $H_k = F_k M_k$ is the total number of information digits transmitted at the kth stage,

$$p_k \sim A e^{-2^{a(\log H_k)^{1/2}}}, \quad A > 0, \quad a > 0. \tag{30}$$

This is a much slower decrease of error probability than Feinstein's result[3] which is

$$p_k \sim B e^{-b \cdot 2^{\log H_k}} = B e^{-b H_k}, \quad B > 0, \ b > 0. \tag{31}$$

A less stringent requirement on the choice of digits checked in a single group is that no two of them have been together in any lower order check group. This requires that there be at least N_k different groups of order k - 1 from which to select digit positions. Thus

$$M_k \geqslant N_k N_{k-1}. \tag{32}$$

If it is possible to approximate equality in (32), and if the statistical dependence so introduced does not seriously weaken inequality (5), then it might be possible to get the result

$$p_k \sim D e^{-2^{\dot{d} \log H_k}}, \quad D > 0, \ d > 0, \tag{33}$$

which is closer to Feinstein's result.

Conclusion

From a practical point of view, this coding procedure has much to recommend it. A question of both theoretical and practical interest is the extent to which the convenience associated with a computable and error-free code is compatible with ideal coding, or the smallest price that must be paid for the convenience if the two are incompatible. No answer to this question is in sight at present. However, the existence of the error-free process, despite its lack of ideality, puts the burden of efficient coding on the first stage of the coding process. For if a coding process succeeds in reducing the equivocation in a received message to some small but positive value E, the remaining errors may always be eliminated at a cost of 4 E (or 3.11 E) in channel capacity: an error-proof termination is available, at a price, to take care of the residual errors left by any other error-correcting scheme.

Acknowledgment

The iterative approach used in this paper was suggested by a comment of Dr. Victor H. Yngve, of the Research Laboratory of Electronics, M.I.T., on the fact that redundancy in language was added at many different levels, a point that he discusses in reference 6.

References

(1) R. W. Hamming, Error Detecting and Error Correcting Codes, Bell System Tech. J. 29, pp. 147-160 (1950).

(2) M. J. E. Golay, Notes on Digital Coding, Proc. I.R.E. 37, p. 657 (1949).

(3) A. Feinstein, Some New Basic Results in Information Theory, these transactions.

(4) I. S. Reed, A Class of Multiple-Error-Correcting Codes and the Decoding Scheme, Technical Report No. 44, Lincoln Laboratory, M.I.T., (1953). See also the paper under this title in these transactions.

(5) D. E. Muller, "Metric Properties of Boolean Algebra and their Applications to Switching Circuits," Report No. 46, Digital Computer Laboratory, University of Illinois (1953).

(6) V. H. Yngve, "Language as an Error-Correcting Code," pp. 73-74, Quarterly Progress Report. Research Laboratory of Electronics, M.I.T., April 15, 1954.

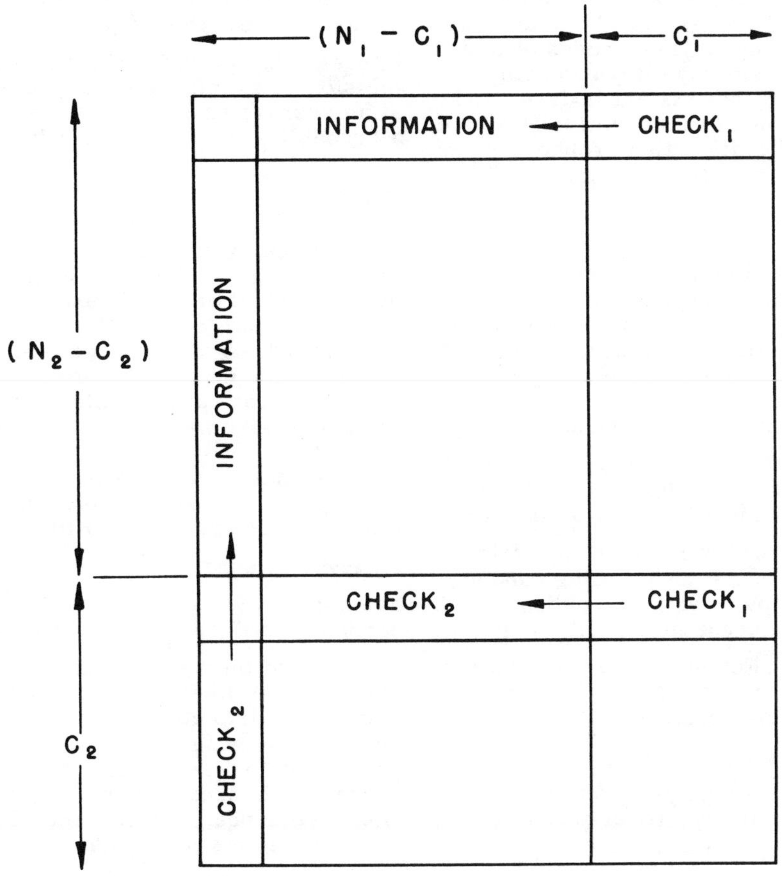

Fig. 1 – Organization of First- and Second-Order Check Digits.

CODING FOR NOISY CHANNELS*

Peter Elias
Department of Electrical Engineering and Research Laboratory of Electronics
Massachusetts Institute of Technology
Cambridge, Massachusetts

Summary: Shannon's and Feinstein's versions of the channel capacity theorem, specialized to the binary symmetric channel, are presented. A much stronger version is proved for this channel. It is shown that the error probability as a function of delay is bounded above and below by exponentials, whose exponents agree for a considerable range of values of the channel and the code parameters. In this range the average behavior of all codes is essentially optimum, but for small transmission rates this is not true. The results of this analysis are shown to apply to check-symbol codes of four kinds which have progressively simpler coding procedures. The last of these is error-free, and makes it possible to transmit information at a rate equal to the channel capacity with a probability one that no decoded symbol will be in error.

Introduction

Since Shannon[1,2] showed that information could be transmitted over a noisy channel at a positive rate with an arbitrarily low probability of error at the receiver, there has been considerable interest in constructing specific transmission schemes that exhibit such behavior.

For a signal transmitted over a channel perturbed by Gaussian noise, Golay[3] and Fano[4] found schemes which in the limit had the desired behavior, but it was a limit of infinite bandwidth or vanishing transmission rate. Rice[5] investigated the characteristics of transmission using randomly selected noise waveforms, and got an indication of exponential decrease in error probability with increasing time delay. Feinstein[6] showed that the same sort of behavior, at least as an upper bound, held true for more general channels.

For the binary channel, Hamming[7], Gilbert[8], Plotkin[9], and Golay[10] investigated a variety of codes, and found some basic properties of the binary symmetric channel. Laemmel[11], Muller[12], and Reed[13] also constructed specific codes and classes of codes. The first constructive code for transmission at a nonzero rate over a noisy binary channel was discovered recently by the author[14]. The investigation reported in the present paper started as a continuation of that work, and an investigation of the rate at which the error probability decreased with delay originally developed from a comparison of check-symbol codes with codes of less restricted types. It seems more sensible to present the results in reverse order. After a definition of the channel and general coding procedures, Shannon's and Feinstein's channel capacity theorems are stated, and a stronger theorem is given for the binary symmetric channel, which shows in considerable detail the behavior of error probability at the receiver as a function of the parameters of the channel and the code, and the delay time. It is then shown that most of these results carry over to a variety of kinds of check-symbol codes. One of these, of primarily academic interest, is error-free[14], and permits the transmission of an infinite sequence of message symbols at an average rate equal to the channel capacity with a probablity one that no decoded digit is in error.

The Channel

The coding problem we will discuss is illustrated in Fig. 1. The problem is to match the output of an ideal binary message source to a binary symmetric noisy channel.

The message source generates a sequence of binary symbols, say the binary digits zero and one. Zeros and ones are selected with equal probability, and successive selections are statistically independent.

The channel accepts binary symbols as an input and produces binary symbols as an output. Each input symbol has a probability $p_o < 1/2$ of being received in error, and a probability $q_o = 1 - p_o$ of being received as transmitted. The transmission error probability p_o is a constant, independent of the value of the symbol being transmitted: the channel is as likely to turn a one into a zero as to turn a zero into a one. The channel, in effect, adds a noise sequence to the input sequence to produce the output sequence; the noise is a random sequence of zeros and ones, synchronous with the signal sequence, in which the ones have probability p_o, and the addition is addition modulo two of each signal digit to the corresponding noise digit ($1+1 = 0+0 = 0$, $0+1 = 1+0 = 1$).

If the message source were connected directly to the channel, a fraction p_o of the received symbols would be in error. A coding procedure for reducing the effect of the errors is shown in Figs. 1 and 2. The output of the message source is segmented into consecutive blocks of length M. There are 2^M such blocks, and they are selected by the source with equal probability. To each input block of M binary symbols is assigned an output block of N binary symbols, N > M.

The input sequences of length M are the messages to be sent; the output sequences of length N are the transmitted signals, and the correspondence between input and output blocks is the code used. The use of the word "code" is justified

*This work was supported in part by the Signal Corps; the Office of Scientific Research, Air Research and Development Command; and the Office of Naval Research.

Reprinted from *IRE Conv. Rec.*, vol. 3, pt. 4, pp. 37–46, 1955.

by Fig. 2, where the correspondence between input and output blocks is given in the form of a codebook. On the left is a column of the 2^M possible messages, listed as M-digit binary numbers in numerical order. Following each message is the N-digit binary number which is the corresponding signal, so that the codebook has 2^M entries, each of which lists a message and the corresponding signal.

The system in operation is shown in Fig. 1. The source selects a message that is coded into a transmitted signal and sent over the noisy channel. The received block of N -- the received, or noisy, signal -- differs from the transmitted signal in about $p_o N$ of its N symbol values. The decoder receives this noisy signal and reproduces one of the 2^M possible messages, with an average probability P_e of making an incorrect choice.

The most general type of decoder is shown in Fig. 3. It is a codebook with 2^N entries, one for each of the possible received signals. The left column is the received sequence, arranged as an N-digit binary number in numerical order. This is followed by the M-symbol message block that will be reproduced when that sequence is received as a noisy signal.

In order to minimize P_e, the codebook must be so constructed that the message that is selected when a given noisy signal is received is the one corresponding to the signal most likely to have been transmitted. For the binary symmetric channel, the signal most likely to have been transmitted is the one that differs from the received signal in the smallest number of symbol positions. This follows from the fact that a particular group of k errors has probability $p_o^k q_o^{N-k}$ of being introduced by the channel; this probability decreases as k increases, for $p_o < 1/2$.

For this channel, the codebook may be simplified. In fact, the transmitter codebook may be used in reverse order. The noisy signal is compared with each of the possible transmitted signals, and the number of positions in which they differ is counted. The signal with the lowest count is assumed to have been transmitted, and the corresponding message block is reproduced as the best guess at the transmitted message. This guess may, of course, be incorrect, and will be if the noise has altered more than half of the positions in which the transmitted signal differs from some other listed signal.

This decoding procedure may be described in a geometrical language introduced by Hamming[7]. Each signal is taken as a point or a vector in an N-dimensional space, with coordinates equal to the values (zero or one) of its N binary symbols. The distance between two points is defined as the number of coordinates in which they differ. In this language, the noisy signal is decoded as the nearest of the signal points, and the corresponding message is chosen.

For given M and N, the error probability P_e depends on the set of points that are used as signals. If these are clustered in a small part of the space, P_e will be large; if they are far from one another, P_e will be small. As specialized to this channel, Shannon's second coding theorem states an asymptotic relationship between M, N, and P_e for a suitable selection of signal points.

Channel Capacity and Error Probability

First, some definitions are required. Given a binary symmetric channel with transmission error probability p_o and $q_o = 1 - p_o$, the equivocation $E_o = E(p_o)$ and the capacity $C_o = C(p_o)$ of the channel, both measured in bits per symbol, are given by

$$E_o = -p_o \log p_o - q_o \log q_o \qquad C_o = 1 - E_o \tag{1}$$

(Here and later, all logarithms are to the base two.)

Given a coding procedure like that illustrated by Figs. 1, 2, and 3, the redundancy E_1 and the transmission rate C_1, also in bits per symbols, are given by

$$E_1 = \frac{N - M}{N} \qquad C_1 = 1 - E_1 = \frac{M}{N} \tag{2}$$

It is convenient to introduce the probability p_1 which is the upper bound of the transmission error probabilities for which this particular code can be expected to work, and $q_1 = 1 - p_1$. These are uniquely defined by

$$p_1 < \frac{1}{2} \qquad E_1 = E(p_1) = -p_1 \log p_1 - q_1 \log q_1 \tag{3}$$

since a plot of E(p) or C(p) is monotonic for $0 \leq p \leq 1/2$.

Finally, the average probability of an error in decoding, which was written as P_e above, will in general be a function of the block length N, the channel capacity C_o or error probability p_o, and the transmission rate C_1 or the probability p_1. It will be written as $P_e(N, p_o, p_1)$.

Shannon's second coding theorem[1], as applied to this channel, follows.

Theorem 1. Given any fixed $C_1 < C_o$ and any fixed $\epsilon > 0$, for all sufficiently large N there are codes which will transmit information at the rate C_1 bits per symbol and will decode it with an error probability per block of N, $P_e(N, p_o, p_1) < \epsilon$. This cannot be done for $C_1 > C_o$.

Shannon's proof of the theorem proves more than the theorem states. A code is a selection of

2^{NC_1} signal sequences from among 2^N possibilities. Including those codes that select the same signal two or more times to represent several different messages, there are $2^{N \cdot 2^{NC_1}}$ different codes. Each of these will have an average decoding error probability (averaged over the different messages, with equal weights). Shannon shows that the average of all of these (averaged over the different codes, with equal weights) is less than ϵ. Since the error probability for each code is positive, it follows that at least one code has an average error probability less than ϵ; and it also follows, as Shannon remarks, that, at most, a fraction f of the codes can have an average error probability as great as ϵ/f, so that almost all of the codes have arbitrarily small error probability; that is, almost all codes are "good" codes, although some "bad" codes do exist. By the same argument, in any one good code the error probability for most of the individual messages is less than ϵ/f, so that by discarding a few of the signal sequences and transmitting at a very slightly slower rate, any good code can be made into a uniformly good code. This result has considerable practical importance, since a uniformly good code will transmit with the specified small error probability, regardless of the probabilities with which message sequences are selected, and there are many information sources whose statistics are not known in detail.

The major question left open by this theorem is how large N must be for given p_o, p_1, and ϵ. Feinstein[6] has proved a stronger version which provides an upper bound for $P_e(N, p_o, p_1)$. As specialized to the binary symmetric channel it may be written as:

Theorem 2. Given any $C_1 < C_o$, an $\epsilon(p_o, p_1) > 0$ can be found. For any sufficiently large N, a code may be constructed which will transmit information at the rate C_1 bits per symbol which can be decoded with $P_e(N, p_o, p_1) < 2^{-\epsilon N}$.

Feinstein's proof consists of the construction of a code that satisfies the requirements of the theorem and is uniform in the sense that all signals are good signals. Some indication of the relation of ϵ to the channel and code parameters is also given.

The next theorem is much stronger than this, but unlike Shannon's and Feinstein's it does not apply to the general discrete noisy channel without memory, but only to the binary symmetric case. Some more definitions are needed. It turns out that the error probability P_e is bounded not only above but below by exponentials in N, and that for a considerable range of channel and code parameters the exponents of the two bounds agree. The error exponent for the best possible code is defined as

$$\alpha_{opt}(N, p_o, p_1) = \frac{-\log P_e(N, p_o, p_1)}{N} \tag{4}$$

and $\alpha_{avg}(N, p_o, p_1)$ is defined as the same function of the average of the error probabilities of all codes.

An additional probability value is also needed, along with the values of α, C, and E which go with it:

$$p_{crit} = \frac{p^{1/2}}{p^{1/2} + q^{1/2}}, \quad q_{crit} = 1 - p_{crit}$$
$$E_{crit} = E(p_{crit}), \quad C_{crit} = 1 - E_{crit} \tag{5}$$
$$\alpha_{crit} = \lim_{N\to\infty} \alpha_{opt}(N, p_o, p_{crit})$$

Finally, the margin in error probability and the margin in channel capacity need labeling:

$$\delta = p_1 - p_o \tag{6}$$
$$\Delta = C_o - C_1$$

For a binary symmetric channel with capacity C_o and transmission rate C_1, the following statements hold.

Theorem 3. (a) For $p_o < p_1 < p_{crit}$, $C_o > C_1 > C_{crit}$, the average code is essentially as good as the best code:

$$\alpha(p_o, p_1) = \lim_{N\to\infty} \alpha_{opt}(N, p_o, p_1) = \lim_{N\to\infty} \alpha_{avg}(N, p_o, p_1)$$
$$= -\Delta - \delta \log \frac{p_o}{q_o} \tag{7}$$

(b) For $p_{crit} < p_1 < 1/2$, the average code is not necessarily optimum; for p_1 near 1/2 it is certainly not. Specifically,

$$\alpha_{avg}(p_o, p_1) = \lim_{N\to\infty} \alpha_{avg}(N, p_o, p_1)$$
$$= \alpha_{crit} + C_{crit} - C_1 \tag{8}$$

where α_{crit} is the $\alpha(p_o, p_1)$ of Eq. (5) with $p_1 = p_{crit}$, while for α_{opt} there are two upper and two lower bounds:

$$\liminf \alpha_{opt}(N, p_o, p_1) \geq \begin{cases} \alpha_{crit} + C_{crit} - C_1 \\ \frac{p_1}{2} \log \frac{1}{4pq} - C_1 \end{cases} \tag{9}$$

$$\limsup \alpha_{opt}(N, p_o, p_1) \leq \begin{cases} -\Delta - \delta \log \frac{p_o}{q_o} \\ \frac{E_1}{4} \log \frac{1}{4pq} \end{cases} \tag{10}$$

As $p_1 \to 1/2$, the second bound in (9) approaches the second bound in (10);

$$\lim_{N\to\infty} \lim_{p_1 \to 1/2} \alpha_{opt}(N, p_o, p_1) = \frac{1}{4} \log \frac{1}{4pq} \quad (11)$$

which is always greater than

$$\alpha_{avg}\left(p_o, \frac{1}{2}\right) = \alpha_{crit} + C_{crit} \quad (12)$$

The content of this theorem is illustrated by Fig. 4. This is a plot of the channel capacity $C(p)$ vs. transmission error probability p for a binary symmetric channel. A dashed line is drawn tangent to the curve at the point given by the channel parameters p_o, C_o. This tangent line has the slope $\log (p_o/q_o)$. The critical point p_{crit}, C_{crit} is the point at which the slope of the curve is $(1/2) \log (p_o/q_o)$. For $p_o < p_1 < p_{crit}$, the $\alpha(p_o, p_1)$ of (7), which is both the average and the optimum error exponent, is the length of a vertical dropped from the channel capacity curve to the tangent line at the ordinate p_1.

At $p_1 = p_{crit}$, the dotted line that determines $\alpha_{avg}(p_o, p_1)$ diverges from the tangent line. For $p_{crit} < p_1 < 1/2$ the exact value of $\alpha_{opt}(N, p_o, p_1)$ is not known, but is given by the length of a vertical at ordinate p_1, dropped from the channel capacity curve and terminating in the shaded region. The upper and lower bounds of this region provide lower and upper bounds, respectively, on the value of α_{opt}. These bounds converge to $(1/4) \log (1/4pq)$ at $p_1 = 1/2$, and near this point α_{opt} is definitely $> \alpha_{avg}$.

The value of α given by the tangent line at $p_1 = 1/2$, although not approached for the transmission of information at any nonzero rate, is the correct value of α_{opt} for transmission of one bit per block of N symbols.

An outline of the proof of Theorem 3 appears in the Appendix. A more detailed presentation, giving bounds on $P_e(N, p_o, p_1)$, as well as on α, will appear elsewhere.

Check-Symbol Codes

The preceding three theorems are interesting in theory but discouraging in practice. They imply that a good code will require a transmitting codebook containing $N \cdot 2^{NC_1}$ binary digits in all, and either a receiver codebook containing $N \cdot 2^N$ binary digits or another copy of the transmitter codebook and 2^{NC_1} comparisons of the received signal with the possible transmitted signals. Since in interesting cases NC_1 may be of the order of 100, the requirements in time and space are unmanageable. Furthermore, it would be quite consistent with these theorems if no code with any simplicity or symmetry properties were a good code.

The theorems that follow show that this is fortunately not the case. Four kinds of codes of increasing simplicity and convenience from the point of view of realization are demonstrated to have essentially the same behavior, from both a channel capacity and an error probability point of view, as the optimum code. The last of the four is of theoretical interest as well, since it permits the receiver to set the decoding error probability arbitrarily low without consulting the transmitter.

A check-symbol code of block length N is a code in which the 2^{NC_1} signal sequences have in their initial NC_1 positions all 2^{NC_1} possible combinations of symbol values. The first NC_1 positions will be called information positions and the last NE_1 will be called check positions. The signal corresponding to a message sequence is that one of the signal sequences whose initial symbols are the message.

A parity check-symbol (pcs) code is a check-symbol code in which the check positions are filled in with digits each of which completes a parity check of some of the information positions. Such codes were discussed in detail first by Hamming[1], who calls them systematic codes. A pcs code is specified by an $NC_1 \times NE_1$ matrix of zeros and ones, the ones in a row giving the locations of the information symbols whose sum modulo two is the check digit corresponding to that row. The process is illustrated in Fig. 5. Such a code requires $NC_1 \times NE_1 = N^2 C_1 E_1 \leq \frac{1}{4} N^2$ binary digits in its codebook, these being the digits in the check-sum matrix.

A sliding pcs code is defined as a pcs code in which the check-sum matrix is constructed from a sequence of N binary symbols by using the first NC_1 of them for the first row, the second to $(NC_1 + 1)$st for the second row ..., the NE_1th to the Nth for the NE_1th row. This code requires only an N-binary-digit codebook.

Finally a convolutional pcs code is defined as one in which check symbols are interspersed with information symbols, and the check symbols check a fixed pattern of the preceding NC_1 information positions if $C_1 \geq 1/2$; if $E_1 > 1/2$, the information symbols add a fixed pattern of zeros and ones to the succeeding NE_1 check positions. Such a code requires $\max(NC_1, NE_1) \leq N$ binary digits in its codebook. It is illustrated by Fig. 6.

<u>Theorem 4</u>. All the results of Theorem 3 apply to check-symbol codes and to pcs codes. The results of part (a) of that theorem apply to sliding pcs codes.

In reading Theorem 3 into Theorem 4, the average involved in α_{avg} is the average of all codes of the appropriate type; that is, all combinations of check symbols for the check-symbol codes, all check-sum matrices for the pcs codes, all sequences of N binary digits for the sliding pcs code.

<u>Theorem 5</u>. The results of part (a) of

Theorem 3 apply to convolutional pcs codes, if $P_e(N, p_o, p_1)$ is interpreted as the error probability per decoded symbol. For infinite memory (each check symbol checking a set of prior information symbols extending back to the start of transmission over the channel) the N in $P_e(N, p_o, p_1)$ for a particular decoded information symbol is the number of symbols which have been received since it was received.

This theorem shows that error-free coding can be attained at no loss either in channel capacity or in error probability, a question raised by the author when the first error-free code was introduced[14]. By waiting long enough, the receiver can obtain as low a probability of error per digit as is desired, without a change of code being necessary. By gradually reducing the ratio of check to information symbols toward E_o/C_o, using the law of the iterated logarithm for binary sequences, it can be shown that in an infinite sequence of message digits transmission is obtained at average rate C_o with probability one of no errors in the decoded message.

Conclusion

An appreciable gain in simplicity has been achieved in going from an arbitrary average code to a convolutional or sliding pcs code. It is possible to encode and decode either of these codes with a codebook of only N or fewer binary digits. However, the decoding operation will require 2^{NC_1} or 2^{NE_1} (whichever is smaller) comparisons, which will take a great deal of time in interesting cases. No decoding procedure that replaces this operation by a small amount of computing has yet been discovered, although the iterated Hamming code, which is error-free[14], gives hope that it may be possible to manage this while still keeping optimum behavior in terms of channel capacity and error probability -- a feature which the iterated Hamming code lacks.

Acknowledgments

After the analytical work reported in this paper was done, but before it had been organized for presentation, I discovered that Dr. Shannon was also working on the problem of error probability, and was to present his results at the same meeting. In discussing the results with Dr. Shannon, he mentioned the geometric relationship between the tangent line and the capacity curve, illustrated in Fig. 4, in the region $p_1 < p_{crit}$. I do not know whether this would have occurred to me in organizing my results, but I do know that it is vital; the information to the right of p_{crit} is my own, but is impossible to present in any other fashion without getting lost in numbers of families of curves.

It is a pleasure to acknowledge my indebtedness to the atmosphere at the Research Laboratory of Electronics, without which this work would not have gotten started; and to my colleagues, Professors Fano, Huffman, and Yngve, who provided that part of the atmosphere most relevent to this specific project.

Appendix

1. Outline Proof of Theorem 3

Using the symbols and definitions of Eqs. (1), (3), (5), and (6), let $k_1 = Np_1$ be an integer. Define $V_N(k)$, the volume of an N-dimensional sphere of radius k, by

$$V_N(k) = \sum_{j=0}^{k} \binom{N}{j} = \sum_{j=0}^{k} \frac{N!}{j!\,(N-j)!} \qquad (A.1)$$

Select $2^N/V_N(k_1)$ sequences as signaling sequences. Then the signaling rate is

$$\frac{1}{N} \log \left\{2^N/V_N(k_1)\right\} = 1 - \frac{1}{N} \log V_N(k_1) \qquad (A.2)$$

If the selection of signal sequences can be made so that every possible received sequence differs from one (and only one) signal sequence in k_1 or fewer positions, then the probability of a detection error will be just the probability $P_I(N, p_o, p_1)$ that more than k_1 out of N errors are made in transmission. This is the tail of the binomial distribution:

$$P_I(N, p_o, p_1) = \sum_{j=k_1+1}^{N} p_o^j q_o^{N-j} \binom{N}{j} \qquad (A.3)$$

Such a selection is not, in general, possible. However, $P_I(N, p_o, p_1)$ of (A.3) is a lower bound to the average decoding error probability $P_e(N, p_o, p_1)$ for any actual selection of signal points: this follows directly from the fact that $p_o^j q_o^{N-j}$ is a monotonically decreasing function of j.

The average of all possible codes is used to provide an upper bound to the decoding error probability of the best code. The average probability of a detection error, $P_{III}(N, p_o, p_1)$, is the probability $P_{II}(N, j, k_1)$ of a decoding error when just j transmission errors have occurred, averaged over the binomial distribution of j. With Eq. (A.3) this gives

$$P_{III}(N,p_o,p_1) = \sum_{j=0}^{N} P_{II}(N,j,k_1)\, p_o^j q_o^{N-j} \binom{N}{j}$$

$$\leq \sum_{j=0}^{k_1} P_{II}(N,j,k_1)\, p_o^j q_o^{N-j} \binom{N}{j} + P_I(N,p_o,p_1) \qquad (A.4)$$

The probability $P_{II}(N, j, k_1)$ of a decoding error when just j transmission errors have occurred is the probability that one of the $\{2^N/V_N(k_1)\} - 1$ incorrect signal sequences differs in j or fewer places from the received sequence. There are a total of $V_N(j)$ sequences which differ from the received sequence in as few as j positions, and the probability of missing all of them in $\{2^N/V_N(k_1)\} - 1$ tries is, for $j < k_1$, bounded by

$$\left(1 - \frac{V_N(j)}{2^N}\right)^{\{2^N/V_N(k_1)\}-1} \geq 1 - \frac{V_N(j)}{V_N(k_1)} \qquad (A.5)$$

Equation (A.5) gives the probability of no decoding error: P_{II} is the probability of a decoding error, so

$$P_{II}(N, j, k_1) \leq \frac{V_N(j)}{V_N(k_1)} \leq \frac{\binom{N}{j}}{\binom{N}{k_1}} \qquad (A.6)$$

Equations (A.4) and (A.6) give

$$P_{III}(N, p_o, p_1) \leq \sum_{j=0}^{k_1} p_o^j q_o^{N-j} \frac{\binom{N}{j}^2}{\binom{N}{k_1}} + P_I(N, p_o, p_1) \qquad (A.7)$$

Now the sums in Eqs. (A.1) and (A.3) are bounded below by the value of their largest term and above by a geometric series multiplied by that term -- the last term in Eq. (A.1), the first in Eq. (A.3). (See Feller[15], p. 126 for the bounds for Eq. (A.3).) The sum in Eq. (A.7) is similarly bounded above and below, if p_1 is less than p_{crit}, which is the condition guaranteeing that the last term in the sum is the largest. Using these results, taking logarithms, and using Stirling's approximation for the binomial coefficients gives, from (A.2),

$$\lim_{N\to\infty} \left\{1 - \frac{1}{N} \log V_N(k_1)\right\} = 1 - E_1 = C_1 \qquad (A.8)$$

from (A.3), for $p_o < p_1 < \frac{1}{2}$,

$$\lim \sup_N \alpha_{opt}(N, p_o, p_1) \leq \lim_{N\to\infty} \frac{-\log P_{II}(N, p_o, p_1)}{N}$$

$$= -\Delta - \delta \log \frac{p_o}{q_o} \qquad (A.9)$$

and from (A.7), for $p_o < p_1 < p_{crit}$,

$$\lim \inf_N \alpha_{opt}(N, p_o, p_1) \geq \lim_{N\to\infty} \frac{-\log P_{III}(N, p_o, p_1)}{N}$$

$$= -\Delta - \delta \log \frac{p_o}{q_o} \qquad (A.10)$$

Together, Eqs. (A.9) and (A.10) prove the first part of the theorem, and cover the region in which the dashed-line and the dotted curves of Fig. 4 coincide.

Since the length represented by α_{opt} in this region is the difference between the curve and its tangent, it is second-order in δ or Δ. In fact, for small δ and Δ,

$$\alpha_{opt}(p_o, p_1) \approx \frac{\delta^2}{2pq} \log e \approx \frac{\Delta^2}{2pq\left(\log \frac{p}{q}\right)^2} \log e$$

For $p_1 > p_{crit}$, the largest term in the sum in Eq. (A.7) is not the last, but is that term for which $j^2/(N-j)^2$ is most nearly equal to p_o/q_o. This term is larger than $P_I(N, p_o, p_1)$ for large N, and the sum is bounded above by k_1, the number of terms, times the largest term. Taking the limit of (1/N) multiplied by the logarithm and using Stirling's approximation gives upper and lower bounds for $\alpha_{avg}(N, p_o, p_1)$ which coincide, giving for $p_{crit} < p_1 < \frac{1}{2}$,

$$\lim \inf_N \alpha_{opt}(N, p_o, p_1) \geq \lim_{N\to\infty} \alpha_{avg}(N, p_o, q_o)$$

$$= \lim_{N\to\infty} \frac{-\log P_{III}(N, p_o, p_1)}{N}$$

$$= C_{crit} + \alpha_{crit} - C \qquad (A.11)$$

This gives the remainder of the dotted curve in Fig. 4.

For p_1 less than p_{crit}, the probability of a detection error as computed above is essentially the probability of escaping from a sphere of radius $k_1 = Np_1$. For p_1 near 1/2, a different point of view is possible and leads to the two solid curves in Fig. 4.

The probability that transmission errors will cause one transmitted signal to be decoded as another is the probability that the noise will alter half or more of the positions in which they differ. If they differ in $Np_1 = k_1$ positions, this probability is just the upper half of the binomial, $P_I\left(Np_1, p_o, \frac{1}{2}\right)$ as given in Eq. (A.3). This is the probability of a particular transition; the total error probability is certainly less than this multiplied by the number of signal sequences. Gilbert[8] has shown that it is always possible to find $2^N/V_N(k_1 - 1)$ signal sequences each of which differs in at least k_1

positions from every other. For large N, by Eq. (A.8), this corresponds to a signaling rate of C_1 bits per symbol, or 2^{NC_1} signal points. Thus

$$P_e(N, p_o, p_1) \leq \frac{2^N}{V_N(k_1 - 1)} P_I\left(Np_1, p_o, \frac{1}{2}\right) \quad (A.12)$$

and asymptotically, from Eqs. (A.8) and (A.3),

$$\lim_{N\to\infty} \frac{-\log P_e(N,p_o,p_1)}{N} = -p_1 \left\{C - 0 + \left(\frac{1}{2} - p_o\right)\log \frac{p_o}{q_o}\right\}$$

$$= p_1 \cdot \frac{1}{2}\left\{\log \frac{1}{2p_o} + \log \frac{1}{2q_o}\right\}$$

$$= \frac{p_1}{2} \log \frac{1}{4pq} \quad (A.13)$$

and

$$\liminf_N \alpha_{opt}(N,p_o,p_1) \geq -C_1 + \frac{p_1}{2} \log \frac{1}{4pq} \quad (A.14)$$

This is the upper solid curve in Fig. 4. For an upper bound to α_{opt} we use a result of Plotkin[9] which shows that there are at most 2N signal points whose mutual minimum distance is as great as N/2. This means that the transmission rate for signal points at this distance is (1 + log N)/N and approaches zero for large N. This result sets a limit to the number of signal points at smaller distances as well. As Plotkin pointed out, if B(N,k) is the number of signal points at mutual distance $\geq$k, then at least half of these agree in their first coordinate. Eliminating the n first coordinates gives

$$B(N-n, k) \geq 2^{-n} B(N, k) \quad (A.15)$$

Using Eqs. (A.14) and (A.15), let n = N - 2k. Then

$$B(N, k) \leq 2^{N-2k} B(2k, k) = 4k \cdot 2^{N-2k} \quad (A.16)$$

For a transmission rate C_1 this determines k:

$$C_1 = 1 - E_1 = \lim_{N\to\infty} \frac{\log B(N,k)}{N} \leq 1 - 2\frac{k}{N},$$

$$\text{or} \quad k \leq \frac{N}{2} E_1 \quad (A.17)$$

Now the error probability for such a set of signal points is certainly greater than the probability of a single transition, which is, in turn, at least as great as the upper half of the binomial

$$P_I\left(\frac{NE_1}{2}, p_o, \frac{1}{2}\right)$$

Thus

$$\limsup_N \alpha_{opt}(N,p_o,p_1) \leq \lim_{N\to\infty} \frac{-\log P_I\left(\frac{NE_1}{2}, p_o, \frac{1}{2}\right)}{N}$$

$$= \frac{E_1}{4} \log \frac{1}{4pq} \quad (A.18)$$

which gives the lower solid curve in Fig. 4. At $p_1 = 1/2$, Eqs. (A.18) and (A.14) give the same value, so that

$$\lim_{N\to\infty} \lim_{p_1\to 1/2} \alpha_{opt}(N, p_o, p_1) = \frac{1}{4} \log \frac{1}{4pq} \quad (A.19)$$

These results prove the remainder of the theorem. It should be noted that Eq. (A.19) does not imply that it is impossible to transmit any information with an error probability less than approximately $2^{-\frac{N}{4}\log\frac{1}{4pq}}$ for finite N. It is only impossible to do so while transmitting at a positive rate in the limit of large N. The transmission of one bit per block of N symbols can be accomplished by picking two signal sequences that differ in every position, with an error probability equal to $P_I(N, p_o, 1/2)$ for which, from Eq. (A.3),

$$\lim_{N\to\infty} \frac{-\log P_I\left(N, p_o, \frac{1}{2}\right)}{N} = \frac{1}{2} \log \frac{1}{4pq} \quad (A.20)$$

This error exponent is twice as great as that for the limit (as $p_1 \to 1/2$) of α_{opt} for positive transmission rates. Other points for the transmission of 2, 3, ... log N bits per block of N symbols fall between the value of Eq. (A.20) and that of Eq. (A.19).

2. Outline Proof of Theorems 4 and 5

A large part of the proof of Theorem 3 carries over directly for Theorems 4 and 5. Any upper bound on α_{opt} for the best possible code is automatically an upper bound for the more restricted class of check-symbol codes. Thus Eqs. (A.9) and (A.18), the tangent line and the lower solid curve in Fig. 4, still apply. To get the upper solid curve, Eq. (A.14), it is necessary to show that Gilbert's result, and thus Eq. (A.12), holds for the kind of code considered. This is obvious for pcs codes; Gilbert's proof requires only trivial modifications in this case. Since the pcs codes are a special case of check-symbol codes, the result follows for these as well. For sliding and convolutional pcs codes Gilbert's result is not obvious, although probably still true; that is why only the first part of Theorem 3 is extended to these cases.

The difficult point in Theorems 4 and 5 is the demonstration that the average of all possible codes, of each of the four types considered, is still given by Eqs. (A.10) and (A.11) and the dotted curve in Fig. 4. This requires a demonstration that the inequality of Eq. (A.6) still applies; that is, that the probability of a decoding error when j transmission errors have occurred is essentially the same, on the average, for the different types of check-symbol codes as for the average of all codes. The remainder of the derivation then follows as before.

This will be done for the pcs codes. When a noisy signal is received, the parity-check sums are recomputed at the receiver and added modulo two per position to the received check symbols, as in the Hamming code[7]. The resulting check-symbol pattern is the pattern caused by the transmission errors alone. The probability that this check-symbol pattern will be misinterpreted when j transmission errors have occurred is the probability that some other collection of j or fewer errors has the same check-symbol pattern. There are $V_N(j) - 1$ other patterns of j or fewer errors, and the probability that one of these has the same check-sum pattern is the probability that one of the $V_N(j) - 1$ differences has a check-sum pattern of all zeros. Now, if the check-sum matrix is filled in at random, any error pattern may produce any check-symbol pattern with equal probability. Therefore the probability of any one error pattern having a check-symbol pattern that vanishes is the reciprocal of the total number of possible check-symbol patterns. This number is $V_N(k_1)$, since $2^N/V_N(k_1)$ messages are being sent, and the total probability of a decoding error when j transmission errors have been made is less than this multiplied by the number of difference patterns:

$$P_{II}(N, j, k_1) \leq \frac{V_N(j) - 1}{V_N(k_1)} \leq \frac{V_N(j)}{V_N(k_1)} \leq \frac{\binom{N}{j}}{\binom{N}{k_1}} \qquad (A.21)$$

which is the inequality of (A.6) obtained by a different route.

The essential point in this argument is that every transmission error pattern, in the ensemble of possible codes, may cause every check-symbol pattern, with equal probability. Given this, the rest of the argument presented above follows. This is easy, but tedious, to show for sliding and convolutional pcs coding; the proofs will be omitted here.

References

1. C. E. Shannon, "A mathematical theory of communication," Bell System Tech. J. 27, 379-423, 623-656 (1948).
2. C. E. Shannon, "Communication in the presence of noise," Proc. I.R.E. 37, 10-21 (1949).
3. M. J. E. Golay, "Note on the theoretical efficiency of information reception with PPM," Proc. I.R.E. 37, 1031 (1949).
4. R. M. Fano, "Communication in the presence of additive Gaussian noise," "Communication Theory," Willis Jackson Ed. (Butterworths, London, 1953) 169-182.
5. S. O. Rice, "Communication in the presence of noise -- probability of error of two encoding schemes," Bell System Tech. J. 29, 60-93 (1950).
6. A. Feinstein, "A new basic theorem of information theory," Trans. I.R.E. (PGIT) 4, 2-22 (1954).
7. R. W. Hamming, "Error detecting and error correcting codes," Bell System Tech. J. 29, 147-160 (1950).
8. E. N. Gilbert, "A comparison of signalling alphabets," Bell System Tech. J. 31, 504-522 (1952).
9. M. Plotkin, "Binary codes with specified minimum distance," Univ. of Penna., Moore School Research Division Report 51-20 (1951).
10. M. J. E. Golay, "Binary coding," Trans. I.R.E. (PGIT) 4, 23-28 (1954).
11. A. E. Laemmel, "Efficiency of noise-reducing codes," pp. 111-118 in "Communication Theory," reference 4 above.
12. D. E. Muller, "Metric properties of Boolean algebra and their application to switching circuits," University of Illinois, Digital Computer Laboratory Report No. 46.
13. I. S. Reed, "A class of multiple error-correcting codes and the decoding scheme," Trans. I.R.E. (PGIT) 4, 38-49 (1954).
14. P. Elias, "Error-free coding," Trans. I.R.E. (PGIT) 4, 30-37 (1954).
15. W. Feller, "An Introduction to Probability Theory and Its Applications" (John Wiley and Sons, Inc., New York, 1950).

A Class of Binary Signaling Alphabets

By DAVID SLEPIAN

(Manuscript received September 27, 1955)

A class of binary signaling alphabets called "group alphabets" is described. The alphabets are generalizations of Hamming's error correcting codes and possess the following special features: (1) all letters are treated alike in transmission; (2) the encoding is simple to instrument; (3) maximum likelihood detection is relatively simple to instrument; and (4) in certain practical cases there exist no better alphabets. A compilation is given of group alphabets of length equal to or less than 10 binary digits.

INTRODUCTION

This paper is concerned with a class of signaling alphabets, called "group alphabets," for use on the symmetric binary channel. The class in question is sufficiently broad to include the error correcting codes of Hamming,[1] the Reed-Muller codes,[2] and all "systematic codes".[3] On the other hand, because they constitute a rather small subclass of the class of all binary alphabets, group alphabets possess many important special features of practical interest.

In particular, (1) all letters of the alphabets are treated alike under transmission; (2) the encoding scheme is particularly simple to instrument; (3) the decoder — a maximum likelihood detector — is the best possible theoretically and is relatively easy to instrument; and (4) in certain cases of practical interest the alphabets are the best possible theoretically.

It has very recently been proved by Peter Elias[4] that there exist group alphabets which signal at a rate arbitarily close to the capacity, C, of the symmetric binary channel with an arbitrarily small probability of error. Elias' demonstration is an existence proof in that it does not show *explicitly* how to construct a group alphabet signaling at a rate greater than $C - \varepsilon$ with a probability of error less than δ for arbitrary positive δ and ε. Unfortunately, in this respect and in many others, our understanding of group alphabets is still fragmentary.

In Part I, group alphabets are defined along with some related concepts necessary for their understanding. The main results obtained up to the present time are stated without proof. Examples of these concepts are given and a compilation of the best group alphabets of small size is presented and explained. This section is intended for the casual reader.

In Part II, proofs of the statements of Part I are given along with such theory as is needed for these proofs.

The reader is assumed to be familiar with the paper of Hamming,[1] the basic papers of Shannon[5] and the most elementary notions of the theory of finite groups.[6]

PART I — GROUP ALPHABETS AND THEIR PROPERTIES

1.1 INTRODUCTION

We shall be concerned in all that follows with communication over the symmetric binary channel shown on Fig. 1. The channel can accept either of the two symbols 0 or 1. A transmitted 0 is received as a 0 with probability q and is received as a 1 with probability $p = 1 - q$: a transmitted 1 is received as a 1 with probability q and is received as a 0 with probability p. We assume $0 \leq p \leq \frac{1}{2}$. The "noise" on the channel operates independently on each symbol presented for transmission. The capacity of this channel is

$$C = 1 + p \log_2 p + q \log_2 q \text{ bits/symbol} \qquad (1)$$

By a *K-letter, n-place binary signaling alphabet* we shall mean a collection of K distinct sequences of n binary digits. An individual sequence of the collection will be referred to as a *letter* of the alphabet. The integer K is called the size of the alphabet. A letter is transmitted over the channel by presenting in order to the channel input the sequence of n zeros and ones that comprise the letter. A *detection scheme* or *detector* for

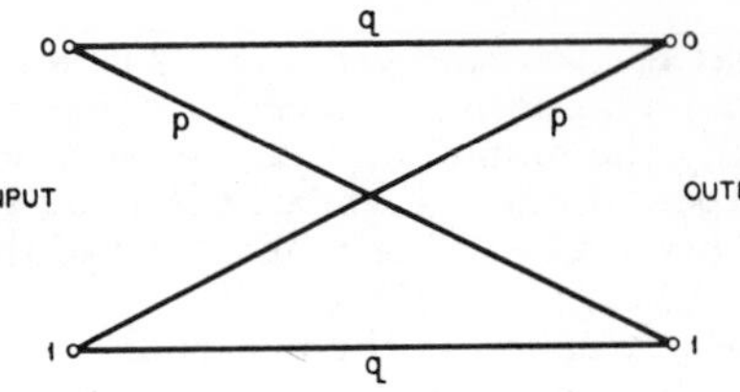

Fig. 1 — The symmetric binary channel.

a given K-letter, n-place alphabet is a procedure for producing a sequence of letters of the alphabet from the channel output.

Throughout this paper we shall assume that signaling is accomplished with a given K-letter, *n-place* alphabet by choosing the letters of the alphabet for transmission independently with equal probability $1/K$.

Shannon[5] has shown that for sufficiently large n, there exist K-letter, n-place alphabets and detection schemes that signal over the symmetric binary channel at a rate $R > C - \varepsilon$ for arbitrary $\varepsilon > 0$ and such that the probability of error in the letters of the detector output is less than any $\delta > 0$. Here C is given by (1) and is shown as a function of p in Fig. 2. No algorithm is known (other than exhaustvie procedures) for the construction of K-letter, n-place alphabets satisfying the above inequalities for arbitrary positive δ and ε except in the trivial cases $C = 0$ and $C = 1$.

1.2 THE GROUP B_n

There are a totality of 2^n different n-place binary sequences. It is frequently convenient to consider these sequences as the vertices of a cube of unit edge in a Euclidean space of n-dimensions. For example the 5-place sequence 0, 1, 0, 0, 1 is associated with the point in 5-space whose

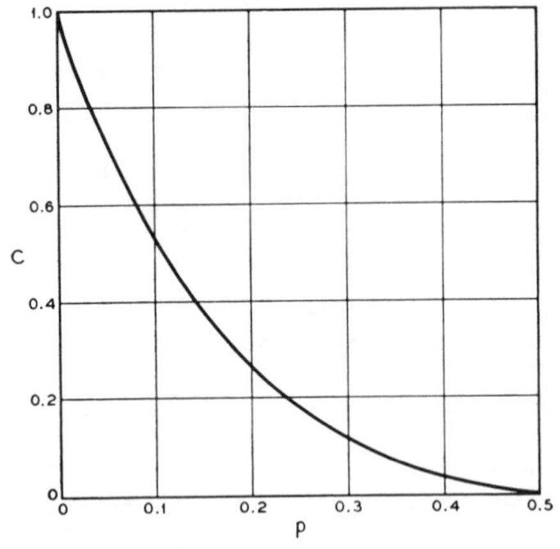

Fig. 2 — The capacity of the symmetric binary channel.
$C = 1 + p \log_2 p + (1 - p) \log_2 (1 - p)$

coordinates are (0, 1, 0, 0, 1). For convenience of notation we shall generally omit commas in writing a sequence. The above 5-place sequence will be written, for example, 01001.

We define the *product of two n-place binary sequences*, $a_1 a_2 \cdots a_n$ and $b_1 b_2 \cdots b_n$ as the n-place binary sequence

$$a_1 \dotplus b_1, \quad a_2 \dotplus b_2, \cdots, a_n \dotplus b_n$$

Here the a's and b's are zero or one and the $\dotplus$ sign means addition modulo 2. (That is $0 \dotplus 0 = 1 \dotplus 1 = 0, \quad 0 \dotplus 1 = 1 \dotplus 0 = 1$) For example, (01101) (00111) = 01010. With this rule of multiplication the 2^n n-place binary sequences form an Abelian group of order 2^n. The elements of the group, denoted by $T_1, T_2, \cdots, T_{2^n}$, say, are the n-place binary sequences; the identity element I is the sequence $000 \cdots 0$ and

$$IT_i = T_iI = T_i; \qquad T_iT_j = T_jT_i; \qquad T_i(T_jT_k) = (T_iT_j)T_k;$$

the product of any number of elements is again an element; every element is its own reciprocal, $T_i = T_i^{-1}$, $T_i^2 = I$. We denote this group by B_n.

All subgroups of B_n are of order 2^k where k is an integer from the set $0, 1, 2, \cdots, n$. There are exactly

$$N(n, k) = \frac{(2^n - 2^0)(2^n - 2^1)(2^n - 2^2) \cdots (2^n - 2^{k-1})}{(2^k - 2^0)(2^k - 2^1)(2^k - 2^2) \cdots (2^k - 2^{k-1})} = N(n, n - k) \tag{2}$$

distinct subgroups of B_n of order 2^k. Some values of $N(n, k)$ are given in Table I.

TABLE I — SOME VALUES OF $N(n, k)$, THE NUMBER OF SUBGROUPS OF B_n OF ORDER 2^k. $N(n, k) = N(n, n - k)$

$n \backslash k$	0	1	2	3	4	5
2	1	3	1			
3	1	7	7	1		
4	1	15	35	15	1	
5	1	31	155	155	31	1
6	1	63	651	1395	651	63
7	1	127	2667	11811	11811	2667
8	1	255	10795	97155	200787	97155
9	1	511	43435	788035	3309747	3309747
10	1	1023	174251	6347715	53743987	109221651

1.3 GROUP ALPHABETS

An n-place *group alphabet* is a K-letter, n-place binary signaling alphabet whose letters form a subgroup of B_n. Of necessity the size of an n-place group alphabet is $K = 2^k$ where k is an integer satisfying $0 \leq k \leq n$. By an (n, k)*-alphabet* we shall mean an n-place group alphabet of size 2^k. Example: the $N(3, 2) = 7$ distinct (3, 2)-alphabets are given by the seven columns

(i)	(ii)	(iii)	(iv)	(v)	(vi)	(vii)	
000	000	000	000	000	000	000	
100	100	100	010	010	001	110	
010	001	011	001	101	110	011	(3)
110	101	111	011	111	111	101	

1.4 STANDARD ARRAYS

Let the letters of a specific (n, k)-alphabet be $A_1 = I = 00 \cdots 0$, $A_2, A_3, \cdots, A_\mu$, where $\mu = 2^k$. The group B_n can be developed according to this subgroup and its cosets:

$$B_n = \begin{matrix} I, & A_2, & A_3, & \cdots, A_\mu \\ S_2, & S_2A_2, & S_2A_3, & \cdots, S_2A_\mu \\ S_3, & S_3A_2, & S_3A_3, & \cdots, S_3A_\mu \\ \cdot & & & \\ \cdot & & & \\ \cdot & & & \\ S_\nu, & S_\nu A_2, & S_\nu A_3, & \cdots, S_\nu A_\mu \end{matrix} \tag{4}$$
$$\mu = 2^k, \qquad \nu = 2^{n-k}.$$

In this array every element of B_n appears once and only once. The collection of elements in any row of this array is called a *coset* of the (n, k)-alphabet. Here S_2 is any element of B_n not in the first row of the array, S_3 is any element of B_n not in the first two rows of the array, etc. The elements $S_2, S_3, \cdots, S_\nu$ appearing under I in such an array will be called the *coset leaders*.

If a coset leader is replaced by any element in the coset, the same coset will result. That is to say the two collections of elements

$$S_i, \quad S_iA_2, \quad S_iS_3, \cdots, S_iA_\mu$$

and

$$S_iA_k, \quad (S_iA_k)A_2, \quad (S_iA_k)A_3, \cdots (S_iA_k)A_\mu$$

are the same.

We define the *weight* $w_i = w(T_i)$ of an element, T_i, of B_n to be the number of ones in the n-place binary sequence T_i.

Henceforth, unless otherwise stated, we agree in dealing with an array such as (4) to adopt the following convention:

the leader of each coset shall be taken to be an element of minimal weight in that coset. (5)

Such a table will be called a *standard array*.

Example: B_4 can be developed according to the (4, 2)-alphabet 0000, 1100, 0011, 1111 as follows

0000	1100	0011	1111	
1010	0110	1001	0101	
1110	0010	1101	0001	(6)
1000	0100	1011	0111	

According to (5), however, we should write, for example

0000	1100	0011	1111	
1010	0110	1001	0101	
0010	1110	0001	1101	(7)
1000	0100	1011	0111	

The coset leader of the second coset of (6) can be taken as any element of that row since all are of weight 2. The leader of the third coset, however, should be either 0010 or 0001 since these are of weight one. The leader of the fourth coset should be either 1000 or 0100.

1.5 THE DETECTION SCHEME

Consider now communicating with an (n, k)-alphabet over the symmetric binary channel. When any letter, say A_j, of the alphabet is transmitted, the received sequence can be of any element of B_n. We agree to use the following detector:

if the received element of B_n lies in column i of the array (4), the detector prints the letter A_i, $i = 1, 2, \cdots, \mu$. The array (4) is to be constructed according to the convention (5). (8)

The following propositions and theorems can be proved concerning signaling with an (n, k)-alphabet and the detection scheme given by (8).

1.6 BEST DETECTOR AND SYMMETRIC SIGNALING

Define the *probability* $\ell_i = \ell(T_i)$ of an element T_i of B_n to be $\ell_i = p^{w_i}q^{n-w_i}$ where p and q are as in (1) and w_i is the weight of T_i. Let Q_i, $i = 1, 2, \cdots, \mu$ be the sum of the probabilities of the elements in the ith column of the standard array (4).

Proposition 1. The probability that any transmitted letter of the (n, k)-alphabet be produced correctly by the detector is Q_1.

Proposition 2. The equivocation[5] per symbol is

$$H_y(x) = -\frac{1}{n}\sum_{i=1}^{\mu} Q_i \log_2 Q_i$$

Theorem 1. The detector (8) is a maximum likelihood detector. That is, for the given alphabet no other detection scheme has a greater average probability that a transmitted letter be produced correctly by the detector.

Let us return to the geometrical picture of n-place binary sequences as vertices of a unit cube in n-space. The choice of a K-letter, n-place alphabet corresponds to designating K particular vertices as letters. Since the binary sequence corresponding to any vertex can be produced by the channel output, any detector must consist of a set of rules that associates various vertices of the cube with the vertices designated as letters of the alphabet. We assume that every vertex is associated with some letter. The vertices of the cube are divided then into disjoint sets, $W_1, W_2, \cdots, W_K$ where W_i is the set of vertices associated with ith letter of the signaling alphabet. A maximum likelihood detector is characterized by the fact that every vertex in W_i is as close to or closer to the ith letter than to any other letter, $i = 1, 2, \cdots, K$. For group alphabets and the detector (8), this means that no element in the ith column of array (4) is closer to any other A than it is to A_i, $i = 1, 2, \cdots, \mu$.

Theorem 2. Associated with each (n, k)-alphabet considered as a point configuration in Euclidean n-space, there is a group of $n \times n$ orthogonal matrices which is transitive on the letters of the alphabet and which leaves the unit cube invariant. The maximum likelihood sets W_1, $W_2, \cdots W_\mu$ are all geometrically similar.

Stated in loose terms, this theorem asserts that in an (n, k)-alphabet every letter is treated the same. Every two letters have the same number of nearest neighbors associated with them, the same number of next nearest neighbors, etc. The disposition of points in any two W regions is the same.

1.7 GROUP ALPHABETS AND PARITY CHECKS

Theorem 3. Every group alphabet is a systematic[3] code: every systematic code is a group alphabet.[7]

We prefer to use the word "alphabet" in place of "code" since the latter has many meanings. In a *systematic alphabet*, the places in any letter can be divided into two classes: the information places — k in number for an (n, k)-alphabet — and the check positions. All letters have the same information places and the same check places. If there are k information places, these may be occupied by any of the 2^k k-place binary sequences. The entries in the $n - k$ check positions are fixed linear (mod 2) combinations of the entries in the information positions. The rules by which the entries in the check places are determined are called *parity checks*. Examples: for the (4, 2)-alphabet of (6), namely 0000, 1100, 0011, 1111, positions 2 and 3 can be regarded as the information positions. If a letter of the alphabet is the sequence $a_1a_2a_3a_4$, then $a_1 = a_2$, $a_4 = a_3$ are the parity checks determining the check places 1 and 4. For the (5, 3)-alphabet 00000, 10001, 01011, 00111, 11010, 10110, 01100, 11101 places 1, 2, and 3 (numbered from the left) can be taken as the information places. If a general letter of the alphabet is $a_1a_2a_3a_4a_5$, then $a_4 = a_2 \dotplus a_3$, $a_5 = a_1 \dotplus a_2 \dotplus a_3$.

Two group alphabets are called *equivalent* if one can be obtained from the other by a permutation of places. Example: the 7 distinct (3, 2)-alphabets given in (3) separate into three equivalence classes. Alphabets (i), (ii), and (iv) are equivalent; alphabets (iii), (v), (vi), are equivalent; (vii) is in a class by itself.

Proposition 3. Equivalent (n, k)-alphabets have the same probability Q_1 of correct transmission for each letter.

Proposition 4. Every (n, k)-alphabet is equivalent to an (n, k)-alphabet whose first k places are information places and whose last $n - k$ places are determined by parity checks over the first k places.

Henceforth we shall be concerned only with (n, k)-alphabets whose first k places are information places. The parity check rules can then be written

$$a_i = \sum_{j=1}^{k} \gamma_{ij} a_j, \qquad i = k + 1, \cdots, n \tag{9}$$

where the sums are of course mod 2. Here, as before, a typical letter of the alphabet is the sequence $a_1a_2 \cdots a_n$. The γ_{ij} are $k(n - k)$ quantities, zero or one, that serve to define the particular (n, k)-alphabet in question.

1.8 MAXIMUM LIKELIHOOD DETECTION BY PARITY CHECKS

For any element, T, of B_n we can form the sum given on the right of (9). This sum may or may not agree with the symbol in the ith place of T. If it does, we say T satisfies the ith-place parity check; otherwise T fails the ith-place parity check. When a set of parity check rules (9) is given, we can associate an $(n - k)$-place binary sequence, $R(T)$, with each element T of B_n. We examine each check place of T in order starting with the $(k + 1)$-*st* place of T. We write a zero if a place of T satisfies the parity check; we write a one if a place fails the parity check. The resultant sequence of zeros and ones, written from left to right is $R(T)$. We call $R(T)$ the *parity check sequence* of T. Example: with the parity rules $a_4 = a_2 \dotplus a_3$, $a_5 = a_1 \dotplus a_2 \dotplus a_3$ used to define the (5, 3)-alphabet in the examples of Theorem 3, we find $R(11000) = 10$ since the sum of the entries in the second and third places of 11001 is not the entry of the fourth place and since the sum of $a_1 = 1$, $a_2 = 1$, and $a_3 = 0$ is $0 = a_5$.

Theorem 4. Let $I, A_2, \cdots A_\mu$ be an (n, k)-alphabet. Let $R(T)$ be the parity check sequence of an element T of B_n formed in accordance with the parity check rules of the (n, k)-alphabet. Then $R(T_1) = R(T_2)$ if and only if T_1 and T_2 lie in the same row of array (4). The coset leaders can be ordered so that $R(S_i)$ is the binary symbol for the integer $i - 1$.

As an example of Theorem 4 consider the (4, 2)-alphabet shown with its cosets below

0000	1011	0101	1110
0100	1111	0001	1010
0010	1001	0111	1100
1000	0011	1101	0110

The parity check rules for this alphabet are $a_3 = a_1$, $a_4 = a_1 \dotplus a_2$. Every element of the second row of this array satisfies the parity check in the third place and fails the parity check in the 4th place. The parity check sequence for the second row is 01. The parity check for the third row is 10, and for the fourth row 11. Since every letter of the alphabet satisfies the parity checks, the parity check sequence for the first row is 00. We therefore make the following association between parity check sequences and coset leaders

$$\begin{aligned} 00 &\rightarrow 0000 = S_1 \\ 01 &\rightarrow 0100 = S_2 \\ 10 &\rightarrow 0010 = S_3 \\ 11 &\rightarrow 1000 = S_4 \end{aligned}$$

1.9 INSTRUMENTING A GROUP ALPHABET

Proposition 4 attests to the ease of the encoding operation involved with the use of an (n, k)-alphabet. If the original message is presented as a long sequence of zeros and ones, the sequence is broken into blocks of length k places. Each block is used as the first k places of a letter of the signaling alphabet. The last n-k places of the letter are determined by fixed parity checks over the first k places.

Theorem 4 demonstrates the relative ease of instrumenting the maximum likelihood detector (8) for use with an (n, k)-alphabet. When an element T of B_n is received at the channel output, it is subjected to the n-k parity checks of the alphabet being used. This results in a parity check sequence $R(T)$. $R(T)$ serves to identify a unique coset leader, say S_i. The product S_iT is then formed and produced as the detector output. The probability that this be the correct letter of the alphabet is Q_1.

1.10 BEST GROUP ALPHABETS

Two important questions regarding (n, k)-alphabets naturally arise. What is the maximum value of Q_1 possible for a given n and k and which of the $N(n, k)$ different subgroups give rise to this maximum Q_1? The answers to these questions for general n and k are not known. For many special values of n and k the answers are known. They are presented in Tables II, III and IV, which are explained below.

The probability Q_1 that a transmitted letter be produced correctly by the detector is the sum, $Q_1 = \sum_1^{\nu} \ell(S_i)$ of the probabilities of the coset leaders. This sum can be rewritten as $Q_1 = \sum_{i=0}^{n} \alpha_i\, p^i q^{n-i}$ where α_i is the number of coset leaders of weight i. One has, of course, $\sum \alpha_i = \nu = 2^{n-k}$ for an (n, k)-alphabet. Also $\alpha_i \leq \binom{n}{i} = \frac{n!}{i!(n-i)}!$ since this is the number of elements of B_n of weight i.

The α_i have a special physical significance. Due to the noise on the channel, a transmitted letter, A_i, of an (n, k)-alphabet will in general be received at the channel output as some element T of B_n different from A_i. If T differs from A_i in s places, i.e., if $w(A_iT) = s$, we say that an s-tuple error has occurred. For a given (n, k)-alphabet, α_i is the number of i-tuple errors which can be corrected by the alphabet in question, $i = 0, 1, 2, \cdots, n$.

Table II gives the α_i corresponding to the largest possible value of Q_1 for a given k and n for $k = 2, 3, \cdots n - 1$, $n = 4 \cdots$, 10 along with a few other scattered values of n and k. For reference the binomial coefficients $\binom{n}{i}$ are also listed. For example, we find from Table II that the best group alphabet with $2^4 = 16$ letters that uses $n = 10$ places has a probability of correct transmission $Q_1 = q^{10} + 10q^9p + 39q^8p^2 + 14q^7p^3$. The alphabet corrects all 10 possible single errors. It corrects 39 of the possible $\binom{10}{2} = 45$ double errors (second column of Table II) and in addition corrects 14 of the 120 possible triple errors. By adding an additional place to the alphabet one obtains with the best (11, 4)-alphabet an alphabet with 16 letters that corrects all 11 possible single errors and all 55 possible double errors as well as 61 triple errors. Such an alphabet might be useful in a computer representing decimal numbers in binary form.

For each set of α's listed in Table II, there is in Table III a set of parity check rules which determines an (n, k)-alphabet having the given α's. The notation used in Table III is best explained by an example. A (10, 4)-alphabet which realizes the α's discussed in the preceding paragraph can be obtained as follows. Places 1, 2, 3, 4 carry the information.

Place 5 is determined to make the mod 2 sum of the entries in places 3, 4, and 5 equal to zero. Place 6 is determined by a similar parity check on places 1, 2, 3, and 6; place 7 by a check on places 1, 2, 4, and 7, etc.

It is a surprising fact that for all cases investigated thus far an (n, k)-alphabet best for a given value of p is uniformly best for all values of $p, 0 \leqq p \leqq \frac{1}{2}$. It is of course conjectured that this is true for all n and k.

It is a further (perhaps) surprising fact that the best (n, k)-alphabets are not necessarily those with greatest nearest neighbor distance between letters when the alphabets are regarded as point configurations on the n-cube. For example, in the best (7, 3)-alphabet as listed in Table III, each letter has two nearest neighbors distant 3 edges away. On the other hand, in the (7, 3)-alphabet given by the parity check rules 413, 512, 623, 7123 each letter has its nearest neighbors 4 edges away. This latter alphabet does not have as large a value of Q_1, however, as does the (7, 3)-alphabet listed on Table III.

The cases $k = 0, 1, n - 1, n$ have not been listed in Tables II and III. The cases $k = 0$ and $k = n$ are completely trivial. For $k = 1$, all $n > 1$ the best alphabet is obtained using the parity rule $a_2 = a_3 = \cdots = a_n = a_1$. If $n = 2j$,

$$Q_1 = \sum_0^{j-1} \binom{n}{i} p^i q^{n-i} + \frac{1}{2} \binom{n}{j} p^j q^j. \text{ If } n = 2j + 1, Q_1 = \sum_0^{j} \binom{n}{i} p^i q^{n-i}.$$

For $k = n - 1$, $n > 1$, the maximum Q_1 is $Q_1 = q^{n-1}$ and a parity rule for an alphabet realizing this Q_1 is $a_n = a_1$.

If the α's of an (n, k)-alphabet are of the form $\alpha_i = \binom{n}{i}$, $i = 0, 1, 2, \cdots, j$, $\alpha_{j+1} = r$ some integer, $\alpha_{j+2} = \alpha_{j+3} = \cdots = \alpha_n = 0$, then there does not exist a 2^k-letter, n-place alphabet of any sort better than the given (n, k)-alphabet. It will be observed that many of the α's of Table II are of this form. It can be shown that

TABLE II — PROBABILITY OF NO ERROR WITH BEST ALPHABETS, $Q_1 = \sum \alpha_i p^i q^{n-i}$

	i	$\binom{n}{i}$	$k = 2$ a_i	$k = 3$ a_i	$k = 4$ a_i	$k = 5$ a_i	$k = 6$ a_i	$k = 7$ a_i	$k = 8$ a_i	$k = 9$ a_i	$k = 10$ a_i
$n = 4$	0	1	1								
	1	4	3								
$n = 5$	0	1	1	1							
	1	5	5	3							
	2	10	2								
$n = 6$	0	1	1	1	1						
	1	6	6	6	3						
	2	15	9	1							
$n = 7$	0	1	1	1	1	1					
	1	7	7	7	7	3					
	2	21	18	8							
	3	25	6								
$n = 8$	0	1	1	1	1	1	1				
	1	8	8	8	8	7	3				
	2	28	28	20	7						
	3	56	27	3							
$n = 9$	0	1	1	1	1	1	1	1			
	1	9	9	9	9	9	7	3			
	2	36	36	33	22	6					
	3	84	64	21							
	4	126	18								
$n = 10$	0	1	1	1	1	1	1	1	1		
	1	10	10	10	10	10	10	7	3		
	2	45	45	45	39	21	5				
	3	120	110	64	14						
	4	210	90	8							
$n = 11$	0	1	1	1	1		1	1	1	1	
	1	11	11	11	11		11	11	7	3	
	2	55	55	55	55		20	4			
	3	165	165	126	61						
	4	330	226	63							
	5	462	54								
$n = 12$	0	1	1	1				1	1	1	1
	1	12	12	12				12	12	7	3
	2	66	66	66				19	3		
	3	220	220	200							
	4	495	425	233							
	5	792	300								

Proposition 5 if $n + \binom{n-k}{2} + \binom{n-k}{3} \geqq 2^{n-k} - 1$ there exists no 2^k-letter, n-place alphabet better than the best (n, k)-alphabet.

When the inequality of proposition 5 holds the α's are either $\alpha_0 = 1$, $\alpha_1 = 2^{n-k} - 1$, all other $\alpha = 0$; or $\alpha_0 = 1$, $\alpha_1 = \binom{n}{1}$, $\alpha_2 = 2^{n-k} - 1 - \binom{n}{1}$ all other $\alpha = 0$; or the trivial $\alpha_0 = 1$ all other $\alpha = 0$ which holds when $k = n$. The region of the $n - k$ plane for which it is known that (n, k)-alphabets cannot be excelled by any other is shown in Table IV.

1.11 A DETAILED EXAMPLE

As an example of the use of (n, k)-alphabets consider the not unrealistic case of a channel with $p = 0.001$, i.e., on the average one binary digit per thousand is received incorrectly. Suppose we wish to transmit messages using 32 different letters. If we encode the letters into the 32 5-place binary sequences and transmit these sequences without further encoding, the probability that a received letter be in error is $1 - (1 - p)^5 = 0.00449$. If the best (10, 5)-alphabet as shown in Tables II and III is used, the probability that a letter be wrong is $1 - Q_1 = 1 - q^{10} - 10q^9p - 21q^8p^2 = 24p^2 - 72p^3 + \cdots = 0.000024$. Thus by reducing the signaling rate by ½, a more than *one hundredfold* reduction in probability of error is accomplished.

A (10, 5)-alphabet to achieve these results is given in Table III. Let a typical letter of the alphabet be the 10-place sequence of binary digits $a_1a_2 \cdots a_9a_{10}$. The symbols $a_1a_2a_3a_4a_5$ carry the information and can be any of 32 different arrangements of zeros and ones. The remaining places are determined by

$$\begin{aligned} a_6 &= a_1 \dotplus a_3 \dotplus a_4 \dotplus a_5 \\ a_7 &= a_1 \dotplus a_2 \dotplus a_4 \dotplus a_5 \\ a_8 &= a_1 \dotplus a_2 \dotplus a_3 \dotplus a_5 \\ a_9 &= a_1 \dotplus a_2 \dotplus a_3 \dotplus a_4 \\ a_{10} &= a_1 \dotplus a_2 \dotplus a_3 \dotplus a_4 \dotplus a_5 \end{aligned}$$

To design the detector for this alphabet, it is first necessary to determine the coset leaders for a standard array (4) formed for this alphabet. This can be done by a variety of special methods which considerably reduce the obvious labor of making such an array. A set of best S's along with their parity check symbols is given in Table V.

A maximum likelihood detector for the (10, 5)-alphabet in question forms from each received sequence $b_1b_2 \cdots b_{10}$ the parity check symbol $c_1c_2c_3c_4c_5$ where

$$\begin{aligned} c_1 &= b_6 \dotplus b_1 \dotplus b_3 \dotplus b_4 \dotplus b_5 \\ c_2 &= b_7 \dotplus b_1 \dotplus b_2 \dotplus b_4 \dotplus b_5 \\ c_3 &= b_8 \dotplus b_1 \dotplus b_2 \dotplus b_3 \dotplus b_5 \\ c_4 &= b_9 \dotplus b_1 \dotplus b_2 \dotplus b_3 \dotplus b_4 \\ c_5 &= b_{10} \dotplus b_1 \dotplus b_2 \dotplus b_3 \dotplus b_4 \dotplus b_5 \end{aligned}$$

According to Table V, if $c_1c_2c_3c_4c_5$ contains less than three ones, the detector should brint $b_1b_2b_3b_4b_5$. The detector should print $(b_1 \dotplus 1)b_2b_3b_4b_5$ if the parity check sequence $c_1c_2c_3c_4c_5$ is either 11111 or 11110; the detector should print $b_1(b_2 \dotplus 1)b_3b_4b_5$ if the parity check sequence is 01111, 00111, 01011, 01101, or 01110; the detector should print $b_1b_2(b_3 \dotplus 1)b_4b_5$ if the parity check sequence is 10111, 10011, 10101, or 10110; the detector should print $b_1b_2b_3(b_4 \dotplus 1)b_5$ if the parity check sequence is 11011, 11001, 11010; and finally the detector should print $b_1b_2b_3b_4(b_5 \dotplus 1)$ if the parity check sequence is 11101 or 11100.

Simpler rules of operation for the detector may possibly be obtained by choice of a different set of S's in Table V. These quantities in general are not unique. Also there may exist non-equivalent alphabets with simpler detector rules that achieve the same probability of error as the alphabet in question.

PART II — ADDITIONAL THEORY AND PROOFS OF THEOREMS OF PART I

2.1 THE ABSTRACT GROUP C_n

It will be helpful here to say a few more words about B_n, the group of n-place binary sequences under the operation of addition mod 2. This group is simply isomorphic with the abstract group C_n generated by n commuting elements of order two, say $a_1, a_2, \cdots, a_n$. Here $a_ia_j = a_ja_i$ and $a_i^2 = I$, $i, j = 1, 2, \cdots, n$, where I is the identity for the group. The eight distinct elements of C_3 are, for example, I, a_1, a_2,

TABLE III — PARITY CHECK RULES FOR BEST ALPHABETS

	$k = 2$	$k = 3$	$k = 4$	$k = 5$	$k = 6$	$k = 7$	$k = 8$	$k = 9$	$k = 10$
$n = 4$	3 2 4 1 2								
$n = 5$	3 1 2 4 2 5 1	4 1 2 5 1 3							
$n = 6$	3 2 4 1 2 5 1 6 1	4 1 2 5 1 3 6 2 3	5 1 2 3 6 1 2 4						
$n = 7$	3 1 4 1 5 1 6 1 2 7 2	4 1 3 5 1 2 6 1 2 3 7 1 2 3	5 1 3 4 6 1 2 4 7 1 2 3	6 1 7 1					
$n = 8$	3 1 4 1 5 2 6 2 7 1 2 8 1 2	4 1 5 1 2 6 1 3 7 2 3 8 1 2 3	5 1 3 4 6 1 2 4 7 1 2 3 8 1 2 3 4	6 1 3 4 7 1 2 4 8 1 2 3	7 1 8 1				
$n = 9$	3 1 4 1 5 1 6 2 7 2 8 1 2 9 1 2	4 1 5 2 6 1 2 7 1 3 8 2 3 9 1 2 3	5 1 3 4 6 1 2 4 7 1 2 3 8 1 2 3 9 1 2 3	6 1 3 4 5 7 1 2 4 5 8 1 2 3 5 9 1 2 3 4	7 1 3 4 8 1 2 4 9 1 2 3	8 1 9 1			
$n = 10$	3 1 4 1 5 1 6 2 7 2 8 1 2 9 1 2 10 1 2	4 1 5 2 6 3 7 1 2 8 1 3 9 2 3 10 1 2 3	5 3 4 6 1 2 3 7 1 2 4 8 1 3 4 9 2 3 4 10 1 2 3 4	6 1 3 4 5 7 1 2 4 5 8 1 2 3 5 9 1 2 3 4 10 1 2 3 4 5	7 1 3 4 5 8 1 2 4 5 9 1 2 3 5 6 10 1 2 3 4 6	8 1 3 4 9 1 2 4 10 1 2 3	9 1 10 1		
$n = 11$	3 1 4 1 5 1 6 2 7 2 8 2 9 1 2 10 1 2 11 1 2	4 3 5 3 6 2 7 1 3 8 1 3 9 1 2 10 1 2 3 11 1 2 3	5 1 3 6 2 4 7 1 4 8 2 3 9 1 3 4 10 2 3 4 11 1 2 3 4		7 1 3 4 5 6 8 1 2 4 5 6 9 1 2 3 5 6 10 1 2 3 4 6 11 1 2 3 4 5	8 1 3 4 5 9 1 2 4 5 7 10 1 2 3 5 6 11 1 2 3 4 6 7	9 1 3 4 10 1 2 4 11 1 2 3	10 1 11 1	
$n = 12$	3 1 4 1 5 1 6 1 7 2 8 2 9 2 10 2 11 1 2 12 1 2	4 1 5 2 6 3 7 1 2 8 1 2 9 1 3 10 2 3 11 1 2 3 12 1 2 3				8 1 3 4 5 6 9 1 2 4 5 6 10 1 2 3 5 6 7 11 1 2 3 4 6 7 12 1 2 3 4 5 7	9 1 2 3 5 6 7 8 10 1 2 3 4 6 11 1 2 4 5 7 12 1 3 4 5 8	10 1 2 3 11 1 2 4 12 1 3 4	11 1 12 1

a_3, a_1a_2, a_1a_3, a_2a_3, $a_1a_2a_3$. The group C_n is easily seen to be isomorphic with the n-fold direct product of the group C_1 with itself.

It is a considerable saving in notation in dealing with C_n to omit the symbol "a" and write only the subscripts. In this notation for example, the elements of C_4 are I, 1, 2, 3, 4, 12, 13, 14, 23, 24, 34, 123, 124, 134, 234, 1234. The product of two or more elements of C_n can readily be written down. Its symbol consists of those numerals that occur an odd number of times in the collection of numerals that comprise the symbols of the factors. Thus, $(12)(234)(123) = 24$.

The isomorphism between C_n and B_n can be established in many ways. The most convenient way, perhaps, is to associate with the element $i_1i_2i_3 \cdots i_k$ of C_n the element of B_n that has ones in places $i_1, i_2, \cdots, i_k$ and zeros in the remaining $n - k$ places. For example, one can associate 124 of C_4 with 1101 of B_4; 14 with 1001, etc. In fact, the numeral notation afforded by this isomorphism is a much neater notation for B_n than is afforded by the awkward strings of zeros and ones. There are, of course, other ways in which elements of C_n can be paired with elements of B_n so that group multiplication is preserved. The collection of all such "pairings" makes up the group of automorphisms of C_n. This group of automorphisms of C_n is isomorphic with the group of non-singular linear homogenous transformations in a field of characteristic 2.

An element T of C_n is said to be *dependent* upon the set of elements $T_1, T_2, \cdots, T_j$ of C_n if T can be expressed as a product of some elements of the set $T_1, T_2, \cdots, T_j$; otherwise, T is said to be *independent* of the set. A set of elements is said to be independent if no member can be expressed solely in terms of the other members of the set. For example, in C_8, 1, 2, 3, 4 form a set of independent elements as do likewise 2357, 12357, 14. However, 135 depends upon 145, 3457, 57 since $135 = (145)(3457)(57)$. Clearly any set of n independent elements of C_n can be taken as generators for the group. For example, all possible products formed of 12, 123, and 23 yield the elements of C_3.

Any k independent elements of C_n serve as generators for a subgroup of order 2^k. The subgroup so generated is clearly isomorphic with C_k. All subgroups of C_n of order 2^k can be obtained in this way.

The number of ways in which k independent elements can be chosen from the 2^n elements of C_n is

$$F(n, k) = (2^n - 2^0)(2^n - 2^1)(2^n - 2^2) \cdots (2^n - 2^{k-1})$$

For, the first element can be chosen in $2^n - 1$ ways (the identity cannot be included in a non-trivial set of independent elements) and the second element can be chosen in $2^n - 2$ ways. These two elements determine a subgroup of order 2^2. The third element can be chosen as any element of the remaining $2^n - 2^2$ elements. The 3 elements chosen determine a subgroup of order 2^3. A fourth independent element can be chosen as any of the remaining $2^n - 2^3$ elements, etc.

Each set of k independent elements serves to generate a subgroup of order 2^k. The quantity $F(n, k)$ is not, however, the number of distinct subgroups of C_n of this order, for, a given subgroup can be obtained

TABLE IV — REGION OF THE n-k PLANE FOR WHICH IT IS KNOWN THAT (n, k)-ALPHABETS CANNOT BE EXCELLED

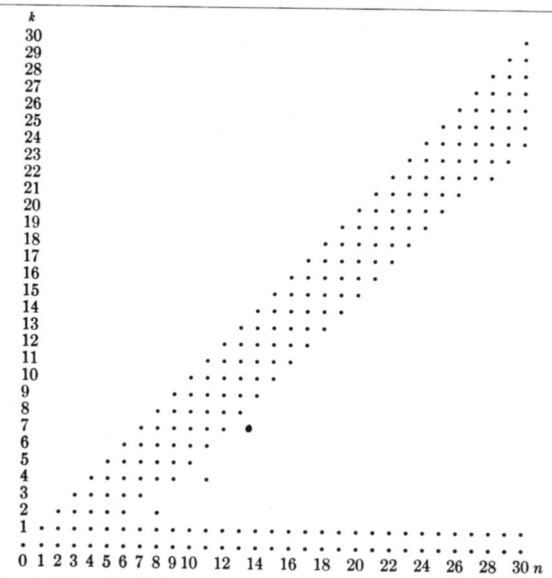

TABLE V — COSET LEADERS AND PARITY CHECK SEQUENCES FOR (10,5)-ALPHABET

$c_1c_2c_3c_4c_5$	$\leftrightarrow$	S	$c_1c_2c_3c_4c_5$	$\leftrightarrow$	S
00000		0000000000	11100		0000100001
10000		0000010000	11010		0001000001
01000		0000001000	11001		0001000010
00100		0000000100	10110		0010000001
00010		0000000010	10101		0010000010
00001		0000000001	10011		0010000100
11000		0000011000	01110		0100000001
10100		0000010100	01101		0100000010
10010		0000010010	01011		0100000100
10001		0000010001	00111		0100001000
01100		0000001100	11110		1000000001
01010		0000001010	11101		0000100000
01001		0000001001	11011		0001000000
00110		0000000110	10111		0010000000
00101		0000000101	01111		0100000000
00011		0000000011	11111		1000000000

from many different sets of generators. Indeed, the number of different sets of generators that can generate a given subgroup of order 2^k of C_n is just $F(k, k)$ since any such subgroup is isomorphic with C_k. Therefore the number of subgroups of C_n of order 2^k is $N(n, k) = F(n, k)/F(k, k)$ which is (2). A simple calculation gives $N(n, k) = N(n, n - k)$.

2.2 PROOF OF PROPOSITIONS 1 AND 2

After an element A of B_n has been presented for transmission over a noisy binary channel, an element T of B_n is produced at the channel output. The element $U = AT$ of B_n serves as a record of the noise during the transmission. U is an n-place binary sequence with a one at each place altered in A by the noise. The channel output, T, is obtained from the input A by multiplication by U: $T = UA$. For channels of the sort under consideration here, the probability that U be any particular element of B_n of weight w is $p^w q^{n-w}$.

Consider now signaling with a particular (n, k)-alphabet and consider the standard array (4) of the alphabet. If the detection scheme (8) is used, a transmitted letter A_i will be produced without error if and only if the received symbol is of the form $S_j A_i$. That is, there will be no error only if the noise in the channel during the transmission of A_i is represented by one of the coset leaders. (This applies for $i = 1, 2, \cdots, \mu = 2^k$). The probability of this event is Q_1 (Proposition 1, Section 1.6). The convention (5) makes Q_1 as large as is possible for the given alphabet.

Let X refer to transmitted letters and let Y refer to letters produced by the detector. We use a vertical bar to denote conditions when writing probabilities. The quantity to the right of the bar is the condition. We suppose the letters of the alphabet to be chosen independently with equal probability 2^{-k}.

The equivocation $h(X \mid Y)$ obtained when using an (n, k)-alphabet with the detector (8) can most easily be computed from the formula

$$h(X \mid Y) = h(X) - h(Y) + h(Y \mid X) \tag{10}$$

The entropy of the source is $h(X) = k/n$ bits per symbol. The probability that the detector produce A_j when A_i was sent is the probability that the noise be represented by $A_i A_j S_\ell$, $\ell = 1, 2, \cdots, \nu$. In symbols,

$$Pr(Y \to A_j \mid X \to A_i) = \sum_\ell Pr(N \to A_i A_j S_\ell) = Q(A_i A_j)$$

where $Q(A_i)$ is the sum of the probabilities of the elements that are in the same column as A_i in the standard array. Therefore

$$Pr(Y \to A_j) = \sum_i Pr(Y \to A_j \mid X \to A_i) Pr(X \to A_i) = \frac{1}{2^k} \sum_i Q(A_i A_j)$$

$$= \frac{1}{2^k}, \qquad \text{since } \sum_i Q(A_i A_j) = \sum_i Q(A_i) = 1.$$

This last follows from the group property of the alphabet. Therefore

$$h(Y) = -\frac{1}{n} \sum Pr(Y \to A_j) \log Pr(Y \to A_j) = \frac{k}{n} \text{ bits/symbol.}$$

It follows then from (10) that

$$h(X \mid Y) = h(Y \mid X)$$

The computation of $h(Y \mid X)$ follows readily from its definition

$$h(Y \mid X) = \sum_i Pr(X \to A_i) h(Y \mid X \to A_i)$$

$$= -\sum_{ij} Pr(X \to A_i) Pr(Y \to A_j \mid X \to A_i) \log Pr(Y \to A_j \mid X \to A_i)$$

$$= -\frac{1}{2^k} \sum_{ij} \sum_\ell Pr(N \to A_i S_\ell A_j) \log \sum_m Pr(N \to A_i S_m A_j)$$

$$= -\frac{1}{2^k} \sum_{ij} Q(A_i A_j) \log Q(A_i A_j)$$

$$= -\sum_i Q(A_i) \log Q(A_i)$$

Each letter is n binary places. Proposition 2, then follows.

2.3 DISTANCE AND THE PROOF OF THEOREM 1

Let A and B be two elements of B_n. We define the *distance*, $d(A, B)$, between A and B to be the weight of their product,

$$d(A, B) = w(AB) \tag{11}$$

The distance between A and B is the number of places in which A and B differ and is just the "Hamming distance."[1] In terms of the n-cube, $d(A, B)$ is the minimum number of edges that must be traversed to go from vertex A to vertex B. The distance so defined is a monotone function of the Euclidean distance between vertices.

It follows from (11) that if C is any element of B_n then

$$d(A, B) = d(AC, BC) \tag{12}$$

This fact shows the detection scheme (8) to be a maximum likelihood detector. By definition of a standard array, one has

$$d(S_i, I) \leqq d(S_i A_j, I) \qquad \text{for all } i \text{ and } j$$

The coset leaders were chosen to make this true. From (12),

$$d(S_i, I) = d(S_i A_m S_i, I A_m S_i) = d(S_i A_m, A_m)$$

$$d(S_i A_j, I) = d(S_i A_j S_i A_m, I S_i A_m) = d(A_j A_m, S_i A_m)$$

$$= d(S_i A_m, A_\ell)$$

where $A_\ell = A_j A_m$. Substituting these expressions in the inequality above yields

$$d(S_i A_m, A_m) \leqq d(S_i A_m, A_\ell) \qquad \text{for all } i, m, \ell$$

This equation says that an arbitrary element in the array (4) is at least as close to the element at the top of its column as it is to any other letter of the alphabet. This is the maximum likelihood property.

2.4 PROOF OF THEOREM 2

Again consider an (n, k)-alphabet as a set of vertices of the unit n-cube. Consider also n mutually perpendicular hyperplanes through the centroid of the cube parallel to the coordinate planes. We call these planes "symmetry planes of the cube" and suppose the planes numbered in accordance with the corresponding parallel coordinate planes.

The reflection of the vertex with coordinates $(a_1, a_2, \cdots, a_i, \cdots, a_n)$ in symmetry plane i yields the vertex of the cube whose coordinates are $(a_1, a_2, \cdots, a_i \dotplus 1, \cdots, a_n)$. More generally, reflecting a given vertex successively in symmetry planes $i, j, k, \cdots$ yields a new vertex whose coordinates differ from the original vertex precisely in places $i, j, k \cdots$. Successive reflections in hyperplanes constitute a transformation that leaves distances between points unaltered and is therefore a "rotation." The rotation obtained by reflecting successively in symmetry planes i, j, k, etc. can be represented by an n-place symbol having a one in places i, j, k, etc. and a zero elsewhere.

We now regard a given (n, k)-alphabet as generated by operating on the vertex $(0, 0, \cdots, 0)$ of the cube with a certain collection of 2^k rotation operators. The symbols for these operators are identical with the sequences of zeros and ones that form the coordinates of the 2^k points. It is readily seen that these rotation operators form a group which is transitive on the letters of the alphabet and which leave the unit cube invariant. Theorem 2 then follows.

Theorem 2 also follows readily from consideration of the array (4). For example, the maximum likelihood region associated with I is the set of points $I, S_2, S_3, \cdots, S_\nu$. The maximum likelihood region associated with A_i is the set of points $A_i, A_iS_2, A_iS_3, \cdots, A_iS_\nu$. The rotation (successive reflections in symmetry planes of the cube) whose symbol is the same as the coordinate sequence of A_i sends the maximum likelihood region of I into the maximum likelihood region of A_i, $i = 1, 2, \cdots, \mu$.

2.5 PROOF OF THEOREM 3

That every systematic alphabet is a group alphabet follows trivially from the fact that the sum mod 2 of two letters satisfying parity checks is again a letter satisfying the parity checks. The totality of letters satisfying given parity checks thus constitutes a finite group.

To prove that every group alphabet is a systematic code, consider the letters of a given (n, k)-alphabet listed in a column. One obtains in this way a matrix with 2^k rows and n columns whose entries are zeros and ones. Because the rows are distinct and form a group isomorphic to C_k, there are k linearly independent rows (mod 2) and no set of more than k independent rows. The rank of the matrix is therefore k. The matrix therefore possesses k linearly independent (mod 2) columns and the remaining $n - k$ columns are linear combinations of these k. Maintaining only these k linearly independent columns, we obtain a matrix of k columns and 2^k rows with rank k. This matrix must, therefore, have k linearly independent rows. The rows, however, form a group under mod 2 addition and hence, since k are linearly independent, all 2^k rows must be distinct. The matrix contains only zeros and ones as entries; it has 2^k distinct rows of k entries each. The matrix must be a listing of the numbers from 0 to $2^k - 1$ in binary notation. The other $n - k$ columns of the original matrix considered are linear combinations of the columns of this matrix. This completes the proof of Theorem 3 and Proposition 4.

2.6 PROOF OF THEOREM 4

To prove Theorem 4 we first note that the parity check sequence of the product of two elements of B_n is the mod 2 sum of their separate parity check sequences. It follows then that all elements in a given coset have the same parity check sequence. For, let the coset be $S_i, S_iA_2, S_iA_3, \cdots S_iA_\mu$. Since the elements $I, A_2, A_3, \cdots, A_\mu$ all have parity check sequence $00 \cdots 0$, all elements of the coset have parity check $R(S_i)$.

In the array (4) there are 2^{n-k} cosets. We observe that there are 2^{n-k} elements of B_n that have zeros in their first k places. These elements have parity check symbols identical with the last $n - k$ places of their symbols. These elements therefore give rise to 2^{n-k} different parity check symbols. The elements must be distributed one per coset. This proves Theorem 4.

2.7 PROOF OF PROPOSITION 5

If

$$n \geqq 2^{n-k} - \binom{n-k}{2} - \binom{n-k}{3} - 1$$

we can explicity exhibit group alphabets having the property mentioned in the paragraph preceding Proposition 5. The notation of the demonstration is cumbersome, but the idea is relatively simple.

We shall use the notation of paragraph 2.1 for elements of B_n, i.e., an element of B_n will be given by a list of integers that specify what places of the sequence for the element contain ones. It will be convenient furthermore to designate the first k places of a sequence by the integers $1, 2, 3, \cdots, k$ and the remaining $n - k$ places by the "integers" $1', 2', 3', \cdots, \ell'$, where $\ell = n - k$. For example, if $n = 8$, $k = 5$, we have

$$10111010 \leftrightarrow 13452'$$
$$10000100 \leftrightarrow 11'$$
$$00000101 \leftrightarrow 1'3'$$

Consider the group generated by the elements $1', 2', 3', \cdots, \ell'$, i.e. the 2^ℓ elements $I, 1', 2', \cdots, \ell', 1'2', 1'3', \cdots, 1'2'3' \cdots \ell'$. Suppose these elements listed according to decreasing weight (say in decreasing order when regarded as numbers in the decimal system) and numbered consecutively. Let B_i be the ith element in the list. Example: if $\ell = 3$, $B_1 = 1'2'3'$, $B_2 = 2'3'$, $B_3 = 1'3'$, $B_4 = 1'2'$, $B_5 = 3'$, $B_6 = 2'$, $B_7 = 1'$.

Consider now the (n, k)-alphabet whose generators are

$$1B_1, 2B_2, 3B_3, \cdots, kB_k$$

We assert that if

$$n \geqq 2^{n-k} - \binom{n-k}{2} - \binom{n-k}{3} - 1$$

this alphabet is as good as any other alphabet of 2^k letters and n places.

In the first place, we observe that every letter of this (n, k)-alphabet (except I) has unprimed numbers in its symbols. It follows that each of the 2^ℓ letters $I, 1', 2', \cdots, \ell', 1'2', \cdots, 1'2' \cdots \ell'$ occurs in a different coset of the given (n, k)-alphabet. For, if two of these letters appeared in the same coset, their product (which contains only primed numbers) would have to be a letter of the (n, k) alphabet. This is impossible since every letter of the (n, k) alphabet has unprimed numbers in its symbol. Since there are precisely 2^ℓ cosets we can designate a coset by the single element of the list $B_1, B_2, \cdots, B_{2^\ell} = I$ which appears in the coset.

We next observe that the condition

$$n \geqq 2^{n-k} - \binom{n-k}{2} - \binom{n-k}{3} - 1$$

guarantees that B_{k+1} is of weight 3 or less. For, the given condition is equivalent to

$$k \geqq 2^\ell - \binom{\ell}{0} - \binom{\ell}{1} - \binom{\ell}{2} - \binom{\ell}{3}$$

We treat several cases depending on the weight of B_{k+1}.

If B_{k+1} is of weight 3, we note that for $i = 1, 2, \cdots, k$, the coset containing B_i also contains an element of weight one, namely the element i obtained as the product of B_i with the letter iB_i of the given (n, k)-alphabet. Of the remaining $(2^\ell - k)$ B's, one is of weight zero, ℓ are of weight one, $\binom{\ell}{2}$ are of weight 2 and the remaining are of weight 3. We have, then $\alpha_0 = 1$, $\alpha_1 = \ell + k = n$. Now every B of weight 4 occurs in the list of generators $1B_1, 2B_2, \cdots, kB_k$. It follows that on multiplying this list of generators by any B of weight 3, at least one element of weight two will result. (E.g., $(1'2'3')(j1'2'3'4') = j4'$) Thus every coset with a B of weight 2 or 3 contains an element of weight 2 and $\alpha_2 = 2^\ell - \alpha_0 - \alpha_1$.

The argument in case B_{k+1} is of weight two or one is similar.

2.8 MODULAR REPRESENTATIONS OF C_n

In order to explain one of the methods used to obtain the best (n, k)-alphabets listed in Tables II and III, it is necessary to digress here to present additional theory.

It has been remarked that every (n, k)-alphabet is isomorphic with C_k. Let us suppose the elements of C_k listed in a column starting with I and proceeding in order $I, 1, 2, 3, \cdots, k, 12, 13, \cdots, (k-1)k, 123, \cdots \cdots, 123 \cdots k$. The elements of a given (n, k)-alphabet can be paired off with these abstract elements so as to preserve group multiplication. This can be done in many different ways. The result is a matrix with elements zero and one with n columns and 2^k rows, these latter being labelled by the symbols I, 1, 2, $\cdots$ etc. What can be said about the columns of this matrix? How many different columns are possible when all (n, k)-alphabets and all methods of establishing isomorphism with C_k are considered?

In a given column, once the entries in rows 1, 2, $\cdots$, k are known, the entire column is determined by the group property. There are therefore only 2^k possible different columns for such a matrix. A table showing these 2^k possible columns of zeros and ones will be called a *modular representation* table for C_k. An example of such a table is shown for $k = 4$ in Table VI.

It is clear that the columns of a modular representation table can also be labelled by the elements of C_k, and that group multiplication of these column labels is isomorphic with mod 2 addition of the columns. The table is a symmetric matrix. The element with row label A and column label B is one if the symbols A and B have an odd number of different numerals in common and is zero otherwise.

Every (n, k)-alphabet can be made from a modular representation table by choosing n columns of the table (with possible repetitions) at least k of which form an independent set.

TABLE VI — MODULAR REPRESENTATION TABLE FOR GROUP C_4

	I	1	2	3	4	12	13	14	23	24	34	123	124	134	234	1234
I	0	0	0	0	0	0	0	0	0	0	0	0	0	0	0	0
1	0	1	0	0	0	1	1	1	0	0	0	1	1	1	0	1
2	0	0	1	0	0	1	0	0	1	1	0	1	1	0	1	1
3	0	0	0	1	0	0	1	0	1	0	1	1	0	1	1	1
4	0	0	0	0	1	0	0	1	0	1	1	0	1	1	1	1
12	0	1	1	0	0	0	1	1	1	1	0	0	0	1	1	0
13	0	1	0	1	0	1	0	1	1	0	1	0	1	0	1	0
14	0	1	0	0	1	1	1	0	0	1	1	1	0	0	1	0
23	0	0	1	1	0	1	1	0	0	1	1	0	1	1	0	0
24	0	0	1	0	1	1	0	1	1	0	1	1	0	1	0	0
34	0	0	0	1	1	0	1	1	1	1	0	1	1	0	0	0
123	0	1	1	1	0	0	0	1	0	1	1	1	0	0	0	1
124	0	1	1	0	1	0	1	0	1	0	1	0	1	0	0	1
134	0	1	0	1	1	1	0	0	1	1	0	0	0	1	0	1
234	0	0	1	1	1	1	1	1	0	0	0	0	0	0	1	1
1234	0	1	1	1	1	0	0	0	0	0	0	1	1	1	1	0

We henceforth exclude consideration of the column I of a modular representation table. Its inclusion in an (n, k)-alphabet is clearly a waste of 1 binary digit.

It is easy to show that every column of a modular representation table for C_k contains exactly 2^{k-1} ones. Since an (n, k)-alphabet is made from n such columns the alphabet contains a total of $n2^{k-1}$ ones and we have

Proposition 6. The weights of an (n, k)-alphabet form a partition of $n2^{k-1}$ into $2^k - 1$ non-zero parts, each part being an integer from the set 1, 2, $\cdots$, n.

The identity element always has weight zero, of course.

It is readily established that the product of two elements of even weight is again an element of even weight as is the product of two elements of odd weight. The product of an element of even weight with an element of odd weight yields an element of odd weight.

The elements of even weight of an (n, k)-alphabet form a subgroup and the preceding argument shows that this subgroup must be of order 2^k or 2^{k-1}. If the group of even elements is of order 2^{k-1}, then the collection of even elements is a possible $(n, k - 1)$-alphabet. This $(n, k - 1)$ alphabet may, however, contain the column I of the modular representation table of C_{k-1}. We therefore have

Proposition 7. The partition of Proposition 6 must be either into $2^k - 1$ even parts or else into 2^{k-1} odd parts and $2^{k-1} - 1$ even parts. In the latter case, the even parts form a partition of $\alpha 2^{k-2}$ where α is some integer of the set $k - 1$, k, $\cdots$, n and each of the parts is an integer from the set 1, 2, $\cdots$, n.

2.9 THE CHARACTERS OF C_k

Let us replace the elements of B_n (each of which is a sequence of zeros and ones) by sequences of $+1$'s and -1's by means of the following substitution

$$\begin{aligned} 0 &\leftrightarrow 1 \\ 1 &\leftrightarrow -1. \end{aligned} \tag{13}$$

The multiplicative properties of elements of B_n can be preserved in this new notation if we define the product of two $+1$, -1 symbols to be the symbol whose ith component is the ordinary product of the ith components of the two factors. For example, 1011 and 0110 become respectively $-11\ -1\ -1$ and $1\ -1\ -11$. We have

$$(-11\ -1\ -1)(1\ -1\ -11) = (-1\ -11\ -1)$$

corresponding to the fact that

$$(1011)\ (0110) = (1101)$$

If the $+1$, -1 symbols are regarded as shorthand for diagonal matrices, so that for example

$$-11\ -1\ -1 \leftrightarrow \begin{vmatrix} -1 & 0 & 0 & 0 \\ 0 & 1 & 0 & 0 \\ 0 & 0 & -1 & 0 \\ 0 & 0 & 0 & -1 \end{vmatrix}$$

then group multiplication corresponds to matrix multiplication.

(While much of what follows here can be established in an elementary way for the simple group at hand, it is convenient to fall back upon the established general theory of group representations[8] for several propositions.

The substitution (13) converts a modular representation table (column I included) into a square array of $+1$'s and -1's. Each column (or row) of this array is clearly an irreducible representation of C_k. Since C_k is Abelian it has precisely 2^k irreducible representations each of degree one. These are furnished by the converted modular table. This table also furnishes then the characters of the irreducible representations of C_k and we refer to it henceforth as a *character table.*

Let $\chi^\alpha(A)$ be the entry of the character table in the row labelled A and column labelled α. The orthogonality relationship for characters gives

$$\sum_{A \subset C_k} \chi^\alpha(A)\chi^\beta(A) = 2^k \delta_{\alpha\beta}$$

$$\sum_{\alpha \subset C_k} \chi^\alpha(A)\chi^\alpha(B) = 2^k \delta_{AB}$$

where δ is the usual Kronecker symbol. In particular

$$\sum_{A \subset C_k} \chi^I(A)\chi^\beta(A) = \sum_{A \subset C_k} \chi^\beta(A) = 0, \qquad \beta \neq I$$

Since each $\chi^\beta(A)$ is $+1$ or -1, these must occur in equal numbers in any column $\beta \neq I$. This implies that each column except I of the modular representation table contains 2^{k-1} ones, a fact used earlier.

Every matrix representation of C_k can be reduced to its irreducible components. If the trace of the matrix representing the element A in an arbitrary matrix representation of C_k is $\chi(A)$, then this representation contains the irreducible representation having label β in the character table d_β times where

$$d_\beta = \frac{1}{2^k} \sum_{A \subset C_k} \chi(A)\chi^\beta(A) \tag{14}$$

Every (n, k)-alphabet furnishes us with a matrix representation of C_k by means of (13) and the procedure outlined below (13). The trace $\chi(A)$ of the matrix representing the element A of C_k is related to the weight of the letter by

$$\chi(A) = n - 2w(A) \tag{15}$$

Equations (14) and (15) permit us to compute from the weights of an (n, k)-alphabet what irreducible representations are present in the alphabet and how many times each is contained. It is assumed here that the given alphabet has been made isomorphic to C_k and that the weights are labelled by elements of C_k.

Consider the converse problem. Given a set of numbers w_1, w_2, $\cdots$, w_{2^k} that satisfy Propositions 6 and 7. From these we can compute quantities $\chi_i = n - 2w_i$ as in (15). It is clear that the given w's will constitute the weights of an (n, k)-alphabet if and only if the 2^k χ_i can be labelled with elements of C_k so that the 2^k sums (14) (β ranges over all elements of C_k) are non-negative integers. The integers d_β tell what representations to choose to construct an (n, k)-alphabet with the given weights w_1.

2.10 CONSTRUCTION OF BEST ALPHABETS

A great many different techniques were used to construct the group alphabets listed in Tables II and III and to show that for each n and k

there are no group alphabets with smaller probability of error. Space prohibits the exhibition of proofs for all the alphabets listed. We content ourselves here with a sample argument and treat the case $n = 10$, $k = 4$ in detail.

According to (2) there are $N(10, 4) = 53{,}743{,}987$ different (10, 4)-alphabets. We now show that none is better than the one given in Table III. The letters of this alphabet and weights of the letters are

I	0
1 6 7 8 10	5
2 6 7 9 10	5
3 5 6 8 9 10	6
4 5 7 8 9 10	6
1 2 8 9	4
1 3 5 7 9	5
1 4 5 6 9	5
2 3 5 7 8	5
2 4 5 6 8	5
3 4 6 7	4
1 2 3 5 7 9	6
1 2 4 5 7 10	6
1 3 4 8 10	5
2 3 4 9 10	5
1 2 3 4 6 7 8 9	8

The notation is that of Section 2.1. By actually forming the standard array of this alphabet, it is verified that

$$\alpha_0 = 1, \qquad \alpha_1 = 10, \qquad \alpha_2 = 39, \qquad \alpha_3 = 14.$$

Table II shows $\binom{10}{2} = 45$, whereas $\alpha_2 = 39$, so the given alphabet does not correct all possible double errors. In the standard array for the alphabet, 39 coset leaders are of weight 2. Of these 39 cosets, 33 have only one element of weight 2; the remaining 6 cosets each contain two elements of weight 2. This is due to the two elements of weight 4 in the given group, namely 1289 and 3467. A portion of the standard array that demonstrates these points is

I	1289	3467
.	.	.
.	.	.
12	89	.
18	29	.
19	28	.
34	.	67
36	.	47
37	.	46
.	.	.
.	.	.

In order to have a smaller probability of error than the exhibited alphabet, it is necessary that a (10, 4)-alphabet have an $\alpha_2 > 39$. We proceed to show that this is impossible by consideration of the weights of the letters of possible (10, 4)-alphabets.

We first show that every (10, 4)-alphabet must have at least one element (other than the identity, I) of weight less than 5. By Propositions 6 and 7, Section 2.8, the weights must form a partition of $10 \cdot 8 = 80$ into 15 positive parts. If the weights are all even, at least two must be less than 6 since $14 \cdot 6 = 84 > 80$. If eight of the weights are odd, we see from $8 \cdot 5 + 7 \cdot 6 = 82 > 80$ that at least one weight must be less than 5.

An alphabet with one or more elements of weight 1 must have an $\alpha_2 \leqq 36$, for there are nine elements of weight 2 which cannot possibly be coset leaders. To see this, suppose (without loss of generality) that the alphabet contains the letter 1. The elements 12, 13, 14, $\cdots$ 1 10 cannot possibly be coset leaders since the product of any one of them with the letter 1 yields an element of weight 1.

An alphabet with one or more elements of weight 2 must have an $\alpha_2 \leqq 37$. Suppose for example, the alphabet contained the letter 12. Then 13 and 23 must be in the same coset, 14 and 24 must be in the same coset, $\cdots$, 1 10 and 2 10 must be in the same coset. There are at least eight elements of weight two which are not coset leaders.

Each element of weight 3 in the alphabet prevents three elements of weight 2 from being coset leaders. For example, if the alphabet contains 123, then 12, 13, and 23 cannot be coset leaders. We say that the three elements of weight 2 are "blocked" by the letter of weight 3. Suppose an alphabet contains at least three letters of weight three. There are several cases: (A) if three letters have no numerals in common, e.g., 123, 456, 789, then nine distinct elements of weight 2 are blocked and $\alpha_2 \leqq 36$; (B) if no two of the letters have more than a single numeral in common, e.g., 123, 345, 789, then again nine elements of weight 2 are blocked and $\alpha_2 \leqq 36$; and (C) if two of the letters of weight 3 have two numerals in common, e.g., 123, 234, then their product is a letter of weight 2 and by the preceding paragraph $\alpha_2 \leqq 37$. If an alphabet contains exactly two elements of weight 3 and no elements of weight 2, the elements of weight 3 block six elements of weight 2 and $\alpha_2 \leqq 39$.

The preceding argument shows that to be better than the exhibited alphabet a (10, 4)-alphabet with letters of weight 3 must have just one such letter. A similar argument (omitted here) shows that to be better than the exhibited alphabet, a (10, 4)-alphabet cannot contain more than one element of weight 4. Furthermore, it is easily seen that an alphabet containing one element of weight 3 and one element of weight 4 must have an $\alpha_2 \leqq 39$.

The only new contenders for best (10, 4)-alphabet are, therefore, alphabets with a single letter other than I of weight less than 5, and this letter must have weight 3 or 4. Application of Propositions 6 and 7 show that the only possible weights for alphabets of this sort are: $35^7 6^7$ and $5^8 4 6^6$ where 5^7 means seven letters of weight 5, etc. We next show that there do not exist (10, 4)-alphabets having these weights.

Consider first the suggested alphabet with weights $35^7 6^7$. As explained in Section 2.9, from such an alphabet we can construct a matrix representation of C_4 having the character $\chi(I) = 10$, one matrix of trace 4, seven of trace 0 and seven of trace -2. The latter seven matrices correspond to elements of even weight and together with I must represent a subgroup of order 8. We associate them with the subgroup generated by the elements 2, 3, and 4. We have therefore

$$\chi(I) = 10, \qquad \chi(2) = \chi(3) = \chi(4) = \chi(23)$$
$$\chi(24) = \chi(34) = \chi(234) = -2.$$

Examination of the symmetries involved shows that it doesn't matter how the remaining χ_i are associated with the remaining group elements. We take, for example

$$\chi(1) = 4, \qquad \chi(12) = \chi(13) = \chi(14) = \chi(123)$$
$$= \chi(124) = \chi(134) = \chi(1234) = 0.$$

Now form the sum shown in equation (14) with $\beta = 1234$ (i.e., with the character χ^{1234} obtained from column 1234 of the Table VI by means of substitution (13). There results $d_{1234} = \frac{1}{2}$ which is impossible. Therefore there does not exist a (10, 4)-alphabet with weights $35^7 6^7$.

The weights $5^8 4 6^6$ correspond to a representation of C_4 with character $\chi(I) = 10, 0^8, 2, (-2)^6$. We take the subgroup of elements of even weight to be generated by 2, 3, and 4. Except for the identity, it is clearly immaterial to which of these elements we assign the character 2. We make the following assignment: $\chi(I) = 10$, $\chi(2) = 2$, $\chi(3) = \chi(4) = \chi(23) = \chi(24) = \chi(34) = \chi(234) = -2$, $\chi(1) = \chi(12) = \chi(13) = \chi(14) = \chi(123) = \chi(124) = \chi(134) = \chi(1234) = 0$. The use of equation (14) shows that $d_2 = \frac{1}{2}$ which is impossible.

It follows that of the 53,743,987 (10, 4)-alphabets, none is better than the one listed on Table III.

Not all the entries of Table III were established in the manner just demonstrated for the (10, 4)-alphabet. In many cases the search for a best alphabet was narrowed down to a few alphabets by simple arguments. The standard arrays for the alphabets were constructed and the best alphabet chosen. For large n the labor in making such a table can be considerable and the operations involved are highly liable to error when performed by hand.

I am deeply indebted to V. M. Wolontis who programmed the IBM CPC computer to determine the α's of a given alphabet and who patiently ran off many such alphabets in course of the construction of Tables II and III. I am also indebted to Mrs. D. R. Fursdon who evaluated many of the smaller alphabets by hand.

REFERENCES

1. R. W. Hamming, B.S.T.J., **29**, pp. 147–160, 1950.
2. I. S. Reed, Transactions of the Professional Group on Information Theory, PGIT-4, pp. 38–49, 1954.
3. See section 7 of R. W. Hamming's paper, loc. cit.
4. I.R.E. Convention Record, Part 4, pp. 37–45, 1955 National Convention, March, 1955.
5. C. E. Shannon. B.S.T.J., **27**, pp. 379–423 and pp. 623–656, 1948.
6. Birkhoff and MacLane, A Survey of Modern Algebra, Macmillan Co., New York, 1941. Van der Waerden, Modern Algebra, Ungar Co., New York, 1953. Miller, Blichfeldt, and Dickson, Finite Groups, Stechert, New York, 1938.
7. This theorem has been previously noted in the literature by Kiyasu-Zen'iti, Research and Development Data No. 4, Ele. Comm. Lab., Nippon Tele. Corp. Tokyo, Aug., 1953.
8. F. D. Murnaghan, Theory of Group Representations, Johns Hopkins Press, Baltimore, 1938. E. Wigner, Gruppentheorie, Edwards Brothers, Ann Arbor, Michigan, 1944.

Part II: Constructions for Block Codes

Many of the papers published in coding theory have introduced new codes of one sort or another. The vast majority of these codes have been designed to correct additive errors on one-way memoryless Hamming-metric channels which are symmetric and have orthogonal input signals. It happens that all of the reprints which I selected for this section are codes designed for channels of this type. Several significant papers about codes for other channels narrowly missed inclusion, and I shall mention a few such papers before discussing those which are included.

Several types of codes to correct synchronization errors have been presented in the papers of Levenstein (in Russian) and in the book by Stiffler [1]. Algebraic codes to correct additive errors on one-way memoryless symmetric channels with certain types of nonorthogonal input signals use the Lee metric instead of the Hamming metric. A number of good Lee-metric codes are presented in Berlekamp's Chapter 9 [2], and in the subsequent papers by Golomb and Welch [3] and Chiang and Wolf [4]. Varshamov [5] recently used some ingenious combinatorial methods to construct a class of interesting codes for asymmetric channels. Many authors have proposed various types of codes for channels with feedback, but such codes are being covered in Slepian's volume [6] rather than in this one. The problem of designing codes to correct arithmetic errors has attracted considerable attention, especially in the U.S.S.R. Finally, there has also been a considerable number of papers devoted to the construction of codes which correct bursts on symmetric Hamming-metric channels with various types of memory. The most successful early burst-correcting codes were due to Fire [7]. The most thorough recent survey of this area is due to Forney [8].

The big breakthrough in the construction of error-correcting codes for one-way memoryless Hamming-metric channels was the advent of the RS (Reed-Solomon) codes and the BCH (Bose-Chaudhuri-Hocquenghem) codes, including the nonbinary BCH codes first introduced by Gorenstein and Zierler. The nonbinary BCH codes contain the RS codes as a proper subset. The Reed-Solomon, Bose-Chaudhuri, Hocquenghem, and Gorenstein-Zierler papers all appeared quite soon after it was realized that good binary codes could be described in terms of finite fields which contain the binary field as a subfield.

The possibility of applying finite field theory to problems in discrete communication was recognized in the later 1950's. E. Prange discovered a number of interesting cyclic binary codes, but his work was never published. The mathematics of finite fields were discussed in considerable detail by Zierler [9], whose list of applications included random number generators for Monte Carlo methods, signal design, anti-multipath communication systems, and radar detection, but *omitted* error-correcting codes! Zierler himself rectified this omission a year later with a paper [10] which introduced methods for decoding the cyclic codes of Prange, but the full power of the abstract approach to coding theory was indicated by Mattson-Solomon [11] only after the appearance of the RS, BCH, and GZ papers. The very notion that the theory of finite fields might be applied to *anything* came as an unpleasant shock to certain pure mathematicians, for whom Levinson wrote a very interesting expository paper [12].

The binary BCH codes include the Hamming and Golay codes, as well as several of Prange's codes. They include some RM codes, and are better than the rest. Nevertheless, it soon was known that, in a certain sense, the BCH codes are asymptotically weak. On the other hand, a certain subset of nonbinary BCH codes, namely the RS codes, are optimal, but one can obtain good long RS codes only by letting the alphabet size grow with the block length. This asymptotic difficulty was first removed by Forney [13], who concatenated short random codes with long RS codes and thereby attained codes which were asymptotically error-free in a stronger sense than the Elias codes. Justesen [14] found a way to circumvent Forney's random choice of inner codes, and thereby obtained the first class of asymptotically good codes which were "constructive" in the strictest sense of the term.

The most significant generalization of the BCH codes yet known is due to Goppa [15]. By working in a larger field, Goppa overcame several of the difficulties which had beset a somewhat similar class of earlier codes due to Srivastava [16]. The Goppa codes include the BCH codes as a proper subclass. They also include numerous binary codes of short to moderate block lengths which are better than any BCH codes, although the improvement is often only a single bit in length for the same distance and redundancy. Like the BCH codes, each Goppa code can be decoded by an algebraic decoding algorithm. But while all long BCH codes are bad, almost all long Goppa codes are good. Asymptotically, the Goppa codes satisfy the Gilbert bound.

Another generalization of BCH codes is the class of "polynomial codes" introduced by Kasami, Lin, and Peterson [17]. These codes include both the BCH codes and a number of codes based on finite geometries. Like the RM codes, of which they are a generalization, the codes based on finite geometries are often proper subcodes of BCH codes of the same length and distance. However, the codes based on finite geometries can be decoded by a threshold decoding algorithm. For certain low-rate codes, the threshold decoding algorithm for finite geometry codes proves even easier to implement than any known algorithm for comparable BCH codes. A more thorough discussion of polynomial codes and codes based on finite geometries may be found in the recent book by Peterson and Weldon [18].

While the above authors have investigated linear codes of increasing generality, others have concentrated their attention on smaller classes of codes with additional structure. The most noteworthy successes in this direction have been new classes of *nonlinear* codes with better distances and rates than comparable linear codes. Nonlinear codes are generally relatively difficult to encode and decode, but many of them have interesting symmetry groups or other properties of mathematical interest.

The close relationship between the best low-rate codes, many of which are nonlinear, and Hadamard matrices was apparently first investigated by Levenshtein [19]. In addition to the equidistant codes, he also constructed a number of other optimal codes of slightly higher rates. Unfortunately, his work is still relatively unknown in the West.

Following Hamming and Golay, the search for perfect codes attracted considerable attention. The only success was Vasiliev's [20] discovery of a class of nonlinear perfect single-error-correcting binary codes with the same parameters as the Hamming codes and the generalization of these codes to nonbinary alphabets [21] and to "mixed" alphabets [22]. Pless [23] showed that there is no nonlinear code comparable to the Golay codes. Because of its uniqueness and remarkable properties, the Golay code became the subject of considerable mathematical interest. Goethals and Siedel [24] and Berlekamp, Siedel, and van Lint [25] used the binary and ternary Golay codes to construct strongly regular graphs with exceptional combinatorial properties. Assmus and Mattson [26] pointed out the close relationship between the extended binary Golay code and certain mathematical objects which had been studied much earlier, namely the Steiner system which consists of the extended Golay codewords of weight 8. This system had been previously studied by Witt [27], who investigated the group of all permutations of the 24 coordinates which preserve the system. Witt showed that this was the remarkable Mathieu group, M_{24}, which has order $24 \cdot 23 \cdot 22 \cdot 21 \cdot 20 \cdot 3 \cdot 16$. This group has a number of quite exceptional properties. It is quintuply transitive (meaning that it contains permutations which permute any five coordinates into any other five, in any order), and it is simple (meaning that it has no normal subgroups). Assmus and Mattson showed that, because of the correspondence between the extended Golay codewords of weight 8 and the Steiner system, M_{24} is also the symmetry group of the Golay code.

In 1964, Leech [28] showed that the Golay code could be embedded in an even more exceptional object: the Leech lattice. This lattice consists of those points in (real) 24-dimensional Euclidean space whose coordinates are integers which satisfy two conditions: 1) modulo 2, each lattice point is congruent to an extended Golay codeword, and 2) modulo 4, the coordinates of each lattice point sum to zero. The distance between any two points in the Leech lattice is at least 8. There are two types of points at distance 8 from the origin: 2^7 points congruent to each of the $3 \cdot 23 \cdot 11$ extended Golay codewords of weight 8, and $4 \cdot \binom{24}{2}$ points which have value ± 2 in two coordinates and value 0 in the other 22. Altogether this gives each lattice point 98256 nearest neighbors. An unusually dense packing of spheres in 24-dimensional Euclidean space may be obtained by placing the spheres' centers at the Leech lattice points. Some of the advantages of this packing for signal design problems in continuous communication theory were calculated by Blake [29].

The symmetry group of the Leech lattice was first investigated by Conway [30], who discovered some quite spectacular results. The group has a central reflection of order 2, which, when factored out, leaves a quotient group which is simple and contains the Mathieu groups and most of the other known sporadic simple groups as proper subgroups. Conway's group provides strong support for an important tacit assumption of many group theorists: any sufficiently "optimum" combinatorial object, by virtue of its optimality, is likely to have a high degree of symmetry, and it should therefore be possible to discover and categorize such objects by studying their symmetries via the theory of finite groups.

Leech and Sloane [31] have used techniques based on codes to construct a large number of dense packings of spheres in Euclidean spaces of various dimensions. Many of these packings are substantially better than any previously known. But just as there is no other code so remarkable as the Golay code, there appears to be no other lattice as remarkable as Leech's.

Because of the sparsity of perfect codes, several authors have weakened the definition of perfection. The definition of "nearly" perfect codes recently introduced by Goethals and Snover [32] appears to be more fruitful than the older definition of "quasi"-perfect codes. It is easily seen that quasi-perfect codes are optimum in the weak sense that, for fixed length and rate, their distance cannot be increased. Soon after the BCH codes were introduced, Gorenstein, Peterson, and Zierler [33] proved that the double-error-correcting primitive binary BCH codes were quasi-perfect, and therefore optimum in this weak sense. Later, it was shown that all long high-rate binary BCH codes are optimum in this same weak sense. Later, a number of people investigating bounds suggested stronger definitions of optimality: for fixed length and distance, maximize the number of codewords or (harder still) for fixed distance and redundancy, maximize the length.

Attempts to strengthen the result of Gorenstein, Peterson, and Zierler met with mixed success. It was shown that no *linear* code of distance 5 has more codewords of length 15 than the double-error-correcting binary BCH code, which has 2^7. The best upper bound that could be obtained on the number of codewords of a nonlinear code of distance 5 and length 15 was Johnson's [34] 2^8. Since this was the simplest example in which the difference between the bounds and the known constructions differed by a full power of two, Robinson chose it as an example of a problem which he posed to high school students in an introductory talk on coding theory. One of them, named Nordstrom, accepted the challenge, and by trial and error, constructed a nonlinear code with 2^8 codewords of length 15 and distance 5, the now-classic Nordstrom-Robinson code [35]. This code was also independently discovered by Zietsiev and Zinoviev [36]. Several previously known nonlinear codes, including the Nadler code [37], were found to be shortened versions of the NR code. Goethals [38] showed how the Nordstrom-Robinson code can be more readily derived as a subcode of the Golay code, and Berlekamp [39] used this observation to explain the surprisingly large symmetry group of the Nordstrom-Robinson code. (It is isomorphic to A_7, the alternating group on 7 letters. The symmetry groups of this code and virtually all good codes of lengths less than 25 are closely related to M_{24} in one way or another.)

The Nordstrom-Robinson code was subsequently generalized to two new infinite classes of codes. First, Preparata [40] constructed nonlinear codes of lengths $4m$ and distance 6, each of which has one more information bit than the extended double-error-correcting binary BCH code of the same length. Later, Kerdock [41] constructed an infinite class of nonlinear low-rate codes. The parameters of the Kerdock codes are dual to the parameters of the Preparata codes. The first Kerdock code, as well as the first Preparata code, is the extended Nordstrom-Robinson code.

Some additional related results are mentioned in the section on weight enumerators and bounds, particularly in the survey paper by Sloane.

References

[1] J. J. Stiffler, *Synchronization in Communication Systems*. Englewood Cliffs, N.J.: Prentice-Hall, 1969.

[2] E. R. Berlekamp, *Algebraic Coding Theory*. New York: McGraw-Hill, 1968.

[3] S. W. Golomb and L. R. Welch, "Algebraic coding and the Lee metric," in *Error-Correcting Codes*, H. B. Mann, Ed. New York: Wiley, 1968.

[4] C.-Y. J. Chiang and J. K. Wolf, "On channels and codes for the Lee metric," *Inform. Contr.*, vol. 19, pp. 159-173, 1971.

[5] R. R. Varshamov, "A class of codes for asymmetric channels and a problem from the additive theory of numbers," *IEEE Trans. Inform. Theory*, vol. IT-19, pp. 92-95, Jan. 1973.

[6] D. Slepian, *Key Papers in the Development of Information Theory*. New York: IEEE Press, 1974.

[7] P. Fire, "A class of multiple-error-correcting codes for non-independent errors," Sylvania Reconnaissance Syst. Lab. Rep. RSL-E-2.

[8] G. D. Forney, Jr., "Burst-correcting codes for the classic bursty channel," *IEEE Trans. Commun. Technol.*, vol. COM-19, pp. 772-781, Oct. 1971.

[9] N. Zierler, "Linear recurring sequences," *SIAM J. Appl. Math.*, vol. 7, pp. 31-48, 1959.

[10] N. Zierler, "On decoding linear error-correcting codes," *IRE Trans. Inform. Theory*, vol. IT-6, pp. 450-459, Sept. 1960.

[11] H. F. Mattson and G. Solomon, "A new treatment of Bose-Chaudhuri codes," *J. Soc. Industr. Appl. Math.*, vol. 9, pp. 654-669; *Math. Rev.*, vol. 24B, p. 1705.

[12] N. Levinson, "Coding theory: A counterexample to G. H. Hardy's conception of applied mathematics," *Amer. Math. Monthly*, vol. 77, pp. 249-258, 1970.

[13] G. D. Forney, Jr., *Concatenated Codes*, Res. Monograph 37. Cambridge, Mass.: MIT Press, 1966.

[14] J. Justesen, "A class of constructive, asymptotically good algebraic codes," *IEEE Trans. Inform. Theory*, vol. IT-18, pp. 652-656, Sept. 1972.

[15] V. D. Goppa, "New class of linear correcting codes," *Probl. Peredach. Inform.*, vol. 6, pp. 24-30, 1970.
—, "Rational presentation of codes and (L, g)-codes," *Probl. Peredach. Inform.*, vol. 7, pp. 41-49, 1971.

[16] J. N. Srivastava, unpublished remarks at the Combinatorial Symp., Univ. North Carolina, Chapel Hill, Apr. 10-14, 1967.

[17] T. Kasami, S. Lin, and W. W. Peterson, "Polynomial codes," *IEEE Trans. Inform. Theory*, vol. IT-14, pp. 807-814, Nov. 1968.

[18] W. W. Peterson and E. J. Weldon, Jr., *Error-Correcting Codes*, 2nd ed. Cambridge, Mass.: MIT Press, 1972.

[19] V. I. Levenshtein, "The application of Hadamard matrices to a coding problem," *Probl. Kibern.*, vol. 5, pp. 123-136 (English translation, pp. 166-184), 1961.

[20] Ju. L. Vasil'ev, "On nongroup close-packed codes," *Probl. Cybern.*, vol. 8, pp. 337-339, 1968; *Math. Rev.*, vol. 29, p. 5661.

[21] J. Schönheim, "On linear and nonlinear single-error-correcting q-nary perfect codes," *Inform. Contr.*, vol. 12, pp. 23-26, 1968.

[22] M. Herzog and J. Schönheim, "Linear and nonlinear single-error-correcting perfect mixed codes," *Inform. Contr.*, vol. 18, pp. 364-368, 1971.

[23] V. Pless, "On the uniqueness of the Golay codes," *J. Combinatorial Theory*, vol. 5, no. 3, pp. 215-228, 1968; *Math. Rev.*, vol. 39, p. 3892.

[24] J. M. Goethals, and J. J. Seidel, "Strongly regular graphs derived from combinatorial designs," *Can. J. Math*, vol. 22, pp. 597-614, 1970.

[25] E. R. Berlekamp, J. J. Seidel, and J. H. van Lint, "A strongly regular graph derived from the perfect ternary Golay code," in *A Survey of Combinatorial Theory*, J. N. Srivastava, Ed. Amsterdam: North-Holland/American Elsevier, 1973, pp. 25-30.

[26] E. F. Assmus, Jr. and H. F. Mattson, "Perfect codes and Mathieu groups," *Arch. Math. Naturvidensk.*, vol. 17, pp. 121-135, 1966.

[27] E. Witt, "Die 5-Fach Transitiven Gruppen von Mathieu," *Abh. Math. Sem. Univ. Hamburg*, vol. 12, pp. 256-264, 1938.

[28] —, "Ueber Steinersche Systeme," *Abh. Math. Sem. Univ. Hamburg*, pp. 265-275, 1938.

[29] J. Leech, "Some sphere packings in higher space," *Can. J. Math.*, vol. 16, pp. 657-682, 1964.

[30] I. F. Blake, "The Leech lattice as a code for the Gaussian channel," *Inform. Contr.*, vol. 19, pp. 66-74, 1971.

[31] J. H. Conway, "A group of order 8, 315, 553, 613, 086, 720," *Bull. London Math. Soc.*, vol. 1, pp. 79-88, 1969.

[32] J. Leech, and N.J.A. Sloane, "Sphere packings and error-correcting codes," *Can. J. Math.*, vol. 23, no. 4, pp. 718-745, 1971.

[33] J. M. Goethals, and S. L. Snover, "Nearly perfect binary codes," *Discrete Math.*, vol. 3, pp. 65-88, 1972.

[34] D. C. Gorenstein, W. W. Peterson, and N. Zierler, "Two-error correcting Bose-Chaudhuri codes are quasi-perfect," *Inform. Contr.*, vol. 3, pp. 291-294, 1960; *Math. Rev.*, vol. 22, p. 9350.

[35] S. M. Johnson," A new upper bound for error-correcting codes," *IRE Trans. Inform. Theory*, vol. IT-18, pp. 203-207, Apr. 1962; *Math. Rev.*, vol. 25, p. 1067.

[36] A. W. Nordstrom and J. P. Robinson, "An optimum nonlinear code," *Inform. Contr.*, vol. 11, pp. 613-616, 1967.

[37] N. V. Semakov and V. A. Zinoviev, "Perfect and quasi-perfect codes of constant weight," *Probl. Peredach. Inform.*, vol. 5, no. 3, pp. 14-18, 1969.

[38] J. H. van Lint, "A new description of the Nadler code," *IEEE Trans. Inform. Theory* (Corresp.), vol. IT-18, pp. 825-826, Nov. 1972.

[39] J. M. Goethals, "On the Golay perfect binary code," *J. Combinatorial Theory*, vol. II, pp. 178-186, 1971.

[40] E. R. Berlekamp, "Coding theory and the Mathieu groups," *Inform. Contr.*, vol. 18, pp. 40-64, 1971.

[41] F. P. Preparata, "A class of optimum nonlinear double-error-correcting codes," *Inform. Contr.*, vol. 13, pp. 378-400, 1968; *Math. Rev.*, vol. 39, p. 3894.

[42] A. M. Kerdock, "A class of low-rate nonlinear binary codes," *Inform. Contr.*, vol. 20, pp. 182-187, 1972.

POLYNOMIAL CODES OVER CERTAIN FINITE FIELDS*†

I. S. REED AND G. SOLOMON‡

Introduction. A code is a mapping from a vector space of dimension m over a finite field K (denoted by $V_m(K)$) into a vector space of higher dimension $n > m$ over the same field ($V_n(K)$). K is usually taken to be the field of two elements Z_2, in which case it is a mapping of m-tuples of binary digits (bits) into n-tuples of binary digits. If one transmits n bits, the additional $n - m$ bits are "redundant" and allow one to recover the original message in the event that noise corrupts the signal during transmission and causes some bits of the code to be in error. A multiple-error-correcting code of order s consists of a code which maps m-tuples of zeros and ones into n-tuples of zeros and ones, where m and n both depend on s, and a decoding procedure which recovers the message completely, assuming no more than s errors occur during transmission in the vector of n bits. The Hamming code [1] is an example of a systematic one bit error-correcting code. We present here a new class of redundant codes along with a decoding procedure.

Let K be a field of degree n over the field of two elements Z_2. K contains 2^n elements. Its multiplicative group is cyclic and is generated by powers of α where α is the root of a suitable irreducible polynomial over Z_2. We discuss here a code E which maps m-tuples of K into 2^n-tuples of K.

Consider the polynomial $P(x)$ of degree $m - 1$

$$P(x) = a_0 + a_1x + \cdots + a_{m-1}x^{m-1},$$

where $a_i \in K$ and $m < 2^n$. Code E is the mapping of the m-tuple $(a_0, a_1, \cdots, a_{m-1})$ into the 2^n-tuple $(P(0), P(\alpha), P(\alpha^2), \cdots, P(1))$; this m-tuple might be some encoded message and the corresponding 2^n-tuple is to be transmitted. This mapping of m symbols into 2^n symbols will be shown to be $(2^n - m)/2$ or $(2^n - m - 1)/2$ symbol correcting, depending on whether m is even or odd.

A natural correspondence is established between the field elements of K and certain binary sequences of length n. Under this correspondence, code E may be regarded as a mapping of binary sequences of mn bits into binary sequences of $n2^n$ bits. Thus code E can be interpreted to be a systematic multiple-error-correcting code of binary sequences.

One should note that the binary representation of code E allows in general for the correction of more than $(2^n - m - 1)/2$ bits since each symbol of the code is represented by n consecutive bits. Hence when the binary errors are strongly correlated or occur in "bursts," this code may be more desirable than other more "efficient" multiple-error-correction codes.

Finally, it should be mentioned that code E may be generalized to polynomials of the mth degree in several variables over K. Evidently, for $K = Z_2$, such codes reduce to Reed-Muller codes [2].

The code E. Consider the field $K = Z_2(\alpha)$. This is the vector space over Z_2 with basis $1, \alpha, \alpha^2, \cdots, \alpha^{n-1}$, where α is the root of a suitable irreducible polynomial over Z_2. The nonzero elements of K form a multiplicative cyclic group. Thus we may represent the elements of K in the order

$$0, \beta, \beta^2, \cdots, \beta^{2^n-2}, \beta^{2^n-1} = 1$$

where β is a generator of the multiplicative cyclic group.

Let $P(x) = a_0 + a_1x + a_2x^2 + \cdots + a_{m-1}x^{m-1}$. The code E sends

$$(a_0, a_1, \cdots, a_{m-2}, a_{m-1}) \rightarrow (P(0), P(\beta), P(\beta^2), \cdots, P(\beta^{2^n-2}), P(1)).$$

Upon receiving the message $(P(0), P(\beta), \cdots, P(1))$, we may decode the message by solving simultaneously any m of the 2^n equations,

$$\begin{aligned} P(0) &= a_0 \\ P(\beta) &= a_0 + a_1\beta + a_2\beta^2 + \cdots + a_{m-1}\beta^{m-1} \\ P(\beta^2) &= a_0 + a_1\beta^2 + a_2\beta^4 + \cdots + a_{m-1}\beta^{2m-2} \\ &\cdot \quad \cdot \quad \cdot \quad \cdot \quad \cdot \quad \cdot \\ P(1) &= a_0 + a_1 + a_2 + \cdots + a_{m-1}. \end{aligned}$$

* Received by the editors January 21, 1959 and in revised form August 26, 1959.

† The work reported here was performed at Lincoln Laboratory, a technical center operated by Massachusetts Institute of Technology with the joint support of the Army, Navy and Air Force, under contract.

‡ Staff members, Lincoln Laboratory, Massachusetts Institute of Technology, Lexington 73, Massachusetts.

We note that any m of these equations are linearly independent since the coefficient determinant for, say, $P(\alpha_1), \cdots, P(\alpha_m)$, is

$$\begin{vmatrix} 1 & \alpha_1 & \alpha_1^2 & \cdots & \alpha_1^{m-1} \\ 1 & \alpha_2 & \alpha_2^2 & \cdots & \alpha_2^{m-1} \\ \cdot & \cdot & \cdot & \cdot & \cdot \\ 1 & \alpha_m & \alpha_m^2 & \cdots & \alpha_m^{m-1} \end{vmatrix}$$

which is a Vandermonde determinant whose value is

$$= \prod_{j<i} (\alpha_i + \alpha_j) \neq 0.$$

Thus in the case of no errors in the received values of $P(\cdot)$, we obtain $\binom{2^n}{m}$ determinations of $(a_0, \cdots, a_{m-1})$.

Any errors occurring in the values of $P(\cdot)$ will immediately disturb the unanimity of the values obtained for the a_n's. Indeed, for sufficiently small numbers of errors, by looking at the largest number of determinations for any $(a_0, \cdots, a_{m-1})$ (the plurality of votes received by any m-tuple) we may detect the order of error made and correct it. We prove the following statement.

Lemma. For s errors we can get at most $\binom{s + m - 1}{m}$ determinations for a wrong m-tuple.

Proof. We look upon the simultaneous solution of m equations as the intersection of m hyperplanes. The linear independence guarantees that they meet at only one point. To obtain more than one solution for any m-tuple, we would need more than m hyperplanes meeting at that point. For a wrong m-tuple, we can have at most $s + m - 1$ hyperplanes intersecting at a single wrong point, where s is the number of mistaken equations and where the remaining $m - 1$ equations are chosen from the $2^n - s$ correct ones. Any more correct hyperplanes would determine the correct solution, i.e., a different point of intersection from the assumed wrong one. Therefore, there are at most $\binom{s + m - 1}{m}$ determinations for any wrong value. Note that we get $\binom{2^n - s}{m}$ determinations for the correct one, and a total of $\binom{2^n}{m} - \binom{2^n - s}{m}$ wrong determinations.

Thus, by examining the vote received by the individual candidates $(a_0, \cdots, a_{m-1})$, we may determine the correct message and the number s.

Note that this is valid only when

$$\binom{2^n - s}{m} > \binom{s + m - 1}{m}$$

or

$$2^n - s > s + m - 1$$

or

$$s < \frac{2^n - m + 1}{2}.$$

The code will thus correct errors of order less than $(2^n - m + 1)/2$. For m odd, we get corrections up to $s = (2^n - m - 1)/2$, and detection at $s = (2^n - m + 1)/2$. For m even, we can correct up to $s = (2^n - m)/2$ and not detect any further errors.

Translation of K into a binary alphabet. We represent the elements of K by n-tuples of zeros and ones, $V_n(Z_2)$, and define a multiplication on $V_n(Z_2)$ corresponding to the multiplication of K. We again note that the multiplicative group of K is generated by powers of β. Let us consider an irreducible polynomial f which generates K over Z_2. Suppose $f(x) = x^n + c_1x^{n-1} + \cdots + c_{n-1}x + c_n = 0$, $c_i \in Z_2$. Following N. Zierler [3], we associate the following finite difference equation

$$a_{n+k} + c_1a_{n-1+k} + c_2a_{n-2+k} + \cdots + c_na_{0+k} = 0$$

where $a_i \in Z_2$.

Thus for any fixed f (giving rise to $(c_1, \cdots, c_n)$) and arbitrary $(a_0, \cdots, a_{n-1})(a_i \neq 0$ for $i = 0, 1, \cdots, n - 1)$ we have a sequence

$$a_0, a_1, \cdots, a_{n-r}, a_n, a_{n+1}, a_{n+2}, \cdots$$

Reprinted with permission from *J. Soc. Ind. Appl. Math.*, vol. 8, pp. 300-304, June 1960.

where the values of a_i for $i \geqq n$ are determined by the above difference equation. Zierler has shown that for suitable irreducible f, the sequence (a_n) is periodic of period $2^n - 1$, i.e., $a_{2^n-1} = a_0$, $a_{2^n+m-1} = a_m$ and the $2^n - 1$ sequences of length n obtained by translating the n-tuple $(a_0, a_1, \cdots, a_{n-1})$ along the derived sequence are all distinct.

Thus if we define

$$\beta = (a_0, \cdots, a_{n-1})$$
$$\beta^2 = (a_1, \cdots, a_n)$$
$$\cdot \quad \cdot \quad \cdot \quad \cdot \quad \cdot$$
$$\beta^m = (a_{m-1}, \cdots, a_{n+m-2})$$

we have a multiplication table for the n-tuples. In other words, multiplication of the elements is simply translation along this periodic sequence generated by f. Note too that the elements β satisfy the algebraic equations satisfied by corresponding elements in K. We have thus defined multiplication on $V_n(Z_2)$ to make this correspond with the multiplication on K.

We remark that the initial choice of $\beta = (a_0, \cdots, a_{n-1})$ is arbitrary and there are $2^n - 1$ such representations. There are of course many other ways of associating vectors with powers of β. The referee has suggested another natural algebraic association of $V_n(Z_2)$ with K.

We identify K with the ring of polynomials in x with coefficients in Z_2, (i.e., $Z_2[x]$) modulo the prime ideal generated by the irreducible $f(x)$. Let $\beta = (a_0, a_1, \cdots, a_{n-1})$ be a nonzero vector of $V_n(Z_2)$. We associate with β the polynomial $\beta(x) = a_0 + a_1x + a_2x^2 + \cdots a_{n-1}x^{n-1} \bmod f(x)$. Consider $\beta(x)^k \bmod f(x)$. This again is a polynomial of formal degree $(n - 1)$. Let β^k be the vector whose components are the n coefficients of this $(n - 1)$-degree polynomial. This establishes a one-one correspondence of $V_n(Z_2)$ with K(if $(0, 0, \cdots, 0)$ is added to correspond to zero in K). While this may be a more natural choice, we prefer our first representation as the more suitable for computability.

Example. Let $n = 3$, $m = 3$. $K = Z_2(\alpha)$ where α is root of $x^3 + x + 1 = 0$. $P(x) = b_0 + b_1x + b_2x^2$.

Code E: $(b_0, b_1, b_2) \rightarrow (P(0), P(\alpha), P(\alpha^2), \cdots, P(\alpha^6), P(1))$.

Binary translation of this code. To $f(x) = x^3 + x + 1$ we associate the difference equation

$$a_n = a_{n-2} + a_{n-3} \qquad (\text{for } n = 3, 4, 5, \cdots).$$

Choose $a_0 = 1$, $a_1 = 1$, $a_2 = 0$. Then

$$\{a_n\} = (1, 1, 0, 0, 1, 0, 1, 1, 1, 0, 0, 1, 0, 1, \cdots).$$

$\{a_n\}$ has period 7, i.e., $a_7 = a_0$, $a_8 = a_1$.

$$0 = (0, 0, 0)$$
$$\alpha = (1, 1, 0)$$
$$\alpha^2 = (1, 0, 0)$$
$$\alpha^3 = (0, 0, 1)$$
$$\alpha^4 = (0, 1, 0)$$
$$\alpha^5 = (1, 0, 1)$$
$$\alpha^6 = (0, 1, 1)$$
$$1 = \alpha^7 = (1, 1, 1).$$

The message $(0, \alpha, \alpha^3) \rightarrow (P(0), P(\alpha), P(\alpha^2), \cdots, P(\alpha^6), P(1))$ translates into (via $P(x) = \alpha x + \alpha^3 x^2$)

$$(\ 0\ 0\ 0\ 1\ 1\ 0\ 0\ 0\ 1\ )$$
$$\rightarrow (\ 0\ 0\ 0,\ 0\ 0\ 1,\ 1\ 1\ 0,\ 1\ 1\ 0,\ 1\ 1\ 1,\ 0\ 0\ 0,\ 0\ 0\ 1,\ 1\ 1\ 1\ ).$$

This code is error correcting up to $(2^3 - 3 - 1)/2 = 2$ symbols.

REFERENCES

1. R. W. Hamming, *Error detecting and error correcting codes*, Bell System Tech. J., 26 (1950), pp. 147–160.
2. I. S. Reed, *A class of multiple-error-correcting codes and the decoding scheme*, Trans. I.R.E., Prof. Group on Information Theory No. 4 (1954), pp. 38–49.
3. N. Zierler, *Linear recurring sequences*, this Journal, 7 (1959), pp. 31–48.

Codes correcteurs d'erreurs

par A. HOCQUENGHEM,

Professeur au Conservatoire des Arts et Métiers,
Ingénieur conseil à la S.E.A.

Généralisant un travail de Hamming, l'auteur construit des codes permettant de corriger k erreurs dans une transmission de digits binaires.

The paper is a generalization of Hamming's work. The author gives a coding system available to correct k errors in a transmission of binary digits.

Eine Arbeit von Hamming verallgemeinernd, entwickelt der Autor Kodes die es ermöglichen bei Ubertragung binärer bits k Fehler zu korrigieren.

Обобщая работу Хамминга, автор предлагает коды, которые дают возможность исправлять k ошибок в передаче двоичных цифр.

1. Introduction.

Introduisons dans un système de transmission un mot, constitué par un nombre de n chiffres binaires :

$$a_1\, a_2 \,\ldots\ldots\ldots\, a_n$$

Le mot reçu peut différer du mot initial par un certain nombre d'erreurs (certains chiffres a_i étant altérés en $1 - a_i$). Pour essayer de détecter et de corriger ces erreurs, on n'utilise que m chiffres du mot comme support de l'information, les chiffres restant appelés chiffres de test devant servir à la vérification du mot après la transmission. Donner une loi de détermination de ces chiffres de test en fonction des m chiffres d'information de façon à pouvoir détecter — ou corriger — un nombre maximum k d'erreurs, c'est former un code détecteur — ou correcteur — de k erreurs.

L'exemple le plus simple est le code détecteur d'une erreur. Dans ce cas $m = n - 1$, et on choisit le chiffre de test de façon que le nombre total de chiffres 1 du mot soit pair. La vérification du mot consiste alors en un test de parité.

Hamming (*Bell System Technical Journal*, 1950) a donné la loi de formation d'un code correcteur d'une erreur. Le nombre de chiffres de test est l'entier N déterminé par les inégalités

$$\text{(11)} \qquad \mathrm{Log}_2(1+n) \leq \mathrm{N} < 1 + \mathrm{Log}_2(1+n)$$

Dans le cas général d'un code correcteur de k erreurs, le nombre de configurations d'erreurs possibles est :

$$\mathrm{H} = 1 + \mathrm{C}_n^1 + \mathrm{C}_n^2 + \ldots + \mathrm{C}_n^k$$

Par suite le code le plus économique utiliserait un nombre de chiffres de test égal à l'entier immédiatement supérieur à $\mathrm{Log}_2\mathrm{H}$. A part le code de Hamming, on n'a pu construire de tels codes. Ceux que nous proposons utilisent un nombre de chiffres de test égal à

$$n - m = kN$$

La différence

$$k\mathrm{N} - \mathrm{Log}_2\mathrm{H}$$

est de l'ordre de $\mathrm{Log}_2(k!)$, donc assez faible pour que ces codes soient satisfaisants.

Après avoir défini un anneau dans lequel nous ferons nos calculs, nous exposerons le code de Hamming sous cette optique, puis les principes de formation des codes qui nous conduiront à une détermination quasi-expérimentale et à une détermination systématique de ces codes. Nous terminerons par un exemple de code correcteur de 2 erreurs.

2. Définition de l'anneau $\mathcal{A}$

Les éléments de l'anneau $\mathcal{A}$ sont les nombres entiers écrits en numération binaire.

A chaque élément de l'anneau $\mathcal{A}$ nous faisons correspondre un polynôme ayant comme coefficients les chiffres de l'élément. Le polynôme est alors défini sur le corps de caractéristique 2.

Toute opération sur les éléments de $\mathcal{A}$ sera faite sur les polynômes correspondants — au cours de ces opérations tout coefficient pair sera remplacé par O, tout coefficient impair par 1. Le résultat sera un polynôme auquel correspondra un élément de l'anneau $\mathcal{A}$.

On a donc toutes les opérations habituelles sur les nombres entiers — afin d'éviter toute ambiguïté, toutes les expressions calculées selon ces règles seront suivies de l'indication $(\mathcal{A})$.

Exemples :

Addition : $101 + 111 = 10 \quad (\mathcal{A})$

Multiplication : $101 \times 111 = 11.011 \quad (\mathcal{A})$

Puissance : $101^2 = 10.001 \quad (\mathcal{A})$

Division : $1.101 = 111 \times 10 + 11 \quad (\mathcal{A})$

En particulier : $p + p = 0,\ (p+q)^2 = p^2 + q^2 \quad (\mathcal{A})$

Lorsque le polynôme sera irréductible sur le corps de caractéristique 2, nous dirons que le nombre correspondant est irréductible (il n'admet pas, dans l'anneau $\mathcal{A}$, d'autre diviseur que lui-même et l'unité).

On peut classer évidemment les nombres dans l'anneau $\mathcal{A}$ par ordre de grandeur, mais beaucoup plus important est le nombre de chiffres. On démontre que parmi les nombres ayant un nombre de chiffres donné, il existe toujours un nombre irréductible.

Etant donné un mot écrit en binaire

$$a_1\, a_2 \ldots a_n$$

nous attacherons à chaque indice i un nombre p_i de l'anneau $\mathcal{A}$) et au mot lui-même nous attacherons le nombre

$$\mathrm{T} = a_1\, p_1 + a_2\, p_2 + \ldots + a_n\, p_n \qquad (\mathcal{A})$$

C'est la considération du nombre T qui, grâce à un choix convenable des nombres p_i nous permettra de corriger les erreurs éventuelles.

3. Code de Hamming.

Nous retrouvons le code de Hamming en faisant

$$p_i = i$$

Les chiffres de test sont les chiffres du mot d'indices

$$1,\ 2,\ 2^2,\ \ldots,\ 2^{\mathrm{N}-1} \text{ (N défini par les inégalités 11).}$$

L'information sera portée par les chiffres

$$a_3\ a_5\ a_6\ a_7\ a_9\ \ldots\ a_n$$

On détermine les chiffres de test par la condition

$$\mathrm{T} = \Sigma\, p_i\, a_i = 0$$

condition qui s'écrit ici

$$a_1 + 2a_2 + 4a_4 + \ldots + 2^{\mathrm{N}-1}\, a_{2^{\mathrm{N}-1}} = 3a_3 + 5a_5 + 6a_6 + \ldots + n$$
$$(\mathcal{A}$$

Le second membre est un nombre binaire connu d'au plus N chiffres. L'égalité détermine donc parfaitement les valeurs des chiffres de test.

Si, après transmission, il n'y a pas d'erreur, on retrouvera $\mathrm{T} = 0$.

S'il y a une erreur, portant par exemple sur le chiffre a_α remplacé par $(1 - a_\alpha)$, le nombre T prendra la valeur :

Reprinted with permission from *Chiffres*, vol. 2, pp. 147-156, 1959.

$$T = a_1 + 2a_2 + \ldots + \alpha(1 - a_\alpha) + \ldots + na_n = \alpha \qquad (\mathcal{A})$$

La valeur de T sera l'indice du chiffre erronné.

S'il y a deux erreurs, portant sur les chiffres d'indice α et β, T prendra la valeur :

$$T = \alpha + \beta \neq 0 \qquad (\mathcal{A})$$

S'il y a plus de deux erreurs, T pourrait être nul. Le code obtenu est donc correcteur d'une erreur, détecteur de deux erreurs.

Il est commode, pour automatiser le contrôle, de supposer les nombres p disposés en matrice. Par exemple pour $n = 7$, on aura la matrice

$$\begin{vmatrix} 0\,0\,0\,1\,1\,1\,1 \\ 0\,1\,1\,0\,0\,1\,1 \\ 1\,0\,1\,0\,1\,0\,1 \end{vmatrix}$$

Aux nombres

0 0 1 0 1 1 0 et 0 0 1 0 0 1 0

correspondront les matrices

$$\begin{vmatrix} 0\,0\,0\,0\,1\,1\,0 \\ 0\,0\,1\,0\,0\,1\,0 \\ 0\,0\,1\,0\,1\,0\,0 \end{vmatrix} \quad \text{et} \quad \begin{vmatrix} 0\,0\,0\,0\,0\,1\,0 \\ 0\,0\,1\,0\,0\,1\,0 \\ 0\,0\,1\,0\,0\,0\,0 \end{vmatrix}$$

Le nombre T s'obtient en faisant suivre chaque ligne de la matrice de son chiffre de parité (§ 1).

On obtient ici :

$$\begin{vmatrix} 0 \\ 0 \\ 0 \end{vmatrix} \quad \text{et} \quad \begin{vmatrix} 1 \\ 0 \\ 1 \end{vmatrix} = 5$$

Le premier nombre est correct, le 5e chiffre du second nombre est faux.

4. Principe d'un code correcteur de k erreurs.

Voyons maintenant à quelles conditions doivent satisfaire les nombres p pour que le calcul de T permette de corriger k erreurs.

Nous supposerons que les chiffres de test sont en nombre suffisant pour que, connaissant les chiffres d'information, on puisse réaliser la condition

(41) $$T = \Sigma a_i p_i = 0 \qquad (\mathcal{A})$$

Si après transmission, les chiffres de rang

$$\alpha_1, \alpha_2 \ldots, \alpha_j \qquad (j \leq k)$$

sont erronés, le nombre T calculé sur le mot déformé prendra la valeur :

$$T = p_{\alpha_1} + p_{\alpha_2} + \ldots + p_{\alpha_j} \qquad (\mathcal{A})$$

Il faut que le nombre ainsi trouvé soit caractéristique des rangs $\alpha_1, \alpha_2, \ldots, \alpha_j$, c'est-à-dire que :

$$p_{\alpha_1} + p_{\alpha_2} + \ldots + p_{\alpha_j} \neq p_{\alpha'_1} + p_{\alpha'_2} + \ldots + p_{\alpha'_{j'}} \qquad (\mathcal{A})$$

lorsque

$$j \leq k, \quad j' \leq k$$

et les deux ensembles

$$(\alpha_1, \alpha_2, \ldots, \alpha_j), \qquad (\alpha'_1, \alpha'_2, \ldots, \alpha'_{j'}),$$

non identiques.

Cette condition peut encore s'écrire :

(42) $$p_{\lambda_1} + p_{\lambda_2} + \ldots + p_{\lambda_l} \neq 0 \qquad (\mathcal{A})$$

lorsque $l \leq 2k$

et les $\lambda_1, \lambda_2, \ldots, \lambda_l$ étant tous différents.

On devra donc choisir les nombres p tels que l'addition, dans l'anneau $\mathcal{A}$, d'au plus $2k$ de ces nombres donne un résultat non nul.

Une fois déterminé un ensemble de n nombres p, il faudra choisir les chiffres de test. Il est commode pour cela de remplacer l'ensemble obtenu par un autre ensemble de n nombres mais contenant les puissances successives de 2 :

$$1, 2, 2^2, \ldots, 2^{K-1}$$

K désignant le nombre de chiffres du plus grand nombre p obtenu.

Disposons pour cela les nombres p en une matrice M de n colonnes et K lignes ($K < n$), chaque nombre p étant donc représenté par une colonne

$$p_i = \begin{vmatrix} \varpi_i^K \\ \vdots \\ \varpi_i^2 \\ \varpi_i^1 \end{vmatrix}$$

Si le rang de cette matrice (dans l'anneau $\mathcal{A}$) est $K' < K$, c'est que $K - K'$ lignes de cette matrice sont des combinaisons linéaires des K' lignes restantes. Si l'on supprime ces $K - K'$ lignes, on obtiendra une matrice M' de nombres p' qui vérifieront encore la condition (42).

Ceci étant, nous pourrons extraire de la matrice M' une matrice carrée Δ de K' lignes dont le déterminant calculé dans l'anneau $\mathcal{A}$ ne sera pas nul. On aura donc

$$\det \Delta = 1$$

puisque les seules valeurs possibles sont 0 ou 1. En multipliant la matrice M' par Δ^{-1}, on obtiendra la matrice M'' formée de nombres p'' tels que

$$p''_i = \begin{vmatrix} \varpi''^{K'}_i \\ \vdots \\ \varpi''^{1}_i \end{vmatrix} = \Delta^{-1} \begin{vmatrix} \varpi'^{K'}_i \\ \vdots \\ \varpi'^{1}_i \end{vmatrix}$$

et par suite les nombres p''_i vérifieront encore la condition (42). De plus, la matrice M'' contiendra à ce moment la matrice $\Delta^{-1} \times \Delta$, c'est-à-dire la matrice unité, donc l'ensemble des p'' contiendra les puissances successives de 2 :

$$1, 2, 2^2, \ldots, 2^{K'-1}$$

Les indices correspondants seront pris comme chiffres de test et la condition (41) déterminera ces chiffres en fonction des chiffres d'information par égalité de deux nombres binaires de K' chiffres.

Tout le problème se ramène donc à construire des ensembles de nombres p satisfaisant à la condition (42).

5. Formation de proche en proche d'une suite de nombres p.

Prenons d'abord :

$$p_1 = 1, \ p_2 = 2, \ p_3 = 2^2, \ \ldots, \ p_{2k} = 2^{2k-1}$$

puis :

$$p_{2k+1} = 2^{2k} - 1$$

$$p_{2k+2} = 2^{2k}$$

Ces nombres satisfont déjà aux conditions (42). Pour prolonger cette suite dans l'ordre des p croissants, supposons être arrivé au nombre p_i de l chiffres. Considérons l'ensemble des nombres p_1 à p_i et de leurs sommes dans l'anneau $\mathcal{A}$ par groupes de 2, 3, ... $(2k - 1)$. Tous les nombres obtenus ont au plus l chiffres.

S'il existe un nombre non contenu dans l'ensemble ainsi formé et compris entre p_i et 2^l, ce nombre sera pris pour valeur de p_{i+1} (s'il y a plusieurs nombres on choisira évidemment le plus petit). Sinon on prendra $p_{i+1} = 2^l$.

On peut ainsi continuer pas à pas jusqu'à l'obtention des n nombres p. Si p_n a K chiffres, les nombres

$$1, 2, 2^2, \ldots, 2^{K-1}$$

seront inclus dans la suite des p. La suite sera donc directement utilisable pour former un code. Il restera à établir le tableau de correspondance entre les H valeurs de la somme

$$p_{\alpha_1} + p_{\alpha_2} + \ldots + p_{\alpha_j} \qquad (\mathcal{A}) \qquad (j \leq k)$$

et la valeur des indices $\alpha_1, \alpha_2, \ldots, \alpha_j$.

Le procédé ainsi défini est assez long à exploiter. Cependant, pour des valeurs raisonnables de n et k, il ne dépasse pas les possibilités d'une calculatrice de moyenne puissance.

La détermination à priori du nombre K de chiffres de test paraît assez difficile. Aussi allons-nous exposer un procédé plus systématique de recherche des nombres p.

6. Formation systématique des nombres p.

La théorie des congruences, si utilisée dans les preuves des opérations arithmétiques, va nous fournir un mode de calcul des nombres p. Désignons par ρ un nombre irréductible de $N + 1$ chiffres et par q' le reste de la division dans l'anneau $\mathcal{A}$ d'un nombre q par ρ. Le nombre q' aura au maximum N chiffres.

Nous poserons alors :

$$p_i = i + 2^N(i^3)' + 2^{2N}(i^5)' + \ldots + 2^{(k-1)N}(i^{2k-1})' \qquad (\mathcal{A})$$

$$(i = 1, 2, \ldots, n)$$

c'est-à-dire que le nombre p_i est formé de la juxtaposition des restes successifs de la division par ϱ des puissances impaires dans l'anneau $\mathcal{A}$ du nombre i. Nous allons montrer que ces nombres p_i satisfont à la condition (42).

En effet, supposons :

$$(61) \qquad p_{\lambda_1} + p_{\lambda_2} + \dots + p_{\lambda_l} = 0 \qquad (\mathcal{A}) \qquad (l \leq 2k)$$

Cela entraînerait :

$$(62) \qquad S_1 = S'_3 = S'_5 = \dots = S'_{2k-1} = 0$$

en posant :

$$S_i = (\lambda_1)^i + (\lambda_2)^i + \dots + (\lambda_l)^i \qquad (\mathcal{A})$$

et

$$S'_i \equiv S_i \pmod{\varrho} \qquad (\mathcal{A})$$

Or, si nous considérons le produit

$$\Pi(\lambda_i + \lambda_j) \quad \begin{matrix} i = 2, 3, \dots, l \\ j = 1, 2, \dots, l-1 \end{matrix} \quad i > j \qquad (\mathcal{A})$$

ce produit peut s'écrire sous forme d'un déterminant de Van der Monde dont le carré contiendra la ligne

$$S_1 \; S_2 \; S_3 \; \dots\dots \; S_l$$

Comme $S_{2i} = S_i^2$, les conditions (62) entraînent

$$\Pi(\lambda_i + \lambda_j) \equiv 0 \pmod{\varrho} \qquad (\mathcal{A})$$

Donc un des facteurs, par exemple $\lambda_i + \lambda_j$, serait divisible par ϱ. Comme la somme dans $\mathcal{A}$ des nombres $\lambda_i + \lambda_j$ a moins de chiffres que le nombre ϱ, il en résulterait

$$\lambda_i + \lambda_j = 0 \qquad \lambda_i = \lambda_j$$

Par suite l'hypothèse (61) ne peut être réalisée que si au moins deux indices étaient égaux.

Donc les nombres p que nous avons formés remplissent la condition (42) et peuvent servir à former un code correcteur de k erreurs. Naturellement on les transformera comme il est indiqué au § 4 pour former une suite contenant des puissances de 2. Dans le cas général, p_i comprenant kN chiffres, il y aura lieu d'utiliser kN chiffres de test.

7. Exemple.

Nous avons formé un code de 15 chiffres correcteur pour 2 erreurs. Ici $N = \mathrm{Log}_2\, 16 = 4$, il y aura 8 chiffres de test.

En prenant $\varrho = 19 = 10.011$, on calcule aisément les nombres p et la matrice M :

$$\begin{vmatrix}
0 & 1 & 1 & 1 & 1 & 0 & 0 & 1 & 1 & 1 & 1 & 1 & 1 & 1 & 1 \\
0 & 0 & 1 & 1 & 0 & 0 & 0 & 0 & 1 & 1 & 1 & 0 & 0 & 0 & 1 \\
0 & 0 & 1 & 0 & 1 & 0 & 0 & 1 & 1 & 1 & 0 & 0 & 1 & 0 & 0 \\
1 & 0 & 1 & 0 & 0 & 1 & 1 & 0 & 1 & 1 & 0 & 0 & 0 & 0 & 0 \\
0 & 0 & 0 & 0 & 0 & 0 & 0 & 1 & 1 & 1 & 1 & 1 & 1 & 1 & 1 \\
0 & 0 & 0 & 1 & 1 & 1 & 1 & 0 & 0 & 0 & 0 & 1 & 1 & 1 & 1 \\
0 & 1 & 1 & 0 & 0 & 1 & 1 & 0 & 0 & 1 & 1 & 0 & 0 & 1 & 1 \\
1 & 0 & 1 & 0 & 1 & 0 & 1 & 0 & 1 & 0 & 1 & 0 & 1 & 0 & 1
\end{vmatrix}$$

$$\begin{matrix} 1 & 2 & 3 & 4 & 5 & 6 & 7 & 8 & 9 & 10 & 11 & 12 & 13 & 14 & 15 \end{matrix}$$

Cette matrice M est de rang 8, car le déterminant formé avec les colonnes 1, 2, 4, 8, 6, 12, 7, 14 (choisies parce qu'elles présentent le plus de zéros) vaut 1.

La matrice Δ (§ 4) sera formée avec ces colonnes. En l'inversant on trouve la matrice :

$$\Delta^{-1} = \begin{vmatrix}
0 & 1 & 1 & 1 & 1 & 1 & 0 & 0 \\
1 & 1 & 0 & 0 & 1 & 0 & 0 & 0 \\
0 & 1 & 0 & 0 & 0 & 0 & 0 & 0 \\
0 & 0 & 1 & 0 & 0 & 0 & 0 & 0 \\
0 & 0 & 0 & 1 & 0 & 0 & 0 & 1 \\
1 & 0 & 0 & 0 & 1 & 1 & 1 & 0 \\
0 & 1 & 1 & 1 & 1 & 1 & 0 & 1 \\
1 & 0 & 1 & 0 & 0 & 1 & 1 & 0
\end{vmatrix}$$

Le produit Δ^{-1}M (dans l'anneau $\mathcal{A}$) donne la matrice définitive :

$$\begin{vmatrix}
1 & 0 & 1 & 0 & 0 & 0 & 0 & 0 & 0 & 0 & 0 & 0 & 1 & 0 & 1 \\
0 & 1 & 0 & 0 & 1 & 0 & 0 & 0 & 1 & 1 & 1 & 0 & 0 & 0 & 1 \\
0 & 0 & 1 & 1 & 0 & 0 & 0 & 0 & 1 & 1 & 1 & 0 & 0 & 0 & 1 \\
0 & 0 & 1 & 0 & 1 & 0 & 0 & 1 & 1 & 1 & 0 & 0 & 1 & 0 & 0 \\
0 & 0 & 0 & 0 & 1 & 1 & 0 & 0 & 0 & 1 & 1 & 0 & 1 & 0 & 1 \\
0 & 0 & 0 & 0 & 0 & 0 & 1 & 0 & 0 & 1 & 1 & 0 & 1 & 0 & 0 \\
0 & 0 & 0 & 0 & 1 & 0 & 0 & 0 & 1 & 0 & 1 & 1 & 0 & 0 & 0 \\
0 & 0 & 1 & 0 & 1 & 0 & 0 & 0 & 0 & 1 & 0 & 0 & 1 & 1 & 1
\end{vmatrix}$$

$$\begin{matrix} 1 & 2 & 3 & 4 & 5 & 6 & 7 & 8 & 9 & 10 & 11 & 12 & 13 & 14 & 15 \end{matrix}$$

Les chiffres de rang

1, 2, 4, 6, 7, 8, 12, 14

serviront de chiffres de test, les 7 autres chiffres seront les supports de l'information.

Pour vérifier et corriger un mot on fera la somme

$$T = \Sigma p_i a_i \qquad (\mathcal{A})$$

Si elle est nulle, il n'y aura pas eu d'altération du mot (ou plus de 4 erreurs). Si T n'est pas nulle, on pourra retrouver les chiffres faux (en admettant qu'il n'y en ait pas plus de 2) en utilisant la table suivante qui donne les valeurs possibles de T suivies entre parenthèses des rangs des chiffres faux.

1 (14) — 2 (12) — 3 (12,14) — 4 (7) — 5 (7,14)
6 (7,12) — 8 (6) — 9 (6,14) — 10 (6,12) — 12 (6,7)
15 (9,10) — 16 (8) — 17 (8,14) — 18 (8,12) — 19 (10,11)
20 (7,8) — 24 (6,8) — 27 (2,5) — 28 (9,11) — 29 (1,13)
32 (4) — 33 (4,14) — 34 (4,12) — 36 (4,7) — 40 (4,6)
41 (5,9) — 42 (5,10) — 44 (3,13) — 46 (2,11) — 48 (4,8)
49 (1,3) — 50 (2,9) — 53 (5,11) — 61 (2,10) — 64 (2)
65 (2,14) — 66 (2,12) — 68 (2,7) — 72 (2,6) — 75 (5,8)
77 (8,13) — 78 (4,11) — 80 (2,8) — 82 (4,9) — 83 (5,6)
85 (6,13) — 88 (3,15) — 89 (5,12) — 90 (5,14) — 91 (5)
93 (4,10) — 96 (2,4) — 98 (8,9) — 102 (6,11) — 103 (5,7)
105 (1,15) — 106 (7,11) — 108 (11,12) — 110 (11) — 111 (11,14)
112 (9,12) — 113 (8,10) — 114 (9) — 115 (9,14) — 116 (13,15)
117 (6,10) — 118 (7,9) — 121 (7,10) — 122 (6,9) — 123 (4,5)
124 (10,14) — 125 (10) — 126 (8,11) — 127 (10,12) — 128 (1)
129 (1,14) — 130 (1,12) — 132 (1,7) — 135 (11,15) — 136 (1,6)
144 (1,8) — 145 (3,4) — 152 (10,15) — 153 (7,13) — 155 (9,15)
156 (13,14) — 157 (13) — 159 (12,13) — 160 (1,4) — 161 (3,8)
169 (2,15) — 176 (3,14) — 177 (3) — 178 (5,15) — 179 (3,12)
181 (3,7) — 185 (3,6) — 189 (4,13) — 192 (1,2) — 195 (3,9)
198 (5,13) — 201 (4,15) — 204 (3,10) — 219 (1,5) — 221 (2,13)
223 (3,11) — 224 (10,13) — 225 (6,15) — 232 (14,15) — 233 (15)
234 (3,5) — 235 (12,15) — 237 (7,15) — 238 (1,11) — 239 (9,13)
241 (2,3) — 242 (1,9) — 243 (11,13) — 249 (8,15) — 253 (1,10)

On remarquera que le nombre T prend 121 valeurs possibles $\left(1 + C_{15}^1 + C_{15}^2\right)$ et qu'on utilise un nombre de 8 chiffres pour l'écrire. Le code utilise un chiffre de test de plus qu'il n'est théoriquement indispensable, mais il n'est pas sûr qu'on puisse construire des codes n'ayant qu'un nombre de chiffres de test strictement égal à l'entier par excès de Log_2 H.

On A Class of Error Correcting Binary Group Codes*

R. C. Bose and D. K. Ray-Chaudhuri

University of North Carolina and Case Institute of Technology

A general method of constructing error correcting binary group codes is obtained. A binary group code with n places, k of which are information places is called an (n,k) code. An explicit method of constructing t-error correcting (n,k) codes is given for $n = 2^m-1$ and $k = 2^m-1-R(m,t) \geqq 2^m-1-mt$ where $R(m,t)$ is a function of m and t which cannot exceed mt. An example is worked out to illustrate the method of construction.

SECTION 1

Consider a binary channel which can transmit either of two symbols 0 or 1. However, due to the presence of "noise" a transmitted zero may sometimes be received as 1, and a transmitted 1 may sometimes be received as 0. When this happens we say that there is an error in transmitting the symbol. The symbols successively presented to the channel for transmission constitute the "input" and the symbols received constitute the "output."

A v-letter n-place binary signalling alphabet A_n may be defined as a set of v distinct sequences $\alpha_0, \alpha_1, \cdots, \alpha_{v-1}$ of n binary digits. The individual sequences may be called the letters of the alphabet. Given a set of v distinct messages, we get an encoder $E_{n,v}$ by setting up a $(1,1)$ correspondence between the messages and the letters of the alphabet. To transmit a message over the channel the n individual symbols of the corresponding letter of the alphabet are presented to the channel in succession. The output is then an n-place binary sequence belonging to the set B_n of all possible binary sequences. A decoder $D_{n,v}$ is obtained by partitioning B_n into v disjoint sets $S_1, S_2, \cdots, S_v$ and setting up a correspondence between these subsets and the letters of the alphabet so that if a sequence belonging to S_i is received as an output, it is read as the letter α_i and interpreted as the corresponding message. The encoder $E_{n,v}$ together with the decoder $D_{n,v}$ constitute a binary n-place code.

Each sequence of B_n can be regarded as an n-vector with elements from the Galois field $GF(2)$. The addition of these vectors may then be defined in the usual manner, the sum of two vectors being obtained by adding the corresponding elements (mod 2). For example, if $n = 6$ and $\gamma_1 = (110011)$ and $\gamma_2 = (101001)$ then $\gamma_1 + \gamma_2 = (011010)$. Clearly the set B_n of all binary n-place sequences forms a group under vector addition. The weight $w(\gamma)$ of any sequence is defined as the number of unities in the sequence. Thus in the example considered $w(\gamma_1) = 4$, $w(\gamma_2) = 3$. The Hamming distance $d(\gamma_1, \gamma_2)$ between two sequences γ_1 and γ_2 is defined as the number of places in which γ_1 and γ_2 do not match (Hamming, 1950). Clearly $d(\gamma_1, \gamma_2) = w(\gamma_1 + \gamma_2)$. In the example $d(\gamma_1, \gamma_2) = 3 = w(\gamma_1 + \gamma_2)$. The Hamming distance satisfies the three conditions for a metric, namely

(a) $d(\gamma_1, \gamma_2) = 0$ if and only if $\gamma_1 = \gamma_2$.

(b) $d(\gamma_1, \gamma_2) = d(\gamma_2, \gamma_1)$,

(c) $d(\gamma_1, \gamma_2) + d(\gamma_2, \gamma_3) \geqq d(\gamma_1, \gamma_3)$.

Let the letter α_i of the alphabet A_n be transmitted over the channel. Let ϵ_i be the vector which has unities in those places, where an error occurs in transmitting a symbol of α_i. Then ϵ_i is the noise vector. The output received is the sequence $\alpha_i + \epsilon_i$, and the number of errors is $w(\epsilon_i)$. The code is said to be t-error correcting if $\alpha_i + \epsilon_i$ belongs to S_i whenever $w(\epsilon_i) \leqq t(i = 0,1 \cdots, v-1)$. It is clear that under these circumstances if there are t or a lesser number of errors in transmitting a letter α_i, the received message will be correctly interpreted.

* This research was supported by the United States Air Force through the Air Force Office of Scientific Research of the Air Research and Development Command, under Contract No. AF 49(638)-213. Reproduction in whole or in part is permitted for any purpose of the United States Government.

A particularly important class of codes has been studied by Slepian (1956). For this class $v = 2^k$ and the letters of the alphabet A_n form a subgroup of B_n. The null sequence is the unit element of B_n, and must also belong to A_n. We shall suppose without loss of generality that $\alpha_0 = (0,0, \cdots, 0)$. Slepian's decoder may be described as follows: If $r = n - k$, then the group B_n can be partitioned into 2^r cosets with respect to the subgroup A_n. The coset containing a particular sequence β consists of the sequences

$$\alpha_0 + \beta, \alpha_1 + \beta, \cdots, \alpha_{v-1} + \beta.$$

In the jth coset we can choose a sequence β_j whose weight does not exceed the weight of any other sequence in the coset, and call it the coset leader. Let $\beta_0, \beta_1, \cdots, \beta_{u-1}$, $(u = 2^r)$ be the coset leaders, where $\beta_0 = \alpha_0$ is the null sequence and leader of the 0th coset A_n. Let S_j be the set of sequences

$$\alpha_j + \beta_0, \alpha_j + \beta_1, \cdots, \alpha_j + \beta_{u-1} \qquad j = 0,1, \cdots, v-1.$$

Then the decoder is obtained by partitioning B_n into $S_0, S_1, \cdots, S_{v-1}$ and setting up the rule that if the sequence received as an output belongs to S_j, it is read as the letter α_j. The code thus obtained may be called an (n,k) binary group code. It is clear that a transmitted message will be correctly interpreted if and only if the error vector happens to be a coset leader. Hence a necessary and sufficient condition for the code to be t-error correcting is that if β is any n-place binary sequence for which $w(\beta) \leq t$, then β is a coset leader. The following lemma, due to Hamming (1950), is then easy to deduce.

Lemma 1. The necessary and sufficient condition for an (n,k) binary group code to be t-error correcting is that each letter of the alphabet except the null letter has weight $2t + 1$ or more.

Since the $v = 2^k$ messages can be transmitted by a k-place binary code if there is no possibility of error, the number $r = n - k$ is called the redundancy for an (n,k) binary group code. In constructing a t-error correcting (n,k) binary group code for given n and t one would like to maximize k (that is, maximize the number of different messages that it is possible to transmit). Varšamov (1957) has shown that if k satisfies the inequality

$$S_r^{2t-1} + \binom{k-1}{1} S_r^{2t-2} + \cdots + \binom{k-1}{2t-2} S_r^{1} + \binom{k-1}{2t-1} < 2^r \quad (1)$$

where

$$S_r^{q} = 1 + \binom{r}{1} + \binom{r}{2} + \cdots + \binom{r}{q} \quad (2)$$

then a t-error correcting (n,k) group code exists.

The main result of the present paper is the following: If $n = 2^m - 1$, then there exists a t-error correcting (n,k) binary group code with $k \geqq 2^m - 1 - mt$.

The method of proof is constructive and is illustrated by considering the case $n = 15$, $t = 3$, for which a 3-error correcting $(15,5)$ binary group code is explicitly obtained.

As an example of comparison between Varšamov's result and our theorem consider the case $n = 31$. Varšamov's result then shows that a 2-error correcting binary group code can be obtained with $k = 18$, and a 3-error correcting binary group code can be obtained with $k = 13$ but is inconclusive for larger values of k. Our method, however, gives an explicit construction for a 2-error correcting binary group code with $k = 21$, and a 3-error correcting binary group code with $k = 16$.

The following table gives some of the values of n, k and t for which a t-error correcting (n,k) binary group code can be constructed by our method. The transmission rate $R = k/n$ is also given.

Reprinted with permission from *Inform. Contr.*, vol. 3, pp. 68-79, Mar. 1960.

TABLE I

t	n	k	R
1	15	11	0.73
2	15	7	0.47
2	31	21	0.68
2	63	51	0.81
2	127	113	0.89
3	15	5	0.33
3	31	16	0.52
3	63	45	0.71
3	127	106	0.83
4	63	39	0.64
4	127	99	0.78
5	127	92	0.72

SECTION 2

We shall now prove a theorem which gives a necessary and sufficient condition for the existence of a t-error correcting (n,k) group code. This theorem appears in a different form in an earlier paper by Bose (1947) but is given here for the sake of completeness.

Theorem 1. The necessary and sufficient condition for the existence of a t-error correcting (n,k) binary group code is the existence of a matrix A of order $n \times r$ and rank $r = n - k$ with elements from $GF(2)$, such that any set of $2t$ row vectors from A are independent.

Proof of Sufficiency. The matrix A has the property (P_{2t}) that any $2t$ row vectors of A are independent. Clearly $r \geqq 2t$. The property (P_{2t}) is invariant under the following operations: (1) interchange of two rows or columns and (2) replacement of the ith column by the sum of ith and jth column, $i \neq j$. By these operations A can be transformed to the matrix.

$$A^* = \left\| \begin{matrix} I_r \\ C \end{matrix} \right\| \tag{3}$$

where A^* has the property (P_{2t}), I_r is the unit matrix of order r, and C is a matrix of order $k \times r$. Consider the matrix

$$C^* = \|C, I_k\|. \tag{4}$$

Then C^* is of order $k \times n$. We shall show that the k rows of C^* (under vector addition (mod 2)) are generators of a group G of order 2^k such that if α is any arbitrary (nonnull) element of G, then $w(\alpha) \geqq 2t + 1$. Let α be the sum of any d row vectors of C^*, $d \leqq k$. We can write $\alpha = (\gamma, \epsilon)$, where γ is the part coming from C and ϵ the part coming from I_k. Now

$$w(\alpha) = w(\gamma) + w(\epsilon) = w(\gamma) + d.$$

Hence

$$w(\alpha) \geqq 2t + 1 \text{ if } d > 2t.$$

Suppose $d \leqq 2t$. If $w(\alpha) < 2t + 1$, then $w(\gamma) \leqq 2t - d$. Let $w(\gamma) = c$. There are exactly c positions in γ which are occupied by unity. Corresponding to each such position we can find a row vector of I_r which has unity in this position (and zero in all other positions). Then these c vectors of I_r together with the d row vectors of C whose sum is γ, constitute a set of $c + d$ vectors which are dependent. Since $c + d \leqq 2t$, this contradicts the fact that A^* has the property (P_{2t}). Thus the weight of any nonnull element of G is greater than or equal to $2t + 1$. It follows from Lemma 1 that the sequences of the subgroup generated by the k rows of C^* form the alphabet of a t-error correcting (n,k) group code.

Proof of Necessity. Suppose there exists a t-error correcting (n,k) binary group code. We can then find a set of k n-place binary sequences, or n-vectors with elements from $GF(2)$, which under addition generate the group of sequences which constitute the letters of the alphabet. By Lemma 1 if α is a sequence of this group $w(\alpha) \geq 2t + 1$. Consider the $k \times n$ matrix C^* whose row vectors are given by these sequences. If we interchange any two rows or columns of C^*, or replace the ith row of C^* by the sum of the ith and the jth row $(i \neq j)$, the transformed matrix still retains the property that its rows generate under addition a group, each sequence of which has weight $2t + 1$ or more. Hence we can without loss of generality take C^* in the canonical form (4) where C is of order $r \times k$ and I_k is the unit matrix of order k. By retracing the arguments used in proving the first part of the theorem, we see that the matrix A^* of order $n \times r$, given by (3), has the property that any two $2t$ row vectors are independent. This proves that the condition of the theorem is necessary.

Corollary 1. The existence of a t-error correcting (n,k) binary group code implies the existence of a t-error correcting $(n - c, k - c)$ binary group code, $0 < c < k$.

If in the matrix C^* given by (4) we delete the last c rows and the last c columns, we get a matrix

$$C_1^* = \|C_1, I_{k-c}\|$$

of order $(k - c) \times (n - c)$, the rows of which generate a group for which each nonnull element is of weight $2t + 1$ or more. The rows of C_1^* generate the alphabet of the required code.

Let V_r denote the vector space of all r-vectors whose elements belong to GF(2). One may then ask the following question. What is the maximum number of vectors in a set Σ chosen from V_r, such that any $2t$ distinct vectors from Σ are independent. This number may be denoted $n_{2t}(r)$, and the problem of finding the set Σ may be called the packing problem (of order $2t$) for V_r. For a given t, $n_{2t}(r)$ is a monotonically increasing function of r.

Let $k = k_t(n)$ denote the maximum value of k such that a t-error correcting (n,k) binary group code for given t and n exists. We can then state the following.

Theorem 2. If $n_{2t}(r) \geq n > n_{2t}(r - 1)$, then $k_t(n) = n - r$.

From Theorem 1 there exists a t-error correcting $[n_{2t}(r), n_{2t}(r) - r]$ binary group code. Taking $c = n_{2t}(r) - n$ in Corollary 1, there exists a t-error correcting $(n, n - r)$ group code. But a t-error correcting $(n, n - r + 1)$ binary group code cannot exist, since from Theorem 1 its existence would imply that $n_{2t}(r - 1) \geq n$. Hence $k_t(n) = n - r$ is the maximum value of k for which a t-error correcting (n,k) binary group code exists.

Thus the problem of finding a t-error correcting n-place binary group code, with the maximum transmission rate k/n, is equivalent to determining the smallest r for which there exists a set of n or more distinct vectors of V_r, such that any $2t$ distinct vectors from the set are independent.

SECTION 3

The theorem to be proved in the next section depends upon the following lemma.

Lemma 2. If $x_1, x_2, \cdots, x_l$ are different nonzero elements of the Galois field $GF(2^m)$, then the equations

$$x_1^{2i-1} + x_2^{2i-1} + \cdots + x_l^{2i-1} = 0, i = 1,2, \cdots, t \tag{5}$$

cannot simultaneously hold if $l \leqq 2t$.

Suppose, if possible, the Eqs. (5) hold simultaneously. Let

$$x^l + p_1 x^{l-1} + p_2 x^{l-2} + \cdots + p_l = 0 \tag{6}$$

be the algebraic equation whose roots are $x_1, x_2, \cdots, x_l$. Then p_j belongs to $GF(2^m)$ and is the sum of the products of the roots taken j at a time $(j = 1,2, \cdots, l)$. We define s_j as the sum of the jth powers of the roots. For a field of characteristic 2 the well-known relations between the symmetric functions s_j and p_j become (Levi, 1942),

$$\begin{aligned} &s_1 + \delta_1 p_1 = 0 \\ &s_2 + p_1 s_1 + \delta_2 p_2 = 0 \\ &s_3 + p_1 s_2 + p_2 s_1 + \delta_3 p_3 = 0 \\ &\vdots \\ &s_l + p_1 s_{l-1} + p_2 s_{l-2} + \cdots + \delta_l p_l = 0 \end{aligned} \tag{7}$$

where $\delta_i = 0$ or 1 according as i is even or odd. From Eqs. (5) $s_j = 0$ when j is odd $(j < 2t)$. It then follows from (7) that $s_j = 0$ if j is even $(j \leqq l)$ and $p_j = 0$ if j is odd $(j \leqq l)$.

Case I. If l is odd then $p_l = x_1 x_2 \cdots x_l \neq 0$, since $x_1, x_2, \cdots, x_l$ are nonzero. This is a contradiction.

Case II. If l is even, say $l = 2c$, the Eq. (6) becomes

$$x^{2c} + p_2 x^{2c-2} + \cdots + p_{2c} = 0 \tag{8}$$

therefore

$$(x^c + q_1 x^{c-1} + \cdots + q_c)^2 = 0 \tag{9}$$

where q_j is the unique square root of p_{2j} in $GF(2^m)$. Hence (6) cannot have more than c distinct roots, which again is a contradiction, since $x_1, x_2, \cdots, x_l$ are distinct by hypothesis.

Hence the lemma is true whether l is odd or even.

SECTION 4

Let V_m be the vector space of m-vectors with elements from $GF(2)$. We can institute a correspondence between the vector $\alpha = (a_0, a_1, \cdots, a_{m-1})$ of $\mathbf{V}_m$ and the element $a_0 + a_1x + \cdots + a_{m-1}x^{m-1}$ of $GF(2^m)$, where x is a given primitive element of the field. This is a (1,1) correspondence in which the null vector α_0 of V_m corresponds to the null element of $GF(2^m)$, and the sum of any two vectors of V_m corresponds to the sum of the corresponding elements of $GF(2^m)$. We can therefore identify the vector α of V_m and the corresponding element of $GF(2^m)$. This in effect defines a multiplication of the vectors of V_m and converts it into a field. In particular we can speak of powers of any vector.

Let V_{mt} be the vector space of all mt-vectors with elements from $GF(2)$. To any vector α_i of V_m there corresponds a unique vector α_i^* of V_{mt} defined by

$$\alpha_i^* = (\alpha_i, \alpha_i^3, \cdots, \alpha_i^{2t-1}) \tag{10}$$

though the converse is not true.

There are $n = 2^m - 1$ distinct nonnull vectors in V_m. Let

$$M^* = \left\| \begin{matrix} \alpha_1, \alpha_1^3, \cdots, \alpha_1^{2t-1} \\ \alpha_2, \alpha_2^3, \cdots, \alpha_2^{2t-1} \\ \vdots \\ \alpha_n, \alpha_n^3, \cdots, \alpha_n^{2t-1} \end{matrix} \right\| \tag{11}$$

be the $n \times mt$ matrix, which has for row vectors the corresponding vectors $\alpha_1^*, \alpha_2^*, \cdots, \alpha_n^*$. We shall show that M^* has the property (P_{2t}) that any of $2t$ distinct row vectors belonging to M^* are independent. For this it is sufficient that the sum of any l row vectors of M^*, $l \leq 2t$, is nonnull. This is ensured by Lemma 2, since α_i can also be regarded as elements of $GF(2^m)$.

Now rank $(M^*) \leq mt$. Since there is essentially only one Galois field $GF(2^m)$, this rank is a definite function of m and t and will be denoted by $R(m,t)$. When $R(m,t) < mt$, we can choose $R(m,t)$ independent columns of M^*, and delete the other columns dependent on them. The matrix A so obtained has still the property (P_{2t}). Using Theorem 1 we have

THEOREM 3. If $n = 2^m - 1$, we can obtain a t-error correcting (n,k) binary group code where

$$k = 2^m - 1 - R(m,t) \geq 2^m - 1 - mt.$$

When n is not of the form $2^m - 1$ t-error correcting (n,k) binary group codes can be deduced from those obtainable from Theorem 3, by using Corollary 1 of Theorem 1. Stronger results than those which can be obtained in this way will be given in a subsequent communication.

SECTION 5

The proofs of the theorems in Sections 2 and 4 are constructive in the sense that they give an actual procedure for obtaining the required codes. We shall illustrate the procedure to be followed by taking the case $m = 4$, $t = 3$. Then $n = 15$ and the rank $R(m,t)$ turns out to be 10. We thus obtain a 3-error correcting (15,5) group code. The roots of the equation

$$x^4 = x + 1 \tag{12}$$

are primitive elements of $GF(2^4)$, that is all the nonzero elements of the field can be expressed as the powers of a root x (Carmichael, 1937, p. 262). Using (12) the powers of n can be expressed alternatively as polynomials in x of degree 3 or less.The 15 nonzero elements or vectors are listed in the following table in two equivalent forms, (1) as powers of the primitive element, and (2) as polynomials of degree 3 or less in the primitive element. We thus have the following table of the 15 nonzero elements or vectors.

$$\begin{aligned}
x^0 &= 1 &&= (1,0,0,0) = \alpha_1 \\
x &= x &&= (0,1,0,0) = \alpha_2 \\
x^2 &= x^2 &&= (0,0,1,0) = \alpha_3 \\
x^3 &= x^3 &&= (0,0,0,1) = \alpha_4 \\
x^4 &= 1 + x &&= (1,1,0,0) = \alpha_5 \\
x^5 &= x + x^2 &&= (0,1,1,0) = \alpha_6 \\
x^6 &= x^2 + x^3 &&= (0,0,1,1) = \alpha_7 \\
x^7 &= 1 + x + x^3 &&= (1,1,0,1) = \alpha_8 \\
x^8 &= 1 + x^2 &&= (1,0,1,0) = \alpha_9 \\
x^9 &= x + x^3 &&= (0,1,0,1) = \alpha_{10} \\
x^{10} &= 1 + x + x^2 &&= (1,1,1,0) = \alpha_{11} \\
x^{11} &= x + x^2 + x^3 &&= (0,1,1,1) = \alpha_{12} \\
x^{12} &= 1 + x + x^2 + x^3 &&= (1,1,1,1) = \alpha_{13} \\
x^{13} &= 1 + x^2 + x^3 &&= (1,0,1,1) = \alpha_{14} \\
x^{14} &= 1 + x^3 &&= (1,0,0,1) = \alpha_{15}
\end{aligned}$$

In obtaining the powers of the elements it should be remembered that each nonzero element of $GF(2^m)$ satisfies $x^{2^m-1} = 1$. Since $m = 4$ we have

$$x^{15} = 1.$$

Thus for example

$$\alpha_7^3 = (x^6)^3 = x^{18} = x^3 = (0,0,0,1) = \alpha_4$$
$$\alpha_7^5 = (x^6)^5 = x^{30} = x^0 = (1,0,0,0) = \alpha_1.$$

It is now easy to calculate the matrix M^* given by (11). For example, the seventh row is $(\alpha_7, \alpha_7^3, \alpha_7^5)$ or (0 0 1 1, 0 0 0 1, 1 0 0 0). Thus

$$M^* = \left\| \begin{matrix}
1\,0\,0\,0 & 1\,0\,0\,0 & 1\,0\,0\,0 \\
0\,1\,0\,0 & 0\,0\,0\,1 & 0\,1\,1\,0 \\
0\,0\,1\,0 & 0\,0\,1\,1 & 1\,1\,1\,0 \\
0\,0\,0\,1 & 0\,1\,0\,1 & 1\,0\,0\,0 \\
1\,1\,0\,0 & 1\,1\,1\,1 & 0\,1\,1\,0 \\
0\,1\,1\,0 & 1\,0\,0\,0 & 1\,1\,1\,0 \\
0\,0\,1\,1 & 0\,0\,0\,1 & 1\,0\,0\,0 \\
1\,1\,0\,1 & 0\,0\,1\,1 & 0\,1\,1\,0 \\
1\,0\,1\,0 & 0\,1\,0\,1 & 1\,1\,1\,0 \\
0\,1\,0\,1 & 1\,1\,1\,1 & 1\,0\,0\,0 \\
1\,1\,1\,0 & 1\,0\,0\,0 & 0\,1\,1\,0 \\
0\,1\,1\,1 & 0\,0\,0\,1 & 1\,1\,1\,0 \\
1\,1\,1\,1 & 0\,0\,1\,1 & 1\,0\,0\,0 \\
1\,0\,1\,1 & 0\,1\,0\,1 & 0\,1\,1\,0 \\
1\,0\,0\,1 & 1\,1\,1\,1 & 1\,1\,1\,0
\end{matrix} \right\|$$

where the vertical divisions separate the parts coming from α, α^3 and α^5. From M^* we can drop the last null column and the 11th column which is identical with the 10th. The 10×15 matrix of rank 10 so obtained we can take as the matrix A of Theorem 1. From what has been shown in Section 4, this matrix has the property $P(6)$ that any 6 row vectors are independent. Using operations (1) and (2) of Section 2, we can then transform A to A^* where

$$A^* = \left\| \begin{matrix} I_{10} \\ C \end{matrix} \right\|$$

and I_{10} is the unit matrix of order 10, and C is the 5×10 matrix given by

$$C = \left\| \begin{matrix}
1\,1\,1\,0\,1\,1\,0\,0\,1\,0 \\
0\,1\,1\,1\,1\,0\,0\,1\,0\,1 \\
1\,1\,0\,1\,1\,0\,1\,1\,1\,0 \\
0\,1\,1\,0\,0\,1\,1\,1\,1\,1 \\
1\,1\,0\,1\,0\,1\,1\,0\,0\,1
\end{matrix} \right\|$$

Taking

$$C^* = \|C, I_5\|$$

we have a matrix of order 5×15 whose rows generate under vector addition (mod 2), the group of 32 sequences which constitute the letters of the alphabet of the required 3-error correcting (15,5) binary group alphabet. It is easy to verify that of the 31 nonnull sequences 15 have weight 7, 15 have weight 8 and one has weight 15, which checks with Lemma 1.

RECEIVED: September 24, 1959; revised October 20, 1959.

REFERENCES

BOSE, R. C. (1947). Mathematical theory of the symmetrical factorial design. *Sankhya* **8**, 107–166.

CARMICHAEL, R. D. (1937). "Introduction to the Theory of Groups of Finite Order." Gin and Co., New York.

DWORK, B. M. AND HELLER, R. M. (1959). Results of a geometric approach to the theory and construction of non-binary, multiple error and failure correcting codes. IRE *Convention Record*. Pt. 4, pp. 123–192.

HAMMING, R. W. (1950). Error detecting and error correcting codes. *Bell System Tech. J.* **29**, 147–160.

LEVI, F. W. (1942). "Algebra," Vol. I p. 147. University of Calcutta.

SACKS, G. E. (1958). Multiple error correction by means of parity checks. IRE *Trans. on Information Theory*, **IT-4**, pp. 145–147.

SLEPIAN, D. (1956). A class of binary signalling alphabets. *Bell System Tech. J.* **35**, 203–234.

VARSAMOV, R. R. (1957). The evaluation of signals in codes with correction of errors. *Doklady Akad. Nauk SSSR* [*N.S.*] **117**, 739–741 (Russian).

Further Results on Error Correcting Binary Group Codes*

R. C. BOSE AND D. K. RAY-CHAUDHURI

University of North Carolina and Case Institute of Technology

The present paper is a sequel to the paper "On a class of error-correcting binary group codes", by R. C. Base and D. K. Ray-Chaudhuri, appearing in *Information and Control* in which an explicit method of constructing a t-error correcting binary group code with $n = 2^m - 1$ places and $k = 2^m - 1 - R(m,t) \geqq 2^m - 1 - mt$ information places is given. The present paper generalizes the methods of the earlier paper and gives a method of constructing a t-error correcting code with n places for any arbitrary n and $k = n - R(m,t) \geqq [(2^m - 1)/c] - mt$ information places where m is the least integer such that $cn = 2^m - 1$ for some integer c. A second method of constructing t-error correcting codes for n places when n is not of the form $2^m - 1$ is also given.

SECTION I

This paper is a continuation of our previous paper, Bose and Ray-Chaudhuri (1960), "On a class of error correcting binary group codes." The notation used there will be followed throughout, with the minimum of explanation.

It was shown that we can obtain a t-error correcting n-place binary group code (n,k) with k information places, if $n = 2^m - 1$ and $k = n - R(m,t)$ where $R(m,t) \leqq mt$ is the rank of a certain matrix whose properties have been investigated. Peterson (1960) has investigated certain interesting properties of these codes, and given the exact value of $R(m,t)$. In Section II, we have generalized our results to the case when $n = (2^m - 1)/c$ where c is the smallest integer for which $cn + 1$ is a power of 2. This generalization enables us to obtain as a special case certain codes with the same values of n and k as those investigated by Prange (1958–1959).

Let V_r denote the vector space of r-vectors with coordinates from $GF(2)$. Following the notation of Bose and Chaudhuri (1960) we shall denote by $n_{2t}(r)$ the maximum number of vectors that it is possible to choose in V_r such that no $2t$ are dependent. A matrix with elements from $GF(2)$ is said to possess the property (P_{2t}) if no set of $2t$ rows are de-dependent. It was shown in our earlier paper that the problem of finding a t-error correcting n-place binary group code (n,k) with k information places, and the maximum transmission rate k/n can be completely solved if we can determine $n_{2t}(r)$ for every value of r and can construct a matrix with r columns and $n_{2t}(r)$ rows, possessing the property (P_{2t}). We constructed a matrix with mt columns and $2^m - 1$ rows, possessing the property (P_{2t}), which establishes the inequality

$$n_{2t}(mt) \geqq 2^m - 1 \tag{1}$$

In Section III, of the present paper we shall find lower bounds for $n_{2t}(r)$ for values of r which are not multiples of t, and construct the corresponding matrix with the property (P_{2t}). This enables us in certain instances to obtain t-error correcting (n,k) binary group codes for which the transmission rate k/n is better than for codes obtainable by using corollary 1, Theorem 1 of our earlier paper.

In Section IV we have given a table which, for given $n \leqq 100$ and $t \leqq 6$, enables us to calculate the best corresponding value of k obtainable by our methods.

SECTION II

For a given positive integer n, let $c = c(n)$ be the smallest integer such that $1 + cn$ is a power of 2. Let this power be denoted by $m = m(n)$. Thus,

$$n = (2^m - 1)/c \tag{2}$$

For example, if $n = 21$, then $c = 3$, $m = 6$; if $n = 31$, $c = 1$, $m = 5$. Again, if $n = 73$, $c = 7$, $m = 9$.

Let x be a primitive element of the Galois field $GF(2^m)$. Then

$$1, x, x^2, \cdots, x^{nc-1}$$

are all the distinct nonzero elements of the field and

$$x^{nc} = 1.$$

Each element of $GF(2^m)$ can be expressed as a polynomial of degree $m - 1$ or less with coefficients from $GF(2)$. Let V_m be the vector-space of m-vectors, with elements from $GF(2)$. Then, as explained in Bose and Ray-Chaudhuri (1960) we can institute a (1,1) correspondence between the vector $\alpha = (a_0, a_1, \cdots, a_{m-1})$ of V_m, and the element

$$a_0 + a_1x + \cdots + a_{m-1}x^{m-1}$$

of $GF(2^m)$. Then the null vector α_0 of V_m corresponds to the null element of $GF(2^m)$, and the sum of any two vectors of V_m corresponds to the sum of the corresponding elements of $GF(2^m)$. We can then identify the vector α of V_m and the corresponding element of $GF(2^m)$. This in effect defines a multiplication of the vectors of V_m and converts it into a field. In particular, we can speak of the powers of any vector. Let us set

$$\alpha_i = x^{ci} = a_{i0} + a_{i1}x + \cdots + a_{i,m-1}x^{m-1} = (a_{i0}, a_{i1}, \cdots, a_{i,m-1}) \tag{3}$$

where $i = 1, 2, \cdots, n$. Then $\alpha_1, \alpha_2, \cdots, \alpha_n$ are all the distinct elements of $GF(2^m)$ which are powers of x^c, that is, α_1. In particular, $\alpha_n = x^{cn} = 1$. Let

$$\alpha_i^* = (\alpha_i, \; \alpha_i^3, \; \cdots, \; \alpha_i^{2t-1}) \tag{4}$$

and

$$M^* = \left\| \begin{matrix} \alpha_1, & \alpha_1^3, & \cdots & \alpha_1^{2t-1} \\ \alpha_2, & \alpha_2^3, & \cdots & \alpha_2^{2t-1} \\ \cdots & \cdots & \cdots & \cdots \\ \alpha_n, & \alpha_n^3, & \cdots & \alpha_n^{2t-1} \end{matrix} \right\| \tag{5}$$

When the α_i's are regarded as m-vectors over $GF(2)$, M^* is a matrix of order $n \times mt$ with elements from $GF(2)$. We shall now prove

LEMMA 1. Any $2t$ row vectors belonging to M^* are independent, i.e., M^* possesses the property (P_{2t}).

This result was proved in our earlier paper by using the properties of

* This research was supported in part by the United States Air Force Office of Scientific Research of the Air Research and Development Command, under Contract No. AF(638)-213.

Reprinted with permission from *Inform. Contr.*, vol. 3, pp. 279–290, Sept. 1960.

power sums. It is possible to generalize this proof. However, we shall give the following alternative proof based on considerations suggested by W. W. Peterson in a private communication.

Let $\beta_1, \beta_2, \cdots, \beta_{2t}$ be any $2t$ elements of $GF(2^m)$ chosen out of $\alpha_1, \alpha_2, \cdots, \alpha_n$. We then have to show that the matrix

$$D = \left\| \begin{matrix} \beta_1 & \beta_1^3 & \cdots & \beta_1^{2t-1} \\ \beta_2 & \beta_2^3 & \cdots & \beta_2^{2t-1} \\ \vdots & \vdots & & \vdots \\ \beta_{2t} & \beta_{2t}^3 & \cdots & \beta_{2t}^{2t-1} \end{matrix} \right\|$$

has rank $2t$. Since $x \to x^2$ is an automorphism of $GF(2^m)$, any linear relation between $\beta_1^u, \beta_2^u, \cdots, \beta_{2t}^u$ implies a corresponding linear relation among $\beta_1^{2u}, \beta_2^{2u}, \cdots, \beta_{2t}^{2u}$ and vice versa. Hence, the rank of

$$D_1 = \left\| \begin{matrix} \beta_1 & \beta_1^2 & \cdots & \beta_1^{2t-1} & \beta_1^{2t} \\ \beta_2 & \beta_2^2 & \cdots & \beta_2^{2t-1} & \beta_2^{2t} \\ \vdots & \vdots & & \vdots & \vdots \\ \beta_{2t} & \beta_{2t}^2 & \cdots & \beta_{2t}^{2t-1} & \beta_{2t}^{2t} \end{matrix} \right\|$$

is the same as the rank of D. However.

$$\det D_1 = \beta_1\beta_2 \cdots \beta_{2t} \prod_{j<i}^{2t} (\beta_i - \beta_j) \neq 0$$

since $\beta_1, \beta_2, \cdots, \beta_{2t}$ are all distinct and nonzero. This shows that rank $(D_1) = 2t$ and completes the proof of the Lemma.

Let $mt < n$. The columns of M^* are not always independent as is clear from the example for the case $n = 15$, $c = 1$, $m = 4$, $t = 3$, discussed in Section 5 of Bose and Chaudhuri (1960). As before, we shall denote the rank of M^* by $R(m,t)$. If $R(m,t) < mt$, then we can choose $R(m,t)$ independent columns of M^* and drop the remaining columns of M^* and thus get a matrix of order $n \times R(m,t)$ with the property (P_{2t}).

LEMMA 2. The rank $R(m,t)$ is the number of distinct residue classes (mod n) among the integers $2^j u (u = 1, 3, \cdots, 2t-1; j \geq 0)$.

This Lemma has been proved by Peterson (1960) for the special case $c = 1$, and his proof can be easily extended to the general case. We shall make a few remarks useful for application of the Lemma.

Denote by $(2^j u)$ the residue class corresponding to the integer $2^j u$. Since

$$\begin{aligned} 2^m u &= (2^m - 1)u + u \\ &= ncu + u \\ &= u (\text{mod } n) \end{aligned}$$

there cannot be more than m distinct residue classes among $(2^j u)$ with a fixed value of u, and in counting the number of residue classes it is sufficient to confine ourselves to values of j in the range $0 \leq j \leq m-1$. Hence,

$$R(m,t) \leq mt.$$

If we arrange the integers $2^j u$ reduced (mod n) in a rectangular scheme, each row corresponding to one value of u, then

(i) If k is the least nonzero positive integer such that

$$u = 2^k u (\text{mod } n)$$

then $k \leq m$. If $k = m$ the residue classes in the corresponding row are all distinct. If $k < m$, then k is a factor of m, and there are k distinct residue classes in the corresponding row.

(ii) If any two rows have one element in common they coincide entirely.

(a) To u we can associate the set of m columns of the submatrix

$$M_u^* = \left\| \begin{matrix} \alpha_u \\ \alpha_u^3 \\ \vdots \\ \alpha_u^n \end{matrix} \right\| \tag{6}$$

of M. The number of independent columns in M_u^* is exactly k. We can therefore delete $m - k$ suitable columns from M_u^* without changing the rank of M^*, or the property (P_{2t}).

(b) When two rows of the scheme corresponding to say u_1 and $u_2 (u_1 < u_2 \leq 2t-1)$ are identical we can delete the submatrix $M_{u_2}^*$ without changing the rank of M^* or the property (P_{2t}).

After the operations (a) and (b) we get from M^* a matrix of order $n \times R(m,t)$ with rank $R(m,t)$ and possessing the property (P_{2t}). Let the matrix so obtained be called A^* which is of order $n \times R(m,t)$ and possesses the property (P_{2t}).

The matrix A^* can serve as the parity check matrix.

Using Theorem 1 of Bose and Chaudhuri (1960) we now get the following results:

THEOREM 1. If n is any integer, and c is the least integer such that $1 + cn = 2^m$, then there exists a t-error correcting binary group code (n,k) for which the number of information places is

$$k = n - R(m,t)$$

where $R(m,t)$ is given by Lemma 2. The letters of the code are binary n-vectors orthogonal to the columns of A^*, i.e., form the left null-space of A^*, where A^* is the matrix defined in the remarks following Lemma 2.

Every n-place binary sequence $(a_0, a_1, \cdots, a_{n-1})$ may be regarded as a polynomial $a_0 + a_1 y + \cdots + a_{n-1} y^{n-1}$ in an indeterminate y. Let R_n denote the set of all such polynomials of degree less than n with coefficients 0 and 1. The addition of polynomials in R_n can be defined in the usual way, i.e., by adding the coefficients mod 2. Let the multiplication be defined mod 2 and mod $(y^n - 1)$. With these operations R_n becomes a ring. Let

$$A^* = \left\| \begin{matrix} b_{10} & b_{20} & \cdots & b_{r0} \\ b_{11} & b_{22} & \cdots & b_{r1} \\ \vdots & & & \\ b_{1,n-1} & b_{2,n-1} & \cdots & b_{r,n-1} \end{matrix} \right\|$$

and let

$$\beta_i' = (b_{i0}, b_{i1}, \cdots, b_{i,n-1}), \qquad i = 1, 2, \cdots, r = R(m,t)$$

Let $\bar{V}(A^*)$ denote the vector space generated by $\beta_1', \beta_2', \cdots, \beta_r'$, and $V(A^*)$ denote the vector space orthogonal to $\bar{V}(A^*)$. If now the vectors of $V(A^*)$ are regarded as polynomials, then for the case $c = 1$, Peterson (1960) has proved that $V(A^*)$ is an ideal in R_n, generated by a certain polynomial $f(y)$ of degree $r = R(m,t)$. Peterson's arguments at once extend to the general case and we have the following:

Let $f_j(x)$ be the minimum polynomial of x^{cj} over $GF(2)$, where x is a primitive element of $GF(2^m)$. Then $V(A^*)$ is the ideal generated by

$$f(y) = \underset{j=1,3,\cdots,(2t-1)}{\text{L.C.M.}} f_j(y)$$

The polynomial $f(y)$ can also be expressed in an alternative form. Let $(p_1, p_2, \cdots, p_r)$ be a set of $r = R(m,t)$ integers containing one integer for each of the $R(m,t)$ distinct residue classes considered in Lemma 2. Then

$$f(y) = (y - x^{cp_1})(y - x^{cp_2}) \cdots (y - x^{cp_r}).$$

For a polynomial $f(y) = a_0 + a_1 y + \cdots a_{n-1} y^{n-1}$ we shall call $(a_{n-1}, a_{n-2}, \cdots, a_0)$ the reversed vector corresponding to $f(y)$. Let

$$\bar{f}(y) = (y^n - 1)/f(y)$$

and let $I[\bar{f}(y)]$ denote the set of 2^r reversed vectors corresponding to the 2^r polynomials of the ideal generated by $\bar{f}(y)$. Peterson's arguments then show that

$$\bar{V}(A^*) = I[\bar{f}(y)]$$

It follows that we can take

$$A^* = [\beta_1, \beta_2, \cdots, \beta_r]$$

where β_i' is the reversed vector corresponding to the polynomial

$$y^{i-1}\bar{f}(y), \qquad (i = 1, 2, \cdots, r),$$

and β_i is the transpose β_i'.

Example 1. Let $n = 21$. Then $c = 3$, $m = 6$. Let $t = 2$. To determine $R(m,t)$, we write the integers $2^j u (u = 1, 3; j = 0,1,2,3,4,5)$ in the following scheme, each row corresponding to one value of u:

$$\begin{matrix} 1, & 2, & 4, & 8, & 16; & 11 \\ 3, & 6, & 12, & 3, & 6, & 12 \end{matrix}$$

We thus get nine distinct residue classes and $R(m,t) = 9$. The number of information places is $k = 21 - 9 = 12$ and we get a 2 error-correcting binary group code (21, 12). To actually construct the code, we have to compute

$$f(y) = (y - x^3)(y - x^6)(y - x^{12})(y - x^{24})$$

$$(y - x^{48})(y - x^{33})(y - x^{9})(y - x^{18})(y - x^{36})$$

where x is a primitive element of $GF(2^6)$. A minimum function of $GF(2^6)$ is $x^6 + x + 1$. Hence, using the relation $x^6 + x + 1 = 0$, the coefficients of the polynomial $f(y)$ will be all reduced to 0 and 1. The 2^{12} message sequences will be the 21-place binary vectors corresponding to the elements of the ideal generated by $f(y)$ in R_{21}. $\bar{V}(M^*)$ is the ideal generated by

$$\bar{f}(y) = (y^{21} - 1)/\bar{f}(y)$$
$$= y^{12} + y^{11} + y^{9} + y^{7} + y^{3} + y^{2} + y + 1.$$

Hence the parity check matrix A^* can be taken as

$$A^* = \|\beta_1 \vdots \beta_2 \vdots \cdots \beta_9\|$$

where β_i' is the (1×21) reversed vector corresponding to $y^{i-1}\bar{f}(y)$, $i = 1, 2, \cdots, 9$. A 2 error-correcting (21, 12) code has also been studied by Prange (1958).

Example 2. Let $n = 73$. Then $c = 7$, $m = 9$. Let $t = 4$. The residue classes corresponding to the integer $2^j u(u = 1,3,5,7; 0 \leqq j \leqq 8)$ can be exhibited as

1,	2,	4,	8,	16,	32,	64,	55,	37
3,	6,	12,	24,	48,	23,	46,	19,	38
5,	10,	20,	40,	7,	14,	28,	56,	39
7,	14,	28,	56,	39,	5,	10,	20,	40.

The third and the fourth rows in this scheme are identical. Hence $R(m,t) = 27$ and $k = 46$. We thus get a 4 error-correcting binary group code (73, 46). This 4 error-correcting (73, 46) group code has also been obtained by Prange (1959).

Example 3. Let $n = 85$. Then $c = 3$, $m = 8$. Let $t = 6$. The residue classes corresponding to the integers $2^j u(u = 1,3,5,7,9,11; 0 \leqq j \leqq 7)$ can be exhibited as

1	2	4	8	16	32	64	43
3	6	12	24	48	11	22	44
5	10	20	40	80	75	65	45
7	14	28	56	27	54	23	46
9	18	36	72	59	33	66	47
11	22	44	3	6	12	24	48

The rows corresponding to $u = 3$ and 11 coincide. Hence $R(m,t) = 40$ and $k = 45$. We thus get a 6 error-correcting binary group code (85, 45).

SECTION III

We shall now discuss a method which enables us to get matrices possessing the property (P_{2t}) by adjoining other matrices. For the purpose of this section the subscripts carried by a matrix will denote the number of rows and columns of the matrix. Thus, $A_{n,r}$ denotes a matrix with n rows and r columns. $O_{n,r}$ will denote a matrix with n rows and r columns, each of whose elements is zero. Also $O_{n,1}$ will denote a column vector with n zero elements, and $O_{1,r}$ a row vector with r zero elements. Finally, $j_{r,1}$ will denote a column vector with r unities as elements. The elements of all the matrices considered belong to $GF(2)$.

LEMMA 3. If $A_{n,r}$ possesses the property (P_{2t}) then the matrix

$$\| A_{n,r},\ B_{n,s} \| \tag{7}$$

obtained from it by adjoining s new columns $(s > 0)$ also possesses the property (P_{2t}).

Proof is obvious.

LEMMA 4. If the matrix $F_{n,r}$ possesses the property (P_{2t}), then the matrix

$$G_{n+1,r+1} = \left\| \begin{array}{c:c} F_{n,r} & j_{n,1} \\ \hdashline O_{1,r} & 1 \end{array} \right\| \tag{8}$$

possesses the property (P_{2t}).

Denote the matrix $\|F_{n,r} \vdots j_{n,1}\|$ formed by the first n rows of $G_{n+1,r+1}$ by $\bar{F}_{n,r+1}$. From Lemma 3, no $2t$ rows of $\bar{F}_{n,r+1}$ can be dependent. Again, consider the $2t$ rows obtained by choosing $2t - 1$ rows from $\bar{F}_{n,r+1}$ and adjoining the last row of $G_{n+1,r+1}$. These cannot be dependent. Otherwise the corresponding $2t - 1$ rows of $F_{n,r}$ which possess the property (P_{2t}) would be dependent. This completes the proof of the Lemma.

THEOREM 2. If the matrix $A_{n,r-r_0}$, $r > r_0$, possesses the property (P_{2t-2}) and the matrices

$$\| A_{n,r-r_0} \vdots T_{n,r_0} \| \quad \text{and} \quad F_{n',r_0+d-1} \tag{9}$$

$d \geqq 1$, possess the property (P_{2t}), then the matrix

$$M_{n+n'+1,r+d} = \left\| \begin{array}{c:c:c:c} A_{n,r-r_0} & T_{n,r_0} & O_{n,d-1} & O_{n,1} \\ \hdashline O_{n',r-r_0} & \multicolumn{2}{c:}{F_{n',r_0+d-1}} & j_{n',1} \\ \hdashline O_{1,r-r_0} & \multicolumn{2}{c:}{O_{1,r_0+d-1}} & 1 \end{array} \right\| \tag{10}$$

also possesses the property (P_{2t}).

Clearly the matrix $\bar{A}_{n,r+d}$ consisting of the first n rows of $M_{n+n'+1,r+d}$ has the property (P_{2t}). Also from Lemma 4, the matrix $\bar{G}_{n'+1,r+d}$ formed by the last $n' + 1$ rows of $M_{n+n'+1,r+d}$ has the property (P_{2t}). To prove the theorem we have to show that the $2t$ rows obtained by choosing any c rows of $\bar{A}_{n,r+d}$ and any $2t - c$ rows of $\bar{G}_{n'+1,r+d}$, $0 \leqq c \leqq 2t$, cannot be dependent. From what has been said this is true for $c = 2t$ or 0. If $c = 2t - 1$, then the last coordinate of the chosen rows adds up to unity. Hence they cannot be dependent because the matrix $A_{n,r-r_0}$ has the property (P_{2t-2}). This completes the proof of the theorem.

As in Section II, let $c = c(n)$ be the smallest integer such that $1 + cn$ is a power of 2, this power being $m = m(n)$. Let $R(m,t)$ be defined as in Lemma 2. We then have

THEOREM 3.

$$n_{2t}[R(m,t) + d] \geqq 1 + n + n_{2t}[R(m,t) - R(m,t-1) + d - 1]$$

where $n_{2t}(r)$ has been defined in the introduction, and

$$1 \leqq d < R(m+1,t) - R(m,t).$$

Let M^* be the matrix given by (2.3). We can then write

$$M^* = \| M_1^*, M_3^*, \cdots, M_u^*, \cdots, M_{2t-1}^* \|$$

where M_u^* is defined by (2.4).

Using the operations (a) and (b) described under Lemma 2, we can drop redundant columns from M^* and arrive at a matrix with n rows and $R(m,t)$ columns. Let the number of columns in the block coming from M_{2t-1}^* be r_0 and the submatrix of these columns be T_{n,r_0}. Let the number of columns coming from the part $\|M_1^*, M_3^*, \cdots, M_{2t-3}^*\|$ be $r - r_0$ and the submatrix of these columns be $A_{n,r-r_0}$. Then $r = R(m,t)$, $r - r_0 = R(m,t-1)$, and the matrices $\|A_{n,r-r_0} \vdots T_{n,r_0}\|$ and $\|A_{n,r-r_0}\|$ possess the properties (P_{2t}) and (P_{2t-2}) respectively. Let

$$r_0 + d - 1 = R(m,t) - R(m,t-1) + d - 1$$

and let

$$n' = n_{2t}(r_0 + d - 1)$$

Then there exists a matrix F_{n',r_0+d-1} with elements from $GF(2)$ and possessing the property (P_{2t}). We can now construct the matrix

$$M_{n+n'+1,r+d}$$

given by (3.4). The required result then follows from Theorem 2.

The most useful case of Theorem 3 is when $c = 1$, $n = 2^m - 1$. For this case we have

COROLLARY (1). $n_{2t}[R(m,t) + d] \geqq 2^m + n_{2t}[R(m,t) - R(m,t-1) + d - 1]$ A less powerful but simpler result is

COROLLARY (2). $n_{2t}(mt + d) \geqq 2^m + n_{2t}(m + d - 1)$

This follows by applying our reasoning to M^* without dropping any redundant column.

Example 4. Let us consider the case $t = 2$, $c = 1$, so that $n = 2^m - 1$. Then $R(m,2) = 2m$ and corollary (2) gives the same result as corollary (1). We know that $n_4(2m) \geqq 2^m - 1$. But one may want to get a bound on $n_4(2m + 1)$. From corollary (2) we have

$$n_4(2m + 1) \geqq 2^m + n_4(m)$$

For example,

(i)
$$n_4(21) \geqq 2^{10} + n_4(10)$$
$$\geqq 2^{10} + 2^5 - 1$$

(ii)
$$n_4(15) \geqq 2^7 + n_4(7)$$
$$\geqq 2^7 + 2^3 + n_4(3)$$
$$\geqq 2^7 + 2^3 + 3$$

SECTION IV

It is easy to see by exhaustive trial that $n_4(m) = m$ for $m = 1,2,3$;

TABLE I

$t = 1$		$t = 2$		$t = 3$		$t = 4$		$t = 5$		$t = 6$	
r	$L_2(r)$	r	$L_4(r)$	r	$L_6(r)$	r	$L_8(r)$	r	$L_{10}(r)$	r	$L_{12}(r)$
2	3	6	7	6	7	14	15	25	31	30	31
3	7	7	11	7	8	15	20	26	37	31	37
4	15	8	15	8	9	16	21	27	63	32	38
5	31	9	21*	9	10	17	22	28	67	33	63
6	63	10	31	10	15	18	23	29	68	34	70
7	127	11	36	11	18	19	24	30	69	35	71
		12	63	12	19	20	31	31	70	36	72
		13	71	13	20	21	37	32	71	37	73
		14	127	14	21	22	38	33	72	38	74
				15	31	23	39	34	73	39	75
				16	37	24	63	35	127	40	85*
				17	38	25	70			41	86
				18	63	26	71			42	127
				19	70	27	73*				
				20	72	28	127				
				21	127						

$n_4(4) = 5$, $n_4(5) = 6$, and $n_4(6) = 8$. Similarly, we can easily see that $n_{2t}(m) = m$ for $m = 1, 2, \cdots 2t$; $n_6(7) = 8$ and $n_{12}(13) = 14$. Using these facts and the results we have obtained, we can construct the following table where $L_{2t}(r)$ denotes the number of vectors in V_r that we can actually obtain such that no $2t$ are dependent. Thus $L_{2t}(r)$ is a lower bound for $n_{2t}(r)$.

The three asterisks in Table I indicate those cases corresponding to the three examples given after Theorem 1. Given n and t, $n \leqq 100$ $t \leqq 6$, we can find out from Table I the maximum possible k for which we can obtain by our methods a t-error correcting (n,k) group code. For this purpose we need to use the fact that if $n_{2t}(r) = n$, then for any positive integer c we have a t-error correcting $(n - c, n - r - c)$ group code. Thus, for instance, if we are seeking the largest value of k for $n = 90$, $t = 4$, we shall note that $L_8(27) < 90 < L_8(28)$ and decide that the required value of k is $90 - 28 = 62$.

RECEIVED: February 23, 1960.

REFERENCES

BOSE, R. C., AND RAY-CHAUDHURI, D. K. (1960). On a class of error correcting binary group codes. *Information and Control* **3**, 68.

PETERSON, W. W. (1960) Encoding and error-correction procedures for Bose-Chaudhuri codes. To appear in *IRE Trans. on Inform. Theory.*

PRANGE, E. (1958). Some cyclic error-correcting codes with simple decoding algorithms. Tech. note AFCRC-TN-58-156, Air Force Cambridge Research Center, Bedford, Massachusetts.

PRANGE, E. (1959). The use of coset equivalence in the analysis and decoding of group codes. Tech. Rept. AFCRC-TR-59-164, Air Force Cambridge Research Center, Bedford, Massachusetts.

A NEW TREATMENT OF BOSE-CHAUDHURI CODES†

H. F. MATTSON* AND G. SOLOMON**

Abstract. Letting A be any (k, n) Bose-Chaudhuri code, we first attach to each a in A (via difference equations over $GF(2)$) a polynomial $g_a(x)$ such that the coordinates of a are the values of $g_a(x)$ on the nth roots of unity. The degree of these polynomials is such that the minimum nonzero weight d of vectors in A is immediately seen to be at least d_0, the usual Bose-Chaudhuri lower bound. This lower bound d_0 is improved over a class of $(h + 1, p)$ codes, where $p = 2h + 1$ has certain prime values, in a number of general theorems. In particular, the (12, 23) Golay code is proved very simply to have $d = 7$; and a (24, 47) code is shown to have $d \geqq 9$, thus improving by 4 the usual lower bound $d_0 = 5$ for that code.

1. Introduction. In constructing an automatic, high-speed communication system, it is usual to take as a "message" an ordered n-tuple in two arbitrary symbols, say 0, 1. Thus a message would consist of $a = (a_0, \cdots, a_{n-1})$ where each a_i is 0 or 1. Since the usual addition and multiplication modulo 2 are convenient to perform on an electronic machine, one is led to consider the set V of all these n-tuples in 0, 1 as a vector space over $F = GF(2)$, the field of two elements, with addition defined component-wise.

In actual systems, errors may occur in transmission of the a_i. In order to cope with this effect, one can decide that only those a in some subset A of V will be transmitted and then try to choose A in such a way that if $a \in A$ is sent and $a^* \in V$ is received, it will then be possible to recover a from a^* provided not too many errors are present. The subset A is usually called a *code*, or *error-correcting code*. Elements of A will be called code-vectors.

In particular, if we choose A as a (linear) subspace of V, it is very simple to describe the number of errors in a single message which one can "correct". For if we define the *weight* $w(a)$ of $a \in V$ as the total number of a_i which equal 1, then the function $\rho(a, b) = w(a + b)$ is a metric on V—the Hamming distance—and is invariant under translation. Therefore, the distance

$$d(a) = \min \{\rho(a, b); b \in A, b \neq a\}$$

is the same for all $a \in A$; in particular $d(a) = d(0) = \min \{w(b); b \in A, b \neq 0\} = d$ is this value. Thus, taking[1] $e = [(d - 1)/2]$, we easily see that if a is sent and a^* received, then the nearest element of A to a^* is a itself whenever the total number of errors present in a^* is at most e.

How to determine an efficient procedure for finding the element of A nearest to a given element of V—the so-called decoding problem—is an important problem in this field. Another important problem is to determine d for a given (k, n) *group code* A, i.e., a k-dimensional subspace of V.

The group codes to which we confine ourselves here are due mainly to Bose and Ray-Chaudhuri [1], who gave a constructive procedure for obtaining a large class of cyclic codes having preassigned error-correcting properties. Their codes can, in many cases, be shown to possess greater error-correcting ability than they were able to demonstrate. These codes have been studied recently by several authors [3, 4, 8, 9]. Their estimates of the error-correcting properties of these codes are based on methods using matrices, linear recursive sequences, and rings of polynomials. We have reproduced their estimates in general and, in a sub-class of Bose-Chaudhuri codes, have improved these estimates. Our methods are based on the treatment of linear recursions as finite-difference equations [7] and illuminate the subject in a particularly simple way. In particular, we change the difficult combinatorial problem of determining the number of 1's in a code-vector into the more tractable algebraic problem of determining the number of zeros of a certain polynomial on a given finite set.

This paper is reasonably self-contained. We do assume, however, certain basic algebraic notions, such as the elementary properties of polynomials and the existence of extension fields. Nevertheless, the reader will be able to understand all of the paper without much trouble even if he has only a nodding acquaintance with algebra.

We give one particular case ($p = 23$) in detail in §5, so as to illustrate the general theorems of the report. For the casual reader this example is probably a good abstract of our work.

Summary. In §2 we present that part of [7] which we need. For completeness, we give the proof (which is quite simple).

In §3 we define group codes obtained from linear recursions and derive a basic result (Lemma 2), which rests on §2. Applying Lemma 2 to the Bose-Chaudhuri codes as defined in [8], we obtain a simple proof of the previously known result (Theorem 1) giving a lower bound on the minimum nonzero weight of code-vectors.

The rest of this report is restricted to a certain class of $(h + 1, p)$ Bose-Chaudhuri codes, defined for certain primes $p = 2h + 1$. In §4 we present general results on these codes, including an apparently strong improvement (Theorem 2) in the lower bound on the above-mentioned minimum weight for the (h, p) subcode of even-weight vectors in the case $p \equiv 1 (\text{mod } 8)$. In Theorem 3 we prove that the previously known lower bound can always be improved by 1, unless that lower bound is quite good already (namely, as big as h). In Theorem 5 we give relations between p and odd weights of code-vectors.

In §5 we give particular results for $p = 23$ (the Golay case [2]) and $p = 47$; our methods give a short and simple proof that the Golay code corrects three errors, and without too much labor, we raise from 5 to 9 the above-mentioned lower bound in the case $p = 47$.

2. Difference equations over $GF(2)$. Throughout this paper, let $F = GF(2)$ denote the field of two elements 0, 1. Let $a_0, a_1, \cdots$ be a linear recursive sequence in F with the a_j determined by the recursion

$$(1) \qquad a_{k+i} + b_1 a_{k+i-1} + b_2 a_{k+i-2} + \cdots + b_k a_i = 0 \qquad (i = 0, 1, 2, \cdots),$$

in which the coefficients $b_1, \cdots, b_k \in F$ are independent of i and the values $a_0, a_1, \cdots, a_{k-1}$ are preassigned.

Equation (1) is simply a linear difference equation with constant coefficients. In order to find a "general solution" of (1), we set $a_j = \beta^j$ for all j in (1), as in the classical case, obtaining

$$\beta^i(\beta^k + b_1\beta^{k-1} + \cdots + b_{k-1}\beta + b_k) = 0,$$

which is satisfied if β is a root of the polynomial $f(x) = x^k + b_1 x^{k-1} + \cdots + b_k$. To obtain a finite field K which contains F and all the roots of $f(x)$ is a standard algebraic procedure, which we assume done without further ado.

We restrict our attention to the case in which the roots $\beta_1, \cdots, \beta_k$ of $f(x)$ are distinct. For any $c_1, \cdots, c_k \in K$, let

$$(2) \qquad a_j = c_1\beta_1{}^j + \cdots + c_k\beta_k{}^j \qquad (j = 0, 1, 2, \cdots).$$

Then this sequence $\{a_j\}$ satisfies (1), except it may not take the preassigned values for $j = 0, 1, \cdots, k - 1$. But there is a unique set of c's in K, obtained as the solution of k linear simultaneous equations having a van der Monde coefficient matrix such that the sequence defined in (2) satisfies (1) for all $j = 0, 1, 2, \cdots$.

Thus every linear recursive sequence $\{a_i\}$ for (1) is obtainable in the form (2). Conversely, every sequence in the form (2) is linear recursive for (1) over K; it lies in F provided only that $a_0, a_1, \cdots, a_{k-1}$ are in F. We state these results in

LEMMA 1. *If $a_0, a_1, \cdots$ is a sequence in F given by the recursion* (1), *and if the polynomial $f(x) = x^k + b_1x^{k-1} + \cdots + b_k$ has no repeated roots, then there exist uniquely determined elements $c_1, \cdots, c_k$ in K such that*

$$a_i = c_1\beta_1{}^i + \cdots + c_k\beta_k{}^i \qquad (i = 0, 1, 2, \cdots),$$

where $\beta_1, \cdots, \beta_k$ are the roots of $f(x)$.

A detailed treatment of this subject, with additional results on linear recursive sequences not needed here, appears in [7].

3. The codes. Let n be an odd integer, and let $f(x) = x^k + b_1x^{k-1} + \cdots + b_k$ be a polynomial with coefficients in F (the set of all such polynomials is denoted by $F[x]$); suppose also that $f(x)$ divides $x^n + 1$. Let K denote the smallest field containing F and all the roots of $x^n + 1$. Let V denote the vector space of all ordered n-tuples $a = (a_0, a_1, \cdots, a_{n-1})$ with each

† Received by the editors March 27, 1961 and in revised form July 24, 1961.

* Sylvania Applied Research Laboratory, Waltham, Massachusetts. This work was supported in part by the Air Force Cambridge Research Laboratory under Contract No. AF 19(604)-6639.

** Staff Member, M.I.T. Lincoln Laboratory (operated with support from the U. S. Army, Navy, and Air Force).

[1] The square brackets here (only) denote the usual greatest-integer function.

[2] The reader may note the inessential distinction between our definition and that of some authors, who "reverse" the code, i.e., who for our $f(x)$ would use $x^k f(1/x)$.

[3] See Corollary 1, (iii), for definition.

[4] See [8]. A is the code W^* of [8, p. 22], with our $f(x)$ equal to the h there.

Reprinted with permission from *J. Soc. Ind. Appl. Math.*, vol. 9, pp. 654–669, Dec. 1961.

a_i in F, with addition of vectors a and $b = (b_0, \cdots, b_{n-1})$ defined by $a + b = (a_0 + b_0, \cdots, a_{n-1} + b_{n-1})$. We define a certain (k, n) group code A as the following subspace of[2] V:

$$A = \{a; a = (a_0, \cdots, a_{n-1}) \in V, a_{i+k} + b_1 a_{i+k-1} + \cdots + b_k a_i = 0, \quad i = 0, 1, \cdots, n-k-1\}. \tag{3}$$

Thus, to form a vector of A, we choose coordinates $a_0, \cdots, a_{k-1}$ arbitrarily from F and determine succeeding coordinates by the linear recursion in (3). A is a cyclic code of dimension k over[3] F. A criterion for the above cyclic code is that if $a(x)$ denotes the polynomial $a_0x^{n-1} + a_1x^{n-2} + \cdots + a_{n-1}$ in $F[x]$ and if $f^\#(x) = (x^n + 1)/f(x)$, then the vector a in V belongs to the cyclic code A if and only if the associated polynomial $a(x)$ is a multiple of $f^\#(x)$. This result is proved in [5] for all n; we essentially prove it in Corollary 2 by our methods for odd n.

Recall that the *weight* $w(a)$ of a vector $a = (a_0, \cdots, a_{n-1}) \in V$ is the total number of a_i which equal 1. Throughout this paper we shall use the notation d for the minimum nonzero weight of vectors a in A:

$$d = \min \{w(a); a \neq (0, 0, \cdots, 0), a \in A\}.$$

We now present the definition of Bose-Chaudhuri codes given in [8]. Let β be any primitive nth root of unity over F, choose $s < n$, and let $f^\#(x)$ be the polynomial over F of least degree which divides $x^n + 1$ and has $\beta, \beta^2, \beta^3, \cdots, \beta^s$ among its roots. Define $f(x) = (x^n + 1)/f^\#(x)$.

DEFINITION. *The Bose-Chaudhuri code for β and s is defined as the code A attached to $f(x)$ by our definition* (3).

It is known that for the above-defined code A, we have[4] $d \geqq d_0 = s + 1$. We shall later prove this result (as Theorem 1) by our methods.

If $\phi(x)$ is any polynomial over F and z any root of $\phi(x)$, then z^2 is also a root of $\phi(x)$; this fact follows from the property that if L is any field containing F, then $(\alpha + \beta)^2 = \alpha^2 + \beta^2$ for all $\alpha, \beta \in L$, which implies $\phi(z)^2 = \phi(z^2)$. If $\phi(x)$ is irreducible over F, then all the roots of $\phi(x)$ are obtainable on repeated squaring of any one of them: if $\phi(z) = 0$, then all the roots of $\phi(x)$ are $z, z^2, z^4, \cdots, z^{2^{j-1}}$, where j is the degree of $\phi(x)$. Also, $z^{2^j} = z$.

In passing, we note that choosing s above to be even is free; for if s is odd, then β^{s+1} is automatically a root of $f^\#(x)$, since $\beta^{(s+1)/2}$ is by definition a root of $f^\#(x)$. Therefore, we shall always take s to be even in the above definition.

We now set up certain definitions basic to this paper. Let n be odd and let $f(x) \in F[x]$ divide $x^n + 1$, as before. Let ζ be a primitive nth root of unity, i.e., an element ζ of K such that $\zeta^n = 1$ and no lower power of ζ is 1. We define

$$E(\zeta) = \{e; 0 \leqq e < n, f(\zeta^e) = 0\}. \tag{4}$$

In other words, if $f(x)$ has degree k, $E(\zeta)$ consists of integers $e_1, \cdots, e_k$ with $0 \leqq e_i < n$ such that $\zeta^{e_1}, \cdots, \zeta^{e_k}$ are all the roots of $f(x)$. For example, when $n = 7$, we have $x^7 + 1 = (x + 1)(x^3 + x + 1)(x^3 + x^2 + 1) = (x + 1)f_0(x)f_1(x)$. Suppose we take $f(x) = (x + 1)f_0(x)$. Since $n = 7$ is prime, any root of $x^7 + 1$ other than 1 is a primitive 7th root of 1. If we take ζ as a root of $f_0(x)$, then $E(\zeta) = \{0, 1, 2, 4\}$. But if we take ζ to be a root of $f_1(x)$, then, since the roots of $f_1(x)$ are ζ, ζ^2, ζ^4, the roots of $f_0(x)$ are now $\zeta^3, \zeta^5, \zeta^6$; thus $E(\zeta) = \{0, 3, 5, 6\}$.

The basis for all the work of this paper is the following result. As above let n be odd, let ζ be a primitive nth root of unity, let $f(x)$ divide $x^n + 1$, and let A be the code attached to $f(x)$ by (3). Then we have

LEMMA 2. *For each $a = (a_0, \cdots, a_{n-1}) \in A$ there is a polynomial $g_a(x)$ with coefficients in K such that $a_i = g_a(\zeta^i)$ for $i = 0, 1, \cdots, n-1$. ζ being fixed, this polynomial is uniquely determined by a. The degree of $g_a(x)$ is at most m, the largest integer in $E(\zeta)$.*

Proof. The roots of $f(x)$ are $\zeta^{e_1}, \cdots, \zeta^{e_k}$, $e_i \in E(\zeta)$, and $f(x)$ has no repeated roots. We may therefore apply Lemma 1 to our vector $a = (a_0, \cdots, a_{n-1}) \in A$, since the recursion (3) defining $a \in A$ is of the type (1). We obtain $c_1, \cdots, c_k \in K$, the c_j's depending on a, of course, such that

$$a_i = c_1(\zeta^{e_1})^i + \cdots + c_k(\zeta^{e_k})^i \tag{5}$$

for $i = 0, 1, \cdots, n-1$. But we can write (5) as

$$a_i = c_1(\zeta^i)^{e_1} + \cdots + c_k(\zeta^i)^{e_k} \qquad (i = 0, 1, \cdots, n-1), \tag{6}$$

and this observation leads us to define the polynomial

$$g_a(x) = c_1x^{e_1} + \cdots + c_kx^{e_k},$$

which, by Lemma 1, is uniquely determined by the vector $a \in A$ (and the choice of ζ). Equation (6) is the statement $a_i = g_a(\zeta^i)$.

The mapping $a \to g_a(x)$ of A into $K[x]$ is linear over F. That is, if $a, b \in A$ then $g_{a+b}(x) = g_a(x) + g_b(x)$.

We emphasize that $g_a(x)$ must take the value 0 or 1 on the group of nth roots of unity $Z = \{1, \zeta, \cdots, \zeta^{n-1}\}$; the weight of a, therefore, is n less the number of zeros of $g_a(x)$ on Z. In this way the combinatorial problem of determining the weight of a vector is transformed into the more algebraic problem of determining the number of zeros of a polynomial on a given set. In particular, we shall next reproduce the Bose-Chaudhuri lower bound d_0 for d simply by proving that we can choose ζ so that the degree of each nonzero $g_a(x)$ is at most $n - d_0$. (The number of zeros of $g_a(x)$ on Z is then at most $n - d_0$, so the weight of a is at least $n - (n - d_0) = d_0$.)

With d_0 as defined by Bose-Chaudhuri via [8], we shall now prove that the minimum nonzero weight d in the code A is at least d_0. We slightly reword the definition of d_0, by singling out $f(x)$ instead of s. As before, let n be odd and let $f(x) \in F[x]$ divide $x^n + 1$ in such a way that there is a primitive nth root β of 1 which is not a root of $f(x)$. Let A be the (Bose-Chaudhuri) code attached by (3) to $f(x)$. Then we have

THEOREM 1. *Let β^{d_0} be the least positive power of β which is a root of $f(x)$. Then d_0 is necessarily odd, and $d \geqq d_0$.*

Proof. As earlier noted, it suffices to prove that for some primitive nth root of unity, ζ, the set $E(\zeta)$ has $n - d_0$ as maximum.

We are given that $\beta, \beta^2, \cdots, \beta^{d_0-1}$ are not roots of $f(x)$ and that β^{d_0} is a root of $f(x)$. It follows immediately that $E(\zeta)$ for $\zeta = \beta^{-1}$ does not contain $n-1, n-2, \cdots, n-(d_0-1)$ but does contain $n - d_0$.

4. The case where n takes certain prime values—general results. For the rest of the paper we restrict our attention to the case where n is an odd prime $p = 2h + 1$ such that $x^p + 1 = (x + 1)f_0(x)f_1(x)$, where $f_0(x)$ and $f_1(x)$ are irreducible over[5] F.

We immediately observe the following three simple consequences:

(i) $x^p + 1$ factors over F as above if and only if 2 has multiplicative order h modulo p. For if n_i is the degree of $f_i(x) (i = 0, 1)$, and if ζ is a root of $f_i(x)$, then we must have $\zeta^{2^{n_i}} = \zeta$, or $2^{n_i} \equiv 1 (\text{mod } p)$. No lower power of 2 can satisfy this congruence, however, or we would not find enough roots of $f_i(x)$. Therefore $n_0 = n_1$, and since $n_0 + n_1 = 2h$, we have $n_0 = n_1 = h$. Conversely, if 2 has multiplicative order h modulo p, then $x^p + 1$ splits over F into three irreducible factors $x + 1, f_0(x)$, and $f_1(x)$. Thus in this case we may, and do, take K as $GF(2^h)$.

(ii) We may, and do, take $f_0(x) = x^h + 0x^{h-1} + \cdots + 1$ and $f_1(x) = x^h + x^{h-1} + \cdots + 1$, since each $f_i(x)$ has degree h and $f_0(x)f_1(x) = x^{p-1} + x^{p-2} + \cdots + x + 1$, the latter implying that the coefficients of x^{h-1} in $f_0(x)$ and $f_1(x)$ are not the same.

(iii) The powers of 2 are precisely the quadratic residues modulo p and $p \equiv \pm 1 (\text{mod } 8)$. For a cyclic group has at most one subgroup of a given order and the subgroup of quadratic residues mod p also has order h. Since, in particular, 2 is a quadratic residue mod p, the law of quadratic reciprocity[6] tells us that $p \equiv \pm 1 (\text{mod } 8.)$[7]. We shall occasionally modify our treatment to deal with one or the other of these two possibilities. Two examples of such primes are $p = 7$ and $p = 17$.

Let us denote by R the set of least positive quadratic residues mod p:

$$R = \{r_i; 0 < r_i < p, r_i \equiv 2^{i-1} (\text{mod } p), i = 1, \cdots, h\}.$$

Let $R' = \{s_1, \cdots, s_h\}$ denote the set of least positive quadratic nonresidues mod p; $R \cup R' = \{1, 2, \cdots, p-1\}$. Let s_1 denote the least member of R' (note that s_1 is thus odd) and choose the notation so that $s_2 \equiv 2s_1, s_3 \equiv 2s_2, \cdots, s_1 \equiv 2s_h (\text{mod } p)$, as we may do since $R' \equiv s_1R$ (mod p). Thus we have chosen $r_1 = 1$; and, as we shall see in a moment, the most advantageous choice of ζ leads to $d_0 = s_1$.

We define our code A as the $(h + 1, p)$ code attached by (3) to $f(x) = (x + 1)f_0(x)$. In this section we present some general results on the minimum nonzero weight d of vectors a in A, and in the next section we shall present further results on d for particular values of p. The code attached by (3) to $(x + 1)f_1(x)$ is equivalent to A, as we shall prove below, after Corollary 2, so we may confine our attention to A. (Two codes are called equivalent if one can be obtained from the other by a permutation of coordinates.)

We now carry out some of the procedures of the previous section for this code A. The distinction between the two cases $p \equiv \pm 1 (\text{mod } 8)$ arises from the fact that for odd primes q, -1 is a quadratic residue (mod q) if and only if $q \equiv 1 (\text{mod } 4)$. In other words

[5] We are grateful to Eugene Prange for suggesting investigation of the Bose-Chaudhuri codes attached to such primes p.

[6] See any book on number-theory, e.g., [10, p. 127].

[7] As a side point which may be of some interest, we note that for primes less than 3,000,000, those congruent to 1 mod 8 are scarce compared with those congruent to -1 mod 8. See Shanks, Daniel, *Quadratic residues and the distribution of primes*, Math. Tables Aids Comput., 13 (1959), pp. 272–284 (Math. Rev., 21 (1960), Rev. No. 7186, p. 1325).

$$p - 1 \in R, \quad \text{if} \quad p \equiv 1 (\bmod 8);$$
$$p - 1 \in R', \quad \text{if} \quad p \equiv -1 (\bmod 8).$$

Let ζ be any primitive pth root of unity. Any choice of ζ leads either to $\{0\} \cup R$ or $\{0\} \cup R'$ as $E(\zeta)$; we always choose ζ so that $p - 1$ is not in $E(\zeta)$. That is,

Choose ζ to be a root of $f_1(x)$ if $p \equiv 1(\bmod 8)$;

Choose ζ to be a root of $f_0(x)$ if $p \equiv -1(\bmod 8)$.

(Recall that in (4) $E(\zeta)$ is defined so that $(\zeta^e + 1)f_0(\zeta^e) = 0$ when e runs through $E(\zeta)$.) In Theorem 1, d_0 is defined as the smallest odd exponent such that β^{d_0} is a root of $f_0(x)$, where β is a root of $f_1(x)$. In other words, since the roots of $f_1(x)$ are $\beta, \beta^2, \beta^4, \cdots, \beta^{2^{h-1}}$, i.e., $\beta^{r_1}, \beta^{r_2}, \cdots, \beta^{r_h}$, we find immediately that $d_0 = s_1$. Thus we have

THEOREM 1'. *Under the restrictions of this section, and letting s_1 be the smallest quadratic nonresidue* $(\bmod p)$ *(s_1 is necessarily odd), we have* $d \geqq s_1 = d_0$.

We summarize the preceding remarks in

LEMMA 3. *The primitive pth root of unity ζ is chosen and the polynomial $g_a(x)$ of Lemma 2 is given as follows: If $a \in A$, then*

$$g_a(x) = \begin{cases} c_0 + c_1 x^{r_1} + c_2 x^{r_2} + \cdots + c_h x^{r_h}, & (f_0(\zeta) = 0) \\ & p \equiv -1(\bmod 8) \\ c_0 + c_1 x^{s_1} + c_2 x^{s_2} + \cdots + c_h x^{s_h}, & (f_1(\zeta) = 0) \\ & p \equiv +1(\bmod 8). \end{cases}$$

The degree of $g_a(x)$ is at most $p - d_0 = p - s_1$ in both cases.

For convenience we shall need a generic set of exponents for $g_a(x)$. Let us say then that $g_a(x) = c_0 + c_1x^{e_1} + c_2x^{e_2} + \cdots + c_hx^{e_h}$, where $e_i = r_i$ for all $i = 1, \cdots, h$ when $p \equiv -1(\bmod 8)$ and $e_i = s_i$ for all i when $p = +1(\bmod 8)$. By our choice of notation we have $e_i \equiv 2^{i-1}e_1(\bmod p)$ for all i; and the roots of $f_0(x)$ are $\zeta^{e_1}, \cdots, \zeta^{e_h}$. Notice that e_h is not necessarily the degree of $g_a(x)$ (even when $c_h \neq 0$).

We now investigate the coefficients c_i of $g_a(x)$. For this we need the following lemma, due to Reed [6].

LEMMA 4. *Let K_0 be any field containing F and $h(x) = \sum c_j'x^j$ any polynomial over K_0. Let β be a primitive mth root of unity in K_0, with m odd, and suppose degree $h(x) < m$. Then we have*

$$c_j' = \textstyle\sum_{i=0}^{m-1} h(\beta^i)\beta^{-ji} \qquad (j = 0, 1, 2, \cdots)$$

Proof. The sum in question is

$$\textstyle\sum_{i=0}^{m-1} \sum_{k \neq j} c_k'\beta^{i(k-j)} + c_j'\sum_{i=0}^{m-1} 1,$$

in which the second part is c_j', since m is odd; the rest is 0 because

$$\sum_{i=0}^{m-1} \beta^{i(k-j)} = \frac{x^m + 1}{x + 1}$$

with $x = \beta^{k-j} \neq 1$.

When we apply Lemma 4 to our polynomial $g_a(x)$ we obtain, recalling Lemma 2,

LEMMA 5. *Let a be in A. The coefficients $c_0, c_1, \cdots, c_h$ of $g_a(x)$ are given by the formulas*

$$c_0 = \textstyle\sum_{i=0}^{p-1} a_i, \qquad (c_0 \in F)$$
$$c_j = \textstyle\sum_{i=0}^{p-1} a_i\zeta^{-ie_j}, \qquad (c_j \in K) \qquad (j = 1, \cdots, h).$$

In particular, $c_0^2 = c_0, c_1^2 = c_2, c_2^2 = c_3, \cdots, c_h^2 = c_1$.

COROLLARY 1.

(i) *The linear mapping from A to $F \times K$ given by $a \to (c_0, c_1)$ (where $a \in A$ and $g_a(x) = c_0 + c_1x^{e_1} + \cdots + c_hx^{e_h}$) is one-one and onto (by equality of dimensions over F).*

(ii) *Furthermore, the polynomial $g_a(x)$ satisfies*

$$g_a(\zeta^i) = c_0 + T(c_1\zeta^{ie_1}), \qquad a \in A,$$

where T denotes the trace from K to F (if $z \in K$, then $T(z)$ is defined as $z + z^2 + z^4 + \cdots + z^{2^{h-1}}$).

(iii) *The code A is cyclic; i.e., for each $a = (a_0, a_1, \cdots, a_{p-1})$ in A, the vector $a' = (a_1, a_2, \cdots, a_{p-1}, a_0)$ is also in A. (For $a_i' = g_a(\zeta^{i+1})$; thus $g_a(\zeta x)$ is the polynomial for the code-vector corresponding to $(c_0, \zeta c_1)$ under the mapping in (i) above. We could have proved this result earlier, of course.)*

(iv) *The polynomial $g_a(x)$, $a \in A$, has degree $p - d_0 = p - s_1$ unless it is constant (since $c_1, \cdots, c_h$ are all 0 or all not 0).*

COROLLARY 2. *Let β be a root of $f_1(x)$; if $a \in A$, then*

$$\textstyle\sum_{i=0}^{p-1} a_i\beta^{-i} = 0.$$

Proof. Since $\beta = \zeta^e$ for some $e \not\equiv 0, e_1, \cdots, e_h(\bmod p)$, the quantity in question is 0 as the coefficient of x^e in $g_a(x)$ when $a \in A$.

Let A_1 be the code attached by (3) to $f(x) = (x + 1)f_1(x)$. We shall show that A and A_1 are equivalent. Let S be any quadratic nonresidue mod p, and consider the permutation of coordinates sending Si to i for each $i = 0, 1, \cdots, p - 1(\bmod p)$. If ζ is chosen according to Lemma 3, then $a \in A$ is mapped by our permutation to $b = (g_a(1), g_a(\beta), g_a(\beta^2), \cdots, g_a(\beta^{p-1}))$, where $\beta = \zeta^S$. That is, $b_i = g_a(\zeta^{Si}) = a_{Si}$. By Corollary 1, (i), the set of $g_a(x)$, $a \in A$, coincides with the corresponding set of polynomials for the code A_1; and by Lemma 3, β is a proper choice of pth root of unity for A_1, since β and ζ are not conjugate. In the case $p \equiv -1$ (mod 8), we may choose $S = -1$ and simply obtain A_1 as the "reverse" of A (followed by one cyclic shift).

We now prove a simple result on the even weights of A when $p \equiv 1$ (mod 8). That is, we investigate d for the code belonging to $f(x) = f_0(x)$ according to the definition in (3). The result is analogous to Theorem 1 in that the argument rests entirely on the degree of the polynomial $g_a(x)$.

THEOREM 2. *Let d' denote the minimum nonzero even weight attained on A. If $p \equiv 1(\bmod 8)$ then $d' \geqq 2 d_0$.*

Proof. Let a be in A, with $w(a)$ even. By Lemmas 3 and 5,

$$g_a(x) = x^{d_0}(c_1 + \cdots + c_hx^{e_h-d_0});$$

and the polynomial in parentheses has degree $p - 2 d_0$. Therefore $w(a) \geqq 2 d_0$.

We now prove

THEOREM 3. *If $d_0 < h$, then $d \geqq d_0 + 1$.*

Proof. By Theorem 1, we know $d \geqq d_0$; and if $d = d_0$, then for some $a \in A$ all roots of $g_a(x)$ are pth roots of unity. Since d_0 is odd, $g_a(0) = c_0 = 1$; therefore, if c denotes the leading coefficient of $g_a(x)$, the constant term $1/c$ (of $(1/c)g_a(x)$) is a pth root of unity. Thus $c^p = 1$, and by Lemma 5 all coefficients c_j in $g_a(x)$ satisfy $c_j^p = 1$. Thus for all i we have $a_i = g_a(\zeta^i) = 1 + T(c\zeta^{e_1i})$; here $c_1\zeta^{e_1i}$ runs through all pth roots of 1 as i goes from 0 to $p - 1$. Thus a_i assumes the value 0 h times (when $c_1\zeta^{e_1i}$ is a root of $f_0(x)$) and the value 1 h times, and also the value $1 + T(1) = 1 + h \cdot 1$. Thus if h is odd, a has weight h, contradicting $w(a) = d_0 < h$; if h is even, then a has weight $h + 1$, also a contradiction.

Note that there is nothing gained in splitting the above hypothesis into "$d_0 < h$(h odd) and $d_0 < h + 1$(h even)" since d_0 is odd. Note also that we have proved that for this class of primes the Bose-Chaudhuri lower bound is never best possible, except perhaps in the case when d_0 is already as big as h, e.g., $p = 7$ (which indeed may be the only such case).

THEOREM 4. *If $d_0 < h$ and $p \equiv 1(\bmod 8)$, then $d \geqq d_0 + 2$.*

Proof. From Theorem 3 we have $d \geqq d_0 + 1$, and if $d = d_0 + 1$, an even number, then Theorem 2 implies $d' = d_0 + 1 \geqq 2 d_0$, or $d_0 \leqq 1$. But $d_0 \geqq 3$ in general. Therefore, $d > d_0 + 1$.

A possible way of showing that $d \geqq d_0 + 2$ when $p \equiv -1(\bmod 8)$ is given by

COROLLARY 3. *Let $p \equiv -1(\bmod 8)$ and let $c_{j+1} = c_1^{2^j}$ be the leading coefficient of $g_a(x)$. If $2^j - 1$ is prime to $2^h - 1$ and if $d_0 < h$, then $d \geqq d_0 + 2$.*

Proof. From Theorem 3 we know that $d \geqq d_0 + 1$. If $a \in A$ has weight $d_0 + 1$, then $c_0 = 0$, by Lemma 5; thus $(1/x)g_a(x) = c_1 + \cdots + c_hx^{r_h-1}$, which has degree $p - d_0 - 1$ by Theorem 1. Thus $w(a) = d_0 + 1$ implies that all roots of $(1/x)g_a(x)$ are pth roots of unity. But $c_{j+1}/c_1 = c_1^{2^j-1}$, by hypothesis, has the same order as c_1; and therefore c_1 is a pth root of unity. This means a has weight $h + 1$ (since $T(1) = 1$), as in the proof of Theorem 3.

The corresponding result for the case $p \equiv 1(\bmod 8)$ is vacuous, since here $j = h/2$.

Our next result is a consequence of the cyclic property of A. As before, let the vector $a = (a_0, \cdots, a_{p-1})$ correspond to the polynomial $a(x) = a_0x^{p-1} + a_1x^{p-2} + \cdots + a_{p-1}$. Then the cyclic shift $(a_1, a_2, \cdots, a_{p-1}, a_0)$ of a corresponds to the polynomial $xa(x)$ reduced modulo $x^p + 1$. If a is in A, then $a(x)$ is a multiple of $f_1(x)$, and conversely, as we remarked in §3. Identifying vectors a in V with their associated polynomials $a(x)$, we shall prove

THEOREM 5. *Let $a \in A$ have odd weight m. Then*

$$m^2 - m + 1 \geqq p \qquad \text{if } p \equiv -1(\bmod 8),$$

and

$$m^2 \geqq p \qquad \text{if } p \equiv +1(\bmod 8).$$

Proof. Let A_1 be the code attached to $(x + 1)f_1(x)$. Let $a \in A$ have odd weight m; since A and A_1 are equivalent, there must be a vector $b \in A_1$

of weight m. Now consider the polynomial $a(x)b(x)$. Since b has weight m, $a(x)b(x)$ is a sum of m cyclic shifts of a, and therefore is an odd-weight vector in A. By the same argument it is in A_1. Therefore $a(x)b(x)$ is a multiple of both $f_0(x)$ and $f_1(x)$, and hence of the product $f_0(x)f_1(x) = x^{p-1} + \cdots + 1$. Since the weight is odd, $a(x)b(x) \equiv x^{p-1} + \cdots + 1 \pmod{x^p + 1}$. But there are at most m^2 terms in $a(x)b(x)$; therefore $m^2 \geqq p$.

In order to refine this estimate in the case $p \equiv -1 \pmod 8$, we use the particular coordinate permutation $i \to -i \pmod p$ known to yield an equivalence between A and A_1 in this case. Then for each coordinate $a_i = 1$ in a, the term x^{p-1-i} in $a(x)$ is matched by the term x^{p-1+i} in $b(x)$; of the m^2 products in $a(x)b(x)$, m are now the same, namely, $x^{p-1-i}x^{p-1+i} = x^{2p-2} (\equiv x^{p-2} \pmod{x^p + 1})$. Thus m of the 1's collapse into a single 1, so that $p \leqq m^2 - (m - 1)$.

5. Particular values of p (23 and 47). We apply our methods to the Golay (12, 23) code $A[2]$, which is known to be three-error-correcting, i.e., $d = \min\{w(a);\, a \neq 0,\, a \in A\} = 7$ [5, pp. 7 ff.].

We factor $x^{23} + 1 = (x + 1)f_0(x)f_1(x)$ into its irreducible parts,[8] where

$$f_0(x) = x^{11} + x^9 + x^7 + x^6 + x^5 + x + 1$$

and

$$f_1(x) = x^{11} + x^{10} + x^6 + x^5 + x^4 + x^2 + 1.$$

If β is a root of f_0, then the roots of f_0 are

$$\beta, \beta^2, \beta^4, \beta^8, \beta^{16}, \beta^9, \beta^{18}, \beta^{13}, \beta^3, \beta^6, \beta^{12},$$

while the roots of f_1 are

$$\beta^5, \beta^{10}, \beta^{20}, \beta^{17}, \beta^{11}, \beta^{22}, \beta^{21}, \beta^{19}, \beta^{15}, \beta^7, \beta^{14}.$$

The code A is the set of all linear recursive sequences in F (of length 23) generated by the difference equation associated with $f_0(x)(x + 1)$, as in §2. For $a = (a_0, a_1, a_2, \cdots, a_{11}, a_{12}, \cdots, a_{22})$ the general term a_k is given by Lemma 1 as

$$(9) \qquad a_k = \textstyle\sum_{i=0}^{11} c_i(\beta^{r_i k}), \qquad r_i \equiv 2^{i-1} \pmod{23}, \qquad 0 < r_i < 23,$$

(and $r_0 = 0$) when the c_i, $i = 0, 1, \cdots, 11$ are determined by the first 12 values $(a_0, a_1, \cdots, a_{11})$. The c_i are in $K = GF(2^{11})$, the smallest field over F containing the 23rd roots of unity.

The code A has dimension 12 in $V_{23}(F)$ and is clearly cyclic.

We may look upon (9) as the value of a polynomial $g_a(x)$ when x runs through the 23rd roots of unity, namely

$$g_a(x) = \textstyle\sum_{i=0}^{11} c_i x^{r_i}.$$

Then

$$a_k = \textstyle\sum_{i=0}^{11} c_i(\beta^{r_i k}) = \sum_{i=0}^{11} c_i(\beta^k)^{r_i}$$

or,

$$a_k = g_a(\beta^k) \qquad (k = 0, \cdots, 22).$$

The coefficients c_i—using Lemma 5—are given very simply by

$$(10) \qquad c_i = (c_1)^{2^{i-1}} \quad (i = 1, \cdots, 11), \qquad c_0 = \textstyle\sum_0^{22} a_i.$$

The exponents r_i of (9) are

$$1, 2, 4, 8, 16, 9, 18, 13, 3, 6, 12.$$

We note that to each code word a is assigned a pair (c_0, c_1). c_0 is either zero or one according as the weight of a is even or odd. c_1 is an element of $K = GF(2^{11})$ and is given very simply by the formula

$$c_1 = \textstyle\sum_{i=0} a_i\beta^{-i},$$

where β is the fixed root of $f_0(x)$ chosen before.

For every c in K we have two code words, one of even and one of odd weight. We thus have a natural mapping of our code A onto $F \times K$ as in Corollary 1.

We note that x^{18} is the highest-degree term of our polynomial, since $\max r_i = 18$, and we see clearly that $g_a(x)$ can have at most 18 roots. Therefore the minimum weight of a in A is the minimum number of 1's that $g_a(x)$ takes on as values over the 23rd roots of unity. We get the Bose-Chaudhuri lower bound immediately, namely $23 - 18 = 5$.

[8] A table of irreducible factors of $x^n + 1$ over F, for odd $n \leqq 35$, appears in [5, pp. 22–23].

We can do better, however, by examining the coefficients of $g_a(x)$. We do so now. If $w(a) = 5$, then $c_0 = 1$; and our polynomial is (let $c = c_1$ in (10))

$$g_a(x) = c^{2^6}x^{18} + c^{2^4}x^{16} + c^{2^7}x^{13} + \cdots + 1.$$

The product of the roots of $g_a(x)$ is of course $1/c^{2^6}$.

If all the 18 roots of $g_a(x)$ are 23rd roots of unity, then their product is a 23rd root of unity—so $(c^{2^6})^{23} = 1$; therefore, $(c^{2^6})^{2^5} = c$ is a 23rd root of 1. In this case, each exponent on c can be reduced mod 23 and we have

$$g_a(x) = c^{18}x^{18} + c^{16}x^{16} + c^{13}x^{13} + \cdots + 1.$$

If we let $y = cx$, we obtain

$$g_a(x) = y^{18} + y^{16} + y^{13} + y^{12} + \cdots y^2 + y + 1.$$

This is a polynomial over F, which if it contains a root β must contain all its conjugates (obtained by squaring). Therefore, $g_a(x)$ can have only 12 roots of unity as zeros; i.e., $w(a)$ is at least 11. We therefore consider only those c's which are not 23rd roots of 1. We conclude $w(a) > 5$.

We now eliminate $w(a) = 6$ very simply. For if $w(a) = 6$, then $c_0 = 0$, and by (10), and

$$g_a(x) = cx + c^2x^2 + \cdots + c^{2^4}x^{16} + c^{2^6}x^{18}$$
$$= x(c + c^2x + \cdots + c^{2^6}x^{17}).$$

The products of the 17 nonzero roots is clearly

$$c/c^{2^6} = (c^{2^6-1})^{-1}.$$

If all these 17 roots are 23rd roots of unity, then

$$(c^{2^6-1})^{23} = 1;$$

but $2^6 - 1$ is prime to $2^{11} - 1$, so c itself must be a 23rd root of unity—again this brings $w(a)$ up to 11.

We get immediately that $w(a) \geqq 7$. We know however of the existence of a vector a with $w(a) = 7$ (see [5]), so that ends the investigation.

Alternatively, we would use our general results to prove $d \geqq 7$ for this code as follows:

We have $d_0 = s_1 = 5$ and $h = 11$; and $c_1^{2^6}$ is the leading coefficient of $g_a(x)$. Now $2^6 - 1$ is prime to $2^{11} - 1$; therefore $d \geqq 7 = d_0 + 2$, by Corollary 3.

Perhaps we should observe that the (12, 23) code A of our definition is not precisely Golay's code but is obtainable from his code on permuting the coordinates suitably.

Some Results for $p = 47$. In this example we rely on the general results of §4. We again have $d_0 = 5$, since $R = \{1, 2, 4, 8, 16, 32, 17, 34, 21, 42, 37, 27, 7, 14, 28, 9, 18, 36, 25, 3, 6, 12, 24\}$. Here $g_a(x)$ has degree $r_{10} = 42$, so that the leading coefficient $c_{10} = c_1^{2^9}$. Now $2^9 - 1$ is prime to $2^{23} - 1$, because $(2^{23} - 1) - 2^5(2^9 + 1)(2^9 - 1) = 31$; and thus the g.c.d. $(2^{23} - 1, 2^9 - 1)$ is either 1 or 31. It cannot be 31, because $2^5 \equiv 1 \pmod{31}$, implying $2^9 \equiv 2^4 \not\equiv 1 \pmod{31}$. Thus by Corollary 3, $d \geqq 7$.

Theorem 5 eliminates vectors of weight 7. We shall eliminate vectors of weight 8 by the following procedure (which also works for vectors of weight 7): Let $a \in A$ have weight 8; then $g_a(x)$ has precisely 39 zeros $\beta_1, \cdots, \beta_{39}$ on the set $Z = \{1, \zeta, \zeta^2, \cdots, \zeta^{46}\}$. Thus we factor $g_a(x)$ (in a possibly larger field than K) as

$$g_a(x) = c_{10}x(x + \gamma_1)(x + \gamma_2)(x + \beta_1) \cdots (x + \beta_{39})$$

where γ_1 and γ_2 are not both 47th roots of 1 (and are not 0).

We define $\alpha_1, \cdots, \alpha_8$ as the 47th roots of 1 other than the β_i's; thus $(x + \alpha_1) \cdots (x + \alpha_8)(x + \beta_1) \cdots (x + \beta_{39}) = x^{47} + 1$.

Our first observation is that $\gamma_1 = \gamma_2 = \gamma$. For $\gamma_1 + \gamma_2 + \beta_1 + \cdots + \beta_{39} = 0$ is the coefficient of x^{41} in $g_a(x)$, and $\alpha_1 + \cdots + \alpha_8 + \beta_1 + \cdots + \beta_{39} = 0$; thus $\gamma_1 + \gamma_2 = \alpha_1 + \cdots + \alpha_8$. Now each $\alpha_i = \zeta^{j_i}$ and the j_ith coordinates of a are precisely the "1-positions" of a. Since $p \equiv -1 \pmod 8$, we have chosen ζ as a root of $f_0(x)$; thus ζ^{-1} is a root β of $f_1(x)$. We apply Corollary 2 to conclude $\alpha_1 + \cdots + \alpha_8 = 0 (= \beta_1 + \cdots + \beta_{39})$.

Now define

$$\textstyle\sum_{i=0}^{39} b_i x^{39-i} = (x + \beta_1) \cdots (x + \beta_{39}).$$

Then $b_0 = 1$ and $b_1 = \beta_1 + \cdots + \beta_{39} = 0$. We now have

$$g_a(x) = c_{10}x(x^2 + \gamma^2)(x^{39} + b_2x^{37} + b_3x^{36} + \cdots + b_{39}).$$

It follows immediately that the coefficient of x^{40-i} in $g_a(x)$ is

$$(11) \qquad c_{10}(\gamma^2 b_i + b_{i+2}) \qquad (i = 0, 1, \cdots, 37).$$

Furthermore,

$$c_{10}\gamma^2\beta_1 \cdots \beta_{39} = c_1;$$

or

$$(12) \qquad (\gamma^2 c_{10}/c_1)^{47} = 1,$$

incidentally proving that γ^2 and hence all conjugates, including γ, are in K.

We finally define

$$\sum_{i=0}^{8} d_i x^{8-i} = (x + \alpha_1) \cdots (x + \alpha_8),$$

in which $d_0 = 1$ and $d_1 = 0$.

We shall now derive various relations between the b_i's, d_i's and γ, which, together with (12), will lead to a contradiction. We first exploit (11) and our list for R. Since 40, 39, 38 $\notin R$, we have

$$(13) \qquad \begin{aligned} b_2 &= \gamma^2 \\ b_3 &= b_1 = 0 \\ b_4 &= \gamma^4. \end{aligned}$$

Continuing this process, and noting that $37 = r_{11}$ and $36 = r_{18}$, etc., we obtain

$$(13) \qquad \begin{aligned} b_5 &= c_{11}/c_{10} = c_{10} \\ b_6 &= c_{18}/c_{10} + \gamma^6 \\ b_7 &= \gamma^2 c_{10}. \end{aligned}$$

We have

$$(14) \qquad \begin{aligned} &(x^8 + d_2 x^6 + d_3 x^5 + \cdots + d_8) \\ &\qquad (x^{39} + b_2 x^{37} + b_4 x^{35} + \cdots + b_{39}) = x^{47} + 1; \end{aligned}$$

we examine the coefficients of $x^{46}, x^{45}, \cdots$ obtained by multiplying out the left-hand side of (14), first obtaining $b_2 + d_2 = 0$ as the coefficient of x^{45}; thus $b_2 = d_2 = \gamma^2$, from (13). Expressing the next few coefficients and using $d_2 = \gamma^2$ and (11) we find $d_3 = 0$

$$(15) \qquad \begin{aligned} 0 &= b_4 + b_2 d_2 + d_4 = d_4 \\ 0 &= b_5 + b_3 d_2 + d_5 = c_{10} + d_5 \\ 0 &= b_6 + b_4 d_2 + b_2 d_4 + d_6 = c_{18}/c_{10} + d_6 \\ 0 &= b_7 + b_5 d_2 + b_2 d_5 + d_7 = \gamma^2 d_5 + d_7 \\ 0 &= b_8 + b_6 d_2 + b_2 d_6 + d_8 = c_8/c_{10} + \gamma^2 d_6 + d_8. \end{aligned}$$

Having now determined all the d_i's in (15), we look for a contradiction in the succeeding equations. We write down the coefficient of x^{35} on the left-hand side of (14), namely

$$(16) \qquad \begin{aligned} 0 &= b_{12} + b_{10} d_2 + b_7 d_5 + b_6 d_6 + b_5 d_7 + b_4 d_8 \\ &= c_{10}(b_7 + d_7) + \gamma^4 c_8/c_{10} + \gamma^6 d_6 + b_6 d_6 \end{aligned}$$

where we have used (13) and (15), which also give $b_7 = d_7$, etc., so that (16) becomes $\gamma^4 c_8/c_{10} + (c_{18}/c_{10})^2 = 0$. By Lemma 5, this equation is equivalent to

$$(17) \qquad \gamma^2 c_7/c_9 = c_{18}/c_{10}$$

We now multiply (17) by $c_9 c_{10}/c_1 c_7$ to obtain $\gamma^2 c_{10}/c_1 = c_9 c_{18}/c_1 c_7$, which, by (12), must be a 47th root of 1. Using Lemma 5 again, we write

$$c_9 c_{18}/c_1 c_7 = c_1^{2^{17}+2^8-2^6-1};$$

we now prove that $c_1^{47} = 1$ by showing $2^{17} + 2^8 - 2^6 - 1$ is prime to $2^{23} - 1$. Let δ denote the greatest common divisor of these two numbers. Then δ divides

$$2^6(2^{17} + 2^8 - 2^6 - 1) - (2^{23} - 1) = 3(2^{12} - 21),$$

and therefore δ divides

$$3 \cdot 2(2^{23} - 1) - (2^{12} + 21)3(2^{12} - 21) = 3 \cdot 439.$$

Now, if q is a prime dividing δ, then $2^{23} \equiv 1 (\text{mod } q)$, so that $q \equiv 1 (\text{mod } 46)$. Thus $3 \nmid \delta$, and since 439 is prime, we need only observe that $439 \equiv -21 \pmod{46}$, so that $439 \nmid \delta$ either. Therefore $\delta = 1$, which implies $c_1^{47} = 1$, or $w(a) = h = 23$ (by the proof of Theorem 3), a contradiction. Thus $d \geqq 9$ for this code.

Since Eugene Prange has found many vectors of weight 11 in A, the results to date are $9 \leqq d \leqq 11$ for this (24, 47) code.

REFERENCES

1. R. C. Bose and D. K. Ray-Chaudhuri, *On a class of error correcting binary group codes*, Information and Control, 3 (1960), pp. 68–79.
2. Marcel J. E. Golay, *Notes on digital coding*, Proc. I.R.E., 37 (1949), p. 657.
3. Daniel Gorenstein and Neal Zierler, *A class of cyclic, linear, error-correcting codes in p^m symbols*, Group Report 55–19, Lincoln Laboratory, (1960); *A class of error-correcting codes in p^m symbols*, this Journal, 9 (1961), pp. 207–214.
4. W. W. Peterson, *Error-Correcting Codes*, Mass. Inst. of Technology and Wiley, New York, 1961.
5. Eugene Prange, *Cyclic error-correcting codes in two symbols*, Report No. AFCRC-TN-57-103, USAF Cambridge Research Laboratory, Bedford, Mass., 1957.
6. Irving Reed and Gustave Solomon, *A decoding procedure for a polynomial code*, Group Report 47-24, Lincoln Laboratory, 1959.
7. Gustave Solomon, *Linear recursive sequences as finite difference equations*, Group Report 47-37, Lincoln Laboratory 1960.
8. Edwin Weiss, *Some connections between linear recursive sequences and error-correcting codes: informal lectures*, Group Report 55-22, Lincoln Laboratory, 1960.
9. ———, *Residue class rings and linear recursive sequences*, Group Report 55-24, Lincoln Laboratory, 1960.
10. Hermann Weyl, *Algebraic Theory of Numbers*, Ann. of Math. Studies No. 1, Princeton Univ. Press, 1940.

A CLASS OF ERROR-CORRECTING CODES IN p^m SYMBOLS*

DANIEL GORENSTEIN† AND NEAL ZIERLER‡

1. Introduction. Bose and Chaudhuri [7] have introduced a class of binary linear error-correcting codes of block length of the form $2^m - 1$ for which W. W. Peterson [8] has devised an economical decoding procedure. The principal purpose of this note is to construct an efficient decoding procedure, in the spirit of Peterson's for the binary case, for the class of codes in p^m symbols and of block length prime to p (where p is any prime and m is any positive integer) obtained in an obvious generalization of the Bose-Chaudhuri codes. It turns out that the "polynomial codes" discussed by Reed and Solomon [3] belong to the general class, and hence may be decoded by our procedure.

There are two areas (at least) of application for codes in $q > 2$ symbols. First, data to be transmitted may appear in such a form and second, although the *B-C* codes tend to be highly efficient for the correction of independent errors, still greater efficiency may be obtained with the general codes when the errors occur in bursts. A code C of the general class is essentially uniquely determined by a triple (q, n, e) where q is the number of symbols, n is the block length (number of symbols per block), and C can correct every instance of e or fewer errors in a block. The parameter k, the number of information symbols per block, is computed as a function of q, n and e (see below). Two examples of the second application are as follows:

Example 1. Suppose we wish to transmit binary data, that errors occur in bursts of length 5 (binary symbols) or less and that acceptable reception results when we are able to correct 2 bursts in a block of length about 60. A *B-C* code for the job has $n = 63$, $e = 10$, $k = 18$ (and, of course, $q = 2$), and so gives a transmission rate $R = k/n = \frac{18}{63} = \frac{2}{7}$. On the other hand, we can take the general code with $q = 2^4$ (so a burst of length 5 causes errors in exactly two symbols, each symbol being encoded in a natural way as a binary 4-tuple) $n = 15$, $e = 4$ and $k = 7$. Then the binary block length is $4 \times 15 = 60$ and the transmission rate is $\frac{7}{15}$.

Example 2. Suppose again that binary data are to be transmitted, that errors occur in bursts of length 9 or less and that we must be able to correct 4 such bursts in a binary block of length around 2050. An appropriate *B-C* code has $n = 2047$, $e = 36$ and k very near 1670, so $R \sim \frac{4}{5}$. A suitable general code has $q = 2^8$, $n = 255$, $e = 8$, $k = 239$ and binary block length $8 \times 255 = 2040$; its rate is then $\frac{239}{255} \sim \frac{15}{16}$.

In §2 some basic facts concerning linear codes over general finite fields are established. In §3 we assemble the necessary machinery from the theory of linear recurring sequences (including several results which do not appear in [6]). The codes are constructed in §4, and the decoding procedure is obtained in §5.

2. Linear codes. Let p be a prime, m and n positive integers, K the field with $q = p^m$ elements and $V = V_n$ the vector space (over K) of n-tuples of elements of K. For u, v in V define the inner product $u \cdot v = \sum_{i=1}^{n} u_i v_i$ and let $\| u \|$, the norm of n, denote the number of nonzero components of u. If A is a subset of V, define $A^{\perp} = \{v \in V: u \cdot v = 0$ for all $u \in A\}$. Suppose now that A is a subspace of V and let $\bar{v}$ denote the A-coset to which $v \in V$ belongs. Define $W(\bar{v})$, the weight of $\bar{v}$, to be min $\| u \|$: $u \in \bar{v}$ and a *leader* of $\bar{v}$ to be an element u of $\bar{v}$ with $\| u \| = W(\bar{v})$. Evidently, the function W may be defined on V by $W(v) = W(\bar{v})$. Taking A as code alphabet, consider the following idealized decoding procedure: assign $v - u$ as (estimate of) transmitted message when v is received where u is an arbitrarily chosen leader of $\bar{v}$. A is said to be *e-error correcting*, e a positive integer, if the foregoing procedure decodes correctly whenever e or fewer errors occur in the transmission of a member of A, i.e., if every element of V of norm $\leqq e$ is the unique leader of the coset to which it belongs. It is convenient to characterize the property of being e-error correcting in the following two ways.

LEMMA 2.1. *A is e-error correcting if and only if $\| u \| > 2e$ for every $u \neq 0 \in A$.*

LEMMA 2.2. *Let L be any n-rowed matrix with coefficients in K whose columns, regarded as members of V, belong to and linearly span $A^{\perp}$. Then A is e-error correcting if and only if every set of $2e$ rows of L is linearly independent.*

Proofs. Suppose $\| u \| > 2e$ for every $u \neq 0$ in A, $v \in A$ is sent and $v + w$ is received where $\| w \| \leqq e$. Then for $u \in A$, $\| v + w - u \| > e$ if $u \neq v$ since $v - u \neq 0 \in A$ and so w is the unique leader of $\overline{v + w} = \bar{w}$. On the other hand, if A contains a nonzero vector u with $\| u \| = t \leqq 2e$, let $s = \min(t, e)$, let $i_1, \cdots, i_s$ be distinct indices of nonzero components of u and let $w \in V$ such that $w_{i_j} = -u_{i_j}$, $j = 1, \cdots, s$ and $w_i = 0$ for $i \notin \{i_1, \cdots, i_s\}$. Then $\| w \| = s \leqq e$; but $\| u + w \| = t - s \leqq s$ (for if $t \geqq e$, then $s = e$ and $t - s = t - e \leqq 2e - e = e = s$ while if $t < e$, $s = t$ and $t - s = 0$) so either w is not a leader of $\bar{w}$ or, if it is a leader, it is not unique.

Lemma 2.2 follows at once from Lemma 2.1 and the fact that $u \in A$ if and only if $uL = 0$.

REMARK. If A is merely an additive subgroup of V, the foregoing remains valid except for Lemma 2.2, and it would be natural to call A a "group code" in this case, as does Slepian [4] for the case $q = 2$. Of course, such an A is always linear over the prime subfield of K.

Define two codes ($=$ subspaces of V) to be *equivalent* if one is mapped on the other by a coordinate permutation. If the k-dimensional subspace A of V has the property that every k-tuple of elements of K appears as the first k components of some member of A, A is said to be a *check code*. As in [4], we note that every linear code is equivalent to a check code; indeed, if A is any k-dimensional subspace, let M be a $k \times n$ matrix whose rows are a basis of A over K. Since M has rank k, we can find k columns, say columns $i_1, \cdots, i_k$, which are linearly independent, and we let π be any coordinate permutation such that $\pi(i_j) = j$, $j = 1, \cdots, k$. Evidently $\pi(A)$ is a check code equivalent to A.

3. Linear recurring sequences. Let K be the q-element field, let d be a positive integer and let $f \in K[x]$: $f(x) = c_d x^d + \cdots + c_0$ with $c_0 c_d \neq 0$. In [6] we study families $G(f) =$ the set of all linear recurring sequences generated by $f =$ the set of all sequences $u_0, u_1, \cdots$ of elements of K satisfying $c_0 u_i + c_1 u_{i-1} + \cdots + c_d u_{i-d} = 0$ for $i = d, d + 1, \cdots$. We show there that if we identify the sequence u with the member $\sum_{i=0}^{\infty} u_i x^i$ of $K(x)$, then $G(f) = \{f_1/f: \deg f_1 < d\}$. For present purposes, however, we wish to consider, rather than sequences of arbitrarily large period, only sequences whose periods divide some fixed $n > 0$; thus, we consider only $G(f)$ where f divides $1 - x^n$ and treat the members of $G(f)$ as n-tuples of elements of K. The embedding of $G(f)$ in $K(x)$ of [6] now goes over into an embedding in the quotient ring $R = K[x]/(1 - x^n)$ as follows. Identify an element $(u_0, \cdots, u_{n-1})$ of V, the space of n-tuples of elements of K, with (the coset modulo $1 - x^n$ in $K[x]$ which contains) $u_0 + \cdots + u_{n-1} x^{n-1}$. Then $G(f) = \{f_1(1 - x^n)/f: \deg f_1 < d\} =$ the ideal in R generated by $(1 - x^n)/f$.

LEMMA 3.1. $G(f) = ((1 - x^n)/f)$.

COROLLARY 3.1. *If $fg \mid 1 - x^n$ and $h \in G(f)$ then $g \mid h$.*

Proof. Since $(1 - x^n)/fg$ is a polynomial g_1 and $h \in G(f)$, there exists a polynomial f_1 (of degree $<$ deg f) such that $h = f_1(1 - x^n)/f = f_1 g_1 g$.

If $f(x) = c_d x^d + \cdots + c_0$, let $f'(x) = c_0 x^d + \cdots + c_d$. Clearly, $f \mid 1 - x^n$ implies $f' \mid 1 - x^n$.

COROLLARY 3.2. *If $fg \mid 1 - x^n$ then $G(f) \perp G(g')$.*

Proof. Let $h \in G(f)$. Since $g \mid h$ by Corollary 3.1, every member v of $G(g)$ satisfies the recurrence relation associated with h by [6, Theorem 1], i.e., $h_0 v_{n-1} + \cdots + h_{n-1} v_0 = 0$. Since $(v_0, \cdots, v_{n-1}) \in G(g)$ if and only if $(v_{n-1}, \cdots, v_0) \in G(g')$, the result follows.

REMARK. It is interesting to note the following special case of Corollary 3.2. Let $q = 2$, $g = x + 1$, f any polynomial which is not a multiple of $x + 1$. Then one period of any sequence generated by f contains an even number of 1's.

By a count of dimensions we have, finally,

COROLLARY 3.3. $G(f)^{\perp} = G((1 - x^n)/f')$.

4. A class of codes in p^m symbols. Let K be the $q = p^m$ element field and let n and e be positive integers with $(p, n) = 1$ and $2e \leqq n$. Then the equation $x^n - 1 = 0$ is separable over K and has a primitive root a in an extension $F = K(a)$ of K whose degree r satisfies: $r = \min\{j: n \mid q^j - 1\}$.

* Received by the editors May 24, 1960 and in revised form November 15, 1960.

† Consultant to Lincoln Laboratory,§ Massachusetts Institute of Technology, Lexington, Massachusetts and Clark University, Worcester, Massachusetts.

‡ Lincoln Laboratory,§ Massachusetts Institute of Technology, Lexington, Massachusetts. Now at Arcon Corporation, Lexington, Massachusetts.

§ Operated with support from the U. S. Army, the U. S. Navy and the U. S. Air Force.

Reprinted with permission from *J. Soc. Ind. Appl. Math.*, vol. 9, pp. 207–214, June 1961.

Let B be the $n \times 2e$ matrix with $B_{ij} = a^{j(i-1)}$, $i = 1, \cdots, n$, $j = 1, \cdots, 2e$; i.e.,

$$B = \begin{bmatrix} 1 & 1 & \cdots & 1 \\ a & a^2 & & a^{2e} \\ a^2 & a^4 & & a^{4e} \\ \cdot & \cdot & & \cdot \\ \cdot & \cdot & & \cdot \\ \cdot & \cdot & & \cdot \\ a^{n-1} & a^{2(n-1)} & \cdots & a^{2e(n-1)} \end{bmatrix}.$$

It follows at once from the fact that a is primitive and $2e \leqq n$ that the $2e$-square matrix formed from any $2e$ rows of B is a Vandermonde matrix in distinct arguments and so is nonsingular.[1] Now choose a basis of F over K and replace each element a^i of B by its r-tuple of coefficients relative to this basis. Then B becomes an $n \times 2re$ matrix L with coefficients in K which clearly also has the property that every set of $2e$ rows of L is linearly independent over K. Let $f_i(x)$ be the minimum polynomial of a^i over K, $i = 1, \cdots, 2e$. Then the ith column of B evidently belongs to $G(f_i')$ over F and it follows at once that column $r(i-1)+j$ of L belongs to $G(f_i')$ over K for $i = 1, \cdots, 2e$; $j = 1, \cdots, r$.

LEMMA 4.1. *Let $1 \leqq i \leqq 2e$. Then columns $r(i-1)+j$, $j = 1, \cdots, r$ of L span $G(f_i')$.*

Proof. Consider the $n \times r$ matrix $L^{(i)}$ consisting of the r columns of L in question. As members of F relative to the chosen basis, its rows are $1, a^i, a^{2i}, \cdots, a^{i(n-1)}$. Hence any d_i consecutive rows are linearly independent over K where $d_i =$ degree f_i so the row rank, and hence the column rank, of $L^{(i)}$ is d_i.

Let $f =$ least common multiple $\{f_1', \cdots, f_{2e}'\}$. Then, from Lemma 4.1 and [6, Theorem 2] we have

LEMMA 4.2. *The linear span over K of the columns of L is $G(f)$.*

The orthogonal complement A of $G(f)$ is an e-error correcting code by Lemma 2.2, and $A = G(g)$ where $g = (1 - x^n)/f'$ by Corollary 3.3.

DEFINITION. *Let q, n and e be as in the first paragraph of this section. A (q, n, e) code is a subspace $G(g)$ of the space V of n-tuples of elements of K $(= GF(q))$ where $g = (1 - x^n)/\text{l.c.m.}\ \{f_1, \cdots, f_{2e}\}$, f_i the minimum polynomial over K of a^i, where a is a primitive nth root of unity.*

We have seen that a (q, n, e) code is an e-error correcting linear code of length n in q symbols containing q^k n-tuples where $k =$ degree g. According to [6, Theorem 3], every (q, n, e) code is closed under coordinate translation (i.e., under $v \to u$ with $u_i = v_{i+j \bmod n}$, $i = 1, \cdots, n$, any fixed j). Furthermore, such a code is essentially unique in the following sense.

LEMMA 4.3. *Two (q, n, e) codes are equivalent under a coordinate permutation.*

Proof. Let A and B be (q, n, e) codes corresponding to primitive nth roots a and b respectively. Then $b = a^t$ for some t prime to n and the coordinate permutation $i \to ti$ modulo n maps A on B.

The "polynomial codes" of Reed and Solomon [3] may be described as follows. Choose parameters q and k where q is a prime power and $0 < k < q - 1$. Let V_{q-1} be the space of $(q-1)$-tuples of elements of the q-element field K, let a be a primitive $(q-1)$th root of unity in K and let T be the mapping from V_k to V_{q-1} defined by:

$$T(u_0, u_1, \cdots, u_{k-1}) = (h(1), h(a), h(a^2), \cdots, h(a^{q-2}))$$

where $h(x) = u_0 + u_1 x + \cdots + u_{k-1}x^{k-1}$. T is obviously a linear mapping of V_k in V_{q-1} and the image of the ith natural basis vector of V_k is $(1, a^i, a^{2i}, \cdots, a^{(q-2)i})$, $i = 0, 1, \cdots, k-1$. But this is a linear recurring sequence generated by the polynomial $f_i' = x - a^i$ so $TV_k = G(g)$ where $g = (x-1)(x-a^{-1}) \cdots (x - a^{-(k-1)})$, which is exactly the $\left(q, q-1, \left[\frac{q-k-1}{2}\right]\right)$ code corresponding to a.

5. Decoding (q, n, e) codes.

Let A be the (q, n, e) code corresponding to the primitive nth root of unity a and let B denote the associated $n \times 2e$ matrix of §4. Suppose the element $u = (u_0, \cdots, u_{n-1})$ of A is transmitted and $u + v$ is received where $v = (v_0, \cdots, v_{n-1})$ is the error n-tuple. Then $(u+v)B = uB + vB = vB$ is a $2e$-tuple $(s_1, \cdots, s_{2e})$ of members of $F = K(a)$ satisfying: $s_j = v_0 + v_1a^t + v_2a^{2j} + \cdots + v_{n-1}a^{(n-1)j}$, $j = 1, \cdots, 2e$. Let $0 \leqq t \leqq e$ and suppose that $\| v \| \leqq t$, i.e., at most t errors have occurred. The decoding procedure consists essentially of computing the s_j (i.e., $(u+v)B$) and then applying the following theorem (which will be proved below).

Let $t_0 = \| v \|$ and let $i_1, \cdots, i_{t_0}$ denote the indices of the nonzero components of v. Let $\sigma_1, \cdots, \sigma_t$ denote the elementary symmetric functions of $a^{i_1}, \cdots, a^{i_{t_0}}$:

$$\sigma_1 = a^{i_1} + \cdots + a^{i_{t_0}},$$

$$\sigma_2 = a^{i_1+i_2} + a^{i_1+i_3} + \cdots + a^{i_1+i_{t_0}} + a^{i_2+i_3} + \cdots + a^{i_{t_0-1}+i_{t_0}},$$

$$\vdots$$

$$\sigma_{t_0} = a^{i_1+i_2+\cdots+i_{t_0}},$$

and if $t_0 < t$,

$$\sigma_{t_0+1} = \cdots = \sigma_t = 0.^2$$

THEOREM 5.1. *Let $0 \leqq t \leqq e$, suppose $\| v \| = t_0 \leqq t$ and let $i_1, \cdots, i_{t_0}$ and $\sigma_1, \cdots, \sigma_t$ be as in the preceding paragraph. Then the following matrix equation holds:*

$$\begin{bmatrix} s_1 & s_2 & \cdots & s_t \\ s_2 & s_3 & \cdots & s_{t+1} \\ s_3 & s_4 & \cdots & s_{t+2} \\ \vdots & \vdots & & \vdots \\ s_t & s_{t+1} & \cdots & s_{2t-1} \end{bmatrix} \begin{bmatrix} (-1)^{t+1}\sigma_t \\ (-1)^t\sigma_{t-1} \\ (-1)^{t-1}\sigma_{t-2} \\ \vdots \\ \sigma_1 \end{bmatrix} = \begin{bmatrix} s_{t+1} \\ s_{t+2} \\ s_{t+3} \\ \vdots \\ s_{2t} \end{bmatrix}.$$

Furthermore, letting M_t denote the foregoing matrix, $| M_t |$ (the determinant of M_t) $= v_{i_1} \cdots v_{i_{t_0}} \sigma_t \prod_{y>z} (a^{i_y} - a^{i_z})^2$. In particular, M_t is nonsingular if and only if $t_0 = t$.

Assuming, as we do, that $\| v \| \leqq e$, we reduce M_e to the form $\begin{bmatrix} I_{t_0} & 0 \\ 0 & 0 \end{bmatrix}$ by a sequence of elementary row operations,[3] where I_j denotes the $j \times j$ identity matrix, simultaneously subjecting the matrix I_e to the same sequence of operations, at the end of which it has the form $\begin{bmatrix} M_{t_0}^{-1} & - \\ - & - \end{bmatrix}$. Having found t_0 and $M_{t_0}^{-1}$ simultaneously in this way, we compute the σ_i: $M_{t_0}^{-1} \begin{bmatrix} s_{t_0+1} \\ \vdots \\ s_{2t_0} \end{bmatrix} = \begin{bmatrix} (-1)^{t_0+1}\sigma_{t_0} \\ \vdots \\ \sigma_1 \end{bmatrix}$. Then the i_j, $j = 1, \cdots, t_0$ are found by substituting successively $1, a, a^2, \cdots$ in $h(x) = x^{t_0} - \sigma_1 x^{t_0-1} + \cdots + (-1)^{t_0}\sigma_{t_0} = (x - a^{i_1}) \cdots (x - a^{i_{t_0}})$. The v_{i_j} are then determined by solving the t_0 simultaneous linear equations

$$v_{i_1}a^{ji_1} + \cdots + v_{i_{t_0}}a^{ji_{t_0}} = s_j \qquad (j = 1, \cdots, t_0).$$

In particular, the coefficient matrix

$$\begin{bmatrix} a^{i_1} & a^{2i_1} & \cdots & a^{t_0 i_1} \\ \vdots & \vdots & & \vdots \\ a^{i_{t_0}} & a^{2i_{t_0}} & \cdots & a^{t_0 i_{t_0}} \end{bmatrix}$$

may be inverted by a sequence of elementary row operations, or explicit formulae may be used.[4]

It remains to prove the assertions of Theorem 5.1, which we do in the following general form. Let F be an arbitrary field, let t be a positive integer and let $y_1, \cdots, y_t, x_1, \cdots, x_t$ be elements of F. Let $s_j = y_1x_1^j + \cdots + y_tx_t^j$, $j = 1, \cdots, 2t$, let $\sigma_1, \cdots, \sigma_t$ be the elementary symmetric functions of $x_1, \cdots, x_t$ and let $h(x) = x^t - \sigma_1 x^{t-1} + \cdots + (-1)^t\sigma_t$. Let W denote the Vandermonde matrix,

$$W = \begin{bmatrix} 1 & \cdots & 1 \\ x_1 & & x_t \\ x_1^2 & & x_t^2 \\ \vdots & & \vdots \\ x_1^{t-1} & \cdots & x_t^{t-1} \end{bmatrix}$$

and let Λ denote the diagonal matrix:

[1] Cf. [1, Ch. X, §5, ex. 6] and [2].

[2] See [5, §26].

[3] [1, Ch. X, § 3, Theorem 8]

[4] See, e.g., [2].

$$\Lambda = \begin{bmatrix} y_1 x_1 & & & \\ & y_2 x_2 & & \\ & & \ddots & \\ & & & y_t x_t \end{bmatrix}.$$

Letting W' denote the transpose of W, direct computation gives

$$W' \begin{bmatrix} (-1)^{t+1}\sigma_t \\ \vdots \\ \sigma_1 \end{bmatrix} = \begin{bmatrix} (-1)^{t+1}\sigma_t + (-1)^t \sigma_{t-1} x_1 + \cdots + \sigma_1 x_1^{t-1} \\ \vdots \\ (-1)^{t+1}\sigma_t + (-1)^t \sigma_{t-1} x_t + \cdots + \sigma_1 x_t^{t-1} \end{bmatrix} = \begin{bmatrix} x_1^t \\ \vdots \\ x_t^t \end{bmatrix}.$$

Multiplying this vector by Λ gives $\begin{bmatrix} y_1 x_1^{t+1} \\ \vdots \\ y_t x_t^{t+1} \end{bmatrix}$ and multiplying this by W gives $\begin{bmatrix} s_{t+1} \\ \vdots \\ s_{2t} \end{bmatrix}$. On the other hand, $W \Lambda W'$ is readily seen to be

$$M_t = \begin{bmatrix} s_1 & s_2 & \cdots & s_t \\ s_2 & s_3 & \cdots & s_{t+1} \\ \vdots & & & \\ s_t & s_{t+1} & \cdots & s_{2t-1} \end{bmatrix}$$

(which proves the matrix equation of Theorem 5.1). Finally $|M_t| = |W\Lambda W'| = |W||\Lambda||W'| = |\Lambda||W|^2 = y_1 \cdots y_t \sigma_t \prod_{y>z}(x_y - x_z)^2$.

NOTE. It should be pointed out that (q, n, e) codes come in compatible families in the following sense. A decoder for a (q, n, e) code A will also decode every (q, n, e') code with $e' < e$ provided only that A and A' correspond to the same primitive nth root of unity. This property could be exploited in a "variable rate" system in which error-correcting capability is matched to time-varying channel statistics.

REFERENCES

1. G. Birkhoff and S. MacLane, *A Survey of Modern Algebra*, Macmillan, New York, 1944.
2. N. Macon and A. Spitzbart, *Inverses of Vandermonde Matrices*, Amer. Math. Monthly, 65 (1958), pp. 95–100.
3. I. S. Reed and G. Solomon, *Polynomial codes*, this Journal, 8 (1960), pp. 300–304.
4. D. Slepian, *A class of binary signaling alphabets*, Bell System Tech. J., 35 (1956), pp. 203–234.
5. B. van der Waerden, *Moderne Algebra*, Ungar, New York, 1943. English Transl., Ungar, 1949.
6. N. Zierler, *Linear recurring sequences*, this Journal, 7 (1959), pp. 31–48.
7. R. C. Bose and D. K. Ray-Chaudhuri, *On a class of error-correcting binary group codes*, Information and Control, 3 (1960), pp. 68–79.
8. W. W. Peterson, *Encoding and error correction procedures for the Bose-Chaudhuri Codes*, IRE Transactions on Information Theory, vol. IT-6 (1960), pp. 459–470.

Performance of Concatenated Codes

G. David Forney, Jr.

1.1 Definitions; the Coding Theorem and Its Deficiencies

Traditionally, one begins a work on coding theory with a discussion of the coding theorem. This custom will be honored here—not for the sake of form, but because the coding theorem has directly motivated much of the research in the field it created. For almost twenty years, since Shannon[3] announced this remarkable result, it has stimulated extraordinary effort. Yet, although our understanding of the coding theorem has been much refined, it cannot be said that we have satisfactory answers to the questions it raises which, from a practical viewpoint, are the most fundamental.

This situation arises from the coding theorem's being an existence theorem. It demonstrates that a certain standard of performance can be obtained by some coding scheme, but fails to specify a particular code which achieves this standard or an encoding and decoding method which is reasonable to consider for implementation. To remedy this deficiency has been the goal of much of the work in coding theory.

In Appendix A we present the coding theorem for block codes on discrete memoryless channels in its most modern form; we shall now briefly summarize the character of this result, after introducing some basic definitions.

By a *channel* is meant some medium which is available to both transmitter and receiver and which is the sole means of communication between them. A single *use* of the channel consists in the transmitter acting on the channel in some way and the receiver observing some characteristic of the channel which reflects the action of the transmitter. In the case of a *discrete* channel, the transmitter is limited to one of a finite set of J actions called the *inputs* x_j, $1 \le j \le J$; correspondingly, the receiver distinguishes only between K classes of observations called *outputs* y_k, $1 \le k \le K$. The relation between inputs and outputs is in general probabilistic; the channel is characterized by specifying $\Pr(\mathbf{y} \mid \mathbf{x})$ for every *input sequence* $\mathbf{x} = \{x_i\} = (x_1, x_2, \cdots)$ and *output sequence* $\mathbf{y} = (y_1, y_2, \cdots)$. The channel is by definition *memoryless* if for every such pair of input and output sequences

$$\Pr[(y_1, y_2, \cdots) \mid (x_1, x_2, \cdots)] = \Pr(y_1 \mid x_1)\Pr(y_2 \mid x_2)\cdots$$

A discrete memoryless channel is then completely characterized by its *transition probability matrix*

$$p_{kj} = \Pr(y_k \mid x_j), \qquad 1 \le j \le J, \quad 1 \le k \le K.$$

By a *block code* of *length* N and *rate* R for such a channel is meant a set of e^{NR} input sequences or *code words* of length N:

$$\mathbf{x}_m = \{x_{mi}\} = (x_{m1}, x_{m2}, \cdots, x_{mN}), \qquad 1 \le m \le e^{NR}.$$

Here we have implicitly defined the units of rate to be *nats*, rather than the more common *bits*. Except where specified, we shall use nats in this work; conversions can be made by use of the relationships

$$1 \text{ nat} = \log_2 e\ (=1.4) \text{ bits};$$
$$1 \text{ bit} = \ln 2\ (=0.69) \text{ nat};$$

where ln indicates the natural (base e) logarithm. Another useful measure of rate comes from observing that, if J is the total number of inputs, J^N is the total number of input sequences of length N, so that the rate is bounded by

$$R \le R_{\max} = \ln J.$$

It is frequently convenient to use the *dimensionless rate* r, defined by

$$r = \frac{R}{R_{\max}};$$

r is seen to lie between zero and one. In terms of r the number of words in the code can be expressed as

$$e^{NR} = e^{NrR_{\max}} = J^{rN}.$$

An *encoder* is a mechanical device which accepts one of e^{NR} commands from a data source and generates the corresponding input sequence for transmission over the channel. Commonly the data source will be a continuous stream of binary data; for every $NR \log_2 e$ bits the encoder generates a code word. A *decoder* is a device which observes an output sequence of length N, processes this sequence, and presents the result to the data sink or user in the form desired. Commonly the user wants to know which code word was transmitted; if the encoder input was $NR \log_2 e$ bits, the decoder output will be $NR \log_2 e$ bits also, called an *estimate*. The event in which the estimate is not identical with the input code word is called an *error*.

The *probability of error* $\Pr(\mathscr{E})$ depends on the code, the channel, and the decoder's processing strategy. If the decoder is deterministic, then its strategy is describable as a mapping from the set of all received sequences $\mathbf{y}$ to the code words $\mathbf{x}_m$, and is specified by listing the sets Y_m of sequences $\mathbf{y}$ which result in a decoder estimate of $\mathbf{x}_m$. If it is assumed that the code, channel, and decoder are as specified and that all code words are equally likely, the probability of error is

$$\Pr(\mathscr{E}) = \sum_{m=1}^{e^{NR}} \Pr(\mathbf{x}_m)\Pr(\mathbf{y} \text{ not in } Y_m \mid \mathbf{x}_m \text{ transmitted})$$

$$= e^{-NR} \sum_{m=1}^{e^{NR}} \sum_{\mathbf{y} \notin Y_m} \Pr(\mathbf{y} \mid \mathbf{x}_m).$$

The probability of error is therefore minimized if the decoder implements the *maximum likelihood decision rule*, for which*

$$\mathbf{y} \in Y_m \text{ iff } \Pr(\mathbf{y} \mid \mathbf{x}_m) > \Pr(\mathbf{y} \mid \mathbf{x}_{m'}), \qquad \text{all } m' \ne m.$$

The stunning result known as the coding theorem is the following. Every discrete memoryless channel has a *capacity* C such that for all rates R less than C there exist codes of rate R which with maximum likelihood decoding have arbitrarily small probability of error. Few readers of this monograph will need to be told of the sensation evoked by the announcement of this theorem nor to be reminded of its implications.

We now know that the probability of error can be made to decrease exponentially with the block length. In Appendix A it is demonstrated that there exists a block code of rate R and length N such that

$$\Pr(\mathscr{E}) \le e^{-NE_L(R)},$$

where $E_L(R)$, called the *lower-bound error exponent*, is a function of R which is specified by the transition probability matrix of the channel and is positive for all rates $R < C$. We also know that for every code of rate R and length N

$$\Pr(\mathscr{E}) \ge e^{-N[E_U(R)+o(1)]},$$

where $o(1)$ is a function which approaches zero as N becomes very large, and $E_U(R)$ is called the *upper-bound error exponent*; $E_U(R) \to 0$ as $R \to C$, so that the channel capacity is precisely known. We discuss the character of $E_L(R)$ and $E_U(R)$ in more detail in Section 4.1.1.

On the basis of the coding theorem one could imagine finding by trial and error a code with performance meeting the standard of the theorem, and implementing it with an encoder which stored the e^{NR} code words in a dictionary and a maximum likelihood decoder capable of computing $\Pr(\mathbf{y} \mid \mathbf{x}_m)$ for each code word. But in such a scheme the size of the encoder and the number of computations required of the decoder would each be increasing exponentially with the block length N, so that, while the probability of error would decrease exponentially with N, it would decrease only weakly algebraically with the complexity. That is, if the complexity $G \sim e^{NR}$, while $\Pr(\mathscr{E}) \sim e^{-NE(R)}$, $\Pr(\mathscr{E}) \sim G^{-E(R)/R}$.

* The question of what to do in the boundary regions where the *a posteriori* probabilities are equal for two code words is of no significance; an error may always be assumed.

One is therefore faced with the problem of finding a code and an associated decoding scheme such that the complexity of the encoder and decoder required to implement the scheme increases much more slowly than exponentially with N. The introduction of linear codes, which we shall discuss in Chapter 2, essentially solves the encoding problem for block codes by yielding a class of codes which can be encoded in encoders of complexity linear in N; furthermore, if there is symmetry between the inputs, as there normally is in channels of practical interest, codes in this class can meet the standard of performance of the coding theorem. Another type of code with these properties is the convolutional or recurrent type; these codes have no block structure, and are largely ignored in this work.

Two approaches to the problem of reducing decoder complexity can be distinguished. The first, sometimes called probabilistic, looks for a more efficient way of implementing something like maximum likelihood decoding than simple comparison; the outstanding result of this approach is sequential decoding, which is suitable for convolutional codes. However, although an eminently practical scheme, sequential decoding is bedeviled by a variability in number of decoding computations which necessitates buffer storage, the probability of whose overflow decreases only weakly algebraically with its size[5]; furthermore, sequential decoding is limited to rates below a certain rate R_{comp} (defined in Section 4.1.1) which is less than capacity. The second approach, sometimes called algebraic, depends on choosing a code based on the structures of modern algebra to facilitate the development of an efficient decoding algorithm; the outstanding result of this approach is the class of BCH codes, which we discuss in the next chapter. However, though the BCH codes of moderate length are useful codes, as the block length becomes large, the BCH codes are, as far as we know, useless, in a sense to be made precise in the last paragraph of Chapter 2.

In this monograph we consider an approach based on the concept of concatenation, to be elucidated in the next section. Concatenation combines the algebraic and probabilistic approaches to get very long codes for which the probability of error on memoryless channels decreases exponentially with block length, while decoder complexity increases only as a small power of the block length. This happy result holds for all rates less than capacity, as we shall show in Chapter 4. In this sense it brings us within hailing distance of the promised land sighted by Shannon so long ago.

1.2 The Idea of Concatenation; Plan of the Monograph

Recall our description of a block coding scheme for a discrete memoryless channel, and imagine how it would appear to an observer who could see only the data source and data sink. For each block, the data source would submit one of e^{NR} commands, and the data sink would be given one of the e^{NR} decoder estimates. Generally the estimate would match the input, but occasional errors would occur. From this viewpoint, the encoder-channel-decoder would appear to be a discrete memoryless *superchannel* with e^{NR} inputs and e^{NR} outputs. One immediately sees that it is possible to design a block code of length n and dimensionless rate r for this superchannel. But now, in terms of uses of the original channel, one has created a block code of over-all length $N_0 = nN$, with e^{nNrR} words, and thus of over-all rate $R_0 = rR$. The encoder for this very long block code happens to consist of two encoders back-to-back and the decoder of two decoders, suitable for much shorter codes. These ideas are illustrated in Figure 1.1, where we have called the two codes *inner* and *outer*.

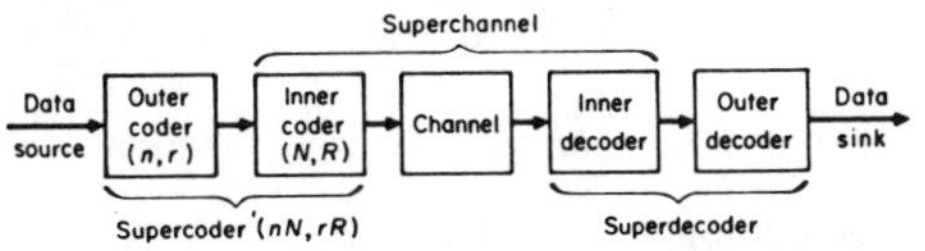

Figure 1.1 Concatenation concepts. n, outer-code length; N, inner-code length; $nN = N_0$ = over-all length; r, outer-code dimensionless rate; R, inner-code rate; $rR = R_0$ = over-all rate.

Although it serves to introduce concatenation, the conceptual separation of the two codes is artificial. As engineers, we want to design both codes simultaneously to get the best over-all performance with the least over-all complexity. We want to know answers to questions like these: What are the best codes to use? Can we reduce the decoding load or increase performance by letting one decoder help the other? Most importantly, what sort of performance is obtainable, at the cost of what complexity?

We find that use of codes of the BCH type is quite satisfactory, though we nowhere prove that they are best. In particular, we always use as an outer code a Reed-Solomon (RS) code. Reed-Solomon codes are nonbinary BCH codes which, whenever they can be used, are optimal, in a sense to be made precise in Chapter 2. Furthermore, RS codes admit an efficient decoding algorithm based on their algebraic structure, which we discuss in Appendix B. Chapter 2 gives a rapid introduction to BCH codes, with emphasis on RS codes.

In Chapter 3 we consider the possibility of the inner decoder passing along something more to the outer decoder than just its best guess. In particular, we let the inner decoder add to its estimate a real number which indicates how reliable it supposes its estimate to be. We show how such information can be efficiently used by the outer decoder in an algebraic decoding scheme in a method we call generalized minimum distance (or GMD) decoding. We also consider other methods of passing along reliability information: erasures, which we treat as highly quantized reliability information, and lists, which we consider only to discard. We develop bounds on performance with these different possibilities, and show that GMD can improve performance without much increase in complexity.

In Chapter 4 we investigate the performance obtainable with concatenation; we find that, with a Reed-Solomon outer code and GMD decoder, we can achieve

$$\Pr(\mathscr{E}) \leq e^{-N_0 E_C(R_0)},$$

where $E_C(R_0)$ is called the *concatenation exponent*, and N_0 and R_0 are the over-all length and rate of the concatenation scheme. It turns out that $E_C(R)$ is less than $E_L(R)$, so that some sacrifice is involved in going to concatenation. On the other hand, $E_C(R)$ is positive for all rates less than capacity, so that the essential feature that makes the coding theorem so provocative—$\Pr(\mathscr{E})$ exponential in the block length for all $R < C$—is preserved. Most important, the complexity of the decoder increases proportionately only to a small power of the block length, N_0.

(*Note:* The text here continues with material from Chapter 4.)

THEOREM 4.3

There is a concatenation scheme with over-all length N_0 and rate R_0 such that

$$\Pr(\mathscr{E}) \leq e^{-N_0 E_{CL}(R_0)}$$

where $E_{CL}(R_0) > 0$ for all $R_0 < C$, the capacity of the original channel.

Therefore whatever sacrifice is involved in using concatenation, it is not in the highest rate for which virtually error-free communication can be obtained.

Let us define the efficiencies

$$\eta(R) = \frac{E_C(R)}{E(R)}$$

and

$$\eta_L(R) = \frac{E_{CL}(R)}{E(R)}.$$

The efficiency is a measure of how much longer the over-all length of a concatenated code must be than that of an unconcatenated code to obtain the same probability of error at the same over-all rate; if the efficiency is 0.1, then the concatenated code must be ten times longer. It is here that one is penalized for going to concatenation; however, the penalty need not be too severe. In fact, at zero rate, $\eta(0) = 1$, and $\eta_L(0) = \frac{1}{2}$, as can be seen from setting $r = R = 0$ in Equations 4.36 and 4.37. At high rates, the efficiency goes to zero, as we see from the following argument. Let the first nonzero derivative of $E(R)$ at $R = C$ be the

nth; normally n will be 2. Near C, $E(R)$ is approximated by the Taylor series expansion

$$E(R) = \frac{E^{(n)}(C)}{n!}(R - C)^n.$$

For R_0 near C,

$$E_C(R_0) = \max_{rR=R_0} (1 - r)E(R)$$

$$= \max_R \left(1 - \frac{R_0}{R}\right)\frac{E^{(n)}(C)}{n!}(R - C)^n.$$

Letting $R_0 = C(1 - \epsilon)$ and $R = C(1 - \delta)$, $0 < \delta < \epsilon \ll 1$, and ignoring all second-order terms in ϵ and δ, we find the optimum δ as

$$\delta = \frac{n}{n+1}\epsilon,$$

whence

$$\eta(R) = \frac{n^n}{(n+1)^{n+1}}\epsilon, \qquad R = C(1 - \epsilon).$$

It follows that while the efficiency approaches zero as $R \to C$, the dropoff in efficiency is not precipitous, being only linear in ϵ. When $E(R)$ has a nonzero second derivative at C, $\eta(R) \simeq 0.15[(C - R)/C]$ for R near C.

Numerical values for the efficiency are given in Tables 4.1 and 4.2 for the two channels we have been using as examples.

TABLE 4.1 Efficiency of Concatenation for the *BSC* with $p = 0.01$ ($C = 0.92$ Bit)

R (bits)	0	0.1	0.2	0.3	0.4	0.5	0.6	0.7	0.8
$\eta(R)$	1.0	0.41	0.30	0.22	0.15	0.10	0.068	0.043	0.022
$\eta_L(R)$	0.5	0.36	0.30	0.22	0.15	0.10	0.068	0.043	0.022

TABLE 4.2 Efficiency of Concatenation for the Unlimited Bandwidth White Gaussian Channel

r_0	0	0.1	0.2	0.3	0.4	0.5	0.6	0.7	0.8	0.9
$\eta(r_0)$	1.0	0.39	0.24	0.18	0.13	0.10	0.076	0.053	0.033	0.016
$\eta_L(r_0)$	0.6	0.37	0.24	0.18	0.13	0.10	0.076	0.053	0.033	0.016

We see that while the efficiency η is 1 at zero rate, it drops rapidly for low rates, and then declines more slowly for higher rates. However, η_L declines less rapidly at low rates, since it starts at $\frac{1}{2}$ at zero rate.

4.2 Performance of RS Codes as Outer Codes

Theorem 4.3 shows that with concatenation one can achieve exponential decrease of probability of error with block length for all rates below capacity. However, derived as the theorem is from coding theorem arguments, it fails to specify a scheme attaining this performance. In this section we give such a scheme.

Let the inner code be the best block code of length N and rate R, whose average probability of error p satisfies

$$p \le \exp\{-N[E(R) + o(1)]\}, \tag{4.38}$$

where $E(R)$ is the true error exponent. Let R be such that e^{NR} is a prime power, and let the outer code be a Reed-Solomon code of length $n = e^{NR}$, minimum distance d, and therefore dimensionless rate $r = (n - d + 1)/n$. Then the concatenated code has over-all length $N_0 = nN = Ne^{NR}$ and over-all rate $R_0 = rR$.

Suppose first that the inner decoder passes on to the Reed-Solomon decoder only an estimate, so that the latter performs errors-only decoding. The RS decoder will fail to decode correctly if and only if the actual number of inner decoder errors t is such that

$$2t \ge d.$$

For a formulation similar to that of Chapter 3, let α now be a random variable which equals 1 when no inner decoding error is made and -1 when an error is made; then the RS decoder will fail to decode correctly if and only if

$$\sum_{i=1}^{n} \alpha_i \le n - d.$$

We can bound the probability that this sum of independent, identically distributed random variables α_i is less than $n - d$ by the Chernov bound techniques of Chapter 3. The random variable α has the moment-generating function

$$g(-s) = \overline{e^{-s\alpha}} = pe^s + (1 - p)e^{-s}.$$

If $\mu(-s) = \ln g(-s)$ and $\delta = d/n$,

$$\Pr(\text{ndc}) = \Pr(\textstyle\sum \alpha_i \le n - d) \le e^{-n[s(\delta-1)-\mu(-s)]}.$$

To maximize the exponent, we set its derivative with respect to s to zero:

$$\delta - 1 = -\mu'(-s) = -\frac{g'(-s)}{g(-s)} = \frac{pe^s - (1-p)e^{-s}}{pe^s + (1-p)e^{-s}},$$

from which we obtain

$$s = \tfrac{1}{2}\ln\frac{(1-p)\delta}{p(2-\delta)}$$

and

$$s(\delta - 1) - \mu(-s)$$
$$= -\frac{\delta}{2}\ln p - \left(1 - \frac{\delta}{2}\right)\ln(1-p) + \frac{\delta}{2}\ln\frac{\delta}{2} + \left(1 - \frac{\delta}{2}\right)\ln\left(1 - \frac{\delta}{2}\right).$$

Now let N become large; then $\delta \to (1 - r)$, $p \to e^{-NE(R)}$, and $(1 - p) \to 1$, so that

$$\Pr(\text{ndc}) \le \exp\left\{-nN\left[\left(\frac{1-r}{2}\right)E(R) - \frac{\mathscr{H}(\delta/2)}{N}\right]\right\},$$

where $\mathscr{H}(x) = -x\ln x - (1 - x)\ln(1 - x)$. For large N the latter term is negligible; using $nN = N_0$, and picking r and R to maximize the exponent subject to $rR = R_0$, we find

$$\Pr(\text{ndc}) \le e^{-N_0E_C(R_0)/2}, \tag{4.39}$$

where

$$E_C(R_0) = \max_{rR=R_0}(1 - r)E(R). \tag{4.40}$$

In other words, with an errors-only RS decoder we can achieve an exponent of one-half the concatenation exponent $E_C(R_0)$ of the previous section. In particular, we can achieve $E_{CL}(0) = \frac{1}{2}E_C(0)$, which is the true zero-rate exponent for equierror channels.

Second, let the inner decoder have the option of putting out an erasure instead of an estimate, so that the RS decoder does erasures-and-errors decoding. We assume that there is some such scheme in which the average probability of erasure is

$$p_x \le Ke^{-NE(R)}, \tag{4.41a}$$

and of error,

$$p_{\mathscr{E}} \le K^{-1}e^{-NE(R)}, \tag{4.41b}$$

for any $K \ge 1$. If $E(R) = E_L(R)$ there is such a scheme, as we stated in Section 4.1.1. Letting α now be the random variable which equals -1 when an error occurs, 0 when an erasure occurs, and 1 otherwise, the RS decoder will fail to decode correctly if and only if

$$\sum_{i=1}^{n} \alpha_i \le n - d.$$

The random variable α has moment-generating function

$$g(-s) = p_{\mathscr{E}}e^s + p_x + (1 - p_{\mathscr{E}} - p_x)e^{-s},$$

which, for large N, and for the worst values of $p_{\mathscr{E}}$ and p_x, equals

$$g(-s) = K^{-1}e^se^{-NE(R)} + Ke^{-NE(R)} + e^{-s}.$$

We first choose K to minimize $g(-s)$:

$$K = e^{s/2}; \qquad g(-s) = 2e^{s/2}e^{-NE(R)} + e^{-s}.$$

With this choice of K we then maximize the exponent with

$$s = \tfrac{2}{3}NE(R) + \tfrac{2}{3}\ln\frac{2\delta}{3-\delta}$$

to get, ignoring all terms which are negligible for large N,

$$\Pr(\text{ndc}) \le e^{-2N_0E_C(R_0)/3}; \tag{4.42}$$

thus with an erasures-and-errors decoder we can achieve an exponent of $\tfrac{2}{3}E_C(R_0)$, assuming that Equations 4.41*a* and 4.41*b* can be satisfied. As one might expect, with this choice of K and s the error probability $p_\mathscr{E}$ is the square of the erasure probability p_x.

Finally, let the inner decoder now put out a weight α, $0 \le \alpha \le 1$, with its estimate, and let the RS decoder do generalized minimum distance decoding. We let the weight α be some function $\alpha(\beta)$ of the criterion function β introduced at the end of Section 4.1.1, for which

$$\Pr(\beta \le x) \le K(x)e^{-NE_L(R)}$$

$$\Pr(\beta \ge x, \text{incorrect estimate}) \le \frac{1}{K(x)}e^{-NE_L(R)}; \tag{4.43}$$

$K(\beta_0) = 1$ and $K(x)$ increases monotonically without limit for $x \ge \beta_0$. We shall set $\alpha = 0$ (erase) whenever $\beta \le \beta_0$; furthermore, if β_1 is such that $K(\beta_1) = e^{NE_L(R)}$, we shall set $\alpha = 1$ whenever $\beta \ge \beta_1$. Define

$$p_c(x)\,dx = \Pr(x \le \beta \le x + dx, \text{correct estimate}),$$

and

$$p_e(x)\,dx = \Pr(x \le \beta \le x + dx, \text{incorrect estimate});$$

then the most unfavorable probability distribution, given Equation 4.43, is

$$\Pr(\beta \le \beta_0) = e^{-NE_L(R)};$$

$$p_c(x)\,dx = e^{-NE_L(R)}K'(x)\,dx, \qquad \beta_0 \le x \le \beta_1;$$

$$p_e(x)\,dx = e^{-NE_L(R)}K^{-2}(x)K'(x)\,dx, \qquad \beta_0 \le x \le \beta_1; \tag{4.44}$$

$$\Pr(\beta \ge \beta_1, \text{correct estimate}) = 1;$$

$$\Pr(\beta \ge \beta_1, \text{incorrect estimate}) = e^{-2NE_L(R)}.$$

The random variable α' which equals $\alpha(\beta)$ when the estimate is correct and which equals $-\alpha(\beta)$ when the estimate is incorrect has moment-generating function

$$g(-s) = \overline{e^{-s\alpha'}} = \Pr(\beta \le \beta_0) + e^{-s}\Pr(\beta \ge \beta_1, c)$$
$$+ e^{s}\Pr(\beta \ge \beta_1, i) + \int_{\beta_0}^{\beta_1} dx[p_c(x)e^{-s\alpha(x)} + p_e(x)e^{s\alpha(x)}]. \tag{4.45}$$

Furthermore, the GMD decoder will fail to decode correctly if and only if

$$\sum_{i=1}^{n} \alpha_i' \le n - d.$$

Wending our way down a familiar path, we choose $\alpha(x)$ to minimize $g(-s)$ by setting the partial derivative

$$\frac{\partial}{\partial\alpha(x)}g(-s) = -sp_c(x)e^{-s\alpha(x)} + sp_e(x)e^{s\alpha(x)}$$

to zero, obtaining

$$\alpha(x) = \frac{1}{2s}\ln\frac{p_c(x)}{p_e(x)} = \frac{1}{s}\ln K(x), \qquad \beta_0 \le x \le \beta_1; \tag{4.46}$$

we shall check later that $\alpha(x) \le 1$. Substituting Equations 4.44 and 4.46 into Equation 4.45, we obtain

$$g(-s) = e^{-NE_L(R)} + e^{-s} + e^{s}e^{-2NE_L(R)} + e^{-NE_L(R)}\int_{\beta_0}^{\beta_1} dx[2K^{-1}(x)K'(x)]. \tag{4.47}$$

Performing the integral,

$$\int_{\beta_0}^{\beta_1} dx\,[2K^{-1}(x)K'(x)] = 2\ln\frac{K(\beta_1)}{K(\beta_0)} = 2NE_L(R),$$

which factor we ignore since

$$2NE_L(R)e^{-NE_L(R)} = e^{-N[E_L(R) + \ln 2NE_L(R)/N]} = e^{-N[E_L(R)+o(1)]}.$$

Maximization over s follows from solving

$$(\delta - 1) = -\mu(-s') = -\frac{e^{s}e^{-2NE_L(R)} - e^{-s}}{e^{s}e^{-2NE_L(R)} + 2e^{-NE_L(R)} + e^{-s}}$$

to get

$$s = NE_L(R) + \ln\frac{\delta}{2-\delta}.$$

(Since $\ln K(x) \le NE_L(R)$, $\beta_0 \le x \le \beta_1$, we now verify that in Equation 4.46 $\alpha(x) \le 1$ when N is large.) Then

$$-s(1-\delta) - \mu(-s) = N[E_L(R) + o(1)],$$

so that, by proper choice of inner and outer code rate, we can attain

$$\Pr(\text{ndc}) \le e^{-N_0E_C(R_0)},$$

where

$$E_C(R_0) = \max_{rR=R_0}(1-r)E_L(R).$$

This will be the same as the concatenation exponent derived earlier if the true error exponent $E(R) = E_L(R)$. Alternatively, if it is found that for some rates $E(R)$ is greater than $E_L(R)$, then it is only necessary that the relatively weak condition 4.43 be true when $E_L(R)$ is replaced by $E(R)$, in order that this exponent equal that found earlier by coding theorem arguments for the best of the GMD channels, the telltale channel. It therefore seems safe to say that with GMD decoding of RS codes we can do as well as can be done with any outer code and any type of decoding, if the inner decoder puts out a GMD-type packet and no distinction is made between errors.

Our story is nearly told; it remains only to ascertain that the complexity of these schemes is reasonable. All schemes require that the inner decoder be a likelihood decoder, which demands a number of comparisons equal to the number of code words $e^{NR} = n$. In a single over-all block the inner decoder must operate n times, so that the total number of computations required of it is proportional to n^2. If the outer decoder is errors-only or erasures-and-errors, the number of computations required is proportional to d^3, or, since r is fixed, to n^3; if it is GMD, then the number required is proportional to n^4. Furthermore, the complexity of an operation is proportional to $(\log n)^2$ or to N^2. We conclude that, for any fixed R_0 less than C, the total complexity of a concatenated decoder is at worst proportional to N_0^4, and without GMD to N_0^3.†

4.3 Discussion

To recapitulate: we have discovered a concatenation exponent

$$E_C(R_0) = \max_{rR=R_0}(1-r)E(R),$$

which is greater than zero wherever the original channel exponent $E(R)$ is greater than zero—for all rates less than capacity—and which equals $E(0)$ at rate zero. Not only have we exhibited a specific scheme—GMD decoding of RS codes—which attains this exponent with reasonable decoder complexity, but we have shown that, under certain symmetry restrictions, no scheme using GMD-type information packets, any outer code, and any decoding method can do better. We can attain qualitatively similar performance with errors-only or erasures-and-errors decoding of RS codes, with some savings in complexity, but with a loss of factors of $\tfrac{1}{2}$ and $\tfrac{2}{3}$ in the exponent.†

We may therefore use concatenated codes at any rate at which we could use unconcatenated. The only loss is in the magnitude of the

† The similarity between the factors ($\tfrac{1}{2}$, $\tfrac{2}{3}$, 1) found here and the factors ($\tfrac{1}{2}$, 0.686, 1) found in Section 3.1.5 is intriguing, both because the factors are so similar and because they are not identical. The discrepancy may well be due to the relative weakness of conditions 4.41*a* and 4.41*b*.

† Recently Ziv[4] has exhibited a scheme involving the interlacing of binary codes, for which the probability of error decreases exponentially as the square root of the over-all block length, while the decoding complexity increases only algebraically, at all rates less than capacity. This work was a stimulus to the investigation reported here. Still more recently Pinsker[3] has reported a scheme capable of achieving negligible probability of error for all $R < C$, in which the outer code is a binary convolutional code suitable for sequential decoding, and in which the average number of decoding computations per bit is bounded. These schemes have in common a relatively short inner block code to be decoded by maximum likelihood decoding, which cleans up the channel to the point that an efficient outer code may be used.

exponent; the efficiency $E_C(R)/E(R)$ declines to zero as R approaches C. Consequently, longer codes are required if concatenation is to be used. Unless the necessity of minimizing decoding delay is overriding, however, the length of a code is less important than the complexity of its implementation, where the concatenated code will generally be superior.

Our formulation permits us to think of concatenating more than two codes, but there seems no good reason to do so. More stages would further reduce the exponent without reducing complexity significantly.

It would be interesting to pin down the reason for the discrepancy between $E(R)$ and $E_C(R)$. It might be, of course, that the concatenated structure of the code itself precludes attaining an exponent better than $E_C(R)$, even if full maximum likelihood decoding were done on the over-all code. If this were so, it would completely account for the discrepancy, but it seems unlikely that such a weak structural requirement as concatenation could affect the exponent markedly. Another possibility is that by including information about more inner code words in the information packet passed to the outer decoder, we could do better; however, we have remarked in Section 4.1.3 that, if all inner code words not on the list are treated as equally likely and the size of the list does not grow exponentially with N, no improvement can be obtained. To let the list size grow exponentially with N is not an attractive alternative. We are left with the possibility of introducing some decoding scheme for the outer decoder which does not treat all code words other than the estimate as equally likely. Indeed, if the inner decoder estimate is $\mathbf{x}_m$, it is much more likely that, if $\mathbf{x}_m$ is wrong, the correct word is among those within distance d of $\mathbf{x}_m$ than that it is among those much farther away. We have been able neither to invent a decoding scheme capable of using such information nor to determine whether such a scheme could offer any improvement.

References

1. R. G. Gallager, "A Simple Derivation of the Coding Theorem and Some Applications," *IEEE Trans. Information Theory*, **IT-11**, 3–17 (1965).
2. G. H. Hardy, J. E. Littlewood, and G. Pólya, *Inequalities*, Cambridge University Press, Cambridge, England, 1952, Chapter 2.
3. M. S. Pinsker, "Some Mathematical Problems of Coding Theory," presented at the 1966 International Symposium on Information Theory, Los Angeles, but unpublished.
4. J. Ziv, "Further Results on the Asymptotic Complexity of an Iterative Coding Scheme, " *IEEE Trans. Information Theory* (Los Angeles Symposium), **IT-12**, 168–171 (1966).

A Class of Constructive Asymptotically Good Algebraic Codes

JØRN JUSTESEN

***Abstract*—For any rate R, $0 < R < 1$, a sequence of specific (n,k) binary codes with rate $R_n > R$ and minimum distance d is constructed such that**

$$\liminf_{n\to\infty} \frac{d}{n} \geq (1 - r^{-1}R)H^{-1}(1 - r) > 0$$

(and hence the codes are asymptotically good), where r is the maximum of $\frac{1}{2}$ and the solution of

$$R = \frac{r^2}{1 + \log_2 [1 - H^{-1}(1 - r)]}.$$

The codes are extensions of the Reed–Solomon codes over $GF(2^m)$ with a simple algebraic description of the added digits. Alternatively, the codes are the concatenation of a Reed–Solomon outer code of length $N = 2^m - 1$ with N distinct inner codes, namely all the codes in Wozencraft's ensemble of randomly shifted codes. A decoding procedure is given that corrects all errors guaranteed correctable by the asymptotic lower bound on d. This procedure can be carried out by a simple decoder which performs approximately $n^2 \log n$ computations.

I. Introduction

THE Gilbert–Varsharmov [1], [2] bound ensures the existence of binary (n,k) codes of arbitrary dimension k and arbitrary block length $n(0 < k < n)$ such that their minimum distance d_n satisfies

$$H(d_n/n) \geq 1 - R_n,$$

where $R_n = k/n$ is the rate and $H(x) = -x \log_2 x - (1 - x) \log_2 (1 - x)$ is the binary entropy function. Thus, at least in principle, it is possible for any rate R, $0 < R < 1$, to select a sequence of linear codes with increasing block length n, minimum distance d_n, and rate $R_n \geq R$ such that

$$\liminf_{n\to\infty} (d_n/n) \dot{\geq} H^{-1}(1 - R), \tag{1}$$

Manuscript received February 7, 1972; revised May 1, 1972. This paper was presented at the IEEE International Symposium on Information Theory, Asilomar, Calif., February 2, 1972.

The author is with the Laboratory for Communication Theory, Technical University of Denmark, Lyngby, Denmark.

where we require $0 < H^{-1}(1 - R) < \frac{1}{2}$ so that the inverse function is uniquely defined.

Hitherto, however, no constructive sequence of binary (n,k) codes with $R_n \geq R$ and

$$\liminf_{n\to\infty} (d_n/n) > 0 \tag{2}$$

had been given for any R, $0 < R < 1$. We shall call a countably infinite sequence of (n,k) codes constructive if it can be specified in terms of entities generally accepted as known in a manner that requires no searching. Thus, in this sense of the word, the cyclic binary Hamming codes are a constructive sequence of codes. We shall call a sequence of codes that satisfy (1) *asymptotically good.*

In Section II of this paper we specify for any rate R, $0 < R < 1$, a constructive countably infinite sequence of (n,k) codes with rates $R_n \geq R$ such that (2) is satisfied. The construction is based on Forney's concept [3] of concatenated codes in which the m information digits of an inner binary code are treated as single digits of an outer Reed–Solomon [4] code over $GF(2^m)$, but we have generalized the concept to allow variation of the inner code. The inner codes are given by a simple algebraic description and shown to be equivalent to the $2^m - 1$ distinct codes in the ensemble of randomly shifted codes described by Massey [5] and attributed to Wozencraft. Alternatively, the codewords of the binary codes we shall construct can be considered as a pair of codewords in different Reed–Solomon codes over $GF(2^m)$.

In Section III we present a decoding method for the codes constructed in Section II that is closely related to Forney's [3] generalized minimum distance decoding as applied by Reddy and Robinson [6] and Weldon [7] to the decoding of iterated codes. It is proved that all error patterns with weight guaranteed correctable by the lower bound on minimum distance in Section II will be corrected by this procedure, and that the number of bit operations required for decoding is proportional to $n^2 \log n$.

Reprinted from *IEEE Trans. Inform. Theory,* vol. IT-18, pp. 652–656, Sept. 1972.

II. The Codes

Let the information sequence of an (N/K) Reed–Solomon (RS) code of block length $N = 2^m - 1$ and rate $r_n = K/N$ over $GF(2^m)$ be

$$\boldsymbol{i} = [i_0,i_1,\cdots,i_{K-1}], \qquad i_j \in GF(2^m)$$

and define the associated polynomial as

$$i(x) = i_0 + i_1x + \cdots + i_{K-1}x^{K-1}.$$

Let α be a primitive element of $GF(2^m)$. The encoded sequence may be written as the N-tuple [8]

$$\boldsymbol{a} = [a_0,a_1,\cdots,a_{N-1}], \qquad a_j \in GF(2^m)$$

or equivalently as the polynomial

$$a(x) = a_0 + a_1x + \cdots + a_{N-1}x^{N-1},$$

where

$$a_j = i(\alpha^j), \qquad 0 \le j \le 2^m - 2.$$

Since $GF(2^m)$ is a vector space over $GF(2)$, the elements of $GF(2^m)$ can, and will hereafter, be taken as binary m-tuples.

Now consider the binary (n,k) code $\mathscr{C}_m$ with codewords

$$\boldsymbol{c} = [c_0,c_1,\cdots,c_{N-1}], \tag{3}$$

where

$$c_j = [a_j,\alpha^j a_j], \qquad 0 \le j \le N - 1. \tag{4}$$

This binary code has length $n = 2mN$, dimension $k = mK$, and rate $R_n = \frac{1}{2}r_N$. $\mathscr{C}_m$ is a linear code over $GF(2^m)$ when the c_j are considered as pairs of elements from $GF(2^m)$, and is consequently also linear over $GF(2)$ when the c_j are expressed as 2^m-tuples over $GF(2)$, as we have done.

The code $\mathscr{C}_m$ may be interpreted as a concatenated code with a Reed–Solomon outer code and a varying inner code. For each j, (4) specifies a $(2m,m)$ binary code; these are all distinct and in fact constitute the $R = \frac{1}{2}$ codes in Wozencraft's ensemble of randomly shifted codes. These codes have the interesting property that any $2m$-tuple a,b with $a \neq 0$ and $b \neq 0$ appears in exactly one code, namely the code for which $\alpha^j = b/a$. This property was used by Massey [5] to show that the ensemble contains codes that meet the Gilbert–Varsharmov bound.

Alternatively the codewords of $\mathscr{C}_m$ may be interpreted as pairs $[\boldsymbol{a},\boldsymbol{b}]$ where $\boldsymbol{a}$ and $\boldsymbol{b} = [b_0,b_1,\cdots,b_{N-1}]$ are both codewords of RS codes. Since the RS codes form a subclass of the Bose–Chaudhuri–Hocquenghem (BCH) codes [8], these codes are cyclic, and the RS code described above is easily seen to have the generator polynomial

$$g(x) = (x + \alpha)(x + \alpha^2)\cdots(x + \alpha^{N-K}).$$

The vector $\boldsymbol{b}$ has the associated polynomial

$$b(x) = a_0 + \alpha a_1x + \cdots + \alpha^{N-1}a_{N-1}x^{N-1} = a(\alpha x)$$

and it is consequently a codeword of the RS code with generator polynomial

$$g_b(x) = (x + 1)(x + \alpha)\cdots(x + \alpha^{N-K-1}).$$

It is customary to accept a sequence of codes specified in terms of primitive field elements of successively larger fields $GF(2^m)$ as constructive. Thus the sequence of codes $\mathscr{C}_m$ as given above is constructive in the same sense as the cyclic Hamming codes or the RS codes. However, following a suggestion by McEliece [9], the specification of a primitive element of $GF(2^m)$, or equivalently a primitive polynomial of degree m over $GF(2)$, can be avoided in the specification of $\mathscr{C}_m$ as follows. Let the $m = 2 \cdot 3^l$-tuple

$$[f_0,f_1,\cdots,f_{m-1}]$$

correspond to

$$f_0 + f_1\sigma + \cdots + f_{m-1}\sigma^{m-1},$$

where σ is a root of the irreducible, but not primitive polynomial [10]

$$F_1(x) = x^{2\cdot 3^l} + x^{3^l} + 1,$$

i.e.,

$$\sigma^{2\cdot 3^l} + \sigma^{3^l} + 1 = 0.$$

If we replace α^j everywhere by the $m = 2\cdot 3^l$-tuple whose digits are the radix-two form of j, $0 \le j < 2^{2\cdot 3^l}$, the codes thus defined for the particular values of $m = 2\cdot 3^l$ are just permutations of the corresponding subsequence of the codes as defined by (3) and (4). This subsequence of codes is, we believe, incontestably constructive.

The codes (without McEliece's modification) are most easily encoded by a RS encoder that forms $a_{N-1},a_{N-2},\cdots,a_0$, followed by an encoder for the inner codes. The inner encoder stores a multiplier that is initially α^{N-1} and is scaled by a factor α^{-1} after each encoding, and so computes the product $b_j = \alpha^j a_j$. These $2N$ $GF(2^m)$ multiplications in the inner encoder are a small addition to the approximately $N(N - K)$ multiplications required by the RS encoder.

Theorem 1: For any given rate R, $0 < R < \frac{1}{2}$, the sequence of binary $(2mN,mK)$, $m = 1,2,\cdots$, codes $\mathscr{C}_m$, with $R_n = \frac{1}{2}r_N = \frac{1}{2}K/N$ chosen as the smallest rate not less than R, satisfy

$$\liminf_{n\to\infty} (d_n/n) \ge (1 - 2R)H^{-1}(\tfrac{1}{2}) \simeq 0.11(1 - 2R), \tag{5}$$

where d_n is the minimum distance of $\mathscr{C}_m$.

The lower bound (5) is plotted in Fig. 1 together with the Gilbert–Varsharmov bound for comparison. For example, with $R = \frac{1}{4}$, the codes of Theorem 1 have minimum distance at least 5.5 percent of the block length, compared to 21.5 percent, which the Gilbert–Varsharmov bound indicates as possible.

In the proof of Theorem 1 we shall make use of the following.

Lemma: Let $o_2(L) \to 0$ as $L \to \infty$. Then for any γ, $0 < \gamma < 1$, and any δ, $0 < \delta < 1$, the total Hamming weight W of $M_L = [\gamma - o_2(L)](2^{L\delta} - 1)$ distinct nonzero binary L-tuples satisfies

$$W \ge \gamma L[H^{-1}(\delta) - o_1(L)](2^{L\delta} - 1). \tag{6}$$

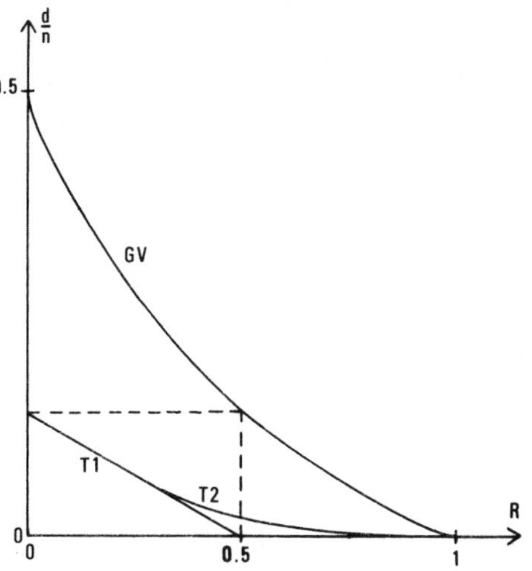

Fig. 1. Comparison of the bounds of Theorem 1 (T1) and Theorem 2 (T2) with the Gilbert–Varsharmov bound (GV).

Proof: We make use of the well-known inequality [8]

$$\sum_{i=1}^{\lambda L} \binom{L}{i} \leq 2^{LH(\lambda)}, \qquad 0 \leq \lambda \leq \tfrac{1}{2}. \tag{7}$$

From this, it follows that the fraction f_L of the M_L specified nonzero L-tuples that have Hamming weight λL or less satisfies

$$f_L \leq M_L^{-1} 2^{LH(\lambda)}, \qquad 0 \leq \lambda \leq \tfrac{1}{2}$$

or

$$f_L \leq [\gamma - o_2(L)]^{-1}(2^{L\delta} - 1)2^{LH(\lambda)}. \tag{8}$$

If we choose

$$\lambda = H^{-1}(\delta - 1/\log L),$$

where L is large enough so that $\delta > 1/\log L$, (8) becomes

$$f_L \leq [\gamma - o_2(L)]^{-1} 2^{-L/\log L}$$

and thus, $f_L = o_3(L)$ where $o_3(L) \to 0$ as $L \to \infty$. Hence the total weight of the M_L L-tuples satisfies

$$W \geq (1 - f_L)M_L H^{-1}(\delta - 1/\log L)L$$

or

$$W \geq [1 - o_3(L)][\gamma - o_2(L)](2^{L\delta} - 1)H^{-1}(\delta)[1 - o_4(L)]L, \tag{9}$$

which by defining

$$\gamma - o_1(L) = [1 - o_3(L)][1 - o_4(1)][\gamma - o_2(L)]$$

gives (6).

We note that since no two randomly shifted codes have any nonzero codewords in common, this lemma may be interpreted as stating that the fraction f_L of M_L randomly shifted codes that have minimum distance $LH^{-1}(\delta - 1/\log L)$ or less vanishes as $L \to \infty$.

Proof of Theorem 1: Consider any nonzero codeword $\boldsymbol{c}$ of $\mathscr{C}_m$. Then $\boldsymbol{a}$ is also nonzero, and, since it is a codeword of an RS code, it contains at least $N - K + 1 > N - K = N(1 - r_N) = N(1 - 2R_n) = (2^m - 1)[1 - 2R - o_2(m)]$ nonzero digits, where $o_2(m) \to 0$ as $m \to \infty$. Since no $2m$-tuple appears in more than one inner code, we may apply the lemma with $L = 2m$, $\delta = \frac{1}{2}$, and $\gamma = 1 - 2R$. The Hamming weight of $\boldsymbol{c}$ is lower bounded by

$$W \geq [H^{-1}(\tfrac{1}{2}) - o_1(2m)](1 - 2R)(2^m - 1)2^m.$$

Consequently the minimum distance d_n of $\mathscr{C}_m$ satisfies

$$d_n/n = d_n/[2m(2^m - 1)] \geq (1 - 2R)[H^{-1}(\tfrac{1}{2}) - o_1(2m)]. \tag{10}$$

Since $m \to \infty$ as $n = 2m(2^m - 1) \to \infty$, (5) follows immediately from (10) and the theorem is proved.

We note that this result could not have been obtained by simply considering the RS code as a binary code. Let d_0 be the minimum distance of the RS code over $GF(2^m)$ and let i be the smallest integer such that $d_0' = 2^i - 1 \geq d_0$. It was proved by Kasami *et al.* [11] that the minimum distance of the BCH code of length $2^m - 1$ and design distance d_0' has minimum distance exactly d_0'. Now as shown in [3] the BCH codes with design distance at least d_0 are subcodes of the RS code with minimum distance d_0, and hence their minimum distance is an upper bound on the minimum distance d_{BRS} of the RS code considered as a binary code of length $n = mN$. Since $d_0' \leq 2d_0'$, we have

$$d_{\mathrm{BRS}}/n = d_{\mathrm{BRS}}/(mN) < 2(N - K)/(mN)$$

and consequently

$$d_{\mathrm{BRS}}/n \to 0, \qquad \text{as } m \to \infty.$$

In order to extend the construction to higher rates, we puncture the codes previously constructed by deleting the last s digits from each inner code. We shall write

$$r_n = m/(2m - s), \qquad 0 \leq s < m$$

for the resulting rate of the inner code.

For a given value of R, $0 < R < 1$, and any choice of r_n, we choose the rate of the RS outer code as the smallest rate such that

$$R_n = r_n r_M \geq R.$$

We again note that any nonzero codeword $\boldsymbol{c}$ in the punctured code $\mathscr{C}_m$ will contain

$$M > (2^m - 1)(1 - r_N) = (2^m - 1)[1 - (R/r_n) - o_2(m)]$$

nonzero a_j. The nonzero (a_j, b_j) are distinct, but after puncturing there may be as many as 2^s identical nonzero c_j, since any c_j is a codeword in exactly 2^s inner codes. Thus at least

$$2^{-s}M > (2^{m-s} - 1)[1 - (R/r_n) - o_3(m)]$$

nonzero $(2m - s)$-tuples are distinct and we can apply the lemma with $L = 2m - s$, $\delta = (m - s)/L = 1 - r_n$, and $\gamma = 1 - (R/r_n)$. We obtain the following lower bound for the Hamming weight of $\boldsymbol{c}$:

$$W \geq [1 - (R/r_n)](2m - s)[H^{-1}(1 - r_n) - o_1(m)](2^{m-s} - 1)2^s. \tag{11}$$

But if r_n approaches r, $\frac{1}{2} \le r < 1$, the ratio s/m is bounded away from 1 and (11) implies

$$\liminf_{n\to\infty} (d_n/n) \ge [1 - (R/r)]H^{-1}(1 - r). \qquad (12)$$

The minimum distance may be maximized by setting the derivative of the right-hand side of (12) with respect to r equal to zero. The condition satisfied by the maximizing choice of r is

$$R = \frac{r^2}{1 + \log_2 [1 - H^{-1}(1 - r)]} \qquad (13)$$

except that $r = \frac{1}{2}$ must be chosen when the solution of (13) would yield $r < \frac{1}{2}$, since our construction permits only inner code rates at least $\frac{1}{2}$.

We summarize the foregoing as Theorem 2.

Theorem 2: For any rate R, $0 < R < 1$, the binary (n,k) codes $\mathscr{C}_m$ punctured so that r_n approaches r, where r is the maximum of $\frac{1}{2}$ and the solution to (13), satisfy

$$\liminf_{n\to\infty} (d_n/n) \ge [1 - (R/r)]H^{-1}(1 - r). \qquad (14)$$

The bound (14) can be viewed as the envelope of a family of straight lines constructed in the following way. The point $[r,H^{-1}(1 - r)]$ on the Gilbert–Varsharmov bound is projected on the axes to give the points $[r,0]$ and $[0,H^{-1} \cdot (1 - r)]$. These two points are then joined by a straight line. Because of the restriction $r \ge \frac{1}{2}$, the bound of Theorem 2 is identical to that of Theorem 1 for $0 < R < 0.30$. For higher rates the bound (14) coincides with the lower bound on the minimum distance for the ensemble of concatenated codes obtained by Zyablov [12]. For rates $r < 0.30$ the bound is inferior to Zyablov's bound because the construction of Section II requires a good ensemble of inner codes with at most $2^m - 1$ codes and we cannot constructively specify such an ensemble for rates less than $\frac{1}{2}$.

We note that, for rates near unity, the maximizing value of r is

$$r \simeq R_n \simeq R^{1/2}.$$

III. A Decoding Procedure

All error patterns of weight less than half the value of our asymptotic lower bound on minimum distance (14) can be corrected by the version of generalized minimum distance decoding which was applied to product codes by Reddy and Robinson [6] and Weldon [7].

It was pointed out earlier that all but a vanishing fraction of the inner codes have minimum distance d_i satisfying

$$\frac{d_i}{2m - s} \ge \frac{D}{2m - s} = H^{-1}(1 - r_n),$$

where D is the lower bound on minimum distance given by (1).

For our decoding procedure, we select the scheme shown in Fig. 2 in which the received block

$$r = [r_0, r_1, \cdots, r_{N-1}],$$

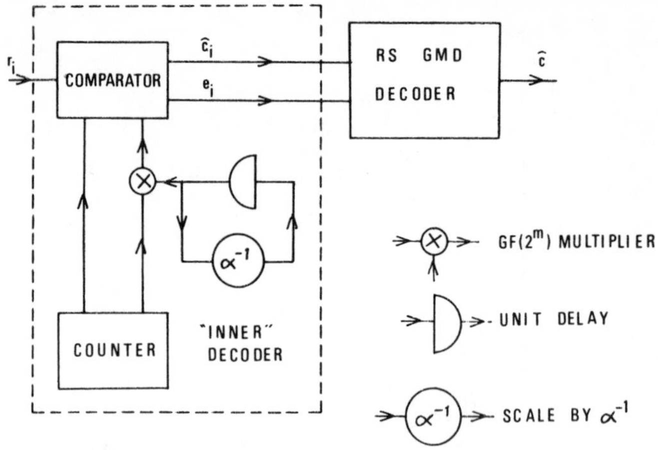

Fig. 2. Block diagram of the decoder for the (n,k) codes of Theorem 2.

where r_i is the received $(2m - s)$-tuple corresponding to the inner codeword c_i, is first decoded by an inner decoder whose decisions are then passed to a generalized minimum distance (GMD) decoder [3] for the RS outer code. The inner decoder may be constructed as shown in Fig. 2 such that, for each received inner word r_i, the decoder (by means of an m-stage binary ring counter) generates all 2^m codewords in the inner code and feeds these to a comparator that also receives r_i. The inner decoder decodes r_i into $\hat{c}_i$ as soon as a codeword $\hat{c}_i$ is found at distance less than $D/2$ from r_i. Since this decoding process must be done $N = 2^m - 1$ times, the inner decoder performs $2^m(2^m - 1)$, or approximately $(n/\log n)^2$ multiplications in all.

Let the output of the comparator be the number e_i together with the estimate $\hat{c}_i$ of c_i, where e_i is defined by

$$e_i = \begin{cases} \text{weight } (\hat{c}_i + r_i), & r_i \text{ decoded} \\ D/2, & \text{otherwise (in this case } \hat{c}_i \text{ may be arbitrary).} \end{cases}$$

Define the normalized weight of the error at position i as

$$\beta_i = \begin{cases} \dfrac{e_i}{D}, & \hat{c}_i = c_i \\ \dfrac{D - e_i}{D}, & \hat{c}_i \ne c_i. \end{cases} \qquad (15)$$

According to the theory of GMD [3], the RS "outer" decoder will decode correctly whenever

$$\sum_{i=0}^{N-1} \beta_i < \frac{N - K + 1}{2}. \qquad (16)$$

Noting that, because of the lemma, the fraction of the r_i yielding errors because $d_i < D$ vanishes as $N \to \infty$ and that the number of errors required to cause a given value of β_i for an inner code with minimum distance at least D is $t_i \ge \beta_i D$, it then follows from (16) that the minimum number of errors that can cause a decoding error asymptotically satisfies

$$t = \sum_{i=0}^{N-1} t_i \ge \sum_{i=0}^{N-1} \beta_i D > (D/2)(N - K + 1) > \tfrac{1}{2}DN(1 - r_N). \qquad (17)$$

But, asymptotically, our lower bound d_B on the minimum distance d is

$$d_B = ND(1 - r_N)$$

so that (17) shows that our decoding procedure corrects all the errors guaranteed correctable by the asymptotic lower bound d_B.

We also note that, to do GMD decoding for the outer RS code the number of trials required equals the number of distinct values of e_i. Since there are $D/2$ allowed values of e_i and since errors-and-erasures decoding of an RS code with Berlekamp's iterative algorithm [13] requires a number of multiplications proportional to 2^{2m}, it follows that the total number of multiplications in our decoding procedure is proportional to $m2^{2m}$. This corresponds to a number of bit operations proportional to $m^3\ 2^{2m}$ or approximately $n^2 \log n$.

IV. Remarks

The code construction given in Theorem 2 for binary codes is easily extended to the construction of asymptotically good codes over any finite field $GF(q)$. In this case, one uses the RS code over the extension field $GF(q^m)$, and chooses the α in (4) as a primitive element of $GF(q^m)$. McEliece's modification cannot be used to remove the last possible objection to constructivity for $q > 2$, however, since we know no expression for irreducible polynomials of arbitrarily large degree over an arbitrary field $GF(q)$.

Finally, we remark that, for any particular block length $n = (2m - s)2^m$, a search for the best fixed inner code requires less than n^2 steps. Even though a good concatenated code with the best fixed inner code can thus be found with a very moderate amount of searching, we have rejected this approach as nonconstructive, since we cannot specify the inner code *a priori*. It is an open question whether the actual distance of our constructive codes with varying inner codes is better or worse than that obtained using the best fixed inner code.

Acknowledgment

The author wishes to thank Prof. J. L. Massey for his help in preparing the final draft of this paper.

References

[1] E. N. Gilbert, "A comparison of signalling alphabets," *Bell Syst. Tech. J.*, vol. 31, pp. 504–522, 1952.
[2] R. R. Varsharmov, "Estimate of the number of signals in error correcting codes," *Dokl. Akad. Nauk. SSSR*, vol. 117, pp. 739–741, 1957.
[3] G. D. Forney, Jr., *Concatenated Codes.* Cambridge, Mass.: M.I.T. Press, 1966.
[4] I. S. Reed and G. Solomon, "Polynomial codes over certain finite fields," *J. Soc. Ind. Appl. Math.*, vol. 8, pp. 300–304, 1960.
[5] J. L. Massey, *Threshold Decoding.* Cambridge, Mass.: M.I.T. Press, p. 21, 1963.
[6] S. M. Reddy and J. P. Robinson, "Random error and burst correction by iterated codes," *IEEE Trans. Inform. Theory*, vol. IT-18, pp. 182–185, Jan. 1972.
[7] E. J. Weldon, Jr., "Decoding binary block codes on *Q*-ary output channels," *IEEE Trans. Inform. Theory*, vol. IT-17, pp. 713–718 (Appendix), Nov. 1971.
[8] W. W. Peterson, *Error Correcting Codes.* Cambridge, Mass.: M.I.T. Press and New York: Wiley, 1961.
[9] R. J. McEliece, private communication, 1972.
[10] S. W. Golomb, *Shift Register Sequences.* San Francisco: Holden-Day, 1967, p. 96.
[11] T. Kasami, S. Lin, and W. W. Peterson, "Some results on weight distributions of BCH codes" (Abstract), *IEEE Trans. Inform. Theory*, vol. IT-12, p. 274, Apr. 1966.
[12] V. V. Zyablov, "On estimation of complexity of construction of binary linear concatenated codes," *Probl. Peredach. Inform.*, vol. 7, pp. 5–13, 1971.
[13] E. R. Berlekamp, *Algebraic Coding Theory.* New York: McGraw-Hill, 1968, p. 184.

О НЕГРУППОВЫХ ПЛОТНО УПАКОВАННЫХ КОДАХ

Ю. Л. ВАСИЛЬЕВ

(НОВОСИБИРСК)

1. В теории корректирующих кодов *) важное место занимает изучение групповых кодов и плотно упакованных кодов (п. у. кодов) [2]. Эти классы кодов позволяют по-разному подойти к решению основной задачи теории корректирующих кодов — построению для данных n и d (n, d) кодов максимальной мощности. Групповые коды обладают некоторой симметрией, что делает их более обозримыми и позволяет в классе этих кодов существенно продвинуть решение основной задачи [3]. В плотно упакованных кодах максимальность предусматривается самим определением.

Отметим два обстоятельства, к которым имеют отношение результаты настоящей заметки.

а) Все известные методы синтеза п. у. кодов [1, 2, 3, 4] дают групповые коды. Тривиально строятся негрупповые п. у. коды, которые являются сдвигами **) групповых п. у. кодов. Однако такие негрупповые коды не представляют самостоятельного интереса, так как их «внутреннее строение» точно такое же, как и у групповых кодов. Код, который не является ни групповым кодом, ни сдвигом группового кода, назовем сильно негрупповым. Возникает вопрос, существуют ли сильно негрупповые коды.

б) В [4] при исследовании симметрии п. у. кодов была доказана теорема: в п. у. $(n, 2l+1)$-коде число кодовых точек, находящихся на расстоянии r, $2l+1 \leqslant r \leqslant n$ от данной кодовой точки z, не зависит ни от выбора точки z, ни от выбора кода. В связи с этой теоремой в [4] было высказано предположение, что при данных n и d имеется по существу самое большее один п. у. (n, d)-код, т. е. что если даны два п. у. (n, d)-кода σ_1 и σ_2, то существует симметрия куба E^n, которая отображает σ_1 в σ_2. Другими словами, в [4] было высказано предположение, что при данных n и d все п. у. коды имеют один и тот же тип [5].

В настоящей заметке предлагается метод построения п. у. $(n, 3)$-кодов. В отличие от других методов [1, 2, 3, 4], позволяющих строить п. у. $(n, 3)$-коды, он дает не только групповые, но и сильно негрупповые коды. Это означает, что ответ на упомянутый выше вопрос положителен и предположение относительно п. у. кодов является, вообще говоря, неверным.

Кроме того, в заметке дана оценка снизу отношения числа типов сильно негрупповых п. у. $(n, 3)$-кодов к числу всех групповых $(n, 3)$-кодов. Эта оценка выражается числом $2^{2^{n\left(\frac{1}{2}-\delta\right)}}$, где $n=2^q-1$, $q=4, 5, 6, \ldots$, $\delta \to 0$ при $n \to \infty$.

2. Пусть $p=2^q-1$, $q=1, 2, \ldots$. Как известно [1], п. у. $(p, 3)$-коды существуют только при указанных значениях p (в п. у. $(p,3)$-коде должно быть $\frac{2^p}{p+1}$ наборов),

Пусть $C^p=\{(\tau_1, \ldots, \tau_p)\}$ — п. у. $(p, 3)$-код, содержащий нулевой набор; E^p — множество всех наборов $(\alpha_1, \ldots, \alpha_p)$ длины p; $\lambda(\tau)$ — произвольная функция, принимающая значения 0 и 1 на наборах $\tau=(\tau_1, \ldots, \tau_p)$ из C^p, причем $\lambda(0, \ldots, 0)=0$. Для набора $\alpha=(\alpha_1, \ldots, \alpha_p)$ положим $|\alpha|=\alpha_1 \oplus \ldots \oplus \alpha_p$.

Паре наборов $\alpha=(\alpha_1, \ldots, \alpha_p)$ и $\tau=(\tau_1, \ldots, \tau_p)$, $\alpha \in E^p$, $\tau \in C^p$; сопоставим набор $(\alpha_1, \ldots, \alpha_p, \alpha_1 \oplus \tau_1, \ldots, \alpha_p \oplus \tau_p, |\alpha| \oplus \lambda(\tau))$, который обозначим через $\alpha \times \tau$. Обозначим через $E^p \times C^p$ множество всех наборов $\alpha \times \tau$, где $\alpha \in E^p$, $\tau \in C^p$.

Т е о р е м а. *Для любой функции* $\lambda(\tau)$ *множество* $E^p \times C^p$, $p=2^q-1$, *является п. у.* $(2^{q+1}-1, 3)$-*кодом, содержащим нулевой набор.*

Д о к а з а т е л ь с т в о. Очевидно, что:

(а) число разрядов в наборе $\alpha \times \tau$ равно $2p+1=2^{q+1}-1$;

(б) число всех наборов $\alpha \times \tau$ во множестве $E^p \times C^p$ равно $\frac{2^{2p}}{p+1}$ — числу наборов в п. у. $(2^{q+1}-1, 3)$-коде;

(в) нулевой набор $(0, \ldots, 0)$ длины $2p+1$ входит в $E^p \times C^p$.

Поэтому остается доказать, что расстояние ϱ между любыми двумя наборами из множества $E^p \times C^p$ не меньше трех. Пусть $\alpha \times \tau$ и $\beta \times \nu$ — произвольные несовпадающие наборы из $E^p \times C^p$, $\alpha=(\alpha_1, \ldots, \alpha_p)$, $\beta=(\beta_1, \ldots, \beta_p)$ $\tau=(\tau_1, \ldots, \tau_p)$, $\nu=(\nu_1, \ldots, \nu_p)$, $\alpha, \beta \in E^p$, $\tau, \nu \in C^p$. По определению

$$\alpha \times \tau=(\alpha_1, \ldots, \alpha_p, \alpha_1 \oplus \tau_1, \ldots, \alpha_p \oplus \tau_p, |\alpha| \oplus \lambda(\tau)),$$
$$\beta \times \nu=(\beta_1, \ldots, \beta_p, \beta_1 \oplus \nu_1, \ldots, \beta_p \oplus \nu_p, |\beta| \oplus \lambda(\nu)).$$

Если $\tau \neq \nu$, то по условию $\varrho(\tau, \nu) \geqslant 3$. Тогда при $\varrho(\alpha, \beta)=0, 1, 2, 3$ имеем соответственно

$$\varrho\left[(\alpha_1 \otimes \tau_1, \ldots, \alpha_p \oplus \tau_p), (\beta_1 \oplus \nu_1, \ldots, \beta_p \oplus \nu_p)\right] \geqslant 3, 2, 1, 0,$$

и следовательно $\varrho(\alpha \times \tau, \beta \times \nu) \geqslant 3$.

Если $\tau=\nu$, то $\alpha \neq \beta$ и $\lambda(\tau)=\lambda(\nu)$. Возможны два случая: $|\alpha| \neq |\beta|$, $|\alpha|=|\beta|$. В первом случае

$$\varrho(\alpha, \beta) \geqslant 1, \varrho\left[(\alpha_1 \oplus \tau_1, \ldots), (\beta_1 \oplus \nu_1, \ldots)\right] \geqslant 1, \quad |\alpha| \oplus \lambda(\tau) \neq |\beta| \oplus \lambda(\nu),$$

откуда $\varrho(\alpha \times \tau, \beta \times \nu) \geqslant 3$. Во втором случае $\varrho(\alpha, \beta) \geqslant 2$ (так как $|\alpha|=|\beta|$, но $\alpha \neq \beta$), $\varrho((\alpha_1 \oplus \tau_1, \ldots), (\beta_1 \oplus \nu_1, \ldots)) \geqslant 2$, откуда $\varrho(\alpha \times \tau, \beta \times \nu) \geqslant 4$. Теорема доказана.

3. Пусть $G^p=\{\tau_1, \ldots, \tau^p)\}$ — п. у. групповой $(p, 3)$-код.

Рассмотрим функцию $\lambda(\tau)$, $\tau \in G^p$, участвующую в построении множества $E^p \times G^p$. Если положить $\lambda(\tau) \equiv 0$ для всех $\tau \in G^p$, то множество $E^p \times G^p$ будет групповым п. у. кодом. Если зафиксировать два каких-либо ненулевых набора

$$\mu=(\mu_1, \ldots, \mu_p) \quad \text{и} \quad \nu=(\nu_1, \ldots, \nu_p), \ \mu \neq \nu, \ \mu, \nu \in G^p,$$

и положить $\lambda(\mu \oplus \nu) \neq \lambda(\mu) \oplus \lambda(\nu)$, то множество $E^p \times G^p$ будет сильно негрупповым п. у. кодом.

Как известно, (3,3)-код и (7,3)-код, являются групповым. Далее, для любого $p=2^q-1$, $q=4, 5, 6, \ldots$, можно указать как групповые, так и негрупповые п. у. $(p, 3)$-коды.

Приведем некоторые оценки. Чтобы по групповому п. у. коду $G=\{(\tau_1, \ldots, \tau_p)\}$ построить сильно негрупповой п. у. код, надо зафиксировать функцию $\lambda(\tau)$, $\tau \in C^p$, на четырех наборах — на нулевом наборе $(\lambda(0, \ldots, 0)=0)$ и на вышеупомянутых наборах μ, ν, $\mu \oplus \nu$.

На остальных наборах из G^p функция $\lambda(\tau)$ может быть произвольной. Число этих наборов равно $\frac{2^p}{p+1}-4$; число различных множеств $E^p \times G^p$, являющихся сильно негрупповыми п. у. $(2p+1,3)$-кодами, будет не менее $2^{\frac{2p}{p+1}-4}$; число типов сильно негрупповых п. у. $(2p+1,3)$-кодов будет не менее $\frac{2^{\frac{2p}{p+1}-4}}{(2p+1)!\,2^{2p+1}}$. Так как каждый групповой $(2p+1,3)$-код полностью определяется своими образующими, число которых не превосходит $2p$, то число групповых $(2p+1,3)$-кодов не превосходит $C_{2^{2p+1}}^{2p} \leqslant 2^{5p^2}$. Сопоставление последних двух оценок дает результат, указанный в самом конце п. 1.

В заключение автор выражает глубокую благодарность В. Глаголеву, давшему ряд ценных советов.

ЛИТЕРАТУРА

1. H a m m i n g R. W., Error detection and error correction codes, BSTJ 29, 1950, 147—160. (Русский перевод: Коды с обнаружением и исправлением ошибок, М., ИЛ, 1956, 7—22).
2. L l o y d S. P., Binary block coding, BSTJ 36, 2, 1957, 517—535. (Русский перевод: Кибернетический сборник 1, М., ИЛ, 1960, 206—226.)
3. B o s e R. C., R a y-C h a u d h u r i, On a class of error correcting binary group codes, Inf. and Control 3, 1, 1960, 68—79. (Русский перевод: Кибернетический сборник 2, М., ИЛ, 1960, 83—94).
4. S h a p i r o H. S., S l o t n i c k D. L., On the theory of errorcorrecting codes, IBM J. Res. and Developm. 3, 1, 1959, 25—34 (Русский перевод: Кибернетический сборник 5, М., ИЛ, 1962.)
5. G. P o l y a, J. Symb. Logik 5, 3, 1940, 98.

Поступило в редакцию 13 XI 1961

*) Напомним основные понятия теории корректирующих кодов. *Расстоянием* между двоичными наборами $(\sigma_1, \ldots, \sigma_n)$ и $(\tau_2, \ldots, \tau_n)$ называется число разрядов, в которых эти наборы не совпадают.

Шаром радиуса l называется множество всех наборов, отстоящих от данного набора на расстоянии не больше l.

Множество $\{(\tau_1, \ldots, \tau_n)\}$ двоичных наборов называется (n, d)-*кодом*, если любые два набора из этого множества находятся друг от друга на расстоянии не меньше d. Если это множество является группой относительно операции

$$(\tau_1, \ldots, \tau_n) \oplus (\nu_1, \ldots, \nu_n)=(\tau_1+\nu_1 \,(\mathrm{mod}\ 2), \ldots, \tau_n+\nu_n \,(\mathrm{mod}\ 2)),$$

то код называется *групповым*. Если $(n, 2l+1)$-код порождает разбиение n-мерного единичного куба E^n на шары радиуса l, которые попарно не пересекаются и объединение которых исчерпывает E^n, то он называется *плотно упакованным* (п. у.).

**) Код $C=\{(\tau_1, \ldots, \tau_n)\}$ назовем сдвигом кода $C=\{(\gamma_1, \ldots, \gamma_n)\}$, если существует такой набор $\xi=(\xi_1, \ldots, \xi_n)$, что $C=\{\gamma_1 \oplus \xi_1, \ldots, \gamma_n \oplus \xi_n)\}$, где через $\oplus$ обозначено сложение по mod 2.

An Optimum Nonlinear Code*

ALAN W. NORDSTROM

United Township High School, East Moline, Illinois 61244

AND

JOHN P. ROBINSON

Department of Electrical Engineering, University of Iowa, Iowa City, Iowa 52240

A systematic nonlinear code having length 15, minimum distance 5, and 256 code words is given in Boolean form. This is the maximum possible number of words for length 15 and distance 5. The distance spectrum of all pairs of code words is an exact multiple of the weight spectrum of the code words.

LIST OF SYMBOLS

$A(n, d)$ = the maximum number of binary n-tuples such that any two n-tuples differ in at least d places.

$B(n, d)$ = the maximum number of code words possible in a linear code of block length n such that the Hamming distance between any two code words is at least d.

$A_0A_1 \cdots B_0B_1 \cdots Z_0Z_1 \cdots$ = Boolean variables.

$\oplus$ = connective symbolizing modulo 2 addition.

I. INTRODUCTION

The optimum code resulted from a study of Nadler's (1962) nonlinear 32-word code and Green's (1966) 64-word nonlinear code. These codes have a minimum Hamming distance of 5, with the former having length 12 and the latter length 13. Both have twice as many words as the best linear code with the same length and minimum distance (Wagner, 1965).

TABLE I

WEIGHT SPECTRA OF THE OPTIMUM CODE AND ITS SHORTENED VERSIONS

Length	0	5	6	7	8	9	10	15
12	1	11	13	2	1	3	1	0
13	1	18	24	4	3	10	4	0
14	1	28	42	8	7	28	14	0
15	1	42	70	15	15	70	42	1

II. THE CODE

The code is systematic with 8 information bits denoted by $X_0, X_1, \cdots, X_7$ and 7 redundant bits denoted by $Y_0, Y_1, \cdots, Y_6$. Each Y, as a Boolean function of the X's, is in the same equivalence class. Y_0 is as follows:

$$Y_0 = X_7 \oplus X_6 \oplus X_0 \oplus X_1 \oplus X_3 \oplus (X_0 \oplus X_4)(X_1 \oplus X_2 \oplus X_3 \oplus X_5) \oplus (X_1 \oplus X_2)(X_3 \oplus X_5), \quad (1)$$

where $\oplus$ denotes modulo 2 addition.

The remaining Y's are found by cyclically shifting X_0 through X_6; i.e., for Y_j substitute $X_{i+j(\mathrm{mod}\ 7)}$ for X_i in (1) where $i = 0, 1, \cdots, 6$ for each j, $j = 0, 1, \cdots, 6$.

Examining (1) we see that all products X_iX_j, $i \neq j$ in $X_0, \cdots, X_5$ appear, except X_0X_4, X_1X_2, X_3X_5. In these 3 missing pairs all 6 variables appear with one linear term in (1) in each pair.

In the general classification method of Roos (1965) this code would be termed a quadratic code.

Shortening this code by dropping one information bit yields a 128-word code having length 13 with minimum distance 5. Deleting two information bits results in Green's (1966) code, three bits yields Nadler's (1962) code. Next we compare these codes from (1) with Johnson's (1962) bound and Wagner's (1965) determination of $B(n, d)$. $B(n, d)$ is $A(n, d)$ restricted to the class of linear codes. We conclude that $A(15, 5) = 256$, using Johnson's upper bound and (1).

$$A(15, 5) = 256 = 2B(15, 5)$$

$$131 \geq A(14, 5) \geq 128 = 2B(14, 5)$$

$$70 \geq A(13, 5) \geq 64 = 2B(13, 5)$$

$$39 \geq A(12, 5) \geq 32 = 2B(12, 5)$$

Surprisingly, these codes have the weight–distance relationship of a linear code. Any given code word can be normalized to the all-zero word by complementing all words in the appropriate positions. The resulting code is the same as the original code with possibly a permutation of the positions. Thus the weight spectrum of the code gives the distance properties. In Table I we list the weight spectra of these codes obtained from (1).

III. METHODOLOGY

In order to determine the distance properties of Nadler's (1962) code and Green's (1965) code, a computer program was written to calculate the distance for all possible combinations of pairs of words. The distance spectra of these two codes were found to be very similar and it was hypothesized that the codes were closely related.

The next step was to put the codes in the same form. Green's code was in a normalized form in that it had one all-zero word. Nadler's code was normalized to this form by changing all the ones in a particular word to zeroes and then changing the corresponding positions in all of the other words.

As the two codes appeared very similar, it was suggested that Nadler's code was contained in Green's code. This was checked by dropping one of the columns of Green's so that it had the same length and form as Nadler's. The shortened 64-word code was then divided into two 32-word codes. By switching certain columns of the first, a 32-word code that was exactly the same as Nadler's was found. By taking the second and normalizing it, and then switching certain columns, Nadler's code was again found in this half of the 64-word code.

Since the column that was dropped contained 32 ones and 32 zeroes, the idea for building a 128-word nonlinear code came to mind. By taking two of Green's codes and adding an extra column of ones to one of them, an extra column of zeroes to the other and then changing all the numbers of certain columns of the first one, a 128-word code could be built.

A suitable program was written which tried all possible combinations of complementing columns. It was successful and a new 128-word nonlinear code was discovered. Extending the technique another step resulted in the optimum code (1) with 256 words of length 15 and distance 5.

IV. CONCLUDING REMARKS

This note presents an optimum nonlinear code with considerable structure and regularity. It is felt that the structure should allow reasonable decoding algorithms. A quasi-algebraic decoding scheme has been proposed (Robinson, 1968). The functional form suggests that a general class of good nonlinear codes may exist.

RECEIVED: August 21, 1967. REVISED: November 24, 1967.

REFERENCES

GREEN, M. W. (1966), Two heuristic techniques for block-code construction *IEEE Trans. Inform. Theory* **IT-12**, 273.

JOHNSON, S. M. (1962), A new upper bound for error-correcting codes. *IRE Trans. Inform. Theory* **IT-8**, 203–207.

NADLER, M. (1962), "Topics in Engineering Logic." Macmillan, New York.

ROBINSON, J. P. (1968), Analysis of Nordstrom's optimum quadratic code. Proceedings of the 1968 Hawaii International Conference on System Sciences Honolulu, Hawaii.

ROOS, JAN-ERIK (1965), An algebraic study of group and nongroup error-correcting codes. *Inform. Control* **8**, 195–214.

WAGNER, T. J. (1965), "Some New Values of $B(n, d)$." University of Texas Report, Austin, Texas.

* The research reported in this paper was supported by the National Science Foundation under the Research Participation Program for Exceptional Secondary Students and Grant GK-816.

Reprinted with permission from *Inform. Contr.*, vol. 11, pp. 613–616, Nov.–Dec. 1967.

НОВЫЙ КЛАСС ЛИНЕЙНЫХ КОРРЕКТИРУЮЩИХ КОДОВ

В. Д. Гоппа

Описан класс двоичных линейных кодов, исправляющих ошибки. Каждый код из этого класса задается некоторым многочленом над $GF(2^m)$. Зная степень t этого многочлена, можно получить следующие оценки для параметров кода: $n \leqslant 2^m$, $k \geqslant n - mt$, $d \geqslant 2t + 1$.

Описанные коды, вообще говоря, нециклические. Единственный циклический код, входящий в рассматриваемый класс,— код Боуза — Чоудхури — Хоквингема (БЧХ). Все основные свойства кода БЧХ определяются, по-видимому, его принадлежностью этому классу кодов, а не классу циклических кодов. Так для всех кодов рассматриваемого класса существует схема декодирования, аналогичная алгоритму Питерсона для кодов БЧХ.

Построение кодов основано на отождествлении исходного пространства двоичных векторов с некоторым множеством рациональных функций.

§ 1. Введение

Линейный код, исправляющий t ошибок, определяется некоторой матрицей с ненулевыми минорами порядка $\leqslant 2t$. В классическом матричном анализе изучаются некоторые специальные типы матриц (над полями характеристики 0) с определенными требованиями к минорам некоторого порядка.

В частности, в книге [1] описаны так называемые вполне положительные матрицы, у которых все миноры порядка $\leqslant r$ положительны. Самой известной вполне положительной матрицей является матрица Вандермонда, и на ее основе построен код Боуза — Чоудхури — Хоквингема (БЧХ). Другая вполне положительная матрица — матрица $\|(x_i - y_j)^{-1}\|$. Эта матрица стала отправным пунктом для создания класса рассматриваемых здесь линейных кодов.

Каждый код из этого класса, так же как и циклический код, задается некоторым порождающим многочленом. Отличие заключается в том, что знание порождающего многочлена циклического кода в общем случае ничего не говорит о корректирующих способностях кода, а по одной только степени порождающего многочлена описываемых кодов можно получить следующие оценки для параметров кода: $n \leqslant 2^m$, $k \geqslant n - m \deg g(z)$, $d \geqslant 2 \deg g(z) + 1$ (здесь $\deg g(z)$ — степень многочлена $g(z)$). Единственный циклический код, входящий в рассматриваемый класс кодов,— это код БЧХ. По-видимому, все основные особенности кода БЧХ объясняются его принадлежностью к построенному классу, а не к классу циклических кодов. Например, для всех описанных в этой статье кодов существует схема декодирования, сводящаяся к решению системы линейных уравнений над конечным полем.

§ 2. Определение класса кодов

Пусть L — некоторое множество элементов поля $GF(2^m)$: $L = \{\alpha_1, \ldots, \alpha_n\}$, $n \leqslant 2^m$, S — векторное пространство размерности n над $GF(2)$.

Поставим в соответствие каждому вектору $x = (a_1, \ldots, a_n)$, $a_i \in GF(2)$, $i = 1, \ldots, n$ рациональную функцию

$$R_x(z) = \sum_{i=1}^{n} \frac{a_i}{z - \alpha_i}.$$

Отображение $x \to R_x(z)$ — гомоморфизм S в аддитивную группу рациональных функций над $GF(2^m)$.

Выберем некоторый многочлен $g(z)$ с коэффициентами из $GF(2^m)$, не имеющий корней в L. Определим линейный код как множество векторов x, для которых $R_x(z) \equiv 0 \bmod g(z)$. Задавая различным образом многочлен $g(z)$ (назовем его порождающим по аналогии с циклическими кодами), можно получать коды с различными свойствами. Вместе с каждым вектором x, имеющим единицы на местах $i_1, i_2, \ldots, i_k$, будем рассматривать многочлен $f(z) = (z - \alpha_{i_1})(z - \alpha_{i_2}) \ldots (z - \alpha_{i_k})$. Очевидно, $R_x(z) = f'(z)/f(z)$, где $f'(z)$ — формальная производная многочлена $f(z)$.

§ 3. Проверочная матрица. Мощность кода

Для кодового вектора $x = (a_1, \ldots, a_n)$ выполняется соотношение

$$R_x(z) = \sum_{i=1}^{n} \frac{a_i}{z - \alpha_i} \equiv 0 \bmod g(z).$$

Это сравнение эквивалентно равенству

$$\sum_{i=1}^{n} a_i \{(z - \alpha_i)^{-1}\}_m = 0,$$

где $\{(z - \alpha_i)^{-1}\}_m$ — элемент, обратный к $(z - \alpha_i)$ в алгебре многочленов по $\bmod g(z)$. Этот элемент находится следующим образом:

$$\{(z - \alpha_i)^{-1}\}_m = \frac{g(z) - g(\alpha_i)}{z - \alpha_i} g^{-1}(\alpha_i),$$

так как в правой части стоит многочлен степени, меньшей чем степень $g(z)$, и

$$\frac{1}{z - \alpha_i} \equiv \frac{g(z) - g(\alpha_i)}{z - \alpha_i} g^{-1}(\alpha_i) \bmod g(z).$$

Поэтому проверочная матрица кода состоит из следующей строки:

$$T = \left(\frac{g(z) - g(\alpha_1)}{z - \alpha_1} g^{-1}(\alpha_1) \ldots \frac{g(z) - g(\alpha_n)}{z - \alpha_n} g^{-1}(\alpha_n) \right).$$

Пусть $g(z) = \sum_{i=0}^{r} b_i z^i$ $(\deg g(z) = r)$.

Тогда матрицу T можно представить так:

$$\begin{pmatrix} b_r g^{-1}(\alpha_1) & \ldots & b_r g^{-1}(\alpha_n) \\ (b_{r-1} + b_r \alpha_1) g^{-1}(\alpha_1) & \ldots & (b_{r-1} + b_r \alpha_1) g^{-1}(\alpha_n) \\ \ldots & \ldots & \ldots \\ (b_1 + b_2 \alpha_1 + \ldots + b_r \alpha_1^{r-1}) g^{-1}(\alpha_1) & \ldots & (b_1 + \ldots + b_r \alpha_n^{r-1}) g^{-1}(\alpha_n) \end{pmatrix}.$$

Отсюда видно, что T — линейное преобразование строк матрицы T^*:

$$T^* = \begin{pmatrix} g^{-1}(\alpha_1) & \ldots & g^{-1}(\alpha_n) \\ \alpha_1 g^{-1}(\alpha_1) & \ldots & \alpha_n g^{-1}(\alpha_n) \\ \ldots & \ldots & \ldots \\ \alpha_1^{r-1} g^{-1}(\alpha_1) & \ldots & \alpha_n^{r-1} g^{-1}(\alpha_n) \end{pmatrix}.$$

Итак, проверочная матрица кода получается умножением матрицы Вандермонда справа на диагональную матрицу

$$T = \begin{pmatrix} 1 & \ldots & 1 \\ \alpha_1 & & \alpha_n \\ \ldots & \ldots & \ldots \\ \alpha_1^{r-1} & \ldots & \alpha_n^{r-1} \end{pmatrix} \begin{pmatrix} g^{-1}(\alpha_1) & & \\ & g^{-1}(\alpha_2) & \\ \ldots & \ldots & \ldots \\ & & g^{-1}(\alpha_n) \end{pmatrix}.$$

В частности, если выбрать $n = 2^k - 1$, в качестве L — все элементы группы порядка n, $g(z) = z^{2r}$, то получается матрица кода БЧХ

$$\begin{pmatrix} 1 & \alpha^{-2r} & \ldots & \alpha^{-(n-1)2r} \\ \ldots & \ldots & \ldots & \ldots \\ 1 & \alpha^{-1} & \ldots & \alpha^{-(n-1)} \end{pmatrix}.$$

Зная проверочную матрицу, легко получить следующую оценку для числа проверочных символов кода.

Код длины $n \leqslant 2^m$ имеет не больше $m \deg g(z)$ проверочных символов.

§ 4. Корректирующая способность кодов

Так как для кодовых многочленов $f'(z) \equiv 0 \bmod g(z)$ и $f'(z) \equiv 0 \bmod \bar{g}(z)$, где $\bar{g}(z)$ — многочлен минимальной степени, являющийся полным квадратом и такой, что $g(z) \mid \bar{g}(z)$, то $\deg f(z) \geqslant \deg \bar{g}(z) + 1$, так что для веса кода получается оценка

$$d \geqslant \deg \bar{g}(z) + 1.$$

Если все корни $g(z)$ различны, то $\bar{g}(z) = g^2(z)$ и

$$d \geqslant 2 \deg g(z) + 1.$$

§ 5. Декодирование

Пусть $y = x + e$; $y, x, e \in S$; x — переданный кодовый вектор, e — вектор ошибки. При переходе к функциям $R_y(z)$, $R_x(z)$ и $R_e(z)$, соответствующим y, x и e, получаем

$$\frac{f_y'(z)}{f_y(z)} = \frac{f_x'(z)}{f_x(z)} + \frac{f_e'(z)}{f_e(z)},$$

а так как $\frac{f_x'(z)}{f_x(z)} \equiv 0 \bmod \bar{g}(z)$, имеем

$$\frac{f_e'}{f_e} \equiv \frac{f_y'}{f_y} \bmod \bar{g}(z).$$

Величина $\theta(z) = \frac{f_y'}{f_y} \bmod \bar{g}(z)$ — синдром, т. е. результат умножения вектора y на проверочную матрицу кода. Если единицы в векторе y расположены на местах $i_1, \ldots, i_k$, то

$$\theta(z) = \frac{\bar{g}(z) - \bar{g}(\alpha_{i_1})}{z - \alpha_{i_1}} \bar{g}^{-1}(\alpha_{i_1}) + \ldots + \frac{\bar{g}(z) - \bar{g}(\alpha_{i_k})}{z - \alpha_{i_k}} \bar{g}^{-1}(\alpha_{i_k}).$$

Искомый многочлен ошибки f_e определяется по синдрому θ из сравнения

$$f_e' \equiv f_e \theta \bmod \bar{g}(z). \qquad (1)$$

Пусть $\deg g(z) = 2t$, $f'/f \equiv \theta \bmod \bar{g}(z)$, $\deg f \leqslant t$ и корни f лежат в L. Если φ — некоторое другое решение сравнения (1), причем $\deg \varphi \leqslant t$, то

$$\varphi' \equiv \varphi\theta \bmod \bar{g}, \quad \varphi' \equiv \varphi \frac{f'}{f} \bmod \bar{g}, \quad (\varphi f)' \equiv 0 \bmod \bar{g}.$$

Так как $(\varphi f)'$ имеет степень $< 2t$, то $(\varphi f)' = 0$, и поскольку f' и f взаимно просты, то $f = \gamma\varphi$, $\gamma \in GF(2^m)$. Таким образом, f_e можно искать по данному θ как единственное (с точностью до постоянного множителя) решение сравнения (1) в виде многочлена степени $\leqslant t$.

Ограничимся рассмотрением случая, когда все корни $g(z)$ различны. Тогда $\bar{g}(z) = g^2(z)$, $\deg g(z) = t$, и f_e можно находить, решая сравнение

$$f_e' \equiv f_e \theta \bmod g. \qquad (2)$$

Это сравнение также имеет единственное решение f_e, $\deg f_e \leqslant t$, так как из $(\varphi f)' \equiv 0 \bmod g$ следует $(\varphi f)' \equiv 0 \bmod g^2$. Ищем решение в виде $f_e = 1 + zu$, $\deg u < t$. Для определения u имеем линейное дифференциальное уравнение в алгебре многочленов по $\bmod g$

Reprinted with permission from *Probl. Peredach. Inform.*, vol. 6, no. 3, pp. 24–30, 1970.

$$u'z+u(1+z\theta)=\theta$$

и. и в операторной записи: $(M+T_\theta)u=\theta$, где M — линейный оператор в алгебре многочленов по $\mathrm{mod}\,g(z)$, проектирующий многочлен на его нечетную часть $Mu=u'z$. В базисе $1, z, \ldots, z^{t-1}$ матрица этого оператора имеет вид

$$M=\begin{pmatrix}0&0&0&0&0\\0&1&0&0&0\\0&0&0&0&0\\0&0&0&1&0\\0&0&0&0&0\end{pmatrix}.$$

T_θ — линейный оператор умножения на элемент $(1+z\theta)$ в той же алгебре. Матрица T_θ имеет вид

$$T_\theta=\begin{pmatrix}c_{00}&\ldots&c_{t-1,0}\\ \cdot&\cdots&\cdot\\ \cdot&\cdots&\cdot\\ \cdot&\cdots&\cdot\\ c_{0,t-1}&\ldots&c_{t-1,t-1}\end{pmatrix},\ \text{где}\ (1+z\theta)z^i\equiv(c_{i0}+\ldots+c_{i,t-1}z^{t-1})\,\mathrm{mod}\,g.$$

Из доказанной единственности решения сравнения (2) следует, что матрица $(M+T_\theta)$ невырождена в случае, когда $f_e(z)$ не имеет нулевого корня. В случае, когда ошибка произошла на месте, соответствующем $\alpha_n=0$, существует решение однородного уравнения $(M+T_\theta)u=0$, т. е. матрица $(M+T_\theta)$ оказывается вырожденной. В этом случае следует исправить одну ошибку (заменить символ на месте α_n), найти вновь синдром θ и решить новую систему с невырожденной матрицей.

Таким образом, получается следующий алгоритм декодирования: 1) найти синдром $\theta(z)$; 2) вычислить $(1+z\theta)z^i$, $i=0, 1, \ldots, t-1$ в алгебре многочленов по $\mathrm{mod}\,g(z)$; 3) построить матрицу $(M+T_\theta)$; 4) если она оказывается вырожденной, то положение одной ошибки известно; исправить ее и перейти к п. 1); 5) в случае невырожденности матрицы решить систему уравнений $(M+T_\theta)u=\theta$; 6) найти корни многочлена $f=1+zu$.

§ 6. Пример кода

Построим код $(16, 8, 5)$, исправляющий все двойные ошибки. В этом случае $m=4$, $t=2$. По таблицам неприводимых многочленов находим, что второй старший коэффициент минимального многочлена для α^3 (α — примитивный элемент $GF(2^4)$) равен 1. Следовательно, $Tr\alpha^3\neq 0$, и многочлен $g(z)=z^2+z+\alpha^3$ неприводим над $GF(2^4)$ [2]. Выберем его в качестве порождающего многочлена кода. Проверочная матрица состоит из двумерных столбцов $\begin{pmatrix}a_{1k}\\a_{0k}\end{pmatrix}$, где

$$a_{0k}+a_{1k}z=\frac{g(z)-g(\alpha_k)}{z-\alpha_k}g^{-1}(\alpha_k)=(z+1+\alpha_k)\frac{1}{\alpha_k^2+\alpha_k+\alpha^3}.$$

Подставляя вместо α_k все элементы $GF(2^4)$, получаем матрицу

γ	α	α^2	α^3	α^4	α^5	α^6	α^7	α^8	α^9	α^{10}	α^{11}	α^{12}	α^{13}	α^{14}	α^{15}	0
a_1	α^4	α^3	α^9	α^4	α	α^8	α^6	α^3	α^6	α	α^2	α^2	α^8	α^9	α^{12}	α^{12}
a_0	α^8	α^{11}	α^8	α^5	α^{11}	α^6	1	α^5	α^{13}	α^6	α^{14}	α^{13}	α^{14}	α^{12}	0	α^{12}.

В таком виде матрицу будем использовать для декодирования. Для кодирования эту матрицу нужно разложить над полем $GF(2)$ и привести к каноническому виду

α	α^2	α^3	α^4	α^5	α^6	α^7	α^8	α^9	α^{10}	α^{11}	α^{12}	α^{13}	α^{14}	α^{15}	0
1	0	0	1	0	1	0	0	0	0	0	0	1	0	1	1
1	0	0	0	0	1	0	0	0	1	1	1	1	1	1	0
0	0	0	1	0	1	1	0	0	0	0	1	1	0	0	0
0	0	0	1	0	1	0	1	0	0	1	0	1	1	1	0
1	0	1	0	0	1	0	0	0	0	0	0	0	1	1	0
0	0	0	0	1	1	0	0	0	0	1	1	0	1	0	0
1	0	0	0	0	1	0	0	1	0	1	0	1	0	0	0
1	1	0	1	0	1	0	0	0	0	0	1	0	1	0	0

С помощью последней матрицы находим кодовый вектор 0 0 1 0 0 0 0 1 0 1 0 0 0 0 1 1 с единицами на позициях 3, 8, 10, 15, 16. Ему соответствует многочлен

$$f(z)=(z-\alpha^3)(z-\alpha^8)(z-\alpha^{10})(z-1)z=z^5+\alpha^7z^4+z^3+\alpha^{10}z^2+\alpha^6z.$$

Производная этого многочлена $f'(z)=z^4+z^2+\alpha^6=g^2(z)$. Следовательно, это действительно кодовый многочлен.

Допустим, что произошла одна ошибка на 5-й позиции, т. е. на позиции, соответствующей α^5. Умножая вектор

1) 0 0 1 0 1 0 0 1 0 1 0 0 0 0 1 1 на проверочную матрицу (над полем $GF(2^4)$), получаем синдром $\theta(z)=\alpha^{11}+\alpha z$.

2) Находим $1+z\theta=1+\alpha^{11}z+\alpha z^2=1+\alpha^{11}z+\alpha(z+\alpha^3)=\alpha+\alpha^6z$; $(1+z\theta)z=\alpha^9+\alpha^{11}z$.

3) Матрицы T_θ, M имеют вид

$$T_\theta=\begin{pmatrix}\alpha&\alpha^9\\\alpha^6&\alpha^{11}\end{pmatrix},\quad M=\begin{pmatrix}0&0\\0&1\end{pmatrix},\quad T_\theta+M=\begin{pmatrix}\alpha&\alpha^9\\\alpha^6&\alpha^{12}\end{pmatrix},\quad \theta=\begin{pmatrix}\alpha^{11}\\\alpha\end{pmatrix}.$$

4) Решая систему уравнений $(T_\theta+M)u=\theta$,

$$\alpha x_1+\alpha^9x_2=\alpha^{11},\quad \alpha^6x_1+\alpha^{12}x_2=\alpha,$$

получаем $x_2=0$, $x_1=\alpha^{10}$, так что искомый многочлен $f_e(z)=1+\alpha^{10}z$.

5) Корень этого многочлена α^5 определяет положение ошибки.

Пусть теперь произошли 2 ошибки на позициях 15 и 16 (т. е. α^{15} и 0). В этом случае $\theta(z)=\alpha^{12}$, $1+z\theta=1+\alpha^{12}z$, $(1+z\theta)z=1+\alpha^{11}z$, и матрица $T_\theta+M=\begin{pmatrix}1&1\\\alpha^{12}&\alpha^{12}\end{pmatrix}$ оказывается вырожденной. Это говорит о том, что f_e имеет нулевой корень, т. е. одна ошибка произошла на 16-й позиции. После исправления этой ошибки получаем синдром $\theta=\alpha^{12}z$ и новую систему

$$\begin{pmatrix}0&1\\\alpha^{12}&\alpha^{11}\end{pmatrix}\begin{pmatrix}x_1\\x_2\end{pmatrix}=\begin{pmatrix}0\\\alpha^{12}\end{pmatrix}.$$

Решая ее, находим $x_2=0$, $x_1=1$, так что $f_e=1+z$.

§ 7. Связь с циклическими кодами

Рассмотрим случай, когда в качестве множества L выбираются все корни n-й степени из 1 над $GF(2)$: $L=\{1, \alpha, \ldots, \alpha^{n-1}\}$, α — первообразный корень уравнения $X^n-1=0$; допустим, что α порождает расширение $GF(2^m)$ поля $GF(2)$. В множестве S всех n-разрядных двоичных слов наряду со структурой аддитивной группы рассмотрим две структуры кольца:

а) с векторным умножением, при котором произведением двух элементов $x=(a_0, \ldots, a_{n-1})$ и $y=(b_0, \ldots, b_{n-1})$ является $z=(a_0b_0, \ldots, a_{n-1}b_{n-1})$ (здесь $a_i, b_i\in GF(2)$), обозначим это кольцо VS;

б) с многочленным умножением по $\mathrm{mod}(X^n-1)$, назовем это кольцо MS.

Х. Матсон и Г. Соломон в своей новой трактовке кодов БЧХ [3] использовали отображение

$$f(X)=a_0+\cdots+a_{n-1}X^{n-1}\to f(\alpha)X^{n-1}+\cdots+f(\alpha^n)=F(X),$$

которое каждому многочлену $f(X)$ над $GF(2)$ ставит в соответствие некоторый многочлен над $GF(2^m)$. Это соответствие взаимно-однозначно, причем обратное отображение совпадает с прямым. Его можно получить, построив, например, интерполяционный многочлен Лагранжа

$$f(X)=\sum_{k=0}^{n-1}\frac{X^n-1}{X-\alpha^k}\frac{f(\alpha^k)}{\alpha^{k(n-1)}}=\sum_{i=0}^{n-1}F(\alpha^{i+1})X^{n-1-i}.$$

В множестве K всех многочленов степени $<n$ над $GF(2^m)$ можно ввести такие же структуры кольца, что и в S — с векторным умножением (VK) и многочленным по mod (X^n-1) (MK). Легко проверяется, что отображение $f(X)\to F(X)$ — гомоморфизм $MS\to VK$ и $VS\to MK$. Так как все элементы VS идемпотентны, то образ множества S при этом отображении состоит из идемпотентов кольца MK. Наоборот, любой идемпотент MK принимает значения 0 или 1 на всех корнях n-й степени из 1. Обозначим через E множество идемпотентов кольца MK. Оно является подкольцом MK (назовем его ME) и в то же время подкольцом VK (назовем его VE). Следовательно, отображение $f(X)\to F(X)$ есть изоморфизм

$$MS\rightleftarrows VE,\quad VS\rightleftarrows ME.$$

Пользуясь этим изоморфизмом, можно определять коды как некоторые подмножества E. Линейные коды — это аддитивные подгруппы E, циклические коды — идеалы кольца VE. Каждый идеал VE — множество многочленов, у которых коэффициенты при некоторых степенях $X^{i_1}, \ldots, X^{i_k}$ равны 0. Например, код БЧХ — это идеал в VE, состоящий из многочленов, у которых или l старших коэффициентов, или l младших равны 0.

Пусть $y=(a_0, \ldots, a_{n-1})\in S$. Справедлива следующая диаграмма, устанавливающая связь между отображением $S\to E$ и $y\to R_y(X)$:

$$y=(a_0\ldots a_{n-1})\begin{matrix}\nearrow\ f(X)=a_0+\ldots+a_{n-1}X^{n-1}\to F(X)=f(\alpha)X^{n-1}+\ldots+f(\alpha^n)\\ \downarrow\\ \searrow\ R_y(X)=\dfrac{a_0}{X-\alpha^0}+\ldots+\dfrac{a_{n-1}}{X-\alpha^{n-1}}\to\{R_y(X)(X^{n+1}+X)\}_m,\end{matrix}$$

где $\{R_y(X)(X^{n+1}+X)\}_m$ означает остаток от деления $R_y(X)(X^{n+1}+X)$ на X^n-1, а вертикальная стрелка — тождественное отображение.

Приведенная диаграмма позволяет установить следующую симметрию между циклическими кодами и кодами, описанными в этой работе.

Циклический код — это множество многочленов, кратных некоторому фиксированному многочлену над $GF(2)$ в пространстве MS. Код, определяемый сравнением $R_x(z)\equiv 0\ \mathrm{mod}\,g(z)$, в случае, когда L является множеством корней многочлена X^n-1, совпадает с множеством многочленов, кратных фиксированному многочлену над $GF(2^m)$ в пространстве ME.

Теорема. *Если код, определяемый условием $R_x(z)\equiv 0\,\mathrm{mod}\,g(z)$ — циклический, то он является кодом БЧХ, т. е. $g(z)=z^l$.*

Доказательство. Допустим, что $g(z)$ имеет ненулевой корень β и порождает циклический код. В пространстве VE этому коду соответствует некоторый идеал C. По определению идеалов кольца VE, если $F(X)\in C$ то и $F(\alpha^iX)\in C$ для всех $i=0, \ldots, n-1$, поэтому $F(\alpha^iX)\equiv 0\,\mathrm{mod}\,g(X)$, $i=0, 1, \ldots, n-1$, откуда $F(X)\equiv 0\,\mathrm{mod}\,g(\alpha^{-i}X)$ и $F(X)$ вместе с корнем $\beta\neq 0$ должен иметь n ненулевых корней, т. е. делиться на X^n-1, что невозможно.

§ 8. Заключение

В настоящей статье описаны только двоичные коды. Обобщение на недвоичный случай и некоторые другие результаты, полученные пока статья находилась в печати, предполагается опубликовать в дальнейшем.

Данная работа обсуждалась на семинаре по теории кодирования при

МГУ. Пользуюсь случаем выразить признательность всем лицам, принявшим участие в обсуждении, в результате которого был устранен ряд неточностей.

ЛИТЕРАТУРА

1. Гантмахер Ф. Р., Крейн М. Г. Осцилляционные матрицы и ядра и малые колебания механических систем. М., Гостехиздат, 1950.
2. Ленг С. Алгебра. М., «Мир», 1968.
3. Матсон Х., Соломон Г. Новая трактовка кодов Боуза — Чоудхури. Сб. «Теория кодирования». М., «Мир», 1964.

Поступила в редакцию
28 апреля 1969 г.

РАЦИОНАЛЬНОЕ ПРЕДСТАВЛЕНИЕ КОДОВ И (L,g)-КОДЫ

В. Д. Гоппа

Описан метод построения корректирующих кодов и класс линейных q-ичных кодов.

§ 1. Введение

Настоящая работа является обобщением результатов, изложенных автором в [1]. В начале статьи описан метод построения корректирующих кодов, позволяющий трактовать линейный q-ичный код веса d как некоторое множество дробно-рациональных функций степени $\geqslant d$. Возможности метода иллюстрируются построением широкого класса кодов, названных (L, g)-кодами. Коды имеют следующие параметры: $n \leqslant q^m$, $k \geqslant n - 2mt$, $d \geqslant 2t+1$. Декодирование всех таких кодов не сложнее, чем у кодов Боуза — Чоудхури — Хоквингема (БЧХ); для некоторых (L, g)-кодов существуют и более простые схемы декодирования. В заключение рассмотрены некоторые частные случаи (L, g)-кодов и исследовано их поведение при $n \to \infty$.

§ 2. Рациональное представление кодов

1. Основной изоморфизм. Пусть $q = p^l$, p — простое число, $L = \{\alpha_1, \alpha_2, \ldots, \alpha_n\}$, $n \leqslant q^m$, $\alpha_i \in GF(q^m)$, $\alpha_i \neq \alpha_j$, причем $GF(q^m)$ — минимальное поле, содержащее L, S — векторное пространство над $GF(q)$ размерности n, S_n — метрическое пространство S с нормой Хемминга, R — векторное пространство рациональных функций над $GF(q^m)$ вида

$$\xi(z) = \sum_{i=1}^{n} \frac{b_i}{z - \alpha_i}, \; b_i \in GF(q), \; \alpha_i \in L. \tag{1}$$

Введем норму в R следующим образом: для любой неприводимой дроби

$$\xi(z) = \frac{\psi(z)}{\varphi(z)} \in R, \; \xi \neq 0, \; \|\xi(z)\| = \deg \varphi(z). \tag{2}$$

Обозначим через R_d пространство R, метризованное с помощью введенной нормы. Пусть $x \in S$, $x = (b_1, b_2, \ldots, b_n)$.

Отображение $x \to \xi_x(z) = \sum_{i=1}^{n} b_i/(z - \alpha_i)$ есть изоморфизм S на R. Так как $\deg \xi_x(z) = \|x\|$, то это отображение — изометрия S_n на R_d, т. е. $S_n \approx \approx R_d$. Данный изоморфизм позволяет определять коды как некоторые подмножества R. В частности, *линейный код веса d — это линейное подпространство R, состоящее из дробей степени не меньше d.*

2. Характеризация R для простого q. Если q — простое число, то $GF(q)$ совпадает с полем вычетов Z_q целых чисел по модулю q, и полезной оказывается следующая характеризация элементов R.

Каждому вектору $x \in S$, $x = (b_1, b_2, \ldots, b_n)$, поставим в соответствие многочлен $f_x(z) = (z - \alpha_1)^{b_1}(z-\alpha_2)^{b_2}\ldots(z-\alpha_n)^{b_n}$. Тогда любая дробь $\xi \in R$ может быть представлена в виде $\xi = f_x' / f_x$, где $x \in S$, f_x' — формальная производная многочлена f_x.

3. Построение двоичных кодов. Для построения двоичных кодов может быть использована более общая конструкция. Пусть S — векторное пространство размерности n над $GF(2)$, K — область целостности с единицей, L — множество n различных элементов K: $L = \{\alpha_1, \alpha_2, \ldots, \alpha_n\}$, D — мультипликативный моноид многочленов над K с корнями из L, D^2 — подмоноид D, состоящий из многочленов с корнями четной кратности.

Два многочлена из D, отличающиеся множителем из D^2, будем считать эквивалентными. Фактор-моноид D / D^2 по этому отношению эквивалентности является мультипликативной группой. В качестве представителей классов эквивалентности будем выбирать многочлены наименьшей степени из класса. Такой представитель может быть получен из любого многочлена $f \in D$ приведением по модулю 2 всех кратностей корней этого многочлена.

Пусть $x \in S$. Соответствие $x \to f_x(z)$ является изоморфизмом аддитивной группы S на мультипликативную группу D / D^2: $S \approx D/D^2$. При этом $\deg f_x(z) = d$, если вес x равен d. Пользуясь этим изоморфизмом, можно определить линейный двоичный код веса d как мультипликативную подгруппу группы D / D^2, состоящую из многочленов степени не меньше d.

Для задания такого кода достаточно выделить такой моноид F, что $D^2 \subset F \subset D$. Тогда линейный код C будет равен F / D^2. Отметим широкие возможности такого метода построения кодов: двоичные коды определяются с помощью многочленов над произвольной областью целостности (в частности, могут быть использованы вещественные многочлены), а вес кодового вектора определяется по степени многочлена.

Если выбрать $K = GF(2^m)$, то $f'(z) = 0$ для любого $f \in D^2$. Так как для любых $\varphi, \psi \in D / D^2$ $(\varphi\psi)' / \varphi\psi = \varphi' / \varphi + \psi' / \psi$, то в этом случае D / D^2 изоморфна аддитивной группе R функций вида φ' / φ, и мы приходим к основному изоморфизму.

§ 3. (L, g)-коды

1. Определение. Оценка веса и мощности. Воспользуемся основным изоморфизмом для определения широкого класса линейных кодов. Пусть $g(z)$ — многочлен над $GF(q^m)$, не имеющий корней в L. Для любого $\xi(z) = \psi(z)/\varphi(z) \in R$ многочлен $\varphi(z)$ взаимно прост с $g(z)$ и, следовательно, обратим в алгебре G многочленов по модулю g. *Определим (L, g)-код как множество элементов $\xi(z) \in R$ таких, что $\xi(z) \equiv 0 \bmod g(z)$.*

Линейность такого кода очевидна. Оценки веса и мощности приведены в следующих теоремах.

Теорема 1. *(L, g)-код имеет не больше $m \cdot \deg g(z)$ проверочных символов.*

Теорема 2. *Вес любого элемента (L, g)-кода не меньше* $\deg g + 1$.

Доказательство. Если $\xi \equiv 0 \bmod g$ и $\xi = \psi / f$, то $\psi \equiv 0 \bmod g$, т. е. $\deg \psi \geqslant \deg g$ и, следовательно, $\deg \xi \geqslant \deg g + 1$, что и требовалось доказать.

При $q \neq 2$ более эффективными являются (L, g)-коды, к проверочной матрице которых добавлена строка из единиц (модифицированные (L, g)-коды).

Теорема 3. *Параметры модифицированного (L, g)-кода удовлетворяют условиям $n \leqslant q^m$, $k \geqslant n - (2t-1)m - 1$, $d \geqslant 2t + 1$.*

Доказательство. Пусть $x \in S$, $x = (b_1, b_2, \ldots, b_n)$, — кодовый вектор. Тогда $\xi = \psi / \varphi \equiv 0 \bmod g$. $\sum_1^n b_i = 0$. Старший коэффициент многочлена ψ равен $\sum_1^n b_i$, поэтому $\deg \varphi \geqslant \deg \psi + 2 \geqslant \deg g + 2$.

Следовательно, для исправления t ошибок достаточно выбрать такой многочлен g, что $\deg g = 2t - 1$.

2. Проверочная матрица. В связи с (L, g)-кодами появляется новая форма проверочной матрицы, при которой проверочная матрица представляет собой строку многочленов (компактная форма проверочной матрицы). Такая форма удобна при анализе свойств кода и при построении схем декодирования. Проверочная матрица (L, g)-кодов может быть представлена в следующих формах:

1-я компактная форма:

$$T_{1\text{к}} = \left\| \frac{g(z) - g(\alpha_1)}{z - \alpha_1} g^{-1}(\alpha_1) \ldots \frac{g(z) - g(\alpha_n)}{z - \alpha_n} g^{-1}(\alpha_n) \right\|;$$

2-я компактная форма:

$$T_{2\text{к}} = \left\| \frac{z^r - \alpha_1^r}{z - \alpha_1} g^{-1}(\alpha_1) \ldots \frac{z^r - \alpha_n^r}{z - \alpha_n} g^{-1}(\alpha_n) \right\|.$$

Здесь $r = \deg g(z)$.

Развернутая форма:

$$T_{\text{р}} = \left\| \begin{matrix} g^{-1}(\alpha_1) & \cdots & g^{-1}(\alpha_n) \\ g^{-1}(\alpha_1)\alpha_1 & \cdots & g^{-1}(\alpha_n)\alpha_n \\ \cdots & \cdots & \cdots \\ g^{-1}(\alpha_1)\alpha_1^{r-1} & \cdots & g^{-1}(\alpha_n)\alpha_n^{r-1} \end{matrix} \right\|.$$

В справедливости указанных форм можно убедиться так же, как это сделано в работе [1].

Проверочная матрица (L, g)-когда получается, таким образом, из «прямоугольной матрицы Вандермонда» умножением столбцов на элементы $g^{-1}(\alpha_i)$. Можно определить более общие коды, выбирая в качестве множителей любые ненулевые элементы, не обязательно равные $g^{-1}(\alpha_i)$. Любые r столбцов полученной таким образом матрицы по-прежнему будут линейно независимы над $GF(q^m)$, поэтому вес кода будет больше r.

Выбор в качестве множителей элементов $g^{-1}(\alpha_i)$ приводит к определению (L, g)-кодов, данному в начале § 3, и позволяет применить для оценки параметров более тонкие методы, не связанные с проверочной матрицей (теоремы 1 и 2). Так, если $q = 2$ и g не имеет кратных корней, то вес любого элемента кода будет больше $2r$; проверочная матрица $T_{\text{р}}$ дает в этом случае оценку всего лишь больше r.

Доказательство. Согласно определению, (L, g)-код является ядром линейного отображения пространства R в алгебру G многочленов по модулю $g(z)$. Число проверочных символов совпадает, таким образом, с размерностью над $GF(q)$ образа R при этом отображении, а эта размерность не может быть больше $m \cdot \deg g$ — размерности самой алгебры.

Reprinted with permission from *Probl. Peredach. Inform.*, vol. 7, no. 3, pp. 41–49, 1971.

3. Декодирование. Пусть $y = x + e$; y, x, $e \in S$, x — переданный кодовый вектор, e — вектор ошибки, имеющий ненулевые координаты b_{i_1}, $b_{i_2}, \ldots, b_{i_k}$, $k \leqslant t$. Умножая вектор y на матрицу T_{1k}' (транспонированную к T_{1k}) получаем синдром $S(z) = \sum_{l=1}^{k} b_{i_l} \frac{g(z) - g(\alpha_{i_l})}{z - \alpha_{i_l}} g^{-1}(\alpha_{i_l})$. Пусть $\psi/f = -\sum_{l=1}^{k} b_{i_l}/(z - \alpha_{i_l})$. Тогда

$$fS \equiv \psi \bmod g, \ \deg f \leqslant t, \ \deg \psi < t, \tag{3}$$

и для декодирования достаточно решить сравнение (3) относительно f и ψ.

Теорема 4. *Если* $\deg g = 2t$, *то для любого многочлена* $S(z)$ *степени* $< 2t$ *множество* D *пар* (u, v) *таких, что* $uS \equiv v \bmod g$, $\deg u \leqslant t$, $\deg v < t$, *содержит ненулевой элемент. Все пары* (u, v) *из* D *имеют одно и то же отношение* $v / u = \eta$.

Доказательство. Пусть A — линейный оператор умножения на S в алгебре вычетов многочленов по модулю $g(z)$. В базисе $1, z, \ldots, z^{2t-1}$ матрица этого оператора имеет вид

$$A = \left\| \begin{matrix} c_{2t-1\,0} & \cdots & c_{2t-1\,2t-1} \\ \cdot & \cdots & \cdot \\ c_{00} & \cdots & c_{0\,2t-1} \end{matrix} \right\|, \quad \text{где } z^j S(z) = \sum_{i=0}^{2t-1} c_{ij} z^i, \ j = 0, 1, \ldots, 2t-1.$$

Обозначим через $A_{t,\,t+1}$ подматрицу матрицы A, составленную из t верхних строк и первых $t+1$ столбцов. Уравнение

$$A_{t,\,t+1} x = 0 \tag{4}$$

всегда имеет решение, откуда следует существование u, удовлетворяющего условию теоремы, и способ его нахождения.

Пусть теперь (u_1, v_1), $(u_2, v_2) \in D$. Тогда $u_1 \bar{S} \equiv \bar{v}_1 \bmod \bar{g}$, $\bar{v}_2 \equiv \bar{S} u_2 \bmod \bar{g}$, где многочлены $\bar{S}$, $\bar{v}_1$, $\bar{v}_2$, $\bar{g}$ получены из S, v_1, v_2, g делением на наибольший общий делитель S и g. Так как $\bar{S}$ и $\bar{g}$ взаимно просты, то $u_1 \bar{v}_2 \equiv \bar{v}_1 u_2 \bmod \bar{g}$, $u_1 \bar{v}_2 - \bar{v}_1 u_2 \equiv 0 \bmod \bar{g}$, и так как степень многочлена в левой части меньше степени $\bar{g}$, то $\bar{v}_1 / u_1 = \bar{v}_2 / u_2$, так что $v_1 / u_1 = v_2 / u_2$, что и требовалось доказать.

Из этой теоремы вытекает следующее правило для декодирования (L, g)-кодов: 1) найти синдром $S(z)$; 2) найти какое-либо решение u системы (4); 3) вычислить $v \equiv u \cdot S \bmod g$; 4) найти $\eta = v / u$ (это отношение будет совпадать в данном случае с $\frac{\psi}{f}$ из сравнения (3), так как f и ψ взаимно просты); 5) разложить η на простые дроби (для этого нужно найти корни $f(z)$ и вычислить $c_{i_l} = \psi(\alpha_{i_l}) / f'(\alpha_{i_l})$); 6) определить значения ошибок: $b_{i_l} = -c_{i_l}$.

Другой способ декодирования связан с использованием проверочной матрицы T_{2k}. При умножении вектора y на матрицу T_{2k}' получаем синдром $S(z) = \sum b_{i_l} \frac{z^r - \alpha_{i_l}^r}{z - \alpha_{i_l}} g^{-1}(\alpha_{i_l})$. Пусть

$$\frac{\psi}{f} = -\sum_{l=1}^{k} b_{i_l} \frac{\alpha_{i_l}^r}{z - \alpha_{i_l}} g^{-1}(\alpha_{i_l}).$$

Тогда декодирование сводится к решению сравнения

$$f \cdot S \equiv \psi \bmod z^r \tag{5}$$

при $\alpha_{i_l} \neq 0$, $l = 1, 2, \ldots, k$, или сравнения

$$f \cdot S \equiv \psi \bmod z^{r-1} \tag{6}$$

при $\alpha_{i_j} = 0$ для некоторого j.

В первом случае $\deg f = k$, во втором — $\deg f = k - 1$, так что при $r = 2t$ согласно теореме 4 существует единственное решение (f, ψ) сравнения (5), где f и ψ взаимно просты, $\deg f \leqslant t$, $\deg \psi < t$, или сравнения (6), где f и ψ взаимно просты, $\deg f \leqslant t - 1$, $\deg \psi < t - 1$.

Берлекемпом [2] была предложена итеративная процедура декодирования кодов БЧХ, которая по существу является методом решения сравнения (5). Процедура заключается в последовательном переходе от сравнения по модулю z^2 к сравнению по модулю z^3, затем к сравнению по модулю z^4 и т. д. Алгоритм Берлекемпа приводит к существенной экономии оборудования, так как не требует запоминания матрицы $A_{t,\,t+1}$.

5. Некоторые специальные (L, g)-коды. Далее будут рассмотрены некоторые частные (L, g)-коды, связанные между собой следующим образом:

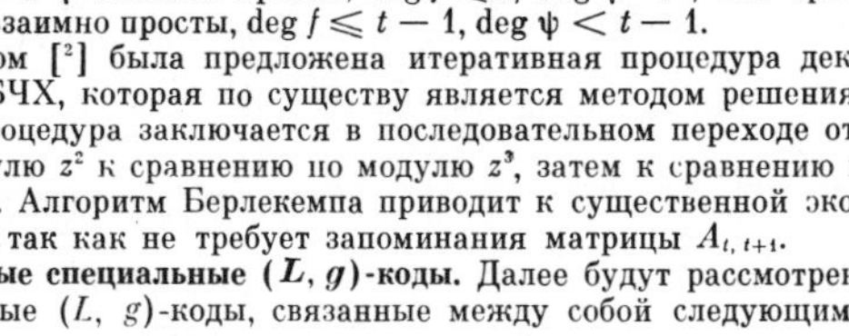

Классификация проводится по виду порождающего многочлена. Кумулятивные коды — это коды типа кодов БЧХ. Сепарабельные коды обладают своеобразной схемой декодирования. Для обоих этих классов кодов улучшаются оценки параметров в двоичном случае: $n \leqslant 2^m$, $k \geqslant n - mt$, $d \geqslant 2t + 1$.

Коды Сриваставы и коды Габидулина были получены в 1967 г. Неприводимые коды обладают самой «сильной» алгебраической структурой, и создается впечатление, что в исследовании свойств этих кодов можно продвинуться достаточно далеко.

§ 4. Кумулятивные коды

Кумулятивные коды — это коды, порождающий многочлен которых имеет один корень: $g(z) = (z - \alpha)^r$.

Максимальное L, которое можно выбрать для таких кодов, $GF(q^m) - \{\alpha\}$, так что их максимальная длина равна $q^m - 1$. Частный случай таких кодов при $g(z) = z^r$ — коды БЧХ (при таком выборе $g(z)$ проверочная матрица T_p превращается в известную проверочную матрицу кода БЧХ).

Теорема 5. *Все кумулятивные коды с одним и тем же* r *имеют одинаковый спектр.*

Доказательство. Если x — элемент $(GF(q^m) - \{\alpha\}, (z - \alpha)^r)$-кода, то $\xi_x(z) = \frac{\psi(z)}{\varphi(z)} \equiv 0 \bmod (z - \alpha)^r$, так что

$$\frac{\psi(z + \alpha)}{\varphi(z + \alpha)} \equiv 0 \bmod z^r.$$

Если $\varphi(z)$ имеет корни $\alpha_{i_1}, \ldots, \alpha_{i_l}$, то $\varphi(z + \alpha)$ имеет корни $\alpha_{i_1} - \alpha, \ldots, \alpha_{i_l} - \alpha$, лежащие в $GF(q^m) - \{0\}$. Таким образом, существует взаимно-однозначное соответствие, сохраняющее вес, между любым кумулятивным кодом и БЧХ-кодом.

Теорема 6. *БЧХ-коды — единственные циклические* (L, g)*-коды с* $L = \{1, \alpha, \alpha^2, \ldots, \alpha^{n-1}\}$, *где* α *— первообразный корень* n*-й степени из единицы.*

Доказательство. При указанном выборе L рациональное представление позволяет следующим образом охарактеризовать циклические коды: циклический код — это множество дробей из R (§ 2), образующих линейное пространство и замкнутое относительно подстановки $z \to \alpha z$, $\alpha \in L$: $C = \{\xi(z) \in R : \xi(\alpha z) \in C, \ a \in L\}$. Предположим, что порождающий многочлен $g(z)$ циклического (L, g)-кода имеет ненулевой корень β. Для любого $\xi(z)$ из этого кода выполняется сравнение

$$\xi(\alpha z) = \frac{\psi(\alpha z)}{\varphi(\alpha z)} \equiv 0 \bmod g(z)$$

для всех $\alpha \in L$; это значит, что $\psi(z)$ вместе с корнем β имеет все корни вида $\alpha\beta$, т. е. $\deg \psi(z) \geqslant n$, что невозможно. Следовательно, $g(z) = z^r$.

§ 5. Сепарабельные коды

1. Определение. Проверочная матрица. *Сепарабельные коды — это коды, порождающий многочлен которых не имеет кратных корней*:

$$g(z) = (z - z_1)(z - z_2) \ldots (z - z_r).$$

Двоичные сепарабельные коды, так же как и коды БЧХ, допускают улучшение оценки своих параметров. Действительно, согласно п. 2 § 2, $f'/f \equiv 0 \bmod g$, и так как производная в поле характеристики 2 является квадратом, то $f'/f \equiv 0 \bmod g^2$. Следовательно, при $\deg g = t$ имеем $n \leqslant 2^m$, $k \geqslant n - mt$, $d \geqslant 2t + 1$. Сепарабельные коды обладают следующей специфической формой проверочной матрицы (матрица Коши): $T_c = \| (z_i - \alpha_j)^{-1} \|$, $i = 1, 2, \ldots, r$, $j = 1, 2, \ldots, n$, где z_i, α_j — различные элементы поля $GF(q^m)$ или некоторого его расширения. Использование проверочной матрицы T_c приводит к эффективному методу декодирования таких кодов, основанному на рациональной интерполяции.

2. Декодирование. Пусть y, x и e — те же, что и в п. 3 § 3. Умножая вектор y на матрицу T_c', получаем синдром $(\xi_1, \xi_2, \ldots, \xi_r)$, где $\xi_i = \xi(z_i)$,

$$\xi(z) = \sum_{l=1}^{k} \frac{b_{i_l}}{z - \alpha_{i_l}}. \tag{7}$$

Декодирование заключается в данном случае в восстановлении дроби $\xi(z)$ по ее значениям в некоторых точках (рациональная интерполяция [3]).

Пусть K — некоторое поле, $\xi(z) = \psi(z) / f(z)$ — неприводимая дробь над полем K, $\deg f = n$, $\deg \psi \neq m$. *Порядком дроби* называется величина $r(\xi)$, равная $2n$ при $m \leqslant n$ и $2n - 1$ при $m > n$. Любая неприводимая рациональная дробь порядка $\leqslant k$ может быть однозначно восстановлена по своим значениям в $k + 1$ точке поля K.

Теорема 7. *Пусть* $z_1, z_2, \ldots, z_{k+1}$ *— различные элементы поля* K. *Каждая неприводимая дробь* $\frac{\psi(z)}{f(z)} = \xi(z) \in K(z)$ *порядка* $r \leqslant k$ *может быть представлена в виде*

$$\xi(z) = \lambda_1 + \cfrac{z - z_{i_1}}{\lambda_2 + \cdots + \cfrac{z - z_{i_{r-1}}}{\lambda_r}}, \tag{8}$$

$$z_{i_1} \neq z_{i_2} \neq \ldots \neq z_{i_{r-1}}.$$

Доказательство. Среди элементов $z_1, z_2, \ldots, z_{k+1}$ найдется элемент z_{i_1} такой, что $f(z_{i_1}) \neq 0$. Положим $\lambda_1 = \xi(z_{i_1})$. Тогда

$$\xi(z) = \lambda_1 + \frac{z - z_{i_1}}{\xi_1(z)}, \ \text{где } \xi_1(z) = \frac{f(z)}{\psi_1(z)}, \ \psi_1(z) = \frac{\psi(z) - \lambda_1 f(z)}{z - z_{i_1}}.$$

Если $m \leqslant n$, то $r(\xi) = 2n$, $\deg \psi < n$, $r(\xi_1) = 2n - 1$. Если $m > n$, то $r(\xi) = 2m - 1$, $\deg \psi_1 = m - 1$, $r(\xi_1) = 2m - 2$. Следовательно, во всех случаях $r(\xi_1) = r(\xi) - 1$. Применяя предыдущие рассуждения к дроби $\xi_1(z)$, затем к дроби $\xi_2(z)$ и т. д., получим разложение (8) ровно за r шагов.

Величины $\lambda_1, \lambda_2, \ldots, \lambda_r$ в разложении (8) определяются последовательной подстановкой $z = z_{i_1}$, $z = z_{i_2}$ и т. д. Вычисления удобно проводить по схеме

$$\begin{matrix} z_1 & y_1^{(1)} & & & & \\ z_2 & y_2^{(1)} & y_1^{(2)} & & & \\ z_3 & y_3^{(1)} & y_2^{(2)} & y_1^{(3)} & & \\ z_4 & y_4^{(1)} & y_3^{(2)} & y_2^{(3)} & y_1^{(4)} & \\ z_5 & y_5^{(1)} & y_4^{(2)} & y_3^{(3)} & y_2^{(4)} & y_1^{(5)} \end{matrix} \tag{9}$$

Здесь

$$y_i^{(1)} = y_i = \xi(z_{i-1}),$$

$$y_i^{(j)} = \frac{z_{i+j-1} - z_{j-1}}{y_{i+1}^{(j-1)} - y_1^{(j-1)}}, \quad j > 1.$$

Если $y_{i+1}^{(j)} = y_1^{(j)}$ при некотором $i \neq 0$, то $(i+1)$-ю строку следует исключить из (9). Величины $\lambda_1, \lambda_2, \ldots, \lambda_r$ — это верхняя диагональ в (9): $\lambda_j = y_1^{(j)}$. Вместо (9) можно построить обратные разности, а затем по ним определить λ_j:

$$\begin{array}{llll} z_1\ y_1 & & & \\ & \rho_1(z_2z_1) & & \\ z_2\ y_2 & & \rho_2(z_3z_2z_1) & \\ & \rho_1(z_3z_2) & & \rho_3(z_4z_3z_2z_1) \\ z_3\ y_3 & & \rho_2(z_4z_3z_2) & \\ & \rho_1(z_4z_3) & & \\ z_4\ y_4 & & & \end{array} \qquad (10)$$

Здесь $\rho_1(z_2z_1) = \dfrac{z_2 - z_1}{y_2 - y_1}$,

$$\rho_2(z_3z_2z_1) = \frac{z_3 - z_1}{\rho_1(z_3z_2) - \rho_1(z_2z_1)} + y_2 \quad \text{и т. д.}$$

По обратным разностям величины λ_j определяются следующим образом:

$$\lambda_j = \rho_j(z_{j+1}z_j \ldots z_1) - \rho_{j-2}(z_{j-1}z_{j-2} \ldots z_1)$$

для $j > 2$ и $\lambda_1 = y_1$, $\lambda_2 = \rho_1(z_2z_1)$. Обратные разности $\rho_j(z_{j+1} \ldots z_1)$ симметричны относительно всех переменных.

Если $q \neq 2$, $\deg g = 2t$, $k \leqslant t$, то дробь $\xi(z)$ (7) имеет порядок $\leqslant 2t$ и однозначно восстанавливается по своим значениям в $2t+1$ узлах. Если $q = 2$, $\deg g = t$ $k \leqslant t$, то следует выполнить такие преобразования:

$$f(z) = u^2(z) + zv^2(z), \quad v^2 = f', \quad f(z_i)\xi_i = f'(z_i),$$

$$u^2(z_i)\xi_i = v^2(z_i)(1 + z_i\xi_i).$$

Если $\xi_i = 0$, то z_i является корнем $v^2(z)$, а если $1 + z_i\xi_i = 0$, то z_i — корень $u^2(z)$. Зная значения $y_i = u(z_i) / v(z_i) = \sqrt{z_i + \xi^{-1}}$ в тех точках z_i, где $\xi_i \neq 0$, можем однозначно восстановить $\dfrac{u(z)}{v(z)}$, представив ее в виде непрерывной дроби (8).

3. Коды Сриваставы — *это коды, порождающий многочлен которых распадается в минимальном поле, содержащем L.* Такие коды описаны в [2]. Там же упоминается схема их декодирования, основанная на рациональной интерполяции. Для кодов Сриваставы справедливы следующие оценки:

$$n \leqslant q^m - 2t, \quad k \geqslant n - 2mt, \quad d \geqslant 2t + 1.$$

4. Коды Габидулина. Частный случай кодов Сриваставы, получающийся при $m = 1$, был независимо получен Э. М. Габидулиным [4], который заметил, кроме того, что если к матрице Коши (в случае, когда $L \subset GF(g)$) присоединить единичную матрицу, то вес кода не уменьшится. Максимальные коды, определяемые такой проверочной матрицей, обладают более высокой скоростью передачи, чем коды Рида — Соломона: $n = q$, $k = n - 2t$, $d = 2t + 1$.

5. Неприводимые коды. *Сепарабельный код назовем неприводимым, если многочлен $g(z)$ неприводим над минимальным полем, содержащим L.*

Проверочная матрица неприводимого кода состоит из строки:

$$T_n = ((z_0 - \alpha_1)^{-1}(z_0 - \alpha_2)^{-1} \ldots (z_0 - \alpha_n)^{-1}),$$

где z_0 — корень $g(z)$. Значение некоторой рациональной функции над $GF(q^m)$ в точке $z_0 \in GF(q^{mr})$, определяет значение этой функции в точках, сопряженных с z_0 над полем $GF(q^m)$:

$$\xi(z_0) = y_0, \quad \xi(\sigma z_0) = \sigma y_0, \ldots, \quad \xi(\sigma^{r-1}z_0) = \sigma^{r-1}y_0.$$

Все обратные разности одного и того же порядка, вычисленные по узлам $z_0, \sigma z_0, \ldots, \sigma^{r-1}z_0$, будут сопряжены:

$$\begin{array}{llll} z_0\ y_0 & & & \\ & \rho_1 & & \\ \sigma z_0\ \sigma y_0 & & \rho_2 & \\ & \sigma\rho_1 & & \rho_3 \\ \sigma^2 z_0\ \sigma^2 y_0 & & \sigma\rho_2 & \\ & \sigma^2\rho_1 & & \\ \sigma^3 z_0\ \sigma^3 y_0 & & & \end{array}$$

Следовательно, для декодирования неприводимых кодов нет необходимости строить всю таблицу обратных разностей. По-видимому, наиболее просто осуществить поэтапное декодирование таких кодов. При этом число ошибок по синдрому y_0 определяется по следующему алгоритму (здесь σ — образующая группы Галуа поля $GF(q^{mr})$ над полем $GF(q^m)$, а обозначение $A := B$ имеет тот же смысл, что и в АЛГОЛе: A присвоить значение, равное B): *при* $q \neq 2$: 1° *положить* $A := z_0$; $B := z_0$; $C := 0$; $D := y_0$; $N := 1$; 2° *вычислить* $E := \sigma D - D$; 3° *если* $E := 0$, *то перейти к* 5°; 4° положить $B := \sigma B$; $E := (B - A)/(E + C)$; $C := \sigma D$; $D := E$; $N := N + 1$; *перейти к* 2°; 5° *выдать значение числа ошибок, равное* $N/2$; *при* $q = 2$: 1° *положить* $A := B := z_0$; $C = 0$; $D := y_0$; $N := 1$; 2° *вычислить* $D := \sqrt{D^{-1} + A}$; 3° *вычислить* $E := \sigma D - D$; 4° *если* $E = 0$, *перейти к* 6°, 5° *положить* $B := \sigma B$; $E := (B - A)/(E + C)$; $C := \sigma D$; $D := E$; $N := N + 1$; *перейти к* 3°; 6° *число ошибок равно* N.

При реализации описанного алгоритма можно производить вычисления или в поле $GF(q^{mr})$, или в поле $GF(q^m)$. Для работы алгоритма необходимо реализовать умножение в конечном поле, нахождение обратной величины и операцию $\sigma x = x^{q^m}$.

§ 6. Поведение (L, g)-кодов при $n \to \infty$

В заключение исследуем поведение (L, g)-кодов, когда $n \to \infty$, а скорость передачи k/n остается фиксированной. Коды, у которых при этом $d/n \to c \neq 0$, принято называть «хорошими», а коды с $d/n \to 0$ — «плохими». Известно, что БЧХ-коды, так же как и все коды, инвариантные относительно аффинной группы преобразований, «плохи» [5], [6]. «Хорошие» коды можно получить, как показал В. Зяблов [7], используя каскадный принцип, открытый Д. Форни [8]. (L, g)-коды также принадлежат к разряду «хороших» кодов. Большинство из них при $n \to \infty$ лежит как угодно близко к границе Варшамова — Гилберта.

Теорема 8. *Пусть $L = GF(q^m)$, $n = q^m$, $H(x)$ — энтропия. Для любого $0 < \lambda < 1$ и $\varepsilon > 0$ вероятность того, что случайно выбранный многочлен $g(z) \in L(z)$ степени $[\lambda n / \log n]$, не имеющий корней в L, порождает (L, g)-код с параметрами $k/n > 1 - H(d/n) - \varepsilon$, стремится к 1 с ростом n.*

Доказательство. Обозначим через M_r сферу радиуса r в пространстве R_d (§ 2), т. е. множество дробей из R степени $\leqslant r$. Пусть N_r — множество всех числителей дробей из M_r, V_r — множество всех нормированных многочленов из $L(z)$ степени $t = [\lambda n / \log n]$, которые не имеют корней в L и делят какой-либо многочлен из N_r, K — множество всех нормированных многочленов из $L(z)$ степени t, не имеющих корней в L. Через $\bar{A}$ будем обозначать, как обычно, мощность множества A.

Любой многочлен из K порождает (L, g)-код $C = \{\xi(z) \in R : \xi(z) \equiv 0 \bmod g(z)\}$ со скоростью передачи $k/n \geqslant 1 - mt/n \geqslant 1 - \lambda$, причем все $g(z) \in V_r$ порождают коды веса $\leqslant r$, а все $g(z) \in K - V_r$ — коды веса $> r$. Оценим мощность всех введенных множеств:

$$\overline{M}_r = \sum_{i=0}^{r} C_n^i (q-1)^i; \quad \overline{N}_r \leqslant \overline{M}_r; \quad \overline{V}_r < \overline{N}_r \cdot C_r^t;$$

$$\overline{K} = \sum_{i=0}^{t} C_n^i (-1)^i n^{t-i} \sim e^{-1} \cdot n^t \sim e^{-1} q^{\lambda n}$$

($X(n) \sim Y(n)$ означает, как обычно, что $X(n) / Y(n) \to 1$).

Выберем μ из условия $H(\mu) = \lambda - \varepsilon$, $0 < \mu < 1/2$ и положим $r = \mu n$. Все $g(z) \in K - V_{\mu n}$ порождают (L, g)-коды с параметрами $H(d/n) > H(\mu) = \lambda - \varepsilon \geqslant 1 - k/n - \varepsilon$, а так как

$$\overline{V}_{\mu n} / \overline{K} < C_{\mu n}^{\lambda n / \log n} \cdot q^{nH(\mu)} / e^{-1} q^{n\lambda} \to 0,$$

то тем самым теорема доказана.

Заметим, что тот же результат справедлив и для более узкого класса (L, g)-кодов — неприводимых кодов.

ЛИТЕРАТУРА

1. Гоппа В. Д. Новый класс линейных корректирующих кодов. Проблемы передачи информации, 1970, **6**, *3*, 24—30.
2. Berlecamp E. Algebraic coding theory. N.-Y., Mc. Grow-Hill, 1968.
3. Милн В. Численный анализ. М., Изд. иностр. лит., 1951.
4. Габидулин Э. М. Декодирование максимальных кодов. Тр. III конф. по теории передачи и кодирования информации, Москва — Ужгород, 1967.
5. Lin S., Weldon E. J. Long BCH-Codes Are Bad. Inform. and Control, 1967, **11**, *10*, 445—451.
6. Kasami T. An Upper Bound on k/n for Affine — Invariant Codes with Fixed d/n. IEEE Trans. Inform. Theory, 1969, **15**, *1*, 174—176.
7. Зяблов В. В. Оценка сложности построения двоичных линейных каскадных кодов. Проблемы передачи информации, 1971, **7**, *1*, 5—13.
8. Forney G. D. Concatenated codes. Cambridge, Massachusetts, The MIT Press, 1966.

Поступила в редакцию
27 февраля 1970 г.

Part III: Decoding Algorithms for Block Codes

The early pioneers recognized that the obvious decoding algorithms are much too cumbersome to be implemented in practice. The first practical idea was the notion of *threshold decoding*, introduced by Reed [1]. The notion was extended to convolutional codes by Massey [2], who found, by trial and error, a much larger set of codes for which threshold decoding could be used. Rudolph [3] applied the tools of finite geometries to construct even more codes of this type. He later showed [4] that, in a certain theoretical sense, the notion of threshold decoding was much less restrictive than had previously been supposed. As mentioned in the section on block code constructions, there are now a large number of known codes based on finite geometries which may be decoded by straightforward threshold decoding algorithms. Research in this direction has yielded many new codes, as well as minor improvements in Reed's original threshold decoding algorithm.

The possibility of decoding data transmitted via linear block codes over the q-ary erasure channel *by solving simultaneous linear equations* over a finite field was first investigated by Epstein [5], who showed that the amount of computation required was only a small power of the block length. Ziv [6] later extended similar techniques to the binary symmetric channel by using a form of concatenated code, the inner code of which effectively converts the channel into an erasure channel with which the outer code can deal by solving simultaneous linear equations.

Except for *sequential decoding*, which is covered in the section on convolutional codes, the next significantly new decoding algorithm was one proposed by Gallager [7] to decode *low-density parity-check codes*. He found that although low-density parity-check matrices yield codes with slightly smaller distances, the relative independence ensured by the low-density constraint makes it possible to decode such codes by relatively straightforward methods whose complexity increases slowly with increasing block length. Recently, the Soviets have shown that, in a certain sense, the complexity of low-density parity check codes is asymptotically minimal.

The most revolutionary decoding algorithm to arise in the past 25 years is *algebraic decoding*. By associating the digits of certain linear cyclic codes with corresponding elements in a finite field, it is possible to define an *error-locator polynomial*, whose roots reveal the locations of the digits which are in error. The decoding problem is thus transformed into two separate computational problems in the finite field: 1) determine the coefficients of the error-locator polynomial, and 2) given the coefficients of the error-locator polynomial, find its roots.

In their original papers, Bose and Ray-Chaudhuri stressed the minimum distance properties of their codes. They were not concerned with decoding them, even though their proof of the lower bound on distance contained the seeds of an algebraic decoding algorithm. The details of such an algorithm were first presented by Peterson [8]. By using the fact that the BCH codes are cyclic, Chien [9] obtained a significantly better algorithm, which was modified and improved by Forney [10]. The Chien algorithm finds the roots of the error-locator polynomial by testing each candidate, but the tests are done sequentially as the about-to-be decoded digits leave the buffer. This method circumvents the computational problem of finding the roots of the error-locator polynomial so that the determination of the coefficients of the error-locator polynomial became the bottleneck of the BCH decoding algorithm. This bottleneck was broken by a new iterative algorithm presented by Berlekamp [11]. Massey [12] pointed out that this same algorithm also solves the linear feedback shift register synthesis problem. Burton [13] showed that by introducing scalar multiples of the reciprocal monic polynomials upon which Berlekamp's algorithm iterates, one can decode all binary BCH codes without doing any Galois field divisions.

In 1971, Berlekamp [14] introduced a different algorithm which simplified the problem of finding the roots in a finite field of a polynomial with given coefficients. However, buffering and timing considerations, combined with the possibility of parallel processing, often make Chien's algorithm more attractive for decoders built as special-purpose hardware.

In the past five years, there have been only minor improvements in BCH decoding algorithms, such as the modifications suggested by Hartmann [15] and van der Horst [16]. The known algorithms are quite good. More general algebraic decoding algorithms can be used to decode arbitrary Goppa codes, but at short to moderate block lengths, the fact that the non-BCH Goppa codes are generally not cyclic seems to cost more in encoding and decoding complexity than the advantages which result from a larger minimum distance. No detailed comparison has yet been published.

The use of the BCH codes on real channels is no longer

limited by the complexity of the decoders so much as by the fact that few real channels are adequately approximated by one-way memoryless Hamming-metric channels for which the BCH codes were designed. Hence, the next major problem for algebraic coding theorists is to generalize the type of channel on which algebraic codes can be effectively used. There has already been some progress in this direction, including the algebraic codes for the Lee metric introduced by Berlekamp [17]. But the most promising real channel, which occurs in deep space communications as well as other places, may be modelled by a one-way symmetric memoryless channel that has binary inputs but real-number outputs. The only progress toward extending the technique of algebraic decoding to channels of this sort is Forney's notion of "generalized minimum-distance decoding." I conclude this section with that paper because I think it likely that one of the forthcoming "classic" papers to be written in algebraic decoding will, in some sense or other, introduce a new approach to this important problem.

References

[1] I. S. Reed, "A class of multiple-error-correcting codes and the decoding scheme," *IRE Trans. Inform. Theory*, vol. IT-4, pp. 38-49, Sept. 1954; *Math. Rev.*, vol. 19, p. 721.

[2] J. L. Massey, *Threshold Decoding*, Res. Monograph 20. Cambridge, Mass.: MIT Press, 1963.

[3] L. D. Rudolph, "Generalized threshold decoding of convolutional codes," *IEEE Trans. Inform. Theory*, vol. IT-16, pp. 739-745, Nov. 1970.

[4] —, "Threshold decoding of cyclic codes," *IEEE Trans. Inform. Theory*, vol. IT-15, pp. 414-418, May 1969.

[5] M. A. Epstein, "Algebraic decoding for a binary erasure channel," M.I.T. Res. Lab. Electron. Rep. 340, 1958.

[6] J. Ziv, "Coding and decoding from time-discrete amplitude continuous memoryless channels," *IEEE Trans. Inform. Theory*, vol. IT-8, pp. S199-S205, Sept. 1962.

—, "Further results on the asymptotic complexity of an iterative coding scheme," *IEEE Trans. Inform. Theory*, vol. IT-12, pp. 168-171, Apr. 1966.

—, "Asymptotic performance and complexity of a coding scheme for memoryless channels," *IEEE Trans. Inform. Theory*, vol. IT-13, pp. 356-359, July 1967.

[7] R. G. Gallager, *Low-Density Parity-Check Codes*, Res. Monograph 21. Cambridge, Mass.: MIT Press, 1963.

[8] W. W. Peterson, "Encoding and error-correction procedures for the Bose-Chaudhuri codes," *IRE Trans. Inform. Theory*, vol. IT-6, pp. 459-470, Sept. 1960.

[9] R. T. Chien, "Cyclic decoding procedures for Bose-Chaudhuri-Hocquenghem codes," *IEEE Trans. Inform. Theory*, vol. IT-10, pp. 357-363, Oct. 1964.

[10] G. D. Forney, Jr., "On decoding BCH codes," *IEEE Trans. Inform. Theory*, vol. IT-11, pp. 549-557, Oct. 1965; *Math. Rev.*, vol. 32, p. 7341.

[11] E. R. Berlekamp, *Algebraic Coding Theory*. New York: McGraw-Hill, 1968.

[12] J. L. Massey, "Shift-register synthesis and BCH decoding," *IEEE Trans. Inform. Theory*, vol. IT-15, pp. 122-127, Jan. 1969; *Math. Rev.*, vol. 39, p. 3887.

[13] H. O. Burton, "Inversionless decoding of binary BCH codes," *IEEE Trans. Inform. Theory*, vol. IT-17, pp. 464-466, July 1971.

[14] E. R. Berlekamp, "Factoring polynomials over large finite fields," *Math. Comput.*, vol. 24, pp. 713-735, 1970.

[15] C. R. P. Hartmann, "Decoding beyond the BCH bound," *IEEE Trans. Inform. Theory* (Corresp.), vol. IT-18, pp. 441-444, May 1972.

[16] J. A. van der Horst, "Complete decoding of some binary BCH codes," Ph.D dissertation, Dep. Elec. Eng., Cornell Univ., Ithaca, N.Y., 1972.

[17] E. R. Berlekamp, "Negacyclic codes for the Lee metric," in R. C. Bose and T. A. Dowling, Ed., *Proc. Conf. Combinatorial Math. and Its Applications*. Chapel Hill, N.C.: Univ. North Carolina Press, 1968, ch. 17. Also see E. R. Berlekamp, *Algebraic Coding Theory*. New York: McGraw-Hill, 1968, ch. 9.

Encoding and Error-Correction Procedures for the Bose-Chaudhuri Codes*

W. W. PETERSON†, MEMBER, IRE

Summary—**Bose and Ray-Chaudhuri have recently described a class of binary codes which for arbitrary m and t are t-error correcting and have length $2^m - 1$ of which no more than mt digits are redundancy. This paper describes a simple error-correction procedure for these codes. Their cyclic structure is demonstrated and methods of exploiting it to implement the coding and correction procedure using shift registers are outlined. Closer bounds on the number of redundancy digits are derived.**

INTRODUCTION

BOSE and Chaudhuri[1] have recently discovered a new class of codes with some remarkable properties. For any positive integers m and t, there is a code in this class that consists of blocks of length $2^m - 1$, that corrects t errors, and that requires no more than mt parity check digits. Thus, the codes cover a wide range in rate and error-correcting ability, unlike most other known classes of codes.[2] These codes are a generalization of the Hamming codes;[3] the case $t = 1$ gives the Hamming code in each case.

In this paper two important properties of these codes are described. First, a method for error correction is described which is a generalization of the simple error-correction procedure that can be used with Hamming codes. The procedure requires a number of operations which increases only as a small power of the length of the codes.

Second, it is shown that these are cyclic codes[4] and,

* Received by the PGIT, December 6, 1959. Part of this work was supported by the U. S. Army Signal Corps, the U. S. Air Force Office of Scientific Research, Air Research and Development Command, and the U. S. Navy Office of Naval Research at the Research Laboratory of Electronics, Mass. Inst. Tech., Cambridge, Mass.; and part of the work was done at the IBM Research Lab., Yorktown, N. Y.

† On leave from the University of Florida, Gainesville. Presently at the Dept. of Elec. Engrg. and Res. Lab. of Electronics, Mass. Inst. Tech., Cambridge, Mass.

[1] R. C. Bose and D. K. Ray-Chaudhuri, "On a class of error-correcting binary group codes," to be published in *Information and Control*.

[2] The only others of which I am aware are I. S. Reed, "A class of multiple-error-correcting codes and decoding scheme," IRE TRANS. ON INFORMATION THEORY, vol. IT-4, pp. 38–49, September, 1954; P. Elias, "Error free coding," IRE TRANS. ON INFORMATION THEORY, vol. IT-4, pp. 29–37, September, 1954; and I. S. Reed and G. Solomon, "Polynomial code," to be published in *J. Soc. Ind. Appl. Math.*

[3] R. W. Hamming, "Error detecting and error correcting codes," *Bell Sys. Tech. J.*, vol. 29, pp. 147–160; April, 1950.

[4] E. Prange, "Some Cyclic Error-Correcting Codes with Simple Decoding Algorithms," Air Force Cambridge Research Center, Bedford, Mass., Tech. Note AFCRC-TN-58-156, April, 1958; "Cyclic Error-Correcting Codes in Two Symbols," Air Force Cambridge Research Center, Bedford, Mass., Tech. Note ARCRC-TN-57-103, September, 1957; "The Use of Coset Equivalence in the Analysis and Decoding of Group Codes," Air Force Cambridge Research Center, Bedford, Mass., Tech. Rept. AFCRC-TR-59-164, June, 1959.

Reprinted from *IRE Trans. Inform. Theory*, vol. IT-6, pp. 459–470, Sept. 1960.

therefore, the encoding can be accomplished very efficiently with a shift register. The theory of the cyclic structure also provides a closer bound on the number of parity checks required to correct a given number of errors.

Construction of the Bose-Chaudhuri Codes

Given an irreducible polynomial $p(X)$ of degree m with 1 and 0 as coefficients, a representation of the Galois Field with 2^m elements $GF(2^m)$ can be formed. It consists of all polynomials of degree $m - 1$ or less. They can be added (modulo 2) term by term in the ordinary way. The rule for multiplication is to multiply in the ordinary way, reducing the answer modulo 2 and modulo $p(X)$ to a polynomial of degree $m - 1$ or less. (That is, consider $p(X) = 0$, and use this equation to eliminate terms of power greater than $m - 1$.) It can be shown then that certain of these polynomials, called primitive elements, have the property that the first $2^m - 1$ powers of such an element are exactly all the $2^m - 1$ nonzero field elements. Also, every nonzero field element is a root of the equation

$$X^{2^m - 1} = 1$$

and conversely. Thus if α is any element of the field, $\alpha^{-1} = \alpha^{2^m - 2}$.

The field elements can also be thought of as vectors whose components are the coefficients of the polynomials. The sum of two vectors corresponds to the sum of the corresponding polynomials.

The Bose-Chaudhuri codes are described by giving the matrix of parity check rules, which is the matrix

$$M = \begin{bmatrix} 1 & 1 & \cdots & 1 \\ \alpha & \alpha^3 & \cdots & \alpha^{2t-1} \\ \alpha^2 & (\alpha^3)^2 & \cdots & (\alpha^{2t-1})^2 \\ \vdots & \vdots & \vdots & \vdots \\ \alpha^{2^m-2} & (\alpha^3)^{2^m-2} & \cdots & (\alpha^{2t-1})^{2^m-2} \end{bmatrix} \qquad (1)$$

where α is a primitive element of the field.

This is a $2^m - 1 \times t$ matrix of $GF(2^m)$ elements, but thinking of each field element as a vector of m binary digits, this is a $2^m - 1 \times mt$ matrix of binary digits. A vector of $2^m - 1$ binary digits is considered a code word if it satisfies the parity check described by each column; *i.e.*, if the product of this vector with the matrix is zero. In other words the set of all code words is the (left) null space of this matrix.

The code that Bose and Ray-Chaudhuri use as an example will be used to illustrate the ideas discussed in this paper. Let α denote a root of the equation $X^4 = X + 1$. This happens to be a primitive element of the field. Then the 15 nonzero field elements are given in Table I.

Taking $t = 3$, the following matrix of parity check rules results:

TABLE I
REPRESENTATION OF $GF(2^4)$

$$\begin{array}{llllllllll}
\alpha^0 & = 1 & & & & & & & = & (1\,0\,0\,0) \\
\alpha^1 & = & & \alpha & & & & & = & (0\,1\,0\,0) \\
\alpha^2 & = & & & & \alpha^2 & & & = & (0\,0\,1\,0) \\
\alpha^3 & = & & & & & & \alpha^3 & = & (0\,0\,0\,1) \\
\alpha^4 & = 1 & + & \alpha & & & & & = & (1\,1\,0\,0) \\
\alpha^5 & = & & \alpha & + & \alpha^2 & & & = & (0\,1\,1\,0) \\
\alpha^6 & = & & & & \alpha^2 & + & \alpha^3 & = & (0\,0\,1\,1) \\
\alpha^7 & = 1 & + & \alpha & & & + & \alpha^3 & = & (1\,1\,0\,1) \\
\alpha^8 & = 1 & & & + & \alpha^2 & & & = & (1\,0\,1\,0) \\
\alpha^9 & = & & \alpha & & & + & \alpha^3 & = & (0\,1\,0\,1) \\
\alpha^{10} & = 1 & + & \alpha & + & \alpha^2 & & & = & (1\,1\,1\,0) \\
\alpha^{11} & = & & \alpha & + & \alpha^2 & + & \alpha^3 & = & (0\,1\,1\,1) \\
\alpha^{12} & = 1 & + & \alpha & + & \alpha^2 & + & \alpha^3 & = & (1\,1\,1\,1) \\
\alpha^{13} & = 1 & & & + & \alpha^2 & + & \alpha^3 & = & (1\,0\,1\,1) \\
\alpha^{14} & = 1 & & & & & + & \alpha^3 & = & (1\,0\,0\,1) \\
\alpha^{15} & = 1 & = & \alpha^0 & & & & & &
\end{array}$$

$$M = \begin{bmatrix}
1\,0\,0\,0 & 1\,0\,0\,0 & 1\,0\,0\,0 \\
0\,1\,0\,0 & 0\,0\,0\,1 & 0\,1\,1\,0 \\
0\,0\,1\,0 & 0\,0\,1\,1 & 1\,1\,1\,0 \\
0\,\,0\,1 & 0\,1\,0\,1 & 1\,0\,0\,0 \\
1\,1\,0\,0 & 1\,1\,1\,1 & 0\,1\,1\,0 \\
0\,1\,1\,0 & 1\,0\,0\,0 & 1\,1\,1\,0 \\
0\,0\,1\,1 & 0\,0\,0\,1 & 1\,0\,0\,0 \\
1\,1\,0\,1 & 0\,0\,1\,1 & 0\,1\,1\,0 \\
1\,0\,1\,0 & 0\,1\,0\,1 & 1\,1\,1\,0 \\
0\,1\,0\,1 & 1\,1\,1\,1 & 1\,0\,0\,0 \\
1\,1\,1\,0 & 1\,0\,0\,0 & 0\,1\,1\,0 \\
0\,1\,1\,1 & 0\,0\,0\,1 & 1\,1\,1\,0 \\
1\,1\,1\,1 & 0\,0\,1\,1 & 1\,0\,0\,0 \\
1\,0\,1\,1 & 0\,1\,0\,1 & 0\,1\,1\,0 \\
1\,0\,0\,1 & 1\,1\,1\,1 & 1\,1\,1\,0
\end{bmatrix} \qquad (2)$$

Of these twelve columns, the last one is trivial and the next to last is a duplicate; these two can be dropped. The rest are independent, and the result is a code with fifteen digit code words of which ten are parity checks and five are information places. The code corrects all triple errors.

An Error-Correction Procedure

Consider the result of multiplying a vector $(r_0, r_1, r_2, \cdots, r_{n-1})$ of $n = 2^m - 1$ components by the matrix M in (1). The result is a vector of t Galois field elements. The first component is

$$r_0 + r_1\alpha + r_2\alpha^2 + \cdots + r_{n-1}\alpha^{n-1} = r(\alpha)$$

where

$$r(X) = r_0 + r_1X + \cdots + r_{n-1}X^{n-1}$$

is the polynomial which corresponds naturally to the given vector. (In what follows no distinction will be made between a vector and the corresponding polynomial.) The other components are clearly $r(\alpha^3), r(\alpha^5), \cdots, r(\alpha^{2t-1})$.

In these terms an equivalent definition of the Bose-Chandhuri codes can be given. A vector is a code word if it is in the left null space of M, *i.e.*, if the parity checks $r(\alpha)$, $r(\alpha^3)$, $r(\alpha^5)$, $\cdots$, $r(\alpha^{2t-1}0)$ are zero. This can be restated as follows:

Definition: A polynomial $s(X)$ is a code vector for a t-error correcting Bose-Chaudhuri code if, and only if, $\alpha, \alpha^3, \cdots, \alpha^{2t-1}$ are roots of $s(X)$.

The first step in devising a decoding method is to characterize the information contained in the parity check calculation for a received vector which may contain errors. Let $e = (e_0, e_1, \cdots, e_{n-1})$ be the vector of errors, *i.e.*, if the errors occur in the positions $i_1, i_2, \cdots, i_v$, then

$$e_i = 1 \text{ for } i = i_1, i_2, \cdots, i_v$$

$$e_i = 0 \text{ otherwise.}$$

There is a one to one correspondence between the elements of the error vector and the elements of $GF(2^m)$ which constitute the first column of the parity check matrix M given by (1), e_i corresponding to the element a^i occurring in the i-th position in the first column of M. The elements $X_1, X_2, \cdots, X_v$ of $GF(2^m)$ which correspond in this way to $e_{i_1}, e_{i_2}, \cdots, e_{i_v}$ may be called the error position numbers. Thus $X_j = a^{i_j}$ $(j = 1, 2, \cdots, v)$.

Lemma 1: If a received vector r has errors in digits numbered $X_1, X_2, \cdots, X_v$, then the parity check vector $r \times M$ is of the form $(S_1, S_3, S_5, \cdots, S_{2t-1})$ where

$$S_j = \sum_{i=1}^{v} X_i^j. \tag{3}$$

Proof: Assume that the vector s was transmitted, and $r = s + e$ received, where e has ones in the positions i_1, $i_2, \cdots, i_v$ and zeros in all other positions. In terms of corresponding polynomials,

$$r(X) = s(X) + e(X)$$

and the result of the parity check calculation is

$$[r(\alpha), r(\alpha^3), \cdots, r(\alpha^{2t-1})].$$

But $s(\alpha) = s(\alpha^3) = \cdots = s(\alpha^2{}_t{}^{-1}) = 0$, so that $r(\alpha) = s(\alpha) + e(\alpha) = e(\alpha)$, $r(\alpha^3) = e(\alpha^3)$, etc. Thus, the result of the parity check calculation is $[e(\alpha), e(\alpha^3), \cdots, e(\alpha^{2t-1})]$. But

$$e(\alpha^j) = e_0 + e_1\alpha^j + e_2\alpha^{2j} + \cdots + e_{n-1}\alpha^{(n-1)j}$$

$$= \sum_{i=1}^{v} \alpha^{i_v j} = \sum_{i=1}^{v} X_i^j \quad \text{Q.E.D.}$$

It is interesting to note that for $t = 1$, if the error occurs, for example, in the component numbered X_1, then the result of the parity check calculation is exactly $S_1 = X_1$ which is the Galois field binary code for the error position number. This is exactly analogous to the method of error-correction for Hamming codes in which the parity check calculation gives the ordinary binary code for the position of the error. In this sense the Bose-Chaudhuri codes for $t = 1$ are equivalent to the Hamming single-error correcting code.

The S_j are the power sum symmetric functions.[5] Thus the parity checks give the first t odd power sum symmetric functions. The first t even ones can be found from the fact that modulo 2, $(a + b)^2 = a^2 + b^2$, and hence

$$S_1^2 = \left[\sum_{i=1}^{v} X_i\right]^2 = \sum_{i=1}^{v} X_i^2 = S_2. \tag{4}$$

Similarly, $S_4 = S_1^4$, $S_6 = S_3^2$, etc.

Suppose that there are t errors. Then the error position numbers $X_1 \cdots X_t$ satisfy the equations

$$S_j = \sum_{i=1}^{t} X_i^j \qquad j = 1, 3, \cdots 2t - 1.$$

This is a set of t equations in t unknowns, the X_i. The solution would tell the positions of the errors. It appears impossible to solve the equations by any direct method, and trying all combinations of t of the $2^m - 1$ field elements would require too many computations. There is, however, an interesting compromise.

The elementary symmetric functions σ_i are related to the power sum symmetric functions S, by Newton's identities:[5]

$$\left.\begin{aligned} &S_1 - \sigma_1 = 0 \\ &S_2 - S_1\sigma_1 + 2\sigma_2 = 0 \\ &S_3 - S_2\sigma_1 + S_1\sigma_2 - 3\sigma_3 = 0 \\ &S_4 - S_3\sigma_1 + S_2\sigma_2 - S_1\sigma_3 + 4\sigma_4 = 0 \\ &S_5 - S_4\sigma_1 + S_3\sigma_2 - S_2\sigma_3 + S_1\sigma_4 - 5\sigma_5 = 0 \\ &\cdots \quad \text{etc.} \end{aligned}\right\} \tag{5}$$

If it is possible to solve Newton's identities for the elementary symmetric functions σ_i, the error position numbers must satisfy the equation

$$X^t - \sigma_1 X^{t-1} + \sigma_2 X^{t-2} \cdots \pm \sigma_t$$
$$= (X - X_1)(X - X_2) \cdots (X - X_t) = 0. \tag{6}$$

Eq. (6) can be solved effectively by merely substituting each of the $n = 2^m - 1$ field elements into the equation. For each digit in the received vector, the corresponding $GF(2^m)$ element is substituted in the equation. If the equation is satisfied, this bit is wrong and must be changed. If the equation is not satisfied, the bit is correct.

[5] See, for example, van der Waerden, footnote 8; J. Riordan, "An Introduction to Combinatorial Analysis," John Wiley and Sons, Inc., New York, N. Y., 1958; T. Muir and W. H. Metzler, "A Treatise on the Theory of Determinants," ch. 21, 1930; or any book on the Theory of Equations.

The proof that it is indeed possible to solve for the ordinary symmetric functions from the power sum symmetric functions is given by the following theorem:[6]

Theorem 1: The $k \times k$ matrix

$$M_k = \begin{bmatrix} 1 & 0 & 0 & 0 & \cdots & 0 \\ S_2 & S_1 & 1 & 0 & \cdots & 0 \\ S_4 & S_3 & S_2 & S_1 & \cdots & 0 \\ \vdots & \vdots & \vdots & \vdots & & \\ S_{2k-4} & S_{2k-5} & S_{2k-6} & S_{2k-7} & & S_{k-3} \\ S_{2k-2} & S_{2k-3} & S_{2k-4} & S_{2k-5} & \cdots & S_{k-1} \end{bmatrix}$$

is nonsingular if power sum symmetric functions S_j are power sums of k or $k - 1$ distinct field elements, and is singular if the S_j are power sums of fewer than $k - 1$ distinct field elements.

The proof requires the following two lemmas:

Lemma 2: If the S_j are power sums of $v \leq k - 2$ distinct field elements, M_k is singular.

Proof:

$$M_k \begin{bmatrix} 0 \\ 1 \\ \sigma_1 \\ \cdot \\ \cdot \\ \cdot \\ \sigma_{k-2} \end{bmatrix} = \begin{bmatrix} 0 \\ 0 \\ 0 \\ 0 \\ 0 \\ 0 \\ 0 \end{bmatrix}$$

by Newton's identities, (5), and thus M_k has a nontrivial null space and must be singular. Q.E.D.

Lemma 3: If the S_j are power sums of k indeterminants $X_1, \cdots, X_k$, then the determinant

$$| M_k | = \prod_{j<i} (X_i + X_j).$$

Proof: If $X_i = X_j$, all of the power sums contain two identical terms, which cancel because the field has characteristic 2 (*i.e.*, $2 = 0$). Then it is just as if there were no more than $k - 2$ distinct elements used in forming the power sums, and, by Lemma 2, the determinant is zero. Therefore, $X_i + X_j$ is a factor of the determinant, for all i and j, and the left-hand side must be divisible by the right-hand side. It is easy to check that the left-hand side is homogeneous of degree $k(k - 1)/2$, the same as the right-hand side, and therefore they must differ at most by a constant factor.

To determine the constant factor, a single special case suffices. If k is odd, let the X_i be the roots of the equation

$$X^k - 1 = 0.$$

Then

$$\sum X_i^j = S_j = 0 \quad \text{if} \quad j \equiv 0 \mod t,$$
$$= 1 \quad \text{if} \quad j \equiv 0 \mod t.$$

There will be exactly one 1 in each row and each column and it follows that $|M_k| = 1$ in this case. For k even, letting the X_i be all of the roots of the equation

$$X^k - X = 0$$

gives the same result. The constant factor, which could be only 0 or 1, must be 1.

Now Theorem 1 follows from the fact that if the determinant $|M_k|$ is zero it must be that some $X_i = X_j$. Since all of the nonzero X_i are distinct, $X_i = X_j = 0$, and there were fewer than $k - 1$ errors. Q.E.D.

If there are actually $t - 1$ errors, it can be seen from Newton's identities, Cramer's Rule and Theorem 1 that the solution for the σ's will yield $\sigma_t = 0$. The corresponding polynomial equation will have zero as one root.

Now let us review the error-correcting procedure. The t-error correcting Bose-Chaudhuri codes give, as the parity checks on received sequences, the odd power-sum symmetric functions up to S_{2t-1} and the intermediate even functions can be calculated simply from these. If it is assumed that no more than t errors occur, then by Theorem 1, with $k = t$, it is either possible to solve for the error position numbers, or there are $t - 2$ or fewer errors. In the latter case, $\sigma_{t-1} = \sigma_t = 0$, and two equations can be dropped, giving a set of $t - 2$ equations in $t - 2$ unknowns to which Theorem 1 can be applied again. Eventually, if there were any errors at all, a set of equations that can be solved for the elementary symmetric functions of the error-position numbers will be found.

The correction procedure consists of three phases:

1) calculate the parity checks and the even numbered S_i;

2) from these, calculate the elementary symmetric functions σ_i; and

3) finally, substitute each field element into the equation

$$X^t + \sigma_1 X^{t-1} + \sigma_2 X^{t-2} \cdots + \sigma_t = 0. \tag{7}$$

Those field elements which satisfy this equation correspond to error positions.

The second step involves a certain amount of trial and error because it is possible to solve the equations and obtain correct solutions only when the number of equations used equals or exceeds by one the number of errors that actually occur. This step might be carried out, as an alternative to the procedure described in the preceding paragraph, by starting with the assumption that two errors occurred, solving, and checking the solution. If the solution doesn't check, four errors would be assumed, and so forth. When a set of answers that checks occurs, it must be the correct solution.

[6] Similar results for a real field appear, for example, in H. O. Faulkes, "Theorems of Kakeya and Polya on Power sums," *Math. Z.*, vol. 65, pp. 345–352; 1956.

If it is assumed that the length n of the code approaches infinity and that the number of errors corrected t is a fixed fraction of n, the number of operations required for error correction can be crudely estimated as follows. The first phase, calculating parity checks, requires a number of operations proportional to the number of digits multiplied by the number of parity checks, or no more than nmt operations. This quantity nmt is proportional to $n^2 \log n$. The second phase requires solving a $t \times t$ set of equations. The number of operations for this task is typically proportional to t^3, but it may have to be done $t/2$ times. This will increase in the limit no faster than n^4. Finally, substituting in a t-degree polynomial requires t multiplications and t additions of m digit numbers, and must be done n times, so that $2\,tmn$ is a rough estimate of the number of operations. This again would vary as $n^2 \log n$. Thus, the total number of operations certainly would increase as a small power of n.

Consider, as an example, the code corresponding to the matrix in (2), which corrects triple errors. The appropriate equations are

$$S_1 + \sigma_1 = 0,$$
$$S_3 + S_2\sigma_1 + S_1\sigma_2 + \sigma_3 = 0, \quad \text{and} \tag{8}$$
$$S_5 + S_4\sigma_1 + S_3\sigma_2 + S_2\sigma_3 = 0.$$

The parity checks for the received vectors give S_1, S_3, and S_5. $S_2 = S_1^2$, and $S_4 = S_1^4$. Solving for the σ's gives

$$\sigma_1 = S_1, \qquad \sigma_2 = (S_1^2S_3 + S_5)/(S_1^3 + S_3) \quad \text{and}$$
$$\sigma_3 = (S_1S_5 + S_3^2 + S_1^3S_3 + S_1^6)/(S_1^3 + S_3), \tag{9}$$

provided that $S_1^3 + S_3 \neq 0$. If there is only one error $S_1^3 + S_3 = 0$. Furthermore, if $S_1^3 + S_3 = 0$, the Newton's identities yield $\sigma_3 = \sigma_1\sigma_2$, and the equation

$$X^3 + \sigma_1X^2 + \sigma_2X + \sigma_3$$
$$= X^3 + \sigma_1X^2 + \sigma_2X + \sigma_1\sigma_2$$
$$= (X + \sigma_1)(X^2 + \sigma_2) = (X + \sigma_1)(X + \sqrt{\sigma_2})^2 = 0$$

has two equal roots, which must be zero, and therefore there is only one error.

As a numerical example, suppose that the vector of all zeros is transmitted, and that errors occur in the 2nd, 5th, and 7th positions. Then

$$r = (0\ 1\ 0\ 0\ 1\ 0\ 1\ 0\ 0\ 0\ 0\ 0\ 0\ 0\ 0)$$
$$r \times M = (1\ 0\ 1\ 1\quad 1\ 1\ 1\ 1\quad 1\ 0\ 0\ 0)$$
$$S_1 = (1\ 0\ 1\ 1) \qquad S_3 = (1\ 1\ 1\ 1) \qquad S_5 = (1\ 0\ 0\ 0).$$

Referring to Table I, one finds

$$S_2 = S_1^2 = (1\ 0\ 1\ 1)^2 = (\alpha^{13})^2 = \alpha^{26} = \alpha^{11} = (0\ 1\ 1\ 1)$$

and

$$S_4 = S_1^4 = \alpha^{52} = \alpha^7 = (1\ 1\ 0\ 1).$$

Then,

$$S_3 + S_1^3 = (1\ 0\ 1\ 0) \neq 0,$$
$$\sigma_1 = S_1 = (1\ 0\ 1\ 1) = \alpha^{13},$$
$$\sigma_2 = (S_1^2S_3 + S_5)/S_3 + S_1^3 = (0\ 0\ 1\ 0)/(1\ 0\ 1\ 0)$$
$$= \alpha^2/\alpha^8 = \alpha^9/\alpha^{15} = \alpha^9.$$

Similarly,

$$\sigma_3 = \alpha^{11}.$$

It is then easy to verify that the equation,

$$X^3 + \alpha^{13}X^2 + \alpha^9X + \alpha^{11} = 0,$$

is satisfied by the three values $X = \alpha$, α^4, and α^6, and only these. These correspond to the errors in r.

Some Properties of Cyclic Codes and Shift Register Generators

Codes for which the code points comprise a cyclic subspace of vectors of zeros and ones have been studied recently by Prange,[4] and, along with theoretical results, he found several efficient codes that can be decoded easily. He has noted that the codes can be coded with the use of a shift-register generator.[7] In this section, some of the theory of cyclic codes and linear recurrent sequences is reviewed briefly from a point of view that is especially well adapted to the study of the Bose-Chaudhuri codes.

A subset C of vectors of n binary digits is called a *cyclic subspace* if it has the following two properties:

1) If v_1 and v_2 are in C, their sum modulo 2 is also in C; that is, C is a subspace, or subgroup; and
2) If $v = (a_0, a_1, \cdots, a_{n-1})$ is in C, the vector $v^1 = (a_{n-1}, a_0, a_1, \cdots, a_{n-2})$ obtained by shifting v cyclically one place is also in C.

Let R_n denote the set of all polynomials

$$a_0 + a_1X \cdots + a_{n-1}X^{n-1}$$

of degree less than n with coefficients 1 and 0. They form a group under modulo 2 addition. Multiplication can be defined modulo $X^n - 1$; that is, these polynomials can be multiplied in the ordinary way, modulo 2, and then reduced again to polynomials of degree less than n by the use of the equation $X^n = 1$. Then R_n is a ring in the mathematical sense. A subset I of R_n is called an *ideal*[8] if it satisfies the following two properties:

1) I is a subgroup of R_n; and
2) if $p(X)$ is in I and $a(X)$ is R_n, then the product $p(X)\ a(X)$ is in I.

---

[7] N. Zierler, "Linear recurring sequences," *J. Soc. Ind. Appl. Math.*, vol. 7, pp. 31–48; March, 1959.

[8] Galois fields and other aspects of algebra used in this paper are treated in many books on modern algebra. See, for example, A. A. Albert, "Fundamental Concepts of Modern Algebra," University of Chicago Press, Chicago, Ill., 1956; G. Birkoff and S. MacLane, "A Survey of Modern Algebra," The Macmillan Co., New York, N. Y., 1953; B. L. van der Waerden, "Modern Algebra," F. Ungar Publishing Co., New York, N. Y., vol. 1 and 2, 1949, 1950.

Considering polynomials $p(X) = a_0 + a_1X \cdots + a_{n-1}X^{n-1}$ to be vectors $(a_0, a_1, \cdots, a_{n-1})$, a cyclic shift is the same as multiplication by X modulo $X^n - 1$. Therefore, *every ideal is a cyclic subspace.* Conversely, if $p(X)$ is in a cyclic subspace C, so is $Xp(X)$. It follows that $X^ip(X)$ must also be in C, and since C is a subspace,

$$\sum_i c_iX^ip(X) = p(X) \sum_i c_iX^i$$

must also be in C. Thus, if $p(X)$ is in C, so is the product of $p(X)$ and any polynomial. Therefore, *every cyclic subspace is an ideal.*

The important but well-known properties of ideals given in the following three lemmas and two theorems are proved here to make the paper self-contained.

Lemma 4: If $p(X)$ and $q(X)$ are in an ideal I, the greatest common divisor (GCD), $d(X)$, of $p(X)$ and $q(X)$ is in I.

This follows directly from the fact that it is always possible to express the $d(X)$ in the form

$$d(X) = a(X)p(X) + b(X)q(X)$$

where $a(X)$ and $b(X)$ are polynomials.

Lemma 5: All polynomials in an ideal I are multiples of the unique polynomial of least degree in I. (That is, every ideal is a principal ideal.)

Proof: Let $p(X)$ be a polynomial of least degree in I. Then, if $q(X)$ is any other polynomial in I, the greatest common divisor of $p(X)$ and $q(X)$ is in I. If $p(X)$ does not divide $q(X)$, then the greatest common divisor of $p(X)$ and $q(X)$ would have lower degree than $p(X)$, which is a contradiction. Therefore, every polynomial in I is divisible by $p(X)$. If $p_1(X)$ and $p_2(X)$ both have minimum degree, each must be divisible by the other, and hence they are equal.

The ideal consisting of all multiples of $p(X)$ is denoted $[p(X)]$. The polynomial of least degree in an ideal is called its generator.

Lemma 6: The generator $p(X)$ of an ideal is a factor of $X^n - 1$.

Proof: The GCD $d(X)$ of $p(X)$ and $X^n - 1$ can be expressed in the form

$$d(X) = a(X)p(X) + b(X)(X^n - 1)$$
$$\equiv a(X)p(X) \mod X^n - 1;$$

hence, $d(X)$ is in the ideal. But $p(X)$ is divisible by $d(X)$, and since $d(X)$ is in the ideal, $d(X)$ is divisible by $p(X)$. Hence, $p(X) = d(X)$.

These results can be summarized as follows:

Theorem 2: A set of polynomials is an ideal in the ring of polynomials modulo $X^n - 1$ if and only if it consists of all multiples of degree less than n of a factor of $X^n - 1$.

Corollary: If $p(X)$ is a polynomial of degree k which divides into $X^n - 1$, $[p(X)]$ is a vector space of dimension $n - k$.

Proof: The elements of $[p(X)]$ are of the form $c(X)p(X)$ where $c(X)$ is an arbitrary polynomial of degree less than $n - k$. Then the $n - k$ coefficients of $c(X)$ are arbitrary.

Theorem 3: If $p(X)\, q(X) = X^n - 1$, the ideals $[p(X)]$ and $[q(X)]$ are null spaces of each other. That is, a polynomial $p_1(X)$ is in $[p(X)]$ if, and only if, $p_1(X)\, q_1(X) = 0$ modulo $(X^n - 1)$ for every polynomial $q_1(X)$ in $[q(X)]$.

Proof: Since $p_1(X)$ is in $[p(X)]$, $p_1(X)$ is a multiple of $p(X)$, for example, $a(X)\, p(X)$. Similarly, $q_1(X) = b(X)\, q(X)$. Then $p_1(X)\, q_1(X) = a(X)\, b(X)\, (X^n - 1) = 0$. Conversely, if $p_1(X)\, q(X) = 0$, then $p_1(X)\, q(X)$ must be a multiple of $X^n - 1$, and $p_1(X)$ must be a multiple of $(X^n - 1)/q(X) = p(X)$.

Note that the fact that the product of two polynomials is zero implies that the dot product of the corresponding two vectors is zero, if in one of them the order of the components is reversed. That is, if

$$(a_0 + a_1X \cdots + a_{n-1}X^{n-1})(b_0 + b_1X \cdots + b_{n-1}X^{n-1}) = 0$$

then

$$(a_0, a_1 \cdots a_{n-1})\cdot(b_{n-1}, b_{n-2}, \cdots b_1, b_0)$$
$$= a_0b_{n-1} + a_1b_{n-2} \cdots + a_{n-1}b_0 = 0,$$

since this is the coefficient of X^{n-1} in the product of the polynomials. Hence, if $[p(x)]$ and $[q(x)]$ are null spaces of each other, the corresponding vector-spaces are null-spaces of each other provided that the order of components in the vectors of one of these is reversed.

Now let us consider a recursion relation (or difference equation) of the form

$$\sum_{j=0}^{k} a_jR_{i-j} = 0, \tag{10a}$$

or

$$R_i = \sum_{j=1}^{k} a_jR_{i-j} \qquad a_0 = a_k = 1. \tag{10b}$$

The solution of these equations for given coefficients a_n will be a sequence of binary digits, $\{R_i\}$. Given the digits $R_0, \cdots, R_{k-1}$, (10) is the rule for calculations R_k, then R_{k+1}, and so forth. Also, the sum of two solutions is again a solution because the equation is linear. Therefore, the solutions form a vector space of dimension k. The solutions are characterized in the following theorem.

Theorem 4: Let $p(X) = \sum_{j=0}^{k} a_jX^j$, $a_0 = a_k = 1$, and let n be the smallest integer for which $X^n - 1$ is divisible by $p(X)$. Let $q(X) = (X^n - 1)/p(X)$. Then the solutions of the difference equation

$$R_i = \sum_{j=1}^{k} a_jR_{i-j}$$

are periodic of period n, and the set made up of the first period of each possible solution, considered as polynomials, is the ideal $[q(X)]$.

Proof: That any vector taken from $[q(X)]$ is a solution can be seen by multiplying a polynomial from $[q(X)]$, for example, $q_1(X)$, by $p(X)$. The digits in the product are formed by the summation in (10a), and, since the product is zero, (10a) is satisfied. Therefore, any sequence formed by repetition of a vector taken from $[q(X)]$ is a solution of (10). Since $q(X) = X^n - 1/p(X)$ has degree $n - k$, then $[q(X)]$ has dimension k, by the corollary to Theorem 2. This is the same as the dimension of the space of solutions, and therefore $[q(X)]$ must include all solutions.

The Cyclic Structure of the Bose-Chaudhuri Codes

It is shown in this section that the Bose-Chaudhuri codes are examples of cyclic codes as studied by Prange.[4] As such they can be generated with very simple equipment, as is illustrated for the (15,5) code in the next section. Out of this theory also comes a better estimate of the number of parity check digits required to correct a given number of errors.

By the alternative definition of the Bose-Chaudhuri codes given in the second section of this paper, a code consists of all polynomials $f(X)$ which have $\alpha, \alpha^3, \cdots, \alpha^{2t-1}$ as roots. Each element α^i of the field is a root of a unique irreducible polynomial $p_i(X)$ of minimum degree. Then $f(X)$ must be divisible by each of the polynomials $p_1(X), p_3(X), \cdots, p_{2t-1}(X)$ and, hence, by their least common multiple:[9]

$$f(X) = \underset{i=1,3,\cdots,2t-1}{\mathrm{LCM}} [p_i(X)]. \tag{11}$$

Since each of the factors $p_i(X)$ is irreducible, the least common multiple of the $p_i(X)$ is simply the product of the polynomials $p_i(X)$, with the duplicates omitted. Duplications are quite possible; they will occur, in fact, for any α^i and α^j that are roots of the same polynomial $p_i(X)$. In other words, should α^i and α^j happen to be roots of the same irreducible polynomial, the columns in the parity check matrix will be dependent, although not necessarily identical. The parity checks produced by the column of powers of α^j will be satisfied if and only if the parity checks produced by the column of powers of α^i are satisfied, and thus one set or the other is unnecessary.

Finally, the set of all sequences that comprise the code can, by Theorem 4, be generated by a recursion relation defined by the polynomial $X^n - 1/f(X)$, and hence by a shift register generator.

At this point it is interesting to study the limiting cases of the minimum and maximum numbers of parity checks. It has already been noted that the nontrivial minimum is the Hamming code. On the other extreme, the last two columns which might be included in the parity check matrix are powers of $\alpha^{2m-2} = \alpha^{-1}$ and $\alpha^{2m-1} = 1$. The last one is a root of the irreducible polynomial $1 + x$ and the resulting code would be the ideal generated by $(1 + x^n)/(1 + x)$. This ideal consists of the zero vector and the vector of all ones, so the code is the trivial repetition of a single information digit $n = 2^m - 1$ times. If α is a primitive element, so is α^{-1}, and therefore the irreducible polynomial of which α^{-1} is a root is primitive. It can be shown then that when only the last two columns, corresponding to α^{-1} and 1, are omitted from the parity check matrix, the resulting code consists of a maximal length sequence, all its shifts, all complements, and a sequence of all 1's, which is then the code studied by San Soucie and Green.[10] This code can also be shown to be equivalent to the Reed-Muller first-order code with any one digit dropped.[11]

It is possible to predict easily which powers of α are roots of the same polynomial, and thus, incidentally, find the degree of the polynomial of which α^i is a root. The method is based on the fact that if a is a root of $f(X)$, then a^2 is also, since $f(a^2) = [f(a)]^2 = 0$. It turns out that $a, a^2, a^4, a^8, \cdots$ are, in fact, all of the roots. In Table II information is given for $m = 4$ and 5. Note that in the first case, $\alpha^{15} = 1$; and in the second, $\alpha^{31} = 1$.

The code for $m = 4$, $t = 3$ has for its generator, by (11),

$$f(X) = p(X)p_3(X)p_5(X)$$

and therefore has $4 + 4 + 2 = 10$ parity checks, and 5 information places. The code for $m = 5$, $t = 5$ has

$$f(X) = p(X)p_3(X)p_5(X)p_7(X)$$

for its generator, and therefore has 20 parity checks. All codes for $m = 4$ and 5 are listed in Table III.

TABLE II
Roots of Polynomials $p_i(X)$

	Polynomial	Roots
$m = 4$	$p(X)$	$\alpha, \alpha^2, \alpha^4, \alpha^8$
	$p_3(X)$	$\alpha^3, \alpha^6, \alpha^{12}, \alpha^9$
	$p_5(X)$	α^5, α^{10} $(\alpha^{20} = \alpha^5)$
	$p_7(X)$	$\alpha^7, \alpha^{14}, \alpha^{13}, \alpha^{11}$
$m = 5$	$p(X)$	$\alpha, \alpha^2, \alpha^4, \alpha^8, \alpha^{16}$
	$p_3(X)$	$\alpha^3, \alpha^6, \alpha^{12}, \alpha^{24}, \alpha^{17}$
	$p_5(X)$	$\alpha^5, \alpha^{10}, \alpha^{20}, \alpha^9, \alpha^{18}$
	$p_7(X)$	$\alpha^7, \alpha^{14}, \alpha^{28}, \alpha^{25}, \alpha^{19}$
	$p_9(X) = p_5(X)$	
	$p_{11}(X)$	$\alpha^{11}, \alpha^{22}, \alpha^{13}, \alpha^{26}, \alpha^{21}$
	$p_{13}(X) = p_{11}(X)$	
	$p_{15}(X)$	$\alpha^{15}, \alpha^{30}, \alpha^{29}, \alpha^{27}, \alpha^{23}$

TABLE III
Rate and Error Correction Ability of Bose-Chaudhuri Codes for $m = 4$ and 5

Length of Code Words	Number of Parity Checks	Number of Information Places	Number of Errors Corrected
n	$n - k$	k	t
15	4	11	1
15	8	7	2
15	10	5	3
31	5	26	1
31	10	21	2
31	15	16	3
31	20	11	5
31	25	6	7

[9] See, for example, Birkhoff and MacLane, *op. cit.*, p. 396.

[10] J. H. Green, Jr. and R. L. San Soucie, "An error-correcting encoder and decoder of high efficiency," Proc. IRE, vol. 46, pp. 1741–1744; October, 1958.

[11] N. Zierler, "On a variation of the first-order Reed-Muller Codes," Lincoln Laboratory Group Rept. 34–80; October, 1958.

Code parameters for some larger codes were calculated on the IBM 704 computer. The results are plotted in Fig. 1. The vertical axis represents rate (percentage of all digits available for information), and the horizontal axis represents the number of errors correctable as a percentage of the total number of digits. The dashed curve represents asymptotic values of a lower bound on the rate of the best code that corrects errors in a given percentage of the digits.[12] The curves drawn for the Bose-Chaudhuri codes for large n fall below the bound for the best code. In fact, it is shown in the Appendix that they approach zero as the length of the code increases indefinitely. This may mean that these codes are truly not optimum, or it may mean that the number of errors correctable by the procedure given in this paper is not the total number of errors correctable by Bose-Chaudhuri codes in the case of very long codes.[13]

The polynomial $p(X)$ can be any primitive polynomial of degree m. The other polynomials $p_i(X)$ are determined

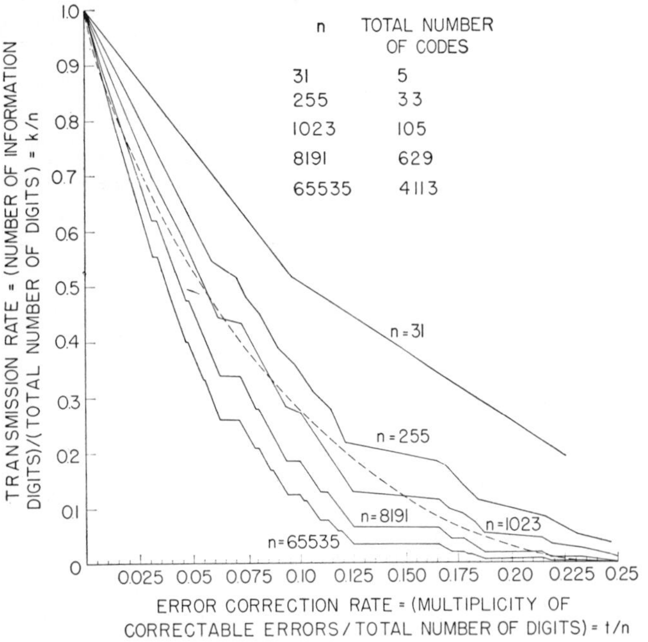

Fig. 1—Error correction and rate for some long Bose-Chaudhuri codes. (Dashed curve is asymptotic lower bound for the rate for the best binary code as given by Gilbert.)

[12] E. N. Gilbert, "A comparison of signaling alphabets," *Bell Sys. Tech. J.*, vol. 31, pp. 504–522; May, 1952.

[13] I have found with the aid of the IBM 704 that the Bose-Chaudhuri two-error correcting codes for $m = 4$ and 5 correct some triple errors and nothing beyond and are therefore optimum. The three-error correcting code for $m = 4$ corrects 420 quadruple and 28 quintuple error patterns and is optimum. The three-error correcting code for $m = S$ corrects 13,020 quadruples and 14,756 quintuples and nothing beyond—this seems good but has not been proved optimum. (See A. B. Fontaine and W. W. Peterson, "Group code equivalence and optimum codes," IRE TRANS. ON INFORMATION THEORY, vol. IT-5, pp. 60–70; May, 1959.) Thus, any non-optimum behavior of these codes occurs only in codes so large that they are difficult to analyze by looking at code words themselves or searching for coset leaders even with the aid of a computer.

by the particular choice of $p(X)$, and the question arises as to how they may be calculated. One simple method is based on the fact that every element of $GF(2^m)$ is a root of the polynomial $X^{2m-1} - 1$. Therefore, each element is a root of one of the factors of $X^{2m-1} - 1$. One needs only to factor this polynomial and test to see which factor has X^i as a root. The following alternative method is useful. It has been noted that the degree m_i of $p_i(X)$ can be easily determined. Then if

$$p_i(X) = a_0 + a_1X + \cdots + a_{m-1}X^{m_i-1} + X^{m_i},$$

since α^i is a root of $p_i(X)$,

$$0 = a_0\alpha^0 + a_1\alpha^1 + \cdots + a_{m-1}\alpha^{m_i-1} + \alpha^{m_i},$$

and if α^i is written as a vector with m components, the resulting set of linear equations can be solved for the coefficients a_i of $p_i(X)$.

There is also an explicit formula

$$p(X^{1/j})p(\alpha X^{1/j}) \cdots p(\alpha^{j-1}X^{1/j})$$

where α is a primitive jth root of unity. It can be shown that when the multiplication is carried out only integral powers of X remain, and these have only ones or zeros as coefficients.

Consider again the sample code discussed by Bose and Chaudhuri. The irreducible factors of $X^{15} - 1$ are

$$X^{15} - 1 = (X - 1)(X^2 + X + 1)(X^4 + X^3 + X^2 + X + 1)(X^4 + X^3 + 1)(X^4 + X + 1).$$

A root of the last factor was taken as α; and thus

$$p(X) = X^4 + X + 1.$$

Then α^3 satisfies the equation $X^5 - 1 = 0$, since $\alpha^{15} = 1$. But $X^5 - 1 = (X - 1)(X^4 + X^3 + X^2 + X + 1)$, and since α^3 is not a root of the first factor, it must be a root of the second. Similarly, α^5 satisfies $X^3 - 1 = 0 = (X - 1)(X^2 + X + 1)$, and so α^5 is a root of $X^2 + X + 1$. The fact that this has degree 2 ties in with the observation that the column of powers of α^5 contained only two independent parity checks.

All code points must be multiples, then, of

$$\begin{aligned} f(X) &= p(X)p_3(X)p_5(X) \\ &= (1 + X + X^4)(1 + X + X^2 + X^3 + X^4) \\ &\qquad \cdot(1 + X + X^2) \\ &= 1 + X + X^2 + X^4 + X^5 + X^8 + X^{10} \\ &= (1\ 1\ 1\ 0\ 1\ 1\ 0\ 0\ 1\ 0\ 1\ 0\ 0\ 0\ 0), \end{aligned} \tag{12}$$

and it can easily be checked that this vector, any cyclic permutation of it, and any sum of permutations, actually do satisfy the parity checks defined by the matrix M in (2).

Mechanizing the Coding and Error-Correction

Shift registers with feedback corrections can be used in a number of ways in mechanizing coding and error-correction procedures. The following uses will be discussed in this section:

1) coding using a shift register with one stage for each information digit in the code,
2) coding using a shift register with one stage for each parity check digit in the code,
3) counting in the Galois field code,
4) multiplying and dividing Galois field elements, and
5) calculating parity checks on received vectors.

Both the methods of coding apply to any cyclic code. The methods will be illustrated using the Bose-Chaudhuri (15, 5) code described by the matrix M in (2).

Every cyclic code is an ideal generated by some polynomial $f(X)$, *i.e.*, a polynomial is a code vector if and only if it is divisible by $f(X)$. This means that, by Theorem 4, a vector is a code vector if and only if it satisfies the recursion relation corresponding to the polynomial $(X^n - 1)/f(X)$. For the code used as an example, by (12),

$$f(X) = 1 + X + X^2 + X^4 + X^5 + X^8 + X^{10}$$
$$(1 - X^{15})/f(X) = 1 + X + X^3 + X^5.$$

Then every sequence satisfying the recursion relation

$$R_j = R_{j-1} + R_{j-3} + R_{j-5}$$

is a code point, and conversely. Such sequences can be generated by putting information digits in the shift register generator shown in Fig. 2 and shifting 15 times. The first five digits coming out will be information digits, and the next ten digits will be a set of parity checks which make the whole sequence a code point. The symbols come out of this encoder low order digits first. The order can be reversed by reversing the order of the shift register feedback connections.

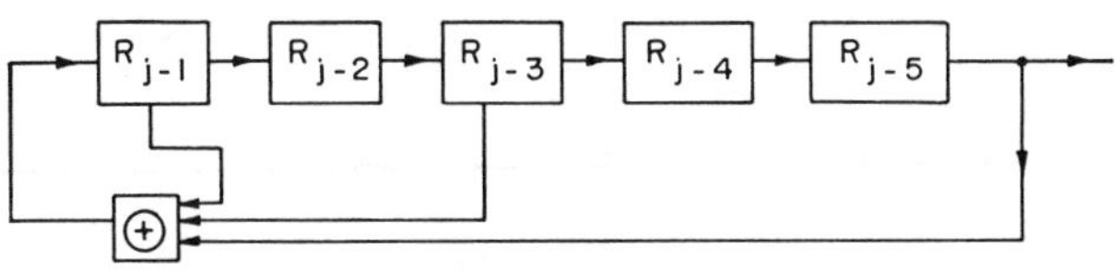

Fig. 2—A shift register for encoding the Bose-Chaudhuri (15,5) code.

A second method of coding is based again on the fact that the coded vector must be, considered as a polynomial, a multiple of $f(X)$. Let $t_0(X)$ be a polynomial in which the k coefficients of the terms involving X^{n-1} through X^{n-k} are arbitrary information digits, and the coefficients of lower order terms are zero. This corresponds to a vector in which the first $n - k$ components are zero, the last k digits arbitrary information digits. Then $t_0(X)$ can be divided by $f(X)$ to produce a quotient and a remainder

$$t_0(X) = f(X)q(X) + r(X),$$

where $r(X)$ has degree less than $(n - k)$, which is the degree of $f(X)$. Then

$$t_0(X) + r(X) = f(X)q(X)$$

and, hence, $t_0(X) + r(X)$ is a code point. But $r(X)$ corresponds to a vector in which all components except the first $n - k$ are zero, since $r(X)$ has degree less than $n - k$. Thus, the sum consists of $n - k$ check digits, the coefficients of $r(X)$, and k information digits, the coefficients of $t_0(X)$.

The next problem is to calculate $r(X)$. In general, the calculation of the remainder after division by a polynomial can be accomplished with a shift register. The method is illustrated in Fig. 3(a). Assuming the divisor is the $f(X)$ for the code used in the example, *i.e.*, $1 + X + X^2 + X^4 + X^5 + X^8 + X^{10}$, the operation of the circuit can be understood as follows: The answer is the same as results from reducing the dividend modulo $f(X)$. This means that the dividend polynomial should be reduced to a polynomial of degree less than 10 using the relation

$$X^{10} = 1 + X + X^2 + X^4 + X^5 + X^8. \qquad (13)$$

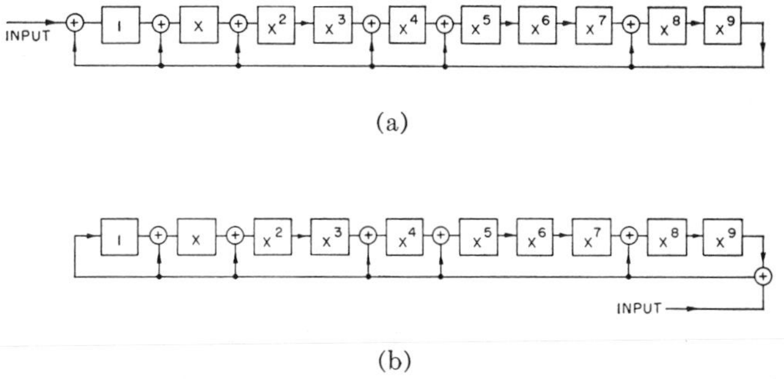

Fig. 3—Shift register for calculating residues modulo $f(X) = 1 + X + X^2 + X^4 + X^5 + X^8 + X^{10}$. (a) Basic circuit; (b) basic circuit with automatic premultiplication by X^{10}.

Now assume that a single one is shifted into the low-order position and then shifted right a number of times. Thinking of the contents of the register as a polynomial with low order digits at the left, each shift corresponds to multiplying by X, at least until a shift out of the high-order position. A one in the high-order position corresponds to X^9, and shifting it out makes it X^{10}. This results in the circuit in adding into the lower order positions the equivalent of X^{10} given in (13), and, hence, in this case the shift still corresponds to multiplying by X and modulo $f(X)$. Thus, successive shifts give successive powers of X modulo $f(X)$.

Now this is a linear device, and a polynomial (which is the sum of powers of X) can be reduced modulo $f(X)$ by shifting it into the device, high power terms first, until the constant term is shifted into the low-order position.

In using this device for calculating the $r(X)$ in (17), the modification shown in Fig. 3(b) can be made to avoid the last $n - k$ shifts which would add $n - k$ zeros into the low-order positions. It amounts to multiplying the input digits by $X^{n-k} = X^{10}$ before adding.

The procedure for coding is then to shift all the information digits into the device in Fig. 3(a) or 3(b). If the device in Fig. 3(a) is used, $n - k$ more shifts must be made with no input. Then the correct check digits remain in the register and should simply follow the information digits, high order digits first, to make a complete code vector. Note that the number of stages in this shift register is $n - k$, while the shift register shown in Fig. 2 has k stages.

A counter which counts in terms of Galois field elements is shown in Fig. 4(a). It works on the same principle as the device shown in Fig. 3(a), but using the primitive polynomial $p(X) = X^4 + X + 1$ of which α is a root. If a 1 is placed in the low-order position, successive shifts give successive powers of α using the relation $\alpha^4 = \alpha + 1$, and these are exactly the representations of $GF(2^4)$ elements given in Table I.

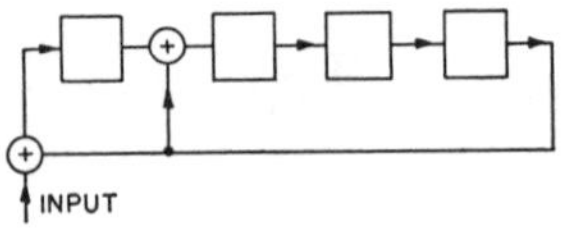

(a)

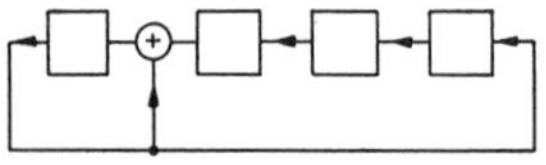

(b)

Fig. 4—Galois field counters for $GF(2^4)$. (a) Increasing powers of α; and (b) decreasing powers of α.

In the device shown in Fig. 3(b), a left shift corresponds to division by α and a 1 shifted out of the low order end α^{-1} is replaced by its equivalent $1 + \alpha^3$. Thus, this device can count down, or give Galois field elements in reverse order. A multiplier can be mechanized by putting one factor in a device A like that shown in Fig. 3(a), the other in a device B like that shown in Fig. 3(b). Then both devices are shifted until the code for 1 appears in device B. The product then appears in A. Division can be done in an anologous manner. Multiplication can also be done in a manner analogous to that used in digital computers with a shift register such as that shown in Fig. 3(a) used in place of an accumulator.

The parity checks corresponding to the first column of Galois field elements in the matrix M of (2) correspond to the Galois field representation of

$$r(\alpha) = r_0 + r_1\alpha + r_2\alpha^2 \cdots r_{2^m-2}\alpha^{2^m-2}.$$

This can be calculated by using the relation $\alpha^4 + \alpha + 1 = 0$ to eliminate terms of degree higher than 3 in α. This, in turn, is exactly what will result if the vector $(r_0, r_1, \cdots, r_2m_2)$ is shifted into the shift register shown in Fig. 3(a) high-order digits first. Note that shifting fifteen times multiplies by α^{15}, but $\alpha^{15} = 1$. Similarly, the device in Fig. 3(b) could be used with the low-order digits entering first.

Calculation of the other parity checks is slightly more complicated. It requires calculating $r(\alpha^j)$ for the first t odd values of j. The first step is to devise a shift register which automatically multiplies by α^j. The example $j = 5$ should make the principles clear. Note that

$$1 \cdot \alpha^5 = \alpha^5 = \alpha + \alpha^2$$
$$\alpha \cdot \alpha^5 = \alpha^6 = \alpha^2 + \alpha^3$$
$$\alpha^2 \cdot \alpha^5 = \alpha^7 = 1 + \alpha + \alpha^3$$
$$\alpha^3 \cdot \alpha^5 = \alpha^8 = 1 + \alpha^2,$$

so that

$$\begin{aligned}\alpha^5(a_0 + a_1\alpha + a_2\alpha^2 + a_3\alpha^3) \\ &= a_0(\alpha + \alpha^2) + a_1(\alpha^2 + \alpha^3) \\ &\quad + a_2(1 + \alpha + \alpha^3) + a_3(1 + \alpha^2) \\ &= (a_2 + a_3) + (a_0 + a_2)\alpha \\ &\quad + (a_0 + a_1 + a_3)\alpha^2 + (a_1 + a_2)\alpha^3.\end{aligned}$$

Thus, the new value of a_0 is the old $a_2 + a_3$, the new a_1 is the old $a_0 + a_2$, etc. A shift register with feedback connections shown in Fig. 5 will give this result. Then, if the received vector $(r_0, r_1, \cdots, r_2m_2)$ is shifted into this device, after fifteen shifts the result $r(\alpha^5)$ will remain in the register.

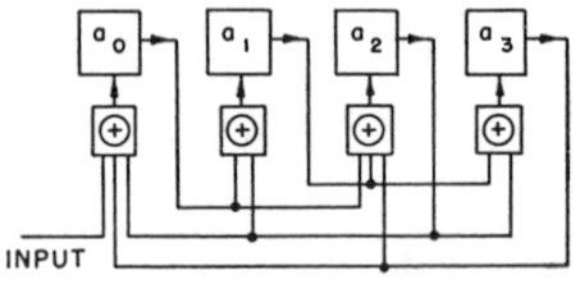

Fig. 5—A circuit for calculating the parity checks $r(\alpha^5)$.

Conclusion

Relatively simple coding and error-correcting methods have been described for the Bose-Chaudhuri codes. The study of coding and error-correction methods for these codes gives additional insight into the remarkable structure of the codes.

Appendix

A bound on the rate of Bose-Chaudhuri codes which correct $t = 2^\lambda$ errors is derived in this Appendix, and it is shown on the basis of this bound that if t is made a fixed fraction of n, the number of digits in the code, the rate must approach zero as n increases indefinitely.

This problem is purely number-theoretic, and can be formulated as follows: The quantity to be studied is the rate, which is the quotient of the number k of information digits and $n = 2^m - 1$, the total number of digits. Since there is one independent parity check for each distinct residue of $j2^i$ for $1 \leq j \leq 2t$, $0 \leq i < m$, the number of such residues in $n - k$. Since $2^m = 2^0$ modulo $2^m - 1$, the condition $0 \leq i < m$ can be replaced by $1 \leq i \leq m$. For convenience in what follows, j will be allowed to take on the value zero also; this adds one distinct residue.

Let $N(s)$ be the number of distinct residues of $j2^i$ for $0 \leq j < 2t = 2^{\lambda+1}$ and $m - s < i \leq m$. Then

$$n - k = N(m) - 1 \geq N(s) - 1 \quad \text{if} \quad s \leq m$$

and

$$k = n - N(m) + 1$$

$$= 2^m - N(m) \leq 2^m - N(s) \quad \text{if} \quad s \leq m. \qquad (14)$$

An equation for $N(s)$, valid only for $s \leq \lambda$, will be derived but this will give an upper bound on k by (14).

Consider first the residues for a particular value of i, $m - \lambda \leq i \leq m$. They can be arranged as follows:

$$\begin{array}{ccccc}
0\cdot 2^i, & 1\cdot 2^i, & 2\cdot 2^i, & \cdots, & (2^{m-i} - 1)2^i \\
(2^{m-i} + 0)2^i, & (2^{m-i} + 1)2^i, & (2^{m-i} + 2)2^i, & \cdots, & (2\cdot 2^{m-i} - 1)2^i \\
(2\cdot 2^{m-i} + 0)2^i, & (2\cdot 2^{m-i} + 1)2^i, & (2\cdot 2^{m-i} + 2)2^i, & \cdots, & (3\cdot 2^{m-i} - 1)2^i \\
\vdots & \vdots & \vdots & & \vdots \\
(2^{\lambda+1} - 2^{m-i} + 0)2^i, & (2^{\lambda+1} - 2^{m-i} + 1)2^i, & (2^{\lambda+1} - 2^{m-i} + 2)2^i, & \cdots, & (2^{\lambda+1} - 1)2^i.
\end{array}$$

In this array there are $2^{\lambda+1-m+i}$ rows. Since $2^m \equiv 1$, the array can be rewritten

$$\begin{array}{ccccc}
0, & 1\cdot 2^i, & 2\cdot 2^i, & \cdots, & (2^{m-i} - 1)2^i \\
1, & 1 + 1\cdot 2^i, & 1 + 2\cdot 2^i, & \cdots, & 1 + (2^{m-i} - 1)2^i \\
2, & 2 + 1\cdot 2^i, & 2 + 2\cdot 2^i, & \cdots, & 2 + (2^{m-i} - 1)2^i \\
\vdots & \vdots & \vdots & & \vdots \\
2^{\lambda+1+i-m} - 1, & (2^{\lambda+1+i-m} - 1) + 1\cdot 2^i, & (2^{\lambda+1+i-m} - 1) + 2\cdot 2^i, & \cdots, & 2^{\lambda+1+i-m} - 1 + (2^{m-i} - 1)2^i.
\end{array}$$

This consists exactly of 2^{m-i} sets of $2^{\lambda+1+i-m}$ successive numbers starting at each multiple of 2^i. The arrangement is shown graphically in Fig. 6.

The important facts can be seen clearly in Fig. 6 but are tedious to prove formally. For each i there are $2^{\lambda+1}$ residues and therefore, in particular, $N(1) = 2^{\lambda+1}$. Two adjacent columns in Fig. 6 have half their residues in common. In particular, $N(2) = 2^{\lambda+1} + 2^\lambda$. Now in adding the contributions to $N(s)$ for larger values of s it is necessary to determine exactly how many residues have occurred in all previous columns combined. There is one other case which must be considered besides the previous adjacent column. Note that the residues and nonresidues of $j\cdot 2^i$ for a particular value of i fall in blocks of $2^{\lambda+1+i-m}$ successive numbers. In determining which residues for a particular value of i, for example, i_0, have occurred before, each block of 2^{i_0+1} successive numbers is treated the same. Each will have two blocks of $2^{\lambda+1+i_0-m}$ residues. The first will already have been counted in the $i_0 + 1^{st}$ column. The fraction of the others to be omitted is the same as the fraction of blocks of length 2^{i_0+1} which were counted as residues for $i \geq i_0 + m - \lambda$, which is the same as $N(\lambda - i_0)/2^m$. Then, since $s = m - i_0$,

$$N(s) = N(s - 1) + 2^\lambda\cdot[1 - 2^{-m}N(\lambda - m + s)] \qquad (15)$$

for $0 < s \leq \lambda$. [$N(s)$ should be considered zero for $s \leq 0$.]

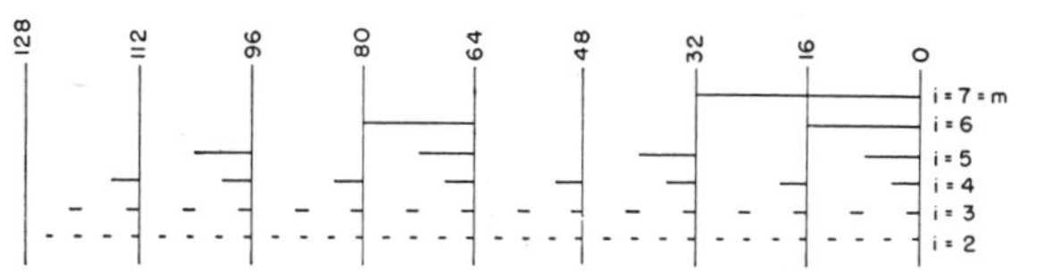

Fig. 6—Distribution of residues of $j2^i$ ($m = 7$, $\lambda = 4$).

Now let

$$R(s) = 1 - N(s)\cdot 2^{-m}.$$

Since $N(s)$ includes the zero residue, the actual number of parity digits is at least $N(s) - 1$. The actual number of information digits is at most $2^m - 1 - N(s) + 1 = 2^m - N(s)$. The actual rate would be at most $[2^m - N(s)]/(2^m - 1)$, but for large m, this is approximately $R(s)$. Then

$$N(s) = 2^m[1 - R(s)],$$

and substitution in (15) results in a difference equation for $R(s)$:

$$R(s) = R(s-1) - 2^{\lambda-m}R(s - m + \lambda) \qquad (16)$$

for $0 < s$. [$R(s)$ should be considered to be 1 for $s \leq 0$.] Clearly,

$$1 \geq R(s) \geq 0 \quad \text{for all} \quad s. \qquad (17)$$

It follows at once from (16) and (17) that $R(s)$ is nonincreasing. Now if there exists $\epsilon > 0$ such that $R(s) > \epsilon$ for all s, choose any $s_0 > m - \lambda + (2^{m-\lambda}/\epsilon)$. Then $R(s_0) = [R(s_0) - R(s_0 - 1)] + [R(s_0 - 1) - R(s_0 - 2)] + \cdots + [R(m - \lambda + 1) - R(m - \lambda)] + R(m - \lambda)$ trivially $= R(m - \lambda) - 2^{\lambda-m}[R(s_0 - m + \lambda) + R(s_0 - m + \lambda - \ ) + \cdots + R(1)]$ by (16) $< R(m - \lambda) - 2^{\lambda-m}(s_0 - m + \lambda)\epsilon$ by hyothesis $< R(m - \lambda) - 1$ by choice of $s_0 < 0$ by half of (17), contradicting the other half, and proving that $R(S) \to 0$ as $s \to \infty$ must hold.

Now suppose that it is required that errors be corrected in a fraction 2^{-v} of the number of digits in a code word. Then

$$2^{-v} = 2^\lambda/2^m - 1 \approx 2^{\lambda-m},$$

so $v \approx m - \lambda$. Then, taking $s = \lambda$, $R(\lambda) = R(m - v)$ is an upper bound on the rate for a code with $2^m - 1$ digits. As m increases this approaches zero. Since rate is a monotone nonincreasing function of the number of errors correctable and the rate approaches zero for arbitrarily small fractions $t/n = 2^{-v}$, it must approach zero for any fraction $t/n > 0$.

Acknowledgment

I have benefited greatly from discussions with many people at the IBM Research Laboratory and at the Research Laboratory of Electronics at Massachusetts Institute of Technology. E. Prange, of the Air Force Cambridge Research Center, J. Griesmer and J. Selfridge of IBM, and M. P. Schutzenberger and S. Golomb at the Research Laboratory of Electronics were especially helpful.

Most of all, I am indebted to R. C. Bose of the University of North Carolina, for lecturing on his and Chaudhuri's fine work so soon after it was done and for the very stimulating discussion we had during his visit to the IBM Research Laboratory in August 1959.

Part of the computation work was done at the M.I.T. Computation Center.

Low-Density Parity-Check Codes*

R. G. GALLAGER†

Summary—A low-density parity-check code is a code specified by a parity-check matrix with the following properties: each column contains a small fixed number $j \geq 3$ of 1's and each row contains a small fixed number $k > j$ of 1's. The typical minimum distance of these codes increases linearly with block length for a fixed rate and fixed j. When used with maximum likelihood decoding on a sufficiently quiet binary-input symmetric channel, the typical probability of decoding error decreases exponentially with block length for a fixed rate and fixed j.

A simple but nonoptimum decoding scheme operating directly from the channel a posteriori probabilities is described. Both the equipment complexity and the data-handling capacity in bits per second of this decoder increase approximately linearly with block length.

For $j > 3$ and a sufficiently low rate, the probability of error using this decoder on a binary symmetric channel is shown to decrease at least exponentially with a root of the block length. Some experimental results show that the actual probability of decoding error is much smaller than this theoretical bound.

Coding for Digital Data Transmission

CODING for error correction is one of the many tools available for achieving reliable data transmission in communication systems. For a wide variety of channels, the Noisy Channel Coding Theorem [1, 6] of Information Theory proves that if properly coded information is transmitted at a rate below channel capacity, then the probability of decoding error can be made to approach zero exponentially with the code length. The theorem does not, however, relate the code length to the computation time or the equipment costs necessary to achieve this low error probability. This paper describes a class of coding and decoding schemes that can utilize the long block lengths necessary for low error probability without requiring excessive equipment or computation.

The codes to be discussed here are special examples of parity-check codes.[1] The code words of a parity-check code are formed by combining a block of binary information digits with a block of check digits. Each check digit is the modulo 2 sum[2] of a prespecified set of information digits. These formation rules for the check digits can be conveniently represented by a parity-check matrix, as in Fig. 1. This matrix represents a set of linear homogeneous modulo 2 equations called parity-check equations, and the set of code words is the set of solutions of these equations. We call the set of digits contained in a parity-check equation a parity-check set. For example, the first parity-check set in Fig. 1 is the set of digits (1, 2, 3, 5).

The use of parity-check codes makes coding (as distinguished from decoding) relatively simple to implement. Also, as Elias [3] has shown, if a typical parity-check code of long block length is used on a binary symmetric channel, and if the code rate is between *critical rate* and channel capacity, then the probability of decoding error will be almost as small as that for the best possible code of that rate and block length.

Unfortunately, the decoding of parity-check codes is not inherently simple to implement, and thus we must look for special classes of parity-check codes, such as described below, for which reasonable decoding procedures exist.

Low-Density Parity-Check Codes

Low-density parity-check codes are codes specified by a matrix containing mostly 0's and only a small number of 1's. In particular, an (n, j, k) low-density code is a code of block length n with a matrix like that of Fig. 2 where each column contains a small fixed number, j, of 1's and each row contains a small fixed number, k, of 1's. Note that this type of matrix does not have the check digits appearing in diagonal form as in Fig. 1. However, for

INFORMATION DIGITS / CHECK DIGITS

x_1	x_2	x_3	x_4	x_5	x_6	x_7
1	1	1	0	1	0	0
1	1	0	1	0	1	0
1	0	1	1	0	0	1

$\longleftrightarrow$

$$x_5 = x_1 \oplus x_2 \oplus x_3$$
$$x_6 = x_1 \oplus x_2 \oplus x_4$$
$$x_7 = x_1 \oplus x_3 \oplus x_4$$

Fig. 1—Example of parity-check matrix.

1	1	1	1	0	0	0	0	0	0	0	0	0	0	0	0	0	0	0	0
0	0	0	0	1	1	1	1	0	0	0	0	0	0	0	0	0	0	0	0
0	0	0	0	0	0	0	0	1	1	1	1	0	0	0	0	0	0	0	0
0	0	0	0	0	0	0	0	0	0	0	0	1	1	1	1	0	0	0	0
0	0	0	0	0	0	0	0	0	0	0	0	0	0	0	0	1	1	1	1
1	0	0	0	1	0	0	0	1	0	0	0	1	0	0	0	0	0	0	0
0	1	0	0	0	1	0	0	0	1	0	0	0	0	0	0	1	0	0	0
0	0	1	0	0	0	1	0	0	0	0	0	0	1	0	0	0	1	0	0
0	0	0	1	0	0	0	0	0	0	1	0	0	0	1	0	0	0	1	0
0	0	0	0	0	0	0	1	0	0	0	1	0	0	0	1	0	0	0	1
1	0	0	0	0	1	0	0	0	0	0	1	0	0	0	0	0	1	0	0
0	1	0	0	0	0	1	0	0	0	1	0	0	0	0	1	0	0	0	0
0	0	1	0	0	0	0	1	0	0	0	0	1	0	0	0	0	0	1	0
0	0	0	1	0	0	0	0	1	0	0	0	0	1	0	0	1	0	0	0
0	0	0	0	1	0	0	0	0	1	0	0	0	0	1	0	0	0	0	1

Fig. 2—Example of a low-density code matrix; $N = 20, j = 3, k = 4$.

* Received by the PGIT, March 15, 1961. Supported in part by the U. S. Army Signal Corps, the AF Office of Scientific Research, the Office of Naval Research, the Mass. Inst. Tech. Computation Center, and the Applied Science Div., Melpar, Inc.

† Elec. Engrg. Dept., Mass. Inst. Tech., Cambridge, Mass.

[1] For a more detailed discussion of parity-check codes, see Slepian [2].

[2] The modulo 2 sum is 1 if the ordinary sum is odd and 0 if the ordinary sum is even.

Reprinted from *IRE Trans. Inform. Theory*, vol. IT-8, pp. 21–28, Jan. 1962.

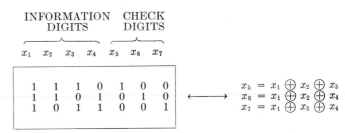

coding purposes, the equations represented by these matrices can always be solved to give the check digits as explicit sums of information digits.

These codes are not optimum in the somewhat artificial sense of minimizing probability of decoding error for a given block length, and it can be shown that the maximum rate at which these codes can be used is bounded below channel capacity. However, a very simple decoding scheme exists for low-density codes, and this compensates for their lack of optimality.

The analysis of a low-density code of long block length is difficult because of the immense number of code words involved. It is simpler to analyze a whole ensemble of such codes because the statistics of an ensemble permit one to average over quantities that are not tractable in individual codes. From the ensemble behavior, one can make statistical statements about the properties of the member codes. Furthermore, one can with high probability find a code with these properties by random selection from the ensemble.

In order to define an ensemble of (n, j, k) low-density codes, consider Fig. 2 again. Note that the matrix is divided into j submatrices, each containing a single 1 in each column. The first of these submatrices contains all its 1's in descending order; *i.e.*, the i'th row contains 1's in columns $(i - 1)k + 1$ to ik. The other submatrices are merely column permutations of the first. We define an ensemble of (n, j, k) codes as the ensemble resulting from random permutation of the columns of each of the bottom $j - 1$ submatrices of a matrix such as Fig. 2, with equal probability assigned to each permutation.[3] There are two interesting results that can be proven using this ensemble, the first concerning the minimum distance of the member codes, and the second concerning the probability of decoding error.

The minimum distance of a code is the number of positions in which the two nearest code words differ. Over the ensemble, the minimum distance of a member code is a random variable, and it can be shown [4] that the distribution function of this random variable can be overbounded by a function such as sketched in Fig. 3. As the block length increases, for fixed $j \geq 3$ and $k > j$, this function approaches a unit step at a fixed fraction δ_{jk} of the block length. Thus, for large n, practically all the codes in the ensemble have a minimum distance of at least $n\delta_{jk}$. In Fig. 4 this ratio of typical minimum distance to block length is compared to that for a parity-check code chosen at random, *i.e.*, with a matrix filled in with equiprobable independent binary digits. It should be noted that for all the specific nonrandom procedures known for constructing codes, the ratio of minimum distance to block length appears to approach 0 with increasing block length.

[3] There is no guarantee that all the rows in such matrices will be linearly independent, and, in fact, all the matrices to be discussed here contain at least j-1 dependent rows. This simply means that the codes have a slightly higher information rate than the matrix indicates.

The probability of error using maximum likelihood decoding for low-density codes clearly depends upon the particular channel on which the code is being used. The results are particularly simple for the case of the BSC, or binary symmetric channel, which is a binary-input, binary-output, memoryless channel with a fixed probability of transition from either input to the opposite output. Here it can be shown [4] that over a reasonable range of channel transition probabilities, the low-density code has a probability of decoding error that decreases exponentially with block length and that the exponent is the same as that for the optimum code of slightly higher rate as given in Fig. 5.

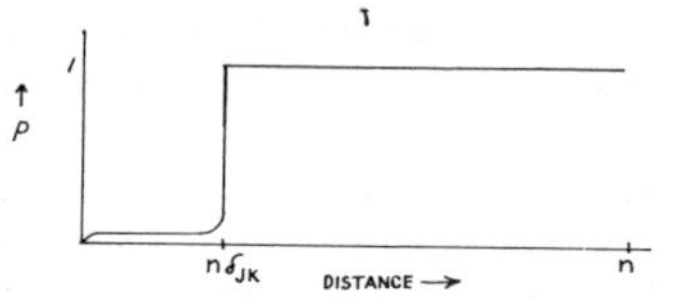

Fig. 3—Sketch of bound to minimum distance distribution function.

j	k	Rate	δ_{JK}	δ
5	6	0.167	0.255	0.263
4	5	0.2	0.210	0.241
3	4	0.25	0.122	0.214
4	6	0.333	0.129	0.173
3	5	0.4	0.044	0.145
3	6	0.5	0.023	0.11

Fig. 4—Comparison of δjk, the ratio of typical minimum distance to block length for an (n, j, k) code, to δ, the same ratio for an ordinary parity-check code of the same rate.

j	k	Rate	RATE FOR EQUIVALENT OPTIMUM CODE
3	6	0.5	0.555
3	5	0.4	0.43
4	6	0.333	0.343
3	4	0.25	0.266

Fig. 5—Loss of rate associated with low-density codes.

Although this result for the BSC shows how closely low-density codes approach the optimum, the codes are not designed primarily for use on this channel. The BSC is an approximation to physical channels only when there is a receiver that makes decisions on the incoming signal on a bit-to-bit basis. Since the decoding procedure to be described later can actually use the channel *a posteriori* probabilities, and since a bit-by-bit decision throws away available information, we are actually interested in the probability of decoding error of a binary-input, continuous-output channel. If the noise affects the input symbols symmetrically, then this probability can again be bounded by an exponentially decreasing function of the block length, but the exponent is a rather complicated function of the channel and code. It is expected that the same type of result holds for a wide class of channels with memory, but no analytical results have yet been derived. For channels with memory, it is clearly advisable, however, to modify the ensemble somewhat, particularly by permuting the first submatrix and possibly by changing the probability measure on the permutations.

Decoding

Introduction

Two decoding schemes will be described here that appear to achieve a reasonable balance between complexity and probability of decoding error. The first is particularly simple but is applicable only to the BSC at rates far below channel capacity. The second scheme, which decodes directly from the *a posteriori* probabilities at the channel output, is more promising but can be understood more easily after the first scheme is described.

In the first decoding scheme, the decoder computes all the parity checks and then changes any digit that is contained in more than some fixed number of unsatisfied parity-check equations. Using these new values, the parity checks are recomputed, and the process is repeated until the parity checks are all satisfied.

If the parity-check sets are small, this decoding procedure is reasonable, since most of the parity-check sets will contain either one transmission error or no transmission errors. Thus when most of the parity-check equations checking on a digit are unsatisfied, there is a strong indication that that digit is in error. For example, suppose a transmission error occurred in the first digit of the code in Fig. 2. Then parity checks 1, 6, and 11 would be violated, and all three parity-check equations checking digit 1 would be violated. On the other hand, at most, one of the three equations checking on any other digit in the block would be violated.

To see how an arbitrary digit d can be corrected even if its parity-check sets contain more than one transmission error, consider the tree structure in Fig. 6. Digit d is represented by the node at the base of the tree, and each line rising from this node represents one of the parity-check sets containing digit d. The other digits in these parity-check sets are represented by the nodes on the first tier of the tree. The lines rising from tier 1 to tier 2 of the tree represent the other parity-check sets containing the digits on tier 1, and the nodes on tier 2 represent the other digits in those parity-check sets. Notice that if such a tree is extended to many tiers, the same digit will appear in more than one place, but this will be discussed.

Assume now that both digit d and several of the digits in the first tier are transmission errors. Then on the first decoding attempt, the error-free digits in the second tier and their parity-check constraints will allow correction of the errors in the first tier. This in turn will allow correction of digit d on the second decoding attempt. Thus digits and parity-check equations can aid in decoding a digit seemingly unconnected with them. The probabilistic decoding scheme to be described next utilizes these extra digits and extra parity-check equations more systematically.

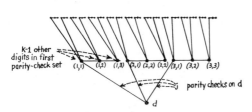

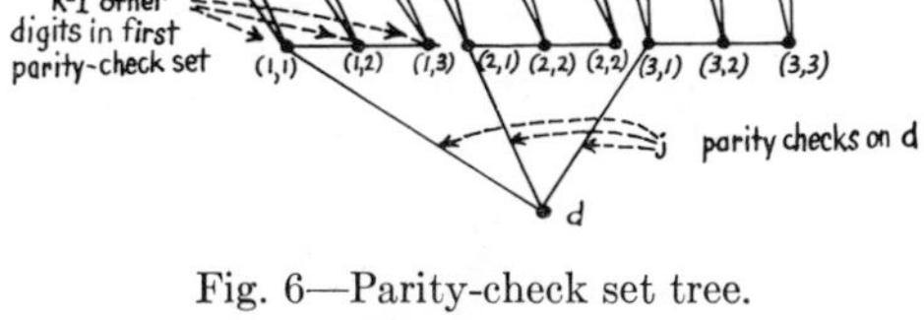

Fig. 6—Parity-check set tree.

Probabilistic Decoding

Assume that the code words from an (n, j, k) code are used with equal probability on an arbitrary binary-input channel. For any digit d, using the notation of Fig. 6, an iteration process will be derived that on the m'th iteration computes the probability that the transmitted digit in position d is a 1 conditional on the received symbols out to and including the m'th tier. For the first iteration, we can consider digit d and the digits in the first tier to form a subcode in which all sets of these digits that satisfy the j parity-check equations in the tree have equal probability of transmission.[4]

Consider the ensemble of events in which the transmitted digits in the positions of d and the first tier are independent equiprobable binary digits, and the probabilities of the received symbols in these positions are determined by the channel transition probabilities $P_x(y)$. In this ensemble the probability of any event conditional on the event that the transmitted digits satisfy the j parity-check equations is the same as the probability of an event in the subcode described above. Thus, *within this ensemble* we want to find the probability that the transmitted digit in position d is a 1 conditional on the set of received symbols $\{y\}$ and on the event S that the transmitted digits satisfy the j parity-check equations on digit d. We write this as

$$Pr\,[x_d = 1 \mid \{y\}, S].$$

Using this ensemble and notation, we can prove the following theorem:

Theorem 1: Let P_d be the probability that the transmitted digit in position d is a 1 conditional on the received digit in position d, and let P_{il} be the same probability for the l'th digit in the i'th parity-check set of the first tier in Fig. 6. Let the digits be statistically independent of each other, and let S be the event that the transmitted digits satisfy the j parity-check constraints on digit d. Then

$$\frac{Pr\,[x_d = 0 \mid \{y\}, S]}{Pr\,[x_d = 1 \mid \{y\}, S]} = \frac{1 - P_d}{P_d} \prod_{i=1}^{j} \left[\frac{1 + \prod_{l=1}^{k-1} (1 - 2P_{il})}{1 - \prod_{l=1}^{k-1} (1 - 2P_{il})} \right].$$

[4] An exception to this statement occurs if some linear combination of those parity-check equations not containing d produces a parity-check set containing only digits in the first tier. This will be discussed later but is not a serious restriction.

In order to prove this theorem, we need the following lemma:

Lemma 1: Consider a sequence of m independent binary digits in which the l'th digit is 1 with probability P_l. Then the probability that an even number of digits are 1 is

$$\frac{1 + \prod_{l=1}^{m} (1 - 2P_l)}{2}.$$

Proof of Lemma: Consider the function

$$\prod_{l=1}^{m} (1 - P_l + P_l t).$$

Observe that if this is expanded into a polynomial in t, the coefficient of t^i is the probability of i1's. The function $\prod_{l=1}^{m} (1 - P_l - P_l t)$ is identical except that all the odd powers of t are negative. Adding these two functions, all the even powers of t are doubled, and the odd terms cancel out. Finally, letting $t = 1$ and dividing by 2, the result is the probability of an even number of ones. But

$$\frac{\prod_{l=1}^{m} (1 - P_l + P_l) + \prod_{l=1}^{m} (1 - P_l - P_l)}{2} = \frac{1 + \prod_{l=1}^{m} (1 - 2P_l)}{2},$$

thus proving the lemma.

Proof of Theorem: By the definition of conditional probabilities,

$$\frac{Pr\,[x_d = 0 \mid \{y\}, S]}{Pr\,[x_d = 1 \mid \{y\}, S]} = \left(\frac{1 - P_d}{P_d}\right)\left(\frac{Pr\,(S \mid x_d = 0, \{y\})}{Pr\,(S \mid x_d = 1, \{y\})}\right).$$

Given that $x_d = 0$, a parity check on d is satisfied if the other $(k - 1)$ positions in the parity-check set contain an even number of 1's. Since all digits in the ensemble are statistically independent, the probability that all j parity checks are satisfied is the product of the probabilities of the individual checks being satisfied. Using Lemma 1 this is

$$Pr\,(S \mid x_d = 0, \{y\}) = \prod_{i=1}^{j} \left[\frac{1 + \prod_{l=1}^{k-1} (1 - 2P_{il})}{2}\right]. \quad (3)$$

Similarly,

$$Pr\,(S \mid x_d = 1, \{y\}) = \prod_{l=1}^{j} \left[\frac{1 - \prod_{l=1}^{k-1} (1 - 2P_{il})}{2}\right]. \quad (4)$$

Substituting (3) and (4) into (2) we get the statement of the theorem; Q.E.D.

Judging from the complexity of this result, it would appear difficult to compute the probability that the transmitted digit in position d is a 1 conditional on the received digits in two or more tiers of the tree in Fig. 6. Fortunately, however, the many-tier case can be solved from the 1-tier case by a simple iterative technique.

Consider first the 2-tier case. We can use Theorem 1 to find the probability that each of the transmitted digits in the first tier of the tree is a 1 conditional on the received digits in the second tier. The only modification of the theorem is that the first product is taken over only $j - 1$ terms, since the parity-check set containing digit d is not included. Now these probabilities can be used in (1) to find the probability that the transmitted digit in position d is 1. The validity of the procedure follows immediately from the independence of the new values of P_{il} in the ensemble used in Theorem 1. By induction, this iteration process can be used to find the probability that the transmitted digit d is 1 given any number of tiers of distinct digits in the tree.

The general decoding procedure for the entire code may now be stated. For each digit and each combination of $j - 1$ parity-check sets containing that digit, use (1) to calculate the probability of a transmitted l conditional on the received symbols in the $j - 1$ parity-check sets. Thus there are j different probabilities associated with each digit, each one omitting 1 parity-check set. Next, these probabilities are used in (1) to compute a second-order set of probabilities. The probability to be associated with one digit in the computation of another digit d is the probability found in the first iteration, omitting the parity-check set containing digit d. If the decoding is successful, then the probabilities associated with each digit approach 0 or 1 (depending on the transmitted digit) as the number of iterations is increased. This procedure is only valid for as many iterations as meet the independence assumption in Theorem 1. This assumption breaks down when the tree closes upon itself. Since each tier of the tree contains $(j - 1)(k - 1)$ more nodes than the previous tier, the independence assumption must break down while m is quite small for any code of reasonable block length. This lack of independence can be ignored, however, on the reasonable assumption that the dependencies have a relatively minor effect and tend to cancel each other out somewhat. Also, even if dependencies occur in the m'th iteration, the first $m - 1$ iterations have reduced the equivocation in each digit. Then we can consider the probabilities after the $m - 1$ iterations to be a new received sequence that should be easier to decode than the original received sequence.

The most significant feature of this decoding scheme is that the computation per digit per iteration is independent of block length. Furthermore it can be shown that the average number of iterations required to decode is bounded by a quantity proportional to the log of the log of the block length.

For the actual computation of the probabilities in Theorem 1, it appears to be more convenient to use (1) in terms of log-likelihood ratios. Let

$$\ln \frac{1 - P_d}{P_d} = \alpha_d \beta_d, \qquad \ln \frac{1 - P_{il}}{P_{il}} = \alpha_{il} \beta_{il}, \quad (5)$$

$$\ln \left[\frac{Pr\,[x_d = 0 \mid \{y\}, S]}{Pr\,[x_d = 1 \mid \{y\}, S]}\right] = \alpha_d' \beta_d',$$

where α is the sign and β the magnitude of the log-likelihood ratio. After some manipulation, (1) becomes

$$\alpha'_d\beta'_d = \alpha_d\beta_d + \sum_{i=1}^{j}\left\{\left(\prod_{l=1}^{k-1}\alpha_{il}\right)f\left[\sum_{l=1}^{k-1}f(\beta_{il})\right]\right\}, \tag{6}$$

where

$$f(\beta) = \ln\frac{e^\beta + 1}{e^\beta - 1}.$$

The calculation of the log-likelihood ratios in (6) for each digit can be performed either serially in time or by parallel computations. The serial computation can be programmed for a general-purpose computer, and the experimental data at the end of this paper was obtained in this manner. For fast decoding, parallel computing is more promising, and Fig. 7 sketches a simplified block diagram showing how this can be done.

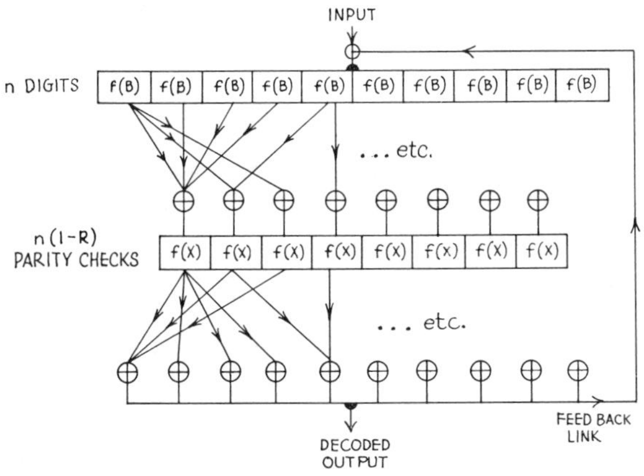

Fig. 7—Decoding apparatus.

If the input to the decoder is in the form of a log-likelihood ratio, the first row of boxes in Fig. 7 computes $f(\beta)$ for each digit, corresponding to the right-most operation in (6). The output from the adders on the next row is $\sum_{l=1}^{k-1} f(\beta_{il})$, corresponding to the two right-most operations in (6). Likewise, successive rows in Fig. 7 correspond to operations in (6) working to the left. Clearly, Fig. 7 omits some details, such as operations on the signs of the log-likelihood ratios and the association of j different log-likelihood ratios with each digit, but these create no essential difficulty.

We see from Fig. 7 that a parallel computer can be simply instrumented requiring principally a number proportional to n of analogue adders, modulo 2 adders, amplifiers, and nonlinear circuits to approximate the function $f(\beta)$. How closely this function must be approximated is a subject for further study, but there are indications that it is not critical.[5]

[5] Some recent experimental work indicates that if computation is strictly digital, 6 significant bits are sufficient to represent $f(\beta)$ without appreciable effect on the probability of decoding error.

Probability of Error Using Probabilistic Decoding

A mathematical analysis of probabilistic decoding is difficult, but a very weak bound on the probability of error can be derived easily.

Assume a BSC with crossover probability p_0 and assume first an (n, j, k) code with $j = 3$ parity-check sets containing each digit. Consider a parity-check set tree, as in Fig. 6, containing m independent tiers, but let the tiers be numbered from top to bottom so that the uppermost tier is the 0 tier and the digit to be decoded is tier m.

Modify the decoding procedure as follows: if both parity checks corresponding to the branches rising from a digit in the first tier are unsatisfied, change the digit; using these changed digits in the first tier, perform the same operation on the second tier, and continue this procedure down to digit d.

The probability of decoding error for digit d after this procedure is an upper bound to that resulting from making a decision after the m'th iteration of the probabilistic decoding scheme. Both procedures base their decision only on the received symbols in the m-tier tree, but the probabilistic scheme always makes the most likely decision from this information.

We now determine the probability that a digit in the first tier is in error after applying the modified decoding procedure described above. If the digit is received in error (an event of probability p_0) then a parity check constraining that digit will be unsatisfied if, and only if, an even number (including zero) of errors occur among the other $k - 1$ digits in the parity-check set. From Lemma 1, the probability of an even number of errors among $k - 1$ digits is

$$\frac{1 + (1 - 2p_0)^{k-1}}{2}. \tag{7}$$

Since an error will be corrected only if both parity checks rising from the digit are unsatisfied, the following expression gives the probability that a digit in the first tier is received in error and then corrected.

$$p_0\left[\frac{1 + (1 - 2p_0)^{k-1}}{2}\right]^2. \tag{8}$$

By the same reasoning, (9) gives the probability that a digit in the first tier is received correctly but then changed because of unsatisfied parity checks.

$$(1 - p_0)\left[\frac{1 - (1 - 2p_0)^{k-1}}{2}\right]^2. \tag{9}$$

Combining (8) and (9), the probability of error of a digit in the first tier after applying this decoding process is

$$p_1 = p_0 - p_0\left[\frac{1 + (1 - 2p_0)^{k-1}}{2}\right]^2 + (1 - p_0)\left[\frac{1 - (1 - 2p_0)^{k-1}}{2}\right]^2. \tag{10}$$

By induction it easily follows that if p_i is the probability of error after processing of a digit in the ith tier, then

$$p_{i+1} = p_0 - p_0\left[\frac{1 + (1 - 2p_i)^{k-1}}{2}\right]^2 + (1 - p_0)\left[\frac{1 - (1 - 2p_i)^{k-1}}{2}\right]^2. \tag{11}$$

We now show that for sufficiently small p_0, the sequence $[p_i]$ converges to 0. Consider Fig. 8 which is a sketch of p_{i+1} as a function of p_i. Since the ordinate for one value of i is the abscissa for the next, the dotted zig-zag line illustrates a convenient graphical method of finding p_i for successive values of i. It can be seen from the figure that if

$$\begin{aligned} 0 < p_{i+1} < p_i &\quad (\text{for } 0 < p_i \le p_0) \\ p_{i+1} = p_i &\quad (\text{for } p_i = 0), \end{aligned} \tag{12}$$

then the sequence $[p_i] \to 0$. It can be seen from (11) that for p_0 sufficiently small, inequality (12) is satisfied. Fig. 9 gives the maximum p_0 for several values of k.

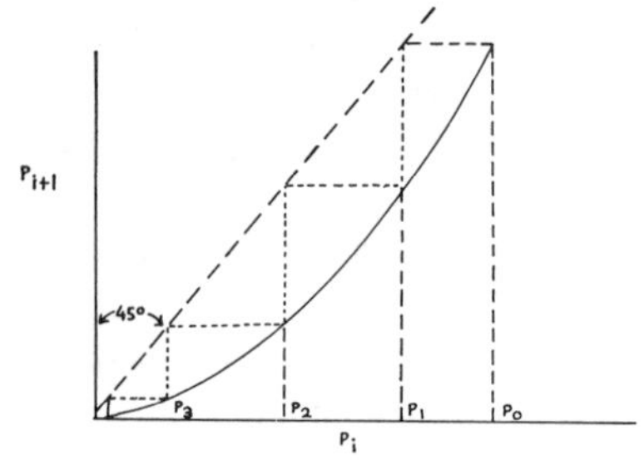

Fig. 8.

j	K	Rate	p_0
3	6	0.5	0.04
3	5	0.4	0.061
4	6	0.333	0.075
3	4	0.25	0.106

Fig. 9—Maximum p_0 for weak bound decoding convergence.

The rate at which $[p_i] \to 0$ may be determined by noting from (11) that for small p_i

$$p_{i+1} \approx p_i 2(k - 1)p_0.$$

From this it is easy to show that for sufficiently large i,

$$p_i \approx c[2(k - 1)p_0]^i,$$

where C is a constant independent of i. Since the number of independent tiers in the tree increases logarithmically with block length, this bound to the probability of decoding error approaches 0 with some small negative power of block length. This slow approach to 0 appears to be a consequence of the modification of the decoding scheme and of the strict independence requirement, rather than of probabilistic decoding as a whole.

This same argument can be applied to codes with more than 3 parity-check sets per digit. Stronger results will be achieved if for some integer b, to be determined later, a digit is changed whenever b or more of the parity-check constraints rising from the digit are violated. Using this criterion and following the reasoning leading to (11) we obtain

$$\begin{aligned} p_{i+1} = p_0 - p_0 \sum_{l=b}^{j-1} \binom{j-1}{l} &\cdot \left[\frac{1 + (1 - 2p_i)^{k-1}}{2}\right]^l \left[\frac{1 - (1 - 2p_i)^{k-1}}{2}\right]^{j-1-l} \\ + (1 - p_0) \sum_{l=b}^{j-1} \binom{j-1}{l} &\cdot \left[\frac{1 - (1 - 2p_i)^{k-1}}{2}\right]^l \left[\frac{1 + (1 - 2p_i)^{k-1}}{2}\right]^{j-1-l}. \end{aligned} \tag{13}$$

The integer b can now be chosen to minimize p_{i+1}. The solution to this minimization is the smallest integer b for which

$$\frac{1 - p_0}{p_0} \le \left[\frac{1 + (1 - 2p_i)^{k-1}}{1 - (1 - 2p_i)^{k-1}}\right]^{2b-j+l}. \tag{14}$$

From this equation, it is seen that as p_i decreases, b also decreases. Fig. 10 sketches p_{i+1} as a function of p_i when b is changed according to (14). The break points in the figure represent changes in b.

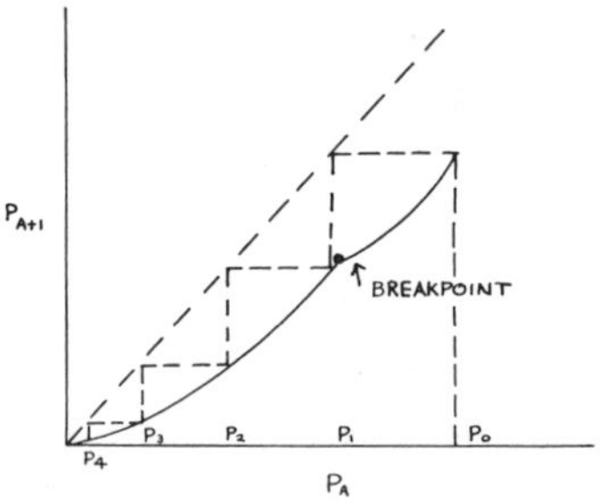

Fig. 10—Behavior of decoding iterations for $j > 3$.

The proof that the probability of decoding error approaches 0 with an increasing number of iterations for sufficiently small cross over probabilities is the same as before. The asymptotic approach of the sequence $[p_i]$ to 0 is different, however. From (14), if p_i is sufficiently small, b takes the value $j/2$ for j even and $j + 1/2$ for j odd. Using these values of b and expanding (13) in a power series in p_i,

$$p_{i+1} = p_0\binom{j-1}{\frac{j-1}{2}}(k - 1)^{(j-1)/2} p_i^{(j-1)/2} + \text{higher order terms} \quad (j \text{ odd}) \tag{15}$$

$$p_{i+1} = \binom{j-1}{j/2}(k - 1)^{j/2} p_i^{j/2} + \text{higher order terms} \quad (j \text{ even}).$$

Using this, it can be shown that for a suitably chosen positive constant C_{jk} and sufficiently large i

$$p_i \leq \exp\left[-C_{jk}\left(\frac{j-1}{2}\right)^i\right] \qquad (j \text{ odd})$$

$$p_i \leq \exp\left[-C_{jk}\left(\frac{j}{2}\right)^i\right] \qquad (j \text{ even}). \tag{16}$$

It is interesting to relate this result to the block length of the code. Since there are $(j-1)^m(k-1)^m$ digits in the m'th tier of a tree, n must be at least this big, giving the left side of (17). On the other hand, a specific procedure can be described [4] for constructing codes satisfying the right side of (17).

$$\frac{\ln(n)}{\ln(j-1)(k-1)} \geq m \geq \frac{\ln\left(\frac{n}{2k} - \frac{n}{2j(k-1)}\right)}{2\ln(k-1)(j-1)}. \tag{17}$$

Combining (16) and (17), the probability of decoding error for a code satisfying (17) is bounded by

$$P_m \leq \exp -C_{jk}\left[\frac{n}{2k} - \frac{n}{2j(k-1)}\right]^{\ln[(j-1)/2]/[2\ln(j-1)(k-1)]} \qquad (j \text{ odd})$$

$$P_m \leq \exp -C_{jk}\left[\frac{n}{2k} - \frac{n}{2j(k-1)}\right]^{\ln[(j/2)]/[2\ln(j-1)(k-1)]} \qquad (j \text{ even}).$$

For $j > 3$, this probability of decoding error bound decreases exponentially with a root of n. Observe that if the number of iterations m which can be made without dependencies were $(2\ln(j-1)(k-1))/(\ln j/2)$ times larger, then the probability of decoding error would decrease exponentially with n. It is hypothesized that using the probabilistic decoding scheme and continuing to iterate after dependencies occur will produce this exponential dependence.

A second way to evaluate the probabilistic decoding scheme is to calculate the probability distributions of the log-likelihood ratios in (6) for a number of iterations. This approach makes it possible to find whether a code of given j and k is capable of achieving arbitrarily small error probability on any given channel. With the aid of the IBM 709 computer, it was found that a code with $j = 3$, $k = 6$ is capable of handling transition probabilities up to 0.07 and with $j = 3$, $k = 4$, transition probabilities up to 0.144 can be handled. These figures are particularly interesting since they disprove the common conjecture that the computational cutoff rate of sequential decoding [7] bounds the rate at which any simple decoding scheme can operate.

Experimental Results

The probability of decoding an error $P(e)$ associated with a coding and decoding scheme can be directly measured by simulating both the scheme and the channel of interest on a computer. Unfortunately, the experiment must be repeated until there are many decoding failures if $P(e)$ is to be evaluated with any accuracy, and thus many times $1/P(e)$ trials are necessary. For block lengths of about 500, an IBM 7090 computer requires about 0.1 seconds per iteration to decode by the probabilistic decoding scheme. Consequently, many hours of computation time are necessary to evaluate even a $P(e)$ of the order of 10^{-4}.

Because of limitations on available computer time, all of the results presented will be for situations in which $P(e)$ is large. Certainly it would be more interesting to have results for small $P(e)$. However, the data presented are at least sufficiently convincing to justify further experimental work.

The first two codes to be discussed were used on the BSC and the last code on a Gaussian noise channel. The BSC was unduly emphasized for the following reasons: first, the effect of channel variations on the BSC can be eliminated by controlling the number of crossovers rather than the crossover probability; next, the BSC is convenient for comparison with other coding and decoding schemes; and finally, it is likely that the operation of the decoding scheme on one channel is typical of its operation on other channels.

A (504, 3, 6) Code on Binary Symmetric Channel

A code of block length 504 with each digit contained in three parity-check sets and each parity-check set containing 6 digits was selected by the IBM 704 computer using a pseudo-random number routine. The only restriction on the code was that no two parity-check sets should contain more than one digit in common. That restriction guaranteed the validity of the first-order iteration in the decoding process and also excluded the remote possibility of choosing a code with minimum distance of 2.

Fig. 11 plots the fraction of times the decoder was unable to decode correctly as a function of the number of crossovers. The number in parentheses beside each point is the number of trials performed with that number of crossovers. In all the trials on this code, the decoder never decoded to the wrong code word; it just failed to find a code word. If a feedback channel is available, this inability to decode troublesome noise patterns is not a serious limitation, since retransmission is possible.

Out of the error patterns correctly decoded, 86 per cent were decoded in between 9 and 19 iterations. The rest were spread out between 20 and 40 iterations. There appeared to be a slight increase in the number of iterations necessary to decode as the number of crossovers was increased from 37 to 41, but not enough to be statistically significant. The other curve drawn in Fig. 11 is the theoretical bound using maximum likelihood decoding.

In a later test made on an IBM 7090 computer, a (504,3,6) code was generated and 1000 sequences of 32 errors each were decoded. The process failed to decode in 26 cases and decoded the other 974 sequences correctly.

These results appear encouraging when we observe that no other known coding and decoding scheme of this rate is able to decode this many errors with a reasonable amount of computation. How well the decoding scheme works with smaller numbers of errors is of greater interest, though. The rate at which the experimental probability of error decreases as the number of crossovers decreases is discouraging, but there is no justification for extrapolating this curve to much smaller numbers of crossovers. Either a great deal of additional experimental data or a new theoretical approach will be necessary for evaluation of smaller numbers of cross overs.

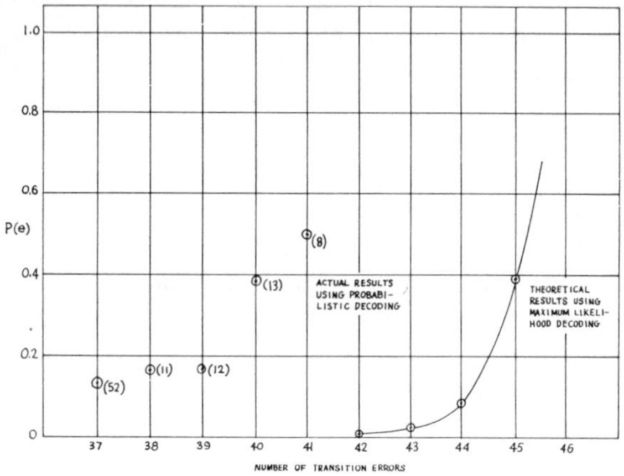

Fig. 11—Experimental results for (504, 3, 6) code as function of number of transition errors.

A (500, 3, 4) Code on the Binary Symmetric Channel

A (500, 3, 4) code, which has a rate of $\frac{1}{4}$, was chosen by the IBM 704 computer in the same way as the (504, 3, 6) code of the last section. Sequences containing from 20 to 77 crossovers were put in to be decoded. There were two sequences for each number of crossovers from 65 to 69 and from 72 to 77 and one sequence for all the other numbers. The decoding was successful for all sequences except one 73-crossover case, one 75-crossover case, and both 77-crossover cases. The theoretical error-correcting breakpoint for the (500, 3, 4) ensemble is 103 errors, and the error-correcting breakpoint for the ensemble of all codes of rate $\frac{1}{4}$ is 108 errors.

A (500, 3, 5) Code on White Gaussian Noise Channel

Assume a channel that accepts inputs of plus or minus 1 and adds a Gaussian random variable of mean 0 and variance 1 to the input to form the output. The log-likelihood ratio of the input conditional on the output is simply twice the received signal. The channel capacity of this channel can be calculated [5] to be 0.5 bits per symbol. However, if the receiver converts the channel into a BSC by making a decision on each symbol and throwing away the probabilities, the probability of crossover becomes 0.16, and the channel capacity is reduced to 0.37 bits per symbol.

In this experiment a (500, 3, 5) code, which has a rate of 0.4 bits per symbol, was simulated on the IBM 704 computer along with the channel just described. Probabilistic decoding was performed using the log-likelihood ratios at the output of the channel. Out of 13 trials, the decoding scheme decoded correctly on 11 trials and failed to decode twice.

This experiment is interesting since it suggests that the loss of rate necessitated by the nonoptimum coding and decoding techniques proposed here is more than compensated for by the opportunity of using the *a posteriori* probabilities at the channel output.

Bibliography

[1] C. E. Shannon, "Certain results in coding theory for noisy channels," *Information and Control*, vol. 1, pp. 6–25; September, 1957.
[2] D. Slepian, "A class of binary signalling alphabets," *Bell Sys. Tech. J.*, vol. 35, pp. 203–234; January, 1956.
[3] P. Elias, "Coding for two noisy channels," in "Information Theory," C. Cherry, Ed., 3rd London Symp., September, 1955; Butterworths Scientific Publications, London, Eng., 1956.
[4] R. G. Gallager, "Low Density Parity Check Codes," Sc.D. thesis, Mass. Inst. Tech., Cambridge; September, 1960.
[5] F. J. Bloom, *et al.*, "Improvement of binary transmission by null-zone reception," Proc. IRE, vol. 45, pp. 963–975; July, 1957.
[6] R. M. Fano, "The Transmission of Information," The Technology Press, Cambridge, Mass.; 1961.
[7] J. M. Wozencraft and B. Reiffen, "Sequential Decoding," The Technology Press, Cambridge, Mass.; 1961.

Cyclic Decoding Procedures for Bose-Chaudhuri-Hocquenghem Codes

R. T. CHIEN, MEMBER, IEEE

Summary—This paper presents new general error-correction procedures for the class of codes known as Bose-Chaudhuri-Hocquenghem codes. It is shown that these procedures are efficient in time required for error-correction, and that they can be implemented with relatively simple electronic circuits. A comparison is also made with existing procedures.

I. Introduction

MANY EFFICIENT cyclic codes with simple implementation are known for the detection and correction of burst-error [1], [2]. For independent errors, however, the only general class of cyclic codes known are the Bose-Chaudhuri-Hocquenghem codes [1] [3] [4]. Since the construction of the codes does not provide means for instrumentation, an outstanding problem was, then, to devise a decoding procedure that lends itself to simple instrumentation. Such a procedure was first given by Peterson [5] for the binary case. It was later modified and generalized by Zieler [6]. Other procedures have also been proposed by a number of people [7] [8].

Recently, Bartee and Schneider [9] constructed an electronic decoder to implement the Peterson procedure for a binary 5-error-correcting Bose-Chaudhuri-Hocquenghem code that has a length of 127 bits. This decoder takes the form of a special purpose computer and is of reasonable complexity.

In comparison to decoders for burst-error-correcting codes, however, the Bartee and Schneider decoder is still an order of magnitude away, both in hardware complexity and in decoding delay. This can be partially attributed to the more complex nature of codes that correct independent errors.[1] Nevertheless, a need exists for more efficient procedures and simpler circuits.

The main purpose of this paper is to report on two new decoding procedures that take advantage of the cyclic property of Bose-Chaudhuri-Hocquenghem codes. The chief advantage of these procedures is in speeding up the decoding process a great deal, although there may also be a saving in hardware. The procedures are perfectly general and can be applied to decode any Bose-Chaudhuri-Hocquenghem code.

Basic knowledge in coding theory and the theory of finite fields is assumed in this paper. Otherwise, the paper is self contained. Readers interested in reviewing the theory of finite field may consult either Peterson [1] or Albert [10].

Manuscript received December 5, 1963; revised April 7, 1964. This work was sponsored by the Rome Air Development Center, Rome, N. Y., under contract AF 30 (602) 2958.

The author is with Thomas J. Watson Research Center, Yorktown Heights, N. Y.

[1] For instance, there are many more different combinations of t-tuple errors than different error bursts of length t.

II. Binary Bose-Chaudhuri-Hocquenghem Codes and the Peterson Decoding Procedure

In its unshortened form, binary Bose Chaudhuri-Hocquenghem codes exist for lengths $n = 2^m - 1$, and with, at most, mt check bits it can correct any set of t independent errors within the block of n bits. m and t are arbitrary positive integers. These codes may be described conveniently with the aid of the theory of finite fields [1].

Let α be a primitive element of the finite field $GF(2^m)$, then the t-error-correcting Bose-Chaudhuri Hocquenghem code may be described as the set of all polynomials $\{a(x)\}$ over $GF(2)$ of degree $n - 1$ or less, such that

$$a(\alpha^i) = 0, \qquad i = 1, 3, 5, \cdots, 2t - 1 \qquad (1)$$

where $a(x) = a_0 + a_1x + a_2x^2 + \cdots + a_{n-1}x^{n-1}$ and $a_i = 0, 1.$ $(i = 0, 1, 2, \cdots, n - 1)$. It is known in coding theory that these polynomials consist of all the multiples of a single polynomial $g(x)$, known as the generator polynomial of the code. $g(x)$ also satisfies the equations

$$g(\alpha^i) \equiv 0 \qquad i = 1, 3, 5, \cdots, 2t - 1. \qquad (2)$$

When the coefficient of $a(x)$ are viewed as a vector, one may equivalently define the code as the set of all n-tuples orthogonal to the parity check matrix

$$M = \begin{bmatrix} 1 & 1 & \cdot & \cdot & 1 \\ \alpha & \alpha^3 & \cdot & \cdot & \alpha^{2t-1} \\ \alpha^2 & \alpha^6 & \cdot & \cdot & \alpha^{2(2t-1)} \\ \alpha^3 & \alpha^9 & \cdot & \cdot & \alpha^{3(2t-1)} \\ \cdot & \cdot & \cdot & \cdot & \cdot \\ \alpha^{n-1} & \alpha^{3(n-1)} & \cdot & \cdot & \alpha^{(n-1)(2t-1)} \end{bmatrix} \qquad (3)$$

that is, the set of all a's for which

$$aM = 0 \qquad (4)$$

where $a = [a_0\ a_1\ a_2\ a_3 \cdots, a_{n-1}]$.

Example: Consider the finite field $GF(2^4)$ defined with the irreducible primitive polynomial $1 + x + x^4$. With $m = 4$, $t = 2$, n is equal to $2^4 - 1 = 15$. $mt = 4 \times 2 = 8$ check bits are required. The polynomials of the (15, 7) code consist of those polynomials of degree 14 or less that

Reprinted from *IEEE Trans. Inform. Theory*, vol. IT-10, pp. 357-363, Oct. 1964.

is a multiple of the generator polynomial

$$g(x) = 1 + x^4 + x^6 + x^7 + x^8 = (1 + x + x^4)(1 + x + x^2 + x^3 + x^4). \quad (5)$$

It may be verified that $g(\alpha) = g(\alpha^3) = 0$. The parity check matrix of the (15, 7) code may be written as

$$M = \begin{bmatrix} 1 & 0 & 0 & 0 & 1 & 0 & 0 & 0 \\ 0 & 1 & 0 & 0 & 0 & 0 & 0 & 1 \\ 0 & 0 & 1 & 0 & 0 & 0 & 1 & 1 \\ 0 & 0 & 0 & 1 & 0 & 1 & 0 & 1 \\ 1 & 1 & 0 & 0 & 1 & 1 & 1 & 1 \\ 0 & 1 & 1 & 0 & 1 & 0 & 0 & 0 \\ 0 & 0 & 1 & 1 & 0 & 0 & 0 & 1 \\ 1 & 1 & 0 & 1 & 0 & 0 & 1 & 1 \\ 1 & 0 & 1 & 0 & 0 & 1 & 0 & 1 \\ 0 & 1 & 0 & 1 & 1 & 1 & 1 & 1 \\ 1 & 1 & 1 & 0 & 1 & 0 & 0 & 0 \\ 0 & 1 & 1 & 1 & 0 & 0 & 0 & 1 \\ 1 & 1 & 1 & 1 & 0 & 0 & 1 & 1 \\ 1 & 0 & 1 & 1 & 0 & 1 & 0 & 1 \\ 1 & 0 & 0 & 1 & 1 & 1 & 1 & 1 \end{bmatrix} \quad (6)$$

Let us denote the received vector by r and its associated polynomial by $r(x)$, then,

$$r = [r_0 r_1 r_2 \cdots r_{n-1}]$$
$$r(x) = r_0 + r_1 x + r_2 x^2 + \cdots + r_{n-1} x^{n-1}. \quad (7)$$

Denote each of the error terms by β_j $(j = 1, 2, \cdots, t)$; it may be deduced that

$$r(\alpha^i) = \sum_{j=1}^{t} \beta_j^i = S_i \qquad (i = 1, 2, 3, \cdots, 2t - 1). \quad (8)$$

The functions S_i are known as power sums.

Example: To illustrate with the code in (6), we assume the code word in question to be

$$a = [1\,0\,0\,0\,1\,0\,1\,1\,1\,0\,0\,0\,0\,0\,0]. \quad (9)$$

With two errors occurring at the fifth and the tenth positions from the left, the received sequence takes the form

$$r = [1\,0\,0\,0\,0\,0\,1\,1\,1\,1\,0\,0\,0\,0\,0]. \quad (10)$$

Computing $r(\alpha^i)$ for $i = 1, 3$, we obtain

$$r(\alpha) = \alpha^4 + \alpha^9 = 1 + \alpha^3$$
$$r(\alpha^3) = \alpha^{12} + \alpha^{27} = \alpha^{12} + \alpha^{12} = 0 \quad (11)$$

as $a(\alpha) = a(\alpha^3) = 0$.

The Peterson procedure consists of three steps [1]:

Step 1: Compute the power sums S_i from the received sequence through the relations

$$S_i = r(\alpha^i) \qquad i = 1, 3, 5, \cdots, 2t - 1. \quad (12)$$

Step 2: Compute the elementary symmetric functions σ_k $(k = 1, 2, \cdots, t)$ from the power sums. The elementary symmetric functions are coefficients of the polynomial $\Sigma(x)$, where

$$\Sigma(x) = x^t + \sigma_1 x^{t-1} + \sigma_2 x^{t-2} + \cdots + \sigma_k x^{t-k} \cdots + \sigma_t$$
$$= (x - \beta_1)(x - \beta_2) \cdots (x - \beta_j) \cdots (x - \beta_t). \quad (13)$$

To obtain the σ_k's from the S_i's, use is made of Newton's identities,

$$S_1 - \sigma_1 = 0$$
$$S_3 - \sigma_1 S_2 + \sigma_2 S_1 - 3\sigma_3 = 0$$
$$S_5 - \sigma_1 S_4 + \sigma_2 S_3 - \sigma_3 S_2 + \sigma_4 S_1 - 5\sigma_5 = 0$$
$$\vdots$$

Step 3: Find the roots β_j $(j = 1, 2, \cdots, t)$ of the polynomial $\Sigma(x)$. These are the error locations.

In determining the efficiency or a decoding procedure, one must pay attention to both hardware complexity and decoding delay.

Peterson [1] has given a rough estimate of the processing time required for each step in his procedure. It is assumed that addition or multiplication in $GF(2^m)$ can be performed within one clock period, and division in a few clock periods. Following this computation, one sees that step 1 takes n clock periods where n is the length of the code. Circuits for accomplishing this are known [1]. Step 2 amounts to solving, at most, a set of t simultaneous linear equations. But one may have to try $t - 1$ times, since the actual number of errors present may be anywhere from one to t. Roughly speaking, step 2 may take as many as $t^4/2$ clock periods. In accomplishing step 3 by the trial and error method, one generates each nonzero element in $GF(2^m)$ in turn and substitutes it in $\Sigma(x)$. One substitution may take $2t$ multiplication and $t - 1$ additions, and this has to be done n times. Step 3, therefore, may take approximately $3tn$ clock periods. Table I contains a comparison of the decoding time in each step. The values are computed for the (127,92) Bose-Chaudhuri-Hocquenghem code. One might conclude that step 3 is the most time consuming of the three steps involved. In the cyclic procedure discussed later, it is shown that step 3 may be accomplished in n clock periods and therefore realizes a great saving in decoding delay. For decoders with serial readout, the error-connection is accomplished during read-out, hence it requires no additional time at all.

TABLE I
DECODING TIME FOR BOSE-CHAUDHURI-HOCQUENGHEM CODES

Decoding Time	Step 1 n	Step 2 $t^4/2$	Step 3 $3tn$
Time Req. for the (127,92) code	127 units	313 units	1905 units

III. Parity Checking and Computing the Polynomial $\Sigma(x)$

The parity-checking operation is to compute the power sums S_i from the received polynomial $r(x)$ with relationships $r(\alpha^i) = S_i$. This can be accomplished in a number of ways. Peterson has suggested the use of a special circuit for each S_i[1]. A universal parity-check circuit can also be designed with a variable frequency clock. The feedback patterns of the shift-register are the same as those of the Peterson circuit for S_1. For the calculation of S_i the clock frequency in the shift-register is i times the frequency at the input.

To convert the power sums S_i $(i = 1, 3, \cdots, 2t - 1)$ into elementary symmetric functions σ_k $(k = 1, 2, \cdots t)$ requires the solution of t simultaneous linear algebraic equations. Well-known procedures may be employed if one is equipped with circuits for carrying out addition, multiplication and division in finite fields. Fast circuits for these operations have been proposed [9] [11]. Where t is small, it may be economical to consider implementation of step 2 by combinational circuits directly. For instance, when $t = 2$,

$$\begin{aligned} \sigma_1 &= S_1 \\ \sigma_2 &= S_1^2 + \frac{S_3}{S_1}. \end{aligned} \tag{15}$$

IV. A Cyclic Error-Location Procedure

In this section we shall describe a new procedure with which we could locate the errors without explicitly solving the polynomial equation $\Sigma(x) = 0$ in $GF(2^m)$. This procedure is based on the cyclic property of the Bose-Chaudhuri-Hocquenghem codes.

First, let us examine the relationship between the coefficients σ_k's of $\Sigma(x)$ and the roots β_j of $\Sigma(x)$. We observe, from the identity

$$x^t - \sigma_1 x^{t-1} + \sigma_{2x}^{t-2} + \cdots + (-1)^t \sigma_t = (x - \beta_1)(x - \beta_2) \cdots (x - \beta_t) \tag{16}$$

that

$$\begin{aligned} \sigma_1 &= \sum_{j=1}^{t} \beta_j \\ \sigma_2 &= \sum_{\substack{j,k=1 \\ j<k}}^{t} \beta_j \beta_k \\ \sigma_3 &= \sum_{\substack{i,j,k=1 \\ i<j<k}}^{t} \beta_i \beta_j \beta_k. \end{aligned} \tag{17}$$

The σ_k's are homogeneous sums of square free products of the roots of order k. This homogeneous property is most useful. Suppose that we transform each β_j $(j = 1, 2, \cdots, t)$ to $\bar{\beta}_j = \alpha\beta_j$ where α is a primitive element of $GF(2^m)$. We may define the new elementary symmetric functions $\bar{\sigma}_k$'s as functions of $\bar{\beta}_j$'s, in the same manner as we did in (17). As a result of the "homogeneous" property, we note the relationship between $\bar{\sigma}_k$'s and σ_ks to be:

$$\bar{\sigma}_k = \alpha^k \sigma_k \qquad k = 1, 2, \cdots, t. \tag{18}$$

In other words, the $\tilde{\beta}_j$'s are the roots of the polynomial

$$\bar{\Sigma}(x) = x^t + \bar{\sigma}_1 x^{t-1} + \bar{\sigma}_2 x^{t-2} + \cdots + \bar{\sigma}_t. \tag{19}$$

In general, after applying such transformation τ times, we obtain

$$\begin{aligned} \tilde{\beta}_j &= \alpha^\tau \beta_j & j &= 1, 2, \cdots, t \\ \tilde{\sigma}_k &= \alpha^{\tau k} \sigma_k & k &= 0, 1, 2, \cdots, t. \end{aligned} \tag{20}$$

Note that $\alpha^n = \alpha^{2^m - 1} = 1$ in $GF(2^m)$; hence, the powers never exceed $n = 2^m - 1$. This may be illustrated by the example where

$$\begin{aligned} \sigma_1 &= 1 + \alpha^3 = \alpha^{14} \\ \sigma_2 &= 1 + \alpha^2 + \alpha^3 = \alpha^{13}. \end{aligned} \tag{21}$$

The polynomials and their roots are illustrated in Fig. 1. for two positions, the initial position and the position after six shift-transformations. It is shown that after six shifts, the root α^9 is transformed into $\alpha^6 \cdot \alpha^9 = \alpha^{15} = 1$. The root α^4 may be transformed into $\alpha^{15} = 1$ after eleven shifts.

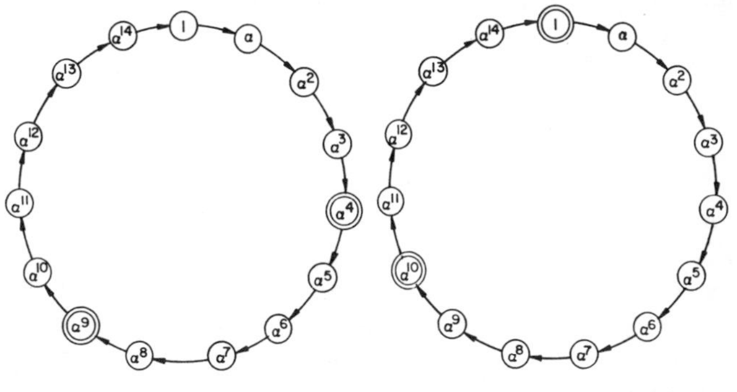

Fig. 1—Roots of $\Sigma(x) = x^2 + \alpha^{14}x + \alpha^{13}$ are shown, in double circles, at left before shifting and at right after six shifts.

Indeed, if we have means for detecting whether a specific element of $GF(2^m)$ is a root of a polynomial $\Sigma(x)$, we may obtain all of its roots by counting and by successive transformations. To simplify circuitry, the element to be detected is chosen to be the unit element of $GF(2^m)$. When $\alpha^0 = 1$ is a root of a polynomial $\Sigma(x)$, we see that $x^k = 1$ $(k = 1, 2, \cdots, t)$ and

$$\Sigma(1) = 1 + \sigma_1 + \sigma_2 + \cdots + \sigma_t = 0. \tag{22}$$

or

$$\sum_{k=1}^{t} \sigma_k = \sigma_1 + \sigma_2 + \cdots + \sigma_t = 1.$$

If $\Sigma\bar{\sigma}_k = 1$ after $\tau_1, \tau_2, \cdots, \tau_t$ shifts, respectively, the roots of the polynomial $\Sigma(x)$ are $\alpha^{n-\tau_1}, \alpha^{n-\tau_2}, \cdots,$ and $\alpha^{n-\tau_t}$. This procedure will be illustrated with our example by listing the $\bar{\sigma}_k$'s $(k = 1, 2)$ after successive transformations. In each shift-transformation the current value of the $\bar{\sigma}_1$ column is multiplied by α and the current value of the $\bar{\sigma}_2$ column by α^2.

	$\bar{\sigma}_1$	$\bar{\sigma}_2$
Initial value	$1 + \alpha^3$	$1 + \alpha^2 + \alpha^3$
After 1 shift	1	1
After 2 shifts	α	α^2
After 3 shifts	α^2	$1 + \alpha$
After 4 shifts	α^3	$\alpha^2 + \alpha^3$
After 5 shifts	$1 + \alpha$	$1 + \alpha^2$
After 6 shifts	$\alpha + \alpha^2$	$1 + \alpha + \alpha^2$.
....	....	....
....	....	....

It is noted that after six transformations $\sum_{k=1}^{t} \bar{\sigma}_k = 1$; hence, $\alpha^{15-6} = \alpha^9$ is a root of the $\Sigma(x)$. By continuing the process one will find that after eleven transformations

$$\bar{\sigma}_1 = 1 + \alpha + \alpha^2$$

$$\bar{\sigma}_2 = \alpha + \alpha^2$$

and again $\sum_{k=1}^{t} \bar{\sigma}_k = 1$. This procedure of obtaining the roots of $\Sigma(x)$ is simple in nature and lends itself to extremely simple implementation, the details of which will be discussed in Section V. Note that for the binary case, information concerning error locations are sufficient for carrying out error correction.

V. Implementation of the Cyclic Error-Location Procedure

The implementation of the error-correction procedure for binary codes as outlined in Section IV follows from theory in a straight forward manner. The entire circuit for error-locating and associated hardware is shown in Fig. 2. The initial values of σ_k's are stored in t σ-registers, and the received sequence is stored in the buffer with high-order bits first.

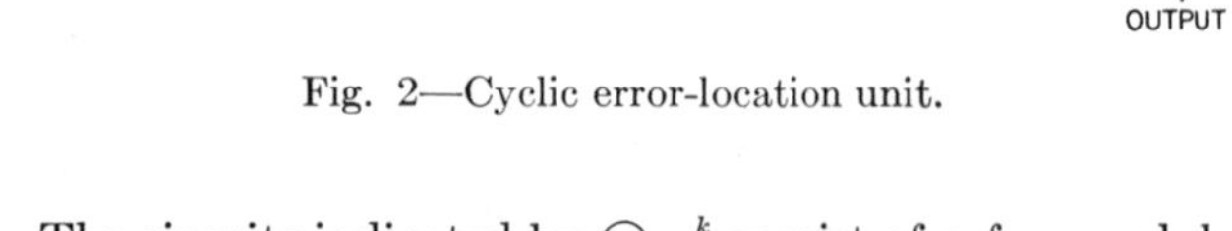

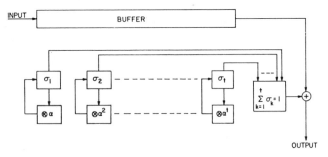

Fig. 2—Cyclic error-location unit.

The circuits indicated by $\otimes$ α^k consist of a few modulo-two adders and a number of feedback connections. The purpose of these α^k-multipliers is to multiply the current contents of the σ_k-register by α^k and subsequently store the product in the σ_k register. These circuits are simply implemented, and they have a structure similar to the circuit for parity checking. Hence, half of all the α^k-multipliers are already part of the system.

The circuit for detecting the condition $\Sigma\sigma_k = 1$ is a simple adder with an OR gate followed by an inverter at the output. The sum of the zeroth order is inverted. The inverter output is a "one" if and only if all inputs to the OR gate are zeros. The circuit shown in Fig. 3 applies to our example. Note that though the σ_k-registers are shown, they are not part of the detection circuit.

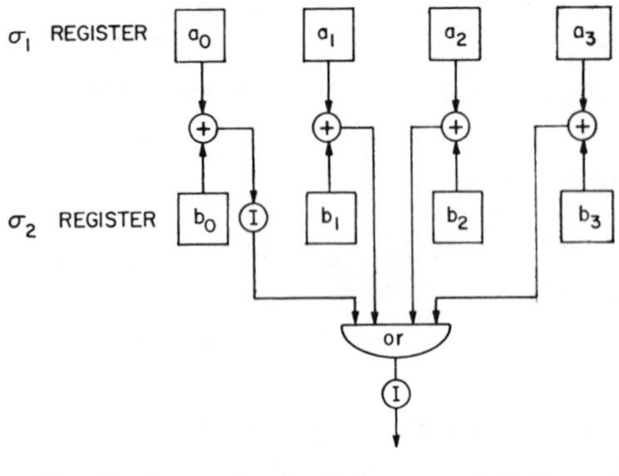

Fig. 3—A sample circuit for error detection.

This error-correction system operates as follows: the received sequences are stored in the buffer with high-order bits first, and the initial values of σ_k's are stored in the σ_k-registers. Since the initial value of σ_ks will only indicate to the detection circuit the presence of errors at the α^0th position, the α^k multipliers are first pulsed once. After this shift transformation, the detection circuit indicates the presence of any error at the leading bit position of the buffer since that position corresponds to α^{n-1}. The bits in the buffer are then shifted out in sequence. Whenever a bit is in error the detection circuit will produce a "one" bit and therefore complement the data bit leaving the buffer at that time. All errors will be corrected with n-shifts provided no more than t-errors are present in the received sequence.

VI. A Direct Method of Implementing the Cyclic Decoding Procedure

The procedure described in Section V is useful in speeding up the decoding process in decoders such as the one constructed at Lincoln Laboratory. It is still necessary to compute the σ_k functions from the S_i power sums before the final step of error correction can be employed. In this section, the concept is carried further to develop a direct method of decoding.

When applying the cyclic decoding procedure, it is only necessary to detect whether "1" is a root of the polynomial $\Sigma(x)$, *i.e.*, when $\sum_{k=1}^{t} \sigma_k = 1$. Now, the S_i's and σ_k's are related by Newton's Identities

$$S_1 - \sigma_1 = 0$$

$$S_3 - \sigma_1 S_2 + \sigma_2 S_1 - 3\sigma_3 = 0$$

$$S_5 - \sigma_1 S_4 + \sigma_2 S_3 - \sigma_3 S_2 + \sigma_4 S_1 - 5\sigma_5 = 0 \quad (23)$$

$$\cdots$$

In Newton's Identities $\sigma_1, \cdots \sigma_t$ are unknowns, and $\sigma_i = 0$ for $i > t + 1$. This set of t linear equations can then be solved, and the unknowns σ_k's can be written as functions of the known quantities S_i's. In matrix notation, this process can be carried out as follows: We write, for the binary case,

$$A\sigma = B \tag{24}$$

where

$$A = \begin{bmatrix} 1 & 0 & 0 & \cdots & 0 \\ S_2 & S_1 & 1 & \cdots & 0 \\ S_4 & S_3 & S_2 & \cdots & 0 \\ & & \vdots & & \\ S_{2t-2} & S_{2t-3} & S_{2t-4} & \cdots & S_{t-1} \end{bmatrix} \tag{25}$$

and

$$B = \begin{bmatrix} S_1 \\ S_3 \\ S_5 \\ \vdots \\ S_{2t-1} \end{bmatrix}. \tag{26}$$

If the determinant $|A|$ is not zero, then the σ_k's may be expressed as

$$\sigma_k = \frac{1}{|A|} \sum_{i=1}^{t} S_{2i-1} A_{i,k} \qquad k = 1, 2, \cdots, t \tag{27}$$

where $A_{i,k}$ ($k = 1, 2, \cdots, t$) are cofactors of the determinant $|A|$.

Now, suppose 1 is a root of the polynomial $\Sigma(x)$, then, $\sum_{k=1}^{t} \sigma_k = 1$. By substituting for σ_k's the expressions shown in (27) for each k, respectively, the following is obtained:

$$\sum_{k=1}^{t} \left\{ \frac{1}{|A|} \sum_{i=1}^{t} S_{2i-1} A_{i,k} \right\} = 1. \tag{28}$$

Since $|A|$ is independent of k, (28) may be written as

$$\sum_{k=1}^{t} \sum_{i=1}^{t} S_{2i-1} A_{i,k} - |A| = 0. \tag{29}$$

Since we are working in a field of characteristic two, (29) may easily be shown to be equivalent to setting the determinant Δ to zero, where

$$\Delta = \begin{vmatrix} 1 & 1 & 1 & \cdots & 1 \\ S_1 & 1 & 0 & \cdots & 0 \\ S_3 & S_2 & 1 & \cdots & 0 \\ S_5 & S_4 & S_3 & \cdots & 0 \\ & \vdots & & & \\ S_{2t-1} & S_{2t-1} & S_{2t-3} & \cdots & S_{t-1} \end{vmatrix} = \begin{bmatrix} 1 & 1 & \cdots & 1 \\ B & & A & \end{bmatrix}. \tag{30}$$

Peterson [1] [3] has shown that $|A| \neq 0$ if the S_i's are power sums of t or $t - 1$ distinct roots, and $|A| = 0$ if the S_i's are power sums of $t - 2$ or fewer distinct roots. In the case where $|A| = 0$, one may delete the last two equations in Newton's identities and end up with $t - 2$ equations and $t - 2$ unknown. The process can then be applied to the reduced set of equations and a new determinant can be obtained in a similar manner.

Thus, the decoder will operate in different modes depending upon the size of the largest nonvanishing determinant. In implementing the procedure, the relations such as (30) are simplified to a set of m binary relations. The circuits are designed for the binary relations and not in $GF(2^m)$.

This error-correction procedure and its implementation are illustrated in detail for two sample cases in Section VII.

VII. Two Sample Decoders

We shall discuss two specific examples in this section and illustrate how one can design circuits to implement the cyclic error-correction procedure described in Section VI.

First, let us consider the case $t = 2$. In this case $|A| \neq 0$ whenever the power sums are not identically zero. One may write (30) in the form

$$\begin{aligned} \Delta &= \begin{vmatrix} 1 & 1 & 1 \\ S_1 & 1 & 0 \\ S_3 & S_2 & S_1 \end{vmatrix} \\ &= S_1(1 + S_1 + S_1^2) + S_3 \\ &= 0. \end{aligned} \tag{31}$$

The relationship (31) is valid for any $GF(2^m)$. Now, let us specifically consider the (15, 7) Bose-Chaudhuri-Hocquenghem code which corrects double errors. Define the field $GF(2^4)$ with the polynomial $x^4 + x + 1$ and substitute into (31), respectively,

$$S_1 = a_0 + a_1\alpha + a_2\alpha^2 + a_3\alpha^3$$

$$S_3 = b_0 + b_1\alpha + b_2\alpha^2 + b_3\alpha^3$$

where α is a primitive element of $GF(2^4)$. One obtains, after some simplification, the following binary relations:

$$\begin{aligned} a_0\bar{a}_2 + a_2\bar{a}_1 + a_1a_3 + b_0 &= 0 \\ (a_1 + a_2)\bar{a}_0 + a_3\bar{a}_2 + b_1 &= 0 \\ (a_0 + a_1)(a_1 + a_2 + a_3) + a_3\bar{a}_2 + b_2 &= 0 \\ (a_1 + a_2)\bar{a}_3 + a_3 + b_3 &= 0. \end{aligned} \tag{32}$$

The complete decoder is shown in Fig. 4 in a block diagram. The circuits for implementing (32) are shown in Fig. 5.

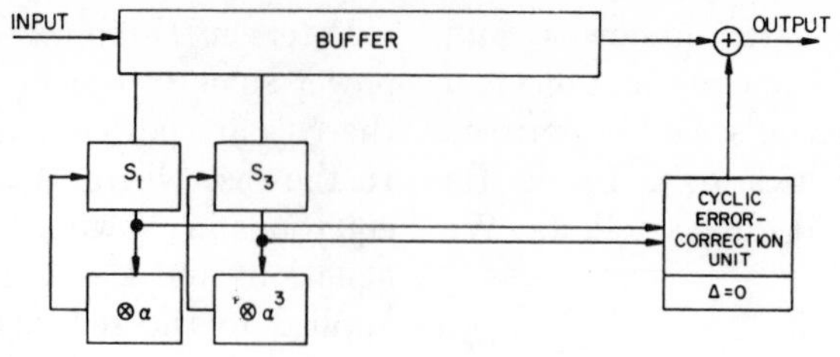

Fig. 4—A cyclic parallel decoder for the BCH (15, 7) code.

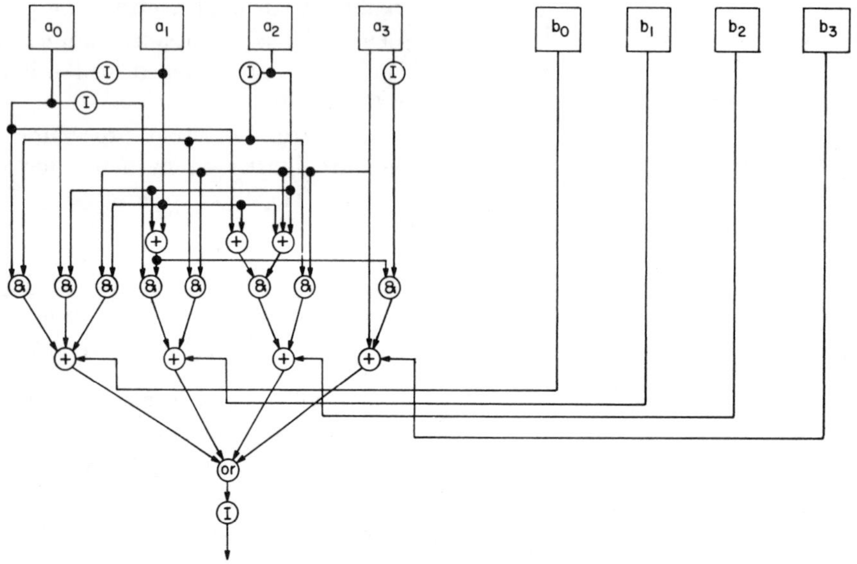

Fig. 5—A sample logical design for the cyclic error-correction unit of the decoder in Fig. 4.

Next, let us consider the case $t = 3$. In this case, the condition is

$$\Delta = \begin{vmatrix} 1 & 1 & 1 & 1 \\ S_1 & 1 & 0 & 0 \\ S_3 & S_2 & S_1 & 1 \\ S_5 & S_4 & S_3 & S_2 \end{vmatrix} \tag{33}$$

$$= S_1^3(1 + S_1 + S_1^3)$$

$$+ S_3(1 + S_1 + S_1^2 + S_1^3 + S_3) + S_5(1 + S_1)$$

$$= 0.$$

When two or three errors occur, the determinant

$$|A| = \begin{vmatrix} 1 & 0 & 0 \\ S_2 & S_1 & 1 \\ S_4 & S_3 & S_2 \end{vmatrix} = S_1^3 + S_3 \tag{34}$$

is nonzero. However, $|A| = 0$ when only one error occurs, as it can easily be seen that

$$S_3 = S_1^3$$

when both are functions of one variable. A system for decoding such a code is shown in Fig. 6.

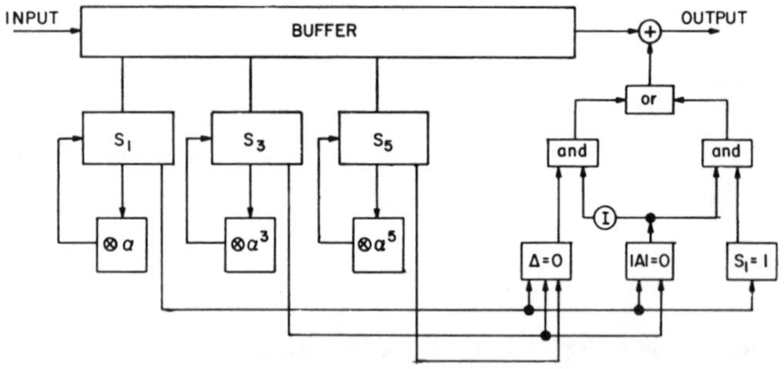

Fig. 6—Systems diagram for a cyclic parallel decoder with $t = 3$.

VIII. Discussion

The decoder suggested by Meggitt [7] utilizes the cyclic property of the code. The procedure suggested by Peterson, *et. al.*, is algebraic. For the method proposed here, both the cyclic and the algebraic properties have been used to advantage.

In the decoding process suggested here, error-correction is carried out automatically as in the case of burst codes. Parity-check circuits are used to advantage in the error-correction phase. Extensions of this technique to non-binary codes is fairly straightforward. Some additional steps will be involved as the errors at any position can occur in more than one way.

Other ways of interpreting the operations of the circuit in Fig. 2 are possible. For instance the circuit may be viewed as one of solving the polynomial equation

$$\sigma_0 + \sigma_1 x + \sigma_2 x^2 + \cdots + \sigma_t x^t = 0$$

with the x's substituted by $1, \alpha, \alpha^2, \cdots, \alpha^{n-1}$, respectively. In fact, an alternative circuit could be designed according to the polynomial

$$\sigma_0 x^t + \sigma_1 x^{t-1} + \cdots + \sigma_t = 0,$$

but it would require either the data in the buffer to be lower-order first, or the α^k-multipliers be replaced by α^k-dividers.

Acknowledgment

The author is indebted to his colleagues Drs. J. E. Meggitt, D. T. Tang, and C. V. Freiman, of Thomas J. Watson Research Center, Yorktown Heights, N. Y., for helpful suggestions in the development of these ideas. Helpful comments were also contributed by Drs. A. H. Frey and F. Corr of IBM Communications Systems Center, Bethesda, Md.

References

[1] W. W. Peterson, "Error-Correcting Codes," John Wiley and Sons, Inc., New York, N. Y.; 1961.

[2] E. Gorog, "Some new classes of cyclic codes used for burst-error correction," *IBM J. Res. Developm.*, vol. 7, pp. 102–111; April, 1963.

[3] R. C. Bose and C. R. Ray-Chaudhuri, "On a class of error-correcting binary group codes," *Information and Control*, vol. 3, pp. 68–79; 1960.

[4] A. Hocquenghem, "Codes correcteurs d'erreurs," *Chiffres*, vol. 2, pp. 147–156; September, 1959.

[5] W. W. Peterson, "Encoding and error-correction procedures for the Bose-Chaudhuri codes," IRE Trans. on Information Theory, vol. IT-6, pp. 459–470; September, 1960.

[6] N. Zieler and D. Gorenstein, "A class of error-correcting codes in p^m symbols," *J. Soc. Ind. Appl. Math.*, vol. 9, pp. 207–214; June, 1961.

[7] J. E. Meggitt, "Error-correcting codes and their implementation for data transmission systems," IRE Trans. on Information Theory, vol. IT-7, pp. 234–244; October, 1961.

[8] R. B. Banerji, "A decoding procedure for double-error correcting Bose-Ray-Chaudhuri codes," Proc. IRE, (Correspondence), vol. 49, p. 1585; October, 1961.

[9] T. C. Bartee and D. I. Schneider, "An electronic decoder for Bose-Chaudhuri-Hocquenghem error-correcting codes," IRE Trans. on Information Theory, vol. IT-8, pp. S 17–24; September, 1962.

[10] A. A. Albert, "*Fundamental Concepts on Higher Algebra,*" University of Chicago Press, Chicago, Ill.; 1956.

[11] T. C. Bartee and D. I. Schneider, "Computation with finite fields," *Information and Control*, vol. 6, pp. 79–98; 1963.

On Decoding BCH Codes

G. DAVID FORNEY, JR., MEMBER, IEEE

Abstract—The Gorenstein-Zierler decoding algorithm for BCH codes is extended, modified, and analyzed; in particular, we show how to correct erasures as well as errors, exhibit improved procedures for finding error and erasure values, and consider in some detail the implementation of these procedures in a special-purpose computer.

I. Introduction

THE DISCOVERY of the binary codes of Bose and Ray-Chaudhuri [1], [2] and (independently) Hocquenghem [3] has been, perhaps, the outstanding success of the search for codes based on algebraic structures. Not the least of their virtues is their capability of being decoded by relatively straightforward algorithms. Peterson [4] was the first to outline an efficient decoding procedure, which was actually realized by Bartee and Schneider [5] in a small special-purpose computer.

Recent work has focused attention on the multisymbol generalizations of these codes. These were first considered by Gorenstein and Zierler [6], who developed an error-correcting algorithm for them.

This paper reports extensions, modifications, and analyses of these decoding procedures. We shall be concerned with the general codes, which we shall call BCH codes, but all of our results apply to the binary special case. In particular, we shall proceed as follows:

Manuscript received January 31, 1965; revised July 1, 1965. The work reported in this paper was supported in part by the Joint Services Electronics Program under Contract DA36-039-AMC-03200(E); and in part by the National Science Foundation (Grant GP-2495), the National Institute of Health (Grant MH-04737-04) and the National Aeronautics and Space Administration (Grants NsG 334 and NsG 496). Portions of this work have appeared in Quarterly Progress Report No. 76, M.I.T. Research Laboratory of Electronics, Cambridge, Mass., p. 236, January 1965.

The author is with the Codex Corporation, Watertown, Mass. He was formerly with the M.I.T. Research Laboratory of Electronics, Cambridge, Mass.

1) to extend the GZ algorithm to correct erasures, as well as errors, by introducing a modified set of parity checks;

2) to improve the final step of the GZ algorithm by giving explicit formulas for error values, thereby eliminating the need for solving simultaneous equations;

3) to introduce an alternative method for determining error values, which has a use when the number of errors to be corrected is small;

4) to note a method for solving for erasures separately from errors;

5) to exhibit a generalization of Chien's [7] 'direct method' of locating errors; and

6) to analyze in some detail the implementation of these procedures in a computer with finite-field arithmetic unit, and determine that the number of operations increases only as a small power of the number of errors to be corrected.

Also, for the reader who likes to confirm his understanding by working out simple examples, we have included a decoding problem, with the aid of which the different procedures can be illustrated.

II. Preliminary Definitions

BCH codes are conveniently described in the language of the theory of finite fields, which has been well developed for this purpose by Peterson [8]. A finite, or Galois, field with p^M elements [written $GF(p^M)$] exists if p is a prime, and M is any integer; p is called the characteristic of the field. In any field there is a zero element 0, a unit element 1, and at least one primitive element α, such that any other nonzero element β can be expressed as a power of α. The order of β is the least integer e such that $\beta^e = 1$; a primitive element α has order $p^M - 1$. If M is a factor of N, the elements of $GF(p^M)$ are included in $GF(p^N)$, and

Reprinted from *IEEE Trans. Inform. Theory*, vol. IT-11, pp. 549–557, Oct. 1965.

the former is said to be a subfield of the latter, or the latter an extension field of the former.

For example, $GF(2^4)$ consists of the elements 0, 1, α, α^2, $\cdots$ α^{14}; 0 and 1 are the subfield $GF(2)$, while 0, 1, α^5, and α^{10} are the subfield $GF(2^2)$. A particular representation for $GF(2^4)$, based on polynomials modulo the irreducible (over $GF(2)$) polynomial $X^4 + X + 1$, is given in Peterson [8] as

0	0000	α^3	0001	α^7	1101	α^{11}	0111
1	1000	α^4	1100	α^8	1010	α^{12}	1111
α	0100	α^5	0110	α^9	0101	α^{13}	1011
α^2	0010	α^6	0011	α^{10}	1110	α^{14}	1001

Since $\alpha^{15} = 1$, $\alpha^i \cdot \alpha^j = \alpha^{(i+j) \bmod 15}$; addition of two elements in the field is achieved by bit-by-bit modulo 2 addition of the two representations.

In this language, code words of length n_0 are represented by sequences of n_0 elements from $GF(p^M)$, which we shall write as a vector

$$\mathbf{f} \equiv (f_1, f_2, \cdots, f_{n_0}).$$

If we define $\mathbf{X}_{(a,b)}$ as the column vector of descending powers of X,

$$\mathbf{X}_{(a,b)} \equiv (X^a, X^{a-1}, \cdots, X^b)^T,$$

where X is an indeterminate and T indicates the transpose, then the dot product

$$\mathbf{f} \cdot \mathbf{X}_{(n_0-1,0)} = \sum_i f_i X^{n_0-i} \equiv f(X)$$

is a polynomial in X of degree $n_0 - 1$, which we define as $f(X)$. Similarly, if β is any element of $GF(p^N)$ or of an extension or subfield thereof, we can define

$$f(\beta^m) \equiv \mathbf{f} \cdot \boldsymbol{\beta}^{\mathbf{m}}_{(n_0-1,0)} = \sum_i f_i \beta^{m(n_0-i)},$$

where

$$\boldsymbol{\beta}^{\mathbf{m}}_{(a,b)} \equiv (\beta^{ma}, \beta^{m(a-1)}, \cdots, \beta^{mb})^T.$$

If the order of β is n_0, BCH codes consist of the set of all $\mathbf{f}$ such that $f(\beta^m) = 0$, for all m in the range $m_0 \leq m \leq m_0 + d - 2$, where m_0 and d are arbitrary integers. It develops [8] that d is a lower bound to the minimum distance of the code. Information on the number of words in some binary BCH codes is given in Peterson [8]. Commonly, β is taken as a primitive element of $GF(p^N)$, and m_0 equals 0 or 1.

In examples, we shall use a Reed-Solomon [8], [9] code of length 15, with elements from $GF(2^4)$, such that $\mathbf{f}$ is in the code if $f(\alpha) = f(\alpha^2) = \cdots = f(\alpha^8) = 0$, where α is a primitive element of $GF(2^4)$. Such a code has $(2^4)^7$ words, or 7 information symbols, and its minimum distance is 9. In our examples with this code, $GF(2^4)$ will be represented as previously.

In an actual communications system, a modulator can transmit one of p^M signals corresponding to the p^M field elements; a code word is physically realized by a sequence of n_0 such signals. At the receiver, the function of the demodulator is to guess which of the p^M signals was sent; n_0 of these guesses constitute a received word, and can be represented by a vector

$$\mathbf{r} \equiv (r_1, r_2, \cdots, r_{n_0})$$

of n_0 elements from $GF(p^M)$. If the ith guess is correct, $r_i = f_i$; if it is incorrect, we say an error has been made; the value of the error is defined as $r_i - f_i$ and the locator of the error as β^{n_0-i}. If there are t errors in all, we shall denote the values by e_j and the locators by X_j, $1 \leq j \leq t$.

From the received word one obtains the parity checks S_m defined by

$$S_m \equiv r(\beta^m) = \sum_{i=1}^{n_0} r_i \beta^{m(n_0-i)} = \mathbf{r} \cdot \boldsymbol{\beta}^{\mathbf{m}}_{(n_0-1,0)}, \qquad m_0 \leq m \leq m_0 + d - 2$$

From the definitions of e_j and X_j,

$$S_m = f(\beta^m) + \sum_{j=1}^{t} e_j X_j^m;$$

but $f(\beta^m) = 0$ for $m_0 \leq m \leq m_0 + d - 2$ and for all code words $\mathbf{f}$, so that

$$S_m = \sum_{j=1}^{t} e_j X_j^m, \qquad m_0 \leq m \leq m_0 + d - 2. \tag{1}$$

The decoding problem is to solve (1), called the parity check equations, for the e_j and X_j. Whenever $2t < d$, the algorithm of Gorenstein and Zierler [6] solves this problem. The algorithm consists of three steps: first the number is found, then the locations, and finally the value of the errors, as we shall explain in more detail later.

In addition to the column vectors $\mathbf{X}_{(a,b)}$ and $\boldsymbol{\beta}^{\mathbf{m}}_{(a,b)}$ already defined, we shall use the column vectors $\mathbf{X}_{j(a,b)}$, $\mathbf{Y}_{k(a,b)}$, and $\mathbf{Z}_{(a,b)}$ of descending powers of X_j, Y_k, and Z, respectively, in which Y_k and Z have yet to be introduced. We shall also use the column vectors

$$\mathbf{S}_{(a,b)} \equiv (S_a, S_{a-1}, \cdots, S_b)^T, \qquad m_0 \leq b \leq a \leq m_0 + d - 2$$

and

$$\mathbf{T}_{(a,b)} \equiv (T_a, T_{a-1}, \cdots, T_b)^T, \qquad 0 \leq b \leq a \leq d - s - 2,$$

where the S_m are the parity checks introduced above, and the T_n are the modified cyclic parity checks to be introduced below. With these expressions and (1) we have, for example,

$$\mathbf{S}_{(a,b)} = \sum_{j=1}^{t} e_j \mathbf{X}_{j(a,b)}.$$

Finally, let us consider the polynomial $\sigma(Z)$ defined by

$$\sigma(Z) \equiv (Z - Z_1)(Z - Z_2) \cdots (Z - Z_L),$$

where Z is an indeterminate, and the Z_l are members of a field. Clearly, $\sigma(Z) = 0$ if and only if Z equals one of the Z_l. Expanding $\sigma(Z)$, we get

$$\sigma(Z) = Z^L - (Z_1 + Z_2 + \cdots + Z_L)Z^{L-1} + \cdots + (-1)^L(Z_1Z_2 \cdots Z_L).$$

The coefficient of $(-1)^{L-l}Z^l$ in this expansion is defined as the $(L - l)$th elementary symmetric function σ_{L-l} of the Z_l; note that σ_0 is always one. We define $\boldsymbol{\sigma}$ as the row vector

$$(\sigma_0, -\sigma_1, \cdots, (-1)^L\sigma_L);$$

then the dot product

$$\boldsymbol{\sigma} \cdot \mathbf{Z}_{(L,0)} = \sigma(Z).$$

III. Decoding Algorithm for Erasures and Errors

Coding is required in communication systems to combat the errors that occur in the guesses of the demodulator, as we have seen. It has long been recognized [10] that there are advantages in allowing the demodulator not to guess at all on certain transmissions when the evidence does not clearly indicate one signal as the most probable; such events are called erasures. It is convenient to imagine that in the event of an erasure the demodulator does make some guess, perhaps arbitrary, but in addition passes on the side information to the decoder that this guess is absolutely unreliable and is to be disregarded.

From information theoretic considerations, allowing the option of erasures is thus a way of passing more information to the decoder about the signal actually received. In the context of the BCH algorithm, an erasure can be regarded as an error whose location is known and therefore does not have to be computed; the use of erasures should allow the shifting of some of the burden of determining error locations from the decoder to the demodulator.

As with errors, if an erasure occurs in the ith place, we say its value is $r_i - f_i$, where r_i represents the arbitrary guess of the demodulator; the value of an erasure may be zero, while that of an error may not. β^{n_0-i} is the locator of the erasure, which is known to the decoder, while that of an error is not. If there are s erasures, we denote their values by d_k and their locators by Y_k, $1 \leq k \leq s$. The values and locators of the t errors which may also occur will continue to be denoted by e_j and X_j, $1 \leq j \leq t$.

We observe that in this case (1), the parity check equations, become

$$S_m = \sum_{j=1}^{t} e_j X_j^m + \sum_{k=1}^{s} d_k Y_k^m, \quad m_0 \leq m \leq m_0 + d - 2. \quad (2)$$

In our vectorial notation we have

$$\mathbf{S}_{(a,b)} = \sum_{j=1}^{t} e_j \mathbf{X}_{j(a,b)} + \sum_{k=1}^{s} d_k \mathbf{Y}_{k(a,b)},$$

$$m_0 \leq b \leq a \leq m_0 + d - 2. \quad (2a)$$

The decoding problem is now to find the e_j, X_j, and d_k, given the S_m and Y_k. The modification of the GZ algorithm now to be presented permits solution of this problem whenever $2t + s < d$. Essentially we derive from the $d - 1$ parity check equations of (2) a set of $d - s - 1$ equations of the same form as (1), to which therefore the GZ algorithm can be applied.

Define

$$\sigma_d(Z) \equiv (Z - Y_1)(Z - Y_2) \cdots (Z - Y_s)$$

and let $\boldsymbol{\sigma}_d$ then be the vector of the symmetric functions σ_{dk} of the erasure locators Y_k, as above. We define the modified cyclic parity checks T_n by

$$T_n \equiv \boldsymbol{\sigma}_d \cdot \mathbf{S}_{(m_0+n+s, m_0+n)}, \qquad 0 \leq n \leq d - s - 2. \quad (3)$$

The range of n is restricted to $0 \leq n \leq d - s - 2$, since we must have $m_0 \leq m_0 + n$ and $m_0 + n + s \leq m_0 + d - 2$. (In the case of no erasures, $T_n = S_{m_0+n}$.)

Combining (2a) and (3), we have

$$\begin{aligned} T_n &\equiv \sum_{j=1}^{t} e_j \boldsymbol{\sigma}_d \cdot \mathbf{X}_{j(m_0+n+s, m_0+n)} + \sum_{k=1}^{s} d_k \boldsymbol{\sigma}_d \cdot \mathbf{Y}_{k(m_0+n+s, m_0+n)} \\ &= \sum_{j=1}^{t} e_j X_j^{m_0} X_j^n \sigma_d(X_j) + \sum_{k=1}^{s} d_k Y_k^{m_0+n} \sigma_d(Y_k) \qquad (4) \\ &= \sum_{j=1}^{t} E_j X_j^n, \qquad 0 \leq n \leq d - s - 2. \end{aligned}$$

Here we have defined

$$E_j = e_j X_j^{m_0} \sigma_d(X_j)$$

and used the fact that $\sigma_d(Y_k) = 0$, since Y_k is one of the erasure locators upon which $\boldsymbol{\sigma}_d$ is defined.

In (4) we now have $d - s - 1$ equations of exactly the form of (1), which are thus soluble for the E_j and X_j by the GZ algorithm whenever $2t < d - s$. In particular, we have the following lemma and theorem whose proofs, being identical to those which appear in Peterson [8], are omitted. We define t_0, the maximum correctable number of errors, as the greatest integer for which $2t_0 < d - s$. Then we have

Lemma: If $t \leq t_0$, then the $t \times t$ matrix M_t has rank t, where

$$M_t \equiv \begin{bmatrix} T_{2t_0-2} & T_{2t_0-3} & \cdots & T_{2t_0-t-1} \\ T_{2t_0-3} & T_{2t_0-4} & \cdots & T_{2t_0-t-2} \\ \vdots & \vdots & & \vdots \\ T_{2t_0-t-1} & T_{2t_0-t-2} & \cdots & T_{2t_0-2t} \end{bmatrix}.$$

Theorem: If $t \leq t_0$, then the $t_0 \times t_0$ matrix M has rank t, where

$$M \equiv \begin{bmatrix} T_{2t_0-2} & T_{2t_0-3} & \cdots & T_{t_0-1} \\ T_{2t_0-3} & T_{2t_0-4} & \cdots & T_{t_0-2} \\ \vdots & \vdots & & \vdots \\ T_{t_0-1} & T_{t_0-2} & \cdots & T_0 \end{bmatrix}.$$

This theorem allows determination of t from the T_n, which is the first step in the algorithm.

Now consider the vector $\boldsymbol{\sigma}_e$ of elementary symmetric functions σ_{ej} of the X_j, and its associated polynomial

$$\sigma_e(X) = \boldsymbol{\sigma}_e \cdot \mathbf{X}_{(t,0)}.$$

We have from (4) and the fact that $\sigma_e(X_j) = 0, 1 \leq j \leq t$,

$$\boldsymbol{\sigma}_e \cdot \mathbf{T}_{(n'+t,n')} = \sum_{j=1}^{t} E_j X_j^{n'} \sigma_e(X_j) = 0,$$

$$0 \leq n' \leq d - s - t - 2.$$

σ_{e0} always equals one; therefore this gives us $d - s - t - 1$ linear equations in t unknowns. Since $2t + s < d$, $t \leq d - s - t - 1$; thus we have sufficient equations to solve for the σ_{ej}, $1 \leq j \leq t$. By defining the vector

$$\boldsymbol{\sigma}_e' \equiv (-\sigma_{e1}, \sigma_{e2}, \cdots, (-1)^t \sigma_{et})$$

the equations specified by $2t_0 - 2t \leq n' \leq 2t_0 - t - 1$ can be expressed in matrix form as

$$-\mathbf{T}_{(2t_0-1,2t_0-t)} = \boldsymbol{\sigma}_e' M_t, \qquad (5)$$

where M_t is as in the lemma and therefore invertible; thus the equations are soluble for $\boldsymbol{\sigma}_e'$ and hence $\boldsymbol{\sigma}_e$. Then since $\sigma_e(\beta^{n_0-i})$ is zero if and only if β^{n_0-i} is an error locator, calculation of $\sigma_e(\beta^{n_0-i})$ for each i will reveal in turn the locations of all t errors, which concludes the second step in the algorithm.

A. Remarks I

In Peterson [8], first the rank of M is found, and then a set of t equations in t unknowns is solved, as in the present work. We remark that with the definition of M_t in the foregoing lemma, these two steps may be combined into one (as is implicit in Gorenstein and Zierler). Consider the equation

$$-\mathbf{T}_{(2t_0-1,t_0)} = \boldsymbol{\sigma}_e'' M, \qquad (6)$$

where $\boldsymbol{\sigma}_e'' \equiv (-\sigma_{e1}, \sigma_{e2}, \cdots, (-1)^t \sigma_{et}, 0, \cdots, 0)$. An efficient way of solving (6) is by a Gauss-Jordan reduction to upper triangular form. Since the rank of M is t, this reduction will leave t nontrivial equations, the last $t_0 - t$ equations being simply $0 = 0$. But now M_t is the upper left-hand corner of M, so that the upper left-hand corner of the reduced M will be the reduced M_t. We can therefore set the last $t_0 - t$ components of $\boldsymbol{\sigma}_e''$ to zero, and get a set of equations equivalent to (5), which can be solved for $\boldsymbol{\sigma}_e'$. Thus we need only one reduction, not two; since Gauss-Jordan reductions are tedious, this may be a significant saving.

This procedure works whenever $t \leq t_0$; that is, whenever the received word lies within distance t_0 of some code word, not counting places in which there are erasures. It will generally be possible to receive words greater than distance t_0 from any code word, and upon such words this procedure must fail. This failure, corresponding to a detectable error, must turn up either in the failure of (6) to be reducible to the form previously described, or in $\sigma_e(X)$ having an insufficient number of nonzero roots of the form β^{n_0-i}; either of these events may occur [11].

Finally, if $d - s$ is even, the preceding algorithm will locate all errors when $t \leq t_0 = (d - s - 2)/2$. Also, if $t = t_0 + 1$, an uncorrectable error can be detected by the nonvanishing of the determinant of the $t \times t$ matrix with T_{d-s-2} in the upper left, T_0 in the lower right. Such an error would be detected at some later stage in the correction process, however.

B. Example 1

Consider the (15, 7), distance 9 Reed-Solomon code introduced earlier. Suppose there occur errors of value α^4 in the first position and α in the fourth position, and erasures of value 1 in the second position and α^7 in the third position.

$$(e_1 = \alpha^4, X_1 = \alpha^{14}, e_2 = \alpha, X_2 = \alpha^{11},$$
$$d_1 = 1, Y_1 = \alpha^{13}, d_2 = \alpha^7, Y_2 = \alpha^{12}).$$

In this case the parity checks S_m will turn out to be

$$S_1 = \alpha^{14}, \quad S_2 = \alpha^{13}, \quad S_3 = \alpha^5, \quad S_4 = \alpha^6,$$
$$S_5 = \alpha^9, \quad S_6 = \alpha^{13}, \quad S_7 = \alpha^{10}, \quad \text{and} \quad S_8 = \alpha^4.$$

With these eight parity checks and two erasure locators, the decoder must find the number and position of the errors. First it forms $\boldsymbol{\sigma}_d = (\sigma_{d0}, \sigma_{d1}, \sigma_{d2})$. (Since we are working in a field of characteristic two, where addition and subtraction are identical, we omit minus signs.)

$$\sigma_{d0} = 1$$
$$\sigma_{d1} = Y_1 + Y_2 = \alpha^{13} + \alpha^{12}$$
$$= (1011) + (1111) = (0100) = \alpha$$
$$\sigma_{d2} = Y_1 Y_2 = \alpha^{13} \cdot \alpha^{12} = \alpha^{10}.$$

Next it forms the six modified cyclic parity checks T_n by (3),

$$T_0 = S_3 + \sigma_{d1} S_2 + \sigma_{d2} S_1$$
$$= \alpha^5 + \alpha \cdot \alpha^{13} + \alpha^{10} \cdot \alpha^{14} = \alpha^5 + \alpha^{14} + \alpha^9$$
$$= (0110) + (1001) + (0101) = (1010) = \alpha^8$$
$$T_1 = S_4 + \sigma_{d1} S_3 + \sigma_{d2} S_2 = \alpha^8$$
$$T_2 = 0, T_3 = \alpha^3, T_4 = \alpha^{13}, T_5 = \alpha^3.$$

Equation (6) now takes the form

$$\begin{aligned} \alpha^3 &= \alpha^{13}\sigma_{e1} + \alpha^3\sigma_{e2} \\ \alpha^{13} &= \alpha^3\sigma_{e1} + \alpha^8\sigma_{e3} \\ \alpha^3 &= \alpha^8\sigma_{e2} + \alpha^8\sigma_{e3}. \end{aligned}$$

With these equations reduced to upper triangular form, the decoder gets

$$\begin{aligned} \alpha^5 &= \sigma_{e1} + \alpha^5\sigma_{e2} \\ \alpha^{10} &= \sigma_{e2} + \sigma_{e3} \\ 0 &= 0. \end{aligned}$$

From the vanishing of the third equation, it learns that only two errors actually occured. Therefore it sets σ_{e3} to zero and solves for σ_{e1} and σ_{e2}, obtaining $\sigma_{e2} = \alpha^{10}$, $\sigma_{e1} = \alpha^{10}$. Finally, it evaluates the polynomial

$$\sigma_e(X) = X^2 + \sigma_{e1} X + \sigma_{e2} = X^2 + \alpha^{10} X + \alpha^{10}$$

for X equal to each of the nonzero elements of $GF(2^4)$; $\sigma_e(X) = 0$ when $X = \alpha^{14}$ and $X = \alpha^{11}$, so that these are the two error locators.

IV. Solving for the Values of the Erased Symbols

Once the errors have been located, they can be treated as erasures. The third and final step of the algorithm is then to determine the values of $s + t$ erased symbols, given that there are no errors in the remaining symbols. To simplify notation, we consider the problem of finding the d_k, given Y_k, $1 \leq k \leq s$, and $t = 0$.

Equation (2) is a set of linear equations in the erasure values, which can be solved by standard techniques. Its particular form gives us another approach, however, which is more efficient.

The derivation of (7), which follows, can be understood by imagining the following. Suppose we wanted to find d_{k_0}. If we continued to treat the remaining $s - 1$ erasures as erasures, but made a stab at guessing d_{k_0}, we would get a word with $s - 1$ erasures and either one or (on the chance of a correct guess) zero errors. The rank of the matrix M_1 would therefore be either zero or one; but M_1 is simply a single modified cyclic parity check, formed from the elementary symmetric functions of the $s - 1$ remaining erasure locators. Its vanishing would therefore tell us when we had guessed d_{k_0} correctly.

Symbolically, let ${}_{k_0}\boldsymbol{\sigma}_d$ be the vector of elementary symmetric functions of the $s - 1$ erasure locators not including Y_{k_0}.

Since $t = 0$, we have from (2a)

$$\mathbf{S}_{(m_0+d-2,m_0+d-s-1)} = \sum_{k=1}^{s} d_k Y_k^{m_0+d-s-1} \mathbf{Y}_{k(s-1,0)}$$

and therefore

$$\begin{aligned} {}_{k_0}T_{d-s-1} &\equiv {}_{k_0}\boldsymbol{\sigma}_d \cdot \mathbf{S}_{(m_0+d-2,m_0+d-s-1)} \\ &= d_{k_0} Y_{k_0}^{m_0+d-s-1} {}_{k_0}\sigma_d(Y_{k_0}) + \sum_{k \neq k_0} d_k Y_k^{m_0+d-s-1} {}_{k_0}\sigma_d(Y_k) \\ &= d_{k_0} Y_{k_0}^{m_0+d-s-1} {}_{k_0}\sigma_d(Y_{k_0}), \end{aligned}$$

since ${}_{k_0}\sigma_d(Y_k) = 0$, $k \neq k_0$. Thus

$$d_{k_0} = \frac{{}_{k_0}T_{d-s-1}}{Y_{k_0}^{m_0+d-s-1} {}_{k_0}\sigma_d(Y_{k_0})}.$$

This yields an explicit formula for d_{k_0} which is valid for any s

$$d_{k_0} = \frac{S_{m_0+d-2} - {}_{k_0}\sigma_{d1} S_{m_0+d-3} + {}_{k_0}\sigma_{d2} S_{m_0+d-4} - \cdots}{Y_{k_0}^{m_0+d-2} - {}_{k_0}\sigma_{d1} Y_{k_0}^{m_0+d-3} + {}_{k_0}\sigma_{d2} Y_{k_0}^{m_0+d-4} - \cdots}. \quad (7)$$

Evidently we can find all erasure values in this way; each requires the calculation of the symmetric functions of a different set of $s - 1$ locators. Alternatively, after finding d_1, we could modify all parity checks to account for this information $[\mathbf{S}'_{(m_0+d-2,m_0)} = \mathbf{S}_{(m_0+d-2,m_0)} - d_1 \mathbf{Y}_{1(m_0+d-2,m_0)}]$, and solve for d_2 in terms of these new parity checks and the remaining $s - 2$ erasure locators, and so forth.

A. Example 2

As a continuation of our previous example, let the decoder solve for e_1. The elementary symmetric functions of X_2, Y_1, and Y_2 are

$$\sigma_3 = X_2 Y_1 Y_2 = \alpha^6,$$

$$\sigma_2 = Y_2 Y_1 + X_2 Y_2 + X_2 Y_1 = \alpha^3,$$

$$\sigma_1 = X_2 + Y_1 + Y_2 = \alpha^6.$$

Therefore

$$e_1 = \frac{\alpha^4 + \alpha^6 \cdot \alpha^{10} + \alpha^3 \cdot \alpha^{13} + \alpha^6 \cdot \alpha^9}{\alpha^7 + \alpha^6 \cdot \alpha^8 + \alpha^3 \cdot \alpha^9 + \alpha^6 \cdot \alpha^{10}} = \frac{\alpha}{\alpha^{12}} = \alpha^4.$$

In a similar manner, e_2 can be found, or the decoder can calculate

$$S'_8 = S_8 + \alpha^4 X_1^8 = \alpha^{13},$$

$$S'_7 = S_7 + \alpha^4 X_1^7 = \alpha^3,$$

$$S'_6 = S_6 + \alpha^4 X_1^6 = 0.$$

Since

$$\sigma'_2 = Y_1 Y_2 = \alpha^{10}, \qquad \sigma'_1 = Y_1 + Y_2 = \alpha,$$

$$e_2 = \frac{\alpha^{13} + \alpha \cdot \alpha^3}{\alpha^{13} + \alpha \cdot \alpha^2 + \alpha^{10} \cdot \alpha^6} = \frac{\alpha^{11}}{\alpha^{10}} = \alpha.$$

Third, $S''_8 = \alpha^2$, $S''_7 = 0$,

$$d_1 = \frac{\alpha^2}{\alpha + \alpha^{12} \cdot \alpha^{13}} = 1$$

and finally, with

$$S'''_8 = \alpha^{13}, \qquad d_2 = \frac{\alpha^{13}}{\alpha^6} = \alpha^7.$$

B. Remarks II

By similar reasoning, we find

$$e_{i_0} = \frac{{}_{i_0}\boldsymbol{\sigma}_e \cdot \mathbf{T}_{(d-s-2,d-s-t-1)}}{X_{i_0}^{m_0+d-s-t-1} {}_{i_0}\sigma_e(X_{i_0}) \sigma_d(X_{i_0})}$$

which gives the error values in terms of the modified cyclic parity checks. We could therefore find all error values by this formula, modify the parity checks S_m accordingly, and then solve for the erasure values by (7).

These results are also obtainable from the first s parity check equations by solving for the s unknown erasure values by Cramer's Rule, with use of explicit formulas for the determinants of van der Monde-like matrices with missing powers.

Reed-Solomon [8], [9] codes are the subclass of BCH codes for which the symbol values and locators are defined on the same field ($M = N$). The number of check symbols in an RS code is equal to $d - 1$, the maximum correctable number of erasures. One way of generating a systematic Reed-Solomon code—that is, a code in which the first k symbols are arbitrary information symbols, while the last $n_0 - k$ are check symbols—would be to use (7) to solve for the last $n_0 - k$ symbols in terms of the parity checks S_m of the first k symbols, as though they were erasures. Since the 'erasures' are always in the last $n_0 - k$ places, each of the final $n_0 - k$ symbols can be expressed as a fixed linear function of the S_m.

V. An Alternative Determination of Error Values

The point of view which led us to the erasure correction procedure just described leads us also to another method of determining the values of the errors. Suppose the number of errors t had been discovered; then the $t \times t$ matrix M_t would have rank t and therefore nonzero determinant. If the decoder now determined the locator X_{i_0} of any error, guessed the corresponding error value e_{i_0}, and modified the T_n accordingly, then the guessed word would either still have t or (on the chance of a correct guess) $t - 1$ errors, and the $t \times t$ matrix M'_t formed from the new T'_n would have zero determinant if and only if the guess were correct.

In general, one would expect this argument to yield a polynomial in e_{i_0} of degree t as the equation of condition, but because of the special form of M_t this equation is only first degree, and an explicit formula for e_{i_0} can be obtained.

Let

$$\mathbf{S}'_{(m_0+n+s,\,m_0+n)} \equiv \mathbf{S}_{(m_0+n+s,\,m_0+n)} - e_{i_0}\mathbf{X}_{i_0(m_0+n+s,\,m_0+n)}.$$

Then

$$T'_n \equiv \boldsymbol{\delta}_d \cdot \mathbf{S}'_{(m_0+n+s,\,m_0+n)}$$

$$= \boldsymbol{\delta}_d \cdot \mathbf{S}_{(m_0+n+s,\,m_0+n)} - e_{i_0}\boldsymbol{\delta}_d \cdot \mathbf{X}_{i_0(m_0+n+s,\,m_0+n)}$$

$$= T_n - e_{i_0}X_{i_0}^{m_0+n}\sigma_d(X_{i_0}) = T_n - E_{i_0}X_{i_0}^{n}.$$

$$M'_t = \begin{bmatrix} T_{2t_0-2} - E_{i_0}X_{i_0}^{2t_0-2} & T_{2t_0-3} - E_{i_0}X_{i_0}^{2t_0-3} & \cdots & T_{2t_0-t-1} - E_{i_0}X_{i_0}^{2t_0-t-1} \\ T_{2t_0-3} - E_{i_0}X_{i_0}^{2t_0-3} & T_{2t_0-4} - E_{i_0}X_{i_0}^{2t_0-4} & \cdots & T_{2t_0-t-2} - E_{i_0}X_{i_0}^{2t_0-t-2} \\ \vdots & \vdots & & \vdots \\ T_{2t_0-t-1} - E_{i_0}X_{i_0}^{2t_0-t-1} & T_{2t_0-t-2} - E_{i_0}X_{i_0}^{2t_0-t-2} & \cdots & T_{2t_0-2t} - E_{i_0}X_{i_0}^{2t_0-2t} \end{bmatrix}$$

Let us expand this determinant into 2^t determinants, using the fact that the determinant of the matrix which has the vector $(\mathbf{a} + \mathbf{b})$ as a row is the sum of the determinants of the two matrices which have $\mathbf{a}$ and $\mathbf{b}$ in that row, respectively. We classify the resulting determinants by the number of rows which have E_{i_0} as a factor.

There is one determinant with no row containing E_{i_0}, which is simply $|M_t|$.

There are t determinants with one row having E_{i_0} as a factor. For example, the first is

$$\begin{vmatrix} -E_{i_0}X_{i_0}^{2t_0-2} & -E_{i_0}X_{i_0}^{2t_0-3} & \cdots & -E_{i_0}X_{i_0}^{2t_0-t-1} \\ T_{2t_0-3} & T_{2t_0-4} & \cdots & T_{2t_0-t-2} \\ \vdots & \vdots & & \vdots \\ T_{2t_0-t-1} & T_{2t_0-t-2} & \cdots & T_{2t_0-2t} \end{vmatrix}.$$

There are $\binom{t}{2}$ determinants with two rows having E_{i_0} as a factor. The first is

$$\begin{vmatrix} -E_{i_0}X_{i_0}^{2t_0-2} & -E_{i_0}X_{i_0}^{2t_0-3} & \cdots & -E_{i_0}X_{i_0}^{2t_0-t-1} \\ -E_{i_0}X_{i_0}^{2t_0-3} & -E_{i_0}X_{i_0}^{2t_0-4} & \cdots & -E_{i_0}X_{i_0}^{2t_0-t-2} \\ T_{2t_0-4} & T_{2t_0-5} & \cdots & T_{2t_0-t-3} \\ \vdots & \vdots & & \vdots \\ T_{2t_0-t-1} & T_{t_0-t-2} & \cdots & T_{2t_0-2t} \end{vmatrix}.$$

But in this determinant the first row is simply X_{i_0} times the second, so that the determinant is zero. Furthermore, in all such determinants with two or more rows having E_{i_0} as a factor, these rows will be some power of X_{i_0} times each other, so that all such determinants are zero.

The t determinants with one row having E_{i_0} as a factor are all linear in E_{i_0}, and contain explicit powers of X_{i_0} between $2t_0 - 2t$ and $2t_0 - 2$; their sum is then

$$-E_{i_0}X_{i_0}^{2t_0-2t}P(X_{i_0})$$

where $P(X_{i_0})$ is a polynomial of degree $2t - 2$, whose coefficients are functions of the original T_n.

Finally, we recall that $E_{i_0} = e_{i_0}X_{i_0}^{m_0}\sigma_d(X_{i_0})$ and that $|M'_t| = 0$ if and only if e_{i_0} is chosen correctly, from which we get the equation of condition

$$0 = |M'_t| = |M_t| - E_{i_0}X_{i_0}^{2t_0-2t}P(X_{i_0})$$

so

$$e_{i_0} = \frac{|M_t|}{X_{i_0}^{m_0+2t_0-2t}\sigma_d(X_{i_0})P(X_{i_0})}. \tag{8}$$

We inquire into the ease of using this formula to compute error values. $|M_t|$ can be obtained as a by-product of the reduction of (6). The only term in the denominator of (8) which is not readily calculable is $P(X_{i_0})$. In general, if A_{ik} is the determinant of the matrix remaining after the ith row and kth column are struck from M_t, then

$$P(X_{i_0}) = \sum_{l=2}^{2t}(-X_{i_0})^{2t-l}\sum_{i+k=l}A_{ik}.$$

A simplification occurs when we are in a field of characteristic two. Note that, because of the diagonal symmetry of M_t, $A_{ik} = A_{ki}$. Any sum $\sum_{i+k=l}A_{ik}$ will consist entirely of pairs $A_{ik} + A_{ki} = 0$, unless l is even, when the entire sum equals A_{jj}, with $j = l/2$. Then

$$P(X_{i_0}) = \sum_{i=1}^{t}X_{i_0}^{2(t-i)}A_{ii}. \tag{9}$$

Evaluation of the coefficients of $P(X)$ in a field of characteristic two therefore involves calculating $t \times (t-1) \times (t-1)$ determinants.

A. Example 3

Let the decoder solve (6) as before, obtaining as a by-product $|M_t| = \alpha^6$. Trivially, $A_{22} = T_4 = \alpha^{13}$, $A_{11} = T_2 = 0$. The first error locator that it will discover is $X_1 = \alpha^{14}$. Then, from (8),

$$e_1 = \frac{|M_2|}{X_1^3(X_1^2 + \sigma_{d1}X_1 + \sigma_{d2})(A_{11}X_1^2 + A_{22})}$$

$$= \frac{\alpha^6}{\alpha^{12}(\alpha^{13} + \alpha \cdot \alpha^{14} + \alpha^{10})\alpha^{13}} = \alpha^4.$$

Similarly, when it discovers $X_2 = \alpha^{11}$,

$$e_2 = \frac{\alpha^6}{\alpha^3(\alpha^7 + \alpha \cdot \alpha^{11} + \alpha^{10})\alpha^{13}} = \alpha.$$

Then it can solve for d_1, d_2 as before

B. Remarks III

The procedure just described for determining error values is clearly applicable in principle to the determination of erasure values. In this case, however, $\boldsymbol{\sigma}_d$ must be replaced by ${}_{k_0}\boldsymbol{\sigma}_d$, the vector of elementary symmetric functions of the $s - 1$ erasures other than the one being considered, and the original modified cyclic parity checks T_n by the modified cyclic parity checks defined on the other $s - 1$ erasure locators. This means that the determinants appearing in (9), as well as $|M_t|$, must be recomputed to solve for each erasure, in contrast to the solution for the error values; this promises to be tedious and to militate against this method in practice. We mention this possibility only because it does allow calculation of the correct value of an erasure, given only the number of errors and the positions of the other erasures, without knowledge of the location or value of the errors, a capability that might be useful in some application.

The erasure-correction scheme with no errors described previously can be seen to be a special case of this algorithm.

Continued development of the point of view in this paper gives us an alternative method of locating the errors. If we tentatively considered a received symbol as an erasure, in a received word with t errors, then the resulting word would have t errors if the trial symbol were correct, and $t - 1$ errors if the trial symbol were in error. The vanishing of the $t \times t$ determinant $|M_t''|$ formed from the T_n'' defined now by $s + 1$ erasure locators would then indicate the error locations. The reader may verify the fact that if X_{j_0} is the locator of the trial symbol, $T_n'' = T_{n+1} - X_{j_0}T_n$, and

$$M_t'' = \begin{bmatrix} T_{2t_0-1} - X_{j_0}T_{2t_0-2} & T_{2t_0-2} - X_{j_0}T_{2t_0-3} & \cdots & T_{2t_0-t} - X_{j_0}T_{2t_0-t-1} \\ T_{2t_0-2} - X_{j_0}T_{2t_0-3} & T_{2t_0-3} - X_{j_0}T_{2t_0-4} & \cdots & T_{2t_0-t-1} - X_{j_0}T_{2t_0-t-2} \\ \vdots & \vdots & & \vdots \\ T_{2t_0-t} - X_{j_0}T_{2t_0-t-1} & T_{2t_0-t-1} - X_{j_0}T_{2t_0-t-2} & \cdots & T_{2t_0-2t+1} - X_{j_0}T_{2t_0-2t} \end{bmatrix}.$$

If we expand $|M_t''|$ by columns, many of the resulting determinants will have one column equal to $-X_{j_0}$ times another. The only ones that will not are

$$D_0 = |\mathbf{T}_{(2t_0-1,2t_0-t)}, \mathbf{T}_{(2t_0-2,2t_0-t-1)}, \cdots, \mathbf{T}_{(2t_0-t,2t_0-2t+1)}|$$

$$-X_{j_0}D_1 = |\mathbf{T}_{(2t_0-1,2t_0-t)}, \cdots, \mathbf{T}_{(2t_0-t+1,2t_0-2t+2)}, -X_{j_0}\mathbf{T}_{(2t_0-t-1,2t_0-2t)}|$$

$$X_{j_0}^2D_2 = |\mathbf{T}_{(2t_0-1,2t_0-t)}, \cdots, \mathbf{T}_{(2t_0-t+2,2t_0-2t+3)}, -X_{j_0}\mathbf{T}_{(2t_0-t,2t_0-2t+1)}, -X_{j_0}\mathbf{T}_{(2t_0-t-1,2t_0-2t)}|,$$

and so forth. Thus if X_{j_0} is a root of the polynomial

$$D(X_{j_0}) = \sum_{j=0}^{t} D_j(-X_{j_0})^j,$$

$|M_t''|$ is zero and X_{j_0} is an error locator. It can be verified that $D_j = \sigma_{e(t-j)}D_t$, so that $D(X) = D_t\sigma_e(X)$, and this method is entirely equivalent to the former one. Furthermore, it is clear that

$$D(X) = \begin{vmatrix} X^t & T_{2t_0-1} & T_{2t_0-2} & \cdots & T_{2t_0-t} \\ X^{t-1} & T_{2t_0-2} & T_{2t_0-3} & \cdots & T_{2t_0-t-1} \\ \vdots & \vdots & \vdots & & \vdots \\ X & T_{2t_0-t} & T_{2t_0-t-1} & \cdots & T_{2t_0-2t+1} \\ 1 & T_{2t_0-t-1} & T_{2t_0-t-2} & \cdots & T_{2t_0-2t} \end{vmatrix}.$$

The condition of the vanishing of this matrix determinant is the generalization to the nonbinary case of the 'direct method' of Chien [7]. It appears to offer no advantages in practice, for to get the coefficients of $D(X)$ one must find the determinants of $t + 1$ $t \times t$ matrices, whereas the coefficients of the equivalent $\sigma_e(X)$ can be obtained as a by-product of the determination of t.

VI. Implementation

We shall now consider in more detail how a BCH decoder might be realized as a special-purpose computer. Bartee and Schneider [12] and Peterson [8] have discussed the construction of finite-field arithmetic units. We shall assume the availability of an arithmetic unit able to realize, in approximate order of complexity, the following functions of finite-field elements: addition ($X = X_1 + X_2$), squaring ($X = X_1^2$), multiplication by β_m, $m_0 \leq m \leq m_0 + d - 2$ ($X = \beta^m X_1$), inversion ($X = X_1^{-1}$), and multiplication ($X = X_1X_2$). Furthermore, because of the bistable nature of common computer elements, we shall assume $p = 2$, so that subtraction is identical to addition and squaring is linear. Finally, if the fields on which the symbol locators and symbol values are defined are different ($M \neq N$), we shall assume that all elements of the smaller field are converted to their representations in the larger field of which the smaller field is a subfield, and that all operations are carried out in the larger field, say, $GF(2^N)$.

We attempt to estimate the approximate complexity of the algorithms described earlier by estimating the number of multiplications required by each and the number of memory registers. All registers will, of course, be N bits long. In a straightforward serial multiplication, the amount of time needed for one multiplication is also proportional to N [8].

During the computation, the received sequence of symbols must be stored in some buffer, awaiting correction. Once the S_m and Y_k have been determined, no further access to this sequence is required.

The calculation of the parity checks $S_m \equiv r(\beta^m) = r_1\beta^{m(n_0-1)} + r_2\beta^{m(n_0-2)} + \cdots + r_{n_0}$ is accomplished by the iteration

$$S_m = ((r_1\beta^m + r_2)\beta^m + r_3)\beta^m + r_4 \cdots$$

which involves $n_0 - 1$ multiplications by β^m. $d - 1$ such parity checks must be formed, requiring $d - 1$ memory registers.

$\boldsymbol{\sigma}_d$ can be calculated at the same time. We note that $\sigma_{dk} = {}_{k_0}\sigma_{dk} + Y_{k_0}{}_{k_0}\sigma_{d(k-1)}$; $\boldsymbol{\sigma}_d$ can be calculated by this recursion relation as each new Y_k is determined. Adding a new Y_k requires s' multiplications when s' are already determined, so that the total number of multiplications, given s erasures, is $s - 1 + s - 2 + \cdots = \binom{s}{2} < \frac{1}{2}d^2$. s memory registers are required ($\sigma_{d0} = 1$).

The modified cyclic parity checks T_n are then calculated by (3). Each requires s multiplications, and there are $d - s - 1$ of them, so that this step requires $s(d - s - 1) < \frac{1}{4}d^2$ multiplications and $d - s - 1$ memory registers.

Equation (6) is then set up in $t_0(t_0 + 1) < \frac{1}{4}d^2$ memory registers. In the worst case, $t = t_0$, the reduction to upper triangular form of these equations will require t_0 inversions and

$$t_0(t_0 + 1) + (t_0 - 1)t_0 + \cdots + (1)(2) = \begin{bmatrix} 2t_0 + 2 \\ 3 \end{bmatrix} + \begin{bmatrix} t_0 + 1 \\ 2 \end{bmatrix} < \frac{(t_0 + 1)^3}{3}$$

multiplications. As d becomes large, this step turns out to be the most lengthy, requiring as it does $\sim d^3/24$ multiplications.

Determination of $\boldsymbol{\sigma}_e$ from these reduced equations involves, in the worst case, a further $\binom{t_0}{2} < d^2/8$ multiplications, and t_0 memory registers.

Finding the roots of $\sigma_e(X) = \sum_{i=0}^{t} \sigma_{e(t-i)}X^i$ is best accomplished by the method of Chien [7]. If $\sum_{i=0}^{t} \sigma_{e(t-i)} = 0$, then 1 is a root of $\sigma_e(X)$. Use the special multipliers by β^m in the arithmetic unit, and let $\sigma'_{e(t-i)} = \beta^{m_0+t-i}\sigma_{e(t-i)}$. Now $\sum_{i=0}^{t} \sigma'_{a(t-i)} = \beta^{m_0+t} \sum_{i=0}^{t} \beta^{-i}\sigma_{e(t-i)}$, which will be zero when $\beta^{-1} = \beta^{n_0-1}$ is a root of $\sigma_e(X)$. All error locators can therefore be found with $n_0 t$ multiplications by β^m, and stored in t memory registers.

At this point we have the option of solving for the error values directly, by (8), or indirectly, by treating the errors as erasures and using (7).

If we choose the former method, we need the t $(t - 1) \times (t - 1)$ determinants A_{ij} of (9). In general this requires $\frac{1}{4}t\binom{2t}{3} < t^4/3$ multiplications, which is rapidly too many as t becomes large. There is a method of calculating all A_{ij} at once which seems feasible for moderate values of t. Let $B_{a_1, a_2, \cdots, a_j}$ be the determinant of the $j \times j$ matrix which remains when all of the rows and columns but the a_1th, a_2th, $\cdots$, a_jth are struck from M_t. In this notation $|M_t| = B_{1,2,\cdots,t}$ and $A_{ij} = B_{1,2,\cdots,j-1,j+1,\cdots,t}$. The reader may verify the fact that

$$B_{a_1,a_2,\cdots,a_j} = T_{2t_0-2a_j}B_{a_1,a_2,\cdots,a_{j-1}} + T^2_{2t_0-2a_j+1}B_{a_1,a_2,\cdots,a_{j-2}} + T^2_{2t_0-2a_j+2}B_{a_1,a_2,\cdots,a_{j-3},a_{j-1}} + \cdots$$

by expanding B in terms of the minors of its last row and cancelling those terms which, because of symmetry, appear twice. The use of this recursion relation allows calculation of all A_{ij} with N_t multiplications (not counting squares), where, for small t, N_t is $N_2 = 0$ (see Example 3), $N_3 = 3$, $N_4 = 15$, $N_5 = 38$, $N_6 = 86$, $N_7 = 172$, $N_8 = 333$, and $N_9 = 616$.

Once the A_{ij} are obtained, the denominator of (8) can be expressed as a single polynomial $E(X)$ by st multiplications; $E(X)$ has terms in X^m for all m such that $m_0 + 2t_0 - 2t \leq m \leq m_0 + 2t_0 + s$, or a total of $2t + s + 1$ terms. The value of $E(X)$ can therefore be obtained for $X = 1, \beta^{-1}, \beta^{-2}, \cdots$, in turn by the Chien method of solving for the roots of $\sigma_e(X)$, and in fact these two calculations may be done simultaneously. Whenever β^{n_0-i} is a root of $\sigma_e(X)$, $E(\beta^{n_0-i})$ will appear as the current value of $E(X)$. Since $|M_t|$ will have been obtained as a by-product of solving for $\sigma_e(X)$, an inversion and multiplication will give the error value corresponding to $X_{i_0} = \beta^{n_0-i}$. An additional $n_0(s + 2t)$ multiplications by β^m are involved here, and $s + 2t$ memory registers.

Finally, we have only the problem of solving for s erasure values, if we have already determined the error values as above, or of $s + t$ erasures, if we know only the error locators. We use (7), which requires the elementary symmetric functions of all erasure locators but one. The reader may verify the fact that ${}_{k_0}\sigma_{dk} = Y_{k_0}^{-1}(\sigma_{d(k+1)} - {}_{k_0}\sigma_{d(k+1)})$. Beginning with ${}_{k_0}\sigma_{d(s-1)} = Y_{k_0}^{-1}\sigma_{ds}$, we can find all ${}_{k_0}\sigma_{dk}$ from the σ_{dk} with $s - 1$ multiplications and an inversion. Then the calculation of (7) requires $2(s - 1)$ multiplications and an inversion. Doing this s times, to find all erasure values requires, therefore, $3s(s - 1)$ multiplications and s inversions. Or we can alter $s - 1$

parity checks after finding the value of the first erasure, and repeat with $s' = s - 1$, and so forth; under the assumption that all $Y_{k_0}^m$ are readily available, this alternative requires only $2s(s - 1)$ multiplications and s inversions. Thus our procedure is simpler than the general methods for solving s linear equations in s unknowns, which require a number of multiplications proportional to s^3.

In order to compare the alternative methods of finding error values, we simply compare the number of multiplications needed in each case, leaving aside all analysis of any other equipment or operations needed to realize either algorithm. For the first method, we need approximately N_t multiplications to find the error values, and $2s(s - 1)$ to find the erasures; for the second, $2(s + t)(s + t - 1)$ to find both the erasures and the errors. Using the values of N_t given here, we find that the first method definitely requires fewer multiplications when $t \leq 7$, which suggests that it ought to be considered whenever the minimum distance of the code is 15 or less.

To summarize, we note that there are two steps with a number of multiplications by β^m proportional to n_0 and a step with a number of memory registers proportional to d^2, and a buffer with memory proportional to n_0; and the complexity of a multiplication and the length of a register are both proportional to N. It follows that if $d \sim \delta n_0$ and $N \sim \log_2 n_0$, we can estimate that the complexity of a BCH decoder will increase with code length n_0 as n_0^b, where b is some small power of the order of 3. Therefore we can expect that implementation of a decoder for a long and complicated BCH code will not be prohibitively more difficult than for a short, simple code.

Acknowledgment

The author is indebted to W. W. Peterson, from whose book [8] comes most of his knowledge of this field. These results appeared in the author's doctoral thesis [13], during the research for which he was supported by a National Science Foundation Graduate Fellowship.

References

[1] R. C. Bose, and D. K. Ray-Chaudhuri, "On a class of error-correcting binary group codes," *Inform. and Contr.*, vol. 3, pp. 68–79, March 1960.
[2] ——, "Further results on error-correcting binary group codes," *Inform. and Contr.*, vol. 3, pp. 279–290, September 1960.
[3] A. Hocquenghem, "Codes correcteurs d'erreurs," *Chiffres*, vol. 2, pp. 147–156, September 1959.
[4] W. W. Peterson, "Encoding and error-correction procedures for the Bose-Chaudhuri codes," *IRE Trans. on Information Theory*, vol. IT-6, pp. 459–470, September 1960.
[5] T. C. Bartee and D. I. Schneider, "An electronic decoder for Bose-Chaudhuri-Hocquenghem error-correcting codes," *IRE Trans. on Information Theory*, vol. IT-8, pp. S17–S24, September 1962.
[6] D. Gorenstein and N. Zierler, "A class of cyclic linear error-correcting codes in p^m symbols," *J. SIAM*, vol. 9, pp. 207–214, June 1961.
[7] R. T. Chien, "Cyclic decoding procedures for Bose-Chaudhuri-Hocquenghem codes," *IEEE Trans. on Information Theory*, vol. IT-10, pp. 357–363, October 1964.
[8] W. W. Peterson, *Error-Correcting Codes*, New York: M.I.T. Press-Wiley, 1961.
[9] I. S. Reed and G. Solomon, "Polynomial codes over certain finite fields," *J. SIAM*, vol. 8, pp. 300–304, June 1960.
[10] P. Elias, "Coding for two noisy channels," in *Information Theory*, C. Cherry, Ed., New York: Academic Press, 1956.
[11] R. T. Chien and D. T. Tang, "On detecting errors after correction," *Proc. IEEE*, vol. 62, p. 974, August 1964.
[12] T. C. Bartee and D. I. Schneider, "Computation with finite fields," *Inform. and Contr.*, vol. 6, pp. 79–98, June 1963.
[13] G. D. Forney, Jr., "Concatenated codes," Sc.D. Thesis, Dept. of Electrical Engineering, M.I.T., Cambridge, Mass., June, 1965. To appear as Technical Report 440, M.I.T. Research Laboratory of Electronics, Cambridge, Mass.

Binary BCH Codes for Correcting Multiple Errors

Elwyn R. Berlekamp

7.2 THE KEY EQUATION FOR DECODING BINARY BCH CODES

If the encoder transmits the binary BCH codeword

$$C(x) = \sum_{i=0}^{n-1} C_i x^i$$

and the channel noise causes additive errors given by the coefficients of the binary polynomial

$$E(x) = \sum_{i=0}^{n-1} E_i x^i$$

then the received word will be given by

$$R(x) = \sum_{i=0}^{n-1} R_i x^i = \sum_{i=0}^{n-1} C_i x^i + \sum_{i=0}^{n-1} E_i x^i$$

For $j = 1, 2, \ldots, 2t$, the codeword is a multiple of the minimal polynomial of α^j, and therefore

$$R(\alpha^j) = 0 + \sum_{i=0}^{n-1} E_i \alpha^{ji} = \sum_{k=1}^{e} X_k{}^j = S_j$$

where we let the Galois field error locations $X_1, X_2, \ldots, X_e$ denote the positions where $E_i = 1$. If $R(x)$ is divided by $M^{(j)}(x)$, the minimal polynomial of α^j $(1 \le j \le 2t)$, then $S_j = R(\alpha^j)$ may be readily computed from the remainder, $r^{(j)}(\alpha^j)$, as in Sec. 5.2. After the decoder has calculated $S_1, S_2, S_3, \ldots, S_{2t}$, the major problem is to find the error locations, $X_1, X_2, \ldots, X_e$, from the equations

$$\sum_{i=1}^{e} X_i{}^j = S_j \qquad j = 1, 2, 3, \ldots, 2t$$

In general, these equations will have many solutions, each corresponding to a different error pattern in the same coset of the additive group of codewords. The decoder must find a solution with as small a value of e as possible.

In order to solve these equations, the decoder first attempts to find the coefficients of the error-locator polynomial, defined by Definition 7.21 [or Eq. (1.46)].

Definition 7.21

$$\sigma(z) = \prod_{i=1}^{e} (1 - X_i z) = 1 + \sum_{j=1}^{e} \sigma_j z^j$$

Once the decoder has determined the error-locator polynomial $\sigma(z)$, the reciprocal roots of $\sigma(z)$ can be found via a Chien search. The errors can then be corrected, as in Fig. 5.14. The hardest part of the decoding procedure is to find the σ's from the S's.

To relate the σ's and the S's, we introduce the generating function

$$S(z) = \sum_{j=1}^{\infty} S_j z^j = \sum_{j=1}^{\infty} \sum_{i=1}^{e} X_i{}^j z^j = \sum_{i=1}^{e} \frac{X_i z}{1 - X_i z}$$

We may eliminate the fractions in this equation by multiplying through by $\sigma(z)$. This gives

$$S(z)\sigma(z) = \sum_{i=1}^{e} \frac{X_i z}{1 - X_i z} \prod_{j=1}^{e} (1 - X_j z) = \sum_{i=1}^{e} X_i z \prod_{j \ne i} (1 - X_j z)$$

Adding $\sigma(z)$ to both sides gives

$$[1 + S(z)]\sigma(z) = \sigma(z) + \sum_{i=1}^{e} X_i z \prod_{j \ne i} (1 - X_j z)$$

We define the polynomial $\omega(z) = \sum_{k=0}^{e} \omega_k z^k$ by the equation

$$\omega(z) = \sigma(z) + \sum_{i=1}^{e} X_i z \prod_{j \ne i} (1 - X_j z) \tag{7.22}$$

We then have the equation

$$[1 + S(z)]\sigma(z) = \omega(z)$$

In general, the decoder knows only the coefficients of the first $2t$ powers of z in $S(z)$; he does not know $S_{2t+1}, S_{2t+2}, S_{2t+3}, \ldots$. In other words, the decoder does not know $S(z)$, but he does know $S(z) \bmod z^{2t+1}$. The relevant equation, which I call the *key equation*, is

$$[1 + S(z)]\sigma(z) \equiv \omega(z) \bmod z^{2t+1} \tag{7.23}$$

Given $S(z)$, we wish to find *both* $\sigma(z)$ and $\omega(z)$ from this equation. The unknown polynomials $\sigma(z)$ and $\omega(z)$ both have degrees $\le e$, the number of errors that actually occurred.

One "physical interpretation" of the key equation has been suggested by Massey (1968). The key equation may be rewritten as

$$S_k + \sum_{i=1}^{k-1} \sigma_i S_{k-i} + \sigma_k = \omega_k$$

or

$$S_k = \omega_k - \sum_{i=1}^{k-1} \sigma_i S_{k-i} - \sigma_k$$

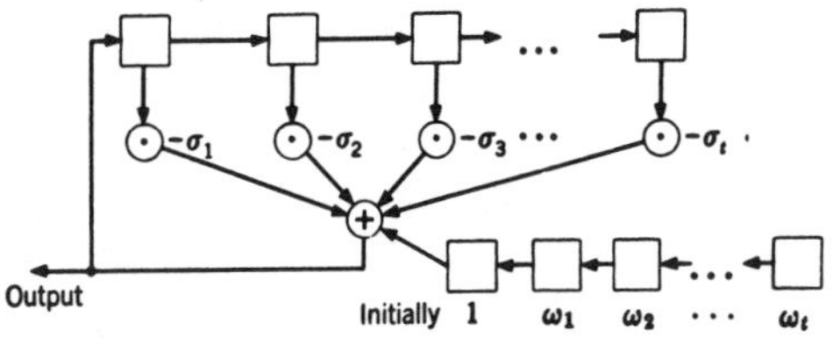

Fig. 7.1 Feedback-shift-register interpretation of the key equation.

This equation gives the kth output of the feedback shift registers of Figs. 7.1 and 7.2, which are wired according to the coefficients of the polynomial $\sigma(z)$ and initially loaded with the coefficients of the polynomial $\omega(z)$. Thus, the key equation may be interpreted as the mathematical formulation of a problem in feedback-shift-register synthesis. Given the output sequence $1 + S(z)$, we wish to determine the connections $\sigma(z)$ and the initial contents $\omega(z)$ of the shortest feedback shift register whose output sequence is $1 + S(z)$.

In our present application, the decoding of binary BCH codes, only $\sigma(z)$ is of any real interest; $\omega(z)$ is just another polynomial of no obvious importance. However, this same key equation arises in several other applications, and we shall encounter it again in Chaps. 9 and 10. In both of these applications of the key equation, as well as in the interpretation as a problem in feedback-shift-register synthesis, the unknown polynomials $\sigma(z)$ and $\omega(z)$ will both have an obvious importance.

Let us now develop an algorithm for solving the key equation over any field.

7.3 HEURISTIC SOLUTION OF THE KEY EQUATION

We wish to solve the key equation

$$(1 + S)\sigma \equiv \omega \bmod z^{2t+1} \tag{7.23}$$

for the polynomials $\sigma(z)$ and $\omega(z)$, given $S(z) \bmod z^{2t+1}$. The problem looks difficult, so we break it up into smaller pieces. We consider the sequence of equations

$$(1 + S)\sigma^{(k)} \equiv \omega^{(k)} \bmod z^{k+1} \tag{7.301}$$

For each $k = 0, 1, 2, \ldots, 2t$, we shall find polynomials

$$\sigma^{(k)} = \sum_i \sigma_i{}^{(k)} z^i$$

and

$$\omega^{(k)} = \sum_i \omega_i{}^{(k)} z^i$$

which solve this equation. In general, these equations may have many solutions. Since the degree of σ is the number of errors, a good decoder

Reprinted with permission from E. R. Berlekamp, *Algebraic Coding Theory.* New York: McGraw-Hill, 1968.

must attempt to find a solution in which degree σ and degree ω are "small."

If we have a solution to (7.301), then we can hope that this same pair of polynomials, $\sigma^{(k)}$ and $\omega^{(k)}$, might also solve the equation

$$(1 + S)\sigma^{(k)} \overset{?}{\equiv} \omega^{(k)} \bmod z^{k+2}$$

In general, we cannot expect to be so lucky. However, we may write

$$(1 + S)\sigma^{(k)} \equiv \omega^{(k)} + \Delta_1^{(k)} z^{k+1} \bmod z^{k+2} \tag{7.302}$$

where we define $\Delta_1^{(k)}$ as the coefficient of z^{k+1} in the product $(1 + S)\sigma^{(k)}$. If $\Delta_1^{(k)} = 0$, then we may evidently proceed by taking $\sigma^{(k+1)} = \sigma^{(k)}$ and $\omega^{(k+1)} = \omega^{(k)}$. In order to define $\sigma^{(k+1)}$ in the case when $\Delta_1^{(k)} \neq 0$, we introduce the auxiliary polynomials $\tau^{(k)}$ and $\gamma^{(k)}$, which will be chosen so as to solve the *auxiliary equation*

$$(1 + S)\tau^{(k)} \equiv \gamma^{(k)} + z^k \bmod z^{k+1} \tag{7.303}$$

Of course we would also like the degrees of $\tau^{(k)}$ and $\gamma^{(k)}$ to be "small." In terms of these auxiliary polynomials, we may define the successive σ's and ω's by

$$\sigma^{(k+1)} = \sigma^{(k)} - \Delta_1^{(k)} z \tau^{(k)} \tag{7.304}$$
$$\omega^{(k+1)} = \omega^{(k)} - \Delta_1^{(k)} z \gamma^{(k)} \tag{7.305}$$

It is readily seen that $\sigma^{(k+1)}$ and $\omega^{(k+1)}$ satisfy the equation

$$(1 + S)\sigma^{(k+1)} \equiv \omega^{(k+1)} \bmod z^{(k+1)+1}$$

if $\sigma^{(k)}$ and $\omega^{(k)}$ satisfy (7.302) and $\tau^{(k)}$ and $\gamma^{(k)}$ satisfy (7.303).

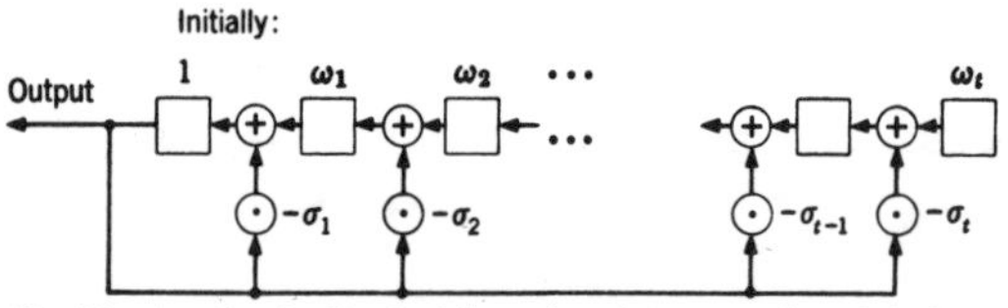

Fig. 7.2 Another feedback-shift-register interpretation of the key equation.

There are two obvious ways to define $\tau^{(k+1)}$ and $\gamma^{(k+1)}$. *Either*

$$\tau^{(k+1)} = z\tau^{(k)} \quad \text{and} \quad \gamma^{(k+1)} = z\gamma^{(k)} \tag{7.306}$$

or

$$\tau^{(k+1)} = \frac{\sigma^{(k)}}{\Delta_1^{(k)}} \quad \text{and} \quad \gamma^{(k+1)} = \frac{\omega^{(k)}}{\Delta_1^{(k)}} \tag{7.307}$$

Either choice will satisfy the equation

$$(1 + S)\tau^{(k+1)} \equiv \gamma^{(k+1)} + z^{k+1} \bmod z^{k+2}$$

if $\sigma^{(k)}$ and $\omega^{(k)}$ satisfy (7.302) and $\tau^{(k)}$ and $\gamma^{(k)}$ satisfy (7.303). If $\Delta_1^{(k)} = 0$, then (7.307) is meaningless, and we are forced to define $\tau^{(k+1)}$ and $\gamma^{(k+1)}$ by (7.306). However, if $\Delta_1^{(k)} \neq 0$, then our choice between (7.306) and (7.307) must be based upon our desire to minimize the degrees of $\tau^{(k+1)}$ and $\gamma^{(k+1)}$. The degrees of $\sigma^{(k+1)}$, $\tau^{(k+1)}$, $\omega^{(k+1)}$, and $\gamma^{(k+1)}$ are given by

$$\deg \sigma^{(k+1)} = \begin{cases} \deg \sigma^{(k)} & \text{if } \Delta_1^{(k)} = 0 \text{ or if } \deg \sigma^{(k)} > \deg \tau^{(k)} + 1 \\ 1 + \deg \tau^{(k)} & \text{if } \Delta_1^{(k)} \neq 0 \text{ and if } \deg \tau^{(k)} > \deg \sigma^{(k)} - 1 \end{cases} \tag{7.308}$$

$$\deg \sigma^{(k+1)} \le \text{either of above} \qquad \text{if } \Delta_1^{(k)} \neq 0 \text{ and if } \deg \sigma^{(k)} = 1 + \deg \tau^{(k)}$$

$$\deg \tau^{(k+1)} = \begin{cases} 1 + \deg \tau^{(k)} & \text{if we use (7.306)} \\ \deg \sigma^{(k)} & \text{if we use (7.307)} \end{cases} \tag{7.309}$$

$$\deg \omega^{(k+1)} = \begin{cases} \deg \omega^{(k)} & \text{if } \Delta_1^{(k)} = 0 \text{ or if } \deg \omega^{(k)} > 1 + \deg \gamma^{(k)} \\ 1 + \deg \gamma^{(k)} & \text{if } \Delta_1^{(k)} \neq 0 \text{ and if } \deg \gamma^{(k)} > \deg \omega^{(k)} - 1 \end{cases} \tag{7.310}$$

$$\deg \omega^{(k+1)} \le \text{either of above} \qquad \text{if } \Delta_1^{(k)} \neq 0 \text{ and if } \deg \omega^{(k)} = 1 + \deg \gamma^{(k)}$$

$$\deg \gamma^{(k+1)} = \begin{cases} 1 + \deg \gamma^{(k)} & \text{if we use (7.306)} \\ \deg \omega^{(k)} & \text{if we use (7.307)} \end{cases} \tag{7.311}$$

The degree of $\sigma^{(k+1)}$ is seen to be subject to an "accidental" decrease if $\deg \sigma^{(k)} = 1 + \deg \tau^{(k)}$ and the leading coefficients of $\sigma^{(k)}$ and $\Delta_1^{(k)}\tau^{(k)}$ happen to be equal. In order to circumcompute such accidents, we base our choice between (7.306) and (7.307) not on the actual degrees of $\sigma^{(k)}$, $\tau^{(k)}$, $\omega^{(k)}$, and $\gamma^{(k)}$ but on an upper bound $D(k)$ which is independent of such vagaries. We shall define this integral function $D(k)$ in such a way that

$$\deg \sigma^{(k)} \le D(k) \tag{7.312}$$
$$\deg \tau^{(k)} \le k - D(k) \tag{7.313}$$

From (7.308), (7.312), and (7.313), we are led to the recursive definition of $D(k)$

$$D(k+1) = \begin{cases} D(k) & \text{if } \Delta_1^{(k)} = 0 \text{ or if } D(k) \ge \dfrac{k+1}{2} \\ k + 1 - D(k) & \text{if } \Delta_1^{(k)} \neq 0 \text{ and } D(k) \le \dfrac{k+1}{2} \end{cases} \tag{7.314}$$

It is readily seen that if $\deg \sigma^{(k)} \le D(k)$ and $\deg \tau^{(k)} \le k - D(k)$, then $\deg \sigma^{(k+1)} \le D(k+1)$. Similarly, if $\deg \omega^{(k)} \le D(k)$ and $\deg \gamma^{(k)} \le k - D(k)$, then $\deg \omega^{(k+1)} \le D(k+1)$.

In order to ensure that $\deg \tau^{(k+1)} \le (k+1) - D(k+1)$ and that $\deg \gamma^{(k+1)} \le (k+1) - D(k+1)$, we must adopt the following rule (7.315) for choosing between (7.306) and (7.307).

$$\text{Use} \begin{cases} (7.306) & \text{if } \Delta_1^{(k)} = 0 \text{ or if } D(k) > \dfrac{k+1}{2} \\ (7.307) & \text{if } \Delta_1^{(k)} \neq 0 \text{ and } D(k) < \dfrac{k+1}{2} \end{cases} \tag{7.315}$$

If $\Delta_1^{(k)} \neq 0$ and $D(k) = (k+1)/2$, then either (7.306) or (7.307) will give us polynomials $\tau^{(k+1)}$ and $\gamma^{(k+1)}$ each having degree $\le k + 1 - D(k+1)$. When in doubt, procrastinate! We postpone the close decision between (7.306) and (7.307) in this case until we have looked at another consideration.

The initial equations are

$$(1 + S)\sigma^{(0)} \equiv \omega^{(0)} \bmod z$$
$$(1 + S)\tau^{(0)} \equiv \gamma^{(0)} + 1 \bmod z$$

The equations may be solved by the obvious initialization

$$\sigma^{(0)} = \tau^{(0)} = \omega^{(0)} = 1 \qquad \gamma^{(0)} = 0 \qquad D(0) = 0 \tag{7.316}$$

We notice that $\deg \sigma^{(0)} = \deg \tau^{(0)} = \deg \omega^{(0)} = 0 = D(0)$ but that† $\deg \gamma^{(0)} = -\infty < D(0)$. Thus, at least initially, we may do even better than the restrictions

$$\deg \omega^{(k)} \le D(k)$$
$$\deg \gamma^{(k)} < k - D(k)$$

We require that at least one of these expressions be satisfied with strict inequality. To this end, we introduce the Boolean function $B(k)$, with initial value $B(0) = 0$. [In general, either $B(k) = 0$ or $B(k) = 1$.] We require that

$$\deg \omega^{(k)} \le D(k) - B(k) \tag{7.317}$$
$$\deg \gamma^{(k)} \le k - D(k) - [1 - B(k)] \tag{7.318}$$

If we take the proper choice between (7.306) and (7.307) in the case when $\Delta_1^{(k)} \neq 0$ and $D(k) = (k+1)/2$, and if we define $B(k)$ carefully, then we can guarantee that (7.317) and (7.318) hold for all k. By examining (7.310) and (7.311), the proper choice is seen to be

$$\text{Use} \begin{cases} (7.306) & \text{if } \Delta_1^{(k)} \neq 0,\ D(k) = \dfrac{k+1}{2}, \text{ and } B(k) = 0 \\ (7.307) & \text{if } \Delta_1^{(k)} \neq 0,\ D(k) = \dfrac{k+1}{2}, \text{ and } B(k) = 1 \end{cases} \tag{7.319}$$

$$B(k+1) = \begin{cases} B(k) & \text{when using (7.306)} \\ 1 - B(k) & \text{when using (7.307)} \end{cases} \tag{7.320}$$

This completes the heuristic derivation of the recursive algorithm. To summarize, we start from the initial conditions (7.316). We then proceed recursively as follows. Define $\Delta_1^{(k)}$ by (7.302), $\sigma^{(k+1)}$ by (7.304), $\omega^{(k+1)}$ by (7.305), and $D(k+1)$ by (7.314). According to (7.315) and (7.319), we then define $\tau^{(k+1)}$ and $\gamma^{(k+1)}$ by (7.306) or (7.307) and $B(k+1)$ according to (7.320). The polynomials defined in this recursive manner are then seen to satisfy Eqs. (7.301), (7.303), (7.312), (7.313), (7.317), and (7.318). We restate the algorithm explicitly as follows.

† Recall that $\deg 0 = -\infty$, as explained in the footnote on page 28.

7.4 AN ALGORITHM FOR SOLVING THE KEY EQUATION OVER ANY FIELD

Initially define $\sigma^{(0)} = 1$, $\tau^{(0)} = 1$, $\omega^{(0)} = 1$, $\gamma^{(0)} = 0$, $D(0) = 0$, $B(0) = 0$. Proceed recursively as follows. If S_{k+1} is unknown, stop; otherwise define $\Delta_1^{(k)}$ as the coefficient of z^{k+1} in the product $(1 + S)\sigma^{(k)}$ and let

$$\sigma^{(k+1)} = \sigma^{(k)} - \Delta_1^{(k)} z\tau^{(k)}$$
$$\omega^{(k+1)} = \omega^{(k)} - \Delta_1^{(k)} z\gamma^{(k)}$$

If $\Delta_1^{(k)} = 0$, or if $D(k) > (k+1)/2$, or if $\Delta_1^{(k)} \neq 0$ and $D(k) = (k+1)/2$ and $B(k) = 0$, set

$$D(k+1) = D(k)$$
$$B(k+1) = B(k)$$
$$\tau^{(k+1)} = z\tau^{(k)}$$
$$\gamma^{(k+1)} = z\gamma^{(k)}$$

But if $\Delta_1^{(k)} \neq 0$ and either $D(k) < (k+1)/2$ or $D(k) = (k+1)/2$ and $B(k) = 1$, set

$$D(k+1) = k + 1 - D(k)$$
$$B(k+1) = 1 - B(k)$$
$$\tau^{(k+1)} = \frac{\sigma^{(k)}}{\Delta_1^{(k)}}$$
$$\gamma^{(k+1)} = \frac{\omega^{(k)}}{\Delta_1^{(k)}}$$

Theorem 7.41 For each k,

7.411. $\sigma^{(k)}(0) = \omega^{(k)}(0) = 1$.
7.412. $(1 + S)\sigma^{(k)} \equiv \omega^{(k)} + \Delta_1^{(k)} z^{k+1} \bmod z^{k+2}$.
7.413. $(1 + S)\tau^{(k)} \equiv \gamma^{(k)} + z^k \bmod z^{k+1}$.
7.414. $\deg \sigma^{(k)} \leq D(k)$ with equality if $B(k) = 1$.
7.415. $\deg \tau^{(k)} \leq k - D(k)$ with equality if $B(k) = 0$.
7.416. $\deg \omega^{(k)} \leq D(k) - B(k)$ with equality if $B(k) = 0$.
7.417. $\deg \gamma^{(k)} \leq k - D(k) - [1 - B(k)]$ with equality if $B(k) = 1$.

Theorem 7.42 For each k,

$$\omega^{(k)}\tau^{(k)} - \sigma^{(k)}\gamma^{(k)} = z^k$$

Theorem 7.43 If σ and ω are any pair of polynomials which satisfy

$$\sigma(0) = 1 \quad \text{and} \quad (1 + S)\sigma \equiv \omega \bmod z^{k+1}$$
$$D = \max\{\deg \sigma, \deg \omega\}$$

then there exist polynomials U and V such that $U(0) = 1$, $V(0) = 0$, $\deg U \leq D - D(k)$, $\deg V \leq D - [k - D(k)]$, and

$$\sigma = U\sigma^{(k)} + V\tau^{(k)}$$
$$\omega = U\omega^{(k)} + V\gamma^{(k)}$$

Theorem 7.44 If σ and ω are relatively prime, and $\sigma(0) = 1$ and $(1 + S)\sigma \equiv \omega \bmod z^{k+1}$, then

7.441. Either $\deg \sigma \geq D(k) + 1 - B(k) \geq D(k)$, or $\deg \omega \geq D(k)$, or both.
7.442. If $\deg \sigma \leq (k+1)/2$ and $\deg \omega \leq k/2$, then $\sigma = \sigma^{(k)}$ and $\omega = \omega^{(k)}$.

Proof of Theorem 7.41, excepting the equality clauses of 7.414 to 7.417 These claims were proved in the heuristic derivation of the algorithm. Readers who prefer a direct proof may verify these claims by a straightforward induction on k.

Proof of Theorem 7.42 According to Theorem 7.41,

$$(1 + S)\sigma^{(k)} \equiv \omega^{(k)} \bmod z^{k+1}$$
$$(1 + S)\tau^{(k)} \equiv \gamma^{(k)} + z^k \bmod z^{k+1}$$

Taking the product of these two congruences gives

$$(1 + S)\tau^{(k)}\omega^{(k)} \equiv (1 + S)\sigma^{(k)}(\gamma^{(k)} + z^k)$$

Dividing by $1 + S$ gives

$$\tau^{(k)}\omega^{(k)} \equiv \sigma^{(k)}\gamma^{(k)} + \sigma^{(k)}z^k$$

Since

$$\sigma^{(k)}z^k \equiv \sigma_0^{(k)}z^k \equiv z^k \bmod z^{k+1}$$

this becomes

$$\tau^{(k)}\omega^{(k)} - \gamma^{(k)}\sigma^{(k)} \equiv z^k \bmod z^{k+1}$$

According to Theorems 7.415 and 7.416, $\deg \omega^{(k)} + \deg \tau^{(k)} \leq k$. According to Theorems 7.414 and 7.417, $\deg \sigma^{(k)} + \deg \gamma^{(k)} \leq k$. Therefore, $\deg\{\tau^{(k)}\omega^{(k)} - \sigma^{(k)}\gamma^{(k)}\} \leq k$ from which we conclude that

$$\boxed{\tau^{(k)}\omega^{(k)} - \sigma^{(k)}\gamma^{(k)} = z^k}$$

Proof of equality clauses of 7.414 to 7.417 If $B(k) = 0$, then

$$\deg \gamma^{(k)} \leq k - D(k) - 1$$
$$\deg \sigma^{(k)} \leq D(k)$$

and

$$\deg\{\sigma^{(k)}\gamma^{(k)}\} \leq k - 1$$

Since

$$\deg\{\omega^{(k)}\tau^{(k)} - \sigma^{(k)}\gamma^{(k)}\} = k$$

it follows that

$$\deg\{\omega^{(k)}\tau^{(k)}\} = k \quad \text{and} \quad \deg \omega^{(k)} = D(k)$$
$$\deg \tau^{(k)} = k - D(k)$$

Similarly, if $B(k) = 1$, then $\deg \omega^{(k)} \leq D(k) - 1$, $\deg \tau^{(k)} \leq k - D(k)$, and $\deg\{\omega^{(k)}\tau^{(k)}\} \leq k - 1$. It follows that $\deg\{\sigma^{(k)}\gamma^{(k)}\} = k$, $\deg \sigma^{(k)} = D(k)$, and $\deg \gamma^{(k)} = k - D(k)$.

Remarks on Theorem 7.43 This theorem gives us the form of the *general solution* (of any degree) of the equations $\sigma(0) = 1$, $(1 + S)\sigma \equiv \omega \bmod z^{k+1}$. In view of Theorems 7.42 and 7.411, it is evident that $\sigma^{(k)}$ and $\tau^{(k)}$ are relatively prime. Hence any polynomial f may be expressed in the form

$$f = U\sigma^{(k)} + V\tau^{(k)}$$

Theorem 7.43 asserts that if σ and ω are a solution to the equations $\sigma(0) = 1$, $(1 + S)\sigma \equiv \omega \bmod z^{k+1}$, then the same U and V hold for both expressions $\sigma = U\sigma^{(k)} + V\tau^{(k)}$ and $\omega = U\omega^{(k)} + V\gamma^{(k)}$. Furthermore, Theorem 7.43 asserts that the degrees of U and V are small.

Proof of Theorem 7.43

$$\omega \equiv (1 + S)\sigma \bmod z^{k+1} \qquad \text{(hypothesis)}$$

Multiplying by $\omega^{(k)}$ and then using Theorem 7.412 gives

$$(1 + S)\sigma^{(k)}\omega \equiv (1 + S)\sigma\omega^{(k)}$$
$$\sigma^{(k)}\omega \equiv \sigma\omega^{(k)}$$
$$\sigma^{(k)}\omega - \sigma\omega^{(k)} = -z^k V \tag{7.45}$$

where $V(0) = 0$ and $\deg V \leq D(k) + D - k$. Multiplying the hypothesis by $\tau^{(k)}$ and then using Theorem 7:413 gives

$$(1 + S)\tau^{(k)}\omega \equiv (1 + S)\sigma(\gamma^{(k)} + z^k)$$
$$\tau^{(k)}\omega \equiv \sigma(\gamma^{(k)} + z^k) \equiv \sigma\gamma^{(k)} + z^k$$
$$\tau^{(k)}\omega - \sigma\gamma^{(k)} = Uz^k \tag{7.46}$$

where $U(0) = 1$ and $\deg U \leq D - D(k)$. Subtracting $\tau^{(k)}$ times (7.45) from $\sigma^{(k)}$ times (7.46) gives

$$(\tau^{(k)}\omega^{(k)} - \sigma^{(k)}\gamma^{(k)})\sigma = z^k(U\sigma^{(k)} + V\tau^{(k)})$$

Using Theorem 7.42, this becomes

$$\boxed{\sigma = U\sigma^{(k)} + V\tau^{(k)}}$$

Similarly, subtracting $\gamma^{(k)}$ times (7.45) from $\omega^{(k)}$ times (7.46) gives

$$(\omega^{(k)}\tau^{(k)} - \sigma^{(k)}\gamma^{(k)})\omega = z^k(U\omega^{(k)} + V\gamma^{(k)})$$

Using Theorem 7.42, this becomes

$$\boxed{\omega = U\omega^{(k)} + V\gamma^{(k)}}$$

Proof of Theorem 7.44 Theorem 7.441 is an immediate consequence of Eq. (7.46) and Theorems 7.415 and 7.417.

Consequently, if $\deg \sigma \leq (k+1)/2$ and $\deg \omega \leq k/2$, then $D(k) \leq (k+1)/2$, with strict inequality if $B(k) = 0$. According to Theorems 7.414 and 7.416, we may deduce that

$$\deg \sigma^{(k)} \leq \frac{k+1}{2} \qquad \deg \omega^{(k)} \leq \frac{k}{2} \qquad \deg (\sigma^{(k)}\omega) \leq k + \tfrac{1}{2} < k+1$$

$$\deg (\sigma\omega^{(k)}) \leq k + \tfrac{1}{2} < k+1$$

$$\deg (-z^k V) = \deg (\sigma^{(k)}\omega - \sigma\omega^{(k)}) < k+1 \qquad \deg V < 1$$

Since $V(0) = 0$, this implies $V = 0$. Theorem 7.43 becomes $\sigma = U\sigma^{(k)}$, $\omega = U\omega^{(k)}$. By hypothesis, σ and ω are relatively prime, so $U = 1$. Q.E.D.

Generalized Minimum Distance Decoding

G. DAVID FORNEY, JR., MEMBER, IEEE

Abstract—**We introduce a new distance measure which permits likelihood information to be used in algebraic minimum distance decoding techniques. We give an efficient decoding algorithm, and develop exponential bounds on the probability of not decoding correctly. In one application, this technique yields the same probability of error as maximum likelihood decoding.**

I. Introduction

COMMUNICATIONS theorists have sought with some success to develop useful codes through the highly developed and well understood tools of modern algebra; the BCH codes [1]–[4] are perhaps the most celebrated result of this effort. The structural properties of algebraic codes frequently permit efficient decoding algorithms whose implementation in digital circuitry is straightforward.

This work was supported in part by the Joint Services Electronic Program under Contract DA36–039–AMC–03200(E), and in part by the National Science Foundation (Grant GP-2495), and the National Aeronautics and Space Administration (Grant NsG-334, Grant NsG-496, and Contract NAS2-2874).

The author is with the Codex Corporation, Watertown, Mass. He was formerly with the M.I.T. Research Laboratory of Electronics, Cambridge, Mass.

One serious deficiency in the algebraic approach is its assumption of a channel model in which in each time period one of q signals is sent and one of q received. This model implies a receiver which makes a hard decision, on the basis of the received signal, which one of the q letters of the transmitter alphabet was most likely to have been sent. Such a receiver discards all information about the reliability of its choice—whether its decision was clearcut or borderline—as well as about the probability of letters other than the one chosen.

One way to ameliorate this deficiency is to allow the receiver the option of not choosing at all when the evidence does not clearly indicate one of the transmissions as the most probable; this non-choice is called an erasure. It can be regarded as a highly quantized reliability indicator, if we artificially assume that the receiver makes some choice and then tags it as absolutely unreliable. Furthermore, algebraic decoding algorithms can be modified to handle erasures efficiently.

In what follows, we investigate an extension of the erasure idea which permits a more flexible use of reliability information, and show that algebraic decoding algorithms can be readily modified to make use of this information.

Reprinted from *IEEE Trans. Inform. Theory,* vol. IT-12, pp. 125–131, Apr. 1966.

Generalized minimum distance decoding, as we call this extension, thus introduces some of the virtues of probabilistic decoding without sacrificing the attractiveness of the algebraic approach.

II. Distance Measure

In algebraic terms, a code word **f** from a code of length n on $GF(q)$ is a vector of length n with elements f_i from the finite field with q elements.

An important figure of merit for a code is its minimum distance d between two code words, where the distance measure is Hamming distance. The Hamming distance [5] $D_H(\mathbf{f}, \mathbf{g})$ between two code words **f** and **g** is simply the number of places in which they differ; more formally, if we define

$$d_H(f_i, g_i) = \begin{cases} 0, & f_i = g_i; \\ 1, & f_i \neq g_i; \end{cases} \tag{1a}$$

then

$$D_H(\mathbf{f}, \mathbf{g}) = \sum_{i=1}^{n} d_H(f_i, g_i) \tag{1b}$$

and

$$d = \min_{\mathbf{f} \neq \mathbf{g}} D_H(\mathbf{f}, \mathbf{g}). \tag{2}$$

Decoding algorithms can generally be formulated in terms of choosing the code word "closest" to the received word, where "closest" is defined by some distance measure. In this paper we consider measures of the following type. We suppose that the receiver decides in each time period which of the q letters was most likely to have been transmitted; its choice, therefore, corresponds to an element of $GF(q)$ which we label r_i. Further, it assigns its choice to some *reliability class* C_j according to how sure it is that its choice is correct. With each class are associated two parameters β_{cj} and β_{ej}, such that

$$0 \leq \beta_{cj} \leq \beta_{ej} \leq 1. \tag{3}$$

It will develop that only the difference of these parameters, $\beta_{ej} - \beta_{cj} = \alpha_j$, which we call the *weight* of class C_j, is significant; clearly $0 \leq \alpha_j \leq 1$.

We then define the *generalized distance* $D_G(\mathbf{r}, \mathbf{f})$ between a received word **r** and a code word **f** as

$$D_G(\mathbf{r}, \mathbf{f}) = \sum_{i=1}^{n} d_G(r_i, f_i) \tag{4a}$$

where

$$d_G(r_i, f_i) = \begin{cases} \beta_{cj}, & r_i = f_i \text{ and } r_i \text{ in } C_j; \\ \beta_{ej}, & r_i \neq f_i \text{ and } r_i \text{ in } C_j. \end{cases} \tag{4b}$$

(Note that this is not a true distance, since it is not symmetric in **r** and **f**.)

In the standard type of decoding, which we shall call errors-only decoding, there is only one reliability class C_1 into which go all outputs, and $\beta_{c1} = 0$, $\beta_{e1} = 1$, so that $\alpha_1 = 1$. In this case, the generalized distance reduces to the Hamming distance.

In this formulation, permitting erasures amounts to introducing a second reliability class C_2 for which $\beta_{c2} = \beta_{e2}$ or $\alpha_2 = 0$; in other words, letters in the erasure class are considered to be the same distance from all transmitted letters.

In general, we will refer to classes for which $\alpha_j = 1$ as *fully reliable* classes, and to those for which $\alpha_j = 0$ as *unreliable*. Intermediate values of α_j will correspond to intermediate levels of reliability. It will be convenient to number the classes in order of decreasing reliability, so that if $j < k$, $\alpha_j \geq \alpha_k$.

III. Minimum Distance Properties

We now suppose that in the course of communicating a word **f** from a code with minimum distance d, the number of letters received correctly ($r_i = f_i$) and put in class C_j is n_{cj}, and the number received incorrectly ($r_i \neq f_i$) and put in class C_j is n_{ej}. We now shall show the following.

Theorem 1

If **f** is sent and n_{cj} and n_{ej} are such that

$$\sum_j [(1 - \alpha_j)n_{cj} + (1 + \alpha_j)n_{ej}] < d, \tag{5}$$

then

$$D_G(\mathbf{r}, \mathbf{f}) < D_G(\mathbf{r}, \mathbf{g}) \text{ for all code words } \mathbf{g} \neq \mathbf{f}.$$

Proof: By the definition of generalized distance of (4),

$$D_G(\mathbf{r}, \mathbf{f}) = \sum_j [\beta_{cj}n_{cj} + \beta_{ej}n_{ej}]. \tag{6}$$

Take any code word $\mathbf{g} \neq \mathbf{f}$, and divide the n places into sets such that

$$i \in \begin{cases} S_0 & \text{if } f_i = g_i; \\ S_{cj} & \text{if } f_i \neq g_i, r_i = f_i, r_i \text{ in } C_j; \\ S_{ej} & \text{if } f_i \neq g_i, r_i \neq f_i, r_i \text{ in } C_j. \end{cases} \tag{7}$$

We note that the numbers of letters $|S_{cj}|$ and $|S_{ej}|$ in the sets S_{cj} and S_{ej} satisfy

$$\begin{aligned} |S_{cj}| &\leq n_{cj} \\ |S_{ej}| &\leq n_{ej}. \end{aligned} \tag{8}$$

Using (1) and (4), we have

$$\begin{aligned} d_G(r_i, g_i) &\geq 0 = d_H(g_i, f_i), \quad i \in S_0; \\ d_G(r_i, g_i) &= \beta_{ej} = d_H(g_i, f_i) - 1 + \beta_{ej}, \quad i \in S_{cj}; \\ d_G(r_i, g_i) &\geq \beta_{cj} = d_H(g_i, f_i) - 1 + \beta_{cj}, \quad i \in S_{ej}, \end{aligned} \tag{9}$$

so that, with use of (2) and (8),

$$\begin{aligned} D_G(\mathbf{r}, \mathbf{g}) &\geq D_H(\mathbf{f}, \mathbf{g}) \\ &\quad - \sum_j [(1 - \beta_{ej})\, |S_{cj}| + (1 - \beta_{cj})\, |S_{ej}|] \\ &\geq d - \sum_j [(1 - \beta_{ej})n_{cj} + (1 - \beta_{cj})n_{ej}]. \end{aligned} \tag{10}$$

But now if

$$d > \sum_j [(1 - \beta_{ej} + \beta_{cj})n_{cj} + (1 - \beta_{cj} + \beta_{ej})n_{ej}],$$

which was assumed in the statement of the theorem (5), we have, with use of (6),

$$D_G(\mathbf{r}, \mathbf{g}) > \sum_j [n_{cj}\beta_{cj} + n_{ej}\beta_{ej}] = D_G(\mathbf{r}, \mathbf{f}),$$

which was to be proved.

In the special case of errors-only decoding, where there is only one class, n_{e1} is the total number of t errors in $\mathbf{r}$, so that the theorem becomes:

$$D_G(\mathbf{r}, \mathbf{f}) < D_G(\mathbf{r}, \mathbf{g}) \quad \text{if} \quad 2t < d.$$

With erasures there is added a second class whose total number $s = n_{c2} + n_{e2}$; since $\alpha_2 = 0$, the theorem becomes:

$$D_G(\mathbf{r}, \mathbf{f}) < D_G(\mathbf{r}, \mathbf{g}) \quad \text{if} \quad 2t + s < d.$$

These are the familiar conditions which insure no error when errors-only or erasures-and-errors decoding is used.

Two conclusions can be drawn from this theorem. First, if generalized distance is used as a decoding criterion, no error will be made when n_{cj} and n_{ej} are such that the inequality of Theorem 1 is satisfied. Second, let us say that a code word is *within the minimum distance* of the received word when the inequality of Theorem 1 is apparently satisfied. The theorem says there can be at most one code word within the minimum distance of any received word. Therefore, if the decoder happens upon some code word within the minimum distance of the received word, it can immediately announce that word as its choice; this property has been the basis for a number [6]–[9] of clever decoding schemes proposed recently, and will be used in the generalized minimum distance decoding subsequently described.

IV. A Generalized Minimum Distance Decoder

In this section we shall introduce a practical decoding algorithm which uses generalized distance as a decoding criterion and works whenever there is a code word within the minimum distance of the received word according to the criterion of Theorem 1.

We assume the existence of a decoding algorithm which is capable of handling erasures and errors, and which works whenever the number of erasures s and of errors t is such that $2t + s < d$. (Elsewhere [10], we have described such an algorithm for BCH codes.)

Suppose, for the purpose of using this erasures-and-errors algorithm, we tentatively consider received letters in some classes as erasures, and the rest as fully reliable. Let E be the set of j's such that the class C_j is considered unreliable, and let R be the complementary set. We then attempt to decode the word by the erasures-and-errors algorithm, which we will be able to do if

$$2 \sum_{j \in R} n_{ej} + \sum_{j \in E} (n_{cj} + n_{ej}) < d. \tag{11}$$

If we succeed, we then check to see whether the code word so generated is within the minimum distance of the received word according to the generalized distance criterion of Theorem 1. If it is, then we have found the unique code word within the minimum distance.

There is no single provisional assignment of erasures for which this method must succeed. However, the following theorem and its corollary show that a small number of such trials must succeed in finding the unique code word within the minimum distance, if there is one.

As above, we let the classes be ordered according to decreasing reliability, so that $\alpha_j \geq \alpha_k$ if $j < k$. We let the total number of classes be J, and define the J-dimensional vector

$$\boldsymbol{\alpha} \equiv (\alpha_1, \alpha_2, \cdots, \alpha_J).$$

Let the sets R_b consist of all $j \leq b$, and E_b of all $j \geq b + 1$, $0 \leq b \leq J$. Let $\boldsymbol{\alpha}_b$ be the J-dimensional vector with ones in the first b places and zeros thereafter, which represents the provisional assignment corresponding to $R = R_b$ and $E = E_b$. The idea of the following theorem is that $\boldsymbol{\alpha}$ is inside the convex hull whose extreme points are the $\boldsymbol{\alpha}_b$, while the expression on the left in (5) is a linear function of $\boldsymbol{\alpha}$, which must take on its minimum value over the convex hull at some extreme point—that is, at one of the provisional assignments $\boldsymbol{\alpha}_b$.

Theorem 2

If

$$\sum_{j=1}^{J} [(1 - \alpha_j)n_{cj} + (1 + \alpha_j)n_{ej}] < d$$

$$\text{and} \quad \alpha_j \geq \alpha_k \quad \text{for} \quad j < k,$$

there is some integer b such that

$$2 \sum_{j=1}^{b} n_{ej} + \sum_{j=b+1}^{J} (n_{cj} + n_{ej}) < d.$$

Proof: Let

$$f(\boldsymbol{\alpha}) = \sum_{j=1}^{J} [(1 - \alpha_j)n_{cj} + (1 + \alpha_j)n_{ej}]. \tag{12}$$

f is clearly a linear function of the J-dimensional vector $\boldsymbol{\alpha}$. Note that

$$f(\boldsymbol{\alpha}_b) = 2 \sum_{j=1}^{b} n_{ej} + \sum_{j=b+1}^{J} (n_{cj} + n_{ej}). \tag{13}$$

We prove the theorem by supposing that $f(\boldsymbol{\alpha}_b) \geq d$, for all b such that $0 \leq b \leq J$, and exhibiting a contradiction. For, let

$$\begin{aligned} \lambda_0 &= 1 - \alpha_1; \\ \lambda_b &= \alpha_b - \alpha_{b+1}, \qquad 1 \leq b \leq J - 1; \\ \lambda_J &= \alpha_J. \end{aligned} \tag{14}$$

We see that

$$0 \leq \lambda_b \leq 1, \quad 0 \leq b \leq J, \quad \text{and} \quad \sum_{b=0}^{J} \lambda_b = 1$$

(so that the λ_b can be regarded as probabilities). But now,

$$\boldsymbol{\alpha} = \sum_{b=0}^{J} \lambda_b \boldsymbol{\alpha}_b.$$

Therefore,

$$f(\boldsymbol{\alpha}) = f\left(\sum_{b=0}^{J} \lambda_b \boldsymbol{\alpha}_b\right) = \sum_{b=0}^{J} \lambda_b f(\boldsymbol{\alpha}_b) \geq d \sum_{b=0}^{J} \lambda_b = d.$$

Thus, if $f(\boldsymbol{\alpha}_b) \geq d$, all b, then $f(\boldsymbol{\alpha}) \geq d$, in contradiction to the given conditions. Therefore, $f(\boldsymbol{\alpha}_b)$ must be less than d for at least one value of b.

The import of this theorem is that if there is some code word which satisfies the generalized distance criterion of Theorem 1, then there must be some provisional assignment in which the least reliable classes are erased and the rest are not which will enable an erasures-and-errors decoder to succeed in finding that code word. But an erasures-and-errors decoder will succeed only if there are apparently no errors and $d - 1$ erasures, or one error and $d - 3$ erasures, and so forth up to t_0 errors and $d - 2t_0 - 1$ erasures, where t_0 is the largest integer such that $2t_0 \leq d - 1$. If by a *trial* we then mean an operation in which the $d - 2i - 1$ least reliable symbols are erased, the resulting provisional word decoded by an erasures-and-errors algorithm, and the resulting code word (if one is found) checked by (5), then we have the following corollary.

Corollary

$t_0 + 1 \leq (d + 1)/2$ trials suffice to decode any received word which is within the minimum distance by the generalized distance criterion of Theorem 1, regardless of how many reliability classes there are.

The maximum number of trials is then proportional only to d. Further, many of the trials—perhaps all—may succeed, so that the average number of trials may be appreciably less than the maximum.

V. Error Bounds for Minimum Distance Decoding

In this section we develop exponentially tight error bounds on the probability of not decoding correctly for minimum distance decoding schemes on memoryless channels. Not decoding correctly means either decoding incorrectly or failing to decode; the latter event will occur when there is no word within the minimum distance of the received word.

Consider the random variable y_i which for each transmitted letter takes on the value $(1 - \alpha_j)$ if the letter is received correctly and put in class C_j, and $(1 + \alpha_j)$ if incorrectly in C_j. Assuming a memoryless channel and unchanging receiver, these random variables will be independent and identically distributed. Let the probability of $(1 - \alpha_j)$ be p_{cj} and that of $(1 + \alpha_j)$ be p_{ej}; then the moment-generating function $g(s)$ of the y_i is

$$g(s) = \overline{e^{sy_i}} = \sum_j [p_{cj}e^{s(1-\alpha_j)} + p_{ej}e^{s(1+\alpha_j)}] \qquad (15)$$

and the semi-invariant moment-generating function is

$$\mu(s) = \ln g(s).$$

We know that no error will be made if n_{cj} and n_{ej} are such that the inequality of Theorem 1 is satisfied, which is the same as requiring that the sum of these n random variables y_i be less than d. The Chernoff bound [11] is an exponentially tight [12] bound on the probability that a sum of independent, identically distributed random variables exceeds a certain number; in this case, it bounds the probability of not decoding correctly by

$$Pr\ (ndc) \leq \exp - n[s\delta - \mu(s)] \qquad (16)$$

for any $s \geq 0$, where we have let $\delta = d/n$.

To get the tightest bound, we maximize the exponent over s and over the choice of weights α_j. Define

$$E(\delta) = \max_{\alpha_j, s} [s\delta - \mu(s)]. \qquad (17)$$

We first maximize over the α_j; since $\mu(s) = \ln g(s)$, this is done by minimizing $g(s)$. Using (15), we have

$$\partial g(s)/\partial \alpha_j = -sp_{cj}e^{s(1-\alpha_j)} + sp_{ej}e^{s(1+\alpha_j)}. \qquad (18)$$

Setting this partial derivative to zero, we find the optimum α_j as

$$\alpha_j = L_j/2s, \qquad 0 \leq L_j \leq 2s, \qquad (19)$$

where L_j is the log likelihood ratio

$$L_j = \ln \frac{p_{cj}}{p_{ej}} = \ln \frac{Pr(\text{received letter correct, given it is in } C_j)}{Pr(\text{received letter incorrect, given it is in } C_j)}.$$

Since α_j must lie between 0 and 1, we must set $\alpha_j = 1$ if $L_j \geq 2s$ or $\alpha_j = 0$ if $L_j \leq 0$.

Thus, the optimum assignment of weights involves erasing any receptions for which the probability of correct choice p is less than $\frac{1}{2}$, considering as fully reliable any receptions for which p is greater than a certain threshold dependent on s (and hence, from (28), on d), and for intermediate p assigning a weight proportional to the log likelihood ratio $\ln p/(1 - p)$. Thus, except for hard limiting at both ends, generalized minimum distance decoding is a method of using the received log likelihood ratio in an algebraic decoding scheme.

With the optimum assignment of weights given by (19),

$$g_{\text{opt}}(s) = e^{2s}p_e(s) + e^s p_x(s) + e^s p_g(s) + p_c(s), \qquad (20)$$

where

$$\begin{aligned} p_e(s) &= \sum_{j \in R} p_{ej} \\ p_x(s) &= \sum_{j \in E} (p_{ej} + p_{cj}) \\ p_g(s) &= \sum_{j \in G} 2\sqrt{p_{ej}p_{cj}} \\ p_c(s) &= \sum_{j \in R} p_{cj} \end{aligned} \qquad (21)$$

and where

$$\begin{aligned} j \in R \quad &\text{if} \quad L_j > 2s \\ j \in E \quad &\text{if} \quad L_j \leq 0 \end{aligned} \qquad (22)$$

and

$$j \, \varepsilon \, G \quad \text{otherwise};$$

and we have used, from (19),

$$e^{s\alpha_j} = \sqrt{p_{cj}/p_{ej}}, \qquad j \, \varepsilon \, G.$$

Specialization to errors-only or erasures-and-errors decoding involves a restriction to a suboptimum set of weights. In the errors-only case all weights must be one; there is only the one set R, and

$$g_{\text{opt}}(s) = e^{2s}p_e + p_c, \tag{23}$$

where

$$p_e = \sum_i p_{ej}$$

and

$$p_c = \sum_i p_{cj}. \tag{24}$$

In the erasures-and-errors case we permit a second class E with $\alpha_E = 0$; the optimum assignment is

$$j \, \varepsilon \, R \quad \text{if} \quad L_j > s;$$
$$j \, \varepsilon \, E \quad \text{if} \quad L_j \leq s. \tag{25}$$

With this assignment

$$g_{\text{opt}}(s) = e^{2s}p_e(s) + e^s p_x(s) + p_c(s), \tag{26}$$

where

$$p_e(s) = \sum_{j\varepsilon R} p_{ej}$$
$$p_x(s) = \sum_{j\varepsilon E} (p_{ej} + p_{cj}), \tag{27}$$
$$p_c(s) = \sum_{j\varepsilon R} p_{cj}.$$

Finally, defining

$$\mu_{\text{opt}}(s) = \ln g_{\text{opt}}(s),$$

we have from (17)

$$E(\delta) = \max_s \, [s\delta - \mu_{\text{opt}}(s)].$$

The optimum s is that for which

$$\delta = \mu'_{\text{opt}}(s) = g'_{\text{opt}}(s)/g_{\text{opt}}(s). \tag{28}$$

It can be shown [12] that $\mu'_{\text{opt}}(s)$ is monotonically increasing with s, and therefore this equation will have a solution with $s \geq 0$ whenever

$$\delta \geq \mu'_{\text{opt}}(0); \tag{29}$$

thus, the smallest minimum distance for which minimum distance decoding can work is $n\mu'_{\text{opt}}(0)$. We see from (22) and (25) that as s approaches zero the assignment of weights for generalized minimum distance decoding and for erasures-and-errors decoding becomes the same, so that $\mu'_{\text{opt}}(0)$, and hence the smallest feasible minimum distance, is the same for either. For errors-only decoding, that will also be the same if there is no possible output for which $L_j < 0$, which will always be the case with binary transmissions. Thus, in very marginal conditions generalized minimum distance decoding offers little or no improvement.

VI. Binary Antipodal Signals on the White Gaussian Channel

As an example, we consider the performance of errors-only, erasures-and-errors, and generalized minimum distance decoding when the channel is white and Gaussian, modulation is binary, and reception is coherent. Because of its tractability, the white Gaussian channel is the most frequently studied continuous memoryless channel; it even appears to be a reasonable model for some space channels.

We suppose that every T seconds the transmitter sends either a given waveform of energy E or its negative, that the noise spectral density is N_0, and that coherent correlation detection is used at the receiver. It is well known that under these conditions the output y of the correlator (matched filter) is a Gaussian random variable which has, under a convenient normalization, mean $\pm E$ and variance $N_0E/2$, where the sign of the mean depends on whether a zero or a one was transmitted. The sign of the correlator output determines whether a zero or a one is chosen, while the bit log likelihood ratio defined above is

$$L = \ln \frac{(\pi N_0 E)^{-1/2} \exp -(|y| - E)^2/N_0E}{(\pi N_0 E)^{-1/2} \exp -(|y| + E)^2/N_0E} = 4\,|y|/N_0. \tag{30}$$

Thus, the log likelihood ratio is proportional to the magnitude of the correlator output, and can be regarded as the magnitude of a Gaussian random variable with mean $4E/N_0$ and variance $8E/N_0$. E/N_0 is called the signal to noise ratio per transmitted bit; it is convenient to define the parameter

$$A = \sqrt{2E/N_0}, \tag{31}$$

so that we have

$$\begin{aligned} Pr_e(x) &= Pr(L = x, \text{incorrect decision}) \\ &= (8\pi A^2)^{-1/2} \exp -(x + 2A^2)/8A^2 \end{aligned} \tag{32a}$$

and

$$\begin{aligned} Pr_c(x) &= Pr(L = x, \text{correct decision}) \\ &= (8\pi A^2)^{-1/2} \exp -(x - 2A^2)/8A^2. \end{aligned} \tag{32b}$$

For errors-only decoding, the likelihood information is thrown away and only a hard decision remains. The probability of bit error p_e is then, using (24),

$$\begin{aligned} p_e &= \int_0^\infty Pr_e(x)\,dx \\ &= \int_A^\infty \frac{1}{\sqrt{2\pi}} e^{-z^2/2}\,dz \\ &= \Phi(A) \end{aligned} \tag{33a}$$

and

$$p_c = 1 - \Phi(A), \tag{33b}$$

where we have implicitly defined $\Phi(x)$ as the error function.

For erasures-and-errors decoding, the optimum strategy is to retain the hard decision if $L > s$, and to erase otherwise. The probability of bit error is then, using (27),

$$p_e(s) = \int_s^\infty Pr_e(x)\,dx$$
$$= \Phi(A + s/2A), \quad (34a)$$

of bit erasure

$$p_x(s) = \int_0^s (Pr_e(x) + Pr_c(x))\,dx$$
$$= \Phi(A - s/2A) - \Phi(A + s/2A), \quad (34b)$$

and of correct decision

$$p_c(s) = 1 - \Phi(A - s/2A). \quad (34c)$$

With generalized minimum distance decoding, the optimum strategy, as we have seen, is to assign a weight $L/2s$ to a reception when $L \leq 2s$, and 1 otherwise. In this case, substituting in (21),

$$p_e(s) = \int_{2s}^\infty Pr_e(x)\,dx$$
$$= \Phi(A + s/A), \quad (35a)$$
$$p_x(s) = Pr\,(L \leq 0) = 0, \quad (35b)$$
$$p_c(s) = \int_{2s}^\infty Pr_c(x)\,dx$$
$$= 1 - \Phi(A - s/A), \quad (35c)$$

and

$$p_g(s) = 2\int_0^{2s} \sqrt{Pr_e(x)\,Pr_c(x)}\,dx$$
$$= e^{-A^2/2}[1 - 2\Phi(s/A)]. \quad (35d)$$

We now suppose that a word is sent from a code of length n with minimum distance d, and investigates the performance of different decoding schemes when the signal-to-noise ratio E/N_0 becomes large. We shall assume for simplicity that the code is a group code, so that from every code word there are the same number n_d of other code words at the minimum distance d, though this restriction can be removed.

The optimum receiving strategy in this or any case is called maximum likelihood decoding. The following brief discussion is intended to orient the reader who is familiar with the geometric representation of signals and noise in n-space; the details may be found, for example, in Wozencraft and Jacobs [13]. We let the n correlator outputs define a point $\mathbf{y}$ in n-space. $\mathbf{y}$ is a random point; given the transmission of a particular word $\mathbf{f}$, the distribution of $\mathbf{y}$ is Gaussian and spherically symmetric about its mean value $\mathbf{y}_f$, where for example for the all-zero word $\mathbf{0}$

$$\mathbf{y}_0 = (-E, -E, \cdots, -E).$$

The projection of $\mathbf{y}$ along any line through $\mathbf{y}_f$ is a one-dimensional Gaussian random variable with mean at $\mathbf{y}_f$ and variance $N_0E/2$. The maximum likelihood rule, given $\mathbf{y}$, is to choose that code word $\mathbf{g}$ for which $\mathbf{y}_g$ is closest in Euclidean distance to $\mathbf{y}$. Take two code words $\mathbf{f}$ and $\mathbf{g}$ such that the Hamming distance between them is d; then the Euclidean distance between $\mathbf{y}_f$ and $\mathbf{y}_g$ is $2E\sqrt{d}$. If one of these is sent, an error will be made if the noise component along the line joining them is greater than $E\sqrt{d}$. Thus,

$$Pr\,(\mathcal{E}) \geq \int_{E\sqrt{d}}^\infty \frac{1}{\sqrt{\pi E N_0}} e^{-y^2/EN_0}\,dy$$
$$= \Phi(A\sqrt{d}), \quad (36)$$

where we have substituted A for $\sqrt{2E/N_0}$. Asymptotically, for large x

$$\Phi(x) \simeq \frac{1}{x\sqrt{2\pi}} e^{-x^2/2} \quad (37)$$

so that the exponential behavior of the probability of decoding error is given by

$$Pr\,(\mathcal{E}) \geq \frac{1}{A\sqrt{2\pi d}} e^{-dA^2/2} \quad (38)$$
$$\simeq e^{-dE/N_0}.$$

Actually, for high signal-to-noise ratios $Pr(\mathcal{E})$ approaches [14]

$$Pr\,(\mathcal{E}) \simeq \frac{n_d}{A\sqrt{2\pi d}} e^{-dE/N_0}$$

but the exponent is the significant quantity.

We now use the Chernoff bounds of the Section V to get comparable exponential bounds for the three types of minimum distance decoding. For errors-only decoding, we have, with use of (23) and (33),

$$g(s) = \Phi(A)e^{2s} + 1 - \Phi(A). \quad (39)$$

As A becomes large, we have, by substitution of the asymptotic form for the error function of (37),

$$g(s) = (2\pi A^2)^{-1/2} \exp\,(2s - A^2/2), \qquad s > A^2/4;$$
$$= 1, \qquad s \leq A^2/4. \quad (40)$$

Thus,

$$E(\delta) = \max_s\,[s\delta - \mu_{\text{opt}}(s)]$$

is maximized by letting $s = A^2/4$, since $\delta < 2$, so that

$$Pr\,(\mathcal{E}) \leq e^{-nE(\delta)}$$
$$= e^{-dA^2/4}$$
$$= e^{-.5dE/N_0}. \quad (41)$$

Thus, the exponent is one half that for maximum likelihood decoding, so that 3 dB more power is required to achieve the same probability of error when E/N_0 is large, as was pointed out by Hackett [14].

For erasures-and-errors decoding, we have, with use of (26) and (34),

$$g(s) = \Phi(A + s/2A)e^{2s}$$
$$+ [\Phi(A - s/2A) - \Phi(A + s/2A)]e^s$$
$$+ 1 - \Phi(A - s/2A) \quad (42)$$

so that as A becomes large

$$g(s) = [(A + s/2A)^{-1} + (A - s/2A)^{-1}](2\pi)^{-1/2} \cdot \exp(-A^2/2 + 3s/2 - s^2/8A^2), \quad s > 2A^2(3 - 2\sqrt{2})$$
$$= 1, \quad s \leq 2A^2(3 - 2\sqrt{2}) \quad (43)$$

where the breakpoint for s is a root of the quadratic equation in the exponent. Again, $E(\delta)$ is maximized by taking s at the breakpoint, so that

$$Pr(\mathcal{E}) \leq \exp - n\delta[2A^2(3 - 2\sqrt{2})]$$
$$= e^{-.686dE/N_0}. \quad (44)$$

Using erasures and errors, therefore, permits a reduction of power over errors-only decoding of 0.686/0.5 or about 1.4 dB.

Finally, for generalized minimum distance decoding, we have, from (20) and (35),

$$g(s) = \Phi(A + s/A)e^{2s} + e^{-A^2/2}[1 - 2\Phi(s/A)]e^{s} + 1 - \Phi(A - s/A), \quad (45)$$

which becomes for large A

$$g(s) = \exp(s - A^2/2), \quad s > A^2/2;$$
$$= 1, \quad s \leq A^2/2. \quad (46)$$

Since $\delta < 1$, $E(\delta)$ is once again maximized at the breakpoint s and

$$Pr(\mathcal{E}) \leq e^{-n\delta A^2/2}$$
$$= e^{-dE/N_0}. \quad (47)$$

It follows that at high signal-to-noise ratios, generalized minimum distance decoding gives effectively the same probability of error as maximum likelihood decoding, even though the former will generally be simpler to implement; in the asymptotic region this essentially algebraic technique achieves the optimum performance.

At the opposite extreme, as we saw at the end of the Section V, neither erasures-and-errors nor generalized minimum distance decoding offers any improvement over errors-only decoding. In fact, for all three schemes

$$\mu'_{\text{opt}}(0) = 2\Phi(A). \quad (48)$$

Since $\Phi(A)$ is the bit probability of error by (33), this means that for a minimum distance scheme to work the signal-to-noise ratio must be high enough that the mean number of errors per block be less than half the minimum distance.

VII. Discussion

We have introduced the distance criterion of Theorem 1, shown that there can be at most one code word within the minimum distance of any received word using this criterion, given an algorithm for finding that code word, and developed error bounds which show that performance is thereby improved, to the point of being equivalent to that obtainable by maximum likelihood decoding on the low-noise white Gaussian channel.

It is clear that the following modification of generalized minimum distance decoding must further improve performance. The $t_0 + 1$ decoding trials of the corollary to Theorem 2 may result in up to $t_0 + 1$ different code words. There is no reason why we could not use the optimum distance measure—likelihood— in comparing these code words to the received word, rather than the generalized distance measure. This must result in decreased probability of error, though it is difficult to measure the amount of improvement analytically.

More generally, this work suggests a class of decoding algorithms in which algebraic techniques would be used to generate a number of code words close in some sense to the received word, and in which these code words would then be compared to the received word by the likelihood measure. It seems plausible that if enough code words were generated, a very close approximation to maximum likelihood decoding would become possible at all noise levels. The central problem is to find an efficient technique of generating a set of code words which will with high probability contain that word which is most likely, given the received word. The use of erasures, as here, is one line of approach; another might be to adapt list decoding, which has found application in sequential decoders, to algebraic algorithms.

Acknowledgment

Most of this work appeared in my doctoral thesis [15], resulting from research for which I was supported by an National Science Foundation Graduate Followship. Conversations with Prof. R. S. Kennedy were particularly helpful.

References

[1] R. C. Bose and D. K. Ray-Chaudhuri, "On a class of error-correcting binary group codes," *Information and Control*, vol. 3, pp. 68–79, March 1960.

[2] ——, "Further results on error-correcting binary group codes," *Information and Control*, vol. 3, pp. 279–290, September 1960.

[3] A. Hocquenghem, "Codes correcteurs d'erreurs," *Chiffres*, vol. 2, pp. 147–156, September 1959.

[4] W. W. Peterson, *Error-Correcting Codes*. Cambridge, Mass.: M.I.T. Press, and New York: Wiley, 1961.

[5] R. W. Hamming, "Error detecting and error correcting codes," *B.S.T.J.*, vol. 29, pp. 147–160, April 1950.

[6] E. Prange, "The use of information sets in decoding cyclic codes," *IRE Trans. on Information Theory*, vol. IT-8, pp. s5–s9, September 1962.

[7] T. Kasami, "A decoding procedure for multiple-error-correcting cyclic codes," *IEEE Trans. on Information Theory*, vol. IT-10, pp. 134–138, April 1964.

[8] J. MacWilliams, "Permutation decoding of systematic codes," *Bell Syst. Tech. J.*, vol. 43, pp. 485–506, January 1964.

[9] L. D. Rudolph and M. E. Mitchell, "Implementation of decoders for cyclic codes," *IEEE Trans. on Information Theory (Correspondence)*, vol. IT-10, pp. 259–260, July 1964.

[10] G. D. Forney, Jr., "On decoding BCH codes," *IEEE Trans. on Information Theory*, vol. IT-11, pp. 549–557, October 1965.

[11] H. Chernoff, "A measure of asymptotic efficiency for tests of a hypothesis based on a sum of observations," *Ann. Math. Stat.*, vol. 23, pp. 493–507, December 1952.

[12] C. E. Shannon and R. G. Gallager, private communications.

[13] J. M. Wozencraft and I. M. Jacobs, *Principles of Communication Engineering*. New York: Wiley, 1965.

[14] C. M. Hackett, Jr., "Word error rate for group codes detected by correlation and other means," *IEEE Trans. on Information Theory*, vol. IT-9, pp. 24–33, January 1963.

[15] G. D. Forney, Jr., "Concatenated Codes," Sc.D. dissertation, Dept. of Elec. Engrg., M.I.T., Cambridge, Mass., June, 1965. To appear as Tech. Rept. 440, M.I.T. Research Lab. of Electronics, Cambridge, Mass.

Part IV: Convolutional Codes and Sequential Decoding

Convolutional codes were first invented by Elias [1], who proved that randomly chosen codes of this type were good. The first practical decoding algorithm for long randomly chosen convolutional codes, called sequential decoding, was invented by Wozencraft and Reiffen [2], and subsequently refined by Fano [3]. It was known that the error probability of sequential decoders could be made to approach zero rapidly [4]. The early workers presented evidence that the average amount of computation was bounded, even for arbitrarily large block lengths, if the rate of the code was less than a certain rate called R_{comp}. For a time, it was generally believed that no decoding algorithm would work very well at rates between R_{comp} and the channel capacity, but this myth was refuted by Pinsker [5]. Good schemes for decoding long codes at rates higher than R_{comp} had also been presented by Ziv [6]-[8], Epstein [9], Gallager [10], Falconner [11], Forney [12], and others.

Some early experimental work on sequential decoding at MIT's Lincoln Laboratory indicated that the distribution of the number of computations in sequential decoding might not be a very well-behaved random variable. This led to a number of theoretical studies on the moments of this random variable, including the works of Savage [13], [14], Jacobs-Berlekamp [15], and Jelinek [16], who showed that the distribution is Paretian. Higher moments of the distribution may be infinite even when the average is finite. These results suggested that the performance of sequential decoders was limited, not by the exponential bounds on the probability of error, but rather by the Paretian bounds on the probability of buffer overflow. This led to a reexamination of the various sequential decoding algorithms, and to some modifications of them. Significant improvements were introduced by Zigangirov [17], [18]. Although Zigangirov's original work was not published in English, some of his algorithms were rediscovered by Jelinek [19]. Zigangirov himself later published a detailed analysis of one of his algorithms for the BSC in English [20]. Viterbi [21] introduced a new algorithm which is particularly effective in decoding convolutional codes of short constraint length. Viterbi's algorithm has found numerous applications.

A number of techniques to construct convolutional codes which are "good" in one sense or another have been presented by Hagelbarger [22], Massey [23], Wyner-Ash [24], Berlekamp [25]-Preparata [26]-Massey [27], Iwadara [28], Robinson-Berstein [29], Piret [30], Larsen [31], and others. The Soviets have recently obtained a number of improved bounds on the complexity of various sequential decoding algorithms. A procedure for constructing asymptotically good time-varying convolutional codes was recently presented by Justesen [32].

Bussgang [33] studied the properties of several particular binary convolutional code generators. Massey and his colleagues [34]-[40] discovered important relationships between error propagation in convolutional decoders and questions concerning the invertibility of linear systems. This work has been extended by Viterbi [41]. It has been shown that, in spite of the difficulties of a Paretian computation distribution for sequential decoders and the possibility of error propagation, convolutional codes are superior to block codes in many respects. The most thorough treatment of convolutional codes, from several different points of view, appears in the recent papers of Forney [42], [43].

References

[1] P. Elias, "Coding for noisy channels," in *IRE Conv. Rec.*, pt. 4, 1955, pp. 37–46.

[2] Wozencraft and B. Reiffen, *Sequential Decoding*, Res. Monograph 10. Cambridge, Mass.: MIT Press, 1961.

[3] R. M. Fano, "A heuristic discussion of probabilistic decoding," *IEEE Trans. Inform. Theory*, vol. IT-9, pp. 64–75, Apr. 1963; *Math. Rev.*, vol. 29, p. 3283.

[4] H. L. Yudkin, "Channel state testing in information decoding," Sc.D. dissertation, Dep. Elec. Eng., Massachusetts Inst. Technol., Cambridge, 1964.

[5] M. S. Pinsker, "On the complexity of decoding," *Probl. Peredach. Inform.*, vol. 1, no. 1, pp. 113–116, 1965.

[6] J. Ziv, "Coding and decoding for time-discrete amplitude continuous memoryless channels," *IEEE Trans. Inform. Theory*, vol. IT-8, pp. S199–S205, Sept. 1962.

[7] ——, "Further results on the asymptotic complexity of an iterative coding scheme," *IEEE Trans. Inform. Theory*, vol. IT-12, pp. 168–171, Apr. 1966.

[8] ——, "Asymptotic performance and complexity of a coding scheme for memoryless channels," *IEEE Trans. Inform. Theory*, vol. IT-13, pp. 356–359, July 1967.

[9] M. A. Epstein, "Algebraic decoding for a binary erasure channel," M.I.T. Res. Lab. Electron. Rep. 340, 1958.

[10] R. G. Gallager, *Low-Density Parity-Check Codes*, Res. Monograph 21. Cambridge, Mass.: MIT Press, 1963.

[11] D. D. Falconer, "A hybrid sequential and algebraic decoding scheme," Ph.D. dissertation, Dep. Elec. Eng., Massachusetts Inst. Technol., Cambridge, 1967.

[12] G. D. Forney, Jr., *Concatenated Codes*, Res. Monograph 37. Cambridge, Mass.: MIT Press, 1966.

[13] J. E. Savage, "The distribution of the sequential decoding computation time," *IEEE Trans. Inform. Theory*, vol. IT-12, pp. 143-147, Apr. 1966.

[14] ——, "Sequential decoding—The computation problem," *Bell Syst. Tech. J.*, vol. 45, pp. 149-175, 1966.

[15] I. M. Jacobs and E. R. Berlekamp, "A lower bound to the distribution of computation for sequential decoding," *IEEE Trans. Inform. Theory*, vol. IT-13, pp. 167-174, Apr. 1967.

[16] F. Jelinek, "An upper bound on moments of sequential decoding effort," *IEEE Trans. Inform. Theory*, vol. IT-15, pp. 140-149, Jan. 1969; *Math. Rev.*, vol. 40, p. 8490.

[17] K. Sh. Zigangirov, "Some sequential decoding procedures," *Probl. Peredach. Inform.*, vol. 2, no. 4, pp. 13-25, 1966.

[18] ——, "An algorithm of sequential decoding in which the error probability decreases in agreement with the random coding bound," *Probl. Peredach. Inform.*, vol. 4, pp. 83-85, 1968.

[19] F. Jelinek, "Fast sequential decoding algorithm using a stack," *IBM J. Res. Develop.*, pp. 675-678, Nov. 1969.

[20] K. Sh. Zigangirov, "On the error probability of sequential decoding on the BSC," *IEEE Trans. Inform. Theory*, vol. IT-18, pp. 199-202, Jan. 1972.

[21] A. J. Viterbi, "Error bounds for convolutional codes and an asymptotically optimum decoding algorithm," *IEEE Trans. Inform. Theory*, vol. IT-13, pp. 260-269, Apr. 1967.

[22] D. W. Hagelbarger, "Recurrent codes: Easily mechanized burst-correcting binary codes," *Bell Syst. Tech. J.*, vol. 38, pp. 969-984, 1959; *Math. Rev.*, vol. 21, p. 7812.

[23] J. L. Massey, *Threshold Decoding*, Res. Monograph 20. Cambridge, Mass.: MIT Press, 1963.

[24] A. D. Wyner and R. B. Ash, "Analysis of recurrent codes," *IEEE Trans. Inform. Theory*, vol. IT-9, pp. 143-156, July 1963; *Math. Rev.*, vol. 29, p. 1088.

[25] E. R. Berlekamp, "Note on recurrent codes," *IEEE Trans. Inform. Theory* (Corresp.), vol. IT-10, pp. 257-258, July 1964.

[26] F. P. Preparata, "Systematic construction of optimal linear recurrent codes for burst error correction," *Calcolo*, vol. 6, pp. 147-153, 1964; *Math. Rev.*, vol. 30, p. 2954.

[27] J. L. Massey, "Implementation of burst-correcting convolutional codes," *IEEE Trans. Inform. Theory* (Corresp.), vol. IT-14, pp. July 1965.

[28] Y. Iwadare, "A class of high-speed decodable burst-correcting codes," *IEEE Trans. Inform. Theory* (Corresp.), vol. IT-14, pp. 817-821, Nov. 1972.

[29] J. P. Robinson and A. J. Bernstein, "A class of binary recurrent codes with limited error propagation," *IEEE Trans. Inform. Theory*, vol. IT-13, pp. 106-113, Jan. 1967.

[30] P. Piret, "Some optimal type B1 convolutional codes," *IEEE Trans. Inform. Theory* (Corresp.), vol. IT-17, pp. 355-356, May 1971.

[31] K. J. Larsen, "Short convolutional codes with maximal free distance for rates 1/2, 1/3, and 1/4," *IEEE Trans. Inform. Theory* (Corresp.), vol. IT-19, pp. 371-372, May 1973.

[32] J. Justesen, "New convolutional code constructions and a class of asymptotically good time-varying codes," *IEEE Trans. Inform. Theory*, vol. IT-19, pp. 220-224, Mar. 1973.

[33] J. J. Bussgang, "Some properties of binary convolutional code generators," *IEEE Trans. Inform. Theory*, vol. IT-11, pp. 90-100, Jan. 1965.

[34] J. L. Massey and M. K. Sain, "Codes, automata, and continuous systems: Explicit interconnections," *IEEE Trans. Automat. Contr.*, vol. AC-12, pp. 644-650, Dec. 1967.

[35] ——, "Inverses of linear sequential circuits," *IEEE Trans. Comput.*, vol. C-17, pp. 330-337, Apr. 1968.

[36] M. K. Sain and J. L. Massey, "Invertibility of linear time-invariant dynamical systems," *IEEE Trans. Automat. Contr.*, vol. AC-14, pp. 141-149, Apr. 1969.

[37] D. D. Sullivan, "Control of error propagation in convolutional codes," Univ. Notre Dame, Notre Dame, Ind., Tech. Rep. EE-667, Nov. 1966.

[38] R. R. Olson, "Note on feedforward inverses for linear sequential circuits," Dep. Elec. Eng., Univ. Notre Dame, Notre Dame, Ind., Tech. Rep. EE-684, Apr. 1, 1968; also *IEEE Trans. Comput.*, to be published.

[39] D. J. Costello, "Construction of convolutional codes for sequential decoding," Dep. Elec. Eng., Univ. Notre Dame, Notre Dame, Ind., Tech. Rep. EE-692, Aug. 1969.

[40] T. N. Morrissey, "A unified analysis of decoders for convolutional codes," Univ. Notre Dame, Notre Dame, Ind., Tech. Rep. EE687, Oct. 1968; also T. N. Morrissey, Jr., "Analysis of decoders for convolutional codes by stochastic sequential machine methods," *IEEE Trans. Inform. Theory*, vol. IT-16, pp. 460-469, July 1970.

[41] A. J. Viterbi, "Convolutional codes and their performance in communication systems," *IEEE Trans. Commun. Technol.*, vol. COM-19, pp. 751-771, Oct. 1971.

[42] G. D. Forney, Jr., "Convolutional codes I: Algebraic structure," *IEEE Trans. Inform. Theory*, vol. IT-16, no. 6, pp. 720-738, Nov. 1970.

[43] ——, "Convolutional codes II: Maximum likelihood decoding," and "Convolutional codes III: Sequential decoding," Inform. Syst. Lab., Stanford Univ., Stanford, Calif., Tech. Rep. 7004-1, 1972.

Sequential Decoding

J. M. Wozencraft and B. Reiffen

1. Concepts

Recognition of the fact that communication can be made as reliable as desired, provided that the transmission rate R_t is less than the channel capacity C, entices the communication engineer to attempt the design of systems that perform this way. We have seen in Chapter 2 that there is no real difficulty in specifying a good code capable of near-ultimate performance. Encoding can be performed with equipment whose complexity grows linearly with n.

However, the decoding problem remains: The general maximum likelihood decoding procedure involves either 2^{nR_t} sequence comparisons, or a dictionary with $2^{n(1-R_t)}$ entries. For many channels of interest, values of nR_t and $n(1-R_t)$ between 30 and at least 60 seem desirable from considerations of error probability and communication efficiency. The numbers involved then range from 10^{10} to 10^{20} : one does not trifle with exponentials.

Fortunately, in our search for good communication techniques, we need not be limited to consideration of ideal maximum-likelihood procedures only. The exponential decrease of $\overline{Pr(\epsilon)}$ with increasing n is a behavior strong enough to overcome minor difficulties with coefficients, at least in the limit of large n. We are motivated to look for other schemes that permit decoding simplification, so long as the exponent in the probability of error is not forfeited.

In particular, it is intuitively clear that almost all numbers of a set of $|S| = 2^{nR_t}$ messages picked at random will differ from a received message $\overline{y}$ to a considerable degree. There may be a subset of messages within S that agree in many positions with $\overline{y}$, but the size of this subset will be small. Even though it may be difficult to select the single most-probable sequence in S, given $\overline{y}$, it should be relatively simple to eliminate most of the possible $\overline{s}_i$ from detailed consideration. Concentration not upon selecting the most probable $\overline{s}_i$, but rather upon discarding from contention all improbable $\overline{s}_i$, is the single key concept in the decoding strategy that we now discuss. The word "improbable" must of course be precisely defined.

Let us assume, rather arbitrarily at first, that "improbable" means "differs from $\overline{y}$ at length n in k_n or more digits." For a given channel transition probability p_o and length n, it is convenient to define this discard function k_n such that the probability that k_n or more transmission errors occur is overbounded by 2^{-K}, where K is some arbitrary positive constant which we call the "probability criterion."

$$\Pr[d_o(n) \geq k_n] \leq 2^{-K} \qquad (3.1)$$

Solving Equation 3.1 precisely for the largest possible function k_n would be tedious. However, we can quite simply obtain a function that is entirely adequate for our purposes from the Chernoff bound.

$$\Pr_o[d(n) \geq np] \leq 2^{-n[T_o(p) - H(p)]}, \quad p \geq p_o \qquad (A.21)$$

We set

$$\frac{K}{n} = T_o(p) - H(p) \qquad (3.2)$$

determine thereby a value of p which we call p_n, and set

$$k_n = np_n \qquad (3.3)$$

Referring to Figure 3.1, we see that p_n may be obtained geometrically by sliding a vertical line segment of length K/n between $T_o(p)$ and $H(p)$, until it just fits. The corresponding abscissa is p_n.

From Figure 3.1 it is clear that K may be so large or n so small that Equation 3.2 has no solution. This simply means that

$$\Pr_o[d(n) = n] = p_o^{\,n} > 2^{-K} \qquad (3.4)$$

There is, in general, some maximum value of n, say n_o, for which Equation 3.4 is true. Then

$$\frac{K}{n_o - 1} \leq -\log p_o < \frac{K}{n_o} \qquad (3.5)$$

Thus, we define

$$k_n = \begin{cases} n & , n \leq n_o \\ np_n \leq n & , n > n_o \end{cases} \qquad (3.6)$$

With k_n specified as in Equation 3.6, our objective is to discard all messages in S that differ from $\overline{y}$ in k_n or more digits out of n. We seek to do so for every value n within the constraint length of the code, which hereafter is designated n_t in order to distinguish it from the variable n. Ultimately, we want to know how many incorrect sequences in S must be retained.

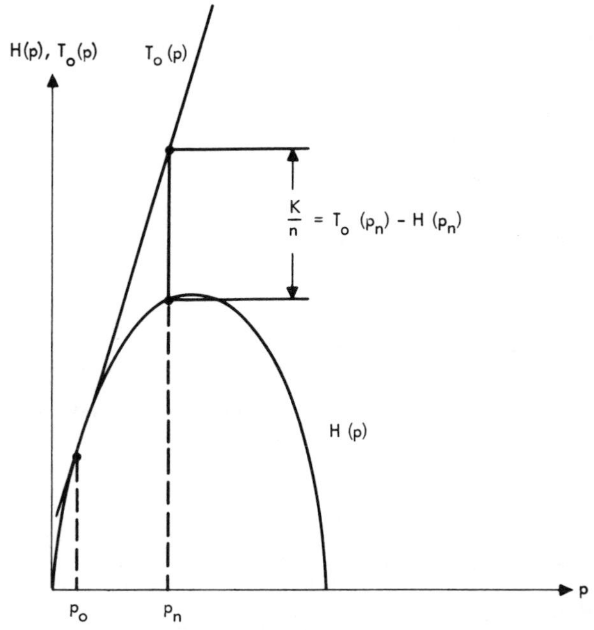

Fig. 3.1. Graphical determination of p_n

It is again convenient to revert to a random-coding analysis. If we assume that each binary digit in the entire code book S is picked independently at random with Pr (1) = 1/2, we can use the Chernoff bound of Equation A.24 to evaluate the probability that any particular incorrect sequence $\overline{s}_i$ is retained at length n

$$\Pr[\overline{s}_i \text{ retained at } n] = \Pr[d_i(n) < k_n] < \Pr_1[d(n) \leq np_n]$$

$$\leq 2^{-n[1 - H(p_n)]}, \quad p_n \leq \frac{1}{2} \qquad (3.7)$$

A sufficiently accurate upper-bound to $\Pr_1[d(n) \leq np_n]$ for $p_n > 1/2$ is unity. There is some minimum value of n, which we call n_1, such that $p_n \leq 1/2$. From Figure 3.1, we see that since H(1/2) = 1,

$$\frac{K}{n_1 + 1} < T_o\left(\frac{1}{2}\right) - 1 \leq \frac{K}{n_1} \qquad (3.8)$$

It is convenient to define the function

$$R_K(n) = \begin{cases} 0 & , n < n_1 \\ 1 - H(p_n) & , n \geq n_1 \end{cases} \qquad (3.9)$$

Thus

$$\Pr[\bar{s}_i \text{ retained at } n] \leq 2^{-nR_K(n)} \tag{3.10}$$

Sketches of the function $R_K(n)$ for several values of K are drawn in Figure 3.2. Of particular interest is the fact that, for any

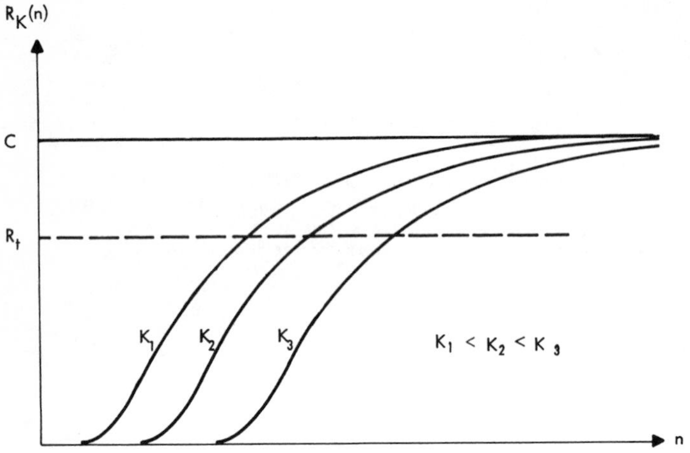

Fig. 3.2. Sketch of $R_K(n)$ vs. n for several values of K

finite K, the limit of $R_K(n)$ as $n \to \infty$ is equal to the channel capacity C. This again is clear from Figure 3.1: As $n \to \infty$, $K/n \to 0$, and the point $p_n \to p_o$. It is this limiting behavior that lends hope to the prospect of curtailing the number of computations necessary for decoding in the limit of small $\Pr(\epsilon)$.

Let us consider a randomly selected code, with constraint length n_t, for which the size $|S(n)|$ of the set of allowable messages is an exponential function of the variable n.

$$|S(n)| \leq A_1 \, 2^{nR_t}, \quad 1 \leq n \leq n_t \tag{3.11}$$

where A_1 is some constant ≥ 1. Clearly, this cannot be a block code; for block codes with constraint length n_t, the number of messages is always $|S| = 2^{n_t R_t}$ regardless of n. A code structure that is consistent with Equation 3.11 is a tree, as shown in Figure 3.3 for the case $R_t = 1/3$, $n_t = 15$. In this case, 2 is clearly a large enough value for the constant A_1 in Equation 3.11. For any reasonably homogeneous tree, A_1 can always be small.[8]

The encoding procedure for a random tree code is parallel to that described in Chapter 2 for random block codes. The information sequence $\bar{x}$ provides instructions that direct the transmitter along a path through S. We adopt the convention that if the ith digit x_i of $\bar{x}$ is zero, we trace the upper branch at the ith node of the tree; if $x_i = 1$, we trace the lower branch. The transmitted sequence $\bar{s}_i$ is the digits encountered along the way. An example is illustrated in the Figure 3.3.

On account of the tree structure of the message set we are considering, we shall seek in decoding to make a decision only about the first information digit x_1. This digit divides the entire tree set S into two subsets: S^o is the subset of all $\bar{s}$ consistent with $x_1 = 0$, and S^1 is the subset of all $\bar{s}$ consistent with $x_1 = 1$. The task of the first-digit decoder is to decide whether a received sequence $\bar{y}$ should be attributed to an $\bar{s}$ in S^o or S^1. In Chapter 4 we discuss the generation of infinite tree codes, and the iteration of the decoding procedure for information digits $x_2, x_3, \ldots$.

The desirability of the random tree code stems from consideration of the average number of incorrect messages $\bar{s}_i$ that differ from the received sequence $\bar{y}$ in fewer than k_n places, and that are therefore not discarded at length n. For any length n, there are $\frac{1}{2}|S(n)|$ possible incorrect sequences $\bar{s}_i$, and for each of these the probability of retention is bounded by $2^{-nR_K(n)}$. Assuming no previous discarding, therefore, we have for the average number $\overline{M}_K(n)$ of messages retained at length n

$$\overline{M}_K(n) \leq \frac{A_1}{2} \, 2^{n[R_t - R_K(n)]} \tag{3.12}$$

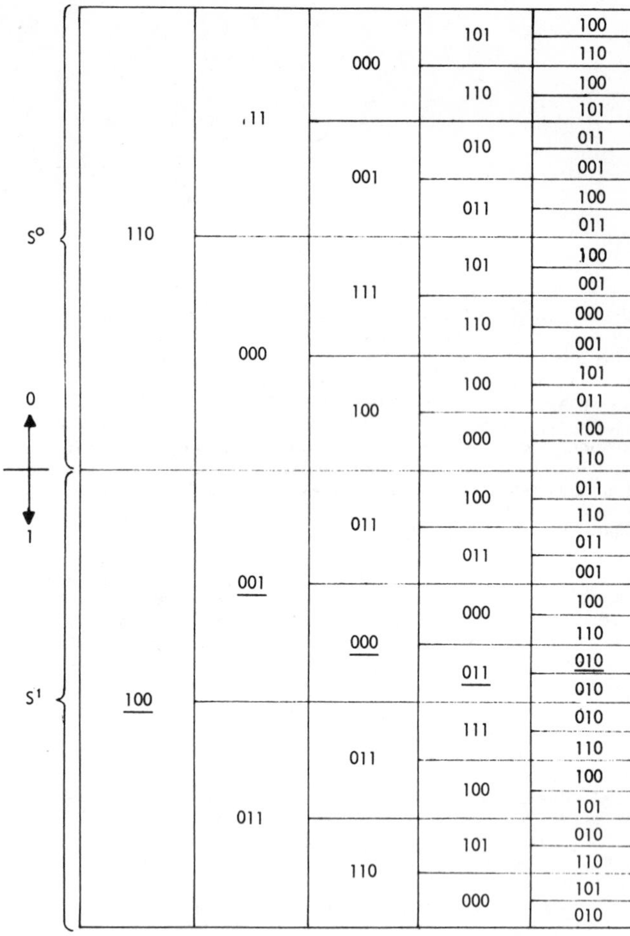

Fig. 3.3. Random tree code. $n_t = 15$, $R_t = \frac{1}{3}$
The underlined digits form the transmitted sequence $\bar{s}$ when the information sequence $\bar{x}$ is 10110

$R_K(n)$ approaches C as $n \to \infty$, and the transmission rate R_t must be less than C. It is evident from Figure 3.2 that for n greater than some n_K, the exponent in Equation 3.12 is negative. Thus we may hope to extend the code constraint length n_t indefinitely beyond n_K, for a fixed probability criterion K, without increasing the number of incorrect messages with which we must cope.

2. Single Criterion Decoding

Let us assume that a transmitted sequence $\bar{s}_o$ belongs to subset S^o of our random tree code. We can correctly decode the first information digit x_1 by using the discard function k_n to eliminate from consideration the entire incorrect subset S^1. Consider a decoding computer that starts at $n = 1$ and attempts to trace every sequence $\bar{s}(n)$ in S. As it proceeds, it compares $\bar{s}(n)$ with the received sequence $\bar{y}(n)$, and counts the number $d(n)$ of 1's in $\bar{y}(n) \oplus \bar{s}(n)$. If $d(n) < k_n$, the computer follows $\bar{s}(n)$ from n to $n+1$. If $d(n) \geq k_n$, the computer discards $\bar{s}(n)$ and considers the next sequence in S that has not yet been discarded at any n. The procedure stops and x_1 is decoded as soon as either subset S^o or S^1 is discarded in toto.

We consider now the problem of bounding the average number of computations that such a decoding computer makes in processing the incorrect subset (by assumption, S^1). We define "one computation" to be the generation of $d(n+1)$ from $d(n)$ and the digits s_{n+1} and y_{n+1}, and the comparison of $d(n+1)$ against k_{n+1}.

$$d(n+1) = d(n) + [s_{n+1} \oplus y_{n+1}] \; (<, =) \; k_{n+1}$$

Let $\overline{N}_K(n+1)$ be the average number of computations at length $n+1$ necessary to check all sequences in $S^1(n)$ which have been accepted out to length n. Since $M_K(n)$ is the average number of retained sequences in $S^1(n)$,

$$\overline{N}_K (n+1) = \Delta(n)\, \overline{M}_K (n) \tag{3.13}$$

$$\overline{N}_K (1) = 1$$

where $\Delta(n) = 1$ or 2 depending on whether or not the sequence $\overline{s}(n)$ being tested divides at length n. If $\overline{N}_K$ is the average number of computations required to process S',

$$\overline{N}_K = \sum_n \overline{N}_K (n) \tag{3.14}$$

Substituting Equations 3.12 and 3.13 into Equation 3.14, we obtain

$$\overline{N}_K < A_1 \sum_n 2^{n[R_t - R_K(n)]} \tag{3.15}$$

where we have overbounded $\Delta(n)$ by 2.

From Figure 3.2 it is clear that in the summation of Equation 3.15 there is a maximum term; let n_{max} be the corresponding value of n. Then from Equations 3.9 and 2.14, for $n \approx n_{max}$ we have

$$n[R_t - R_K(n)] = n[H(p_n) - H(p_t)] \tag{3.16}$$

Considering n as a continuous variable and differentiating, we find the maximum corresponds to

$$n H'(p_n) \frac{dp_n}{dn} + H(p_n) - H(p_t) = 0 \tag{3.17}$$

But from Equations 3.2 and 2.19,

$$-\frac{K}{n^2} \frac{dn}{dp_n} = H'(p_o) - H'(p_n) \tag{3.18}$$

Combining Equations 3.17 and 3.18, we have

$$\frac{K}{n} = \frac{[H(p_n) - H(p_t)]\,[H'(p_o) - H'(p_n)]}{H'(p_n)} \tag{3.19}$$

Explicit solution for n_{max} and the corresponding p_{max} cannot be accomplished due to the transcendental nature of the $H(p)$ function. However, the form of Equation 3.19 is instructive. Given values for K, p_o, and p_t, we can solve the transcendental Equations 3.19 and 3.2 simultaneously for n_{max} and p_{max}. If, having done so, we hold p_o and p_t constant and double K, the solution for p_{max} will be unchanged and the solution for n_{max} will be doubled. Whatever value p_{max} takes is clearly invariant to the ratio K/n_{max}. Thus, using Equation 3.2, we may write

$$n_{max} [H(p_{max}) - H(p_t)] = K \frac{[H(p_{max}) - H(p_t)]}{[T_o(p_{max}) - H(p_{max})]} \tag{3.20}$$

We see from Equation 3.20 that the maximum term in the summation of Equation 3.15 can be expressed in the form $2^{\alpha K}$, where $\alpha > 0$ is a constant independent of K and n_t. Since the summation is clearly finite (n cannot exceed the code length n_t), we have

$$\overline{N}_K \leq n_t 2^{\alpha K} \tag{3.21}$$

The fact that $\overline{N}_K$ is bounded by a number that grows only linearly with n_t (given K) might tempt us to employ a decoding procedure using a single criterion K. That doing so is not completely satisfactory can be seen as follows. We have defined the discard criterion k_n in such a way that the probability of discarding the correct sequence $\overline{s}_o$ at any length n is

$$Pr_o [d(n) \geq k_n] \approx 2^{-K} \tag{3.22}$$

For the single criterion decoding procedure, the probability of error is substantially the probability that the correct subset, as well as the incorrect subset, is discarded.

In Chapter 2 it has been shown that the attainable probability of error decreases exponentially with code length n_t; we desire not only to curtail the computational problem in decoding, but also to preserve this exponential error behavior. But if we make our probability criterion K proportional to n_t, then the bound on computations $\overline{N}_K$ (Equation 3.21) grows exponentially with n_t. This growth can be avoided by using an ascending sequence of probability criteria K_1 K_2 K_3 . . . and a corresponding sequence of discard functions $k_n(1)$, $k_n(2)$, $k_n(3)$, - - - .

3. Multiple Criteria Decoding

We consider next a multi-criteria decoding computer that searches through the message tree set S, in accordance with the following rules.

1. The computer begins with the smallest criterion K_1, and starts out to generate sequentially the entire set S. As the computer proceeds, it discards any sequence of length n that differs from the received sequence of length n in $k_n(1)$ or more places.
2. As soon as the computer discovers any sequence in S that is retained through length n_t, it prints out the corresponding first information digit.
3. If the complete set S is discarded, the computer adopts the next larger criterion K_2, and its corresponding discard function $k_n(2)$.
4. The computer continues this procedure until some sequence in S is retained through length n_t. It then prints the corresponding first information digit.

When these rules are adopted, the computer never uses a criterion K_j unless the correct subset S^o (and, hence, the correct sequence $\overline{s}_o$) is discarded for K_{j-1}. The probability that $\overline{s}_o$ is discarded depends only on the channel noise. Accordingly, by averaging over all channel noise sequences, we can bound the average number of computations required to eliminate the incorrect subset S'.

Roughly speaking, only when the channel is especially noisy will we require a large criterion before accepting some sequence in S. Most frequently, we will be able to accept an element of S ($\overline{s}_o$ in particular) with a small criterion.

Let $\overline{N}_j$ represent the value of $\overline{N}_K$ when $K = K_j$, and let $Pr(j)$ be the probability that the decoder uses the jth criterion. If $\overline{N}$ is the average number of computations required to process S', then

$$\overline{N} = \sum_j Pr(j)\, \overline{N}_j \tag{3.23}$$

Criterion K_j is used only when $\overline{s}_o$ fails to satisfy criterion K_{j-1}. The correct sequence $\overline{s}_o$ is subjected to at most n_t tests at criterion K_{j-1}. For each test the probability that $\overline{s}_o$ is rejected is $\leq 2^{-K_{j-1}}$. Since the probability of a union of events is upper-bounded by the sum of the probabilities of the individual events,

$$Pr(j) \leq Pr[\overline{s}_o \text{ fails to satisfy } K_{j-1}] \leq n_t 2^{-K_{j-1}} \tag{3.24}$$

It is convenient to let

$$K_j = K_1 + (j-1)\,\Delta K \tag{3.25}$$

Then we may write

$$Pr(j) \leq \begin{cases} 1, & j = 1 \\ n_t 2^{\Delta K} 2^{-K_j}, & j > 1 \end{cases} \tag{3.26}$$

Substituting Equations 3.15 and 3.26 into Equation 3.23, we obtain

$$\overline{N} \leq A_1 \sum_n 2^{n[R_t - R_1(n)]} + A_1 n_t 2^{\Delta K} \sum_n \sum_{j \geq 2} 2^{-K_j + n[R_t - R_j(n)]} \tag{3.27}$$

where we use the notation $R_j(n) = R_K(n)$ for $K = K_j$.

In order to motivate the derivation of the bound on $\overline{N}$ which follows, we first consider the exponent in the double summation of Equation 3.27. For each j, Equation 3.9 states that there is a range of n for which

$$R_j(n) = 1 - H(p_n^{(j)}) \tag{3.28}$$

where, when $K = K_j$, $p_n^{(j)}$ is the solution to the equation

$$K_j = n[T_o(p_n^{(j)}) - H(p_n^{(j)})] \tag{3.29}$$

Over this range of n, the exponent in the double summation of Equation 3.27 becomes

$$-n[T_o(p_n^{(j)}) - 2H(p_n^{(j)}) + 1 - R_t] \, .$$

The function $T_o(p) - 2H(p) + 1$ has its minimum value at $p = p_{crit}$, as defined in Equation 2.35. Furthermore, $p_n^{(j)}$ is a monotonically decreasing function of n for fixed j. It follows that, given j, there exists some n for which $p_n^{(j)}$ (very nearly) equals p_{crit}.

We therefore define

$$R_{comp} = T_o(p_{crit}) - 2H(p_{crit}) + 1 \tag{3.30}$$

After some algebraic manipulation, it can be shown that

$$1 - R_{comp} = \log[1 + \sqrt{4 p_o q_o}] = \log \frac{q_o}{(q_{crit})^2} \tag{3.31}$$

We shall see presently that for $R_t < R_{comp}$, $\overline{N}$ is upper-bounded by a quantity that varies slowly with n_t; for $R_t \geq R_{comp}$, the bound diverges. This divergence follows from the fact that for each j, there exists a term in the summation on n of Equation 3.27 which is no smaller than $2^{-n[R_{comp} - R_t]}$. Figure 3.4 is a plot of R_{comp}/C versus C, together with a plot of R_{crit}/C versus C for comparison.

We now proceed to bound $\overline{N}$ for $R_t < R_{comp}$. First we shall obtain a bound to $\overline{N}_K$, the average number of computations required to process the incorrect subset for a fixed criterion K. We have already observed

$$N_K \leq A_1 \sum_n 2^{n[R_t - R_K(n)]} \tag{3.15}$$

where

$$R_K(n) = \begin{cases} 0, & n < n_1 \\ 1 - H(p_n), & n \geq n_1 \end{cases} \tag{3.9}$$

The quantity n_1 has been defined by

$$\frac{K}{n_1 + 1} < T_o\left(\frac{1}{2}\right) - 1 \leq \frac{K}{n_1} \tag{3.8}$$

Thus we may rewrite Equation 3.15 as

$$\overline{N}_K \leq A_1 \sum_{n=1}^{n_1 - 1} 2^{nR_t} + A_1 \sum_{n=n_1}^{n_t} 2^{n[H(p_n) - H(p_t)]} \tag{3.15a}$$

Furthermore, for $n \geq n_1$,

$$\frac{K}{n} = T_o(p_n) - H(p_n) \tag{3.2}$$

Thus Equation 3.15a may be rewritten as

$$\overline{N}_K \leq A_1 \sum_{n=1}^{n_1 - 1} 2^{nR_t} + A_1 2^{-K} \sum_{n=n_1}^{n_t} 2^{n[T_o(p_n) - H(p_t)]} \tag{3.15b}$$

Due to the fact that p_n is a transcendental function of n, an exact evaluation of the right-hand side of Equation 3.15b is not feasible. We have observed, however, in the discussion following Equation 3.27 that a large contribution to $\overline{N}$ arises when $p_n = p_{crit}$. Since $T_o(p_n)$ increases monotonically with p_n, we see from Equation 3.15b that we can over-estimate $\overline{N}_K$ by over-estimating p_n by any quantity p_n' which varies with n in a simple fashion. Furthermore, at that n where $p_n = p_{crit}$, it is desirable to make $p_n' = p_n$.

This overestimation is most simply accomplished by constructing a tangent to the H(p) curve at $p = p_{crit}$. The equation of this tangent is

$$T_{crit}(p) = -p \log p_{crit} - q \log q_{crit} \tag{3.32}$$

If p_n' is defined as the solution to

$$\frac{K}{n} = T_o(p_n') - T_{crit}(p_n') \tag{3.33}$$

then it is obvious from Figure 3.5 that $p_n' \geq p_n$. Alternatively, if p_n' is defined as the solution to

$$\frac{K}{n} = T_o(p_n') - 1 \tag{3.34}$$

then Figure 3.5 clearly shows that $p_n' \geq p_n$.

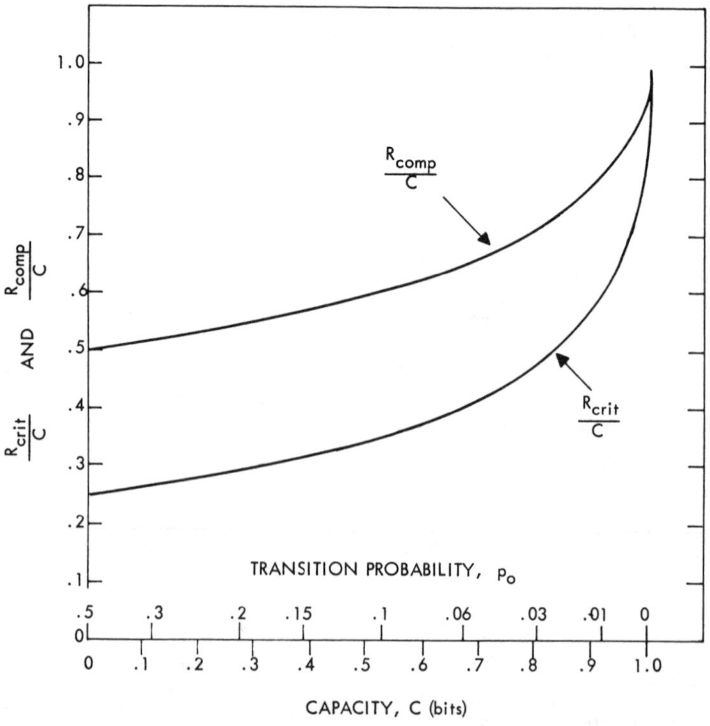

Fig. 3.4. $\frac{R_{crit}}{C}$ and $\frac{R_{comp}}{C}$ vs. C and p_o for the binary symmetric channel

We adopt for p'_n the smaller of the two values implied by Equation 3.33 and 3.34. Accordingly, we define p_2 as that value of p for which

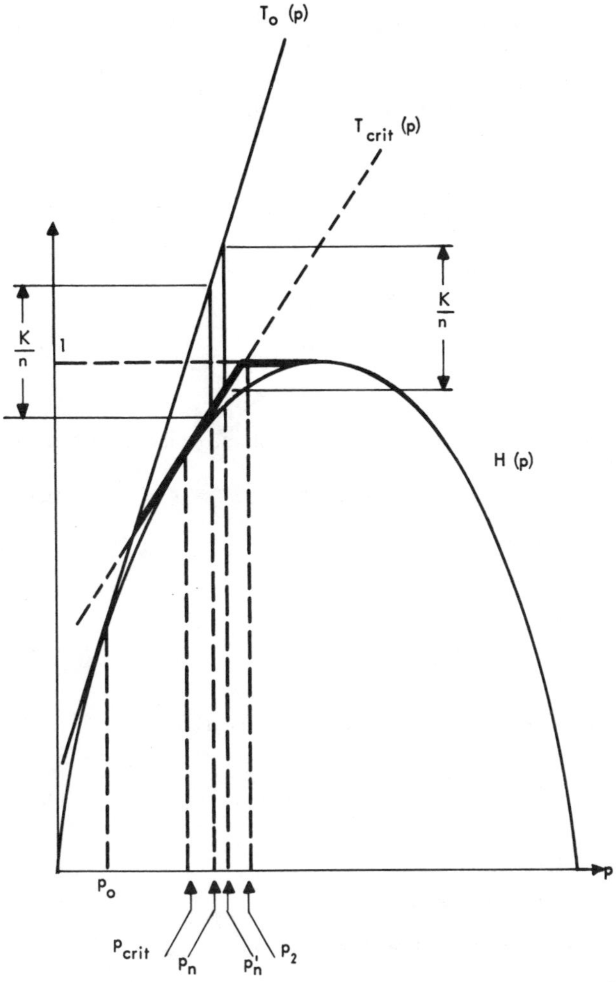

Fig. 3.5. Piecewise linear approximation to H (p) and overestimation of p_n by p'_n

$$T_{crit}(p) = 1 \tag{3.35}$$

After some manipulation, it can be shown that

$$p_2 = \frac{1 + \log q_{crit}}{\log \dfrac{q_{crit}}{p_{crit}}}, \quad q_{crit} = 1 - p_{crit} \tag{3.36}$$

Next, define n_2 such that

$$\frac{K}{n_2 + 1} \leq T_o(p_2) - 1 < \frac{K}{n_2} \tag{3.37}$$

Substituting Equation 3.36 into 3.37, we obtain

$$\frac{K}{n_2 + 1} \leq R_{comp} < \frac{K}{n_2} \tag{3.38}$$

Equation 3.38 follows after considerable algebra from the defining equation of $T_o(p)$ and the relation

$$\log \frac{q_o}{p_o} = 2 \log \frac{q_{crit}}{p_{crit}} \tag{2.34}$$

Finally, we define p'_n by

$$\frac{K}{n} = \begin{cases} T_o(p'_n) - 1, & n \leq n_2 \quad (3.39a) \\ T_o(p'_n) - T_{crit}(p'_n), & n > n_2 \quad (3.39b) \end{cases}$$

Then, solving Equation 3.39b, we have

$$p'_n = \frac{\dfrac{K}{n} + \log \dfrac{q_o}{q_{crit}}}{\log \dfrac{q_{crit}}{p_{crit}}}, \quad n > n_2 \tag{3.40}$$

The desired bound on $\overline{N}_K$ now follows from a succession of substitutions. Substituting Equation 3.40 into Equation 3.33, and using Equations 3.31 and 3.32, we obtain

$$T_o(p'_n) = \frac{2K}{n} + 1 - R_{comp}, \quad n > n_2 \tag{3.41}$$

Substituting Equation 3.39a and 3.41 into 3.15 b, we obtain

$$\overline{N}_K \leq A_1 \sum_{n=1}^{n_2} 2^{nR_t} + A_1 2^K \sum_{n=n_2+1}^{n_t} 2^{-n[R_{comp} - R_t]} \tag{3.42}$$

The inequality in Equation 3.42 is preserved, since $p'_n \geq p_n$ implies $T_o(p'_n) \geq T_o(p_n)$

Now,

$$\sum_{n=1}^{n_2} 2^{nR_t} < 2^{n_2 R_t} \sum_{n=0}^{\infty} 2^{-nR_t} = \frac{2^{n_2 R_t}}{1 - 2^{-R_t}} \tag{3.43}$$

From Equation 3.38,

$$n_2 < \frac{K}{R_{comp}} \tag{3.44}$$

Substituting Equation 3.44 into 3.43, we obtain

$$\sum_{n=1}^{n_2} 2^{nR_t} < \frac{2^{KB}}{1 - 2^{-R_t}} \tag{3.45}$$

where

$$B = \frac{R_t}{R_{comp}} \tag{3.46}$$

Similarly, for $R_t < R_{comp}$

$$\sum_{n=n_2+1}^{n_t} 2^{-n[R_{comp} - R_t]} < 2^{-(n_2+1) R_{comp}(1-B)} \sum_{n=0}^{\infty} 2^{-nR_{comp}(1-B)}$$

$$< \frac{2^{-(n_2+1) R_{comp}(1-B)}}{1 - 2^{-R_{comp}(1-B)}} \tag{3.47}$$

Equation 3.38 states that

$$n_2 + 1 \geq \frac{K}{R_{comp}} \tag{3.48}$$

Substituting Equation 3.48 into 3.47, we obtain

$$\sum_{n=n_2+1}^{n_t} 2^{-n[R_{comp} - R_t]} < \frac{2^{-K(1-B)}}{1 - 2^{-R_{comp}(1-B)}}, \quad R_t < R_{comp} \tag{3.49}$$

Substitution of Equation 3.45 and 3.49 into 3.42 finally yields the desired bound on $\overline{N}_K$.

$$\overline{N}_K \leq A_2 \, 2^{KB}, \quad R_t < R_{comp} \tag{3.50}$$

where

$$A_2 = A_1 \left[\frac{1}{1 - 2^{-R_t}} + \frac{1}{1 - 2^{-R_{comp}(1-B)}} \right] \tag{3.51}$$

We now proceed to bound $\overline{N}$. Substituting Equations 3.26 and 3.50 into 3.23, we obtain

$$\frac{\overline{N}}{A_2} \leq 2^{BK_1} + n_t 2^{\Delta K} \sum_{j \geq 2} 2^{K_j (B-1)} \tag{3.52}$$

Substituting Equation 3.25 into 3.52, noting that $B < 1$ for $R_t < R_{comp}$, and bounding the summation by the sum of a geometric series, we obtain

$$\frac{\overline{N}}{A_2} \leq 2^{BK_1} + n_t 2^{(B-1)K_1} \cdot \frac{2^{B\Delta K}}{1 - 2^{(B-1)\Delta K}} \tag{3.53}$$

We desire to select K_1 and ΔK to minimize the bound on $\overline{N}$. Differentiating with respect to these variables, and setting the resulting equations equal to zero, we obtain

$$\Delta K = \frac{\log B}{B - 1} \tag{3.54}$$

$$K_1 = \log n_t + \Delta K \tag{3.55}$$

Substitution of Equations 3.54 and 3.55 in Equation 3.53 then yields the result

$$N \leq \frac{A_2 B^{\frac{B}{B-1}}}{1 - B} n_t^B, \quad B < 1 \tag{3.56}$$

Equation 3.56 is sufficient to show that the average number of computations required to process the incorrect subset varies less rapidly than n_t. In order to exhibit more explicitly the behavior with respect to B, we note first that

$$1 \leq B^{\frac{B}{B-1}} \leq e, \quad 0 \leq B \leq 1 \tag{3.57}$$

Furthermore, for $0 < R_t < R_{comp}$ $(0 < B < 1)$,

$$1 - 2^{-(R_{comp} - R_t)} = 1 - 2^{-R_{comp}(1-B)} \geq (1 - B)\left(1 - 2^{-R_{comp}}\right)$$

Thus (3.58)

$$A_2 < \frac{A_1}{1 - B} \left[\frac{1}{1 - 2^{-R_t}} + \frac{1}{1 - 2^{-R_{comp}}} \right] \tag{3.59}$$

Using Equations 3.57 and 3.59 to overbound Equation 3.56 we have * the final result

$$\overline{N} < \frac{e A_1}{(1 - B)^2} n_t^B \left[\frac{1}{1 - 2^{-R_t}} + \frac{1}{1 - 2^{-R_{comp}}} \right], \quad B < 1 \tag{3.60}$$

We see that the bound diverges as $B \to 1$, and grows less rapidly than linearly with n_t for $B<1$.

4. Probability of Error

Thus far we have considered a procedure for decoding the first digit of a random tree code S of length n_t and transmission rate R_t. It remains to be shown that this procedure is effective not only in curtailing the average number of computations on the incorrect subset S', but also in achieving a small probability of error.

We have established a set of successive probability criteria K_j and a corresponding set of discard criterion functions $k_n(j)$. The first information digit corresponding to the transmitted sequence $\overline{s}_0$ is decoded when first we discover in our search procedure a sequence $\overline{s}$ in S that satisfies

$$d(n) < k_n(j) \tag{3.61}$$

for all n in the range $1 \leq n \leq n_t$. We commit a decoding error if and only if this sequence $\overline{s}$ is in the incorrect subset S'. Let us abbreviate $k_{n_t}(j)$ as $k_t(j)$. Since Equation 3.61 holds for all n, it holds for $n = n_t$.

$$d(n_t) < k_t(j) \tag{3.62}$$

It follows that

$$\Pr(\epsilon \mid \text{decode on } j) \leq \Pr[\text{any } d_i(n_t) \leq k_t(j)] \tag{3.63}$$

Furthermore, it follows from the random construction of the tree code S that, by the same arguments used in Chapter 2

$$\Pr[\text{any } d_i(n_t) \leq k_t(j)] \leq \begin{cases} \frac{1}{2} |S| \Pr_1[d(n_t) \leq k_t(j)] \\ 1 \end{cases} \tag{3.64}$$

Next let us consider the probability $\Pr(j)$ that the jth criterion is actually used. We have already established that

$$\Pr(j) \leq \begin{cases} 1, & j = 1 \\ n_t 2^{\Delta K} 2^{-K_j}, & j > 1 \end{cases} \tag{3.26}$$

For K_1 and ΔK chosen as in Equations 3.54 and 3.55,

$$n_t 2^{\Delta K} 2^{-K_1} = 1$$

Thus,

$$\Pr(j) \leq n_t 2^{\Delta K} 2^{-K_j}, \quad \text{all } j \tag{3.65}$$

In accordance with the definition of our discard functions (Equation 3.2)

$$K_j = n_t [T_o(p_j) - H(p_j)] \tag{3.66}$$

where we define $p_j = p_{n_t}$ for $K = K_j$

Thus,

$$\Pr(j) \leq n_t 2^{\Delta K} 2^{-n_t [T_o(p_j) - H(p_j)]}, \quad \text{all } j \tag{3.67}$$

The average probability of error $\overline{\Pr(\epsilon)}$ is given by

$$\overline{\Pr(\epsilon)} = \sum_j \Pr(j) \Pr(\epsilon \mid \text{decode on } j) \tag{3.68}$$

Define $\hat{j}$ such that

$$p_{\hat{j}} \leq p_t < p_{\hat{j}+1} \tag{3.69}$$

where p_t is the solution of $R_t = 1 - H(p_t)$

Then

$$\sum_{j > \hat{j}} \Pr(j) \Pr(\epsilon \mid \text{decode on } j) \leq \sum_{j > \hat{j}} \Pr(j) =$$

$\Pr$ [smallest criterion satisfied by $\overline{s}_o$ is greater than $K_{\hat{j}}$]

$$\leq n_t 2^{-K_{\hat{j}}} \leq n_t \frac{2^{\Delta K}}{2} 2^{-n_t [T_o(p_{\hat{j}}) - H(p_{\hat{j}})]} \tag{3.70}$$

The last inequality in Equation 3.70 follows from the fact that $2^{\Delta K}/2 > 1$ for ΔK selected in accordance with Equation 3.54. Furthermore, for $j \leq \hat{j}$,

$$\frac{1}{2} |S| \Pr_1 [d(n_t) \leq k_t(j)] \leq \frac{1}{2} A_1 2^{n_t [H(p_j) - H(p_t)]} \tag{3.71}$$

This follows from Equations 3.11 and A.24. Substituting Equations 3.64, 3.67, 3.70, and 3.71 into Equation 3.68, we obtain

$$\overline{\Pr(\epsilon)} \leq \frac{1}{2} A_1 n_t 2^{\Delta K} \sum_{j=1}^{\hat{j}} 2^{-n_t [H(p_t) + T_o(p_j) - 2H(p_j)]}$$

$$+ \frac{1}{2} A_1 n_t 2^{\Delta K} 2^{-n_t [T_o(p_{\hat{j}}) - H(p_{\hat{j}})]} \tag{3.72}$$

We are concerned with the asymptotic behavior of Equation 3.72 for large n_t. Since ΔK is independent of n_t, $K_{j+1}/n_t - K_j/n_t = \Delta K/n_t \to 0$ as $n_t \to \infty$. Thus, from Equation 3.69, $p_{\hat{j}} \to p_t$ as $n_t \to \infty$. Applying this result to Equation 3.72, we obtain

*The summations on n and j each produced a factor $1/1 - B$ in the bound of Equation 3.60. The summation on n produced this factor from the geometric series implied by the straight line approximation to $H(p)$. We conjecture that a more refined analysis will show that the summation on n converges for $B = 1$. In this case, the bound on $\overline{N}$ would be proportional to $n_t^B / 1 - B$.

$$\frac{\overline{\Pr(\epsilon)}}{A_1 n_t 2^{\Delta K}} \leq \sum_{j=1}^{\hat{j}} 2^{-n_t [H(p_t) + T_o(p_j) - 2H(p_j)]} \qquad (3.73)$$

We have shown in Chapter 2 that for random block codes of length n_t the probability of error is upper-bounded by

$$\overline{[\Pr(\epsilon)]}_{block} \leq \sum_{i=n_t p_o}^{n_t p_t} 2^{-n_t [H(p_t) + T_o(p_i) - 2H(p_i)]} \qquad (2.33)$$

where $p_i = i/n_t$.

The summation of Equation 2.33 clearly over-bounds the summation of Equation 3.73. This is true since the index i of Equation 2.33 steps through all integers between $n_t p_o$ and $n_t p_t$, whereas the range of the index j in Equation 3.73 can only imply a subset of these integers.

If we therefore use the previous results given in Equations 2.36 and 2.37, and substitute for ΔK the minimizing value of Equation 3.54, we have for the first digit in a random tree code,

$$\overline{\Pr(\epsilon)} \leq \frac{A_1 n_t^2}{B^{1/1-B}} 2^{-n_t [T_o(p_t) - H(p_t)]}$$

$$(\text{for } R_{comp} > R_t \geq R_{crit}) \qquad (3.74)$$

and

$$\overline{\Pr(\epsilon)} \leq \frac{A_1 n_t^2}{B^{1/1-B}} 2^{-n_t [H(p_t) + T_o(p_{crit}) - 2H(p_{crit})]}$$

$$(\text{for } R_t < R_{crit}) \qquad (3.75)$$

The exponential behavior of the ensemble average probability of error for sequential decoding is the same as for block codes.

5. Synopsis

For convenience of reference, we reproduce here the major results of the chapter.

Encoding. $\overline{x}$, a binary sequence of length $n_t R_t$, is transformed into $\overline{s}$, a binary sequence of length n_t, which is sent over a binary symmetric channel. The binary sequence $\overline{y}$ of length n_t is available at the output of the channel. The set S of all $\overline{s}$ sequences has $2^{n_t R_t}$ elements. The transformation is such that the elements of S can be represented as a tree (see Figure 3.3). We call S a tree code. This code has the crucial property that, if truncated at length $n < n_t$, the remainder contains approximately 2^{nR_t} sequences.

Decoding. Associated with an increasing sequence of criteria $K_1, K_2, \ldots, K_j \ldots$, the decoder has in storage (or is able to compute) a matrix

$$\begin{matrix} k_1(1) & k_2(1) \ldots & k_n(1) \ldots & k_{n_t}(1) \\ k_1(2) & k_2(2) \ldots & k_n(2) \ldots & k_{n_t}(2) \\ \cdot & & & \\ \cdot & & & \\ k_1(j) & k_2(j) \ldots & k_n(j) \ldots & k_{n_t}(j) \\ \cdot & \cdot & \cdot & \cdot \\ \cdot & \cdot & \cdot & \cdot \end{matrix}$$

The jth row in the matrix corresponds to criterion K_j. The following algorithm is employed to decode x_1, the first digit in $\overline{x}$.

1. The decoder starts with the smallest criterion K_1 and proceeds as if to generate sequentially the entire set S. The decoder discards any sequence that differs from the received sequence in $k_n(1)$ or more digits out of n.
2. As soon as the decoder discovers any sequence in S that is retained through length n_t, it decodes x_1 as the first digit in the $\overline{x}$ that implied the acceptable $\overline{s}$.
3. If no sequence in S is retained through length n_t, the decoder starts again with criterion K_2. Whenever all sequences in S fail to satisfy criterion K_j, the procedure starts again with criterion K_{j+1} until a sequence is found that does satisfy a criterion.

Number of Computations. We define S^1 as the subset of all elements of S that are not consistent with the correct x_1. For a suitably chosen $k_n(j)$ matrix, the average number of computations required to process S^1 is upper-bounded by a quantity proportional to $n_t^B/(1-B)^2$.* This result holds for $B = R_t/R_{comp} < 1$. The quantity R_{comp} is dependent on the channel transition probability; this dependence is shown on figure 3.4.

Probability of Error. With the decoding algorithm described above, and for that choice of $k_n(j)$ matrix that gives the cited bound on average number of computations, the average probability that x_1 will be decoded incorrectly has the same exponent as does the $\overline{\Pr(\epsilon)}$ for random block codes. This result holds for $0 < R_t < R_{comp}$

*This average is computed over the ensemble of all tree codes with given dimensions.

A Heuristic Discussion of Probabilistic Decoding*

ROBERT M. FANO†, FELLOW, IEEE

This is another in a series of invited tutorial, status and survey papers that are being regularly solicited by the PTGIT Committee on Special Papers. We invited Professor Fano to commit to paper his elegant but unelaborate explanation of the principles of sequential decoding, a scheme which is currently contending for a position as the most practical implementation to date of Shannon's theory of noisy communication channels.
—*Special Papers Committee.*

THE PURPOSE of this paper is to present a heuristic discussion of the probabilistic decoding of digital messages after transmission through a randomly disturbed channel. The adjective "probabilistic" is used to distinguish the decoding procedures discussed here from algebraic procedures[1] based on special structural properties of the set of code words employed for transmission.

In order to discuss probabilistic decoding in its proper frame of reference, we must first outline the more general problem of transmitting digital information through randomly disturbed channels, and review briefly some of the key concepts and results pertaining to it.[2] These key concepts and results were first presented by C. E. Shannon, in 1948,[3] and later sharpened and extended by Shannon and others. The first probabilistic decoding procedure of practical interest was presented by J. M. Wozencraft,[4] in 1957, and extended soon thereafter by B. Reiffen.[5] Equipment implementing this procedure has been built at Lincoln Laboratory[6] and is at present being tested in conjunction with telephone lines.

I. The Encoding Operation

We shall assume, for the sake of simplicity, that the information to be transmitted consists of a sequence of equiprobable and statistically independent binary digits. We shall refer to these digits as information digits, and to their rate, R, measured in digits per second, as the information transmission rate.

The complex of available communication facilities will be referred to as the transmission channel. We shall assume that the channel can accept as input any time function whose spectrum lies within some specified frequency band, and whose rms value and/or peak value are within some specified limits.

The information digits are to be transformed into an appropriate channel input, and must be recovered from the channel output with as small a probability of error as possible. We shall refer to the device that transforms the information digits into the channel input as the encoder, and to the device that recovers them from the channel output as the decoder.

The encoder may be regarded, without any loss of generality, as a finite-state device whose state depends, at any given time, on the last ν information digits input to it. This does not imply that the state of the device is uniquely specified by the last ν digits. It may depend on time as well, provided that such a time dependence is established beforehand and built into the decoder, as well as into the encoder. The encoder output is uniquely specified by the current state, and therefore is a function of the last ν information digits. We shall see that the integer ν, representing the number of digits on which the encoder output depends at any given time, is a critical parameter of the transmission process.

The encoder may operate in a variety of ways that depend on how often new digits are fed to it. The digits may be fed one at a time every $1/R$ seconds, or two at a time every $2/R$ seconds, and so forth. The limiting case in which the information digits are fed to the encoder in blocks of ν every ν/R seconds is of special interest and corresponds to the mode of operation known as block encoding. In fact, if each successive block of ν digits is fed to the encoder in a time that is short compared with $1/R$, the decoder output depends only on the digits of the last block, and is totally independent of the digits of the preceding blocks. Thus, the encoder output during each

* Received November 24, 1962. The work of this laboratory is supported in part by the U. S. Army, the Air Force Office of Scientific Research, and the Office of Naval Research.

†Department of Electrical Engineering and Research Laboratory of Electronics, M.I.T., Cambridge, Mass.

[1] W. W. Peterson, "Error-Correcting Codes," The M.I.T. Press, Cambridge, Mass., and John Wiley and Sons, Inc., New York, N. Y.; 1961.

[2] R. M. Fano, "Transmission of Information, The M.I.T. Press, Cambridge, Mass., and John Wiley and Sons, Inc., New York, N. Y.; 1961.

[3] C. E. Shannon, "A mathematical theory of communication," *Bell Sys. Tech. J.*, vol. 27, pp. 379–623; July, October, 1948.

[4] J. M. Wozencraft, "Sequential Decoding for Reliable Communications," Res. Lab. of Electronics, M.I.T., Cambridge, Mass., Technical Rept. 325; 1957. See also J. M. Wozencraft and B. Reiffen, "Sequential Decoding," The M.I.T. Press, Cambridge, Mass., and John Wiley and Sons, Inc., New York, N. Y.; 1961.

[5] B. Reiffen, "Sequential Encoding and Decoding for the Discrete Memoryless Channel," Res. Lab. of Electronics, M.I.T., Cambridge, Mass., Technical Rept. 374; 1960.

[6] K. M. Perry and J. M. Wozencraft, "SECO: A self-regulating error correcting coder-decoder," IRE Trans. on Information Theory, vol. IT-8, pp. 128–125; September, 1962.

Reprinted from *IEEE Trans. Inform. Theory*, vol. IT-9, pp. 64–74, Apr. 1963.

time interval of length ν/R, corresponding to the transmission of one particular block of digits, is completely independent of the output during the time intervals corresponding to preceding blocks of digits. In other words, each block of ν digits is transmitted independently of all preceding blocks.

The situation is quite different when the information digits are fed to the encoder in blocks of size $\nu_0 < \nu$. Then the encoder output depends not only on the digits of the last block fed to the encoder, but also on $\nu - \nu_0$ digits of preceding blocks. Therefore, it is not independent of the output during the time interval corresponding to preceding blocks. As a matter of fact a little thought will indicate that the dependence of the decoder output on his own past extends to infinity in spite of the fact that its dependence on the input digits is limited to the last ν. For this reason, the mode of operation corresponding to $\nu_0 < \nu$ is known as sequential encoding. The distinction between block encoding and sequential encoding is basic to our discussion of probabilistic decoding.

The encoding operation, whether of the block or sequential type, is best performed in two steps, as illustrated in Fig. 1. The first step is performed by a binary encoder that generates n_0 binary digits per input information digit, where the integer n_0 is a design parameter to be selected in view of the rest of the encoding operation, and of the channel characteristics. The binary encoder is a finite-state device whose state depends on the last ν information digits fed to it, and possibly on time as discussed above. The dependence of the state on the information digits is illustrated in Fig. 1, by showing the ν information digits as stored in a shift register with serial input and parallel output. It can be shown that the operation of the finite-state encoder need not be more complex than a modulo-2 convolution of the input digits with a periodic sequence of binary digits of period equal to $n_0\nu$. A suitable periodic sequence can be constructed simply by selecting the $n_0\nu$ digits equiprobably and independently at random. Thus, the complexity of the binary encoder grows linearly with ν, and its design depends on the transmission channel only through the selection of the integers n_0 and ν.

The second part of the encoding operation is a straightforward transformation of the sequence of binary digits generated by the binary encoder into a time function that is acceptable by the channel. Because of the finite-state character of the encoding operation, the resulting time function must necessarily be a sequence of elementary time functions selected from a finite set. The elementary time functions are indicated in Fig. 1 as $S_1(t)$, $S_2(t)$, $\cdots$, $S_M(t)$, where M is the number of distinct elementary time functions, and T is their common duration. The generation of these elementary time functions may be thought of as being controlled by a switch, whose position is in turn set by the digits stored in a μ-stage binary register. The digits generated by the binary encoder are fed to this register μ at a time, so that each successive group of μ digits is transformed into one of the elementary

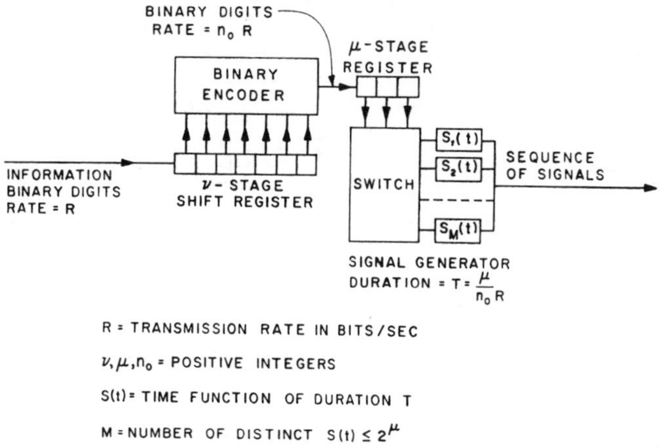

Fig. 1—The encoding operation.

signals. The number of distinct elementary signals M cannot exceed 2^μ, but it may be smaller. A value of M substantially smaller than 2^μ is used when some of the elementary signals are to be employed more often than others. For instance, with $M = 2$ and $\mu = 2$ we could make one of the 2 elementary signals occur 3 times as often as the other, by connecting 3 of the switch positions to one signal and the remaining one to the other.

While the character of the transformation of binary digits into signals envisioned in Fig. 1 is quite general, the range of the parameters involved is limited by practical considerations. The number of distinct elementary signals M must be relatively small, and so must be the integer n_0. The values of M and n_0, as well as the forms of the elementary signals, must be selected with great care in view of the characteristics of the transmission channel. In fact, their selection results in the definition of the class of time functions that may be fed to the channel, and therefore, in effect, to a redefinition of the channel.[7] Thus, one faces a compromise between equipment complexity and degradation of channel characteristics.

Fig. 2 illustrates two choices of parameters and of elementary signals, which would be generally appropriate when no bandwidth restriction is placed on the signal and thermal agitation noise is the only disturbance present in the channel. In case a) each digit generated by the binary encoder is transformed into a binary pulse, while in case b) each successive block of 4 digits is transformed into a sinusoidal pulse 4 times as long, and of frequency proportional to the binary number spelled by the group of 4 digits. The example illustrated in Fig. 3 pertains instead to the case in which the signal bandwidth is so limited that the shortest pulse duration permitted is equal to the time interval corresponding to the transmission of 2 information digits. In this case the elementary signals are pulses of the shortest permissible duration, with 16 different amplitudes.

[7] J. Ziv, "Coding and decoding for time-discrete amplitude continuous memoryless channels," IRE Trans. on Information Theory, vol. IT-8, pp. 199–205; September, 1962.

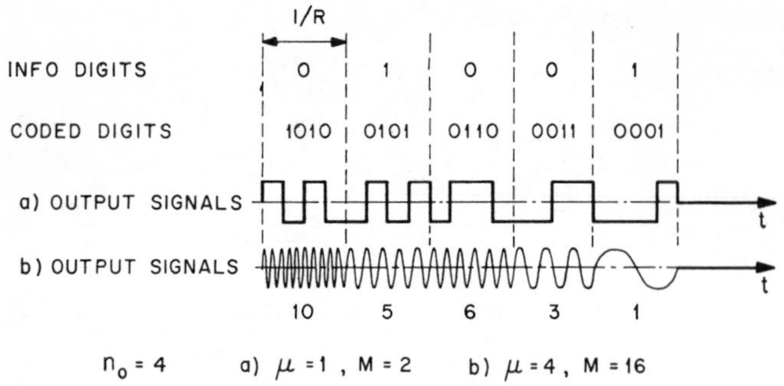

Fig. 2—Examples of encoding for a channel with unlimited band.

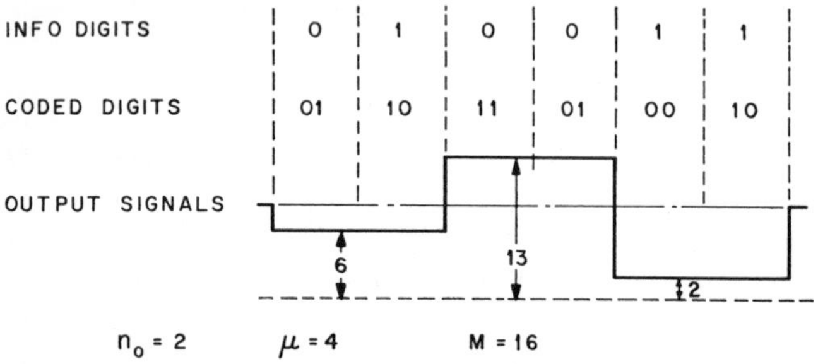

Fig. 3—Example of encoding for a band-limited channel.

These examples should make it clear that the encoding process illustrated in Fig. 1 includes, as special cases, the traditional forms of modulation employed in digital communication. What distinguishes the forms of encoding envisioned here from the traditional forms of modulation is the order of magnitude of the integer ν. In the traditional forms of modulation the value of ν is very small, often equal to 1 and very seldom greater than 5. Here instead we envision values of ν of the order of 50 or more. The reason for using large values of ν will become evident later on.

II. Channel Quantization

Let us suppose that the encoding operation has been fixed to the extent of having selected the duration and identities of the elementary signals. We must consider next how to represent the effect of the channel disturbances on these signals. Since most of our present detailed theoretical knowledge is limited to channels without memory, we shall limit our discussion to such channels. A channel without memory can be defined for our purpose as one whose output during each time interval of length T, corresponding to the transmission of an elementary signal, is independent of the channel input and output during preceding time intervals. This implies that the operation of the channel can be described within any such time interval without reference to the past or the transmission. We shall also assume that the channel is stationary in the sense that its properties do not change with time.

Let us suppose that the elementary signals are transmitted with probabilities $P(S_1)$, $P(S_2)$, $\cdots$, $P(S_M)$, and indicate by $S'(t)$ the channel output during the time interval corresponding to the transmission of a particular signal. The observation of the output $S'(t)$ changes the probability distribution over the ensemble of elementary signals, from the *a priori* distribution $P(S)$ to the *a posteriori* conditional distribution $P(S \mid S')$. The latter distribution can be computed, at least in principle, from the *a priori* distribution and the statistical characteristics of the channel disturbances. More precisely, we may regard the output $S'(t)$ as a point S' in a continuous space of suitable dimensionality. Then, if we indicate by $p(S' \mid S_k)$ the conditional probability density (assumed to exist) of the output S' for a particular input S_k, and have

$$p(S') = \sum_{k=1}^{M} P(S_k)p(S' \mid S_k) \tag{1}$$

the probability density of S' over all input signals, we obtain

$$P(S \mid S') = \frac{P(S)p(S' \mid S)}{p(S')}. \tag{2}$$

Knowing the *a posteriori* probability distribution $P(S \mid S')$ is equivalent, for our purposes, to knowing the output signal S'. In turn, this probability distribution depends on S' only through the ratios of the M probability densities $p(S' \mid S)$. Furthermore, these probability densities cannot be determined, in practice, with infinite precision. Thus, we must decide, either implicitly or explicitly, the tolerance within which the ratios of these probability densities are to be determined.

The effect of introducing such a tolerance is to lump together the output signals S' for which the ratios of the probability densities remain within the prescribed tolerance. Thus, we might as well divide the S' space into regions in which the ratios of the densities remain within the prescribed tolerance, and record only the identity of the particular region to which the output signal S' belongs.

Such a quantization of the output space S' is governed by considerations similar to those governing the choice of the input elementary signals, namely, equipment complexity and channel degradation. We shall not discuss this matter further, except for stressing again that such quantizations are unavoidable in practice; their net result is to substitute for the original transmission channel a new channel with discrete sets of possible inputs and outputs, and a correspondingly reduced transmission capability.[7]

III. Channel Capacity

It is convenient at this point to change our terminology to that commonly employed in connection with discrete channels. We shall refer to the set of elementary input signals as the input alphabet, and to the individual signals as input symbols. Similarly, we shall refer to the set of regions in which the channel output space has been divided as the output alphabet, and to the individual regions as output symbols. The input and output alphabets will be indicated by X and Y, respectively, and particular symbols belonging to them will be indicated by x and y. Thus, the transmission channel is completely described by the alphabets X and Y, and by the set of conditional probability distributions $P(y \mid x)$.

We have seen that the net effect of the reception of a symbol y is to change the *a priori* probability distribution $P(x)$ into the *a posteriori* probability distribution

$$P(x \mid y) = \frac{P(x)P(y \mid x)}{P(y)} = \frac{P(x, y)}{P(y)}, \tag{3}$$

where $P(x, y)$ is the joint probability distribution of input and output symbols. Thus, the information provided by a particular output symbol y about a particular input symbol x is defined as

$$I(x; y) = \log \frac{P(x \mid y)}{P(x)} = \log \frac{P(y \mid x)}{P(y)} = \log \frac{P(x, y)}{P(x)P(y)}. \tag{4}$$

We shall see that this measure of information and its average value over the input and/or output alphabets play a central role in the problem under discussion.

It is interesting to note that $I(x; y)$ is a symmetrical function of x and y, so that the information provided by a particular y about a particular x is the same as the information provided by x about y. In order to stress this symmetry property, $I(x; y) = I(y; x)$ is often referred to as the mutual information between x and y. In contrast,

$$I(x) = \log \frac{1}{P(x)} \tag{5}$$

is referred to as the self-information of x. This name follows from the fact that, for a particular symbol pair $x = x_k$, $y = y_i$, $I(x_k; y_i)$ becomes equal to $I(x_k)$ when $P(x_k \mid y_i) = 1$, that is, when the output symbol y_i uniquely identifies x_k as the input symbol. Thus, $I(x_k)$ is the amount of information that must be provided about x_k in order to uniquely identify it, and as such is an upper bound to the value of $I(x_k; y)$.

In the particular case of an alphabet with L equiprobable symbols, the self-information of each symbol is equal to $\log L$. The information is measured in bits when base-2 logarithms are used in the expressions above. Thus the self-information of the symbols of a binary equiprobable alphabet is equal to 1 bit.

Let us suppose that the input symbol is selected from the alphabet X with probability $P(x)$. The average, or expected value, of the mutual information between input and output symbols is, then,

$$I(X; Y) = \sum_{XY} P(x, y) I(x; y). \tag{6}$$

This quantity depends on the input probability distribution $P(x)$ and on the characteristics of the channel represented by the conditional probability distributions $P(y \mid x)$. Thus, its value for a given channel depends on the probability distribution $P(x)$ alone.

The channel capacity is defined as the maximum value of $I(X; Y)$ with respect to $P(x)$, that is,

$$C = \underset{P(x)}{\text{Max}}\, I(X; Y). \tag{7}$$

It can be shown[8] that if a source that generates sequences of x symbols is connected to the channel input, the average amount of information per symbol provided by the channel output about the channel input cannot exceed C, regardless of the statistical characteristics of the source.

IV. Error Probability for Block Encoding

Let us consider now the special case of block encoding, and suppose that a block of ν information digits is transformed by the encoder into a sequence of N elementary signals, that is, into a sequence of N input symbols. Since the information digits are, by assumption, equiprobable and independent of one another, it takes an amount of information equal to log 2 (1 bit) to identify each of them. Thus, the information transmission rate per channel symbol is given by

$$R = \frac{\nu}{N} \log 2. \tag{8}$$

(Note that the same symbol is used to indicate the information transmission rate, whether per channel symbol or per unit time.)

The maximum amount of information per symbol which the channel output can provide about the channel input is equal to C, the channel capacity. It follows that we cannot expect to be able to transmit the information digits with any reasonable degree of accuracy at any rate $R > C$. Shannon's fundamental theorem asserts, furthermore, that for any $R < C$ the probability of erroneous decoding of a block of ν digits can be made as small as desired by employing a sufficiently large value of ν and a correspondingly large value of N. More precisely, it is possible[9] to achieve a probability of error per block bounded by

$$P_e < 2^{-\nu(\alpha/R)+1}, \tag{9}$$

where α is independent of ν and varies with R as illustrated schematically in Fig. 4. Thus, for any $R < C$, the probability of error decreases exponentially with increasing ν.

It is clear from (9) that the probability of error is controlled primarily by the product of ν and α/R, the latter quantity being a function of R alone for a given channel. Thus, the same probability of error can be obtained with a small value of ν and relatively small value of R, or with a value of R close to C and a correspondingly larger value of ν. In the first situation, which corresponds to the traditional forms of modulation, the encoding and decoding equipment is relatively simple because of the small value of ν, but the channel is not utilized efficiently. In the second situation, on the contrary, the channel is efficiently utilized, but the relatively large value of ν implies that the terminal equipment must be substantially more complex. Thus, we are faced with a compromise between efficiency of channel utilization and complexity of terminal equipment.

[8] Fano, *op. cit.*, see Sec. 5.2.
[9] Fano, *op. cit.*, see ch. 9.

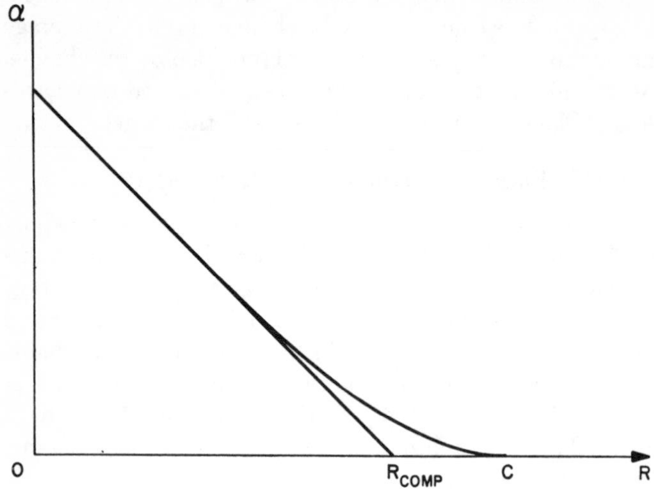

Fig. 4—Relation between the exponential coefficient α and the information transmission rate R in (9).

It was pointed out in Section I that the operation to be performed by the binary encoder is relatively simple, namely, the convolution of the input information digits with a periodic sequence of binary digits of period equal to $n_0\nu$. Thus, roughly speaking, the complexity of the encoding equipment grows linearly with ν. On the other hand, the decoding operation is substantially more complex, both conceptually and in terms of the equipment required to perform it. The rest of this paper is devoted to it.

V. Probabilistic Block Decoding

We have seen that in the process of block encoding each particular sequence of ν information digits is transformed by the encoder into a particular sequence of N channel-input symbols. We shall refer to any such sequence of input symbols as a code word, and we shall indicate by u_k the code word corresponding to the sequence of information digits which spells k in the binary number system. The sequence of N output symbols resulting from an input code word will be indicated by v.

The probability that a particular code word u will result in a particular output sequence v is given by

$$P(v \mid u) = \prod_{j=1}^{N} [P(y \mid x)]_j, \qquad (10)$$

where the subscript j indicates that the value of the conditional probability is evaluated for the input and output symbols that occupy the jth positions in u and v. On the other hand, since all sequences of information digits are transmitted with the same probability, the *a posteriori* probability of any particular code word u after the reception of a particular output sequence v is given by

$$P(u \mid v) = \frac{P(v \mid u)P(u)}{\sum_U P(v \mid u)P(u)} = 2^{-\nu} \frac{P(v \mid u)}{P(v)}. \qquad (11)$$

Thus, the code word that *a posteriori* is most probable for a particular output v is the one that maximizes the conditional probability $P(v \mid u)$ given by (10). We can conclude that in order to minimize the probability of error the decoder should select the code word with the largest probability $P(v \mid u)$ of generating the sequences v output from the channel.

While the specification of the optimum decoding procedure is straightforward, its implementation presents very serious difficulties for any sizable value of ν. In fact, there is no general procedure for determining the code word corresponding to the largest value of $P(v \mid u)$ without having to evaluate this probability for most of the 2^ν possible code words. Clearly, the necessary amount of computation grows exponentially with ν and very quickly becomes prohibitively large. However, if we do not insist on minimizing the probability of error, we may take advantage of the fact that, if the probability of error is to be very small, the *a posteriori* most probable code word must be almost always substantially more probable than all other code words. Thus, it may be sufficient to search for a code word with a value of $P(v \mid u)$ larger than some appropriate threshold, and take a chance on the possibility that there be other code words with even larger values, or that the value for the correct code word be smaller than the threshold.

Let us consider, then, what might be an appropriate threshold. Let us suppose that for a given received sequence v there exists a code word u_k for which

$$P(u_k \mid v) \geq \sum_{i \neq k} P(u_i \mid v), \qquad (12)$$

where the summation extends over all the other $2^\nu - 1$ code words. Then u_k must be the *a posteriori* most probable code word. The condition expressed by (12) can be rewritten, with the help of (11), as

$$P(v \mid u_k) \geq \sum_{i \neq k} P(v \mid u_i). \qquad (13)$$

The value of $P(v \mid u_k)$ can be readily computed with the help of (10). However, we are still faced with the problem of evaluating the same conditional probability for all of the other code words. This difficulty can be circumvented by using an approximation related to the random-coding procedure employed in deriving (9).

In the process of random coding each code word is constructed by selecting its symbols independently at random according to some appropriate probability distribution $P_0(x)$. The right-hand side of (9) is actually the average value of the probability of error over the ensemble of code-word sets so constructed. This implies, incidentally, that satisfactory code words can be obtained in practice by following such a random construction procedure.

Let us assume that the code words under consideration have been constructed by selecting the symbols independently at random according to some appropriate probability distribution $P_0(x)$. It would seem reasonable then to substitute for the right-hand side of (13) its average value over the ensemble of code-word sets constructed in the same random manner. In such an ensemble of code-word sets, the probability $P_0(u)$ that any particular input sequence u be chosen as a code word is

$$P_0(u) = \prod_{j=1}^{N} [P_0(x)]_j, \tag{14}$$

where the subscript j indicates that $P_0(x)$ is evaluated for the jth symbol of the sequence u. Thus, the average value of the right-hand side of (12), with the help of (10), is

$$(2^\nu - 1) \sum_U P_0(u)P(v \mid u) = (2^\nu - 1) \prod_{j=1}^{N} [P_0(y)]_j, \tag{15}$$

where U is the set of all possible input sequences, and

$$P_0(y) = \sum_X P_0(x)P(y \mid x) \tag{16}$$

is the probability distribution of the output symbols when the input symbols are transmitted independently with probability $P_0(x)$. Then substituting the right-hand side of (15) for the right-hand side of (13) and expressing $P(v \mid u_k)$ as in (10) yields

$$\prod_{j=1}^{N} \left[\frac{P(y \mid x)}{P_0(y)}\right]_j \geq 2^\nu - 1. \tag{17}$$

Finally, approximating $2^\nu - 1$ by 2^ν and taking the logarithm of both sides yields

$$\sum_{j=1}^{N} \left[\log \frac{P(y \mid x)}{P_0(y)}\right]_j \geq NR, \tag{18}$$

where R is the transmission rate per channel symbol defined by (8).

The threshold condition expressed by (18) can be given a very interesting interpretation. The jth term of the summation is the mutual information between the jth output symbol and the jth input symbol, with the input symbols assumed to occur with probability $P_0(x)$. If the input symbols were statistically independent of one another, the sum of these mutual informations would be equal to the mutual information between the output sequence and the input sequence. Thus, (18) states that the channel output can be safely decoded into a particular code word if the mutual information that it provides about the code word, evaluated as if the N input symbols were selected independently with probability $P_0(x)$, exceeds the amount of information transmitted per code word.

It turns out that the threshold value on the right-hand side of (18) is not only a reasonable one, as indicated by our heuristic arguments, but the one that minimizes the average probability of error for threshold decoding over the ensemble of randomly constructed code-word sets. This has been shown by C. E. Shannon in an unpublished memorandum. The bound on the probability of error obtained by Shannon is of the form of (9); however, the value of α is somewhat smaller than that obtained for optimum decoding. Shannon assumes in his derivation that an error occurs whenever (17) either is satisfied for any code word other than the correct one or it is not satisfied for the correct code word.

The probability of error for threshold decoding, although larger than for optimum decoding, is still bounded as in (9). This fact encourages us to look for a search procedure that will quickly reject any code word for which (17) is not satisfied, and thus converge relatively quickly on the code word actually transmitted. We observe, on the other hand, that, even if we could reject an incorrect code word after evaluating (17) over some small but finite fraction of the N symbols, we would still be faced with an amount of computation that would grow exponentially with ν. In order to avoid this exponential growth, we must arrange matters in such a way as to be able to eliminate large subsets of code words, by evaluating the left-hand side of (17) over some fraction of a single code word. This implies that the code words must possess the kind of tree structure which results from sequential encoding, as discussed in the next section.

It is just the realization of this fact that led J. M. Wozencraft to the development of his sequential decoding procedure in 1957. Other decoding procedures, both algebraic[1] and probabilistic,[10] have been developed since, which are of practical value in certain special cases. However, sequential decoding remains the only known procedure that is applicable to all channels without memory. As a matter of fact, there is reason[11] to believe that some modified form of sequential decoding may yield satisfactory results in conjunction with a much broader class of channels.

VI. Sequential Decoding

The rest of this paper is devoted to a heuristic discussion of a sequential decoding procedure recently developed by the author. This procedure is similar in many respects to that of Wozencraft,[4–6] but it is conceptually simpler and therefore it can be more readily explained and evaluated. An experimental comparison of the two procedures is in progress at Lincoln Laboratory, M. I. T., Lexington, Mass. A detailed analysis of the newer procedure will be presented in a forthcoming paper.

Let us reconsider in greater detail the structure of the encoder output in the case of sequential encoding, that is, when the information digits are fed to the encoder in blocks of size ν_0 (in practice ν_0 is seldom larger than 3 or 4). The encoder output, during the time interval corresponding to a particular block, is selected by the digits of the block from a set of 2^{ν_0} distinct sequences of channel input symbols. The particular set of sequences from which the output is selected is specified, in turn, by the $\nu - \nu_0$

[10] R. G. Gallager, "Low density parity-check codes," IRE Trans. on Information Theory, vol. IT-8, pp. 21–28; January, 1962.

[11] R. G. Gallager, "Sequential Decoding for Binary Channels with Noise and Synchronization Errors," Lincoln Lab., M.I.T., Lexington, Mass., Rept. No. 25G-2; 1961.

information digits preceding the block in question. Thus, the set of possible outputs from the encoder can be represented by means of a tree with 2^{ν_0} branches stemming from each node. Each successive block of ν_0 information digits causes the encoder to move from one node to the next one, along the branch specified by the digits of the block.

The two trees shown in Fig. 5 correspond to the two examples illustrated in Figs. 2(b) and 3. The first example yields a binary tree ($\nu_0 = 1$), while the second example yields a quaternary tree ($\nu_0 = 4$).

In summary, the encoding operation can be represented in terms of a tree in which the information digits select at each node the branch to be followed. The path in the tree resulting from the successive selections constitutes the encoder output. This is equivalent to saying that each block of ν_0 digits fed to the encoder is represented for transmission by a sequence of symbols selected from a set of 2^{ν_0} distinct sequences, but the particular set from which the sequence is selected depends on the preceding $\nu - \nu_0$ information digits. Thus, the channel output, during the time interval corresponding to a block of ν_0 information digits, provides information not only about these digits but also about the preceding $\nu - \nu_0$ digits.

The decoding operation may be regarded as the process of determining, from the channel output, the path in the tree followed by the encoder. Suppose, to start with, that the decoder selects at each node the branch which *a posteriori* is most probable, on the basis of the channel output during the time interval corresponding to the transmission of the branch. If the channel disturbance is such that the branch actually transmitted does not turn out to be the most probable one, the decoder will make an error, thereby reaching a node that does not lie on the path followed by the encoder. Thus, none of the branches stemming from it will appear as a likely channel input. If by accident one branch does appear as a likely input, the same situation will arise with respect to the branches stemming from the node in which it terminates, and so forth and so on. This rough notion can be made more precise.

Let us suppose that the branches of the tree are constructed, as in the case of block encoding, by selecting symbols independently at random according to some appropriate probability distribution $P_0(x)$. This is accomplished in practice by selecting equiprobably at random the $n_0\nu$ binary digits specifying the periodic sequence with which the sequence of information digits is convolved, and by properly arranging the connections of switch positions to the elementary signals in Fig. 1. Then, as in the case of threshold block decoding, the decoder, as it moves along a path in the tree, evaluates the quantity

$$I_N = \sum_{j=1}^{N} \left[\log \frac{P(y \mid x)}{P_0(y)} \right]_j, \tag{19}$$

where y in the jth term of the summation is the jth symbol

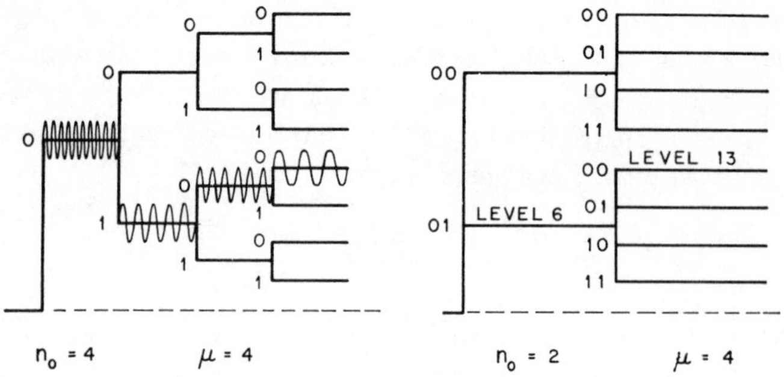

Fig. 5—Encoding trees corresponding to the examples of Figs. 2(b) and 3.

output from the channel, and x in the same term is the jth symbol along the path followed by the decoder.

As long as the path followed by the decoder coincides with that followed by the encoder I_N can be expected to remain greater than NR (R, the information transmission rate per channel symbol, is still equal to the number of channel symbols divided by the number of corresponding information digits, but it is no longer given by (8)). However, once the decoder has made a mistake and has thereby arrived at a node that does not lie on the path followed by the encoder, the terms of I_N corresponding to branches beyond that node are very likely to be smaller than R. Thus I_N must eventually become smaller than NR, thereby indicating that an error must have occurred at some preceding node. It is clear that in such a situation the decoder should try to find the place where the mistake has occurred in order to get back on the correct path. It would be desirable, therefore, to evaluate for each node the relative probability that a mistake has occurred there.

VII. Probability of Error along a Path

Let us indicate by N the order number of the symbol preceding some particular node, and by N_0 the order number of the last output symbol. Since all paths in the tree are *a priori* equiprobable, their *a posteriori* probabilities are proportional to the conditional probabilities $P(v \mid u)$, where u is the sequence of symbols corresponding to a particular path, and v is the resulting sequence of output symbols. This conditional probability can be written in the form

$$P(v \mid u) = \prod_{j=1}^{N} [P(y \mid x)]_j \prod_{j=N+1}^{N_0} [P(y \mid x)]_j. \tag{20}$$

The first factor on the right-hand side of (20) has the same value for all the paths that coincide over the first N symbols. The number of such paths, which differ in some of the remaining $N_0 - N$ symbols, is

$$m = 2^{(N_0-N)R/\log 2}. \tag{21}$$

As in the case of block decoding, it is impractical to compute the second factor on the right-hand side of (20) for each of these paths. We shall again circumvent this difficulty by averaging over the ensemble of randomly constructed trees. By analogy with the case of threshold block decoding, we obtain

$$P_0(v \mid u) = \prod_{j=1}^{N} [P(y \mid x)]_j \prod_{j=N+1}^{N_0} [P_0(y)]_j, \qquad (22)$$

where $P_0(y)$ is given by (16).

Let P_N be the probability that the path followed by the encoder is one of the $m - 1$ paths that coincide with the one followed by the decoder over the first N symbols, but differ from it thereafter. By approximating $m - 1$ with m, we have

$$P_N = K_1 2^{(N_0 - N)R/\log 2} \prod_{j=1}^{N} [P(y \mid x)]_j \prod_{j=N+1}^{N_0} [P_0(y)]_j \qquad (23)$$

$$= K_2 2^{-NR/\log 2} \prod_{j=1}^{N} \left[\frac{P(y \mid x)}{P_0(y)}\right]_j,$$

where K_1 and K_2 are proportionality constants. Finally, taking the logarithm of both sides of (23) yields

$$\log P_N = \log K_2 + \sum_{j=1}^{N} [\log \frac{P(y \mid x)}{P_0(y)} - R]_j. \qquad (24)$$

The significance of (24) is best discussed after rewriting it in terms of the order number of the nodes along the path followed by the decoder. Let us indicate by N_b the number of channel symbols per branch (assumed for the sake of simplicity to be the same for all branches) and by n the order number of the node following the Nth symbol. Then (24) can be rewritten in the form

$$\log P_n = \log K_2 + \sum_{k=1}^{n} \lambda_k, \qquad (25)$$

where

$$\lambda_k = \sum_{j=(k-1)N_b+1}^{kN_b} [\log \frac{P(y \mid x)}{P_0(y)} - R]_j \qquad (26)$$

is the contribution to the summation in (24) of the kth branch examined by the decoder. Finally, we can drop the constant from (25) and focus our attention on the sum

$$L_n = \sum_{k=1}^{n} \lambda_k, \qquad (27)$$

which increases monotonically with the probability P_n.

A typical behavior of L_n as a function of n is illustrated in Fig. 6. The value of λ_k is normally positive, in which case the probability that an error has been committed at some particular node is greater than the probability that an error has been committed at the preceding node. Let us suppose that the decoder has reached the nth node, and the value of λ_{n+1} corresponding to the *a posteriori* most probable branch stemming from it is positive. Then, the decoder should proceed to examine the branches

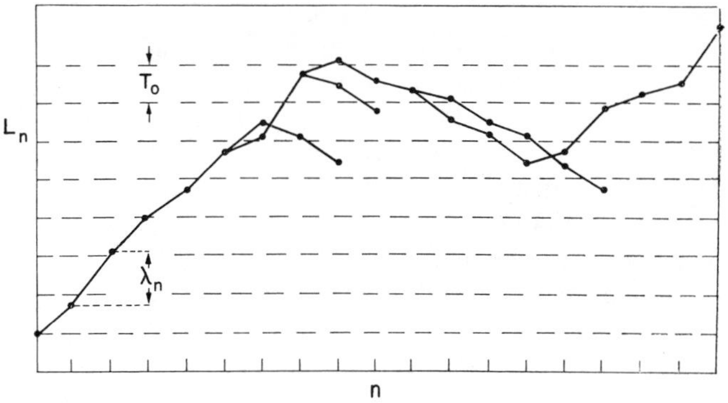

Fig. 6—Behavior of the likelihood function L_n along various tree paths. The continuous curve corresponds to the correct path.

stemming from the following node, on the assumption that the path is correct up to that point. On the other hand, if the value of λ_{n+1} is negative, the decoder should assume that an error has occurred and examine other branches stemming from preceding nodes in order of relative probability.

VIII. A Specific Decoding Procedure

It turns out that the process of searching other branches can be considerably simplified if we do not insist on searching them in exact order of probability. A procedure is described below in which the decoder moves forward or backward from node to node depending on whether the value of L at the node in question is larger or smaller than a threshold T. The value of T is increased or decreased in steps of some appropriate magnitude T_0 as follows. Let us suppose that the decoder is at some node of order n, and that it attempts to move forward by selecting the most probable branch among those still untried. If the resulting value of L_{n+1} exceeds the threshold T, the branch is accepted and T is reset to the largest possible value not exceeding L_{n+1}. If, instead, L_{n+1} is smaller than T, the decoder rejects the branch and moves back to the node of order $n - 1$. If $L_{n-1} \geq T$, the decoder attempts again to move forward by selecting the most probable branch among those not yet tried, or, if all branches stemming from that node have already been tried, it moves back to the node of order $n - 2$. The decoder moves forward and backwards in this manner until it is forced back to a node for which the value L is smaller than the current threshold T.

When the decoder is forced back to a node for which L is smaller than the current threshold, all of the paths stemming from that node must contain at least a node for which L falls below the threshold. This situation may arise because of a mistake at that node or at some preceding node, as illustrated in Fig. 6 by the first curve branching off above the correct curve. It may also result from the fact that, because of unusually severe channel disturbances, the values of L along the correct path reach a maximum and then decrease to a minimum before rising again, as illustrated by the main curve in Fig. 6. In either case, the threshold must be reduced by T_0 in order to allow the decoder to proceed.

After the threshold has been reduced, the decoder attempts again to move forward by selecting the most probable branch, just as if it had never gone beyond the node at which the threshold had to be reduced. This leads the decoder to retrace all of the paths previously examined to see whether L remains above the new threshold along any one of them. Of course, T cannot be allowed to increase while the decoder is retracing any one of these paths, until it reaches a previously unexplored branch. Otherwise, the decoder would keep retracing the same path over and over again.

If L remains above the new threshold along the correct path, the decoder will be able to continue beyond the point at which it was previously forced back, and the threshold will be permitted to rise again, as discussed above. If, instead, L still falls below the reduced threshold at some node of the correct path, or an error has occurred at some preceding node for which L is smaller than the reduced threshold, the threshold will have to be further reduced by T_0. This process is continued until the threshold becomes smaller than the smallest value of L along the correct path, or smaller than the value of L at the node at which the mistake has taken place.

The flow chart of Fig. 7 describes the procedure more precisely than can be done in words. Let us suppose that the decoder is at some node of order n. The box at the extreme left of the chart examines the branches stemming from that node and selects the one that ranks ith in order of decreasing *a posteriori* probability. (The value of λ for this branch is indicated in Fig. 7 by the subscript $i(n)$, and the integer $i(n)$ is assumed to be stored for future use for each value of n. The number of branches is $b = 2^{\nu_0}$. Thus $1 \leq i(n) \leq b$.) Next, the value L_{n+1} is computed by adding L_n and $\lambda_{i(n)}$. The value of L_n may be needed later, if the decoder is forced back to the nth node, and therefore it must be stored or recomputed when needed. For the sake of simplicity, the chart assumes that L_n is stored for each value of n.

The chart is self-explanatory beyond this point except for the function of the binary variable F. This variable is used to control a gate that allows or prevents the threshold from increasing, the choice depending on whether $F = 0$ or $F = 1$, respectively. Thus, F must be set equal to 0 when the decoder selects a branch for the first time, and equal to 1 when the branch is being retraced after a reduction of threshold. The value of F is set equal to 1 each time a branch is rejected; it is reset equal to 0 before a new branch is selected only if $T \leq L_n < T + T_0$ for the node to which the decoder is forced back. The value F is reset equal to 0 after a branch is accepted if $T \leq L_{n+1} < T + T_0$ for the node at which the branch terminates. It can be checked that, after a reduction of threshold, F remains equal to 1 while a path is being retraced,

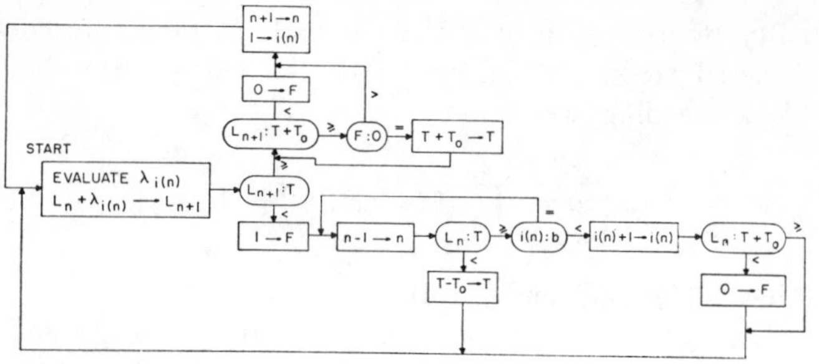

Fig. 7—Flow charge of the sequential-decoding procedure.
$1 \to F$ stands for: set F equal to 1
$L_n + \lambda_{i(n)} \to L_{n+1}$ stands for: set L_{n+1} equal to $L_n + \lambda_{i(n)}$
$n + 1 \to n$ stands for: substitute $n + 1$ for n(increase n by one)
$i(n) + 1 \to i(n)$ stands for: substitute $i(n) + 1$ for $i(n)$ (increase $i(n)$ by one)
$T + T_0 \to T$ stands for: substitute $T + T_0$ for T (increase T by T_0)
L_{n+1}: T stands for: compare L_{n+1} and T; follow path marked $\geq$, if $L_{n+1} \geq T$.

and it is reset equal to 0 at the node at which the value of L falls below the previous threshold.

IX. Evaluation of the Procedure

The performance of the sequential decoding procedure outlined in the preceding section has been evaluated analytically for all discrete memoryless channels. The details of the evaluation and the results will be presented in a forthcoming paper. The general character of these results and their implications are discussed below. The most important characteristics of a decoding procedure are its complexity, the resulting probability of error per digit, and the probability of decoding failure. We shall define and describe these characteristics in order.

The notion of complexity actually consists of two related but separate notions: the amount of equipment required to carry out the decoding operation, and the speed at which the equipment must operate. Inspection of the flow chart shown in Fig. 7 indicates that the necessary equipment consists primarily of that required to generate the possible channel inputs, namely, a replica of the encoder, and that required to store the channel output and the information digits decoded. All other quantities required in the decoding operation can be either computed from the channel output and the information digits decoded, or stored in addition to them, if this turns out to be more practical. In Section I we found that the complexity of the encoding equipment increases linearly with the encoder memory ν, since the binary encoder must convolve two binary sequences of lengths proportional to ν. The storage requirements will be discussed in conjunction with the decoding failures.

The speed at which the decoding equipment must operate is not the same for all of its parts. However, it seems reasonable to measure the required speed in terms of the average number, $\bar{n}$, of branches which the decoder must examine per branch transmitted. A very conservative upper bound to $\bar{n}$ has been obtained which has the following properties. For any given discrete channel without memory, there exists a maximum information transmission rate for which the bound to $\bar{n}$ remains finite for messages of unlimited length. This maximum rate is given by

$$R_{comp} = \underset{P_0(x)}{\text{Max}} \{-\log \sum_{Y} [\sum_{X} P_0(x) \sqrt{P(y \mid x)}]^2\}. \quad (28)$$

Then, for any transmission rate $R < R_{comp}$, the bound on $\bar{n}$ is not only finite but also independent of ν. This implies that the average speed at which the decoding equipment has to operate is independent of ν.

The maximum rate given by (28) bears an interesting relation to the exponential factor α in the bound, given by (9), to the error probability for optimum block decoding. As shown in Fig. 4, the curve of α vs R, for small values of R, coincides with a straight line of slope -1. This straight line intersects the R axis at the point $R = R_{comp}$. Clearly, $R_{comp} < C$. The author does not know of any channel for which R_{comp} is smaller than $\frac{1}{2}$ C, but no definite lower bound to R_{comp} has yet been found.

Next, let us turn our attention to the two ways in which the decoder may fail to reproduce the information digits transmitted. In the decoding procedure outlined above no limit is set on how far back the decoder may go in order to correct an error. In practice, however, a limit is set by the available storage capacity. Thus, decoding failures will occur whenever the decoder proceeds so far along an incorrect path that, by the time it gets back to the node where the error was committed, the necessary information has already been dropped from storage. Any such failure is immediately recognized by the decoder because it is unable to perform the next operation specified by the procedure.

The manner in which such failures are handled in practice depends on whether or not a return channel is available. If a return channel is available, the decoder can automatically ask for a repeat.[12] If no return channel is available, the stream of information digits must be broken into segments of appropriate length and a fixed sequence of $\nu - \nu_0$ digits must be inserted between segments. In this manner, if a decoding failure occurs during one segment, the rest of the segment will be lost but the decoder will start operating again at the beginning of the next segment.

The other type of decoding failure consists of digits erroneously decoded which cannot be corrected, regardless of the amount of storage available to the decoder. These errors are inherently undetectable by the decoder, and therefore do not stop the decoding operation. They arise in the following way.

The decoder, in order to generate the branches that must be examined, feeds the information digits decoded to a replica of the encoder. As discussed in Section VI, the set of branches stemming from a particular node is specified by the last $\nu - \nu_0$ information digits. Then, let us suppose that the decoder is moving forward along an incorrect path and that it generates, after a few incorrect digits, a sequence of $\nu - \nu_0$ information digits that happen to coincide with those transmitted. This is a very improbable event because the decoder is usually forced back long before it can generate that many digits. However, it can indeed happen if the channel disturbance is sufficiently severe during the time interval involved. After such an event, the replica of the encoder (which generates the branches to be examined) becomes completely free of incorrect digits, and therefore the decoding operation proceeds just as if the correct path had been followed all the time. Thus, the intervening errors will not be corrected. As a matter of fact, if the decoder were forced back to the node where the first error was committed, it would eventually take again the same incorrect path.

The resulting probability of error per digit decoded is bounded by an expression similar to (9). However, the exponential factor α is larger than for block encoding, although, of course, it vanishes for $R = C$. This fact may be explained heuristically by noting that the dependence of the encoder output on its own past extends beyond the symbols corresponding to the last ν information digits. Thus, we might say that, for the same value of ν, the effective constraint length is larger for sequential encoding than for block encoding.

Finally, let us consider further the decoding failures mentioned above. Since these decoding failures result from insufficient storage capacity, we must specify more precisely the character of the storage device to be employed. Suppose that the storage device is capable of storing the channel output corresponding to the last n branches transmitted. Then a decoding failure occurs whenever the decoder is forced back n nodes behind the branch being currently transmitted. This is equivalent to saying that the decoder is forced to make a final decision on each information digit within a fixed time after its transmission. Any error in this final decision, other than errors of the type discussed above, will stop the entire decoding operation. No useful bound could be obtained to the probability of occurrence of the decoding failures resulting from this particular storage arrangement.

Next, let us suppose that the channel output is stored on a magnetic tape, or similar buffer device, from which the segments corresponding to successive branches can be individually transferred to the decoder upon request. Suppose also that the internal memory of the decoder is limited to n branches. Then, a decoding failure occurs whenever the decoder is forced back n branches from the farthest one ever examined, regardless of how far back this branch is from the one being currently transmitted.

Let us indicate by k the order number of the last branch dropped from the decoder's internal memory. There are two distinct situations in which the decoder may be

[12] J. M. Wozencraft and M. Horstein, "Coding for two-way channels," in "Information Theory, Fourth London Symposium," C. Cherry, Ed., Butterworths Scientific Publications, London, England, p. 11; 1961

forced back to this branch after having examined a branch of order $k + n$. The value of L along the correct path falls below L_k at some node of order equal to, or larger than, $k + n$; or it falls below some threshold $T \leq L_k$ at some earlier node, and there exists an incorrect path, stemming from the node of order k, over which the value L remains above T up to the node of order $k + n$.

An upper bound to the probability of occurrence of these events can be readily found. It is similar to (9), with $\nu = n\nu_0$, and with a value of α approximately equal to that obtained for threshold block decoding.

X. Conclusion

The main characteristic of sequential decoding that makes it particularly attractive in practice is that the complexity of the necessary equipment grows only linearly with ν, while the required speed of operation is independent of ν. Thus, it is economically feasible to use values of ν sufficiently large to yield a negligibly small probability of error for transmission rates relatively close to channel capacity.[6]

Another important feature of sequential decoding is that its mode of operation depends very little on the channel characteristics, and therefore most of the equipment can be used in conjunction with a large variety of channels.

Finally, it should be stressed that sequential decoding is in essence a search procedure of the hill-climbing type. It can be used, in principle, to search any set of alternatives represented by a tree in which the branches stemming from different nodes of the same order are substantially different from one another.

О СЛОЖНОСТИ ДЕКОДИРОВАНИЯ

М. С. Пинскер

В работе показывается, что для весьма общего класса каналов при любых скоростях передачи, меньших пропускной способности канала, можно построить методы кодирования так, что число операций при декодировании меньше некоторой константы, не зависящей от вероятности ошибки.

Настоящая заметка представляет собой предварительную публикацию. В ней показывается, что для весьма общего класса каналов, при любых скоростях передачи R, меньших пропускной способности канала C, можно построить методы группового кодирования и декодирования так, что число операций при декодировании, приходящееся на один информационный символ, будет меньше некоторой константы, не зависящей от вероятности ошибки декодирования ε (однако зависящей от R, и с временем задержки ν, зависящим от ε). При этом здесь под числом операций при декодировании понимается среднее число сравнений последовательности символов на выходе канала с кодовыми комбинациями на входе канала, приходящееся на декодирование одного информационного символа.

Вопрос о построении кодов, допускающих простое декодирование, ввиду его важности как для теории, так и для практики, привлекает внимание многочисленных исследователей. Весьма интересны публикации по последовательному декодированию [1—4] (и по низкоплотностным кодам [5]). В этих работах было указано, что для скоростей передачи R, меньших некоторого $R_{\text{выч}}$, сверточное кодирование и последовательное декодирование позволяют вести передачу так, что вероятность ошибки экспоненциально убывает с увеличением времени задержки ν, в то время как число операций при декодировании растет как некоторая степень ν или даже не зависит от ν.

Для дискретного канала без памяти значения пороговой скорости $R_{\text{выч}}$, рассматриваемые в указанных работах, оцениваются сверху выражением

$$R_{\text{выч}} \leqslant \overline{R}_{\text{выч}} = \max_{p_i} \left\{- \log \sum_j \left[\sum_i p_i \sqrt{p_{j/i}}\right]^2\right\}, \qquad (1)$$

где $p_{j/i}$ — переходные вероятности канала, а p_i — вероятности символов на входе канала. Следует заметить, что $\overline{R}_{\text{выч}}$, найденное по этой формуле, может быть немногим больше $\frac{1}{2}C$.

Перейдем к построению кодирования и декодирования, приводящих к указанному выше результату. При этом мы можем для простоты считать, что рассматриваемый канал является двоичным симметричным каналом баз памяти, так как в общем случае рассуждения совершенно аналогичны. Для такого канала формула (1) принимает вид:

$$\overline{R}_{\text{выч}} = \overline{R}_{\text{выч}}(p) = -\log\left(\frac{1}{2}\left(1 + 2\sqrt{p(1-p)}\right),\right), \qquad (2)$$

где p — вероятность искажения входного символа, а пропускная способность канала равна

$$C = 1 + p \log p + (1-p) \log (1-p). \qquad (3)$$

Прежде всего мы остановимся на доказательстве того, что если какой-либо код позволяет вести передачу с любыми скоростями, меньшими $\overline{R}_{\text{выч}}$, при вероятности ошибки на символ, меньшей $\varepsilon > 0$, и числе операций, не зависящем от ε (но зависящем от R), то итерация этого кода и блочного кода позволяет вести передачу со скоростями, меньшими пропускной способности C, при вероятности ошибки, меньшей ε, и числе операций при декодировании, не зависящем от ε. Итерационные коды рассматривались Элайесом [6] (см. также [7]). Для удобства изложения мы предположим, что оба кода групповые. Заметим (см. [7]), что если один из кодов позволяет вести передачу со скоростью R_1, второй — со скоростью R_2, то скорость передачи при итерации этих кодов равна

$$R = R_1 R_2. \qquad (4)$$

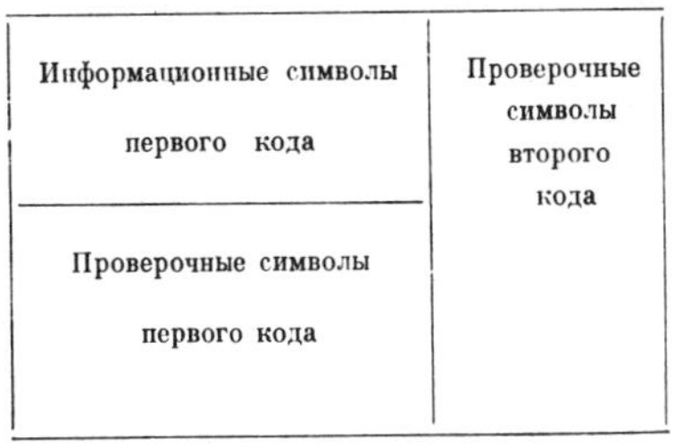
Информационные символы первого кода
Проверочные символы второго кода
Проверочные символы первого кода

Рис. 1

Итерацию кодов можно пояснить рис. 1 (см. [7], § 5, 8). Столбцы на рис. 1 образуют кодовые слова одного кода, строки — кодовые слова другого; пусть число столбцов n_2, а строк n_1; информационных элементов в строке k_2, а в столбце k_1, очевидно, $n = n_1 n_2$,

$$k = k_1 k_2, \quad R = \frac{k}{n} = \frac{k_1 k_2}{n_1 n_2} = R_1 R_2.$$

Если первый код сверточный, в котором информационные символы перемежаются проверочными, то вместо рис. 1 удобно пользоваться рис. 2.

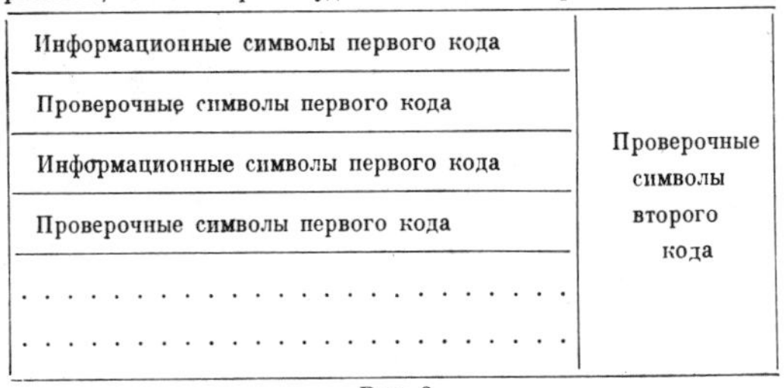
Информационные символы первого кода
Проверочные символы первого кода
Информационные символы первого кода
Проверочные символы первого кода
Проверочные символы второго кода

Рис. 2

Рассмотрим отношение

$$\overline{R}_{\text{выч}}(p)/C = \frac{-\log\left(\frac{1}{2}\left(1 + 2\sqrt{p(1-p)}\right)\right)}{1 + p \log p + (1-p)\log(1-p)},$$

график зависимости этого отношения от p имеет вид, указанный на рис. 3. Легко видеть, что

$$\lim_{p \to 0} \overline{R}_{\text{выч}}(p)/C = 1, \qquad \lim_{p \to \frac{1}{2}} \overline{R}_{\text{выч}}(p)/C = \frac{1}{2}. \qquad (5)$$

Пусть теперь задана скорость $R < C$. Выберем $R_1 < 1$, $R_2 < C$ так, чтобы $R_1 R_2 = R$, и целое число n_2 столь большим, чтобы блочный код длины n_2 с $k_2 = [n_2 R_2] + 1$ информационными символами обеспечивал при декодировании вероятность ошибки, меньшую некоторого числа p'. Далее выберем p' столь малым, чтобы для двоичного симметричного канала с вероятностью ошибки p' скорость R_1 была меньше $\overline{R}_{\text{выч}}(p')$; поскольку $R_1 < 1$ и $\overline{R}_{\text{выч}}(p') \to 1$ при $p' \to 0$, то это всегда можно осуществить.

В соответствии с рис. 2 построим итерационный код так, чтобы кодовые слова блочного кода, образованные строками матрицы рис. 2, содержали n_2 элементов, из которых $k_2 = [n_2 R_2] + 1$ информационные, и обеспечивали вероятность ошибки на символ меньшую p', а столбцы матрицы образовывались элементами первого кода (в частности бесконечного сверточного кода, используемого в последовательном декодировании) со скоростью передачи

$$R_1' = \frac{[n_1 R_1]}{n_1}.$$

Для построенного кода

$$R = \lim_{n_1 \to \infty} \frac{k_1 k_2}{n_1 n_2} = \lim_{n_1 \to \infty} \frac{([n_2 R_2] + 1)[n_1 R_1]}{n_1 n_2} \geqslant R_1 R_2. \qquad (6)$$

Декодирование можно осуществить по этапам. Сначала исходя из блочного кода производится декодирование по строкам, так что информационные и проверочные символы первого кода с вероятностью p' восстанавливаются без ошибки. При этом ошибки в разных символах одного столбца будут независимы.

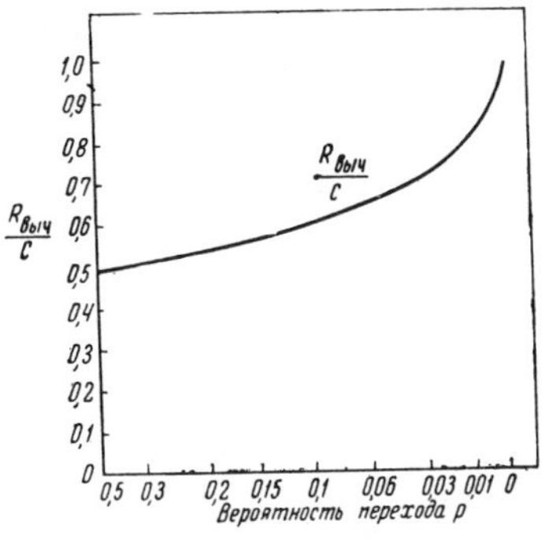

Рис. 3

Далее, исходя из первого кода производится декодирование по столбцам; восстанавливаются информационные символы. Поскольку при $R < \overline{R}_{\text{выч}}(p')$ существует модификация последовательного декодирования, при которой число операций не зависит от вероятности ошибки ε [3], то можно считать, что число операций при декодировании на втором этапе N_1 не зависит от ε. Первый этап (декодирование блочного кода) требует не более чем $N_2 = 2^{k_2}$ операций (N_2 — мощность блочного кода), N_2 также не зависит от ε. Таким образом, общее число операций при декодировании на обоих этапах не зависит от ε и не превосходит числа $N = N_1 N_2$, где N_1 — величина, ограничивающая число операций в первом коде при скорости R_1 и вероятности искажения p'.

За искомый мы можем принять код, являющийся итерацией сверточного и блочного кодов.

Нетрудно вывести асимпототическую оценку порядка дополнительного числа операций N_2, вносимых блочным кодом при $R \to C$. Имеем при $R \to C$

$$R_1 \to 1, \quad R_2 \to C; \quad R_1 \leqslant \overline{R}_{\text{выч}}(p') = -\log \frac{1}{2}\left(1 + 2\sqrt{p'(1-p')}\right) \approx$$

$$\approx 1 - 2\sqrt{p'}; \qquad C > R_2 = \frac{R}{R_1} \geqslant \frac{R}{1 - 2\sqrt{p'}} \approx R\left(1 + 2\sqrt{p'}\right). \qquad (7)$$

Как известно [8], в двоичных симметричных каналах существуют блочные групповые коды такие, что вероятности ошибки удовлетворяют неравенству

$$p' \leqslant 2^{-k_2 \varkappa (C - R_2)^2}, \qquad (8)$$

где $\varkappa$ — некоторая константа.
Отсюда

$$\log p' \leqslant -k_2 \varkappa (C - R_2)^2, \qquad (9)$$

следовательно, в соответствии с (7),

$$N_2 = 2^{k_2} \leqslant 2^{-\frac{\log p'}{\varkappa (C - R_2)^2}} \approx 2^{-\frac{\log p'}{\varkappa \left(C - (R + 2\sqrt{p'})\right)^2}}. \qquad (10)$$

Параметр $2\sqrt{p'} < C - R$ в нашем распоряжении. Положим $2\sqrt{p'} = \beta (C - R)$, $0 < \beta < 1$. Из (10) получаем

$$N_2 \approx 2^{-\frac{\log \left[\frac{\beta (C-R)}{2}\right]^2}{\varkappa [(1-\beta)(C-R)]^2}} = \frac{2^{-\frac{1}{(1-\beta)^2 (C-R)^2}}}{\beta^2 (C-R)^2 / 4}. \qquad (11)$$

Надо выбрать $0 < \beta < 1$ так, чтобы правая часть (11) была минимальна.

В заключение отметим, что итерация кодов привлечена здесь только в силу простоты использования известных фактов для доказательства ограниченности числа операций при декодировании $R > \overline{R}_{\text{выч}}$. Можно указать некоторые модификации последовательного декодирования, для которых указанная ограниченность числа операций имеет место для $R > \overline{R}_{\text{выч}}$.

Исследования по построению кодов с простым декодированием, в частности по последовательному декодированию, находятся, по-видимому, в начальной фазе. Проблема математически точного и полного определения сложности декодирования, а также получения точных оценок констант, выражающих сложность декодирования, требует дальнейших исследований.

В последующих публикациях будут приведены оценки числа операций при декодировании в принятом выше смысле.

Reprinted with permission from *Probl. Peredach. Inform.*, vol. 1, no. 1, pp. 113–116, 1965.

ЛИТЕРАТУРА

1. Возенкрафт Дж. М., Рейффен Б. Последовательное декодирование, М., Изд. иностр. лит., 1963.
2. Ziv J. Coding and decoding for time — discrete amplitude continuons memoryless channels. IRE Trans. Inform. Theory, 1962, 8, *5*, 199—205.
3. Fano R. M. A heuristic discussion of probilistic decoding. IEEE Trans. Inform Theory, 1963, 9, *2*, 64—74 (Рус. пер.: Фано Р. М. Эвристическое обсуждение вероя тностного декодирования. Теория кодирования, сб., М., Изд. «Мир», 1964 166—198.)
4. Pèrry K. M., Wozencraft J. M. SECO: A self —regulating error correcting, coder — decoder. IRE Trans. Inform. Theory, 1962, 8, *5*, 125—135. (Рус. пер.: Перри К., Возенкрафт Дж. М. SECO, саморегулирующееся кодирующее устройство с исправлением ошибок. Зарубежная радиоэлектроника, 1964, *9*, 3—16.)
5. Gallager R. G. Low density parity — check codes. IRE Trans. Inform. Theory, 1962, 8, *1*, 21—28. (Рус. пер.: Галлагер Р. Г. Коды с малой плотностью проверок на четность. Теория кодирования, сб., М., Изд. «Мир», 1964, 139—166.)
6. Elias P. Error-free coding. IRE Trans., PGIT, 1954, 4, *1*, 29—31. (Рус. пер.: Элайес П. Безошибочное декодирование. Коды, с обнаружением и исправлением ошибок, сб., М., Изд. иностр. лит., 1956, 51—79.)
7. Питерсон У. Коды, исправляющие ошибки, М., «Мир», 1964.
8. Fano R. M. Transmission of information, M. I. T. Press, Cambridge, Mass., John Wiley and Sons, Inc., New York, 1961.

Поступила в редакцию
14 ноября 1964 г.

A Lower Bound to the Distribution of Computation for Sequential Decoding

IRWIN MARK JACOBS, MEMBER, IEEE, AND ELWYN R. BERLEKAMP, MEMBER, IEEE

Abstract—In sequential decoding, the number of computations which the decoder must perform to decode the received digits is a random variable. In this paper, we derive a Paretian lower bound to the distribution of this random variable. We show that $P[C > L] \approx L^{-\rho}$, where C is the number of computations which the sequential decoder must perform to decode a block of Λ transmitted bits, and ρ is a parameter which depends on the channel and the rate of the code. Our bound is valid for all sequential decoding schemes and all discrete memoryless channels.

In Section II we give an example of a special channel for which a Paretian bound can be easily derived. In Sections III and IV we treat the general channel. In Section V we relate this bound to the memory buffer requirements of real-time sequential decoders. In Section VI, we show that this bound implies that certain moments of the distribution of the computation per digit are infinite, and we determine lower bounds to the rates above which these moments diverge. In most cases, our bounds coincide with previously known upper bounds to rates above which the moments converge.

We conclude that the performance of systems using sequential decoding is limited by the computational and buffer capabilities of the decoder, not by the probability of making a decoding error. We further note that our bound applies only to sequential decoding, and that, in certain special cases (Section II), algebraic decoding methods prove superior.

Manuscript received August 25, 1965; revised June 3, 1966. This paper presents the results of one phase of research carried out at the Jet Propulsion Laboratory, California Institute of Technology, Pasadena, under Contract NAS 7–100, sponsored by the National Aeronautics and Space Administration.

I. M. Jacobs is with the Department of Applied Electrophysics, University of California, San Diego, La Jolla, Calif. He was formerly with the Massachuetts Institute of Technology, Cambridge, and the Jet Propulsion Laboratory, Pasadena, Calif.

E. R. Berlekamp is with Bell Telephone Laboratories, Inc., Murray Hill, N. J. He was formerly with the Department of Electrical Engineering, and Electronics Research Laboratory, University of California, Berkeley, Calif., and the Jet Propulsion Laboratory, Pasadena, Calif.

I. Introduction

A SEQUENTIAL decoder is an algorithm for decoding tree codes based on knowledge of the code and the channel. The decoder starts from the

Reprinted from *IEEE Trans. Inform. Theory*, vol. IT-13, pp. 167–174, Apr. 1967.

origin of the tree. Branch by branch, it then selects a path into the tree. Whenever a suitably defined distance between the transmitted letters associated with the path being followed and the received sequence of channel output letters becomes too large, the decoder retraces its steps and searches for a more acceptable path. Details of the search procedure depend upon the particular sequential decoding algorithm. Readers unfamiliar with sequential decoding should begin by studying the Fano Algorithm, which is discussed in Chapter 6 of Wozencraft and Jacobs.[12]

The essential features of sequential decoding which we require here are that 1) the branches are examined sequentially, so that at any node of the tree the decoder's choice among a set of previously unexplored branches does not depend on received branches deeper in the tree, and 2) the decoder performs at least one computation for each node of every examined path. Algorithms which do not have these two properties are not considered to be sequential decoding algorithms.

We shall show that *any* sequential decoding algorithm operating over a discrete memoryless channel must occasionally perform a certain large number of computations before it can find the correct path. More specifically, we obtain a lower bound to the distribution of computation that is essentially Paretian.

Specifically, let C be the number of branches which the sequential decoder investigates before it correctly decodes the first Λ transmitted information bits. C is a random variable, because the behavior of the decoder depends on the received sequence, which in turn depends on the channel noise. We wish to show that there is a substantial probability that C assumes unpleasantly large values. In particular, we would like to show that

$$P[C > L] \approx \Lambda L^{-\alpha}$$

where α is a number depending only on the channel and the rate of the code (and not on L).

Our argument is based on proving two things. First, we show that, with sufficiently high probability, the channel noise will be extremely severe for a certain initial interval, which we shall call the burst length N. Second, we shall show that no sequential decoder can be expected to find the correct path through such a noise burst without doing more than L computations. In general, we will choose both the burst length N and the severity of the channel noise during this burst in a manner which depends on L.

II. An Example

Before giving the general derivation, we first consider a special example that illustrates the principal ideas with a minimum of obscuring technicalities. Suppose that our channel is a discrete memoryless channel with A input letters and $A + 1$ output letters. One output letter, the erasure, has the same nonzero probability W of being received, no matter which input letter is transmitted. A block of information bits is to be communicated over this channel, using a homogeneous tree code which has u branches exiting from each node and for which V channel symbols are transmitted per branch. We define the rate of this code as

$$R = \frac{1}{V} \ln u \quad \text{nats per channel symbol.}$$

Now suppose that the first N received letters are erasures. This erasure burst will occur with probability W^N. These N letters then correspond to the transmission of N/V branches. To simplify the argument in this example, we consider only values of N which are integer multiples of V. After observing these N erasures, the decoder still has no information about which of the N-letter paths diverging from the origin is the initial segment of the transmitted path. The number of possible N-letter transmitted paths is $u^{N/V} = e^{NR}$. By the rules of the game, the sequential decoder cannot use received letters beyond the first N to order its initial search among these e^{NR} paths. No matter which order it first elects to examine these paths, there is probability $\frac{1}{2}$ that it must examine at least $\frac{1}{2}$ of them before locating the correct path. These examinations require at least one computation per path, and hence at least

$$L = \tfrac{1}{2}e^{NR}$$

total computations if the correct path is eventually to be followed. (Counting only 1 computation per path rather than one computation per branch does not significantly weaken the bound, since paths may contain many common branches.)

We therefore observe that the number of computations necessary to find the correct path through the burst increases exponentially in N. However, the probability of a severe noise lasting for a burst of length N decreases exponentially in N. We shall see that the Paretian distribution of computation is the combined result of these two opposing effects. (A similar observation was made by Savage.[10])

For given $L \leq 2^{(\Lambda-1)}$, let us choose $N = \ln (2L)/R$. To avoid the complicating Diophantine constraints, in this example we assume that $\ln (2L)/R$ turns out to be an integer and a multiple of V. By the above arguments, if the first N digits are erasures, L or more computations occur with probability at least $\frac{1}{2}$. Hence,

$$\begin{aligned} P[C > L] &> \tfrac{1}{2}W^N \\ &= \tfrac{1}{2}e^{(NR \ln W)/R} \\ &= \tfrac{1}{2}(2L)^{(\ln W)/R} \\ &= aL^{-\alpha} = (a/\Lambda)\Lambda L^{-\alpha}. \end{aligned}$$

Here, $\alpha = -\ln W/R$ and $a = 2^{-(1+\alpha)}$. Both of these quantities depend on the channel and the rate of the code, but not on L. A distribution of the form $aL^{-\alpha}$ is called a Paretian distribution with exponent α. For large L, the behavior of the distribution is dominated by the behavior of the factor $L^{-\alpha}$. For this reason, there is no essential difference between the distribution for the number of computations necessary to decode the first Λ digits and the distribution for the number of computations necessary to decode the first digit.

The Paretian distribution has certain essential properties which place some rather strong limitations on sequential decoding. For example, if $1 < \alpha < 2$, then the Paretian distribution has finite mean but infinite variance. Under these circumstances, the *average* number of computations per decoded digit may be bounded, but the variance of the distribution of the computation will be infinite. (For further discussions and applications, see Gnedenko and Kolmogorov[7] and Mandelbrot.[14]) The decoder will require many more than the average number of computations all too often, as wel shall show more conclusively in Section VI.

The above example demonstrates the inherent inability of sequential decoding to deal effectively with severe bursts of noise. The difficulties arise from the essential nature of the sequential decoding algorithms, *not* from any fundamental limitations on convolutional codes or from any basic difficulties arising from bursts. In fact, Epstein[3] has demonstrated algebraic schemes which perform extremely well on the erasure channel just considered. These schemes permit reliable communication at any rate up to capacity (which on erasure channels may be much larger than R_{comp}). The number of computations performed by the algebraic decoder is never greater than a fixed small power of the block length, no matter how severe the channel noise. Berlekamp[1] has exhibited convolutional codes which are uniquely optimum in their ability to correct erasure bursts, and more recent schemes have been devised whereby these erasure burst convolutional codes may be implemented at a cost proportional to the constraint length and a decoding time per digit which is independent of the constraint length. All of these algebraic schemes succeed by violating both of the defining constraints on sequential decoding. The algebraic schemes search many branches of the tree simultaneously by dealing with equations which must be satisfied by the values of the erased digits, and they violate the sequential requirement by skipping over the erased positions to begin their computations on the unerased digits following the burst.

III. Lower Bound to Error Probability for Block Codes with List Decoding

Consider a general discrete memoryless channel with an A-letter input alphabet and a Q-letter output alphabet, and the transition probabilities $\{q_{i,j}\}$; $i = 1, 2, \cdots, A$; $j = 1, 2, \cdots, Q$. Following Gallager,[6] we define the function

$$E_0(\rho) = \max_{\mathbf{p}} \left\{ -\ln \sum_{j=1}^{Q} \left(\sum_{i=1}^{A} p_i q_{i,j}^{1/(1+\rho)} \right)^{(1+\rho)} \right\}; \qquad 0 < \rho < \infty \qquad (1)$$

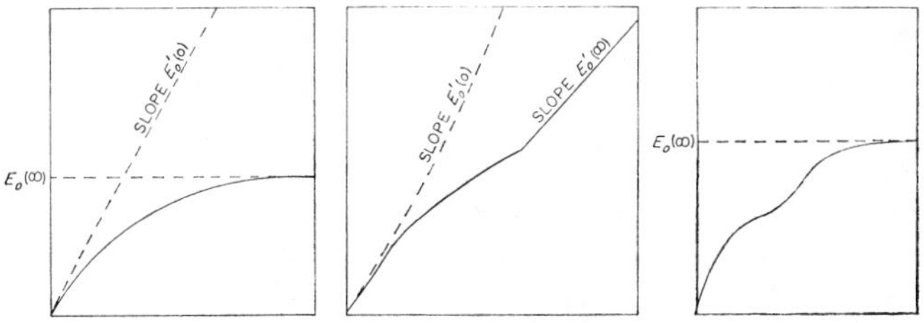

Fig. 1. Several plots of $E_0(\rho)$ vs. ρ.

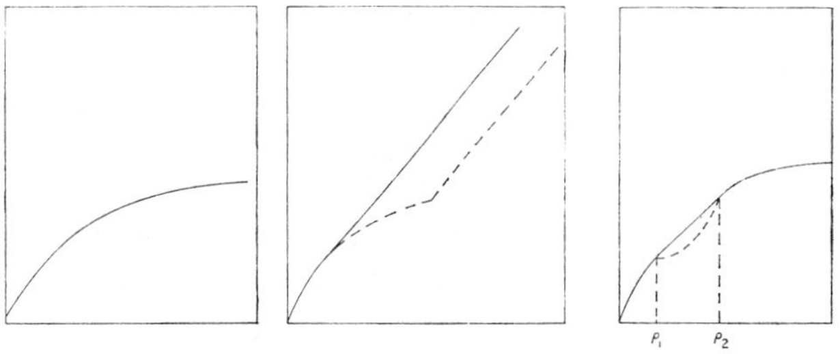

Fig. 2. Corresponding plots of $\hat{E}_0(\rho)$ vs. ρ.

where $\mathbf{p} = (p_1, p_2, \cdots, p_A)$ is a vector of non-negative components which sum to one. Several plots of $E_0(\rho)$ vs. ρ are shown in Fig. 1. Various general properties of this function are discussed by Gallager[6] and by Shannon, Gallager, and Berlekamp.[11] We define $\hat{E}_0(\rho)$ as the concave hull of $E_0(\rho)$ as shown in Fig. 2. For many common channels, $E_0(\rho)$ is itself concave, and $\hat{E}_0(\rho) = E_0(\rho)$. For channels for which $E_0(\rho)$ is not concave, certain coding theorem lower bounds are stated more simply in terms of the function $\hat{E}_0(\rho)$. Certain other coding theorems may be equivalently expressed in terms of either $E_0(\rho)$ or $\hat{E}_0(\rho)$. For example, the sphere packing exponent may be defined as

$$E_{sp}(R) = \underset{0<\rho<\infty}{\text{l.u.b.}} (E_0(\rho) - \rho R) = \underset{0<\rho<\infty}{\text{l.u.b.}} (\hat{E}_0(\rho) - \rho R).$$

These two definitions are equivalent, since the maximizing value of ρ must occur at a point $\rho_{\max}$ where $E_0(\rho_{\max}) = \hat{E}_0(\rho_{\max})$. Part of the advantage of using the definition in terms of $\hat{E}_0(\rho)$ stems from the fact that $d\hat{E}_0(\rho)/d\rho$ is a monotonic nonincreasing function of ρ. Thus the maximum of the second expression occurs at $\rho_{\max}$ if $d\hat{E}_0(\rho_{\max})/d\rho = R$.

The coding theorem used here may be stated as follows. Let one of M equiprobable messages be transmitted over a discrete memoryless channel by means of a block code containing M codewords. Each codeword is an ordered sequence of N letters of the input alphabet. The receiver processes the N-letter channel output sequence and produces a list of L messages. We define $P_e(L)$ as the probability that the transmitted message is not a member of this list.

Shannon, Gallager, and Berlekamp[11] have obtained the following lower bound to $P_e(L)$. If, for any $\rho > 0$, the code size M and the list size L satisfy

$$M/L \geq \exp\{N\hat{E}_0'(\rho) + O_1(\sqrt{N})\}, \qquad (2)$$

then

$$P_e(L) \geq \exp\{-N(\hat{E}_0(\rho) - \rho\hat{E}_0'(\rho)) - O_2(\sqrt{N})\}. \quad (3)$$

Here $\hat{E}_0' = d\hat{E}_0/d\rho$ and $O_1(\sqrt{N})$ and $O_2(\sqrt{N})$ are terms which are of the order of $\sqrt{N}$.[1] The above result applies to any block code of length N and to any list-of-L receiver. In fact, it even applies to any feedback coding and decoding scheme which transmits no more than N letters across the given forward channel, even though the decoder may be permitted to send back anything he wishes across a noiseless, delayless reverse channel of unlimited capacity.

Part of the importance of this theorem stems from its similarity to upper-bound random coding theorems, such as those given by Gallager.[6] For $L \geq \rho$, this theorem differs from lower bounds to $P_e(L)$ only in the asymptotically unimportant terms $O_1(\sqrt{N})$ and $O_2(\sqrt{N})$. The theorem therefore gives an exponentially tight expression for the probability of list decoding error of the best block code, if one uses a sufficiently large list. Under such circumstances, the rate is most conveniently defined as $R = \ln(M/L)/N$, or as a function of ρ by $R(\rho) = \hat{E}_0'(\rho)$.

For purposes of bounding the distribution of computation in sequential decoding, however, it is convenient to use instead the definitions

$$R \triangleq \frac{\ln M}{N}; \qquad M = e^{NR} \quad (4)$$

and

$$R_\rho \triangleq \hat{E}_0(\rho)/\rho. \quad (5)$$

Suppose that we have a given block code with rate $R \geq R_\rho$. Given L, we may satisfy (2) by choosing N sufficiently large so that

$$L \leq \exp\{N[R_\rho - \hat{E}_0'(\rho)] - O_1(\sqrt{N})\}. \quad (6)$$

The most economical choice of N is the smallest integral multiple of V which satisfies (6). For this choice of N,

$$L > \exp\{(N - V)(R_\rho - \hat{E}_0'(\rho)) - O_1(\sqrt{N - V})\}. \quad (7)$$

Since (6) is satisfied, so is (2); the bound of (3) is therefore valid. Substituting (5) into (3) gives

$$P_e(L) \geq \exp\{-N\rho(R_\rho - \hat{E}_0'(\rho)) - O_2(\sqrt{N})\}. \quad (8)$$

With the help of (7), (8) may be written as

$$P_e(L) \geq (L \exp O_1(\sqrt{N - V}))^{-N\rho/(N-V)} \exp -O_2(\sqrt{N})$$
$$= L^{-\rho} \exp\left\{\frac{-\rho V}{N - V} \ln L - \frac{N\rho}{N - V} O_1(\sqrt{N - V}) - O_2(\sqrt{N})\right\}. \quad (9)$$

We estimate the bracketed quantity by observing from (6) and (7) that N is of the order of $\ln L$. More precisely,

$$N = \frac{\ln L}{R_\rho - \hat{E}_0'(\rho)} + O_3(\sqrt{\ln L}). \quad (10)$$

We conclude that, for given L and ρ, there exists a block length N such that for every list-of-L receiver and every block code of length N and rate $R \geq R_\rho$,

$$P_e(L) \geq L^{-\rho} \exp\{-O_4(\sqrt{\ln L})\}. \quad (11)$$

The error probability of a list-of-L receiver for a code of rate $R \geq R_\rho$ is thus seen to be lower-bounded by a quantity which is essentially Paretian in the variable L. This bound is obtained by choosing N in such a way that the exponential decrease in error probability (with N) is balanced by the exponential increase in list size.

In the next section we shall show how this result for block codes may be applied to tree codes. Given ρ and L, and $R \geq R_\rho$, then, with probability given by (11), the decoder will be subjected to an initial channel noise burst of length N given by (6) and (7), and this noise burst will be sufficiently severe that L wrong paths of length N will appear more probable than the transmitted path of length N. We will claim that the sequential decoder can do no better than to examine each of these more probable paths first, and that this requires L computations.

IV. Application to Sequential Decoding

Let C denote the number of computations required to decode the first Λ information bits. We are interested in $P[C > L]$ for a tree code of rate R. Choose ρ such that $R \geq R_\rho$, where R_ρ is defined by (4). This constraint merely places a lower bound on ρ, so we may select ρ as the smallest value for which this constraint is satisfied. We next select the burst length N to be that integral multiple of V which satisfies (6) and (7). Next, let $\{\mathbf{y}_i\}$ denote the set of transmitted sequences of length N extending out from the tree origin. Notice that there are $M = e^{RN}$ such sequences. This set $\{y_i\}$ can therefore be considered to be a block code of length N and rate R.

We are now in a position to apply the result (11) on block codes. Let $\mathbf{r}$ denote the first N digits of the received sequence. Arrange the M probabilities $P(\mathbf{r}/\mathbf{y}_i)$ in order of decreasing size. Let $I_B(\mathbf{r})$ denote the set of subscripts of the last $M - L$ entries on this list. When $\mathbf{r}$ is received, the set $I_B(\mathbf{r})$ contains the indexes of the $M - L$ least-likely messages. These are precisely the messages which would not be on a maximum likelihood receiver's list of length L. The joint probability of transmitting a $\mathbf{y}_i$ with an index in $I_B(\mathbf{r})$ and receiving $\mathbf{r}$ is given by

$$\sum_{i \in I_B(\mathbf{r})} \frac{1}{M} P[\mathbf{r}/\mathbf{y}_i].$$

The error probability for a maximum likelihood list decoder may therefore be written

$$P_e(L) = \sum_{\text{all } \mathbf{r}} \sum_{i \in I_B(\mathbf{r})} \frac{1}{M} P[\mathbf{r}/y_i]. \quad (12)$$

$P_e(L)$ is lower-bounded in accordance with (11).

[1] We say that a_n is of the order of b_n, denoted by $a_n = 0(b_n)$, if there exists a constant K such that, for all indexes n, $a_n < K\, b_n$.

Assuming that $\mathbf{r}$ is received, we next consider the behavior of any sequential decoder searching out from the tree origin. Eventually, some one of the $\{\mathbf{y}_i\}$ is examined to its full length N, and the decoder proceeds on beyond the Nth letter. Since our initial bound involves only the first N letters of the tree, we may expedite the decoder's search by stationing a benevolent, omniscient genie at the $N + 1$st letter. This genie tells the decoder whether or not the decoder has selected the correct $\mathbf{y}_i$. Suppose that the first $\mathbf{y}_i$ examined is wrong. Then, the decoder will retrace its steps in the order directed by its specific search algorithm and eventually examine another sequence $\mathbf{y}_i$ to length N. If this sequence is incorrect, the genie will again notify the decoder of his error. The decoder will then examine a third sequence to length N, and so on. In the rare case that the transmitted path is not among the first $M - 1$ of the $\{\mathbf{y}_i\}$ reached by the decoder, the genie-aided algorithm examines all M sequences $\{\mathbf{y}_i\}$. For each received sequence $\mathbf{r}$ of length N, we thus have another ordering of the $\{\mathbf{y}_i\}$, namely, the ordering in which the sequential decoder examines these sequences. This second ordering depends on $\mathbf{r}$, the code sequences $\{\mathbf{y}_i\}$, and the algorithm. By our definition of a sequential decoding algorithm, however, this ordering is independent of the transmitted and received letters at lengths greater than N from the tree origin. (In the absence of the genie, if the last code word on the list is transmitted, and if the code and noise at depth greater than N are such that no error is made, the $\{\mathbf{y}_i\}$ of length N will be examined in the same order that they appear on the list.)

Notice that we are interested only in the order in which the decoder examines *new* members of the set $\{\mathbf{y}_i\}$. Repeated visits to any $\mathbf{y}_i$ are not entered on our second list. After ordering the sequences $\{\mathbf{y}_i\}$ according to the order in which they are visited, we may define the set $I_s(\mathbf{r})$ as the subscripts of the last $M - L$ sequences $\{\mathbf{y}_i\}$ examined by the sequential algorithm.

We next use the second property of sequential decoding algorithms and insist that any node which is examined by the algorithm require at least one computation. Therefore, if any member of $I_s(\mathbf{r})$ is transmitted and if $\mathbf{r}$ is received, at least L computations must be performed by the genie-aided decoder. These computations correspond to nodes at depth N into the tree. Let $F_N(L)$ denote the (unconditioned) probability that at least L computations are performed by the genie-aided decoder at a distance of N from the origin of the tree. Then

$$F_N(L) \geq \sum_{\text{all } \mathbf{r}} \sum_{i \in I_s(\mathbf{r})} \frac{1}{M} P[\mathbf{r}/\mathbf{y}_i]. \tag{13}$$

Now, for each $\mathbf{r}$, $I_B(\mathbf{r})$ and $I_s(\mathbf{r})$ each specify a list of $M - L$ codewords $\{\mathbf{y}_i\}$. Since $I_B(\mathbf{r})$ specifies the $M - L$ least likely codewords, we have

$$\sum_{i \in I_s(\mathbf{r})} P[\mathbf{r}/\mathbf{y}_i] \geq \sum_{i \in I_B(\mathbf{r})} P[\mathbf{r}/\mathbf{y}_i]. \tag{14}$$

Consequently,

$$\begin{aligned} F_N(L) &\geq \sum_{\text{all } \mathbf{r}} \sum_{i \in I_B(\mathbf{r})} \frac{1}{M} P[\mathbf{r}/\mathbf{y}_i] \\ &= P_e(L) \\ &\geq L^{-\rho} \exp\{-O_4(\sqrt{\ln L})\}. \end{aligned} \tag{15}$$

This expression places in evidence the Paretian behavior of the distribution of computation of a sequential decoder and provides a stronger, more generalized version of the result obtained for the erasure channel of Section II.

The expression (15) provides the key to obtaining a variety of useful lower bounds. In one case, for example, a system may be designed to permit L or fewer computations on each block of Λ consecutive letters before moving on to the next block. Then $P[C > L]$ is the probability that a block remains undecoded. If blocks that cannot be decoded in real time are set aside to be decoded at leisure, then $P[C > L]$ is the fraction of blocks that cannot be decoded in real time.

An alternate system provides a fixed buffer to hold incoming data and permits computation to proceed, without any fixed upper limit L on the number of computations permitted for each Λ consecutive letters. This system will fail when the buffer overflows. The probability of this occurrence is analyzed in detail in the next section.

Finally, analysis of the queuing of the data in the buffer generally requires information about the moments of the distribution of the number of computations per information bit (the "service time" per bit). Moments of this distribution are treated in Section VI.

V. Probability of Memory Overflow or Error

In this section, we lower-bound the probability $P_0(\Lambda)$ that a sequential decoder with a fixed buffer size will either overflow its memory buffer or make one or more errors before completely decoding a block of Λ information digits. It is assumed that the digits arrive periodically from the channel and that the decoder includes a memory capable of storing data associated with Γ information bits. We further assume that the received data for each information bit must remain in storage from the time of arrival of that data until the bit is processed; that is, until a final decision is made on the bit.

If at least k information bits are not decoded during any time interval in which data for $k + \Gamma$ information bits are received from the channel, the decoder must be more than Γ information bits behind, and the memory must overflow. Denote by μ the maximum number of computations which the decoder can perform in a time interval equal to the interarrival time of the digits in the information sequence. The decoder cannot perform more than $\mu(k + \Gamma)$ computations in a time interval during which the $k + \Gamma$ information bits come out of the channel. Hence, whenever the algorithm requires more than $\mu(k + \Gamma)$ computations to decode k information bits, buffer overflow occurs.

We now show that the lower bound (15) on the number of computations performed by the genie-aided decoder can be used to lower-bound $P_0(\Lambda)$, the probability of overflow or error achieved by any unaided sequential decoder with memory size Γ while decoding a block of Λ information bits.

We first must choose an argument L and the appropriate N for the probability function $F_N(L)$. We set

$$L = \mu(k + \Gamma) \tag{16}$$

in which k is the number of information bits corresponding to N channel symbols. At rate $R \geq R_\rho$ nats per channel symbol,[2]

$$k = [NR/\ln 2]^+. \tag{17}$$

Here N is that integral multiple of V for which (6) and (7) are satisfied. Note that in this case both the left and right sides of (6) and (7) vary with N.

We next divide the block of Λ bits into $[\Lambda/k]^-$ k-bit (N channel-symbol) sub-blocks. Assume first that the friendly genie is again available and that he informs the decoder whenever it is about to exit from a sub-block on the wrong path. The decoder then returns into the sub-block, searching for the next acceptable path, and eventually finding the correct one. Thus the genie-aided decoder invariably enters each of the $[\Lambda/k]^-$ sub-blocks on the correct path. Next observe that the assumption of an empty memory at the beginning of each sub-block reduces the probability of overflow. We conclude that, for the genie-aided decoder, the probability of memory overflow while decoding any one of the $[\Lambda/k]^-$ sub-blocks, say λ, is under-bounded by the probability that $\mu(\Gamma + k)$ or more computations are required to locate the correct path through the sub-block. Thus,

$$\lambda \geq F_N(L) \tag{18}$$

where L and N are defined (together with k) in the discussion encompassing (16) and (17).

We next note that, since the channel was assumed memoryless, the noise disturbing any sub-block is independent of the noise disturbing all other sub-blocks. We conclude that, for the genie-aided decoder operating on a block of Λ bits, the probability of error is zero, and the probability of memory overflow in at least one sub-block $P_G(\Lambda)$ is

$$\begin{aligned} P_G(\Lambda) &= 1 - (1 - \lambda)^{[\Lambda/k]^-} \\ &\geq [\Lambda/k]^-\lambda - \binom{[\Lambda/k]^-}{2}\lambda^2 \\ &\geq \left(\frac{\Lambda}{k} - 1\right)\lambda\left(1 - \frac{\Lambda\lambda}{2k}\right). \end{aligned} \tag{19}$$

[2] Note that $NR/\ln 2 = N/V \log_2 u$. Since N is chosen to be an integral multiple of V, the notation $[\;]^+$ in (17) is necessary only if u, the number of branches per node, is not a power of 2.

It remains to exorcise the genie. Observe that whenever the genie-aided decoder has a memory overflow, so must the unaided decoder, unless it makes an error. (In the absence of error, the unaided decoder performs all of the computations required of the genie-aided decoder, and in addition it must perform computations in subsequent sub-blocks whenever it exits from a given sub-block along an incorrect path. These additional computations serve the same function as the genie, ultimately informing the decoder, in the absence of error, that it exited from the sub-block on the wrong path.) Thus the probability of overflow for the genie-aided decoder $P_G(\Lambda)$ is a lower bound to the probability of overflow or error for the unaided decoder $P_0(\Lambda)$,

$$P_0(\Lambda) \geq P_G(\Lambda). \tag{20}$$

The rest is algebraic manipulation. From (7), (16), and (17), we observe that

$$\ln L > (N - V)(R_\rho - \hat{E}_0'(\rho)) - O_1(\sqrt{N - V}) \tag{21}$$

and

$$\begin{aligned} \ln L &= \ln \{\mu\Gamma + \mu[NR/\ln 2]^+\} \\ &\leq \ln \mu\Gamma + \ln\left\{1 + \frac{1}{\Gamma}\left(\frac{NR}{\ln 2} + 1\right)\right\} \\ &\leq \ln \mu\Gamma + \frac{NR}{\Gamma \ln 2} + 1. \end{aligned} \tag{22}$$

Combining (21) and (22) and grouping terms involving N, we obtain

$$N\left(R_\rho - \hat{E}_0'(\rho) - \frac{R}{\Gamma \ln 2}\right) - O_1(\sqrt{N - V}) \leq \ln \mu\Gamma + V(R_\rho - E_0'(\rho)) + 1$$

so that

$$N = O_5(\ln \mu\Gamma) \tag{23}$$

whenever

$$\Gamma > \frac{R}{(R_\rho - \hat{E}_0'(\rho)) \ln 2}, \tag{24}$$

a condition which is satisfied in all cases of interest. From (17) and (23), the number of information bits in each sub-block satisfies

$$k = \left[\frac{NR}{\ln 2}\right]^+ = O_6(\ln \mu\Gamma). \tag{25}$$

From (15), (16), (18), and (25) we obtain for the probability of overflow in any one sub-block

$$\begin{aligned} \lambda &\geq L^{-\rho} \exp\{-O_4(\sqrt{\ln L})\} \\ &= (\mu\Gamma)^{-\rho}\left(1 + \frac{k}{\Gamma}\right)^{-\rho} \\ &\quad \cdot \exp\left\{-O_4\left(\sqrt{\ln \mu\Gamma + \ln\left[1 + \frac{k}{\Gamma}\right]}\right)\right\} \end{aligned} \tag{26}$$

$$= (\mu\Gamma)^{-\rho} \exp\left\{-\rho \ln\left(1 + \frac{O_6(\ln \mu\Gamma)}{\Gamma}\right) - O_4\left(\sqrt{\ln \mu\Gamma + \ln\left(1 + \frac{O_6(\ln \mu\Gamma)}{\Gamma}\right)}\right)\right\}$$

$$\geq (\mu\Gamma)^{-\rho} \exp\{-O_7(\sqrt{\ln \mu\Gamma})\}.$$

Let us assume a minimum size for the block length Λ

$$\Lambda > 2k = 2\, O_6(\ln \mu\Gamma) \tag{27}$$

and a (nontrivial) value for λ such that

$$\frac{\lambda\Lambda}{2k} \leq \frac{1}{2}. \tag{28}$$

From (19), (20), (25)–(28) we obtain our desired result

$$\begin{aligned} P_0(\Lambda) &\geq P_G(\Lambda) \\ &\geq \left(\frac{\Lambda}{k} - 1\right)\lambda\left(1 - \frac{\Lambda\lambda}{2k}\right) \\ &\geq \frac{\lambda\Lambda}{2k}\left(1 - \frac{\Lambda\lambda}{2k}\right) \geq \frac{\lambda\Lambda}{4k} \\ &\geq \Lambda(\mu\Gamma)^{-\rho}\exp\{-O_7(\sqrt{\ln \mu\Gamma}) - \ln(4O_6(\ln \mu\Gamma))\} \\ &\geq \Lambda(\mu\Gamma)^{-\rho}\exp\{-O_8(\sqrt{\ln \mu\Gamma})\}. \end{aligned} \tag{29}$$

We thus find that the probability of error or overflow for a sequential decoder must increase at least linearly with block length for a fixed $R \geq R_\rho$, memory size Γ, and computer speed μ.

VI. Moments of Computation

The average number of computations c required to decode a block of length Λ bits is defined as the total number of computations used in decoding the block divided by Λ. Exploiting (15), we obtain lower bounds to the tth moment of computation $\overline{c^t}$.

First observe that, if the number of computations performed on the first k bits for any $k \leq \Lambda$ is at least $l\Lambda$, then $c \geq l$. Letting $L = l\Lambda$ and choosing N to satisfy (6) and (7), we have

$$P[c > l] \geq F_N(l\Lambda) \tag{30}$$

whenever

$$k = \frac{RN}{\ln 2} \leq \Lambda. \tag{31}$$

Let $\hat{l}$ be the smallest value of l for which (31) is violated. Recalling from (10) that N is of the order of $\ln L$, we have

$$\begin{aligned} N = O(\ln l\Lambda) &\leq \Lambda \quad \text{for} \quad 1 \leq l < \hat{l} \\ &> \Lambda \quad \text{for} \quad l = \hat{l}. \end{aligned} \tag{32}$$

For later use, we observe that, for some constant $\beta > 0$,

$$\hat{l} > e^{\beta\Lambda}. \tag{33}$$

We next write the tth moment of c as

$$\begin{aligned} \overline{c^t} &= \sum_{l=1}^{\infty} l^t P[c = l] \\ &= \sum_{l=1}^{\infty} [l^t - (l-1)^t] P[c \geq l] \\ &= \sum_{l=1}^{\infty} l\left[1 - \left(1 - \frac{1}{l}\right)^t\right] l^{t-1} P[c \geq l]. \end{aligned}$$

Let $x = 1/l$. For $0 \leq x \leq 1$,

$$\begin{aligned} 1 - x \leq (1-x)^t \leq 1 - tx \quad &\text{if} \quad 0 < t \leq 1 \\ 1 - tx \leq (1-x)^t \leq 1 - x \quad &\text{if} \quad 1 \leq t. \end{aligned}$$

Consequently, for $t > 0$,

$$\overline{c^t} \geq b_t \sum_{l=1}^{\infty} l^{t-1} P[c \geq l] \tag{34}$$

in which

$$b_t = \begin{cases} t & \text{if} \quad 0 < t \leq 1 \\ 1 & \text{if} \quad t \geq 1 \end{cases}. \tag{35}$$

If $R \geq R_\rho$, (30), (34), and (15) yield

$$\begin{aligned} \overline{c^t} &\geq b_t \sum_{l=1}^{\hat{l}-1} l^{t-1} F_N(l\Lambda) \\ &\geq b_t \sum_{l=1}^{\hat{l}-1} l^{t-1} (l\Lambda)^{-\rho} \exp -O_4(\sqrt{\ln l\Lambda}). \end{aligned} \tag{36}$$

Using this inequality, we may prove the following theorem.

Theorem

If $R > R_t$, then $\lim_{\Lambda\to\infty} \overline{c^t} = \infty$.

Remarks

Using random coding, Savage,[10] Yudkin,[13] and Falconner[4] have shown that, for all values of $t \geq 0$ and for the Fano algorithm as described in Chapter 6 of Wozencraft and Jacobs,[12] the tth moment of computation $\overline{c^t}$ is bounded above by a constant independent of Λ whenever $R < \tilde{R}_t$. Here, $\tilde{R}_\rho \triangleq E_0(\rho)/\rho$. Thus the rate above which the tth moment of computation diverges and below which it can be bounded is now uniquely determined as $R = R_t = \tilde{R}_t$ for all memoryless channels for which $\hat{E}_0(\rho) = E_0(\rho)$. When these two functions differ, we conjecture that the correct answer is given by $\hat{E}_0(\rho)/\rho$, but this has not yet been conclusively demonstrated.

In particular, the "computational cutoff rate" R_{comp}, defined as the supremum of rates for which the first moment of computation is bounded independent of Λ, is bounded by $E_0(1) \leq R_{\text{comp}} \leq \hat{E}_0(1)$.

Proof: Select ρ such that $R \geq R_\rho > R_t$, which implies that

$$t - \rho > 0. \tag{37}$$

For some constants γ' and γ (independent of l and Λ), the exponential factor in (36) is bounded by

$$\exp -O(\sqrt{\ln l\Lambda}) > \exp\{-\gamma'\sqrt{\ln l\Lambda}\} > \gamma(l\Lambda)^{-(t-\rho)/2}. \tag{38}$$

By (38), the bound (36) on the tth moment of computation becomes

$$\overline{c^t} > b_t\gamma\Lambda^{-(t+\rho)/2}\sum_{l=1}^{\hat{l}-1} l^{-1+(t-\rho)/2}. \tag{39}$$

However,

$$\sum_{l=1}^{\hat{l}-1} l^{-1+\alpha} \geq \int_1^{\hat{l}} x^{-1+\alpha}\,dx = (\hat{l}^\alpha - 1^\alpha)/\alpha. \tag{40}$$

Together, (39) and (40) yield

$$\overline{c^t} > b_t\gamma\left(\frac{2}{t-\rho}\right)\Lambda^{-(t+\rho)/2}(\hat{l}^{(t-\rho)/2} - 1) \tag{41}$$

which, in view of the exponential growth of $\hat{l}$ with Λ evidenced in (33), completes the proof of the theorem.

VII. Conclusion

We conclude that the behavior of systems using sequential decoding is limited by the Paretian distribution of computation as given in (15). In Section V we used this bound to calculate lower bounds to the probability of buffer overflow, and in Section VI we used this bound to show that, if $R > R_\rho = E_0(\rho)/\rho$, then the ρth moment of the distribution of computation/decoded digit (as well as the ρth moment of the distribution of the computation/decoded block) diverges.

These results show that the distribution of computation, rather than the probability of decoding error, is the limiting factor in the performance of systems using sequential decoding. These results also show that the average number of computations per decoded digit may be a misleading figure, because higher moments of the distribution typically diverge, and the normal law of large numbers may not apply. The implications of this phenomenon for systems using sequential decoding are discussed in Wozencraft and Jacobs.[12]

This unfortunate behavior of the distribution of computation is apparently peculiar to sequential decoding. For certain special channels, there are known algebraic decoding methods which do not suffer from these restrictions. Some of these methods were noted in Section II; others are described by Berlekamp.[2] For these reasons, one is led to investigate hybrid decoding schemes which combine algebraic decoding's well-behaved distribution of computation with sequential decoding's exponentially decreasing error probability and independence of the detailed properties of the channel. Such schemes have been suggested by Pinsker[9] and by Falconer,[5] and there is reason to hope that further research may provide even more attractive decoding methods.

References

[1] E. R. Berlekamp, "A class of convolution codes," *Information and Control*, vol. 6, pp. 1–13, 1963.

[2] ——, "Practical BCH decoders" (to be published).

[3] M. A. Epstein, "Algebraic decoding for a binary erasure channel," M.I.T. Research Lab. of Electronics, Cambridge, Mass., Tech. Rept. 340, 1958.

[4] D. D. Falconer, "An upper bound on the distribution of computation for sequential decoding with rate above R_{comp}," M.I.T. Research Lab. of Electronics, Cambridge, Mass., Quart. Prog. Rept. 81, pp. 174–179, 1966.

[5] ——, "A hybrid sequential and algebraic decoding scheme," Ph.D. dissertation, Massachusetts Institute of Technology, Cambridge, February 1967.

[6] R. G. Gallager, "A simple derivation of the coding theorem and some applications," *IEEE Trans. on Information Theory*, vol. IT-11, pp. 3–18, January 1965.

[7] B. V. Gnedenko and A. N. Kolmogorov, *Limit Distributions for Sums of Independent Random Variables.* London: Addison-Wesley, 1954.

[8] I. M. Jacobs, "Probabilities of overflow and error in a coded PSK system with sequential decoding," Jet Propulsion Lab., Pasadena, Calif., SPS 37–33, vol. 4, 1965.

[9] M. Pinsker, paper presented at *1966 Internat'l Symp. on Information Theory*, Los Angeles, Calif., February 2, 1966.

[10] J. E. Savage, "The distribution of the sequential decoding computation time," *IEEE Trans. on Information Theory*, vol. IT-12, pp. 143–147, April 1966.

[11] C. E. Shannon, R. G. Gallager, and E. R. Berlekamp, "Lower bounds to error probability for coding on discrete memoryless channels," *Information and Control* (to be published).

[12] J. M. Wozencraft and I. M. Jacobs, *Principles of Communication Engineering.* New York: Wiley, 1965.

[13] H. L. Yudkin, unpublished correspondence, 1965.

[14] B. Mandelbrot, "Self-similar error clusters in communication systems and the concept of conditional stationarity," *IEEE Trans. on Communication Technology*, vol. COM-13, pp. 71–90, March 1965.

НЕКОТОРЫЕ ПОСЛЕДОВАТЕЛЬНЫЕ ПРОЦЕДУРЫ ДЕКОДИРОВАНИЯ

К. Ш. Зигангиров

Предлагается несколько последовательных алгоритмов декодирования случайных древовидных кодов; сообщения передаются по бинарному симметричному каналу. Получены приближенные выражения для вероятности неправильного декодирования и среднего числа операций, затраченных на декодирование одного символа.

1. Как известно, общая процедура декодирования по методу максимума правдоподобия требует числа операций такого же порядка, что и число сообщений, что делает практически невозможной ее техническую реализацию. Предложено несколько модификаций последовательного декодирования [1—4], которые могут быть реализованы на несложных специализированных вычислительных машинах.

Хотя предлагаемые ниже четыре процедуры построены на других принципах (ближе всего к ним процедура Фано [2, 3]), мы сохраним за ними название — последовательное декодирование. Эти процедуры позволяют при всех скоростях, меньших $R_{\text{выч}}$, получать в среднем конечное число операций на декодированный символ. Применяемый здесь математический аппарат конечноразностных уравнений позволяет получить довольно точные оценки для характеристик процедур.

Пусть информационные сообщения представляют собой последовательности равновероятных и статистически независимых (информационных) символов 0 и 1. Сообщения закодированы случайным древовидным кодом [1], скорость передачи — R. Для простоты ниже рассматривается только случай, когда $1/R = m$ — целое. В общем случае получаются аналогичные результаты. Сигналы, представляющие собой также последовательности 0 и 1, передаются по бинарному симметричному каналу без памяти: вероятность правильной передачи символа по каналу $q(>0{,}5)$, вероятность неправильной передачи $p = 1 - q$.

Введем некоторые определения. Ветвью кодового дерева длины n будем называть последовательность n символов в канале, соответствующую некоторому сообщению длины Rn. Ребром кодирующего дерева называется совокупность m символов в канале, соответствующих одному информационному символу. Узлами кодирующего дерева будем называть конечные символы в ребрах. Правильным поддеревом будем называть совокупность ветвей, у которых первый информационный символ совпадает с переданным, неправильным поддеревом — совокупность ветвей, у которых первый информационный символ отличен от переданного.

Одной операцией при декодировании будем считать комплекс вычислений, связанных со сравнением одного ребра кодирующего дерева с соответствующим отрезком принятого сигнала. Такое определение операции носит условный характер. Более правильно число операций при декодировании определять числом сдвигов регистра в декодирующем устройстве. Однако следует отметить, что обработка одного ребра кодирующего дерева, как правило, связана с одним и реже с несколькими сдвигами регистра.

Среди характеристик процедуры декодирования наибольший интерес представляют вероятность ошибочного декодирования символа $P(\varepsilon)$ и среднее число операций на декодированный информационный символ N (если оно существует). Целесообразно минимизировать одну из этих величин при условии, что другая задана. Необходимо также учитывать, чтобы емкость памяти декодера M и задержка декодирования τ (число просмотренных принятых канальных символов, следующих за последним декодированным ребром в момент декодирования следующего ребра) не вышла за разумные пределы.

В предлагаемых ниже процедурах большую роль играет величина T, которую мы назовем «функцией правдоподобия». Определим функцию правдоподобия T_i для i-й ветви кодирующего дерева длины n_i следующим образом:

$$T_i = \frac{q^{k_i} p^{n_i - k_i}}{\left(\frac{1}{2}\right)^{n_i}} \left(\frac{1}{2}\right)^{n_i R}, \qquad (1)$$

где k_i — число символов принятого сигнала, совпадающих с символами i-й ветви. Обозначим $z_i = \log_2 T_i = \alpha k_i + \beta(n_i - k_i)$, где $\alpha = \log_2 q + 1 - R$, $\beta = \log_2 p + 1 - R$. Отметим, что функция z с точностью до знака совпадает со скошенным расстоянием («tilted» distance), используемым в процедуре Фано [3]. Совокупность статистик T_j для всех просмотренных ветвей при заданных априорных вероятностях позволяет однозначно вычислить апостериорные вероятности того, что передана одна из этих ветвей. Вероятность π_i передачи i-й ветви при априори равновероятных сообщениях равна $\pi_i = T_i / \sum_j T_j$, где суммирование в знаменателе производится по всем обработанным ветвям.

В дальнейшем будем рассматривать только величины z, относящиеся к узлам кодирующего дерева. Для определенности будем считать, что переданному сообщению соответствуют ветви (разной длины) с индексами 0. Общий принцип, объединяющий рассматриваемые процедуры, состоит в том, что в любой момент производится сопоставление (оно состоит в вычислении z) принятого сообщения и той ветви, для которой значение z максимально среди всех, хранящихся в памяти декодера. В случае, если несколько ветвей имеют максимальное z, обрабатываемая ветвь выбирается среди них случайно с одинаковой вероятностью для каждой. Различными являются лишь правила окончания декодирования символа и объем запоминаемых данных. Процедуры построены на тех же принципах, что и последовательный анализ Вальда [5], и сходны с рассмотренными в [6] процедурами поиска при обнаружении сигналов.

Будут рассмотрены следующие процедуры последовательного декодирования: декодирование с фиксированной вероятностью ошибки и декодирование с фиксированной задержкой без ограничений на емкость памяти декодера и те же процедуры с ограничениями на емкость памяти декодера.

2. Рассмотрим первую процедуру — декодирование с фиксированной вероятностью ошибки без ограничений на емкость памяти декодера. В этом случае декодирование первого символа заканчивается, как только для одной из ветвей значение z превзойдет порог a. Предполагается, что в процессе декодирования декодер запоминает величины z для всех обработанных ветвей — ограничения на емкость памяти декодера не налагаются. После того как произошло превышение порога для одной из ветвей, оставляются лишь ветви, принадлежащие поддереву, к которому относится эта ветвь. Остальные ветви отбрасываются. Второй символ декодируется аналогично.

Опишем процедуру подробнее. В начальный момент всем ветвям приписывается значение логарифма «функции правдоподобия», равное нулю. Обрабатываемое ребро выбирается среди двух начальных случайно с равными вероятностями. Пусть для определенности выбрано ребро, соответствующее первому информационному символу 1. Если после обработки соответствующее значение z окажется меньше нуля, то производится обработка ребра, соответствующего первому информационному символу 0. Если величина z, соответствующая информационному символу 0, окажется больше величины z, соответствующей информационному символу 1, то для обработки выбираются либо ветви, которые имеют первые символы 00, либо ветви с первыми символами 01. Если между полученными z соблюдается противоположное неравенство, то для обработки выбираются ветви, начинающиеся символами 10, 11, если значения z для выбранных ветвей равны, то ветвь для обработки выбирается случайно среди ветвей, начинающихся символами 00, 01, 10, 11.

Рассмотрим второй случай, когда в результате обработки ребра, соответствующего первому информационному символу 1, соответствующее ему z окажется больше нуля. В этом случае производится обработка второго ребра у сообщений, которые начинаются либо символами 10, либо символами 11. Дальнейшая обработка производится таким образом, чтобы обрабатываемой ветви соответствовало значение z, максимальное среди всех, хранящихся в памяти декодера.

Отметим, что в памяти машины хранятся величины z, соответствующие тем обработанным узлам кодирующего дерева, для которых либо не производилась обработка выходящих из них ребер, либо производилась обработка только одного ребра. Логарифмы «функции правдоподобия» узлов, у которых произведена обработка двух исходящих из них ребер, «сбрасываются» и в дальнейшем сравнении z не участвует.

3. Найдем верхнюю границу для вероятности ошибочного декодирования первого символа. При этом будем использовать конечноразностные уравнения, аналогичные тем, которые используются при получении характеристик последовательного анализа.

Рассмотрим неправильное поддерево. Пусть ξ — значение z, запомненное декодером для одного из узлов поддерева, $L(\xi)$ — вероятность того, что хотя бы для одной из ветвей поддерева, проходящих через данный узел, при дальнейшей обработке z превзойдет порог a ранее, чем достигнет некоторого значения $b(<0)$. Так как кодирование носит чисто случайный характер и передача сигналов производится по каналу без памяти, L зависит только от ξ.

Тогда, рассмотрев соотношение между ξ и значением z в следующих узлах поддерева, получим

$$1 - L(\xi) = \sum_{k=0}^{m} C_m^k (1/2)^m \{1 - L[\xi + k\alpha + (m-k)\beta]\}^2. \qquad (2)$$

При решении классических конечноразностных уравнений такого типа необходимо иметь не менее $2m$ граничных условий. Легко получить два следующих граничных условия:

$$M_1 L(\xi) = 1, \quad M_2 L(\xi) = 0, \qquad (3)$$

где $M_1 f(\xi)$ означает математическое ожидание функции f от тех значений величины z, которые она принимает после превышения порога a, а $M_2 f(\xi)$ — математическое ожидание f от значений величины z, которые она принимает после выхода за порог b.

Reprinted with permission from *Probl. Peredach. Inform.*, vol. 2, no. 4, pp. 13–25, 1966.

Остальные условия являются следствием следующих свойств функции $L(\xi)$: функция $L(\xi)$ монотонно возрастает при возрастании ξ; решение уравнения (2) при фиксированном ξ монотонно возрастает, если a или b убывает.

После выхода за верхний порог величина z принимает значение на отрезке $[a, a + m\alpha)$, после выхода за нижний порог — на отрезке $(b + m\beta, b]$. Допустим, что в момент перехода верхней границы z принимает значение, равное a, а в момент перехода нижней границы — значение, равное $b + m\beta$. Тогда условия (3) примут вид

$$L(a) = 1, \quad L(b + m\beta) = 0. \tag{4}$$

Поскольку a и $b + m\beta$ заведомо не превышают значений z, принимаемых после выхода за верхнюю и нижнюю границы, соответственно, решение уравнения (2) с граничными условиями (4) мажорируют сверху решение уравнения (2) с граничными условиями (3).

Уравнение (2) удобно переписать в виде

$$L(\xi) = 2\sum_{k=0}^{m} C_m^k (1/2)^m L[\xi + k\alpha + (m+k)\beta] -$$
$$- \sum_{k=0}^{m} C_m^k (1/2)^m L^2[\xi + k\alpha + (m-k)\beta].$$

Отсюда

$$L(\xi) \leqslant 2\sum_{k=0}^{m} C_m^k (1/2)^m L[\xi + k\alpha + (m-k)\beta]. \tag{5}$$

Рассмотрим сначала решение уравнения

$$L_0(\xi) = 2\sum_{k=0}^{m} C_m^k (1/2)^m L_0[\xi + k\alpha + (m-k)\beta], \tag{6}$$

$$L_0(a) = 1, \quad L_0(b + m\beta) = 0. \tag{7}$$

Общее решение уравнения (6) запишется в виде

$$L_0(\xi) = \sum_i A_i 2^{\lambda_i \xi},$$

где A_i — произвольные постоянные, а λ_i — действительные (в силу монотонности функции L) корни характеристического уравнения

$$1 = 2^{1-m}(2^{\lambda\alpha} + 2^{\lambda\beta})^m.$$

Два единственных действительных корня λ удовлетворяют уравнению (при $R = 1/m$)

$$2^{1-R} = 2^{\lambda\alpha} + 2^{\lambda\beta}, \tag{8}$$

где один из корней $\lambda_0 = 1$, другой корень при $R < C$ удовлетворяет неравенству $1 > \lambda_1 > 0$.

Решение уравнения (6) с граничным условием (7) имеет вид

$$L_0(\xi) = \frac{2^\xi - 2^{b+m\beta} + 2^{\lambda_1\xi} - 2^{\lambda_1(b+m\beta)}}{2^a - 2^{b+m\beta} + 2^{\lambda_1 a} - 2^{\lambda_1(b+m\beta)}}. \tag{9}$$

Непосредственная проверка показывает, что при любых $\xi \in [b, a]$

$$L_0(\xi) \leqslant 2^{\xi - a}. \tag{10}$$

Докажем, что $L(\xi)$ мажорируется сверху функцией $L_0(\xi)$ *. Вычитая (6) из (5) и заменяя $u(\xi) = L(\xi) - L_0(\xi)$, получим

$$u(\xi) \leqslant 2\sum_{k=0}^{m} C_m (1/2)^m u[\xi + k\alpha + (m-k)\beta], \tag{11}$$
$$u(a) = 0, \quad u(b + m\beta) = 0.$$

Сделаем замену $u(\xi) = 2^{\lambda\xi} v(\xi)$, где λ удовлетворяет неравенству $\lambda_1 < \lambda < 1$. Тогда

$$v(\xi) \leqslant 2\sum_{k=0}^{m} C_m^k \left(\frac{2^{\lambda\alpha}}{2}\right)^k \left(\frac{2^{\lambda\beta}}{2}\right)^{m-k} v[\xi + k\alpha + (m-k)\beta], \tag{12}$$

$$v(a) = 0, \quad v(b + m\beta) = 0.$$

Можно показать, что при $\lambda \in (\lambda_1, 1)$

$$2\sum_{k=0}^{m} C_m^k \left(\frac{2^{\lambda\alpha}}{2}\right)^k \left(\frac{2^{\lambda\beta}}{2}\right)^{m-k} < 1. \tag{13}$$

Допустим, что существует $\xi \in (b + m\beta, a)$ такое, что $v(\xi) > 0$. Тогда $\max\limits_{\xi \in [b+m\beta, a]} v(\xi) = v(\xi_0) > 0$. Из (12) и (13) получим

$$v(\xi_0) < \max_{\xi \in [b+m\beta, a]} v(\xi) = v(\xi_0).$$

* Идея доказательства принадлежит Р. З. Хасьминскому.

Полученное противоречие доказывает, что $v(\xi) \leqslant 0$ при $\xi \in [b + m\beta, a]$. Отсюда $L(\xi) \leqslant L_0(\xi) \leqslant 2^{\xi - a}$.

Таким образом, вероятность того, что при обработке хотя бы для одной ветви неправильного поддерева z достигнет порогового значения a ранее, чем некоторого значения b при любых $b(<0)$, не превышает $L_0(0)$, а следовательно, 2^{-a}. Отсюда следует, что вероятность ошибочного декодирования первого символа, также не превышает 2^{-a}. Аналогично, вероятность ошибочного декодирования второго символа, при условии правильного декодирования первого символа, не превышает 2^{-a}. Таким образом,

$$P(\varepsilon) < 2^{-a}. \tag{14}$$

4. Найдем теперь оценки для среднего числа операций, затраченных на обработку неправильного поддерева при условии правильного декодирования первого символа.

Рассмотрим совокупность значений логарифмов «функции правдоподобия» для ветвей, соответствующих переданному сообщению, длины $0, m, 2m, \ldots, -z_0(0), z_0(m), z_0(2m), \ldots$. Обозначим $\eta = \min z_0(im)$. Естественно, что обработка неправильного поддерева прекращается, как только все значения z, соответствующие ветвям неправильного поддерева, окажутся меньше η.

Пользуясь разностными уравнениями для функции распределения минимальной координаты случайно блуждающей величины, приращение которой принимает с вероятностью q значение α и с вероятностью p — значение β [7], легко показать, что функция распределения величины η ограничена сверху (при $R < C$)

$$P(\eta < x) \leqslant \begin{cases} 1 & \text{при } x \geqslant 0, \\ 2^{-h_1 x} & \text{при } x < 0, \end{cases} \tag{15}$$

где h_1 — корень (< 0) уравнения

$$1 = q2^{h\alpha} + p2^{h\beta}. \tag{16}$$

Интересно отметить, что корни характеристических уравнений (8) и (16) связаны соотношением: $\lambda = 1 + h$.

Рассмотрим неправильное поддерево. Пусть ξ — значение z, запомненное декодером для одного из узлов поддерева; $N_\eta(\xi)$ — среднее число операций, затраченное на анализ участков ветвей, проходящих через данный узел и начинающихся данным ребром до выхода на один из порогов a или η. Тогда

$$N_\eta(\xi) = 2\sum_{k=0}^{m} C_m (1/2)^m N_\eta[\xi + k\alpha + (m-k)\beta] + 1, \tag{17}$$

$$N_\eta(a) = 0, \quad N_\eta(\eta) = 0. \tag{18}$$

Здесь, мы пренебрегли «перескоком» через границу. Учет явлений «перескока» увеличивает среднее число операций, даваемое решением данного уравнения. Оценка сверху для среднего числа операций с учетом «перескока» рассмотрена ниже.

Решая уравнение (17), при $2^a \gg 1$ получим

$$N_\eta(\xi) \approx 2^{-\lambda_1(\eta - \xi)} - 1. \tag{19}$$

Усредняя по возможным положениям η, согласно (15), получим

$$N_\eta(0) \approx \frac{h_1}{1 + 2h_1} - 1 \quad \left(\text{при } h_1 < \frac{1}{2}\right). \tag{20}$$

Если доопределить процедуру, прекращая декодирование символа не только тогда, когда хотя бы одно из значений z станет больше a, но и когда все значения z станут меньше a/h_1 (при этом, по-прежнему, $P(\varepsilon) = O(2^{-a})$), получим

$$N(0) \approx a \ln 2 - 1 \quad \left(\text{при } h_1 = -\frac{1}{2}\right) \tag{21}$$

и

$$N(0) \approx \frac{-h_1}{1 + 2h_1}\left[2^{-\frac{1+2h_1}{h_1}a} - 1\right] - 1 \quad \left(\text{при } h_1 > -\frac{1}{2}\right). \tag{22}$$

Из уравнения (16) найдем значение R, соответствующее $h_1 = -1/2$. Оно равно: $R_{\text{выч}} = 1 - \log_2 [1 + \sqrt{4pq}]$.

Рассмотрим ветви, у которых первые $(k-1)$ информационных символов совпадают с переданными, а k-й символ отличен от переданного. Легко показать, что среднее число операций, затраченных на анализ отрезков этих ветвей, начинающихся k-м ребром (при условии правильного декодирования), также приближенно определяется формулами (20) — (22).

Учитывая сказанное, нетрудно показать, что среднее число операций, затрачиваемое на декодирование символа, асимптотически определяемое как отношение среднего числа операций, затраченных на декодирование первых k информационных символов к k, при больших k равно (при $2^{-a} \ll 1$):

$$N \approx \frac{h_1}{1 + 2h_1} \text{ при } R < R_{\text{выч}}, \tag{23}$$

$$N \approx a \ln 2 \text{ при } R = R_{\text{выч}}, \tag{24}$$

$$N \approx \frac{-h_1}{1 + 2h_1} 2^{-\frac{1+2h_1}{h_1}a} \text{ при } R > R_{\text{выч}}. \tag{25}$$

Таким образом, при $R < R_{\text{выч}}$ среднее число операций на декодированный символ, независимо от $P(\varepsilon)$, приближенно определяется формулой (23)

Как отмечалось выше, уравнение (17) не учитывает явлений «перескока» через границу в момент окончания обработки ветви. Поэтому среднее число операций, определяемое (23)—(25), фактически является оценкой снизу (довольно близкой к истинному значению) для среднего числа операций, затраченных на декодирование символа.

Используя стандартную методику [5], можно получить оценку сверху для среднего числа операций, затраченных на декодирование символа. При выводе используется то обстоятельство, что в момент окончания обработки ветви z принимает значение, принадлежащее отрезку $[\eta + m\beta, \eta]$. Отсюда, например,

$$N < 2^{-\frac{1}{R}\beta(1+h_1)} \frac{h_1}{1+2h_1} \text{ при } R < R_{\text{выч}}. \qquad (26)$$

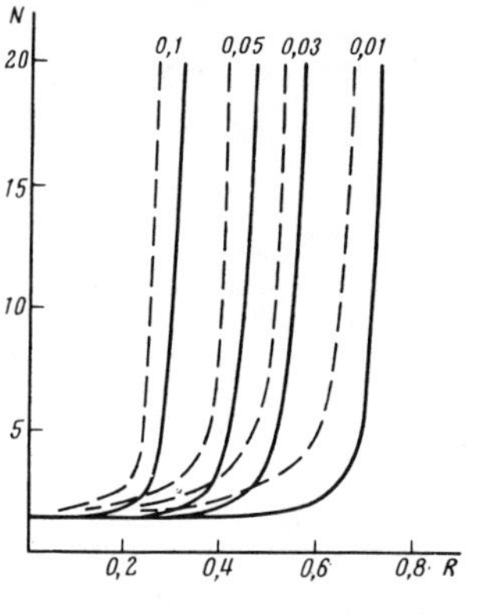

Рис. 1. График зависимости среднего числа операций на декодированный символ N от скорости передачи R для каналов с вероятностью искажения символа p = 0,1; 0,05; 0,03; 0,01. Сплошные кривые — для первой и второй процедур, пунктирные — для третьей и четвертой

График среднего числа операций на декодированный символ, вычисленного по формуле (23), (в предположении, что при $R \to 0$ $N \to 1{,}5$) как функция скорости передачи R для каналов с $p = 0{,}1$; 0,05; 0,03; 0,01 представлен на рис. 1 (сплошные кривые).

5. Рассмотрим неправильное поддерево. Пусть ξ — значение z, запомненное декодером для одного из узлов этого поддерева, случайная величина $n(\xi)$ — число операций, производимых при анализе участков ветвей, начинающихся данным ребром и проходящих через данный узел, $F(s, \xi) = M[e^{sn(\xi)}]$ — производящая функция $n(\xi)$. Тогда

$$F(s, \xi) = e^s \sum_{k=0}^{m} C_m^k (1/2)^m F^2 [s, \xi + k\alpha + (m-k)\beta], \qquad (27)$$

$$F(s, \eta) = 1, \quad F(s, a) = 1.$$

Дифференцируя (27) по s и полагая $s = 0$, нетрудно получить уравнения для моментов числа операций. В частности, уравнение для второго момента $K_\eta(\xi)$ имеет вид

$$K_\eta(\xi) = 2 \sum_{k=0}^{m} C_m^k (1/2)^m K_\eta [\xi + k\alpha + (m-k)\beta] + A 2^{2\lambda_1(\xi - \eta)} - 1, \qquad (28)$$

$$K_\eta(\eta) = 0, \quad K_\eta(a) = 0. \qquad (29)$$

Здесь $A = 2^{1-m}[2^{2\lambda_1\alpha} + 2^{2\lambda_1\beta}]^m$.

Решение (28) при $a \to \infty$ существует, если $R < R_{\text{выч}}$, и имеет вид ($\xi = 0$)

$$K_\eta(0) \approx \frac{A}{1-A}[2^{-2\lambda_1\eta} - 2^{-\lambda_1\eta}] - (2^{-\lambda_1\eta} - 1). \qquad (30)$$

Усредняя по возможным положениям нижнего порога η, получим

$$K_\eta(0) \approx \frac{A}{1-A}\left[\frac{h_1}{2+3h_1} - \frac{h_1}{1+2h_1}\right] - \left[\frac{h_1}{1+2h_1} - 1\right]$$

$$\left(\text{при } h_1 < -\frac{2}{3}\right).$$

Значение R, соответствующее корню $h_1 = -2/3$,

$$R_2 = 1 - \frac{3}{2}\log_2(\sqrt[3]{p} + \sqrt[3]{q}). \qquad (31)$$

Таким образом, при $R < R_2$ дисперсия числа операций на декодированный символ конечна и равна

$$D \approx \frac{A}{1-A}\left[\frac{h_1}{3+2h_1} - \frac{h_1}{1+2h_1}\right] - \left[\frac{h_1}{1+2h_1} - 1\right] - \left[\frac{h_1+1}{1+2h_1}\right]^2 \qquad (32)$$

Из уравнения (27) следует, что при фиксированном положении нижнего порога η и $a = \infty$ момент k-го порядка ($k = 0, 1, \ldots$) для числа операций существует при скоростях $R < R_{k-1}$, где

$$R_k = 1 - \frac{k+1}{k}\log_2[p^{\frac{1}{k+1}} + q^{\frac{1}{k+1}}], \quad k = 1, 2, \ldots,$$

$$R_0 = C = 1 + p\log_2 p + q\log_2 q$$

и не существует при $R \geqslant R_{k-1}$. При этом момент k-го порядка при $\eta \to -\infty$ возрастает как $O(2^{-k\lambda_1\eta})$, причем существует B_k такое, что $B_k 2^{-k\lambda_1\eta}$ мажорирует k-й момент сверху равномерно по η.

Используя неравенства, которые можно получить, например, из неравенства Гельдера и Минковского [8],

$$\mathbf{M}(x^\gamma) \geqslant [\mathbf{M}x]^\gamma \quad \text{при } \gamma > 1,$$

$$\mathbf{M}(x^\gamma) \leqslant [\mathbf{M}x]^\gamma \quad \text{при } \gamma < 1,$$

можно показать, что момент δ-го порядка, где δ — любое положительное, возрастает при $\eta \to -\infty$ как $O(2^{-\delta\lambda_1\eta})$ и мажорируется сверху величиной $B_\delta 2^{-\delta\lambda_1\eta}$, где B_δ — постоянная. Отсюда следует, что при усреднении по возможным положениям нижнего порога η момент δ-го порядка для числа операций существует при $R < R_\delta$, где

$$R_\delta = 1 - \frac{\delta+1}{\delta}\log_2(p^{\frac{1}{\delta+1}} + q^{\frac{1}{\delta+1}}), \qquad (33)$$

и не существует при $R \geqslant R_\delta$. С учетом (16), это утверждение можно видоизменить. При любой скорости R существуют моменты порядка ниже, чем $-h_1/(1+h_1)$, и не существуют моменты порядка выше или равного $-h_1/(1+h_1)$. Из существования моментов порядка $-h_1/(1+h_1) - \varepsilon$, согласно неравенству Чебышева, следует, что

$$P(n > x) \leqslant C_1(\varepsilon) x^{\frac{h}{1+h_1} + \varepsilon}, \qquad (34)$$

где $C_1(\varepsilon) = O\left(\frac{1}{\varepsilon}\right)$ при $\varepsilon \to 0$.

6. Рассмотрим вторую процедуру декодирования с фиксированной задержкой без ограничения на емкость памяти декодера. Эта процедура применима и для сверточных кодов.

Правило окончания декодирования символа следующее: декодирование символа заканчивается, как только длина одной из ветвей достигнет величины τ. В этом случае часть ветвей отбрасывается и оставляется лишь поддерево, к которому принадлежит указанная ветвь. Аналогично декодируется второй символ. Как и в предыдущем случае, ограничений на емкость памяти декодера не налагается.

Среднее число операций, затраченных на декодирование одного символа, так же как и в предыдущей процедуре, при $P(\varepsilon) \ll 1$ приближенно определяется формулой (23).

Найдем вероятность неправильного декодирования. Вероятность того, что логарифм «функции правдоподобия» определенной ветви длины τ, принадлежащей неправильному поддереву, превзойдет a, ограничена сверху (аналогично неравенству Чернова)

$$P_i(\varepsilon) \leqslant 2^{\tau u(s)} \frac{h_1}{h_1 + s} \quad \text{при} \quad s < -h. \qquad (35)$$

Здесь $u(s) = \log_2 1/2(2^{s\alpha} + 2^{s\beta})$. Множитель $h_1/(s+h_1)$ появился вследствие того, что характеристическая функция величины η имеет вид

$$f(s) = M(2^{-s\eta}) = \int_{-\infty} 2^{-s\eta}(-h_1 2^{-h_1\eta}\ln 2)\,d\eta = \frac{h_1}{h_1+s}. \qquad (36)$$

Поскольку при $R < R_{\text{выч}}$ $h_1 < -1/2$, из (35) получим

$$P_i(\varepsilon) \leqslant C_1 2^{\tau u(1/2)} = C_1 2^{-\frac{R_{\text{выч}}+R}{2}\tau}. \qquad (37)$$

Отсюда вероятность неправильного декодирования при $R < R_{\text{выч}}$

$$P(\varepsilon) \leqslant P_i(\varepsilon) 2^{\tau R} \leqslant C_1 2^{-\frac{R_{\text{выч}}-R}{2}\tau}. \qquad (38)$$

Аналогичная оценка для вероятности ошибки приведена в [3] для алгоритма Фано.

Для рассматриваемого алгоритма можно получить и более точную оценку. Действительно,

$$P_i(\varepsilon) \leqslant \begin{cases} C_2 2^{\tau u(s_0)} \text{ при } R \leqslant R_0', \\ C_3 2^{\tau u(-h_1-\varepsilon)} \text{ при } R \geqslant R_0'. \end{cases}$$

Здесь $C_2 = \text{const}$, $C_3 = O\left(\frac{1}{\varepsilon}\right)$ при $\varepsilon \to 0$,

$$s_0 = \frac{\log_2 p'/q'}{\log_2 p/q}, \quad q' = \frac{-\beta}{\log_2 q/p}, \quad p' = \frac{\alpha}{\log_2 q/p}.$$

Скорость передачи R_0', при которой $s_0 = -h$, определяется из системы уравнений

$$\begin{cases} \alpha 2^{s\alpha} + \beta 2^{s\beta} = 0, \\ q 2^{-s\alpha} + p 2^{-s\beta} = 1. \end{cases} \qquad (39)$$

Таким образом,

$$P(\varepsilon) \leqslant \begin{cases} C_2 2^{\tau[u(s_0)+R]} \text{ при } R \leqslant R_0' \\ C_3 2^{\tau[u(-h_1-\varepsilon)+R]} \text{ при } R \geqslant R_0'. \end{cases} \qquad (40)$$

Правая часть неравенства (40) при $R < R_0'$ может быть также записана в виде $C_2 2^{-\tau[1-H(p')-R]}$.

На рис. 2 приведен график коэффициента при τ в экспоненте верхней границы вероятности ошибки, вычисленной по формуле (40), в зависимости от R для каналов с $p = 0{,}1$; 0,05; 0,03; 0,01.

7. Обе рассмотренные выше процедуры обладают одним существенным недостатком, который делает невозможной их практическую реали-

зацию. Он состоит в том, что память декодера должна быть очень большой, порядка $[P(\varepsilon)]^{\frac{1+h}{h_1}}$.

Нетрудно видеть однако, что большинство хранящихся в памяти декодера логарифмов «функций правдоподобия» с большой вероятностью не потребуется при дальнейшем декодировании. Поэтому при переполнении памяти машины можно производить сброс некоторых величин z, а если в дальнейшем их знание потребуется, произвести вычисление этих статистик заново.

Рассмотрим две процедуры, при выполнении которых требуемый объем памяти декодера растет только линейно с ростом $-\log P(\varepsilon)$.

Пусть в любой момент в памяти системы хранится $2nR$ чисел, где n — длина обрабатываемой ветви (nR — число узлов). Половина этих чисел представляет собой значения z в узлах обрабатываемой ветви. Остальные nR чисел также соответствуют пройденным узлам обрабатываемой ветви. Каждое из них представляет собой минимальный логарифм «функции правдоподобия» обработанных ветвей, отошедших от рассматриваемой в соответствующем узле.

Возвращение к обработке ветвей, выходящих из пройденного узла, производится в случае, если величина z обрабатываемой ветви становится меньше минимального логарифма «функции правдоподобия» ветвей, отошедших от обрабатываемой в этом узле.

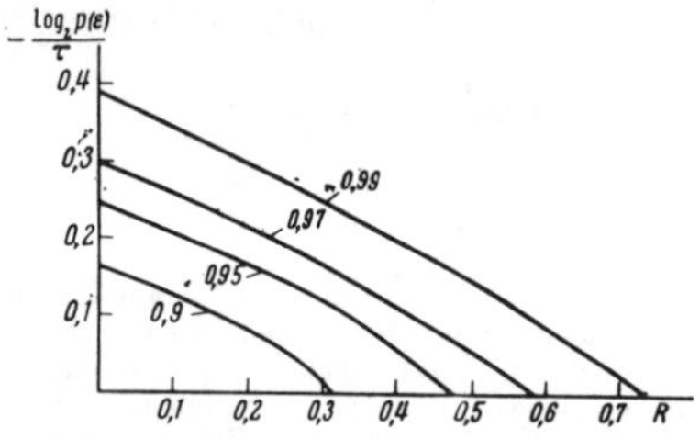

Рис. 2. График зависимости коэффициента при задержке τ в экспоненте вероятности ошибки от скорости передачи R для каналов с вероятностью правильной передачи $q = 0{,}9$; $0{,}95$; $0{,}97$; $0{,}99$

Естественно рассмотреть две схемы декодирования, аналогичные двум рассмотренным выше. Декодирование символа в одной из них — третьей схеме — прекращается, как только величина z для обрабатываемой ветви превысит порог a, декодирование символа в другой — четвертой прекращается, как только длина обрабатываемой ветви достигнет τ. Легко видеть, что формулы (4) и (40) для вероятности ошибки $P(\varepsilon)$ сохраняют свой вид.

Найдем среднее число операций на декодированный символ. Отметим, что при возвращении к обработке ветвей предварительно пройденного узла каждая из этих ветвей наращивается по крайней мере на одно ребро (не меньше). Будем полагать, что при возвращении к обработке ветвей пройденного узла наращивание производится ровно на одно ребро (это увеличивает число операций). «Перескоком» через порог будем пренебрегать.

Рассмотрим неправильное поддерево. Пусть ξ — значение z, запомненное декодером для одного из узлов поддерева, n — число ребер, отделяющих данный узел от начала дерева, $N_\eta(\xi, n)$ — среднее число операций, затраченных на анализ участков ветвей, проходящих через данный узел и начинающихся определенным ребром. Тогда

$$N_\eta(\xi, n) = 2\sum_{k=0}^{m} C_m^k (1/2)^m N_\eta[\xi + k\alpha + (m-k)\beta] + n + 1,$$
$$N_\eta(\eta, n) = 0, \quad N_\eta(a, n) = 0, \tag{41}$$
$$N_\eta(\xi, n) = N_\eta(\xi, 0) + n[2^{\lambda_1(\xi-\eta)} - 1].$$

Решение уравнения (41) (при $2^a \gg 1$) имеет вид

$$N_\eta(\xi, n) \approx B(\xi - \eta)2^{\lambda_1(\xi-\eta)} + (n-1)[2^{\lambda_1(\xi-\eta)} - 1],$$

где

$$B = \frac{-R2^{-R}}{\alpha 2^{\lambda_1\alpha} + \beta 2^{\lambda_1\beta}}.$$

Усредняя $N_\eta(\xi, n)$ по возможным положениям η, получим (при $\xi = 0$, $n = 0$, $h_1 < -1/2$).

$$N(0,0) = -B\frac{h_1}{(1+2h_1)^2} - \left(\frac{h}{1+2h_1} - 1\right).$$

Число операций, затраченных на обработку первого ребра правильного поддерева в принятых предположениях, равно $\frac{h_1}{1+2h_1} - 1$.

Таким образом, среднее число операций, затраченных на декодирование одного символа, при $R < R_{\text{выч}}$ и любых вероятностях, $P(\varepsilon)$, в принятых предположениях, приближенно равно

$$N \approx -B\frac{h_1}{(1+2h_1)^2}. \tag{42}$$

График N в зависимости от скорости передачи R, рассчитанный по формуле (42) для различных вероятностей искажения символа p, представлен на рис. 1 (пунктирные кривые).

Оценка сверху для среднего числа операций с учетом «перескока»

$$N < B2^{-(1+h_1)\frac{2\beta}{R}}\left[\frac{h_1}{(1+2h_1)^2} - \frac{2\beta}{R}\frac{h_1}{1+2h_1}\right]. \tag{43}$$

Аналогично можно получить приближенные выражения для моментов числа операций высших порядков. Интересно отметить, что момент δ-го порядка, так же как и для первых двух процедур, существует при $R < R_\delta$ (см. формулу (33)) и не существует при $R \geqslant R_\delta$. Как следствие этого получим, что асимптотика распределения числа операций, необходимых для декодирования одного символа, подчиняется неравенству (34), коэффициент $C_1(\varepsilon)$ имеет более высокий порядок стремления к бесконечности при $\varepsilon \to 0$, чем для первых двух процедур.

8. Для рассмотренных в предыдущем пункте процедур среднее число операций на символ оказывается несколько большим, чем для первой и второй процедур. Однако путем увеличения емкости памяти декодера на постоянную величину можно модифицировать третью и четвертую процедуры таким образом, чтобы среднее число операций на символ, требуемое при их выполнении, было как угодно близко к среднему числу операций на символ первой процедуры.

Модификация заключается в добавлении памяти, способной хранить 2^l чисел. Хранящиеся в памяти числа представляют собой логарифмы «функций правдоподобия» ветвей, расстояние которых до обрабатываемой ветви не превышает l. (Под «расстоянием» между двумя ветвями здесь понимается сумма числа ребер, отделяющих последние узлы этих ветвей от ближайшего общего узла.) Логарифмы «функций правдоподобия ветвей, расстояние которых превышает l, сбрасываются.

Аналогично (38), можно показать, что вероятность повторения проделанных операций есть $O(2^{-l\frac{R_{\text{выч}}-R}{2}})$ при $l \to \infty$. Тогда среднее число операций на декодированный символ имеет вид

$$N \approx \frac{h_1}{1+2h_1} + O(2^{-l\frac{R_{\text{выч}}-R}{2}}). \tag{44}$$

По-видимому, качественное поведение предложенных другими авторами алгоритмов последовательного декодирования (Фано [2], Кошелев [4]), обеспечивающих постоянное в среднем число операций при $R < R_{\text{выч}}$, аналогично поведению рассмотренных здесь процедур. В частности, для процедуры Фано показано, что асимптотика распределения числа операций на символ подчиняется распределению Парето (Pareto) [9].

Выражаю благодарность Р. Л. Добрушину, В. Н. Кошелеву, М. С. Пинскеру, Р. З. Хасьминскому и Б. С. Цыбакову за советы и замечания, высказанные при обсуждении работы.

ЛИТЕРАТУРА

1. Возенкрафт Дж. М., Рейффен Б. Последовательное декодирование, М., Изд-во иностр. лит., 1963.
2. Fano R. M. A heuristic duscussion of probabilistic decoding. IEEE Trans. Inform. Theory, 1963, 9, 2, 64—74. (Рус. пер.: Фано Р. М. Эвристическое обсуждение вероятностного декодирования. Сб. «Теория кодирования», М., «Мир», 1964, 166—198).
3. Wozencraft J. M. and Jacobs I. M. Principles of communication engineering. New York, John Wiley, 1965.
4. Кошелев В. Н. Оценка сложности последовательного декодирования случайных древовидных кодов. Проблемы передачи информации, 1966, 2, 2, 12—28.
5. Вальд А. Последовательный анализ. М., Физматгиз, 1960.
6. Зигангиров К. Ш. Задача поиска в системе с конечным числом позиций. Радиотехника и электроника, 1963, 8, 1, 16—23.
7. Феллер В. Введение в теорию вероятностей и ее приложения. М., «Мир», 1964.
8. Беккенбах Э., Беллман Р. Неравенства. М., «Мир», 1965.
9. Savage J. E. Sequential Decoding — The Computation Problem. Bell System Techn J., 1966, 45, 1, 149—175.

Поступила в редакцию
9 марта 1966 г.

F. Jelinek

Fast Sequential Decoding Algorithm Using a Stack*

Abstract: In this paper a new sequential decoding algorithm is introduced that uses stack storage at the receiver. It is much simpler to describe and analyze than the Fano algorithm, and is about six times faster than the latter at transmission rates equal to R_{comp}, the rate below which the average number of decoding steps is bounded by a constant. Practical problems connected with implementing the stack algorithm are discussed and a scheme is described that facilitates satisfactory performance even with limited stack storage capacity. Preliminary simulation results estimating the decoding effort and the needed stack size are presented.

1. Introduction

Sequential decoding is a method of communication through noisy channels that uses tree codes (see Fig. 2). Several decoding algorithms have been suggested,[1–3] the one due to Fano being universally acknowledged as the best (see also Sec. 10.4 of Jelinek[4]). The first two algorithms have the common characteristic that, when used with appropriate tree codes signaling over memoryless channels, their probability of error decreases exponentially to zero at all transmission rates less than capacity C, while for all rates less than a value called R_{comp}, the average amount of decoding effort per decoded information bit is bounded above by a constant. R_{comp} is a function of the channel transmission probabilities only, and exceeds $C/2$ for all binary symmetric channels. Figure 1 contains the plot of R_{comp}/C as a function of the crossover probability p. In contrast, all widely used methods of algebraic coding achieve arbitrarily reliable performance only when the transmission rate is sufficiently reduced toward zero. (It ought to be added, however, that algebraic schemes are much simpler than sequential ones.) Furthermore, these methods work only for symmetrical channels with the same number of outputs as inputs.

In this paper we will introduce a new sequential algorithm that is faster than all competing ones, and that is very simple to describe and analyze. To realize its speed advantage without an increase in error probability, it is necessary to increase by a considerable amount the memory of the decoder. However, in a suitable environment (e.g., when a general purpose computer is used for decoding as is done in the Pioneer space program) the increase in speed will be well worth the added cost. The price of memories will continue to drop rapidly in the foreseeable future, thereby widening the applicability of our method. It is hoped that this paper will stimulate further research into tree decoding algorithms. In particular, it would be interesting to know what compromises between the present and Fano's algorithm are possible, and what the trade-off is between finite memory size and speed.

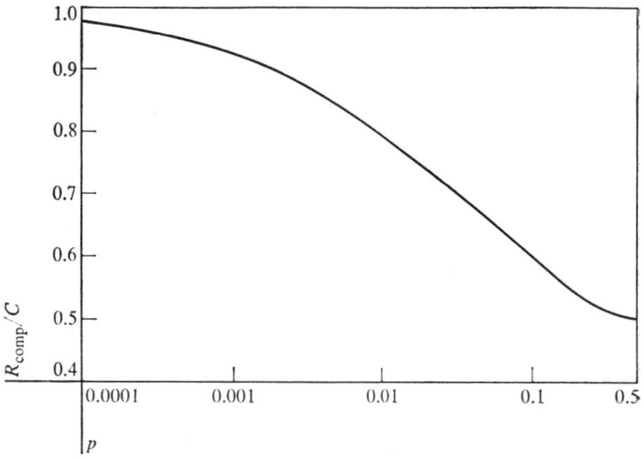

Figure 1 The graph of R_{comp}/C as a function of the crossover probability p of a binary symmetric channel.

Since no previous knowledge of sequential decoding is assumed (the exception being Section 7), we start by describing tree encoding. Then in Section 3 we give our stack decoding algorithm and provide an example. Section 4 contains an outline of the analysis of the present scheme that leads to previously known results about the average number of decoding steps and the probability of error.

The author's current address is School of Electrical Engineering, Cornell University, Ithaca, New York 14850.

* A skeletal version of this paper was presented at the Third Princeton Conference on System Science, March 1969.

Reprinted with permission from *IBM J. Res. Develop.*, vol. 13, pp. 675–678, Nov. 1969.

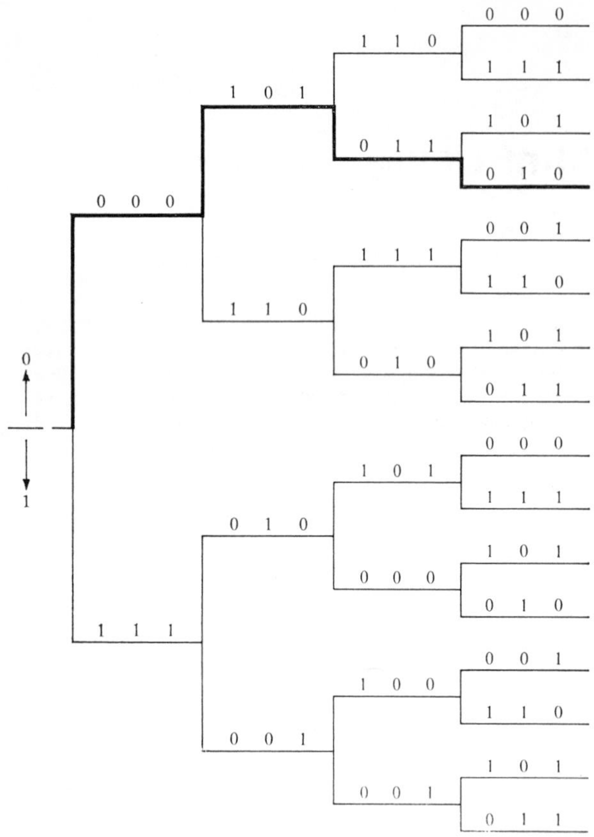

Figure 2 Example of a binary tree code of rate $R = 1/3$.

We next describe the outcome of simulations that enable us to compare the stack and Fano's algorithm (Fig. 6). In Section 6 we modify the algorithm to limit the growth of the stack, and show how effectively this can be done (Fig. 7). Section 7 contains a procedure that handles stack overflows and thereby prevents catastrophic terminations. Finally, Section 8 describes how a stack is actually implemented and in what manner it specifies the required information.

2. Tree encoding

Let us assume that we desire to communicate over a *discrete memoryless channel* characterized by a transmission probability matrix $[w_0(\eta/\xi)]$, where $\xi \in \{0, 1, \cdots, \alpha - 1\}$ are the channel inputs and $\eta \in \{0, 1, \cdots, \beta - 1\}$ are the corresponding channel outputs. Thus, for any integer n, the probability that an arbitrary output sequence $\eta_1, \eta_2, \cdots, \eta_n$ is received, given that an arbitrary input sequence $\xi_1, \xi_2, \cdots, \xi_n$ was transmitted, is equal to

$$\prod_{i=1}^{n} w_0(\eta_i/\xi_i).$$

Let us further assume that the information source generates outputs $s \in \{0, 1, \cdots, d - 1\}$ that are to be communicated to the user located at the other end of the channel. An appropriate tree code of rate $R \equiv 1/(n_0 \log_2 d)$ bits per channel used will have d branches leaving every node, each branch being associated with a sequence of n_0 channel input digits. An example of a tree code appropriate for a binary source ($d = 2$) and a channel with binary inputs ($\alpha = 2$) is given in Fig. 2. Since $n_0 = 3$, the code has rate $R = 1/3$. The correspondence between source outputs and channel inputs (i.e., the encoding rule) is as follows.

At the start of the encoding process the encoder is "located" at the root of the tree. If the first digit s_1 generated by the source is a 0, the encoder follows the upper branch out of the root node to the next node and puts out (transmits through the channel) the sequence associated with that branch (in this case 000). If $s_1 = 1$, the encoder follows the lower branch to the next node and puts out the corresponding sequence (in this case 111). In general, just before the source generates the ith digit s_i, the encoder is located at some node $i - 1$ branches deep in the tree. If $s_i = 0$ the encoder leaves this node along the top branch and, if $s_i = 1$, it leaves the node along the bottom branch; in both cases it transmits the sequence corresponding to the branch along which it traveled. In this way a message sequence $s_1, s_2, \cdots$ traces a path through the tree and the tree sequence corresponding to that path is then transmitted. (The generalization to a d-nary tree is obvious: At time i the encoder leaves the current node along the $(s_i + 1)$th branch of the fan-out and transmits the corresponding sequence, where $s_i \in \{0, 1, \cdots, d - 1\}$.) Thus in Fig. 2 the message sequence 0011 determines the path indicated by the thick line and causes the sequence 000101011010 to be sent through the channel.

In principle, the encoding tree can be continued indefinitely and thus the total number Γ of levels it can have (the tree displayed in Fig. 2 has four levels) is arbitrary. Since a d-nary tree with Γ levels has d^{Γ} paths, there is a problem of how the encoder can store its tree code. We will not concern ourselves here with that issue. Let the reader be assured that no difficulty arises. The tree is never actually stored; all that is stored is a very simple algorithm that can generate the digits associated with the tree branches whenever the former are required. The usual tree codes are called *convolutional* and their description can be found on pp. 377 to 383 of Ref. 4. It will be easier for us to continue to act as if the encoder stored the entire tree.

3. The decoding algorithm

From the preceding description of tree encoding it follows that the natural transmission units we are dealing with are not channel digits themselves, but sequences of n_0 of these that correspond to the tree branches. Accordingly, it will simplify further discussion if we henceforth restrict our attention to the *n_0-product channel* $[w(y/x)]$ whose input symbols x and output symbols y are strings of n_0 inputs and outputs of the underlying channel $[w_0(\eta/\xi)]$. From this point of view, the product channel input alphabet corresponding to the tree code of Fig. 2 is octal, and the message sequence 0011 causes 0532 to be transmitted. Let x^* represent some sequence $\xi_1^*, \xi_2^*, \cdots, \xi_{n_0}^*$ and let y^* represent $\eta_1^*, \eta_2^*, \cdots, \eta_{n_0}^*$. Then

$$w(y^*/x^*) \equiv \prod_{i=1}^{n_0} w_0(\eta_i^*/\xi_i^*). \tag{1}$$

Thus the product channel has inputs $x \in \{0, 1, \cdots, a-1\}$ and outputs $y \in \{0, 1, \cdots, b-1\}$, where $a = \alpha^n$, $b = \beta^n$, and α and β are the sizes of the input and output alphabets of the underlying channel, respectively. A tree path of length i is specified by the vector $\mathbf{s}^i \equiv (s_1, s_2, \cdots, s_i)$ (we will use boldface for vectors and superscripts will indicate their length) formed from the corresponding message digits. We will denote by $x_j(\mathbf{s}^i), j \leq i$, the transmitted symbol associated with the jth branch of the path $\mathbf{s}^i$. Thus in Fig. 2 $x_3(010s_4) = 7$ for all s_4, and $x_2(10s_3s_4) = 2$ for all s_3s_4. Let us now assume that a sequence $\mathbf{y}^\Gamma$ was received through the channel (Γ is the number of levels in the tree) and that we wish to *decode* this sequence, i.e., to determine the identity of the message sequence $\mathbf{s}^\Gamma$ put out by the source. We will denote by $\hat{\mathbf{s}}^\Gamma$ the receiver's estimate of $\mathbf{s}^\Gamma$ and, of course, we aim at having $\hat{\mathbf{s}}^\Gamma$ equal to $\mathbf{s}^\Gamma$. We recall that, when $\mathbf{s}^\Gamma$ was inserted into the encoder, the latter produced the channel input sequence $\mathbf{x}^\Gamma(\mathbf{s}^\Gamma) = x_1(\mathbf{s}^\Gamma), x_2(\mathbf{s}^\Gamma), \cdots, x_\Gamma(\mathbf{s}^\Gamma)$ in the way described in the preceding section. Our problem is to specify the operation of the *decoder*.

Let $r(x), x \in \{0, 1, \cdots, a-1\}$ be a suitable probability distribution (its choice will be clarified below) over the input symbols of the product channel, and define the output distribution

$$w_r(y) \equiv \sum_x w(y \mid x)r(x). \tag{2}$$

If the sequence $\mathbf{y}^\Gamma$ was received, we will be interested in the *likelihoods*

$$L(\mathbf{s}^j) \equiv \sum_{i=1}^{j} \lambda_i(\mathbf{s}^i) \tag{3}$$

of the various paths $\mathbf{s}^j = (\mathbf{s}^i, s_{i+1}, \cdots, s_j)$, where the *branch likelihood function* $\lambda_i(\mathbf{s}^i)$ of the branch leading from node $\mathbf{s}^{i-1}$ to node $\mathbf{s}^i$ (note that a path uniquely defines the tree node on which it terminates, and vice versa) is defined by

$$\lambda_i(\mathbf{s}^i) \equiv \log_2 \frac{w[y_i \mid x_i(\mathbf{s}^i)]}{w_r(y_i)} - n_0 R. \tag{4}$$

From the decoder's point of view, $L(\mathbf{s}^j)$ and $\lambda_i(\mathbf{s}^i)$ are functions of the paths only, since the received sequence $\mathbf{y}^\Gamma$ is fixed throughout the decoding process and the branch symbols $x_i(\mathbf{s}^i)$ are determined by the tree code that is known in advance.

We are now ready to describe the decoding algorithm for a binary tree ($d = 2$). This restriction will make the explanation easier to follow, but will be subsequently removed.

(1) Compute $L(0) = \lambda_1(0)$ and $L(1) = \lambda_1(1)$, the likelihoods of the two branches' leaving the root node (see Fig. 2), and place them into the decoder's memory.
(2) If $L(0) \geq L(1)$, eliminate $L(0)$ from the decoder's memory and compute $L(00) = L(0) + \lambda_2(00)$ and $L(01) = L(0) + \lambda_2(01)$. Otherwise, eliminate $L(1)$ and compute $L(10) = L(1) + \lambda_2(10)$ and $L(11) = L(1) + \lambda_2(11)$. Therefore we end with the likelihoods of three paths in the decoder's memory, two paths of length 2 and one of length 1.
(3) Arrange the likelihoods of the three paths in decreasing order. Take the path corresponding to the topmost likelihood, compute the likelihoods of its two possible one-branch extensions, and eliminate from memory the likelihood of the just extended path [e.g., if $L(0)$, $L(10)$, $L(11)$ are in the memory and, say, $L(10) \geq L(0) \geq L(11)$, then $L(10)$ is replaced in the memory by the newly computed values of $L(100) = L(10) + \lambda_3(100)$ and $L(101) = L(10) + \lambda_3(101)$. In case, say, $L(0) \geq L(11) \geq L(10)$, then $L(0)$ is replaced by $L(00)$ and $L(01)$]. At the end of this step the decoder's memory will contain the likelihoods of four paths, and either two of these will be of length 3 and one each of lengths 1 and 2, or all four paths will be of length 2.
(4) The search pattern is now clear. After the kth step, the decoder's memory will contain exactly $k + 1$ likelihoods corresponding to paths of various lengths and *different* end nodes. The $(k + 1)$th step will consist of finding the largest likelihood in the memory, determining the path to which it corresponds, and replacing that likelihood with the likelihoods corresponding to the two one-branch extensions of that path.
(5) The decoding process terminates when the path to be extended next is of length Γ, i.e., leads from the root node to the last level of the tree.

Because of the ordered nature of the decoder's memory, we refer to it as a *stack*. We next illustrate the *stack decoding algorithm* by an example. Consider the binary tree of Fig. 3 with nodes as numbered. Let paths be associated with their terminal nodes. Let the numbers written on top of the branches represent the values of the corresponding branch likelihood function for some received sequence $\mathbf{y}^3$. Thus, the likelihood of path 8 is equal to $-6 + 1 + 1 = -4$. The state of the stack

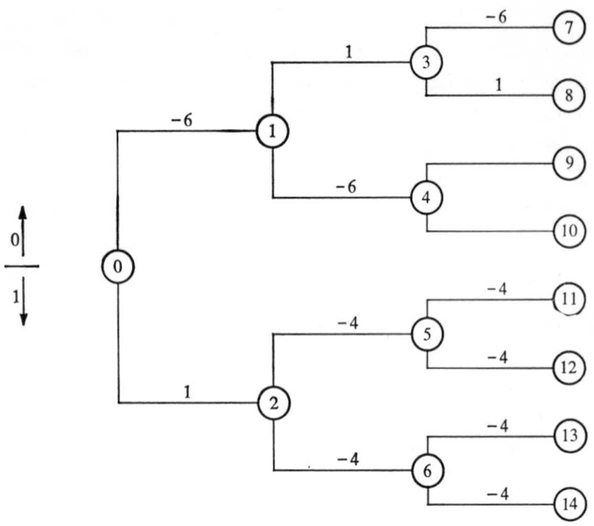

Figure 3 An example of likelihood-value assignment to tree branches induced by a code and a received sequence.

Figure 4 The $\mathfrak{D}_i$ classification of nodes of a binary tree relative to the transmitted message $(s_1, s_2, s_3, s_4, \cdots)$.

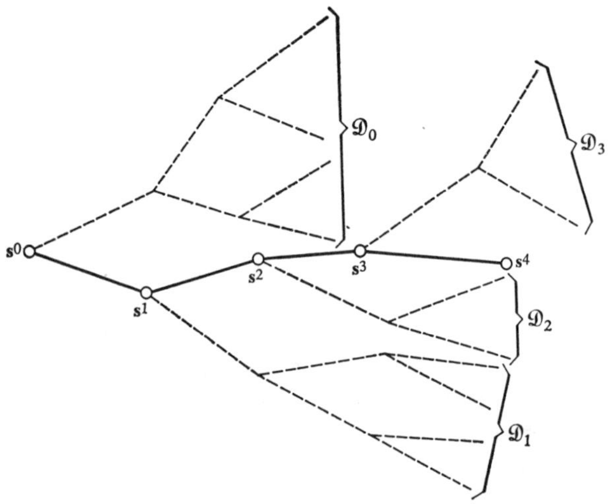

during decoding would then be as follows (topmost path on the left).

1st state:	0
2nd state:	2, 1
3rd state:	5, 6, 1
4th state:	6, 1, 11, 12
5th state:	1, 11, 12, 13, 14
6th state:	3, 11, 12, 13, 14, 4
7th state:	8, 11, 12, 13, 14, 7, 4.

When trying to perform the eighth step, the decoder would find path 8 on top of the stack and, since the latter has length 3 ($=\Gamma$), decoding would terminate.

It remains to generalize our algorithm to d-nary trees. The notion of a stack will facilitate a concise statement of the procedure.

(1) At the beginning of the decoding process, the stack contains only the root node of the tree (i.e., the empty path) with its likelihood set arbitrarily to zero.
(2) A decoding operation consists of ascertaining the path $\mathbf{s}^i$ that corresponds to the likelihood L_i at the top of the stack (i.e., the largest of the likelihoods in the stack), eliminating L_i from the stack, computing the likelihoods $\lambda_{i+1}^1, \cdots, \lambda_{i+1}^d$ of the branches that leave the end node of the path $\mathbf{s}^i$, and inserting the new path likelihoods $L_{i+1}^i = L_i + \lambda_{i+1}^i$, $i = 1, 2, \cdots, d$, into their proper positions according to size (clearly the stack contains path identifications and their likelihood values).
(3) The search ends when the decoder finds at the top of the stack a path whose length is Γ. That path is then considered to have been taken by the encoder.

In our algorithm, the likelihoods are used as a distance measure between the received sequence and the code word sequences on the various paths of the tree. The heuristic reasons for this choice of the measure were made clear in an IBM research report by the author[6] in which the stack algorithm was developed as an intuitively natural way of taking advantage of the tree structure of the code. This aspect makes the algorithm very attractive from a pedagogical point of view. The Fano method[3] can then be considered to be a particular implementation of the stack search in which the decoder's memory is eliminated for the price of an increase in the number of steps necessary for decoding (see Fig. 6). We will see in the next section that the stack algorithm lends itself to a relatively simple analysis.

4. Probabilty of error and average number of decoding steps

We now wish to compute upper bounds to the probability of decoding error and to the average number of decoding steps. Much of the argument is identical to that which applies to the Fano algorithm[4,5] and we will therefore limit ourselves to the differences between the two analyses and to a statement of results. A complete treatment can be found in our earlier IBM research report.[6]

It will be convenient to partition the nodes of the tree into sets $\mathfrak{D}_i$, $i = 0, 1, \cdots, \Gamma$, defined relative to the path $\mathbf{s}^\Gamma$ actually taken by the encoder.

Definition 1: The incorrect subset $\mathfrak{D}_i$ consists of the end node of the initial segment $\mathbf{s}^i$ of the true path $\mathbf{s}^\Gamma = (\mathbf{s}^i, s_{i+1}, \cdots, s_\Gamma)$, of the d-1 terminal nodes of the incorrect branches $s_{i+1}^* \neq s_{i+1}$ stemming from the end node of $\mathbf{s}^i$, and of all nodes lying on paths leaving these $d-1$ nodes.

Error Bounds for Convolutional Codes and an Asymptotically Optimum Decoding Algorithm

ANDREW J. VITERBI, SENIOR MEMBER, IEEE

Abstract—The probability of error in decoding an optimal convolutional code transmitted over a memoryless channel is bounded from above and below as a function of the constraint length of the code. For all but pathological channels the bounds are asymptotically (exponentially) tight for rates above R_0, the computational cutoff rate of sequential decoding. As a function of constraint length the performance of optimal convolutional codes is shown to be superior to that of block codes of the same length, the relative improvement increasing with rate. The upper bound is obtained for a specific probabilistic nonsequential decoding algorithm which is shown to be asymptotically optimum for rates above R_0 and whose performance bears certain similarities to that of sequential decoding algorithms.

Manuscript received May 20, 1966; revised November 14, 1966. The research for this work was sponsored by Applied Mathematics Division, Office of Aerospace Research, U. S. Air Force, Grant AFOSR-700-65.

The author is with the Department of Engineering, University of California, Los Angeles, Calif.

I. Summary of Results

SINCE Elias[1] first proposed the use of convolutional (tree) codes for the discrete memoryless channel, it has been conjectured that the performance of this class of codes is potentially superior to that of block codes of the same length. The first quantitative verification of this conjecture was due to Yudkin[2] who obtained

Reprinted from *IEEE Trans. Inform. Theory,* vol. IT-13, pp. 260–269, Apr. 1967.

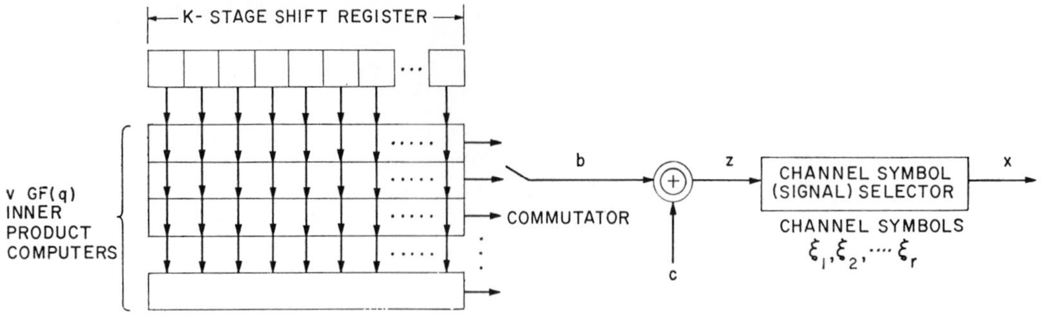

Fig. 1. Encoder for q-ary convolutional (tree) code.

an upper bound on the error probability of an optimal convolutional code as a function of its constraint length, which is achieved when the Fano sequential decoding algorithm[3] is employed.

In this paper, we obtain a lower bound on the error probability of an optimal convolutional code independent of the decoding algorithm, which for all but pathological channels is asymptotically (exponentially) equal to the upper bound for rates above R_0, the computational cutoff rate of sequential decoding. Also, a new probabilistic nonsequential decoding algorithm is described, which exhibits and exploits a fundamental property of convolutional codes. An upper bound on error probability utilizing this decoding algorithm is derived by random coding arguments, which coincides with the upper bound of Yudkin.[2] In the limit of very noisy channels, upper and lower bounds are shown to coincide asymptotically (exponentially) for all rates, and the negative exponent of the error probability, also known as the reliability, is shown to be

$$\lim_{N\to\infty} \frac{1}{N} \ln (1/P_E) = \begin{cases} C/2 & 0 \le R \le C/2 \\ C - R & C/2 \le R < C \end{cases}$$

where N is the code constraint length (in channel symbols), R is the transmission rate and C is channel capacity. This represents a considerable improvement over block codes for the same channels. Also, it is shown that in general in the neighborhood of capacity, the negative exponent is linear in $(C - R)$ rather than quadratic, as is the case for block codes.

Finally, a semisequential modification of the decoding algorithm is described which has several of the basic properties of sequential decoding methods.[3],[4]

II. Description and Properties of the Encoder

The message to be transmitted is assumed to be encoded into the data sequence **a** whose components are elements of the finite field of q elements, $GF(q)$, where q is a prime or a power of a prime. All messages are assumed equally likely; hence all sequences **a** of a fixed number of symbols are equally probable. The encoder consists of a K-stage shift register, v inner-product computers, and an adder, all operating over $GF(q)$, together with a channel symbol selector connected as shown in Fig. 1. After each q-ary symbol of the sequence is shifted into the shift register, the uth computer ($u = 1, 2, \cdots v$) forms the inner product of the vector in the shift register, which is a subsequence of **a**, with some fixed K-dimensional vector $\mathbf{g}_u$, whose components are also elements of $GF(q)$. The result is a matrix multiplication of the K symbol subsequence of **a** (as a row vector) with a Kxv matrix G (whose uth column is $\mathbf{g}_u$) to produce v symbols of the sequence **b**. This is added to v symbols of a previously stored (or generated) q-ary sequence **c**, whose total length is $(L + K - 1)v$ symbols. The v symbol subsequence of **z** thus generated can be any one of q^v v-component vectors. By properly selecting the matrix G and subsequence of **c** [or by selecting them at random with uniform probability from among the ensemble of all q^{Kv} matrices and q^v vectors with components in $GF(q)$], all possible v symbol subsequences of **z** can be made to occur with equal probability. Finally the channel symbol selection (or signal selection in the case of continuous channels) consists of a mapping of each q-ary symbol of **z** onto an r-ary channel symbol x_i of the channel input sequence **x** (where $r \le q$), as follows: let n_1 of the q-ary symbols be mapped into ξ_1, n_2 into ξ_2, etc., such that

$$\sum_{i=1}^{r} n_i = q.$$

Thus if each symbol of **z** is with uniform probability any element of $GF(q)$, the probability distribution of the jth channel input symbol x_j is

$$P(x_j = \xi_i) = \frac{n_i}{q} \qquad (i = 1, 2, \cdots r) \quad \text{for all } j$$

and by proper choice of q and r any rational channel input distribution can be attained. Furthermore, since one q-ary data symbol thus produces v channel symbols, the transmission rate of the system is

$$R = \frac{\ln q}{v} \frac{\text{nats}}{\text{channel symbol}} \tag{1}$$

and thus, by proper choice of q (which must be a prime or the power of a prime) and v, any rate can be closely approximated.

We note also that the encoder thus produces a tree code with q branches, each containing v channel symbols, emanating from each branching node since for every

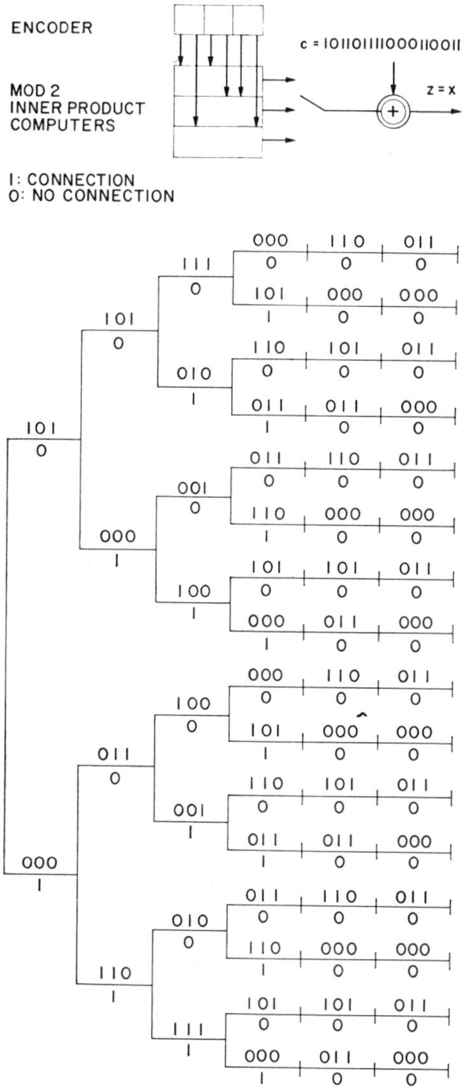

Fig. 2. Tree code for $q = 2$, $v = 3$, $r = 2$, $L = 4$, $K = 3$.

shift of the register a potentially different set of v channel symbols is generated for each of the q possible values of the data symbol. An example is shown in Fig. 2 for $q = 2$, $v = 3$, $r = 2$, $K = 3$. The data symbol a_i is indicated below each branch while the channel symbols x_i are indicated above each branch.

The procedure continues until L data symbols are fed into the shift register followed by a sequence of $K - 1$ zeros. L is known as the (*branch*) *tree length*, and $N = Kv$ as the (*symbol*) *constraint length* of the code. The overall encoding algorithm thus produces a tree code with L branching levels. All branches contain v channel symbols except for the q^L final branches which contain $N = Kv$ channel symbols. The example of Fig. 2 shows such a tree code for $L = 4$ and $K = 3$.

A basic property of the convolutional code thus generated by the K-stage shift register is the following.

A) Two divergent paths of the tree code will converge (i.e., produce the same channel symbols) after the data symbols corresponding to the two paths have been identical for K consecutive branches. Two paths are said to be *totally distinct* over any sequence of branches for which this event does not occur.

We now proceed to derive the lower bound on error probability for an optimal convolutional code using property A) and lower bound results for optimal block codes.

III. The Lower Bound

Suppose a magic genie informs the decoder as to the exact state of each branch data symbol a_i for all branches i ($i = 1, 2, \cdots L + K - 1$) except for the m consecutive branches $j + 1, j + 2, \cdots j + m (0 \leq j \leq L - m)$. Thus to decode the tree the decoder must decide upon which of the q^m possible m-symbol q-ary data sequences corresponding to these m branches actually occurred, or equivalently he must decide among the corresponding q^m alternate paths through the tree. To do this he has available the $(L + K - 1)v$ symbol received tree sequence $\mathbf{y} = (\mathbf{y}_1, \mathbf{y}_2, \cdots \mathbf{y}_{L+K-1})$ where $\mathbf{y}_i$ is the received symbol sequence for the ith branch. Actually since the a_i are known for all $i \leq j$, he needs only examine $\mathbf{y}_i$ for $i \geq j + 1$. Furthermore, the q^m alternate paths in question, which diverge at the $(j + 1)$th branch must converge again at the $(j + m + K)$th branch, for since all the corresponding branch data symbols a_i are identical for $i \geq j + m + 1$, by the $(j + m + K)$th branch the data symbols in the shift register will be identical for all paths in question. Thus the q^m paths are totally distinct over at most $m + K - 1$ branches. Now letting

$$\mu = \frac{m}{K} \tag{2}$$

and having denoted the constraint length in channel symbols by

$$N = Kv \tag{3}$$

we obtain from (1), (2), and (3)

$$m \ln q = \mu N R. \tag{4}$$

The optimal decoder for paths which are a priori equally likely must compute the $q^m = e^{\mu NR}$ likelihood functions $p(\mathbf{y} \mid \mathbf{a})$, where $\mathbf{a} = (a_{j+1}, \cdots a_{j+m})$ is an m-component q-ary vector which specifies the path, and $\mathbf{y} = (\mathbf{y}_{j+1}, \cdots \mathbf{y}_{j+m+K-1})$ is an $(m + K - 1)v = (\mu + 1)N - v$ component vector, and select the path corresponding to the greatest. The resulting error probability is lower bounded by the lower bound[5]–[7] for the best block code with $e^{\mu NR}$ words of length $(\mu + 1)N - v$ channel symbols transmitted over a memoryless channel with discrete input space:

$$P_E(\mu, N, R) > \exp\{-N(\mu + 1)[E_L(R, \mu) + o(\mu N)]\} \tag{5}$$

where

$$o(\mu N) \to 0 \quad \text{linearly} \quad \mu N \to \infty$$
$$E_L(R, \mu) = \operatorname*{l.u.b.}_{0 \leq \rho \leq \infty} \left[\hat{E}_0(\rho) - \rho \frac{\mu}{\mu + 1} R \right] \tag{6}$$

and $\hat{E}_0(\rho)$ is the concave hull of the function

$$E_0(\rho) = \max_{p(x)} \{-\ln \sum_Y [\sum_X p(x)p(y \mid x)^{1/1+\rho}]^{1+\rho}\} \quad (7)$$

where X and Y are the channel input and output spaces, respectively, $p(y \mid x)$ is the channel transition probability distribution, and $p(x)$ is an arbitrary probability distribution on the input space. Furthermore, the function $E_0(\rho)$ has the following basic properties which are proved in Gallager:[5]

a) $E_0(0) = 0$ and $E_0(\rho) > 0$ for all $\rho > 0$,

b) $E_0'(\rho) > 0$ for all finite ρ, and $\lim_{\rho \to 0} E_0'(\rho) = C$ which is the channel capacity.

For most channels of interest $E_0(\rho)$ is itself a concave function. When this is not the case the channel is said to be pathological.[5]

This bound, known as the sphere-packing bound, is the tightest exponential bound for high rates. For low rates a tighter bound, which has been recently derived,[7] is considered below. $E_L(R, \mu)$ can be obtained by solving the parametric equations

$$E_L(R, \mu) = \hat{E}_0(\rho) - \rho\hat{E}_0'(\rho) \quad (8a)$$

$$R = \frac{\mu + 1}{\mu} \hat{E}_0'(\rho). \quad (8b)$$

But $\mu = m/K$ can be any multiple of $1/K$ up to L/K, since m cannot exceed L. Hence, since no particular demands can be made on the magic genie,

$$P_E(N, R) \geq \max_{(1/K) \leq \mu \leq (L/K)} P_E(\mu, N, R)$$

$$> \exp \{-N \min_{(1/K) \leq \mu \leq (L/K)} (\mu + 1)[E_L(R, \mu) + o(\mu N)]\} \quad (9)$$

corresponding to the least obliging genie for the particular R.

Thus we seek the lower envelope

$$E_L(R) = \min_{(1/K) \leq \mu \leq (L/K)} (\mu + 1)E_L(R, \mu). \quad (10)$$

It follows from (6) and (7) and property b) that

$$\lim_{\mu \to 0} (\mu + 1)E_L(R, \mu) = \underset{0 \leq \rho \leq \infty}{\text{l.u.b.}}\ \hat{E}_0(\rho) = \hat{E}_0(\infty)$$

$$\lim_{\mu \to \infty} (\mu + 1)E_L(R, \mu) = \infty \quad \text{for} \quad R < C.$$

The family of functions $(\mu + 1)E_L(R, \mu)$ is sketched in Fig. 3. To find the lower envelope we must minimize $E_L(R, \mu)$ over the set of possible μ for each R. For the purposes of the lower bound we shall let L/K be as large as required for the minimization. First, let us minimize over all positive real μ and then restrict μ to be a multiple of $1/K$. Thus from (8a) we have

$$\frac{\partial[(\mu + 1)E_L(R, \mu)]}{\partial\mu}$$

$$= \hat{E}_0(\rho) - \rho\hat{E}_0'(\rho) + (\mu + 1)[-\rho\hat{E}_0''(\rho)] \frac{\partial\rho}{\partial\mu} \quad (11)$$

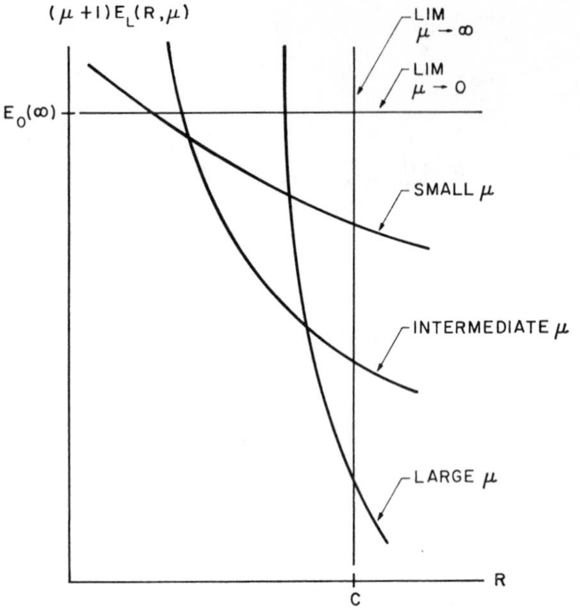

Fig. 3. Family of functions $(\mu + 1)\ E_L\ (R, \mu)$.

while from (8b) we have

$$\hat{E}_0''(\rho) \frac{\partial\rho}{\partial\mu} = \frac{R}{(\mu + 1)^2}. \quad (12)$$

Combining (11) and (12) and setting the former equal to zero, we find that the function has a stationary point at

$$\mu = \frac{\rho R}{\hat{E}_0(\rho) - \rho\hat{E}_0'(\rho)} - 1. \quad (13)$$

Furthermore, differentiating (11) and using (12), we find that

$$\frac{\partial^2[(\mu + 1)E_L(R, \mu)]}{\partial\mu^2} = -\frac{R^2}{(\mu + 1)^3\hat{E}_0''(\rho)} \geq 0$$

so that (13) corresponds to an absolute minimum. Inserting (13) in (8b) yields

$$R = \frac{\hat{E}_0(\rho)}{\rho} \quad (14)$$

and since $\hat{E}_0(\rho)$ is concave it follows that $R = \hat{E}_0(\rho)/\rho \geq \hat{E}_0'(\rho)$ which implies that the solution (13) for μ is nonnegative. From (8a), (13), and (14) we obtain

$$\min_{0 \leq \mu < \infty} (\mu + 1)E_L(R, \mu) = \rho R = \hat{E}_0(\rho). \quad (15)$$

Now, since μ is restricted to be a multiple of $1/K$, let us consider altering (13) by adding a positive real number δ large enough to make μ an element of this set. In any case $\delta < 1/K$. But changing μ by this amount in (9) alters the exponent by an amount proportional to $N/K = v$, which is a constant parameter of the encoder and hence, normalized by N, is $o(N)$. The rate is also altered by an amount of the order of $1/K$ by this change in μ, but if we adjust for this change by returning R to its original value (14), we again alter P_E by an amount of magnitude $o(N)$. Thus from (9), (10), (14), and (15) we obtain

Theorem 1

The probability of error in decoding an arbitrarily long convolutional code tree of constraint length N (channel symbols) transmitted over a memoryless channel is bounded by

$$P_E > \exp\{-N[E_L(R) + o(N)]\}$$

where

$$E_L(R) = \hat{E}_0(\rho) \qquad (0 \le \rho < \infty) \tag{16a}$$

and

$$R = \hat{E}_0(\rho)/\rho. \tag{16b}$$

Taking the derivative of (14) we find

$$\frac{\partial R}{\partial \rho} = \frac{\hat{E}_0'(\rho) - \hat{E}_0(\rho)/\rho}{\rho} \le 0 \quad \text{for all} \quad \rho > 0$$

where we have made use of the fact that $\hat{E}_0(\rho)$ is concave. Also, from property b) we have $\lim_{\rho\to 0} \hat{E}_0(\rho)/\rho = \hat{E}'(0) = C$. Thus we obtain

Corollary 1

The exponent $E_L(R)$ in the lower bound is a positive monotone decreasing continuous function of R for all $0 \le R < C$.

A graphical construction of the exponent-rate curve from a plot of the function $E_0(\rho)$ is shown in Fig. 4. We defer further consideration of the properties of (16) until after an upper bound is obtained.

A tighter lower bound on error probability for low rates is obtained by replacing the sphere packing bound of (6) by the tighter lower bound for low rates recently obtained by Shannon, Gallager, and Berlekamp.[7] For this bound (6) is replaced by

$$E_L(R, \mu) = E_x - \frac{\tilde{\rho}\mu R}{\mu + 1} \quad \left(0 \le R \le \frac{\mu+1}{\mu}\hat{E}_0'(\tilde{\rho})\right) \tag{17a}$$

where

$$E_x = \max_{p(x)} \{-\lim_{\rho\to\infty} [\rho \ln \sum_X \sum_X p(x)p(x') \cdot(\sum_Y \sqrt{p(y \mid x)p(y \mid x')})^{1/\rho}]\} = \hat{E}_0(\tilde{\rho}). \tag{17b}$$

The straight line of (17a) is tangent to the curve of (6) at $R = [(\mu+1)/\mu]\hat{E}_0'(\tilde{\rho})$. Repeating the minimization with respect to μ we find

$$E_L(R) = \min_\mu [(\mu+1)E_x - \tilde{\rho}\mu R]$$

$$= E_x, \qquad 0 \le R \le \frac{\hat{E}_0(\tilde{\rho})}{\tilde{\rho}}.$$

Thus, we have

Corollary 2

For low rates a tighter lower bound than that of Theorem 1 is:

$$P_E > \exp\{-N[E_L(R) + o(N)]\}$$

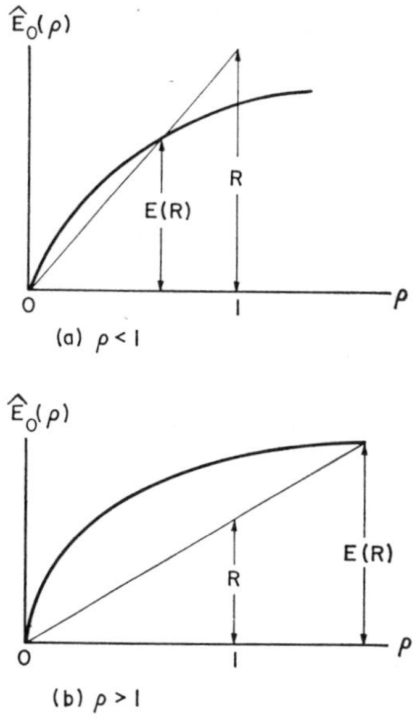

Fig. 4. Graphical construction of $E_L(R)$ from $\hat{E}_0(P)$.

where

$$E_L(R) = E_x, \qquad 0 \le R \le \frac{\hat{E}_0(\tilde{\rho})}{\tilde{\rho}}, \tag{18}$$

$\tilde{\rho}$ is the solution to the equation $\hat{E}_0(\tilde{\rho}) = E_x$, and E_x is given by (17b).

IV. A Probabilistic Nonsequential Decoding Algorithm

We now describe a new probabilistic nonsequential decoding algorithm which, as we shall show in the next section, is asymptotically optimum for rates $R > R_0 = E_0(1)$. The algorithm decodes an L-branch tree by performing L repetitions of one basic step. We adopt the convention of denoting each branch of a given path by its data symbol a_i, an element of $GF(q)$. Also, although $GF(q)$ is isomorphic to the integers modulo q only when q is a prime, for the sake of compact notation, we shall use the integer r to denote the rth element of the field.

In *Step 1* the decoder considers all q^K paths for the first K branches (where K is the branch constraint length of the code) and computes all q^K likelihood functions $\prod_{i=1}^K p(\mathbf{y}_i \mid a_i)$. The decoder then compares the likelihood function for the q paths:

$$(0, a_2, a_3, \cdots a_K),$$
$$(1, a_2, a_3, \cdots a_K),$$
$$\cdots\cdots\cdots\cdots\cdots$$
$$(q-1, a_2, a_3, \cdots a_K)$$

for each of the q^{K-1} possible vectors $(a_2, a_3 \cdots a_K)$. It thus performs q^{K-1} comparisons each among q path likelihood functions. Let the path corresponding to the greatest likelihood function in each comparison be denoted the *survivor*. Only the q^{K-1} survivors of as many comparisons are preserved for further consideration; the remaining paths are discarded. Among the q^{K-1} survivors each of the q^{K-1} vectors $(a_2, a_3, \cdots a_K)$ is represented uniquely, since by the nature of the comparisons no two survivors can agree in this entire subsequence.

Step 2 begins with the computation for each survivor of Step 1 of the likelihood functions of the q branches emanating from the $(K + 1)$th branching node and multiplication of each of these functions by the likelihood function for the previous K branches of the particular path. This produces q^K functions for as many paths of length $K + 1$ branches, and each of the subsequences $a_2, a_3, \cdots a_{K+1}$ are represented uniquely. Again the q^K functions are compared in groups of q, each comparison being among the set of paths:

$$(\alpha_{11}^{(1)}, 0, a_3, a_4 \cdots a_{K+1})$$
$$(\alpha_{21}^{(1)}, 1, a_3, a_4 \cdots a_{K+1})$$
$$\cdots\cdots\cdots$$
$$(\alpha_{q1}^{(1)}, q - 1, a_3, a_4 \cdots a_{K+1})$$

where $\alpha_{k1}^{(1)}$ corresponds to the first branch of the survivor of a comparison performed at the first step. Again only the survivors of the set of q^{K-1} comparisons are preserved and the remaining paths are discarded. The algorithm proceeds in this way, at each step increasing the population by a factor of q by considering the set of q branches emanating from each surviving path and then reducing again by this factor by performing a new set of comparisons and excluding all but the survivors.

In particular, at *Step* $j + 1$ the decoder performs q^{K-1} sets of comparisons among groups of q paths, which we denote

$$(\alpha_{11}^{(j)}, \alpha_{12}^{(j)}, \cdots \alpha_{1j}^{(j)}, 0, a_{j+2}, a_{j+3}, \cdots a_{j+K}),$$
$$(\alpha_{21}^{(j)}, \alpha_{22}^{(j)}, \cdots \alpha_{2j}^{(j)}, 1, a_{j+2}, a_{j+3}, \cdots a_{j+K}),$$
$$\cdots\cdots\cdots$$
$$(\alpha_{q1}^{(j)}, \alpha_{q2}^{(j)} \cdots \alpha_{q,j}^{(j)}, q - 1, a_{j+2}, a_{j+3}, \cdots a_{j+K})$$

where the vectors $(\alpha_{k1}^{(j)}, \alpha_{k2}^{(j)}, \cdots \alpha_{kj}^{(j)})$ depend on the outcome of the previous set of comparisons. Again by the nature of the comparisons no two survivors can agree in all of the last $K - 1$ branches and there is a one-to-one correspondence between each of the q^{K-1} survivors and the subsequences $(a_{j+2}, \cdots a_{j+K})$.

This procedure is repeated through the $(L - K + 1)$th step. Beyond this point branching ceases because only zeros are fed into the shift register. Thus at step $L - K + 2$ the decoder compares the likelihood functions for the q paths:

$$(\alpha_{11}^{(L-K+1)}, \alpha_{12}^{(L-K+1)}, \cdots \alpha_{1,L-K+1}^{(L-K+1)}, 0, a_{L-K+3} \cdots a_L, 0),$$
$$(\alpha_{21}^{(L-K+1)}, \alpha_{22}^{(L-K+1)}, \cdots \alpha_{2,L-K+1}^{(L-K+1)}, 1, a_{L-K+3} \cdots a_L, 0),$$
$$\cdots\cdots\cdots$$
$$(\alpha_{q1}^{(L-K+1)}, \alpha_{q2}^{(L-K+1)}, \cdots \alpha_{q,L-K+1}^{(L-K+1)}, q - 1, a_{L-K+3} \cdots a_L, 0)$$

for each of the q^{K-2} possible vectors $(a_{L-K+3} \cdots a_L)$ resulting in q^{K-2} survivors. Thus, for this and all succeeding steps the population fails to grow since all further branches correspond only to zeros entering the shift register, and it is reduced by a factor of q by the comparisons. Thus, just after the $(L - 1)$th step there are only q survivors:

$$(\alpha_{11}^{(L-1)}, \cdots \alpha_{1,L-1}^{(L-1)}, 000 \cdots 0),$$
$$(\alpha_{21}^{(L-1)}, \cdots \alpha_{2,L-1}^{(L-1)}, 100 \cdots 0),$$
$$\cdots\cdots\cdots$$
$$(\alpha_{q1}^{(L-1)}, \cdots \alpha_{q,L-1}^{(L-1)}, q - 1, 00 \cdots 0).$$

At *Step L*, therefore, there remains a single comparison among q paths, whose survivor will be accepted as the correct path. While this decoding algorithm is clearly suboptimal, the optimal being a comparison of the likelihood functions of all q^L paths at the end of the tree based on $(L + K - 1)v$ received channels symbols, we shall show in the next section that the algorithm is asymptotically optimum for $R > R_0 = E_0(1)$ for all but pathological channels.

V. Random Coding Upper Bound

If we now assume that the matrix G is randomly selected with a uniform distribution from the ensemble of q^{vK} matrices of elements in $GF(q)$ and the sequence $\mathbf{c}$ is also randomly selected from among all possible $(L + K - 1)v$-dimensional vectors with components in the same field, the channel symbols along a given path regarded as random variables have the following properties[8] in addition to A):

B) The probability distribution of the jth channel symbol for any path is the same for all j, and for all paths

$$P(x_j = \xi_i) = P_i \qquad (i = 1, 2, \cdots r).$$

C) Successive channel symbols along a given path are statistically independent

$$P(x_1 = \xi_{i_1}, x_2 = \xi_{i_2}, \cdots x_{(L+K-1)v} = \xi_{i_{(L+K-1)v}}) = \prod_{j=1}^{(L+K-1)v} P(x_j = \xi_{i_j}).$$

We shall need one more property before we can proceed, which requires a modification of the encoder:

D) Symbols along arbitrary subsequences of any two totally distinct paths are independent.

Reiffen[8] proved property D) for the present encoder but only within the first K-branch constraint length. To ensure that D) is satisfied over the entire L-branch tree, we must modify the encoder. One obvious way is to randomly select a new Kxv generator matrix G after each new data symbol a_j is shifted into the register. However, Massey[9] has recently shown that it is possible to ensure D) by introducing only $2v$ new components into the first two rows of the generator matrix for each new data symbol, and simply shifting all the rows of the previous generator matrix two places downward and discarding the last two rows.

We now proceed to obtain an upper bound on the error probability for the class of convolutional codes which possess the above properties, by analyzing the performance of the decoding algorithm of the previous section. We recall that the correct path is eliminated if it fails to have the largest likelihood function in any one of the L comparisons among q alternatives in which it is involved.

In particular, let us consider the situation at the $(j + 1)$th step. Without loss of generality, we may assume that the correct path corresponds to the all zeros data sequence. Although the comparison at this step is with only $q - 1$ other paths, there is a multitude of potential adversaries. Thus, with the first $j + K$ branches of the correct path denoted by the vector $\mathbf{0} = (00 \cdots 0)$, consider all the paths of the form $\alpha_{21}^{(i)}, \alpha_{22}^{(i)}, \cdots \alpha_{2j}^{(i)} 100 \cdots 0$. There is only one such path which diverged from the correct path K branches back: namely, the one for which $\alpha_{21}^{(i)} \cdots \alpha_{2j}^{(i)} = 00 \cdots 0$. But there are $q - 1$ potential adversaries of this form which diverged from the correct path $K + 1$ branches back: namely, those for which $\alpha_{21}^{(i)} \cdots \alpha_{2j-1}^{(i)} = 00 \cdots 0$ and $\alpha_{2j}^{(i)}$ is any element of $GF(q)$ except 0. Similarly, there are $(q - 1)q$ potential adversaries of this form which diverged from the correct path $K + 2$ branches back: namely, those for which $\alpha_{21}^{(i)} \cdots \alpha_{2,j-2}^{(i)} = 00 \cdots 0$, $\alpha_{2,j-1}^{(i)}$ is any element except 0, and $\alpha_{2j}^{(i)}$ is any element of $GF(q)$. Continuing in this way, we find that there are $(q - 1)q^{l-1}$ potential adversaries of this form which diverged $K + l$ branches back. However, there are exactly as many potential adversaries for which $a_{j+1} = 2$, as these are adversaries for which $a_{j+1} = 1$, and similarly for $a_{j+1} = 3, 4, \cdots q - 1$. Thus, the total number of potential adversaries which diverged from the correct path $K + l$ branches back $(l = 1, 2, \cdots)$ is $(q - 1)^2 q^{l-1}$, while $q - 1$ paths diverged K branches back.

Before we can proceed to bound the error probability, we must establish that of all the potential adversaries which diverged from the correct path $K + l$ branches back only those that are totally distinct from it can actually be adversaries in the comparison of likelihood functions. We recall from property A) that two paths which diverge at a given branch will converge again after K branches if all of the next K data symbols are identical. Furthermore, any pair of paths having data symbols which are never identical for K consecutive branches remain totally distinct from the initial divergent branch. We now observe that by the nature of the decoding algorithm no two adversaries in any comparison can agree in K (or more) consecutive branch data symbols beyond their point of initial divergence, for at the outcome of each preceding set of comparisons there was one and only one surviving path with a particular sequence of K data symbols.

Thus, all the actual adversaries to the correct path at step $j + 1$ are totally distinct from it and consequently the branch channel symbols are statistically independent [Property D]. Further, we have no more than $q - 1$ possible adversaries to the correct path which diverged K branches (or N channel symbols) back and no more than $(q - 1)^2 q^{l-1}$ possible adversaries to the correct path which diverged $K + l$ branches (or $(K + l)v = N + (\ln q/R)l$ channel symbols) back, where $l = 1, 2, \cdots$.

Thus, the expected probability of an error in the comparison at the $(j + 1)$th step is bounded by the union bound,

$$\overline{P(j+1)} < \sum_{l=0}^{j} \overline{\Pr \text{ (error caused by a possible adversary which diverged } K + l \text{ branches back)}}. \tag{19}$$

The zeroth term of this sum is bounded by the probability of error for a block code of $(q - 1)$ words (the maximum number of possible adversaries) each of length N channel symbols, while the lth term $(l \geq 1)$ is bounded by the error probability for a block code of $(q - 1)^2 q^{l-1}$ words each of length $N + (\ln q/R)l$ channel symbols. Since all symbols of each codeword are mutually independent and symbols of the correct codeword are independent of symbols of any other codeword, we may use the random coding upper bound on block codes[5],[1] for the lth term. Thus, if for the given transmission rate the convolutional encoder is mechanized, as described above, so that the input symbol distribution is that which achieves the maximum of (7), we have,

$$\begin{aligned}\overline{P(j+1)} &< (q-1)^{\rho} \exp[-NE_0(\rho)] + \sum_{l=1}^{j} [(q-1)^2 q^{l-1}]^{\rho} \\ &\quad \cdot \exp\left[-\left(N + \frac{\ln q}{R} l\right) E_0(\rho)\right] \\ &< (q-1) \exp[-NE_0(\rho)] \sum_{l=0}^{\infty} q^{l[\rho - E_0(\rho)/R]} \\ &= \frac{q-1}{1 - q^{-\epsilon/R}} \exp[-NE_0(\rho)] \quad (0 < \rho \leq 1)\end{aligned} \tag{20}$$

where $\epsilon = E_0(\rho) - \rho R > 0$. This bound is independent of j. We again use a union bound to express the error probability in decoding the L branch tree in terms of (20) and thus obtain

$$\begin{aligned}\overline{P_E} &< \sum_{j=0}^{L-1} \overline{P(j+1)} \\ &< \frac{L(q-1)}{1 - q^{-\epsilon/R}} \exp[-NE_0(\rho)] \quad (0 < \rho \leq 1)\end{aligned} \tag{21}$$

where $\epsilon = E_0(\rho) - \rho R > 0$ and since at least one code in the ensemble must have $P_E < \overline{P_E}$, and $E_0(\rho)$ is a monotonically increasing function of ρ, we have

Theorem 2

The probability of error in decoding an L-branch q-ary tree code transmitted over a memoryless channel is bounded by

$$P_E < \frac{L(q-1)}{1 - q^{-\epsilon/R}} \exp[-NE(R)]$$

[1] Note that Gallager's proof of the upper bound for block codes[5] requires only that the correct word symbols be independent of the symbols of any incorrect word, and not that incorrect words be mutually independent.

where[2]

$$E(R) = \begin{cases} R_0, & 0 \leq R = R_0 - \epsilon < R_0 \quad \text{(22a)} \\ E_0(\rho), & R_0 - \epsilon \leq R = \dfrac{E_0(\rho) - \epsilon}{\rho} < C \\ & (0 < \rho \leq 1) \quad \text{(22b)} \end{cases}$$

and

$$R_0 = E_0(1) = \max_{p(x)} \{-\ln \sum_Y [\sum_X p(x)\sqrt{p(y \mid x)}]^2\}.$$

Since the bound was shown for the specific probabilistic decoding algorithm described above, and $\epsilon > 0$ can be made arbitrarily small for N arbitrarily large, we have comparing (16) and (22), whenever $E_0(\rho)$ is concave,

$$\lim_{N\to\infty} \frac{\ln (1/P_E)}{N} = E(R) = E_L(R) \quad \text{for } R_0 \leq R < C \quad (23)$$

and consequently

Corollary 1

For all but pathological channels the specific probabilistic decoding algorithm described in Section IV is asymptotically (exponentially) optimum for $R \geq R_0$.

Yudkin[2] has obtained an upper bound with the exponent of (22) for the undetectable error probability of the Fano sequential decoding algorithm.[3] Thus the Fano algorithm is also asymptotically optimum in this sense for $R \geq R_0$. However, the average number of computations per branch is unbounded for $R > R_0$ in the latter, while for the nonsequential algorithm considered here the number of computations per branch is proportional to q^K independent of rate. Also, as we shall show below, the number of computations required with this algorithm for a convolutional code of constraint length N is essentially the same as the number required by a maximum likelihood decoder for a block code of block length N, all the other parameters being the same.

The random coding upper bound exponent (with $\epsilon = 0$) is greater than the random coding exponent for block codes for all rates ($0 < R < C$), as is seen by comparing (22) with the exponent for block codes[5] of length N:

$$E(R) = \begin{cases} R_0 - R, & 0 \leq R \leq E_0'(1) \quad \text{(24a)} \\ E_0(\rho) - \rho E_0'(\rho), & E_0'(1) \leq R = E_0'(\rho) < C \\ & (0 < \rho \leq 1). \quad \text{(24b)} \end{cases}$$

From property b) of $E_0(\rho)$, we have $E_0'(\rho) > 0$. Also, from (24b) we have $E_0(\rho)/\rho \geq E_0'(\rho)$, and the conclusion follows.

The same is true also for the lower bound. For $R > E_0'(\tilde{\rho})$, the best known lower bound for block codes[5]–[7] coincides with the sphere packing bound, which is the same as (24b) for nonpathological channels but with ρ extended to $\tilde{\rho} \geq 1$. Thus for this range the lower bound on convolutional codes (16) exceeds this for the reasons just stated. For $R < E_0'(\tilde{\rho})$, the best known bound for block codes[7] is $E_L(R) = E_x - \tilde{\rho}R$ ($\rho \geq 1$), while from (18) for convolutional codes we have $E_L(R) = E_x$ for $0 < R \leq E_0(\tilde{\rho})/\tilde{\rho} > E_0'(\tilde{\rho})$ which therefore exceeds the lower bound for block codes in this region also. For pathological channels the same argument applies to $\hat{E}_0(\rho)$.

[2] If $E_0''(\rho) > 0$ for some ρ on the unit interval, (22b) may specify more than one value of $E(R)$ for a given R. In this case we should choose the greater, with the result that $E(R)$ is a discontinuous function.

VI. Limiting Cases and Comparisons with Block Codes

Of particular interest is the behavior of the exponent in the neighborhood of capacity. We have from the properties a), b), and equation (7)

$$\hat{E}_0(0) = 0, \qquad \hat{E}_0'(0) = C, \qquad E_0''(0) \leq 0.$$

We must solve the parametric equations

$$E_L(R) = \hat{E}_0(\rho) \quad (25a)$$

$$R = \frac{\hat{E}_0(\rho)}{\rho} \qquad (0 \leq \rho \leq 1) \quad (25b)$$

for R in the neighborhood of C, which corresponds to ρ in the neighborhood of 0. Thus, excluding for this purpose the case in which $E_0''(0) = 0$, and expanding $\hat{E}_0(\rho)$ in a Taylor series about $\rho = 0$ neglecting terms higher than quadratic, we obtain

$$\hat{E}_0(\rho) \approx \rho C + \frac{\rho^2}{2} E_0''(0) \approx E_0(\rho). \quad (26)$$

Then from (25b) and (26) we have

$$\rho = \frac{2(C - R)}{-E_0''(0)}.$$

Substituting in (26) and neglecting terms higher than linear in $C - R$ we obtain (setting $\epsilon \approx 0$ in the upper bound)

$$E(R) \approx E_L(R) = \hat{E}_0(\rho) \approx \frac{2C}{-E_0''(0)} (C - R).$$

In contrast, for block codes the exponent for rates in the neighborhood of $C(\rho = 0)$, as obtained by repeating the above argument in connection with (24b), is

$$E(R) = E_L(R) \approx \frac{1}{-2E_0''(0)} (C - R)^2.$$

Another interesting limiting case is that of "very noisy" channels which includes the time-discrete white Gaussian channel. A memoryless channel is said to be very noisy if $p(y \mid x) = p(y)(1 + \epsilon_{xy})$ where $|\epsilon_{xy}| \ll 1$ for all x and y in the channel input and output spaces X and Y. For this class of channels it has been shown[5] that when the input distribution is optimized so that $I(X; Y) = C$, then

$$\hat{E}_0(\rho) = E_0(\rho) \approx \frac{\rho C}{1 + \rho} \quad (27)$$

where

$$C \approx \max_{p(x)} \sum_X \sum_Y p(x)p(y) \frac{\epsilon_{xy}^2}{2}.$$

Also

$$R_0 = E_0(1) \approx \frac{C}{2} \approx E_x$$

and from (17b), it follows that $\tilde{\rho} = 1$. Thus, with $\epsilon = 0$ we find from (18), (22), and (27)

$$E(R) \approx E_L(R) \approx C/2, \qquad 0 \leq R \leq C/2. \qquad \text{(28a)}$$

For rates above $C/2$ we have from (16), (22), and (27)

$$R = \frac{E_0(\rho)}{\rho} \approx \frac{C}{1+\rho}.$$

Solving for ρ in terms of R, and substituting in (27), we obtain from (16) and (22):

$$E(R) \approx E_L(R) \approx C - R, \qquad \frac{C}{2} \leq R < C. \qquad \text{(28b)}$$

From (28a) and (28b) we note that for very noisy channels the upper and lower bounds are exponentially equal for all rates, that they remain at the zero rate level of $C/2$ up to $R = C/2$ and then decrease linearly for rates up to C. This is to be compared with the corresponding result for block codes:[5]

$$E(R) \approx E_L(R) \qquad \text{(29)}$$

$$\approx \begin{cases} \frac{C}{2} - R, & 0 \leq R \leq C/4 \\ (\sqrt{C} - \sqrt{R})^2, & C/4 \leq R < C. \end{cases}$$

The two exponents for very noisy channels (28) and (29) are plotted in Fig. 5. The relative improvement increases with rate. For $R = R_0 = C/2$, the exponent for convolutional codes is almost six times that for block codes.

While the upper and lower bound exponents are identical in the limiting case, we see from the example of the error-bound exponents for three binary symmetric channels (with $p = 0.01$, $p = 0.1$, and $p = 0.4$), shown normalized by C in Fig. 6, that as the channel becomes less noisy the upper and lower bounds diverge for $R < R_0$. In fact, if for all ρ, $E_0''(\rho) \equiv 0$, then $E_0(\rho) = \rho C$, so that $R_0 = C$. Thus, the upper bound exponent equals R_0 for all $R < C$.

There remains to show that this significant improvement over the performance of block codes is achievable without additional decoding complexity. But we observe that in decoding L branches or $L \ln q$ nats the decoding algorithm considered makes slightly less than Lq^K branch likelihood function computations or $Lvq^K = (L/K)Nq^K$ symbol likelihood function computations. Now the equivalent block code transmits $L \ln q$ nats in blocks of $K \ln q$ nats at a rate $R = \ln q/v = K \ln q/N$ nats/symbol, which corresponds to transmitting one of q^K words of length N symbols. Thus, the decoder must perform Nq^K symbol likelihood function computations per block and

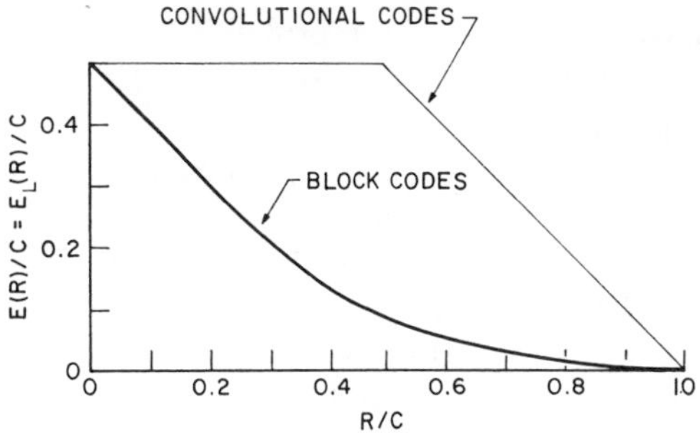

Fig. 5. $E(R)$ for very noisy channels with convolutional and block codes.

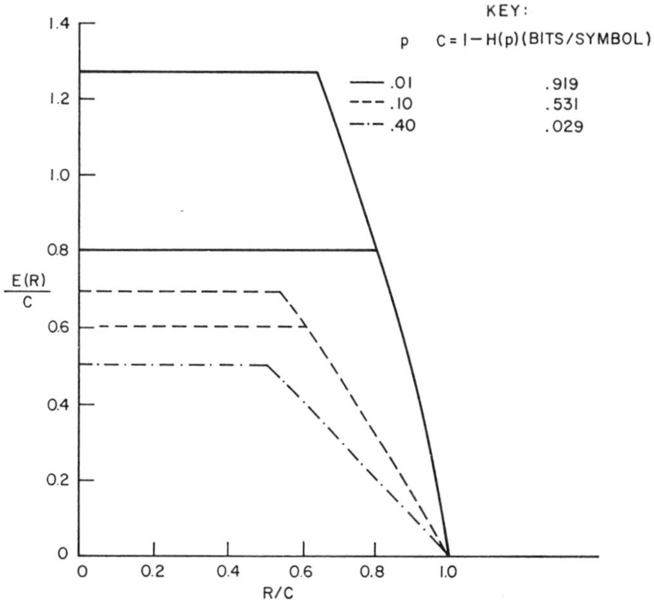

Fig. 6. $E(R)$ and $E_L(R)$ for the binary symmetric channels with evolutional codes ($p = 0.01$, $p = 0.1$, $p = 0.4$).

repeat this L/K times. Consequently, the number of computations is essentially the same for the convolutional code decoding algorithm described as is required for maximum likelihood decoding of the equivalent block code.

We should note, however, that since $K - 1$ zeros are inserted between trees of L branches, the actual rate for convolution codes is reduced by a factor of $L/(L + K - 1)$ from that of block codes, a minor loss since, because of the greatly increased exponent, we can afford to increase L (which affects P_E only linearly) enough to make this factor insignificant.

VII. A Semi-Sequential Modification of the Decoding Algorithm

We observe from (22) with the substitution $N = Kv = K \ln q/R$, that

$$P_E < \frac{L(q-1)}{1 - q^{-\epsilon/R}} (q^K)^{-R_0/R} \quad \text{for } 0 \leq R = R_0 - \epsilon < R_0 \quad (30)$$

for the specific decoding algorithm considered. However, as we have just noted, the number of likelihood function computations per decoded branch is slightly less that q^K, which means that the error probability decreases more than linearly with computational complexity for rates in this region.

Now let us consider an iterated version of the previous algorithm. At first we shall employ the aid of a magic genie. It is clear that the nonsequential decoding algorithm can be modified to make decisions based on k branches where $k < K$, the constraint length, and that the resulting error probability is the same as (30) with K replaced by k. Thus suppose the decoder attempts to decode the L-branch tree using $k = 1$ and at the end of the tree the genie either tells him he is correct or requires him to start over with $k = 2$ and that he proceeds in this way each time increasing k by 1 until he is either told he is correct or he reaches the constraint length K. Then, since at each iteration the number of computations is increased by a factor q, the number of computations per branch performed by the end of the kth iteration is $q + q^2 + \cdots + q^k = [q(q^k - 1)/(q - 1)] < 2q^k$. Thus, denoting the total number of computations per branch by γ, we have using (30),

$$\text{Prob}\,(\gamma > 2q^k) < \frac{L(q-1)}{1 - q^{-\epsilon/R}}\,(q^k)^{-R_0/R}, \qquad 0 \le R = R_0 - \epsilon < R_0$$

or

$$\text{Prob}\,(\gamma > \Gamma) < \frac{L(q-1)}{1 - q^{-\epsilon/R}}\left(\frac{\Gamma}{2}\right)^{-R_0/R}, \qquad 0 \le R = R_0 - \epsilon < R_0 \tag{31}$$

which is known as a Pareto distribution. Also, we have for the expected number of computations per branch

$$\bar{\gamma} < \sum_{k=1}^{K} q^k P_E(k-1) < \frac{L(q-1)}{1 - q^{-\epsilon/R}} \sum_{k=1}^{\infty} q^{-[(\epsilon k - R_0)/R]} = \frac{L(q-1)q}{(1 - q^{-\epsilon/R})^2}, \qquad 0 \le R = R_0 - \epsilon < R_0. \tag{32}$$

Thus, the expected number of computations per branch increases no more rapidly than the tree length for $R < R_0$, a feature of sequential decoding. Actually the Fano algorithm has been shown[10] to have a Pareto distribution on the number of computations with a higher exponent than R_0/R for $R < R_0$ and an expected number of computations which is independent of the tree or constraint length. However, with the Wozencraft algorithm[4] $\bar{\gamma}$ increases linearly with constraint length. The major drawback of this scheme, besides the genie which we shall dispose of presently, is that the number of storage registers required at the kth iteration is q^k and consequently the required storage capacity also has a Pareto distribution.

To avoid employing the genie, the decoder must have some other way to decide whether or not the kth iteration produces the correct path. One way to achieve this is to compare the likelihood function for the last N symbols of the decoded path with a threshold. If it exceeds this threshold the total path is accepted as correct; otherwise the algorithm is repeated with k increased by 1. Since the last N symbols occur after the tree has stopped branching, these can be affected by the last K branches only since no more than K data symbols are in the coder shift register when these channel symbols are being generated. Thus, there are only q^K possible combinations of channel symbols for the final branches which are of length N channel symbols. The upper bound on the probability of error for a threshold decision involving q^K code words of block length N selected independently is[11]

$$P_T < 2\exp\left[-NE_T(R)\right]$$

where

$$E_T(R) = \max_{p(x)} \Big\{\max_{0 \le \rho \le 1} \cdot \Big[-\ln \sum_X \sum_Y p(x)p(y \mid x)^{1-\rho}p(y)^\rho - \rho R\Big]\Big\} > 0, \qquad 0 \le R < C$$

and

$$R = \frac{K \ln q}{N} = \frac{\ln q}{v} \quad \text{as before.}$$

By choosing N or K large enough, P_T can be made sufficiently small, although clearly it can not be as small as P_E of (22), which results from use of the nonsequential algorithm.

Although this algorithm is rendered impractical by the excessive storage requirements, it contributes to a general understanding of convolutional codes and sequential decoding through its simplicity of mechanization and analysis.

Acknowledgment

The author gratefully acknowledges the helpful suggestions and patience of Dr. L. Kleinrock during numerous discussions.

References

[1] P. Elias, "Coding for noisy channels," *IRE Conv. Rec.*, pt. IV, pp. 37–46, 1955.
[2] H. L. Yudkin, "Channel state testing in information decoding," Ph.D. dissertation, Dept. of Elec. Engrg., M.I.T., Cambridge, Mass., September 1964.
[3] R. M. Fano, "A heuristic discussion of probabilistic decoding," *IEEE Trans. on Information Theory*, vol. IT-9, pp. 64–76, April 1963.
[4] J. M. Wozencraft and B. Reiffen, *Sequential Decoding*. Cambridge, Mass.: M.I.T. Press, and New York: Wiley, 1961.
[5] R. G. Gallager, "A simple derivation of the coding theorem and some applications," *IEEE Trans. on Information Theory*, vol. IT-11, pp. 3–18, January 1965.
[6] R. M. Fano, *Transmission of Information*. Cambridge, Mass.: M.I.T. Press, and New York: Wiley, 1961.
[7] C. E. Shannon, R. G. Gallager, and E. R. Berlekamp, "Lower bounds to error probability for coding on discrete memoryless channels," *Information and Control* (to be published).
[8] B. Reiffen, "Sequential encoding and decoding for the discrete memoryless channel," M.I.T. Lincoln Laboratory, Lexington, Mass., Rept. 25, G-0018, August 1960.
[9] J. L. Massey, private communication.
[10] J. E. Savage, "Sequential decoding—The computation problem," *Bell Sys. Tech. J.*, vol. 45, pp. 149–175, January 1966.
[11] C. E. Shannon, unpublished notes.

Inverses of Linear Sequential Circuits

JAMES L. MASSEY, MEMBER, IEEE, AND MICHAEL K. SAIN, MEMBER, IEEE

Abstract—**This paper states the necessary and sufficient conditions for the existence of a feedforward inverse for a feedforward linear sequential circuit and gives an implicit procedure for constructing such inverses. It then goes on to give the necessary and sufficient conditions for the existence of general inverses with finite delay and gives procedures for constructing a class of such inverses. The discussion considers both the transfer function matrix description and the structural matrix description of the linear sequential circuit, together with the complementary nature of the results obtained from these two viewpoints. Finally, a large part of the work is motivated by results and techniques which have been applied in the study of continuous-time linear dynamical systems and thus serves to point out the advantages which may accrue through simultaneous study of both continuous-time systems and linear sequential circuits.**

Index Terms—**Convolution codes, feedforward sequential circuits, information lossless, inverses, linear sequential circuits.**

I. Introduction

THE STUDY of linear sequential circuits is well established as a "classical" subfield of automata theory by virtue of the depth and breadth of known results. This paper considers an aspect of the theory of linear sequential machines that seems to have received little attention, namely, the existence and construction of inverse circuits. Intuitively, a sequential circuit with N inputs and K outputs is an "inverse with delay L" for a linear sequential circuit with K inputs and N outputs if the former circuit, when cascaded with the latter, yields an overall sequential circuit which acts as a pure delay of L time units. An inverse with delay $L=0$ is called an "instantaneous" inverse.

In Section II, it is shown that a practical problem in error-correcting codes raises an interesting theoretical question concerning inverses, namely, when does a feedforward linear sequential circuit possess a feedforward inverse? A complete answer is provided in Sections III and IV, including an implicit construction technique when the inverse exists.

In Section V, the existence of inverses is treated in a more formal manner. A necessary and sufficient condition is given for the existence of an inverse with delay L in terms of the structural matrices which characterize the linear sequential circuit.

Finally, in Section VI, a construction procedure for inverses is formulated by analogy to techniques used in the study of linear dynamical systems. This construction technique proceeds from the structural matrices of the original linear sequential circuit to the structural matrices of the inverse.

Manuscript received August 28, 1967; revised January 4, 1968. This work was supported in part by National Science Foundation Grant GK 804. A preliminary version of these results was presented at the 8th Annual Symposium on Switching and Automata Theory, Austin, Tex., October 1967.

The authors are with the Department of Electrical Engineering, University of Notre Dame, Notre Dame, Ind. 46556

II. Convolutional Codes and Feedforward Inverses

The class of error-correcting codes known as convolutional codes is becoming increasingly important in coding theory because of ease in implementation, suitability for sequential decoding, and certain inherent superiorities relative to block codes.[1] Convolutional codes may be conveniently described from a sequential circuit point of view,[2] and this description leads to an interesting problem concerning the existence of a special kind of inverse.

Let $i_j(0)$, $i_j(1)$, $i_j(2)$, $\cdots$, be a sequence of digits from a finite field $GF(q)$. The D transform of such a sequence will be denoted as

$$I_j(D) = i_j(0) + i_j(1)D + i_j(2)D^2 + \cdots.$$

Let $I_j(D)$, $j=1, 2, \cdots, K$ be considered as the transforms of K sequences of information digits to be encoded in a convolutional code of rate $R=K/N$. Such a code is defined by a set of polynomials, $G_{ij}(D)$, $i=1, 2, \cdots, N$, $j=1, 2, \cdots, K$, over $GF(q)$, called the "code-generating polynomials,"[3] such that the transforms of the N sequences of encoded digits $T_i(D)$, $i=1, 2, \cdots, N$ may be written as

$$T_i(D) = \sum_{j=1}^{K} G_{ij}(D)I_j(D). \tag{1}$$

Equation (1) may be written in matrix form as

$$\boldsymbol{T}(D) = \overline{\boldsymbol{G}}(D)\boldsymbol{I}(D) \tag{2}$$

where $\overline{\boldsymbol{G}}(D)$ is the $N\times K$ matrix whose ijth entry is $G_{ij}(D)$, and where $\boldsymbol{T}(D)$ and $\boldsymbol{I}(D)$ are vectors whose components are $T_j(D)$ and $I_j(D)$, respectively.

From a sequential circuit viewpoint, $\overline{\boldsymbol{G}}(D)$ can be interpreted as the transfer function matrix of a K-input N-output "linear sequential circuit" (LSC). The fact that these transfer functions are all polynomials specifies the LSC as having finite input-memory, that is, as having a "feedforward" (FF) realization in terms of delay units, $GF(q)$ adders, and $GF(q)$ scalors. From the coding standpoint, it is essential that this LSC should also possess an FF inverse, either instantaneous or with delay, so that decoding errors made on the corrupted encoded

Reprinted from *IEEE Trans. Comput.*, vol. C-17, pp. 330–337, Apr. 1968.

sequences received at the decoder do not interfere indefinitely with the recovery of the information sequences.

As an illustration of the necessity of this latter condition, consider the $K=1$, $N=2$ binary code with $G_{11}(D)=1+D$ and $G_{21}(D)=1+D^2$. The all-one information sequence with transform $I_1(D)=1/(1+D)$ then results in the encoded sequences having the transforms $T_1(D)=1$ and $T_2(D)=1+D$ which contain only three nonzero digits. If these digits are changed to zeros by the channel, the decoder will estimate $T_1(D)=T_2(D)=0$, and hence $I_1(D)=0$, since the actual received sequence is a valid encoded sequence. Thus, although only three errors have been made in decoding the received sequences, every information digit is decided incorrectly; that is, the decoding errors are never forgotten in passing from the encoded sequences to the information sequence. However, for the code $G_{11}(D)=1+D$ and $G_{21}(D)=1+D+D^2$, it is seen immediately that $DT_1(D)+T_2(D)=(D+D^2+1+D+D^2)I_1(D)=I_1(D)$, so that an FF inverse exists for the encoding LSC and decoding errors are forgotten after one time unit in recovering the information sequence from the encoded sequences.

The discussion above has singled out the following problem in automata theory as being of practical import in connection with convolutional coding: under what conditions does an FF LSC possess an FF inverse, either instantaneous or with delay? The solution to this problem is given in Sections III and IV.

In conclusion of Section II, it is convenient to prove a lemma which is useful in demonstrating when an FF inverse does not exist.

Lemma 1: If for a given LSC there exists an input sequence with infinitely many nonzero digits such that the corresponding output sequence has only finitely many nonzero digits, then the LSC has no FF inverse, with delay or without delay.

Proof: Note that the term sequence is used in the vector sense. It is assumed that no digits are stored in the encoder at the start of the information sequence. Accordingly, the all-zero information sequence yields the all-zero encoded sequence. Therefore, the all-zero encoded sequence must always be inverted as the all-zero information sequence. Hence if the inverse has input-memory M, then M time units after its last nonzero input it must commence to produce only zeros. Thus, the information sequence hypothesized in the lemma could not be recovered, and no finite-input-memory inverse can exist.

III. Single-Input LSCs and FF Inverses

Let $p_1(D)$, $p_2(D)$, $\cdots$, $p_n(D)$ be polynomials over some field. Their greatest common divisor, denoted

$$\text{GCD}\,[p_1(D), p_2(D), \cdots, p_n(D)],$$

is the monic (highest coefficient unity) polynomial of greatest degree which divides each polynomial. It is known[4] that there exist polynomials $a_1(D), a_2(D), \cdots, a_n(D)$ such that

$$a_1(D)p_1(D) + \cdots + a_n(D)p_n(D) = \text{GCD}\,[p_1(D), \cdots, p_n(D)]. \quad (3)$$

With this preliminary, the problem posed in the previous section is readily solved for the single-input ($K=1$) case.

Theorem 1: A single-input N-output FF LSC has an FF inverse, either with delay or without delay, if and only if

$$\text{GCD}\,[G_{11}(D), G_{21}(D), \cdots, G_{N1}(D)] = D^L \quad (4)$$

for some $L \geq 0$. Moreover, there exists an FF inverse with delay exactly L, and no inverse of any kind exists with smaller delay.

Proof: Suppose first that (4) is satisfied for some $L \geq 0$. Then by (3) there exist polynomials $Q_1(D)$, $Q_2(D)$, $\cdots$, $Q_N(D)$ such that

$$Q_1(D)G_{11}(D) + Q_2(D)G_{21}(D) + \cdots + Q_N(D)G_{N1}(D) = D^L. \quad (5)$$

When multiplied by $I_1(D)$, (5) becomes

$$Q_1(D)T_1(D) + Q_2(D)T_2(D) + \cdots + Q_N(D)T_N(D) = D^L I_1(D), \quad (6)$$

and hence the $Q_i(D)$ are the transfer functions of an N-input single-output FF LSC which recovers $I_1(D)$ with delay exactly L.

Conversely, suppose that (4) is violated, that is, there exists no non-negative integer L such that (4) is satisfied. Then there exists a polynomial $P(D)$ with $P(0)\neq 0$ and of degree at least one such that $P(D)$ divides $G_{j1}(D)$ for $j=1, 2, \cdots, N$. The input sequence whose transform is $I_1(D)=1/P(D)$ then contains infinitely many nonzero digits, but each output sequence $T_j(D)=G_{j1}(D)/P(D)$ is a polynomial and hence has only finitely many nonzero digits. From Lemma 1, it follows that no FF inverse exists.

Finally, if (4) is satisfied, each $G_{j1}(D)$ has D^L as a factor, and thus there is no nonzero response until L time units after an input is applied. Thus no inverse can have delay less than L.

The simplicity of the single-input case does not extend to the multiple-input situation which is treated in Section IV. It is important to remark, however, that the single-input problems are of greatest current interest in the theory of convolutional codes.

IV. Multiple-Input LSCs and FF Inverses

In the general case, $\bar{G}(D)$ is an $N\times K$ matrix having $\binom{N}{K}$ distinct $K\times K$ submatrices. Let $\Delta_i(D)$, $i=1, 2, \cdots, \binom{N}{K}$ denote the determinants of these submatrices. In terms of these quantities, the main result of this section can be stated as follows.

Theorem 2: A K-input N-output FF LSC has an FF inverse, either with delay or without delay, if and only if

$$\text{GCD}\left[\Delta_i(D), i = 1, 2, \cdots, \binom{N}{K}\right] = D^L \quad (7)$$

for some $L \geq 0$. Moreover, there exists an FF inverse with delay exactly L. (But, for $K > 1$, there may exist FF inverses with delay less than L.)

Proof of Sufficiency of Condition (7): Let $\overline{G}(D)_n$ denote the nth $K \times K$ submatrix of $\overline{G}(D)$, and let $\overline{C}(D)_n$ be the matrix whose ijth element is the cofactor of the jith element of $\overline{G}(D)_n$. Note that the entries in $\overline{C}(D)_n$ are polynomials. Also, let $\boldsymbol{T}(D)_n$ be the vector of K output transforms corresponding to the rows used to form $\overline{G}(D)_n$. Clearly,

$$\overline{C}(D)_n \overline{G}(D)_n = \Delta_n(D) \bar{I}_K, \quad (8)$$

where $\bar{I}_K$ is the $K \times K$ identity matrix. Upon postmultiplying both sides of (8) by $\boldsymbol{I}(D)$, it follows that

$$\overline{C}(D)_n \boldsymbol{T}(D)_n = \Delta_n(D) \boldsymbol{I}(D). \quad (9)$$

From (9), it is seen that FF LSC's with transfer function matrices $\overline{C}(D)_n$ can be used to produce $\Delta_n(D) I_j(D)$, for each j and for $n = 1, 2, \cdots, \binom{N}{K}$. In addition, if condition (7) is satisfied, it is possible to find a further FF circuit to act on these $\binom{N}{K}$ outputs $\Delta_n(D) I_j(D)$, in accordance with Theorem 1, to produce the final output sequence $D^L I_j(D)$. The cascade of these FF LSCs is then an FF inverse with delay exactly L.

Before proceeding to the proof of necessity in Theorem 2, it is useful first to obtain certain ancillary results. Recall that a polynomial is called "irreducible" if it has degree at least one and cannot be expressed as the product of polynomials of a smaller degree. An irreducible factor $m(D)$ of a polynomial $P(D)$ has "multiplicity" e if $[m(D)]^e$ divides $P(D)$ but $[m(D)]^{e+1}$ does not.

Lemma 2: If $\Delta_n(D)$, the determinant of $\overline{G}(D)_n$, is nonzero and has an irreducible factor $m(D)$ of multiplicity e, then there is some element of $\overline{G}(D)_n$ whose cofactor is not divisible by $[m(D)]^e$.

Proof: From (8), it is seen that

$$[\Delta_n(D)] \times \det [\overline{C}(D)_n] = [\Delta_n(D)]^K, \quad (10)$$

or equivalently

$$\det [\overline{C}(D)_n] = [\Delta_n(D)]^{K-1}.$$

But if every cofactor has $[m(D)]^e$ as a factor, then the left-hand side of (10) has $m(D)$ as an irreducible factor of multiplicity at least Ke. The right member of (10), however, has $m(D)$ as an irreducible factor of multiplicity only $(K-1)e$. The lemma follows, therefore, by contradiction.

Lemma 3: Under the hypothesis of Lemma 2, there exists a vector of polynomials $\boldsymbol{P}(D) = (P_1(D), \cdots, P_K(D))$ such that for at least one index r, $[m(D)]^e$ does not divide $P_r(D)$, and such that

$$\overline{G}(D)_n \boldsymbol{P}(D) = \Delta_n(D) \boldsymbol{Q}(D), \quad (11)$$

where $\boldsymbol{Q}(D) = (Q_1(D), \cdots, Q_K(D))$ is also a vector of polynomials.

Proof: By Lemma 2, there is some row of $\overline{C}(D)_n$, say the rth, whose greatest common divisor $P_r(D)$ is not divisible by $[m(D)]^e$. Further, there exist polynomials $Q_1(D), \cdots, Q_K(D)$ such that

$$P_r(D) = Q_1(D) C_{r1}(D)_n + \cdots + Q_K(D) C_{rK}(D)_n,$$

where $C_{ij}(D)_n$ denotes the ijth element of $\overline{C}(D)_n$. Define $P_i(D)$, $i \neq r$, from the relation

$$\boldsymbol{P}(D) = \overline{C}(D)_n \boldsymbol{Q}(D). \quad (12)$$

Multiplying both sides in (12) by $\overline{G}(D)_n$, (11) follows, and the lemma is proved.

Lemma 4: For any polynomial vector $\boldsymbol{P}(D)$ satisfying (11), and for any index j, $1 \leq j \leq N$, the jth row of $\overline{G}(D)$ satisfies

$$\sum_{i=1}^{K} G_{ji}(D) P_i(D) = \Delta(D) A_j(D), \quad (13)$$

where $\Delta(D) = \text{GCD}\ [\Delta_i(D),\ i = 1,\ 2, \cdots, \binom{N}{K}]$ and $A_j(D)$ is some polynomial.

Proof: If j corresponds to one of the rows of $\overline{G}(D)_n$, then (13) follows directly from (11). For other j, define the polynomials $B_1(D), \cdots, B_K(D)$ as the solutions of

$$[B_1(D), \cdots, B_K(D)] \overline{G}(D)_n = \Delta_n(D) [G_{j1}(D), \cdots, G_{jk}(D)]. \quad (14)$$

Postmultiplying by $\overline{C}(D)_n$ in (14), it follows that

$$[B_1(D), \cdots, B_K(D)] = [G_{j1}(D), \cdots, G_{jK}(D)] \overline{C}(D)_n \quad (15)$$

from which it is seen that each $B_i(D)$ is the determinant of some $K \times K$ submatrix of $\overline{G}(D)$ and hence is divisible by $\Delta(D)$, the greatest common divisor of these determinants. Moreover, upon postmultiplying by $\boldsymbol{P}(D)$ in (14), (11) implies

$$[B_1(D), \cdots, B_K(D)] \boldsymbol{Q}(D) = \sum_{i=1}^{K} G_{ji}(D) P_i(D). \quad (16)$$

Defining the left member of (16) as $A_j(D)\ \Delta(D)$ gives (13).

With the completion of these preliminaries, it is possible to return to the question of necessity in Theorem 2.

Proof of the Necessity of Condition (7): Suppose that the greatest common divisor $\Delta(D)$ of the $\Delta_i(D)$ is not D^j for some j and further suppose that $\Delta(D) \neq 0$. Then let $m(D) \neq D$ be an irreducible factor of $\Delta(D)$ of multiplicity e. There is then some index, say n, such that

$\Delta_n(D)$ has $m(D)$ as an irreducible factor of multiplicity exactly e. For this n, define $\boldsymbol{P}(D)$ as the polynomial vector given in Lemma 3 and define the input sequences by the transforms

$$I_i(D) = P_i(D)/[m(D)]^e, \quad i = 1, 2, \cdots, K. \quad (17)$$

From Lemma 3, it follows that at least one of these input sequences has infinitely many nonzero digits. The corresponding output sequences have transforms given by

$$T_j(D) = \sum_{i=1}^{K} G_{ji}(D)I_i(D) = [\Delta(D)A_j(D)]/[m(D)]^e,$$
$$j = 1, 2, \cdots, N, \quad (18)$$

where (13) from Lemma 4 has been employed. But the right member of (18) is a polynomial since $[m(D)]^e$ divides $\Delta(D)$, so that there are only finitely many nonzero digits in each output sequence. By Lemma 1, no FF inverse exists. Hence, the necessity of condition (7) is established.

In coding theory, attention is often restricted to "systematic" codes in which $\bar{G}(D)_1 = \bar{I}_K$, so that condition (7) is automatically satisfied with $L=0$. However, nonsystematic codes are of special importance in connection with sequential decoding, so that condition (7) is one that should always be considered when choosing a nonsystematic convolutional code. It is interesting to note that if the $G_{ji}(D)$ are selected randomly, as is often considered in coding theory, then there is a significant probability that condition (7) will not be satisfied. For instance, with $K=1$ and $N=2$, the probability is 1/2 in the binary case that $G_{j1}(D), j=1, 2$ will have an even number of terms and hence be divisible by $(1+D)$. Thus, the probability is 1/4 that $\Delta(D)$ will have $(1+D)$ as a factor.

Theorems 1 and 2 are not restricted to systems defined over finite fields, but apply just as well to sampled-data circuits in which the field in question is that of the real numbers.

As an example of a case in which L in (7) is not the least delay for an FF inverse system, consider the K-input K-output LSC with $\bar{G}(D) = D\bar{I}_K$, where $\bar{I}_K$ is the $K \times K$ identity matrix. Clearly, there is an FF inverse with delay unity, but $\Delta(D) = D^K$ so that $L > 1$ if $K > 1$.

V. The Existence of Inverses with Delay L

In the preceding sections, LSCs have been characterized by their transfer function matrices. The alternative formulation in terms of the structural matrices $\bar{A}$, $\bar{B}$, $\bar{C}$, and $\bar{E}$ will be exploited in the sequel.

Let $\boldsymbol{i}(k)$ denote the K vector whose components are the input digits at time unit k, $i_j(k), j=1, 2, \cdots, K$. Similarly let $\boldsymbol{t}(k)$ denote the N vector of output digits at time unit k. Then a "state vector" $\boldsymbol{x}(k)$ having S components, and matrices $\bar{A}$, $\bar{B}$, $\bar{C}$, and $\bar{E}$ can be de-

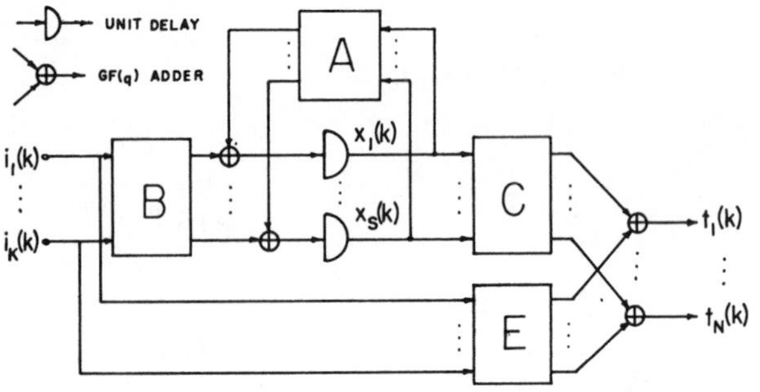

Fig. 1. Canonical realization of an LSC in terms of its structural matrices.

fined so that operation of the LSC may be expressed as[5]

$$\boldsymbol{x}(k+1) = \bar{A}\boldsymbol{x}(k) + \bar{B}\boldsymbol{i}(k) \quad (19a)$$
$$\boldsymbol{t}(k) = \bar{C}\boldsymbol{x}(k) + \bar{E}\boldsymbol{i}(k) \quad (19b)$$

for $k=0, 1, 2, \cdots$. It is assumed throughout the remainder of the paper that $\boldsymbol{x}(0) = \boldsymbol{0}$, where $\boldsymbol{0}$ denotes the all-zero vector; that is, it is assumed that the LSC is initially at rest.

A canonical realization of the LSC in terms of its structural matrices is shown in Fig. 1. The state vector is taken as the outputs of a set of S unit delay stages. The blocks in the figure which are labelled with the structural matrices denote the necessary scalors (constant multipliers) and $GF(q)$ adders so that the vector output of the block is the associated matrix applied to the vector of inputs. Conversely, any LSC constructed from unit delays, scalors, and adders may be considered as in the canonical realization of Fig. 1, and its structural matrices may be readily determined. When desired, the transfer function matrix can always be found from the relation

$$\bar{G}(D) = \bar{C}[D^{-1}\bar{I}_S - \bar{A}]^{-1}\bar{B} + \bar{E}, \quad (20)$$

where the entries in (20) are, in general, rational functions in D rather than simply polynomials as was the case for the FF LSCs considered in Sections II through IV.

Under the assumption that $\boldsymbol{x}(0) = \boldsymbol{0}$, (19) can be solved to obtain

$$\boldsymbol{t}(0) = \bar{E}\boldsymbol{i}(0) \quad (21a)$$
$$\boldsymbol{t}(k) = \sum_{j=1}^{k} \bar{C}\bar{A}^{j-1}\bar{B}\boldsymbol{i}(k-j) + \bar{E}\boldsymbol{i}(k) \quad (21b)$$

for $k=1, 2, \cdots$. For convenience, introduce the notation

$$\bar{J}_0 = \bar{E}$$
$$\bar{J}_j = \bar{C}\bar{A}^{j-1}\bar{B}, \quad j = 1, 2, \cdots,$$

so that (21) can be rewritten as

$$\boldsymbol{t}(k) = \sum_{j=0}^{k} \bar{J}_j\boldsymbol{i}(k-j). \quad (22)$$

From (22), it follows that the response over the first $L+1$ time units is given by the matrix equation

$$\begin{bmatrix} t(0) \\ t(1) \\ \cdot \\ \cdot \\ \cdot \\ t(L) \end{bmatrix} = \begin{bmatrix} \bar{J}_0 & \bar{0} & \cdots & \bar{0} \\ \bar{J}_1 & \bar{J}_0 & \cdots & \bar{0} \\ \cdot & \cdot & \cdots & \cdot \\ \cdot & \cdot & \cdots & \cdot \\ \bar{J}_L & \bar{J}_{L-1} & \cdots & \bar{J}_0 \end{bmatrix} \begin{bmatrix} i(0) \\ i(1) \\ \cdot \\ \cdot \\ \cdot \\ i(L) \end{bmatrix} \tag{23}$$

which will be abbreviated to

$$T_L = \bar{M}_L I_L. \tag{24}$$

T_L, the left member of (23), is the "response segment" over time units 0 through L. Similarly, I_L is the "input segment" over the same period. $\bar{M}_L$ is the matrix which via (23) connects the output and input segments.

The following formal definition may now be stated.

Definition 1: An LSC "has an inverse with delay L" if for every non-negative integer k, the input segment I_k is uniquely determined by the response segment T_{k+L}.

From the linearity of an LSC, Theorem 3 readily follows.

Theorem 3: An LSC has an inverse with delay L if and only if I_0 is uniquely determined by T_L, that is, if and only if (24) may be uniquely solved for $i(0)$.

Proof: Necessity follows directly from Definition 1. Sufficiency may be shown as follows. Suppose $i(0)$ can be found from T_L. Then the effect of $i(0)$ on the entire output sequence can be subtracted out. The remaining modified output sequence, omitting the time-unit-0 portion, is the same as if $i(1)$ were the first input to the LSC. Hence $i(1)$ can also be found from its first $L+1$-time-unit portion. Thus I_1 can be found from T_{L+1}. By a similar argument I_k can be found from T_{L+k} for every k.

Now the rank of the matrix $\bar{M}_L$, denoted rank $(\bar{M}_L)$, is the maximum number of linearly independent columns (or rows) that can be found in $\bar{M}_L$. Alternatively, rank $(\bar{M}_L)$ is the dimension of the column space (or row space) of $\bar{M}_L$, that is, the dimension of the vector space consisting of all distinct linear combinations of columns (or rows) of $\bar{M}_L$.

It is now possible to state the main result of this section. By way of convention, rank $(\bar{M}_{-1})$ is taken to be 0.

Theorem 4: An LSC has an inverse with delay L if and only if

$$\text{rank}\,(\bar{M}_L) = \text{rank}\,(\bar{M}_{L-1}) + K, \tag{25}$$

where K is the number of inputs for the LSC.

As a simple example of this result, consider the binary LSC with $K=2$, $N=2$, and $S=1$, defined by

$$\bar{A} = [0], \quad \bar{B} = [0 \quad 1], \quad \bar{C} = [0 \quad 1]', \quad \bar{E} = \begin{bmatrix} 1 & 0 \\ 0 & 0 \end{bmatrix},$$

where the superscript (′) denotes transposition. A simple calculation yields

$$\bar{M}_0 = \begin{bmatrix} 1 & 0 \\ 0 & 0 \end{bmatrix}, \quad \text{rank}\,(\bar{M}_0) = 1,$$

$$\bar{M}_1 = \left[\begin{array}{cc|cc} 1 & 0 & 0 & 0 \\ 0 & 0 & 0 & 0 \\ \hline 0 & 0 & 1 & 0 \\ 0 & 1 & 0 & 0 \end{array}\right], \quad \text{rank}\,(\bar{M}_1) = 3,$$

so that an inverse with delay $L=1$ does exist but there is no inverse with delay $L=0$.

Before proving Theorem 4, it is convenient first to establish the following.

Lemma 5: An LSC has an inverse with delay L if and only if $T_L = \mathbf{0}$ implies that $i(0) = \mathbf{0}$.

Proof: From Theorem 3, it follows that an inverse with delay L exists if and only if the same T_L cannot result from two input segments I_L with differing values of $i(0)$. This latter condition, by linearity, is equivalent to the condition that the difference of these two input segments, which is an I_L with $i(0) \neq \mathbf{0}$, must not result in the response segment $T_L = \mathbf{0}$. But this is the condition in Lemma 5, which is thereby established.

Proof of Theorem 4: From (24) and Lemma 5, it follows that an inverse with delay L exists if and only if both of the following conditions are satisfied.

1) The first K columns of $\bar{M}_L$ are linearly independent, for otherwise and only then will there be an $i(0) \neq \mathbf{0}$, with $i(1) = i(2) = \cdots = i(L) = \mathbf{0}$, which results in $T_L = \mathbf{0}$.

2) The column space of the first K columns of $\bar{M}_L$ and the column space of the remaining KL columns of $\bar{M}_L$ have only $\mathbf{0}$ in common, for otherwise and only then will there be an $i(0) \neq \mathbf{0}$, together with values of $i(1)$, $i(2), \cdots, i(L)$ not all zero, such that $T_L = \mathbf{0}$.

Next note directly from (24) that the column space of the last KL columns of $\bar{M}_L$ has dimension just rank $(\bar{M}_{L-1})$. Moreover, conditions 1) and 2) above will both be satisfied if and only if adding the first K columns increases the dimension of the column space by K. Hence the theorem follows.

Similar arguments can be used to show that rank $(\bar{M}_L) -$ rank $(\bar{M}_{L-1})$ is a monotonically nondecreasing function of L. Since this difference is at most K, it follows that the following limit always exists

$$\lim_{L\to\infty} \frac{1}{K} [\text{rank}\,(\bar{M}_L) - \text{rank}\,(\bar{M}_{L-1})]$$

and equals unity when and only when there is some finite L such that an inverse with delay L exists.

In the proof of Theorem 4, condition 1), namely that the first K columns of $\bar{M}_L$ are linearly independent, has an interesting interpretation. These K columns are partitioned by rows into the blocks $\bar{J}_0, \bar{J}_1, \cdots, \bar{J}_L$, which is evident in (23). Condition 1) can then be restated as a corollary.

Corollary 1: An LSC has an inverse with delay L only if

$$\text{rank}\,(\bar{J}_0'\bar{J}_1' \cdots \bar{J}_L') = K. \tag{26}$$

By the Cayley–Hamilton result and Corollary 1 the necessary condition (26) can be strengthened to the following corollary.

Corollary 2: If condition (26) is not satisfied for $L=S$, then it is never satisfied, and no inverse with finite delay exists.

Consider now the input sequence $\boldsymbol{i}(0)$, **0**, **0**, **0**, $\cdots$. For this situation, (24) becomes

$$T_L = [\bar{J}_0'\bar{J}_1'\bar{J}_2' \cdots \bar{J}_L']'\boldsymbol{i}(0), \tag{27}$$

which by the linearity of the LSC is the relationship governing the determination of the input at a single nonzero point in time from the observed outputs. This suggests the following definition.

Definition 2: An LSC is "pointwise input observable" if every $\boldsymbol{i}(0)$ in the input sequence $\boldsymbol{i}(0)$, **0**, **0**, **0**, $\cdots$, is uniquely determined from the output sequence $\boldsymbol{t}(0)$, $\boldsymbol{t}(1)$, $\boldsymbol{t}(2)$, $\cdots$.

From (27) and Corollary 2, it is clear that an LSC is pointwise input observable if and only if

$$\text{rank}\,(\bar{J}_0'\bar{J}_1'\bar{J}_2' \cdots \bar{J}_S') = K. \tag{28}$$

Moreover, pointwise input observability is then a necessary condition for an LSC to have an inverse with finite delay L.

VI. The Construction of Inverses with Finite Delay

The present section continues the assumptions of Section V, namely, that the LSC is described in the manner of (19) by a four-tuple of matrices $(\bar{A}, \bar{B}, \bar{C}, \bar{E})$ and that it is initially at rest. The results are motivated by the work of Youla and Dorato[6] on continuous-time, linear dynamical systems for the case $N=K$, and extensions are made herein for the case $N>K$. The development serves as an explicit example of the applicability of certain algebraic approaches in both LSCs and continuous-time dynamical systems.

Denote the first nonzero matrix in the sequence $\bar{J}_0, \bar{J}_1, \bar{J}_2, \cdots$, as $\bar{J}_I$. The discussion divides naturally into two parts, according to whether the rank $(\bar{J}_I)$ is equal to K or is not equal to K.

Theorem 5: Let rank $(\bar{J}_I)=K$. Then the LSC has an inverse with delay I which can be realized in the form

$$\begin{aligned} \boldsymbol{i}(k) &= -(\bar{F}\bar{J}_I)^{-1}\bar{F}\bar{C}\bar{A}^I\mathbf{x}(k) \\ &\quad + (\bar{F}\bar{J}_I)^{-1}\bar{F}\boldsymbol{t}(k+I) \end{aligned} \tag{29a}$$

$$\begin{aligned} \mathbf{x}(k+1) &= [\bar{A} - \bar{B}(\bar{F}\bar{J}_I)^{-1}\bar{F}\bar{C}\bar{A}^I]\mathbf{x}(k) \\ &\quad + \bar{B}(\bar{F}\bar{J}_I)^{-1}\bar{F}\boldsymbol{t}(k+I), \end{aligned} \tag{29b}$$

where (29) is initiated with $\mathbf{x}(0)=\mathbf{0}$ just I time units after (19) is initiated and where $\bar{F}$ is any $K\times N$ matrix such that $\bar{F}\bar{J}^I$ has an inverse. Moreover, there is no inverse with less delay.

Proof: Under the assumptions of the theorem, namely, $\bar{J}_0=\bar{J}_1= \cdots =\bar{J}_{I-1}=\bar{0}$, it is always possible to write

$$\boldsymbol{t}(j) = \mathbf{0}, \quad j = 0, 1, 2, \cdots, I-1, \tag{30a}$$

$$\boldsymbol{t}(k+I) = \bar{C}\bar{A}^I\mathbf{x}(k) + \bar{J}_I\boldsymbol{i}(k), \quad k = 0, 1, 2, \cdots. \tag{30b}$$

That there is no inverse with delay less than I follows immediately from (30a). Moreover, (30b) can be viewed as N equations in K unknowns (which are the K components of $\boldsymbol{i}(k)$). That these equations have a solution follows from construction; that the solution is unique follows from the fact that rank $(\bar{J}_I)=K$. Premultiplication of both members of (30b) by a matrix $\bar{F}$ which selects a set of K linearly independent equations from the N equations, which are given, yields

$$\bar{F}\boldsymbol{t}(k+I) = \bar{F}\bar{C}\bar{A}^I\mathbf{x}(k) + \bar{F}\bar{J}_I\boldsymbol{i}(k),$$

where $\bar{F}\bar{J}_I$ has an inverse. Inversion gives (29a) and substitution of (29a) into (19a) gives (29b), and thus the theorem is established. Although in Theorem 5 the existence of the inverse with delay I was established by construction, it should be observed that Theorem 4 predicts the existence of such a construction, since

$$\text{rank}\,(\bar{J}_I) = \text{rank}\,(\bar{J}_{I-1}) + K = 0 + K = K.$$

As an example illustrating the difference between the inverses of Theorems 1 and 5, consider the $K=1$, $N=2$ binary LSC with $G_{11}(D)=D$ and $G_{21}(D)=1+D+D^2$. By Theorem 1 this FF LSC must have an instantaneous FF inverse. This is indeed the case, for $(1+D)G_{11}(D)+G_{21}(D)=1$. The realization (19) can be given in the form

$$\bar{A} = \begin{bmatrix} 0 & 1 \\ 0 & 0 \end{bmatrix}, \qquad \bar{B} = \begin{bmatrix} 0 \\ 1 \end{bmatrix}$$

$$\bar{C} = \begin{bmatrix} 0 & 1 \\ 1 & 1 \end{bmatrix}, \qquad \bar{E} = \begin{bmatrix} 0 \\ 1 \end{bmatrix}.$$

Since $\bar{J}_0=\bar{E}$ has rank one, Theorem 4 shows the existence of the instantaneous inverse, and Theorem 5 (with $\bar{F}=[0\ 1]$) constructs such an inverse having the transfer functions $H_{11}(D)=0$ and $H_{12}(D)=1/(1+D+D^2)$ which is clearly not FF.

More generally, the transfer function matrix $\bar{H}(D)$ associated with (29) by means of (20) is given by

$$\begin{aligned} \bar{H}(D) &= [-(\bar{F}\bar{J}_I)^{-1}\bar{F}\bar{C}\bar{A}^I[D^{-1}\bar{I}_S - \bar{A} \\ &\quad + \bar{B}(\bar{F}\bar{J}_I)^{-1}\bar{F}\bar{C}\bar{A}^I]^{-1}\bar{B}(\bar{F}\bar{J}_I)^{-1}\bar{F} + (\bar{F}\bar{J}_I)^{-1}\bar{F}]D^{-I} \\ &= [D^{-I}(\bar{F}\bar{J}_I)^{-1}\bar{F}][\bar{I}_N - \bar{C}\bar{A}^I(D^{-1}\bar{I}_S - \bar{A})^{-1} \\ &\quad \cdot[\bar{I}_S + \bar{B}(\bar{F}\bar{J}_I)^{-1}\bar{F}\bar{C}\bar{A}^I(D^{-1}\bar{I}_S - \bar{A})^{-1}]^{-1} \\ &\quad \cdot\bar{B}(\bar{F}\bar{J}_I)^{-1}\bar{F}] \\ &= D^{-I}(\bar{F}\bar{J}_I)^{-1}\bar{F}[\bar{I}_N \\ &\quad + \bar{C}\bar{A}^I(D^{-1}\bar{I}_S - \bar{A})^{-1}\bar{B}(\bar{F}\bar{J}_I)^{-1}\bar{F}]^{-1}, \end{aligned} \tag{31}$$

where the last step follows from a well-known matrix inverse identity.[7] In order to make (31) causal, and to conform with Theorem 5 in initiating (29) with an I-time-unit lag, the inverse transfer function matrix will be written $D^I\bar{H}(D)$. The original LSC transfer function matrix $\bar{G}(D)$ obtained from (19) and (30b) is

$$\begin{aligned}\bar{G}(D) &= D^I[\bar{J}_I + \bar{C}\bar{A}^I(D^{-1}\bar{I}_S - \bar{A})^{-1}\bar{B}] \\ &= D^I[\bar{I}_N + \bar{C}\bar{A}^I(D^{-1}\bar{I}_S - \bar{A})^{-1}\bar{B}(\bar{F}\bar{J}_I)^{-1}\bar{F}]\bar{J}_I. \quad (32)\end{aligned}$$

From (31) and (32) it is seen that $D^I\bar{H}(D)\bar{G}(D) = D^I\bar{I}_K$ as desired. Moreover, it is also apparent that if $\bar{G}(D)$ is a matrix of polynomials corresponding to an FF LSC, then $D^I\bar{H}(D)$ will, in general, have elements which are rational functions of D as a result of the inversion in (31). As an example of this assertion, it is instructive to consider the case $K=1$.

For cases such as this one, in which $K \leq N$, it is usually more convenient to invert a $K\times K$ matrix rather than an $N\times N$ matrix. Therefore, rewrite (31) in the manner

$$\begin{aligned}D^I\bar{H}(D) &= (\bar{F}\bar{J}_I)^{-1}\bar{F}[\bar{I}_N - \bar{C}\bar{A}^I(D^{-1}\bar{I}_S - \bar{A})^{-1}B[\bar{I}_K \\ &\quad + (\bar{F}\bar{J}_I)^{-1}\bar{F}\bar{C}\bar{A}^I(D^{-1}\bar{I}_S - \bar{A})^{-1}\bar{B}]^{-1}(\bar{F}\bar{J}_I)^{-1}\bar{F}]. \quad (33)\end{aligned}$$

Equation (33) has a special advantage over (31) in this instance, when $K=1$, for then (33) simplifies to

$$D^I\bar{H}(D) = \bar{F}/[(\bar{F}\bar{J}_I) + \bar{F}\bar{C}\bar{A}^I(D^{-1}\bar{I}_S - \bar{A})^{-1}\bar{B}]. \quad (34)$$

But the denominator of (34) is simply related to $\bar{G}(D)$ by (32), so that

$$\bar{H}(D) = \bar{F}/\bar{F}\bar{G}(D), \quad (35)$$

from which it is clear that $\bar{H}(D)$ has elements which are reciprocals of polynomials when $\bar{G}(D)$ has elements which are polynomials.

The remaining development of this section is concerned with displaying an inverse construction procedure for the case in which rank $(\bar{J}_I)\neq K$. It is assumed that the LSC has an inverse, that is, the set of determinants $\Delta_i(D)$, $i=1, 2, \cdots, \binom{N}{K}$, are not all identically zero. Consider now the matrix $D^{-I}\bar{G}(D)$, denoted $\bar{G}^1(D)$, and note that $\bar{J}_I=\bar{G}^1(0)$. Let the $\binom{N}{K}$ distinct determinants which can be formed from $K\times K$ minors of $\bar{G}^1(D)$ be denoted $\Delta_i^1(D)$, $i=1, 2, \cdots, \binom{N}{K}$. By assumption, they are not all identically zero. Since rank $(\bar{J}_I)\neq K$, then there exists a positive integer m such that

$$\text{GCF}\left[\overset{1}{\Delta_i}(D),\ i = 1, 2, \cdots, \binom{N}{K}\right] = D^mR_1(D),^1 \quad (36)$$

where $R_1(D)$ is a rational function satisfying $R_1(0)\neq 0$. Let $\Delta_n^1(D)$ be any of the $\binom{N}{K}$ determinants satisfying $\Delta_n^1(D) = D^mR(D)$ where $R(0)\neq 0$, and let $\bar{G}^1(D)_n$ be the submatrix of $\bar{G}^1(D)$ with the corresponding K rows.

[1] The greatest common factor (GCF) of the rational functions $P_i(D)/Q_i(D)$, $i=1, 2, \cdots, n$, where GCD $[P_i(D), Q_i(D)]=1$, is here defined as GCD $[P_i(D), \cdots, P_n(D)]$/GCD $[Q_i(D), \cdots, Q_n(D)]$.

Then there exists a nonsingular $K\times K$ matrix $\bar{W}^1(D)$, such that $D\bar{W}^1(D)$ is a matrix of polynomials in D, and such that

$$\text{GCF}\left[\overset{2}{\Delta_i}(D),\ i=1, 2, \cdots, \binom{N}{K}\right] = D^{m-1}R_2(\text{D}), \quad (37)$$

where $R_2(0)\neq 0$ and the $\Delta_i^2(D)$ are the determinants formed from $K\times K$ minors of

$$\bar{G}^2(D) = \bar{G}^1(D)\bar{W}^1(D). \quad (38)$$

Moreover, $\bar{G}^2(D)$ has a structural realization in the form $(\bar{A},\ \bar{B}\bar{W}_2^1+\bar{A}\bar{B}\bar{W}_1^1,\ \bar{C}\bar{A}^I,\ \bar{J}_I\bar{W}_2^1+\bar{C}\bar{A}^I\bar{B}\bar{W}_1^1)$, where $\bar{W}_2^1$ contains the elements of $\bar{W}^1(D)$ which are independent of D and $\bar{W}_1^1=D(\bar{W}^1(D)-\bar{W}_2^1)$.

In order to verify the assertions above, note that there exists at least one K vector $\boldsymbol{a}^1\neq\boldsymbol{0}$ such that $\bar{J}_I\boldsymbol{a}^1=\boldsymbol{0}$. Since $\boldsymbol{a}^1\neq\boldsymbol{0}$, it has at least one nonzero component, say the jth component a_{1j}. Define $\bar{W}^1(D)$ to be the matrix with jth column $\boldsymbol{a}^1D^{-1}$ and all other columns equal to those of the identity matrix $\bar{I}_K$. Observe that the determinant of $\bar{W}^1(D)=D^{-1}a_{1j}$. Equation (37) is then established by noting that the determinant of $\bar{G}^2(D)_i$ is just the product of the determinants of $\bar{G}^1(D)_i$ and $\bar{W}^1(D)$. In order to examine the question of a structural realization for $\bar{G}^2(D)$, consider the formal series expansion

$$\begin{aligned}\bar{G}^1(D) &= \bar{J}_I + (\bar{C}\bar{A}^I)\bar{B}D + (\bar{C}\bar{A}^I)\bar{A}\bar{B}D^2 \\ &\quad + (\bar{C}\bar{A}^I)\bar{A}^2\bar{B}D^3 + \cdots. \quad (39)\end{aligned}$$

By virtue of the fact that $\bar{W}_1^1$ has zeros everywhere, except in the jth column, where it has $\boldsymbol{a}^1$, and since $\bar{W}_2^1$ is independent of D, it is possible to write

$$\begin{aligned}\bar{G}^2(D) &= \bar{G}^1(D)[\bar{W}_1^1D^{-1} + \bar{W}_2^1] \\ &= [\bar{J}_I\bar{W}_2^1 + (\bar{C}\bar{A}^I)\bar{B}\bar{W}_1^1] \\ &\quad + [(\bar{C}\bar{A}^I)\bar{B}\bar{W}_2^1 + (\bar{C}\bar{A}^I)\bar{A}\bar{B}\bar{W}_1^1]D + \cdots \quad (40)\end{aligned}$$

which can be identified with the structural matrices asserted above.

The discussion above can be generalized and stated as a lemma.

Lemma 6: Suppose that rank $(\bar{J}_I)\neq K$. Let m be the positive integer defined in (36). Then there exists a sequence of LSCs $\bar{G}^i(D)$, $i=1, 2, \cdots, m+1$, such that rank $(\bar{G}^{m+1}(0))=K$. Moreover, the $\bar{G}^i(D)$ can be defined recursively in the manner

$$\begin{aligned}\bar{G}^{i+1}(D) &= \bar{G}^i(D)\bar{W}^i(D), \\ &\quad i = 1, 2, \cdots, m,\ \bar{G}^1(D) = D^{-I}\bar{G}(D), \quad (41)\end{aligned}$$

where each $\bar{G}^i(D)$ has the structural realization $\bar{J}_I^i+\bar{C}\bar{A}^I(D^{-1}\bar{I}_S-\bar{A})^{-1}\bar{B}^i$ and where $\bar{W}^i(D)$ is a $K\times K$ matrix which is equal to $\bar{I}_K$ in all columns except one, say the jth, in which it has a K vector $\boldsymbol{a}^i$ (satisfying $\bar{J}_I^i\boldsymbol{a}^i=\boldsymbol{0}$, $a_{ij}\neq 0$) multiplied by D^{-1}. Finally, the structural matrices $\bar{J}_I{}^i$ and $\bar{B}^i$ are defined by the recursions $\bar{J}_I{}^{i+1} = \bar{J}_I{}^i \bar{W}_2{}^i + \bar{C}\bar{A}^I \bar{B}^i \bar{W}_1{}^i$ and $\bar{B}^{i+1} = \bar{B}^i \bar{W}_2{}^i + \bar{A}\bar{B}^i \bar{W}_1{}^i$, under the initial conditions $\bar{J}_I{}^1 = \bar{J}_I$ and $\bar{B}^1 = \bar{B}$, for $i = 1, 2, \cdots, m$. The nonzero elements of $\bar{W}_2{}^i$ are those of $\bar{W}^i(D)$ which do not depend on D, and $\bar{W}_1{}^i = D(\bar{W}^i(D) - \bar{W}_2{}^i$.)

Proof: The proof of Lemma 6 follows by induction on the case $i = 1$ discussed in preceding paragraphs. It is important to point out, nonetheless, that the existence of a^i, $i = 1, 2, \cdots, m$, is assured by (36) and the construction (41) which together imply that rank $(\bar{J}_I{}^i) \neq K$, $i = 1, 2, \cdots, m$.

Lemma 6 is indicative of a series of nonsingular transformations $\bar{W}^i(D)$, $i = 1, 2, \cdots, m$, by means of which the transfer function matrix $\bar{G}^1(D)$, to which Theorem 5 does not apply, is converted to a transfer function matrix $\bar{G}^{m+1}(D)$, to which Theorem 5 does apply. By means of the relations (41) and the definition of $\bar{G}(D)$, it follows that

$$\begin{aligned}\bar{G}(D) &= D^I \bar{G}^{m+1}(D)[\bar{W}^m(D)]^{-1}[\bar{W}^{m-1}(D)]^{-1} \cdots \\ &\quad \cdot [\bar{W}^1(D)]^{-1} \\ &= [D^I \bar{G}^{m+1}(D)][D\bar{W}^m(D)]^{-1} \cdots [D\bar{W}^1(D)]^{-1} D^m, \qquad (42)\end{aligned}$$

which permits the statement of the main result of the section.

Theorem 6: Under the hypotheses of Lemma 6, the LSC has an inverse which can be constructed as the cascade of the inverse of $D^I \bar{G}^{m+1}(D)$, according to Theorem 5, with the unit-delay LSCs $D\bar{W}^m(D)$, $D\bar{W}^{m-1}(D)$, $\cdots$, $D\bar{W}^1(D)$. The inverse has finite delay $L = I + m$.

Proof: Theorem 6 follows directly from (42) and Lemma 6.

VII. Conclusions

The question of the existence of and realizations for inverse linear sequential circuits is of considerable interest in the theory of information processing. For example, certain desirable properties of convolutional error-correcting code schemes suggest the study of feedforward inverses for feedforward linear sequential circuits.

This paper establishes the necessary and sufficient conditions for a feedforward linear sequential circuit to have a feedforward inverse and gives an implicit procedure for its construction. The concept of an inverse with finite delay is defined, and necessary and sufficient conditions are given for the existence of such an inverse. Finally, construction techniques for inverses with finite delay are given in recursive form by analogy with known results in continuous-time linear dynamical systems.

The first part of the paper approaches the linear sequential circuit from the viewpoint of its transfer function matrix, whereas the second part describes the circuit in terms of state variables and structural matrices. The results are complementary in nature and serve to emphasize the need of further investigations to interrelate methods associated with the two descriptions. Further incentive for such studies is derived from the fact that the second part of the paper is motivated largely by studies in continuous-time systems.

Finally, we remark that our Definition 1 for the existence of an inverse with delay L for an LSC is equivalent for LSCs to Huffman's[8] general concept of "information lossless of finite order L" for a sequential circuit.

Acknowledgment

Prof. Sain wishes to acknowledge beneficial conversations with Prof. R. J. Leake and Prof. C. R. Souza at the University of Notre Dame. He is also indebted to Prof. W. R. Perkins of the University of Illinois for calling the results of Youla and Dorato to his attention.

References

[1] A. J. Viterbi, "Error bounds for convolutional codes and an asymptotically optimum decoding algorithm," *IEEE Trans. Information Theory*, vol. IT-13, pp. 260–269, April 1967.

[2] J. L. Massey and M. K. Sain, "Codes, automata, and continuous systems: explicit interconnections," *IEEE Trans. Automatic Control*, vol. AC-12, pp. 645–650, December 1967.

[3] J. L. Massey, *Threshold Decoding*. Cambridge, Mass.: M.I.T. Press, 1963.

[4] G. Birkhoff and S. Maclane, *A Survey of Modern Algebra*, 3rd ed. New York: Macmillan, 1965, p. 71.

[5] A. Gill, *Linear Sequential Circuits*. New York: McGraw-Hill, 1966, pp. 21–26.

[6] D. C. Youla and P. Dorato, "On the inverse of linear dynamical systems," Polytechnic Institute of Brooklyn, N. Y., Electrophysics Memorandum PIBMRI-1319-66, March 1966.

[7] M. K. Sain, "On 'A useful matrix inversion formula and its applications,' " *Proc. IEEE* (*Letters*), vol. 55, p. 1753, October 1967.

[8] D. A. Huffman, "Canonical forms for information-lossless finite-state logical machines," *IRE Trans. Circuit Theory*, vol. CT-6, pp. 41–59, March 1959.

Convolutional Codes I: Algebraic Structure

G. DAVID FORNEY, JR., MEMBER, IEEE

Abstract—A convolutional encoder is defined as any constant linear sequential circuit. The associated code is the set of all output sequences resulting from any set of input sequences beginning at any time. Encoders are called equivalent if they generate the same code. The invariant factor theorem is used to determine when a convolutional encoder has a feedback-free inverse, and the minimum delay of any inverse. All encoders are shown to be equivalent to minimal encoders, which are feedback-free encoders with feedback-free delay-free inverses, and which can be realized in the conventional manner with as few memory elements as any equivalent encoder. Minimal encoders are shown to be immune to catastrophic error propagation and, in fact, to lead in a certain sense to the shortest decoded error sequences possible per error event. In two appendices, we introduce dual codes and syndromes, and show that a minimal encoder for a dual code has exactly the complexity of the original encoder; we show that systematic encoders with feedback form a canonical class, and compare this class to the minimal class.

I. Introduction

Block codes were the earliest type of codes to be investigated, and remain the subject of the overwhelming bulk of the coding literature. On the other hand, convolutional codes have proved to be equal or superior to block codes in performance in nearly every type of practical application, and are generally simpler than comparable block codes in implementation. This anomaly is due largely to the difficulty of analyzing convolutional codes, as compared to block codes. It is the intent of this series to stimulate increased theoretical interest in convolutional codes by review and clarification of known results and introduction of new ones. We hope, first, to advance the understanding of convolutional codes and tools useful in their analysis; second, to motivate further work by showing that in every case in which block codes and convolutional codes can be directly compared theoretically, the convolutional are as good or better.

Two converging lines of development have generated interest in an algebraic approach to convolutional codes. On the one hand, the success of algebraic methods in generating classes of good block codes suggests that constructive methods of generating good convolutional codes might be developed through use of algebraic structures. Correspondingly one might expect that powerful decoding techniques based on such structures might be discovered. (Fortunately, good codes and decoding methods not relying on such constructions are already known.) On the other hand, the usefulness of regarding convolutional encoders as linear sequential circuits has begun to become evident, as in the observation of Omura [1] and others that the Viterbi [2] maximum-likelihood decoding algorithm is the dynamic programming solution to a certain control problem, and in the observation of Massey and his colleagues [3]–[8] that certain questions concerning error propagation are related to questions concerning the invertibility of linear systems. As the theory of finite-dimensional linear systems is seen increasingly as essentially algebraic, we have another motive for examining convolutional encoders in an algebraic context.

Our result is a series of structure theorems that dissect the structure of convolutional codes rather completely, mainly through use of the invariant factor theorem. We arrive at a class of canonical encoders capable of generating any convolutional code, and endowed with all the desirable properties one might wish, except that in general they are not systematic. (The alternate canonical class of systematic encoders with feedback is discussed in Appendix II.) The results do not seem to suggest any constructive methods of generating good codes, and say little new in particular about the important class of rate-$1/n$ codes, except for putting known results in a more general context. It appears that the results obtained here for convolutional codes correspond to block-code results ([9], ch. 3).

II. Problem Formulation

We are purposefully going to take a rather long time getting to our main results. Most of this time will be spent in definitions and statements of fundamental results in convolutional coding theory, linear system theory, and algebra. It is a truism that when dealing with fundamentals, once the problem is stated correctly, the results are easy. We feel it is important that the right formulation of the problem (like justice) not only be done, but be seen to be done, in the eyes of readers who may have backgrounds in any of the three areas noted.

After exhibiting a simple convolutional encoder for motivation, we move to a general definition of convolutional encoders, which we see amount to general finite-state time-invariant linear circuits. We discuss the decoding problem, which leads to a definition of convolutional encoder equivalence and to certain desirable code properties. Along the way certain algebraic artifacts will intrude into the discussion; in the final introductory section we collect the algebra we need, which is centered on the invariant factor theorem.

Convolutional Encoder

Fig. 1(a) shows a simple binary systematic rate-1/2 convolutional encoder of constraint length 2. The input to this encoder is a binary sequence

Manuscript received December 18, 1969. Part of this paper was presented at the Internatl. Information Theory Symposium, Ellenville, N. Y., January 27–31, 1969.
The author is with Codex Corporation, Newton, Mass. 02158.

Reprinted from *IEEE Trans. Inform. Theory,* vol. IT-16, pp. 720–738, Nov. 1970 and vol. IT-17, p. 360, May 1971.

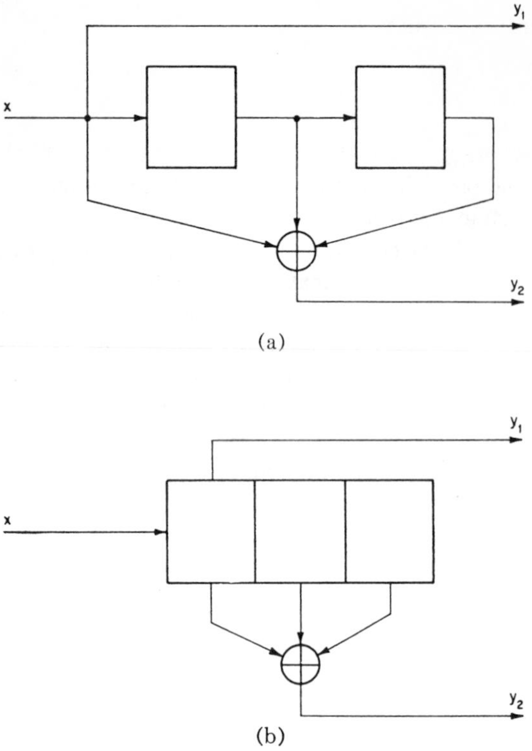

Fig. 1. (a) A rate-1/2 systematic convolutional encoder. (b) Alternate representation.

$$x = (\cdots, x_{-1}, x_0, x_1, \cdots).$$

The outputs are two binary sequences y_1 and y_2 (hence the rate is 1/2). The first output sequence y_1 is simply equal to the input x (hence the code is systematic). The elements y_{2i} of the second sequence y_2 are given by

$$y_{2i} = x_i \oplus x_{i-1} \oplus x_{i-2},$$

where $\oplus$ denotes modulo 2 addition. Therefore, the encoder must save 2 past information bits, so we say the constraint length is 2.

Sometimes it is convenient to draw the encoder as in Fig. 1(b), with the output a function of the memory contents only. (This corresponds to the distinction in automata theory between Moore and Mealy machines.) Some authors would therefore say this code had constraint length 3, since the outputs are a function of a span of 3 input bits. Others measure constraint length in terms of output bits and would assign this code a constraint length of 6. Our definition of constraint length is chosen to coincide with the number of memory elements in a minimal realization.

The term "convolutional" comes from the observation that the output sequences can be regarded as a convolution of the input sequence with certain generator sequences. With the input and output sequences, we associate sequences in the delay operator D (D transforms):

$$x(D) = \cdots + x_{-1}D^{-1} + x_0 + x_1D + x_2D^2 + \cdots$$

$$y_1(D) = \cdots + y_{1,-1}D^{-1} + y_{10} + y_{11}D + y_{12}D^2 + \cdots$$

$$y_2(D) = \cdots + y_{2,-1}D^{-1} + y_{20} + y_{21}D + y_{22}D^2 + \cdots$$

(The delay operator D corresponds to z^{-1} in sampled-data theory, but is purely an indeterminate or place-holder, whereas z is a complex variable.) Now the input/output relationships are expressed concisely as

$$y_1(D) = g_1(D)x(D)$$

$$y_2(D) = g_2(D)x(D),$$

where the generator polynomials $g_1(D)$ and $g_2(D)$ are

$$g_1(D) = 1$$

$$g_2(D) = 1 + D + D^2,$$

and ordinary sequence multiplication with coefficient operations modulo 2 and collection of like powers of D is implied.

Similarly, we can define a general (n, k) conventional convolutional encoder by a matrix of generator polynomials $g_{ij}(D)$, $1 \le i \le k$, $1 \le j \le n$, with coefficients in some finite field F. There are k-input sequences $x_i(D)$, n-output sequences $y_j(D)$, each a sequence of symbols from F, with input/output relations given by

$$y_j(D) = \sum_{i=1}^{k} x_i(D)g_{ij}(D),$$

again with all operations in F. If we define the constraint length for input i as

$$\nu_i = \max_{1 \le j \le n} [\deg g_{ij}(D)],$$

then the general conventional encoder can be realized by k shift registers, the ith of length ν_i, with the outputs formed as linear combinations in F on the appropriate shift register contents. We call this the obvious realization, and note that the number of memory elements required is equal to the overall constraint length, defined as the sum of constraint lengths

$$\nu = \sum_{i=1}^{k} \nu_i.$$

For notational convenience we shall generally suppress the parenthetical D in our subsequent references to sequences; thus x_i means $x_i(D)$, $y_j = y_j(D)$, and so forth, where the fact that a letter represents a sequence (transform) should be clear from the context.

Convolutional Encoder—General Definition

The encoders of the previous section are linear sequential circuits. We now consider all finite-state time-invariant linear sequential circuits as candidates for convolutional encoders.

Definition 1: An (n, k) convolutional encoder over a finite field F is a k-input n-output constant linear causal finite-state sequential circuit.

Let us dissect this definition.

K-input: There are k discrete-time input sequences x_i, each with elements from F. We write the inputs as the row vector $\mathbf{x}$. Sequences must start at some finite time and may or may not end. (Then we can represent a sequence such as $1 + D + D^2 + \cdots$ by a ratio of polynomials such as $1/(1 + D)$, without encountering the ambiguity $1/(1 + D) = D^{-1} + D^{-2} + \cdots$.) If a sequence x_i "starts" at time d (if the first nonzero element is x_{id}), we say it has delay d, del $x_i = d$, and if it ends at time d', we say it has degree d', deg $x_i = d'$, in analogy to the degree of a polynomial. Similarly, we define the delay and degree of a vector of sequences as the minimum delay or maximum degree of the component sequences: del $\boldsymbol{x}$ = min del x_i, deg $\boldsymbol{x}$ = max deg x_i. A finite sequence has both a beginning and an end. (Note that most authors consider only sequences starting at time 0 or later. It turns out that this assumption clutters up the analysis without compensating benefits.)

N-output: There are n-output sequences y_i, each with elements from F, which we write as the row vector $\boldsymbol{y}$. The encoder is characterized by the map G, which maps any vector of input sequences $\boldsymbol{x}$ into some output vector $\boldsymbol{y}$, which we can write in the functional form

$$\boldsymbol{y} = G(\boldsymbol{x}).$$

Constant (Time-Invariant): If all input sequences are shifted in time, all output sequences are correspondingly shifted. In delay-operator notation,

$$G(D^n\boldsymbol{x}) = D^nG(\boldsymbol{x}) \qquad n \text{ any integer.}$$

(Note that most probabilistic analyses of convolutional codes have been forced to assume nonconstancy to obtain ensembles of encoders with enough randomness to prove theorems.)

Linear: The output resulting from the superposition of two inputs is the superposition of the two outputs that would result from the inputs separately, and the output "scales" with the input. That is,

$$G(\boldsymbol{x}_1 + \boldsymbol{x}_2) = G(\boldsymbol{x}_1) + G(\boldsymbol{x}_2)$$

$$G(\alpha\boldsymbol{x}_1) = \alpha G(\boldsymbol{x}_1) \qquad \alpha \in F.$$

It is easy to see that constancy and linearity together give the broader linearity condition

$$G(\alpha\boldsymbol{x}_1) = \alpha G(\boldsymbol{x}_1),$$

where α is any sequence of elements of F in the delay operator D. Furthermore, they imply a transfer-function representation for the encoder. For let $\boldsymbol{\varepsilon}_i$, $1 \leq i \leq k$, be the unit inputs in which the ith input at time 0 is 1, and all other inputs are 0. Let the generators $\boldsymbol{g}_i$ be defined as the corresponding outputs (impulse responses):

$$\boldsymbol{g}_i = G(\boldsymbol{\varepsilon}_i) \qquad 1 \leq i \leq k.$$

Then since any input $\boldsymbol{x}$ can be written

$$\boldsymbol{x} = \sum_{i=1}^{k} x_i\boldsymbol{\varepsilon}_i,$$

we have by linearity

$$\boldsymbol{y} = G(\boldsymbol{x}) = \sum_{i=1}^{k} x_i\boldsymbol{g}_i.$$

Thus we can define a $k \times n$ transfer function matrix G, whose rows are the generators $\boldsymbol{g}_i$, such that the input/output relationship is

$$\boldsymbol{y} = \boldsymbol{x}G.$$

Therefore from this point we use the matrix notation $\boldsymbol{y} = \boldsymbol{x}G$ in preference to the functional notation $\boldsymbol{y} = G(\boldsymbol{x})$.

Finally, this definition implies that a zero input gives a zero output so that the encoder may not have any transient response (nonzero starting state).

Causal: If the nonzero inputs start at time t, then the nonzero outputs start at time $t' \geq t$. Since the unit inputs start at time 0, this implies that all generators start at time 0 or later. As sequences, all generators must therefore have all negative coefficients equal to 0, or del $\boldsymbol{g}_i \geq 0$. Conversely, the condition that all generators satisfy del $\boldsymbol{g}_i \geq 0$ implies

$$\begin{aligned}\text{del } \boldsymbol{y} &= \text{del} \sum_{i=1}^{k} x_i\boldsymbol{g}_i \\ &\geq \min_{1\leq i\leq k} [\text{del } x_i + \text{del } \boldsymbol{g}_i] \\ &\geq \text{del } \boldsymbol{x},\end{aligned}$$

causality.

Finite-state: The encoder shall have only a finite number of memory elements, each capable of assuming a finite number of values. The physical state of an encoder at any time is the contents of its memory elements; thus there are only a finite number of physical states. A more abstract definition of the state of an encoder at any time is the following: the state $\boldsymbol{s}$ of an encoder at time t^- is the sequence of outputs at time t and later if there are no nonzero inputs at time t or later. Clearly the number of states so defined for any fixed t is less than or equal to the number of physical states, since causality implies that each physical state gives some definite sequence, perhaps not unique. Thus an encoder with a finite number of physical states must have a finite number of abstract states as well.

By studying the abstract states, we develop further restrictions on the generators $\boldsymbol{g}_i$. Let us examine the set of possible states at time 1^-, which we call the state space Σ. (By constancy the state spaces at all times are isomorphic.) Let P be the projection operator that truncates sequences to end at time 0, and Q the complementary projection operator $1 - P$ that truncates sequences to start at time 1:

$$xP = x_dD^d + \cdots + x_{-1}D^{-1} + x_0$$

$$xQ = x_1D^2 + x_2D^2 + \cdots .$$

Then any input $\boldsymbol{x}$ is associated with a state at time 1^- given by

$$\boldsymbol{s} = \boldsymbol{x}PGQ;$$

conversely, any state in Σ can be so expressed using any input giving that state. Now the state space Σ is seen to satisfy the conditions to be a vector space over the field F, for P, G, and Q are all linear over F; that is, if

$$s_1 = x_1 PGQ,$$

and

$$s_2 = x_2 PGQ,$$

then the combinations

$$s_1 + s_2 = (x_1 + x_2)PGQ$$

$$\alpha s_1 = (\alpha x_1)PGQ \qquad \alpha \in F,$$

are also states. As a vector space, Σ therefore has a finite dimension dim Σ, or else the number of states $q^{\dim \Sigma}$, where q is the number of elements in the finite field, would be infinite.

Consider now for simplicity a single-input single-output linear sequential circuit with transfer function g, so $y = xg$. Let s_d be the state obtained from a unit input at time $-d$:

$$s_d = D^{-d} g Q \qquad d \geq 0.$$

If the state space has finite dimension dim Σ, then at most dim Σ of these states can be linearly independent, and in particular there is some linear dependence relation (with coefficients ψ_d in F) between the first dim $\Sigma + 1$ of these states:

$$0 = \sum_{d=0}^{\dim \Sigma} \psi_d s_d$$

$$= \psi(D^{-1}) g Q,$$

where

$$\psi(D^{-1}) = \sum_{d=0}^{\dim \Sigma} \psi_d D^{-d}$$

is some nonzero polynomial over F of degree dim Σ or less in the indeterminate D^{-1}. In order for $\psi(D^{-1})g$ to be 0 at time 1 and later, g itself must be equal to a ratio of polynomials $h(D^{-1})/\psi(D^{-1})$, with the degree of the numerator $h(D^{-1})$ not greater than that of the denominator $\psi(D^{-1})$ in order that g be causal, del $g \geq 0$. Any sequence g that can be so expressed will be called a realizable function, realizable meaning both causal and finite. Clearly a realizable function can also be expressed (by multiplying numerator and denominator by $D_{\deg \psi}$) as a ratio of polynomials in D rather than D^{-1}, in which case the condition deg $h \leq$ deg ψ is transformed to the condition that D not be a factor of the denominator after reduction to lowest terms.

We shall always assume that the expression $h(D^{-1})/\psi(D^{-1})$ of a realizable function g has been reduced to lowest terms and has a monic denominator ($\psi_{\deg \psi} = 1$). A canonical realization of such a realizable function is shown in Fig. 2. The realization involves a feedback shift register with deg ψ memory elements storing elements of F, which essentially performs long division; if h/ψ is in lowest terms then the dimension of the state-space dim Σ is also equal to deg ψ.

A convolutional encoder is then any linear sequential circuit whose transfer-function matrix G is realizable, that is, has components that are realizable functions. It is clear that a brute-force finite-state realization of such an encoder always exists, since one can simply realize the kn components and sum their outputs; in Appendix II we discuss the result of realization theory that shows how to construct a canonical realization, namely, one with a number of memory elements equal to the dimension of the abstract state space. We see that the only respect in which this definition is more general than that of a conventional convolutional encoder is that realizable generators in general involve feedback and, as sequences, have infinite length. It might seem that getting an infinite-length generator sequence from a finite-state encoder is a good bargain, but we shall see in the sequel that in fact feedback buys nothing.

Communications Context

So far we have defined a general convolutional encoder merely as a general realizable linear sequential circuit, and developed basic concepts in linear system theory. It is only when we put the encoder in a communications context that questions unlike those posed in linear system theory arise. The most important new concept involves a definition of encoder equivalence very different from the linear-system-theoretic one.

Consider then the use of a convolutional encoder in a communications system, shown in Fig. 3. From the k-input sequences x, called information sequences, the encoder G generates a set of n-output sequences y, called a codeword, which is transmitted over some noisy channel. The received data, whatever their form, are denoted by r; a decoder operates on r in some way to produce k decoded sequences $\hat{x}$, preferably not too different from x.

We see immediately that all is lost if the encoder map G is not one-to-one, for then even if there were no noise in the channel the decoder would make mistakes. In fact, by linearity, if $y = x_1G = x_2G$ for two different inputs, then for any output y' there are at least two inputs x' and $x' + x_1 - x_2$ such that

$$y' = x'G = (x' + x_1 - x_2)G;$$

and by constancy the difference between the inputs may be made to extend over all time by concatenating them if they are not infinite already, so that the probability of decoding error would be at least $\frac{1}{2}$. We therefore require that the encoder map be one-to-one in any useful convolutional encoder. There is therefore some *inverse* map G^{-1}, obviously linear and constant, which takes outputs y back to the unique corresponding input x,

$$xGG^{-1} = x \qquad \text{for all } x;$$

$$GG^{-1} = I_k,$$

where I_k is the identity transformation. (Here we have used an $n \times k$ transfer-function matrix representation for G^{-1}, as we may, since G^{-1} is constant and linear; in matrix terminology G^{-1} is a right inverse.) Of course this inverse map may not be realizable, but we shall show below that when there is any inverse there is always a

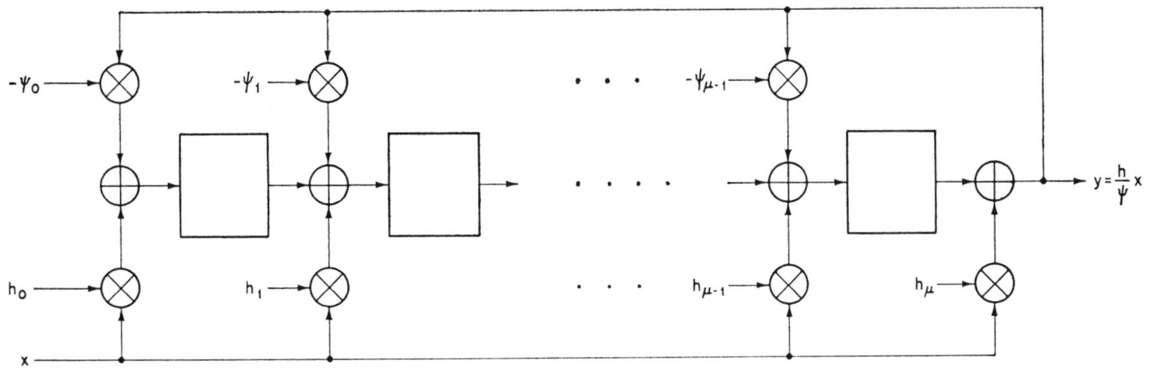

Fig. 2. Canonical realization of the transfer function $h(D^{-1})/\psi(D^{-1})$.

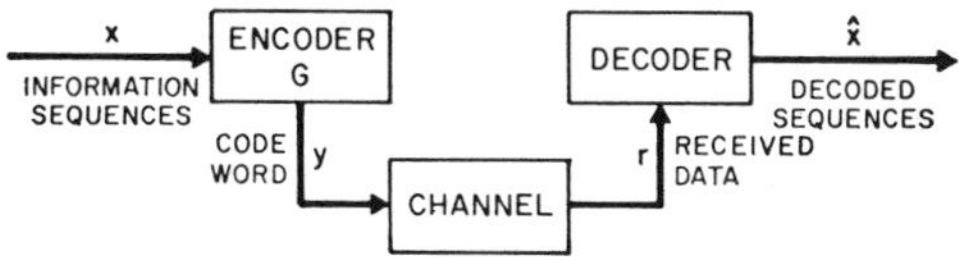

Fig. 3. Communications system using convolutional coding.

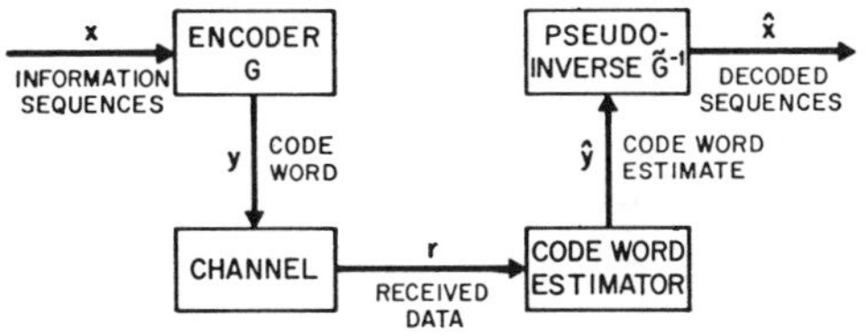

Fig. 4. Same system with decoder in two parts.

realizable pseudo-inverse $\tilde{G}^{-1}$ such that $G\tilde{G}^{-1} = I_k D^d$ for some $d \geq 0$. $\tilde{G}^{-1}$ is also called an inverse with delay d, and G is sometimes called information-lossless of order d.

Returning to Fig. 3, we now split the decoder conceptually into two parts, a codeword estimator that produces from the received data r some codeword estimate $\hat{\boldsymbol{y}}$, followed by a realizable pseudo-inverse $\tilde{G}^{-1}$ that assigns to the codeword estimate the appropriate information sequences $\hat{\boldsymbol{x}}$ (see Fig. 4). In practice a decoder is usually not realized in these two pieces, but it is clear that since all the information about the information sequences $\boldsymbol{x}$ comes through $\boldsymbol{y}$, the decoder can do no better estimating $\boldsymbol{x}$ directly than by estimating $\boldsymbol{y}$ and making the one-to-one correspondence to $\boldsymbol{x}$. (In fact, any decoder can be made into a codeword estimator by appending an encoder G.)

When G is one-to-one, as long as the codeword estimator makes no errors, there will be no error in the decoded sequences. However, even in the best-regulated communications systems, decoding errors will sometimes occur. We define the error sequences as the difference between the estimated codeword $\hat{\boldsymbol{y}}$ and the codeword $\boldsymbol{y}$ actually sent:

$$\boldsymbol{e} = \hat{\boldsymbol{y}} - \boldsymbol{y}.$$

Correspondingly the information errors $\boldsymbol{e}_x$ are defined as the difference between the decoded sequences $\hat{\boldsymbol{x}}$ and the information sequences $\boldsymbol{x}$, with allowance for the pseudo-inverse delay d:

$$\begin{aligned} \boldsymbol{e}_x &= \hat{\boldsymbol{x}} D^{-d} - \boldsymbol{x} \\ &= \hat{\boldsymbol{y}} \tilde{G}^{-1} D^{-d} - \boldsymbol{y} G^{-1} \\ &= \boldsymbol{e} G^{-1}. \end{aligned}$$

The one-to-one correspondence between these two definitions is exhibited explicitly through the inverse G^{-1}. As is implicit in our terminology, we consider the error sequences $\boldsymbol{e}$ as the more basic of the two.

Since the error sequences are the difference between two codewords, $\boldsymbol{e}$ is itself a codeword, in fact the codeword generated by $\boldsymbol{e}_x$:

$$\boldsymbol{e} = \boldsymbol{e}_x G.$$

We make a decomposition of $\boldsymbol{e}$ into short codewords, called error events, as follows. Start at some time when the codeword estimator has been decoding correctly. An error event starts at the first subsequent time at which $\boldsymbol{e}$ is nonzero. The error event stops at the first subsequent time when the error sequences in the event form a finite codeword, after which the decoder will be decoding correctly again. Thus we express $\boldsymbol{e}$ as a sum of nonoverlapping finite codewords, the error events.

Implicit in the above analysis is the assumption that infinite error events do not occur. Such a possibility is not excluded in principle, but can generally be disregarded, on the basis of the following plausibility argument. If two codewords differ in an infinite number of places, then as time goes on, we can expect the evidence in the received data r to build up in favor of the correct word, with probability approaching 1. A well-designed codeword estimator will use this information efficiently enough that very long error events have very small probabilities and infinite error events have probability 0. More precisely, the average error-event length should be small, at least for channel noise, which is in some sense small, so that most of the time the decoder is decoding correctly. We say that any codeword estimator or decoder not satisfying this assumption is subject to ordinary error propagation, and exclude it from the class of useful decoders. Note that any decoder that is capable of starting to decode correctly sooner or later, regardless of what time it is turned on, can be made immune to ordinary error propagation simply by restarting whenever it detects it is in trouble, or even periodically.

We owe to Massey and Sain [4] the observation that if there is any infinite information sequence $\boldsymbol{x}_0$ such that the corresponding codeword $\boldsymbol{y}_0 = \boldsymbol{x}_0 G$ is finite, then even in the absence of ordinary error propagation decoding catastrophes can occur. For there will generally be a nonzero probability for the finite error event $\boldsymbol{e} = \boldsymbol{y}_0$ to occur, which will lead to infinite information errors $\boldsymbol{e}_x = \boldsymbol{x}_0$. Thus $\hat{\boldsymbol{x}}$ will differ from the original information sequences $\boldsymbol{x}$ in an infinite number of places, even if no further decoding errors are made by the codeword estimator. (In fact, the only chance to stop this propagation is to make another error.) Massey and Sain call this catastrophic error propagation. Since $\tilde{G}^{-1}$ must supply the infinite output $\boldsymbol{x}_0$ in response to the finite input $\boldsymbol{y}_0$, it must have internal feedback for the above situation to occur. We therefore require that any useful encoder must not only have a realizable pseudo-inverse $\tilde{G}^{-1}$, but one that is feedback-free. (We see later that if G has no such inverse, then there is indeed an infinite input leading to a finite output.)

We come at last to our most important observations (see also [3]). The codeword estimator dominates both complexity and performance of the system: complexity, because both G and $\tilde{G}^{-1}$ represent simple one-to-one (and in fact linear) maps, while the codeword estimator map from received data r to codeword $\hat{\boldsymbol{y}}$ is many-to-one; performance, because in the absence of catastrophic error propagation the probability of decoding error is equal to the probability of an error event multiplied by the average decoding errors per error event, with the former generally dominant and the latter only changed by small factors for reasonable pseudo-inverses $\tilde{G}^{-1}$ (in fact, in many practical applications the probability of error event is the significant quantity, rather than the probability of decoding error). But the performance and complexity of the codeword estimator depend only on the set of codewords $\boldsymbol{y}$, not on the input/output relationship specified by G (assuming that all $\boldsymbol{x}$ and hence $\boldsymbol{y}$ are equally likely, or at least that the codeword estimator does not use codeword probabilities, as is true, for example, in maximum-likelihood estimation). Therefore it is natural to assert that two encoders generaating the same set of codewords are essentially equivalent in a communications context. So we give the following definitions.

Definition 2: The *code* generated by a convolutional encoder G is the set of all codewords $\boldsymbol{y} = \boldsymbol{x}G$, where the k inputs $\boldsymbol{x}$ are any sequences.

Definition 3: Two encoders are *equivalent* if they generate the same code.

These definitions free us to seek out the encoder in any equivalence class of encoders that has the most desirable properties. We have already seen the desirability of having a feedback-free realizable pseudo-inverse $\tilde{G}^{-1}$. Our main result is that any code can be generated by an encoder G with such a $\tilde{G}^{-1}$, and in fact with the following properties.

1) G has a realizable feedback-free zero-delay inverse G^{-1}.
2) G is itself conventional (feedback-free).
3) The obvious realization of G requires as few memory elements as any equivalent encoder.
4) Short codewords are associated with short information sequences, in a sense to be made precise later.

In the context of linear system theory, the study of convolutional encoders under equivalence can be viewed as the study of those properties of linear systems that belong to the output space alone, or of the invariants over the class of all systems with the same output spaces.

Algebra

We assume that the reader has an algebraic background roughly at the level of the introductory chapters of Peterson [9]. He will therefore be familiar with the notion of a field as a collection of objects that can be added, subtracted, multiplied, and divided with the usual associative, distributive, and commutative rules; he will also know what is meant by a vector space over a field. Further, he will understand that a (commutative) ring is a collection of objects with all the properties of fields except division. He should also recall that the set of all polynomials in D with coefficients in a field F, written conventionally as $F[D]$, is a ring.

The polynomial ring $F[D]$ is actually an example of the very best kind of ring, a principal ideal domain. The set of integers is another such example. Without giving the technical definition of such a ring, we can describe some of its more convenient properties. In a principal ideal domain, certain elements r, including 1, have inverses r^{-1} such that $rr^{-1} = 1$; such elements are called units. The unit integers are ± 1, and the unit polynomials are those of degree 0, namely, the nonzero elements of F. Those elements that are not units can be uniquely factored into products of primes, up to units; a prime is an element that has no factor but itself, up to units. (The ambiguity induced by the units is eliminated by some convention: the prime integers are taken to be the positive primes, while the prime polynominals are taken to be *monic* irreducible polynomials, where monic means having highest order coefficient equal to 1.) It follows that we can cancel: if $ab = ac$, then $b = c$; this is almost as good as division. Further, we have the notion of the greatest common divisor of a set of elements as the greatest product of primes that divides all elements of the set, again made unique by the conventional designation of primes.

Other principal ideal domains have already occurred in our discussions. In general, if R is any principal ideal domain and S is any multiplicative subset—namely, a group of elements containing 1 but not 0 such that if $a \in S$, $b \in S$, then $ab \in S$—then the ring of fractions $S^{-1}R$ consisting of the elements r/s where $r \in R$ and $s \in S$ is a principal ideal domain. (All elements of R that are in S are thereby given inverses and become units.) Letting R be the polynomials $F[D]$, we have the following examples.

1) Let S consist of all nonzero polynomials. Then $S^{-1}R$ consists of all ratios of polynomials with the denominator nonzero, which are called the rational functions, written conventionally $F(D)$. Obviously, in $F(D)$ all nonzero elements are invertible, so $F(D)$ is actually a field, called the field of quotients of $F[D]$.

2) Let S consist of all nonnegative powers of D, including $D^0 = 1$. Then $S^{-1}R$ consists of elements $D^{-n}f(D)$, where $f(D)$ is a polynomial; in other words, $S^{-1}R$ is the set of finite sequences $F_{ft}(D)$. Clearly all the irreducible polynomials except D remain as primes in $F_{ft}(D)$.

3) Let S consist of all polynomials with nonzero constant term; that is, not divisible by D. Then $S^{-1}R$ consists of ratios of polynomials in D with a nonzero constant term in the denominator. We saw earlier that these are precisely the generators realizable by causal finite-state systems; these are therefore called the realizable functions $F_{rz}(D)$. Note that in $F_{rz}(D)$, D is the only prime element.

In addition to the above, we shall be considering the ring of polynomials in D^{-1}, $F[D^{-1}]$. We originally obtained the realizable functions as ratios of polynomials in D^{-1} with the degree of the numerator less than or equal to the degree of the denominator; the realizable functions thus form a ring of fractions of $F[D^{-1}]$, but not of the type $S^{-1}R$. In this ring the only prime is expressed as $(1/D^{-1})$.

We also use the ring containing all sequences x such that del $x \geq 0$, which in algebra is called the formal power series in D and denoted conventionally as $F[[D]]$; $F[[D]]$ is also a principal ideal domain, whose only prime is D.

If G is a matrix of field elements, then it generates a vector space over the field. If it is a matrix of ring elements, then it generates a module over the ring. (A module is defined precisely like a vector space, except the scalars are in a ring rather than a field. This is the difference between block and convolutional codes.) The main theorem concerning modules over a principal ideal domain—some would say the only theorem—is a structure theorem, which, when applied to matrices G, is called the invariant-factor theorem. This theorem alone, when extended and applied to different rings, yields most of our results.

Invariant-Factor Theorem: Let R be a principal ideal domain and let G be a $k \times n$ R-matrix. Then G has an invariant-factor decomposition

$$G = A\Gamma B,$$

where A is a square $k \times k$ R-matrix with unit determinant, hence with an R-matrix inverse A^{-1}; B is a square $n \times n$ R-matrix with R-matrix inverse B^{-1}; and Γ is a $k \times n$ diagonal matrix, whose diagonal elements γ_i, $1 \leq i \leq k$, are called the invariant factors of G with respect to R. The invariant factors are unique, and are computable as follows: let Δ_i be the greatest common divisor of the $i \times i$ subdeterminants (minors) of G, with $\Delta_0 = 1$ by convention; then $\gamma_i = \Delta_i/\Delta_{i-1}$. We have that γ_i divides γ_{i+1} if γ_{i+1} is not zero, $1 \leq i \leq k - 1$. The matrices A and B can be obtained by a computational algorithm (sketched below); they are not in general unique. Finally, if there is any decomposition $G = A\Gamma B$ such that A and B are invertible R-matrices and Γ is a diagonal matrix with $\gamma_i \mid \gamma_{i+1}$ or $\gamma_{i+1} = 0$, then the γ_i are the invariant factors of G with respect to R.

Sketch of Proof [10]: G is said (in this context only) to be equivalent to G' if $G = AG'B$, where A and B are square $k \times k$ and $n \times n$ R-matrices with unit determinants; the assertion of the theorem is that G is equivalent to a diagonal matrix Γ that is unique under the specified conditions on the γ_i. Since any such A can be represented as the product of elementary row operations, and B of elementary column operations (interchange of rows (columns), multiplication of any row (column) by a unit in R, addition of any R-multiple of any row (column) to another), it can be shown that the Δ_i are preserved under equivalence. In particular, therefore, Δ_1 divides all elements of all equivalent matrices. We will now show that there exists an equivalent matrix in which some element divides all other elements, hence is equal to Δ_1 up to units. Let G not be already of this form, and let α and β be two nonzero elements in the same row or column such that α does not divide β. (If there is no element in the same row or column as α not divisible by α, there is some such β in some other column, and this column can be added to the column containing α to give an equivalent matrix for which the prescription above can be satisfied.) By row or column permutations α may be placed in the upper-left corner and β in the second entry of the first row or column; we assume column for definiteness. Now there exist x and y such that $\alpha x + \rho y = \delta$, where δ is the greatest common divisor of α and β, and has fewer prime factors than α since $\alpha \nmid \beta$, $\delta \neq \alpha$. The row transformation below then preserves equivalence while replacing α by δ:

$$\left[\begin{array}{cc|c} x & y & 0 \\ -\beta/\delta & \alpha/\delta & \\ \hline & 0 & I_{k-2} \end{array}\right] \cdot \left[\begin{array}{c|c} \alpha \ \cdot & \cdot \\ \beta \ \cdot & \\ \hline \cdot & \cdot \end{array}\right] = \left[\begin{array}{c|c} \delta \ \cdot & \cdot \\ 0 \ \cdot & \\ \hline \cdot & \cdot \end{array}\right].$$

If δ does not now divide all elements of the equivalent matrix, the construction can be repeated and δ replaced by some δ' with fewer prime factors. This descending chain can therefore terminate only with a δ that does divide all elements of the equivalent matrix. Since $\delta = \Delta_1 = \gamma_1$, up to units, multiplication of the top row by a unit puts γ_1 in the upper-left corner, and the whole first row and column can be cleared to zero by transformations of the above type (with $x = 1$, $y = 0$), giving the equivalent matrix

$$G' = \left[\begin{array}{c|c} \gamma_1 & 0 \\ \hline 0 & G_1 \end{array}\right]$$

where γ_1 divides every element of G_1. Similarly, G_1 is equivalent to a matrix G_1' of the same form, so

$$G \cong \left[\begin{array}{cc|c} \gamma_1 & 0 & 0 \\ 0 & \gamma_2 & \\ \hline & 0 & G_2 \end{array}\right]$$

where γ_1 divides γ_2 and γ_2 divides all elements of G_2. Continuing in this way, we arrive at a diagonal matrix Γ meeting the conditions of the theorem. Its uniqueness and the formula for the γ_i are obtained from the relationship $\Delta_i = \prod_{i' \leq i} \gamma_{i'}$. Q.E.D.

The invariant-factor decomposition involves a similarity transformation in some respects reminiscent of diagonalizing transformations of square matrices over a field; the invariant factors have some of the character of eigenvalues. The analogy cannot be pressed very far however.

The extension of the invariant-factor theorem to rings of fractions is immediate.

Invariant-Factor Theorem (Extension): Let R be a principal ideal domain and let Q be a ring of fractions of R. Let G be a $k \times n$ Q-matrix. Let ψ be the least common multiple of all denominators in G; then ψG is an R-matrix. Consequently ψG has an invariant-factor decomposition

$$\psi G = A\Gamma' B.$$

Dividing through by ψ, we obtain an invariant-factor decomposition of the Q-matrix G with respect to R

$$G = A\Gamma B$$

where

$$\Gamma = \Gamma'/\psi.$$

Here A and B are R-matrices with R-matrix inverses A^{-1} and B^{-1}. The diagonal elements γ_i of Γ are elements of Q uniquely determined as $\gamma_i = \gamma'_i/\psi = \alpha_i/\beta_i$, where α_i and β_i are obtained by canceling common factors in γ'_i and ψ, gcd $(\alpha_i, \beta_i) = 1$. Since $\gamma'_i \mid \gamma'_{i+1}$ if $\gamma'_{i+1} \neq 0$, we have that $\alpha_i \mid \alpha_{i+1}$ if $\alpha_{i+1} \neq 0$ and $\beta_{i+1} \mid \beta_i$, $1 \leq i \leq k - 1$. Explicitly, if ψ_i is the least common multiple of the denominators of the $i \times i$ subdeterminants of G, if θ_i is the greatest common divisor of the numerators, and $\Delta_i = \theta_i/\psi_i$ with $\Delta_0 = 1$ by convention, then

$$\gamma_i = \alpha_i/\beta_i = \Delta_i/\Delta_{i-1} \qquad 1 \leq i \leq k.$$

The γ_i are called the invariant factors of the Q-matrix G with respect to R. Finally, if there exists any $G = A\Gamma B$ satisfying the above conditions, then the diagonal terms of Γ are the invariant factors of G with respect to R.

We may picture an invariant factor decomposition more concretely as follows. Let a $k \times k$ scrambler A be defined as a $k \times k$ R-matrix with an R-matrix inverse A^{-1}. We call it a scrambler because the map $\mathbf{x}' = \mathbf{x}A$ is a one-to-one permutation of all the k-dimensional R-vectors $\mathbf{x}$. For example,

$$A = \begin{bmatrix} 1 & D+1 \\ 1 & D \end{bmatrix}$$

$$A^{-1} = \begin{bmatrix} D & D+1 \\ 1 & 1 \end{bmatrix}$$

are two inverse binary polynomial 2×2 scramblers. (The reader may at first be surprised, as was the author, by the existence of nontrivial pairs of scramblers that are feedback-free and thus not subject to infinite error propagation.) The only 1×1 scramblers are the trivial ones consisting of units of R.

Now we illustrate an invariant factor decomposition of G with respect to R by the block diagram of Fig. 5. Input sequences are scrambled in the $k \times k$ R-scrambler A; the outputs are then operated on individually by the invariant factors γ_i; finally, the k outputs plus $n - k$ dummy zeroes are scrambled in an $n \times n$ R-scrambler B to give the output sequences.

The invariant-factor theorem and its extension are well known in linear system theory, particularly in the work of Kalman [11], [12], who attributes the first engineering use of the extension above to McMillan [13]. As far as the author is aware, its use has generally been confined to the rings of polynomials and rational functions. The utility of considering additional rings will become clear in the sections to follow.

III. Structural Theorems

Our principal results are presented in three sections. In the first, we show how to determine whether G has inverses of various kinds. In the second, we show that every encoder G is equivalent to a so-called minimal encoder, which is a conventional convolutional encoder with a feedback-free inverse. In the final section we point out other desirable properties of minimal encoders: they require as few memory elements to realize as any equivalent encoder, they allow easy enumeration of error events by length, and they ensure that short codewords correspond to short information sequences in a way not shared by nonminimal encoders. We conclude that a minimal encoder is the natural choice to generate any desired code. In two appendices, we discuss dual codes and systematic encoders; the latter may also be taken as canonical encoders for any code.

Inverses

In this section we shall determine when a $k \times n$ Q-matrix G has an R-matrix right inverse G^{-1}, where Q is a ring of fractions of R and $k \leq n$. The results are stated in terms of the invariant factors γ_i of G with respect to R. We assume $\gamma_k \neq 0$, otherwise G has rank less than k and thus no inverse of any kind.

Consider the invariant-factor decomposition of G with respect to R, as illustrated in Fig. 5. The outputs of the scrambler A when its inputs range over all possible sequences are simply all possible sequences in some different

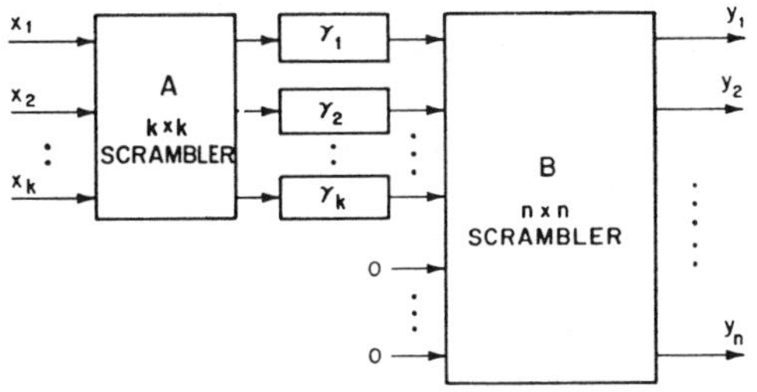

Fig. 5. Invariant-factor decomposition of (n, k) encoder.

order, since A is invertible. Moreover, if the inputs to A range over all vectors of k elements of R, then the outputs are all vectors of k elements of R in a different order, since A and A^{-1} are R-matrices. In particular, there is some input R-vector $\mathbf{x}_k$ to A such that the output $\mathbf{x}_kA$ is $\boldsymbol{\varepsilon}_k$, the kth unit input, namely, $\mathbf{x}_k = \boldsymbol{\varepsilon}_kA^{-1}$. Now the input vector $\gamma_k^{-1}\mathbf{x}_k$ gives $\gamma_k^{-1}\boldsymbol{\varepsilon}_k$ at the output of A, and $\boldsymbol{\varepsilon}_k$ at the output of Γ, hence the R-vector $\boldsymbol{\varepsilon}_kB$ at the matrix output. But $\gamma_k^{-1}\boldsymbol{\varepsilon}_k$, hence $\gamma_k^{-1}\mathbf{x}_k$, is an R-vector if and only if γ_k^{-1} is an element of R. Continuing with this argument, we have the following.

Lemma 1: Let R be a principal ideal domain and Q a ring of fractions of R. Let G be a Q-matrix with invariant-factor decomposition $G = A\Gamma B$ with respect to R, and invariant factors $\gamma_i = \alpha_i/\beta_i$, $1 \leq i \leq k$. If $\alpha_k \neq 1$, then there is a vector $\gamma_k^{-1}\boldsymbol{\varepsilon}_kA^{-1}$ that is not an R-vector but which gives an R-vector output.

Proof: If $\alpha_k \neq 1$, then $\gamma_k^{-1} = \beta_k/\alpha_k$ is not an element of R, hence $\gamma_k^{-1}\boldsymbol{\varepsilon}_k$ is not an R-vector, hence $\gamma_k^{-1}\boldsymbol{\varepsilon}_kA^{-1}$ is not an R-vector, since if it were $(\gamma_k^{-1}\boldsymbol{\varepsilon}_kA^{-1})\,A$ would be an R-vector. But $\gamma_k^{-1}\boldsymbol{\varepsilon}_kA^{-1}G = \gamma_k^{-1}\boldsymbol{\varepsilon}_k\Gamma B = \boldsymbol{\varepsilon}_kB$ is an R-vector since $\boldsymbol{\varepsilon}_k$ and B are in R. Q.E.D.

We call the numerator α_k of the kth invariant factor that appears in Lemma 1 the minimum factor of G with respect to R; this designation will be justified by Theorem 2.

From Lemma 1 we obtain a general theorem on inverses, application of which to particular rings R will settle many questions concerning inverses. We note that if $G = A\Gamma B$, then $G^{-1} = B^{-1}\Gamma^{-1}A^{-1}$ is an inverse for G, where Γ^{-1} is the $n \times k$ matrix with diagonal elements γ_i^{-1}. Since A^{-1} and B^{-1} are R-matrices, G^{-1} is certainly an R-matrix if all γ_i^{-1} are elements of R. The following theorem says that if some γ_i^{-1} is not an element of R, then G has no R-matrix inverse.

Theorem 1: Let R be a principal ideal domain and Q a ring of fractions of R. Let G be a Q-matrix whose invariant factors with respect to R are $\gamma_i = \alpha_i/\beta_i$, $1 \leq i \leq k$. Then the following statements are equivalent.

1) G has an inverse G^{-1}, which is an R-matrix.
2) There is no $\mathbf{x}$ that is not an R-vector such that $\mathbf{y} = \mathbf{x}G$ is an R-vector, or $\mathbf{y} \in R$-implies $\mathbf{x} \in R$.
3) $\alpha_k = 1$.

Proof: We shall show $1 \Rightarrow 2 \Rightarrow 3 \Rightarrow 1$.

($1 \Rightarrow 2$). If $\mathbf{y} = \mathbf{x}G$ is an R-vector, then $\mathbf{x} = \mathbf{y}G^{-1}$ is an R-vector since G^{-1} is an R-matrix.

($2 \Rightarrow 3$). By Lemma 1, if $\alpha_k \neq 1$, then $\mathbf{x} = \gamma_k^{-1}\boldsymbol{\varepsilon}_kA^{-1}$ is a vector not an R-vector such that $\mathbf{x}G$ is an R-vector.

($3 \Rightarrow 1$). If $\alpha_k = 1$, then $\alpha_i = 1$, $1 \leq i \leq k$, since $\alpha_i \mid \alpha_k$. Hence the inverse invariant factors $\gamma_i^{-1} = \beta_i/\alpha_i = \beta_i$ are all elements of R. If $G = A\Gamma B$ is an invariant-factor decomposition of G with respect to R, then A^{-1} and B^{-1} are R-matrices, so that $G^{-1} = B^{-1}\Gamma^{-1}A^{-1}$ is an R-matrix that serves as the desired inverse. Q.E.D.

We then obtain the results we need as special cases.

Corollary 1: An encoder G has a feedback-free inverse iff its minimum factor with respect to the polynomials $F[D]$ is 1.

Corollary 2: An encoder G has a feedback-free pseudo-inverse iff its minimum factor with respect to the finite sequences $F_{ft}(D)$ is 1. Furthermore, in this case and only in this case is there no infinite $\mathbf{x}$ such that $\mathbf{y} = \mathbf{x}G$ is finite.

Corollary 3: An encoder G has a realizable inverse iff its minimum factor with respect to the realizable functions $F_{rz}(D)$ is 1.

Corollary 4: An encoder G has a realizable pseudo-inverse iff its minimum factor with respect to the rational functions $F(D)$ is 1; that is, $\alpha_k \neq 0$, or the rank of G is k.

Here we have used the obvious facts that a feedback-free pseudo-inverse implies and is implied by a finite-sequence inverse, and similarly with a realizable pseudo-inverse and a rational inverse, where one is obtained from the other in both cases by multiplication by $D^{\pm d}$, d being the delay of the pseudo-inverse. We also note that since G is both an $F_{rz}(D)$ and an $F(D)$ matrix, the invariant factors with respect to these rings cannot have denominator terms; further, since $F(D)$ is a field, the only greatest common divisors are 1 and 0, and the rank of G equals the rank of Γ since A and B are invertible.

With equal ease, we can obtain a sharper result on pseudo-inverses $\tilde{G}^{-1}$ in the cases where G has no inverse. We make the following general definition of a pseudo-inverse.

Definition: $\tilde{G}^{-1}$ is an R-matrix pseudo-inverse for G with factor ψ if $\tilde{G}^{-1}$ is an R-matrix and $G\tilde{G}^{-1} = \psi I_k$.

If $G^{-1} = B^{-1}\Gamma^{-1}A^{-1}$ is not an R-matrix inverse for G, it is because Γ^{-1} is not an R-matrix. Since $\gamma_i^{-1} = \beta_i/\alpha_i$ and $\alpha_i \mid \alpha_k$, $1 \leq i \leq k$, $\tilde{G}^{-1} = \alpha_kG^{-1} = B^{-1}\,(\alpha_k\Gamma^{-1})A^{-1}$ is an R-matrix pseudo-inverse for G with factor α_k. Theorem 2 shows that this is the minimum such factor.

Theorem 2: Let R be a principal ideal domain and Q a ring of fractions of R. Let G be a Q-matrix whose invariant factors with respect to R are $\gamma_i = \alpha_i/\beta_i$, $1 \leq i \leq k$. Then G has an R-matrix pseudo-inverse $\tilde{G}^{-1}$ with factor α_k; further, all R-matrix pseudo-inverses have factors ψ such that α_k divides ψ.

Proof: The discussion preceding the theorem shows how to construct a pseudo-inverse with factor α_k. Therefore let $\tilde{G}^{-1}$ be any R-matrix pseudo-inverse, and consider the input $\mathbf{x} = \gamma_k^{-1}\boldsymbol{\varepsilon}_kA^{-1}$. By Lemma 1, $\mathbf{x}G$ is an R-vector, hence $\mathbf{x}G\tilde{G}^{-1}$ is an R-vector, but

$$\mathbf{x}G\tilde{G}^{-1} = \psi\mathbf{x}$$
$$= \psi\gamma_k^{-1}\boldsymbol{\varepsilon}_kA^{-1}.$$

Hence $(\psi\gamma_k^{-1}\varepsilon_k A^{-1})A = \psi\gamma_k^{-1}\varepsilon_k$ is an R-vector, which is to say $\psi\gamma_k^{-1} = \psi\beta_k/\alpha_k$ is an element of R. But gcd $(\alpha_k, \beta_k) = 1$; hence α_k must divide ψ in R. Q.E.D.

We note that we could have obtained Theorem 1 as a consequence of Theorem 2 and Lemma 1.

The pseudo-inverses we are interested in are realizable pseudo-inverses with delay d, or, in the terminology introduced above, $F_{rz}(D)$-matrix pseudo-inverses with factor D^d. Since the only prime in the ring of realizable functions $F_{rz}(D)$ is D, and since G is itself realizable, the invariant factors of G with respect to $F_{rz}(D)$ are $\gamma_i = D^{d_i}$ (or zero); and in particular the minimum factor is D^{d_k}. We then define the delay d of a realizable matrix as $d = d_k$, so $\alpha_k = D^d$ and $d_i \leq d, 1 \leq i \leq k$. Then Theorem 2 answers a problem stated by Kalman ([11], 10.10e) with the following corollary.

Corollary 1: Let G be realizable and let the minimum factor with respect to the realizable functions $F_{rz}(D)$ be $\alpha_k = D^d$. Then G has a realizable pseudo-inverse $\tilde{G}^{-1}$ with delay d, and no realizable pseudo-inverse with delay less than d.

Similarly, the question of the delay of a feedback-free inverse, which was investigated in [4] and [7], is answered by Corollary 2.

Corollary 2: Let G be realizable and let the minimum factor with respect to the polynomials $F[D]$ be α_k. Then G has a feedback-free pseudo-inverse $\tilde{G}^{-1}$ with delay d' if and only if $\alpha_k = D^d$ for $d' \geq d$.

For computation, it is convenient not to have to compute invariant factors repeatedly, so we use the following lemma.

Lemma 2: Let G have invariant factors γ_i with respect to R; then the invariant factors with respect to Q are γ_i', where $\gamma_i' = \gamma_i$ up to units in Q and γ_i' is a product of primes in Q.

Proof: Let $\gamma_i = \gamma_i' \gamma''$, with γ_i'' a unit in Q. Let $G = A\Gamma B$ be an invariant factor decomposition of G with respect to R. Let B' be the Q matrix obtained from B by multiplying the ith row by γ_i''; then det $B' =$ (det B) $\pi_i\gamma_i''$ is a unit in Q, since det B is a unit in R. Hence $G = A\Gamma'B'$ is an invariant-factor decomposition of G with respect to Q, so the γ_i' are invariant factors of G with respect to Q. Q.E.D.

Now we have the following recipe for deciding whether G has inverses of various types. Let $G = \{\boldsymbol{h}_i/\psi_i\}$; multiply each row through by its denominator to obtain the polynomial matrix $G' = \{\boldsymbol{h}_i\}$. Compute all $k \times k$ subdeterminants, and find their greatest common divisor θ_k. Let $\Delta_k = \theta_k/\pi_i\psi_i$ and reduce to lowest terms. Now $\Delta_k = \pi_i\gamma_i = \pi\alpha_i/\pi\beta_i = A_k/B_k$. If $A_k = 1$, G has a polynomial inverse. If $A_k = D^d$, G has a polynomial pseudo-inverse. If A_k is a polynomial not divisible by D, G has a realizable inverse. Finally, if $A_k \neq 0$, G has a realizable pseudo-inverse. The minimum pseudo-inverse delay d is the greatest common delay of the $k \times k$ subdeterminants minus the greatest common delay of the $(k-1) \times (k-1)$ subdeterminants.

In [7], Olson gives a test for the existence and minimum delay of any feedforward inverse that is equivalent to the above; although more cumbersome, Olson's result and proof are remarkable for being carried through successfully without the aid of the powerful algebraic tools used here.

Canonical Encoders Under Equivalence

The theorems of this section are aimed at the determination of a canonical encoder in the equivalence class of encoders generating a given code.

The idea of the first theorem of this section is as follows. Any encoder G has the invariant factor decomposition $A\Gamma B$ with respect to the polynomials $F[D]$ as is illustrated in Fig. 5, where A is a $k \times k$ polynomial scrambler, B is an $n \times n$ polynomial scrambler, and Γ is a set of generally nonpolynomial transfer functions. If the inputs to A are all the k-tuples of sequences; then since A is invertible the outputs are also all the k-tuples in some different order. If none of the γ_i is zero, then the outputs of Γ are also all k-tuples of sequences, if we allow many-to-one encoders, so that G may have rank $r < k$ and $\gamma_{r+1} = \cdots = \gamma_k$ may be zero, then the outputs of Γ are all k-tuples of sequences in which the last $k - r$ components are zero. But the outputs of B with these inputs are the code generated by G; G is therefore equivalent to the encoder G_0 represented by the first r rows of B. Now since B is polynomial and has a polynomial inverse, G_0 is also polynomial and has a polynomial right inverse G_0^{-1} consisting of the first r columns of B^{-1}. These observations are made precise in the following theorem and proof.

Theorem 3: Every encoder G is equivalent to a conventional convolutional encoder G_0 that has a feedback-free delay-free inverse G_0^{-1}.

Remark: In other words, G_0 and G_0^{-1} are polynomial and $G_0G_0^{-1} = I_r$.

Proof: Let G have invariant-factor decomposition $G = A\Gamma B$ with respect to $F[D]$. Let G_0 be the first r rows of B,

$$G_0 = \{\boldsymbol{b}_i\} \qquad 1 \leq i \leq r.$$

G_0 is polynomial since B is, and has a polynomial inverse G_0^{-1} equal to the first r columns of B^{-1}. To show equivalence, let $\boldsymbol{y}_0$ be any codeword in the code generated by G; then

$$\begin{aligned} \boldsymbol{y}_0 &= \boldsymbol{x}_0 G \\ &= (\boldsymbol{x}_0 A\Gamma)(B) \\ &= \sum_{i=1}^{r} (\boldsymbol{x}_0 A\Gamma)_i \boldsymbol{b}_i \\ &= \boldsymbol{x}_1 G_0, \end{aligned}$$

where $\boldsymbol{x}_1$ is the vector consisting of the first r sequences of $\boldsymbol{x}_0 A\Gamma$; hence any codeword in the code generated by G is also in the code generated by G_0. Conversely, let $\boldsymbol{y}_1$ be any word in the code generated by G_0; then

$$\begin{aligned} y_1 &= x_1 G_0 \\ &= x_1' B \\ &= (x_0' A \Gamma) B \\ &= x_0' G, \end{aligned}$$

where x_1' is the k-dimensional vector equal to x_1 in the first r positions and to zero thereafter, while $x_0' = x_1' \Gamma^{-1} A^{-1}$, where Γ^{-1} has diagonal elements γ_i^{-1} for $\gamma_i \neq 0$ and 0 for $\gamma_i = 0$; hence any codeword in the code generated by G_0 is also in the code generated by G. Q.E.D.

At this point we remark that all (n, n) encoders are obviously uninteresting: for if G has full rank, $\gamma_n \neq 0$, then the code consists of all n-tuples of sequences, hence G is equivalent to the identity encoder I_n; while if $\gamma_n = 0$, then G has rank $r < n$ and is equivalent to some (n, r) code.

Let us define any encoder meeting the conditions on G_0 in Theorem 3 as basic (from the fact that the set of generators in such an encoder is a basis for the $F[D]$-module of polynomial codewords).

Definition 4: A *basic* encoder G is a conventional convolutional encoder with a feedback-free inverse G^{-1}; that is, G is polynomial, G^{-1} is polynomial, and $GG^{-1} = I_k$.

In virtue of Corollary 1 to Theorem 1, an encoder is basic if and only if it is polynomial and has minimum factor with respect to $F[D]$ of 1, so that all invariant factors are equal to 1.

Basic encoders are not in general unique; the following theorem characterizes the class of basic encoders generating any code.

Theorem 4: If G is a basic encoder, G' is an equivalent basic encoder if and only if $G' = TG$, where T is a $k \times k$ polynomial scrambler; that is, T is a square polynomial matrix with polynomial inverse T^{-1}.

Proof: Let $G' = TG$; then both G' and $G'^{-1} = G^{-1}T^{-1}$ are polynomial since T, G, G^{-1}, and T^{-1} are all polynomial, so G' is basic. G' is equivalent to G since if $y_0 = x_0 G$, then $y_0 = x_1 G'$ where $x_1 = x_0 T^{-1}$, while if $y_0 = x_0 G'$, then $y_0 = x_1 G$ where $x_1 = x_0 T$. Conversely, let G' be a basic encoder equivalent to G; then the generators g_i of G are in both codes, being generated by ε_i in the latter case and by input vectors

$$t_i = g_i G'^{-1}$$

in the former, since $t_i G' = g_i$. The t_i are polynomial since g_i and G'^{-1} are. Let T then be the matrix that has the t_i for rows; then $G = TG'$. We can repeat the argument with G and G' reversed to obtain $G' = SG$ for some polynomial matrix S. Since both G and G' have full rank, T and S must be invertible, and since $G = TSG$, S must be the inverse of T and vice versa. Q.E.D.

In the special case $k = 1$ (a rate-$1/n$ code), the basic encoder generating any code is uniquely defined up to units (and in the binary case, uniquely defined period, since the only unit is 1), since the only 1×1 scramblers are the units of $F[D]$, or the nonzero elements of F. Given any $(n, 1)$ code with generator $g = h/\psi$, where h is a vector of n polynomials, we can find the equivalent basic encoder by multiplying through by the denominator ψ and then canceling the greatest common divisor of the elements of h. These facts have been known for some time [4], and represent all that this paper has to say about rate-$1/n$ codes. Since it is generally expected that there are good codes in the $(n, 1)$ class, in fact the best codes for most practical purposes, it is questionable whether the results of this paper will assist in the search for constructive methods of obtaining good codes.

For (n, k) encoders with $k \geq 2$, we can distinguish a subclass of basic encoders with further useful properties. By appealing to realizability theory (see Appendix II), one can show that a basic encoder G that has maximum degree μ among its $k \times k$ subdeterminants can be realized with μ and no fewer than μ memory elements. However, the obvious realization in general requires $\Sigma \nu_i = \nu$ memory elements. We therefore define the following.

Definition 5: A basic encoder G is *minimal* if its overall constraint length ν in the obvious realization is equal to the maximum degree μ of its $k \times k$ subdeterminants.

We note first that μ is invariant over all equivalent basic encoders; for if G and G' are two such encoders, by Theorem 4 $G = TG'$ where det T is a unit in $F[D]$, so that the $k \times k$ subdeterminants of G are those of G' up to units. We now proceed to show that among all equivalent basic encoders there is at least one that is minimal. It is helpful to consider the backwards encoder $\mathbf{G}$ associated with any basic encoder, defined as the encoder with generators $D^{-\nu_i} g_i$. As thus defined, $\mathbf{G}$ has elements that are polynomials in D^{-1}, hence is anticausal rather than causal. Another way of stating this definition is to let V be the $k \times k$ diagonal matrix with diagonal elements $D^{-\nu_i}$; then

$$\mathbf{G} = VG.$$

We are interested in $\mathbf{G}$ because of the following lemma.

Lemma 3: If G is basic, G is minimal if and only if $\mathbf{G}$ has an anticausal inverse.

Remark: Recall that since we allow only sequences that "start" at some time, all anticausal sequences are actually polynomials in D^{-1} (elements of $F[D^{-1}]$), with finite negative delay and degree less than or equal to 0. Hence an anticausal inverse for $\mathbf{G}$ must in fact be an $F[D^{-1}]$-inverse.

Proof: If $\mathbf{G}$ has an anticausal inverse, then $G = I_k V^{-1} \mathbf{G}$ is (loosely speaking) an invariant-factor decomposition for G with respect to $F[D^{-1}]$, so that G has invariant factors $\gamma_i = (1/D^{-\nu_i})$ with respect to $F[D^{-1}]$ and $\pi\gamma_i = (1/D^{-\nu})$. From the extended invariant factor theorem, $\pi\gamma_i = \theta_k/\psi_k$, where θ_k is the greatest common divisor of the numerators and ψ_k the least common multiple of the divisors of the $k \times k$ subdeterminants of G as ratios of polynomials in D^{-1}. But since G is basic, these subdeterminants are polynomials in D with no common prime divisor, all of which can be expressed as ratios of polynomials in D^{-1} as $\Delta = \Delta'/D^{-\delta}$, where Δ' is a polynomial in D^{-1} and where δ is the degree of Δ as a polynomial in D. Hence $\theta_k = 1$ and $\psi_k = D^{-\mu}$, where μ is the maximum degree of any $k \times k$ subdeterminant of G. Hence $D^{\nu} = \theta_k/\psi_k = D^{\mu}$, so $\nu = \mu$ and G is minimal.

Conversely, if G is minimal, then at least one of the $k \times k$ subdeterminants of $\mathbf{G}$ has a nonzero constant term, since the $k \times k$ subdeterminants of $\mathbf{G}$ are those of G times the determinant of V, which is $D^{-\nu} = D^{-\mu}$, and at least one of the subdeterminants of G has degree μ. Thus D^{-1} is not a common factor of the $k \times k$ subdeterminants of $\mathbf{G}$; further, there can be no other common factor since such a factor would imply a common factor for G, but G is basic. Hence the $k \times k$ subdeterminants of $\mathbf{G}$ are relatively prime as polynomials in D^{-1}, so that $\mathbf{G}$ has invariant factors equal to 1 with respect to $F[D^{-1}]$ by the invariant-factor theorem, and consequently an anticausal inverse by Theorem 1. Q.E.D.

The desired result is then obtained with a constructive proof.

Theorem 5: Every encoder G is equivalent to a minimal encoder.

Proof: From Theorem 3 it is sufficient to prove the theorem when G is basic. We shall show that whenever G is basic but not minimal, there exists an equivalent basic encoder with reduced constraint length. By Lemma 3, $\mathbf{G}$, the backwards encoder associated with G, has no anticausal inverse, which in turn implies a factor of D^{-1} in all $k \times k$ subdeterminants of $\mathbf{G}$. In other words, the matrix $\mathbf{G}$ modulo D^{-1} (consisting of all constant coefficients of elements of $\mathbf{G}$) is singular. (Since these are the high-order (ν_ith-order) coefficients of G, the high-order coefficients equally form a singular matrix.) There is therefore a linear combination of the generators of $\mathbf{G}$ with coefficients in F that is equal to a vector divisible by D^{-1}, thus with all-zero constant terms; that is, for some nonzero vector $\boldsymbol{f}$ with elements in F,

$$\boldsymbol{f}\mathbf{G} = 0 \bmod D^{-1}.$$

It follows that

$$\boldsymbol{f}V\mathbf{G} = 0 \bmod D^{-1};$$

or deg $(\boldsymbol{f}VG) < 0$, or consequently deg $(\boldsymbol{f}VGD^{\nu_0}) < \nu_0$, where ν_0 is any integer. Let us choose ν_0 equal to the largest constraint length of any generator $\boldsymbol{g}_i$ such that f_i is nonzero; then $\boldsymbol{x} = \boldsymbol{f}VD^{\nu_0}$ is polynomial (in D), so $\boldsymbol{y} = \boldsymbol{x}G$ is polynomial, but deg $\boldsymbol{y} < \nu_0 = \deg \boldsymbol{g}_i$ for some $\boldsymbol{g}_i$ upon which $\boldsymbol{y}$ is linearly dependent. Hence we can replace $\boldsymbol{g}_i$ by $\boldsymbol{y}$ to get an equivalent generator matrix with shorter constraint length. Q.E.D.

If we are actually interested in computing a minimal encoder equivalent to some given encoder G, we can proceed as follows. First we multiply through by denominator terms to obtain a polynomial matrix, and compute $k \times k$ subdeterminants. If ψ is some polynomial common to all such subdeterminants, then G mod ψ is singular since all $k \times k$ subdeterminants are zero mod ψ. Thus there is some linear combination of generators that equals zero mod ψ, which combination can be determined by inspection or systematically by reduction to triangular form, mod ψ. This same linear combination of generators not mod ψ will give a polynomial vector divisible by ψ; we cancel the ψ and use the result to replace one of the generators entering into the linear combination. Thus we will eventually arrive at a basic encoder. If the basic encoder is not minimal, there is some linear combination of the generators of the associated backwards encoder that is equal to 0 mod D^{-1}, which can be used to replace one of the backwards-encoder generators. Since the mod D^{-1} leaves only the zero-order terms of the backward generators, we can work instead with the ν_ith-order terms of the basic encoder, and continue until the matrix of high-order coefficients is nonsingular over F, when we will have our minimal encoder.

Properties of Minimal Encoders

The most important property of minimal encoders is that they are also minimal in the sense of requiring as few memory elements as any equivalent encoder. They also have a property, called the predictable degree property, useful in analyzing error events.

The proof of the former result depends on state-space arguments. We recall our definition of the projection operators P and $Q = 1 - P$ as the operators that truncate to times ≤ 0 and ≥ 1, respectively, and the abstract definition of the states as the set of outputs at time 1 and later for inputs zero at time 1 and later; for any input $\boldsymbol{x}$ the associated state is

$$\boldsymbol{s} = \boldsymbol{x}PGQ.$$

We shall show that the state space Σ_m of a minimal encoder G_m has no greater dimension than the state space Σ_G of any other equivalent encoder G. The result will then follow from the following lemma.

Lemma 4: If an encoder G can be realized with ν memory elements, dim $\Sigma_G \geq \nu$. Equality holds for minimal encoders in the obvious realization.

Proof: If F has q elements, there are q^{ν} physical states, and $q^{\dim \Sigma_G}$ abstract states. Since G is causal the output at time 1 and later when the inputs stop at time 0 is uniquely defined by the physical state. Hence the number of physical states is not less than the number of abstract states.

For minimal encoders in the obvious realization, the physical states correspond to the q^{ν} states

$$\boldsymbol{s} = \left[\sum_{i=1}^{k} \sum_{j=0}^{\nu_i - 1} f_{ij} D^{-j} \boldsymbol{g}_i\right] Q \qquad f_{ij} \in F.$$

The claim is that all such states are different, or, in view of linearity, that no state is equal to zero unless all $f_{ij} = 0$. Consider the input $\boldsymbol{x}$ to the backwards encoder $\mathbf{G}$ defined by

$$\boldsymbol{x} = \sum_{i=1}^{k} \sum_{j=0}^{\nu_i - 1} f_{ij} D^{\nu_i - j} \boldsymbol{\varepsilon}_i;$$

if any $f_{ij} \neq 0$ then deg $\boldsymbol{x} > 0$. The corresponding output is

$$y = xG$$
$$= \sum_{i=1}^{k} \sum_{j=0}^{\nu_i - 1} f_{ij} D^{-j} g_i,$$

which gives one of the states $s = yQ$ defined above. Now if $s = 0$, then deg $y \leq 0$, which is impossible since G has an anticausal inverse, so deg $y \leq 0$ implies deg $x \leq 0$ from Theorem 1. Hence the physical state space and the abstract state space are isomorphic. Q.E.D.

Realizability theory (see Appendix II) shows that any linear circuit can actually be realized with dim Σ_G memory elements; such a realization of a particular G is called canonical in linear system theory.

In the present context, where we are concerned with equivalence classes of encoders G, it seems appropriate to call an encoder canonical if the dimension of its state space is minimum over all equivalent encoders. The next lemma shows that canonical encoders are those with polynomial inverses, which is curiously close to the feedback-free inverse condition that eliminates catastrophic error propagation. The idea of the lemma is to look at the set CQ of all codewords truncated to $t \geq 1$. Some of these truncated codewords are still equal to codewords, namely, the codewords in the set C_{G_m} that do not actually "start" until time 1 or later. The remainder must be equal to codewords plus a state, regardless of the encoder that generates the code. Each of the equivalence classes of CQ modulo C_{G_m} must therefore contain at least one state. But such equivalence classes S_{G_m} form a vector space of dimension dim S_{G_m} so that the minimum state-space dimension is dim S_{G_m}. The remainder of the proof shows that Σ_G is isomorphic to S_{G_m} if and only if G has a polynomial inverse.

Lemma 5: Let Σ_m be the state space of a minimal encoder G_m, and Σ_G the state space of any equivalent encoder G. Then dim $\Sigma_m \leq$ dim Σ_G, with equality if and only if G has a polynomial inverse.

Proof: Let CQ be the space of all yQ, y any codeword in C. Let C_{G_m} and C_G be the spaces of all $y = xG_m$ or xG such that del $x \geq 1$. It is easy to verify that these are all vector spaces over F, and to see that C_{G_m} and C_G are subspaces of CQ. Now by Theorem 1 with $R = F[[D]]$, the ring of formal power series in D, del $y \geq 0$ implies del $x \geq 0$ if and only if G has a causal inverse, which by constancy is the same as del $y \geq 1$ implies del $x \geq 1$. Since G_m has a causal inverse, C_{G_m} consists of all y with del $y \geq 1$. All elements in C_G are therefore in C_{G_m}, with equality iff C_G has a causal (hence realizable) inverse.

Let $S_G = CQ/C_G$ be the equivalence classes of CQ modulo C_G; that is, for any $yQ \in CQ$, an element in S_G is the set of all $y'Q$ such that

$$yQ = y'Q + xG \qquad \text{del } x \geq 1.$$

From its definition, it is immediate that S_G is a vector space over F, a subspace of CQ, and that CQ has the direct sum decomposition

$$CQ = C_G \oplus S_G.$$

Similarly, we can define $S_{G_m} = CQ/C_{G_m}$, and obtain

$$CQ = C_{G_m} \oplus S_{G_m}.$$

Since $C_G \leq C_{G_m}$, $S_{G_m} \leq S_G$.

Now each such equivalence class contains a state. For any yQ can be expressed as

$$yQ = xGQ$$
$$= xPGQ + xQGQ$$
$$= s + x'G \qquad s \text{ a state, del } x' \geq 1,$$

so yQ is equivalent to s modulo C_G. Hence $S_G \leq \Sigma_G$ in the sense of isomorphism, and $S_{G_m} \leq \Sigma_m$. Equality holds if each equivalence class contains only one state, so that if s_1 and s_2 are any different states,

$$s_1 \neq s_2 \bmod C_G.$$

If s is the state $s_1 - s_2$, this means

$$s \neq y = xG \qquad \text{del } x \geq 1.$$

Let $s = x_1GQ$ for some x_1 with deg $x_1 \leq 0$; then

$$s - y = x_2GQ \neq 0,$$

where $x_2 = x_1 - x$, so deg $x_2 \geq 1$. Now deg $x_2G \leq 1$ implies deg $x_2 \leq 0$ for both x_2g and x_2 finite if and only if G has an $F[D^{-1}]$inverse, by Theorem 1 with $R = F[D^{-1}]$. The backwards encoder $\mathbf{G}$ corresponding to any minimal encoder has such an inverse, hence a fortiori so does G, so $S_{G_m} = \Sigma_m$. In summary, in the sense of isomorphism,

$$\Sigma_m = S_{G_m} \leq S_G \leq \Sigma_G,$$

where the second equality holds iff G has a causal inverse, and the third iff G has an $F[D^{-1}]$ inverse. Since the invariant factors of G with respect to $F[D^{-1}]$ are the same as those with respect to $F[D]$ except for powers of D, and the invariant factors of $G = A\Gamma B$ with respect to $F[D^{-1}]$ do not contain negative powers of D, since they are realizable, we see that the condition that G have a polynomial inverse is necessary and sufficient for equality. Q.E.D.

In view of Lemma 4, we now have the following required theorem.

Theorem 6: Let G be an encoder realizable with ν memory elements and let G_m be an equivalent minimal encoder with overall constraint length μ, hence μ memory elements in the obvious realization. Then $\nu \geq \mu$.

Proof:

$$\nu \geq \dim \Sigma_G \geq \dim \Sigma_m = \mu. \qquad \text{Q.E.D.}$$

From Lemma 5, the class of encoders that can be realized with dim Σ_m memory elements is quite broad, and includes basic encoders. Minimal encoders are unique, however, in being realizable with dim Σ_m memory elements as conventional encoders in the obvious realization. It is of course possible for a nonminimal encoder to be less complex than an equivalent minimal encoder by virtue of having fewer adders, multipliers, interconnections, and so forth; furthermore, system complexity is usually dominated by the codeword estimator, not the encoder. However, Theorem 6 does tend to discourage spending much time on looking for great encoder simplifications through unconventional approaches, such as the use of feedback.

Another property of minimal encoders, called the predictable degree property, is a useful analytical tool. Note that in general for any conventional encoder with constraint lengths ν_i, and any codeword $\boldsymbol{y} = \boldsymbol{x}G$,

$$\deg \boldsymbol{y} \leq \max_{1 \leq i \leq k} (\deg x_i \boldsymbol{g}_i) = \max_{1 \leq i \leq k} (\deg x_i + \nu_i).$$

If equality holds for all $\boldsymbol{x}$, we say G has the predictable degree property. Now we have the following.

Lemma 6: Let G be a basic encoder; then G has the predictable degree property if and only if G is minimal.

Proof: By Lemma 3, G is minimal iff its backwards encoder $\mathbf{G}$ has an anticausal inverse. By Theorem 1 with $R = F[D^{-1}]$, $\mathbf{G}$ has an anticausal inverse iff $\deg \boldsymbol{x}\mathbf{G} \leq 0$ implies $\deg \boldsymbol{x} \leq 0$, or equivalently iff $\deg \boldsymbol{x} \geq 1$ implies $\deg \boldsymbol{x}\mathbf{G} \geq 1$, or by constancy $\deg \boldsymbol{x} \geq d$ implies $\deg \boldsymbol{x}\mathbf{G} \geq d$. But

$$\boldsymbol{y} = \boldsymbol{x}\mathbf{G} = \sum_{i=1}^{k} x_i D^{-\nu_i} \boldsymbol{g}_i = \boldsymbol{x}'G$$

where the $\boldsymbol{x}'$ corresponding to $\boldsymbol{x}$ is given by

$$\boldsymbol{x}' = \sum_{i=1}^{k} x_i D^{-\nu_i} \boldsymbol{\varepsilon}_i.$$

Now $\deg \boldsymbol{x} \geq d$ iff $\max \deg x_i \geq d$ iff $\max (\deg x_i' + \nu_i) \geq d$. Hence $\deg \boldsymbol{x} \geq d$ implies $\deg \boldsymbol{x}\mathbf{G} \geq d$ iff $\max (\deg x_i' + \nu_i) \geq d$ implies $\deg \boldsymbol{y} = \deg \boldsymbol{x}'G \geq d$, which is the same as the predictable degree property. Q.E.D.

In our earlier discussion, we asserted that the error events of interest are the finite codewords. Let us normalize all such words $\boldsymbol{y}$ to start at time zero, del $\boldsymbol{y} = 0$. When G has a zero-delay feedback-free inverse, these are precisely the words generated by inputs $\boldsymbol{x}$ that are finite and start at zero, del $\boldsymbol{k} = 0$. When G has the predictable degree property as well, we can easily enumerate the finite codewords by degree, since for each possible input $\boldsymbol{x}$ we can compute the degree of the output knowing only the constraint lengths ν_i. In fact the number of codewords of degree $\leq d$ is equal to the number of ways of choosing k polynomials x_i such that $\deg x_i \leq d - \nu_i$ or $x_i = 0$. For example, if an $(n, 2)$ binary code has $\nu_1 = 2$, $\nu_2 = 4$, then there is one codeword of degree less than 2 (the all-zero word); there are two of degree ≤ 2, 4 of degree ≤ 3, 16 of degree ≤ 4, 64 of degree ≤ 5, and so forth. Of course, all equivalent encoders have the same codewords and hence the same distribution of codeword lengths.

The predictable degree property also guarantees that in some sense short codewords will be associated with short information sequences. Let us establish a partial ordering of information sequences such that $\boldsymbol{x} < \boldsymbol{x}'$ if $\deg x_i \leq \deg x_i'$ for all i, with at least one strict inequality. Codewords $\boldsymbol{y}$ can be ordered by their degrees $\deg \boldsymbol{y}$, namely, the maximum degree of all components y_i. Now we have the following:

Lemma 7: $\boldsymbol{x} < \boldsymbol{x}'$ implies $\deg \boldsymbol{y} \leq \deg \boldsymbol{y}'$ if and only if G is minimal, where $\boldsymbol{y} = \boldsymbol{x}G$, $\boldsymbol{y}' = \boldsymbol{x}'G$.

Proof: If G is minimal, and $\boldsymbol{x} < \boldsymbol{x}'$, then

$$\deg \boldsymbol{y} = \max (\deg x_i + \nu_i) \leq \max (\deg x_i' + \nu_i) = \deg \boldsymbol{y}'.$$

If G is not polynomial, then it has an infinite generator $\boldsymbol{g}_i = \boldsymbol{h}_i/\psi_i$ where $\boldsymbol{h}_i$ and ψ_i are polynomial. Let $\boldsymbol{x} = \boldsymbol{\varepsilon}_i$, $\boldsymbol{x}' = \psi_i \boldsymbol{\varepsilon}_i$; then $\boldsymbol{x} < \boldsymbol{x}'$ but $\boldsymbol{y} = \boldsymbol{g}_i$ is infinite whereas $\boldsymbol{y}' = \boldsymbol{h}_i$ is finite, hence $\deg \boldsymbol{y}' < \deg \boldsymbol{y} = \infty$.

If G is polynomial but not basic, then by Lemma 1 there is an infinite input $\boldsymbol{x} = \gamma_k^{-1} \boldsymbol{\varepsilon}_k A^{-1}$ that gives a finite output $\boldsymbol{y}$, but $\boldsymbol{x}' = \beta_k \boldsymbol{\varepsilon}_k A^{-1}$ is finite and gives finite output $\boldsymbol{y}' = \alpha_k \boldsymbol{y}$ where $\gamma_k = \alpha_k/\beta_k$, $\deg \alpha_k \geq 1$. Hence $\boldsymbol{x}' < \boldsymbol{x}$, but $\deg \boldsymbol{y}' = \deg \alpha_k + \deg \boldsymbol{y} > \deg \boldsymbol{y}$.

If G is basic but does not have the predictable degree property, then for some $\boldsymbol{x}$

$$\deg \boldsymbol{y} < \max (\deg x_i + \nu_i).$$

Let $\boldsymbol{x}' = x_i \boldsymbol{\varepsilon}_i$ for some i for which the maximum on the right is attained; then $\boldsymbol{x}' < \boldsymbol{x}$, but

$$\deg \boldsymbol{y}' = \deg x_i + \nu_i > \deg \boldsymbol{y}.$$

Q.E.D.

This lemma is not quite as sharp as one would like, since the ordering of the inputs $\boldsymbol{x}$ is only partial, so that $\deg \boldsymbol{y} > \deg \boldsymbol{y}'$ does not imply $\boldsymbol{x} > \boldsymbol{x}'$ but only $\boldsymbol{x} \nless \boldsymbol{x}'$. Also, the ordering of codewords $\boldsymbol{y}$ by degree does not take into account the lengths of the individual sequences y_i. However, Lemma 7 reassures us that by choosing a minimal encoder we will not get an excessive number of information errors out per error event.

IV. Conclusions

All our results tend to the same conclusion: regardless of what convolutional code we want to use, we *can* and *should* use a minimal encoder to generate it. Minimal encoders are therefore to be considered a canonical class, like systematic encoders for block codes. (In Appendix II we consider systematic encoders as a canonical class for convolutional codes.)

It should be noted that none of our results depends on the finiteness of F, so that all apply to sampled-data filters, with F the field of real or complex numbers. (In Lemma 4 we do need some continuity restriction, such as that the output be a linear function of the state, when F is infinite; otherwise a multidimensional abstract state space could be mapped into a single real physical memory element, for example, by a Cantor mapping.) In this context polynomial generators correspond to tapped delay line filters. The results on inverses are of clear interest here, but whether the remaining results are or not depends on whether our definition of equivalence is germane to some sampled-data problem.

There are several obvious directions for future work. First, the essential similarity in statement and proof of Theorems 3 and 5 suggests that there ought to be some way of setting up the problem so that the equivalence of any encoder to a minimal encoder could be shown without the intermediary of basic encoders. Second, there ought to be some way of treating the constraint lengths μ_j referenced to the output ($\mu_j = \max_{1 \le i \le k} \deg g_{ij}$) comparable in simplicity to our treatment of the constraint lengths ν_i referenced to the inputs. Third, at least on memoryless channels, permutations of the transmitted sequences or shifts of one relative to the others do not result in essentially different codes; it might be interesting to study encoders and codes under such a more general definition of equivalence (i.e., including column permutations of G and multiplication of columns by D^n). Finally, of course, the problem of constructing classes of "good" codes remains outstanding.

Appendix I

Dual Codes and Syndromes

We saw in Theorem 3 that all encoders were equivalent to an encoder G whose rows were the first k rows of a polynomial matrix B; G had a polynomial inverse G^{-1} that consisted of the first k columns of B^{-1}. It may have seemed that the last $n - k$ rows of B and columns of B^{-1} had no purpose; here we show how we may derive dual codes and form syndromes from them.

Dual Codes

Let B be an $n \times n$ polynomial matrix with polynomial inverse B^{-1}. Any k rows of B can be taken as the generators G of an (n, k) code; the remaining $n - k$ rows, which we call $(H^{-1})^T$, where T means transpose, can be taken as the generators of another $(n, n - k)$ code. G and $(H^{-1})^T$ have polynomial inverses G^{-1} and H^T formed from the corresponding columns of B^{-1}. Clearly if $\boldsymbol{y}$ is any codeword in the first code, say $\boldsymbol{y} = xG$, then $\boldsymbol{y}H^T = 0$, since $\boldsymbol{y}$ is a linear combination of generators $\boldsymbol{g}_i$ all of which satisfy $\boldsymbol{g}_i H^T = 0$, by virtue of $\boldsymbol{g}_i$ being a row in B and H^T consisting of noncorresponding columns of B^{-1}. The code generated by G can be defined equally well as the set of row vectors $\boldsymbol{y}$ such that $\boldsymbol{y}H^T = 0$, or the null space of H^T.

The set of row vectors $\boldsymbol{z} = xH$ for x any $(n - k)$-dimensional row vector of sequences can be considered to be a code generated by H, called the dual code, since if $\boldsymbol{y}$ is generated by G, $\boldsymbol{z}$ by H, then the inner product $\boldsymbol{y}\boldsymbol{z}^T$ is zero by virtue of $GH^T = 0$. The following lemma leads to an interesting theorem relating a code to its dual code.

Lemma 8: The $(n - k) \times (n - k)$ subdeterminants of H are equal up to units to the $k \times k$ subdeterminants of G.

Proof: Recall that G is the first k rows of B, while H^T is the last $(n - k)$ columns of B^{-1}, and $\det B =$ a unit in $F[D]$. We shall show that the upper left $k \times k$ subdeterminant of B is equal to the lower right $(n - k) \times (n - k)$ subdeterminant of B^{-1}; the same proof carries through for any other selection of columns in B and the corresponding rows in B^{-1} by transposition.[1] Write

$$B = \left[\begin{array}{c|c} B_{11} & B_{12} \\ \hline B_{21} & B_{22} \end{array}\right] \qquad B^{-1} = \left[\begin{array}{c|c} B'_{11} & B'_{12} \\ \hline B'_{21} & B'_{22} \end{array}\right],$$

where the dotted lines separate rows and columns into groups of k and $n - k$. Consider the matrix product

$$\left[\begin{array}{c|c} B_{11} & B_{12} \\ \hline 0 & I_{n-k} \end{array}\right] \left[\begin{array}{c|c} B'_{11} & B'_{12} \\ \hline B'_{21} & B'_{22} \end{array}\right] = \left[\begin{array}{c|c} I_k & 0 \\ \hline B'_{21} & B'_{22} \end{array}\right];$$

taking determinants we have

$$|B_{11}|\,|B^{-1}| = |B'_{22}|,$$

but $|B^{-1}|$ is a unit in $F[D]$, hence $|B_{11}| = |B'_{22}|$ up to units. Q.E.D.

Now we have the following.

Theorem 7: If G is equivalent to a minimal encoder with overall constraint length μ, then the dual code can also be generated by a minimal encoder with overall constraint length μ.

Proof: Let $G = A\Gamma B$ be an invariant-factor decomposition of G with respect to the polynomials $F[D]$; the first k rows of B represent an equivalent basic encoder G'. The $k \times k$ subdeterminants of G' therefore have greatest common divisor 1. Furthermore, if μ is the maximum degree of any $k \times k$ subdeterminant of G', then by the discussion leading to Theorem 5 any minimal encoder equivalent to G' has overall constraint length μ. The transpose of the last $n - k$ columns of B^{-1} represents a polynomial encoder H for the code that is dual to that generated by G' and hence G. By Lemma 8, the $(n - k) \times (n - k)$ subdeterminants of H, being the same as those of G, have greatest common divisor equal to 1 and greatest degree μ. Hence H is basic, and is equivalent to a minimal encoder with overall constraint length μ. Q.E.D.

Syndromes

To use syndromes, the receiver must be able to make tentative individual symbol decisions on the elements of the codeword $\boldsymbol{y}$. These we call the received word $\boldsymbol{y}_r$; unlike the output $\hat{\mathbf{Y}}$ of a codeword estimator, $\boldsymbol{y}_r$ is not required to be a codeword. The received errors $\boldsymbol{e}_r$ are defined by

$$\boldsymbol{e}_r = \boldsymbol{y}_r - \boldsymbol{y},$$

where $\boldsymbol{y}$ is the codeword actually sent.

[1] We are indebted to Prof. R. G. Gallager for the main idea of the proof.

Let G be basic and equal to the first k rows of a polynomial invertible matrix B; the polynomial inverse B^{-1} is then made up of two parts, the inverse G^{-1} and the dual encoder transpose H^T. If we pass the received word $\mathbf{y}_r$ through B, the output may be correspondingly divided into two parts, the noisy estimates defined by

$$\begin{aligned} \mathbf{x}_r &= \mathbf{y}_r G^{-1} \\ &= (\mathbf{y} + \mathbf{e}_r)G^{-1} \\ &= \mathbf{x} + \mathbf{e}_r G^{-1}, \end{aligned}$$

and the syndromes defined by

$$\begin{aligned} \mathbf{s}_r &= \mathbf{y}_r H^T \\ &= (\mathbf{y} + \mathbf{e}_r)H^T \\ &= \mathbf{e}_r H^T. \end{aligned}$$

Note that since both G^{-1} and H^T are polynomial, received errors propagate in the noisy estimates and syndromes for only a finite time. If received errors are sparse, the noisy estimates $\mathbf{x}_r$ correspond quite closely to the actual information sequences $\mathbf{x}$, each received error generally giving a finite number of errors in $\mathbf{x}_r$. (As Sullivan [6] has pointed out, the number of errors in $\mathbf{x}_r$ is sometimes minimized over all inverses by using a suitable pseudo-inverse $\tilde{G}^{-1}$. See the example in Appendix II.) As for the syndromes, by virtue of Theorem 7, absence of errors in the received word from time $t - \mu$ to $t - 1$ is sufficient to guarantee that syndromes at time t and later are due only to errors at time t and later. Since the probability of such an event must be nonzero, any decoder using a feedback-free syndrome generator will eventually be able to get started or resynchronized by repeatedly resetting its internal state to that which would have existed had all past syndromes been zero (the 'syndrome reset' technique).

As with block codes, the syndromes are in one-to-one correspondence with the classes of apparent errors $\mathbf{e}_a$ (defined as $\mathbf{e}_a = \mathbf{y}_r - \mathbf{y}$) between which the codeword estimator must choose, since $\mathbf{e}_a H^T = \mathbf{s}_r$ if and only if $\mathbf{y} = \mathbf{y}_r - \mathbf{e}_a$ is a codeword, which is a necessary condition for a codeword estimator output. Another way of saying this is that H^T is a many-to-one homomorphism from n-tuples to $(n - k)$-tuples whose kernel is the code generated by G. (As a dual encoder, H executes a one-to-one map in the reverse direction.) Syndromes are especially useful when the channel is such that the received errors $\mathbf{e}_r$ are independent of which codeword was actually sent, because then the decoding rule can be simply a map from syndromes $\mathbf{s}_r$ to apparent errors $\mathbf{e}_a$.

Finally, we observe that for any code a syndrome former H^T can be realized with the same number of memory elements as a minimal encoder since H^T is just the dual encoder H with inputs and outputs exchanged. (Theorem 8 can be used to prove this.)

Appendix II

Systematic Encoders

We have settled on minimal encoders as the canonical encoders for convolutional codes. In general a minimal encoder is nonsystematic; that is, the information sequences do not in general form part of the codeword. On the other hand, convolutional coding theory and practice have been almost exclusively concerned with systematic encoders. We believe that this historical happenstance is due mostly to misconceptions: false analogies with block codes, apprehensions about error propagation, feelings of insecurity about not having the true information bits buried somewhere in the received data. However, there are some situations in which a systematic code is definitely to be preferred.

Costello [8] was apparently the first to notice that every convolutional encoder is equivalent to a systematic encoder that may itself contain feedback, but which has a delay-free feedback-free inverse. In this Appendix we extend Costello's result to show that there is such an encoder that is realizable with the same number of memory elements as the minimal encoder, and thus also merits the label 'canonical.' The proof requires an appeal to realizability theory, whose main theorem we state, with a sketch of its proof. As a corollary we obtain a generalization of the result of Bussgang that every convolutional encoder is equivalent to a feedback-free systematic encoder over a decoding constraint length. We conclude with a comparison of our two classes of canonical encoders, minimal and systematic.

Realizability Theory: Main Theorem

We will first make a slight digression into realizability theory to pick up its main theorem [11].

Theorem 8: Let G have invariant factors γ_i with respect to $F[D^{-1}]$. Assume $\gamma_i \neq 0$, $1 \leq i \leq k$, and let $\gamma_i = \alpha_i/\beta_i$, where α_i and β_i are polynomials in D^{-1} with no common factors and β_i is monic; let μ_i be the degree of β_i. Then the state space Σ_G has dimension $\mu = \Sigma \mu_i$, and G is realizable with μ memory elements, but with no less than μ.

Sketch of Proof: That G needs at least μ memory elements for realization can be shown by exhibiting μ independent states and thus proving that the state space has dimension of at least μ. Let G have an invariant factor decomposition $G = A\Gamma B$ with respect to $F[D^{-1}]$, and consider the μ anticausal inputs

$$D^{-j}\boldsymbol{\varepsilon}_i A^{-1} \qquad 1 \leq i \leq k, 0 \leq j \leq \mu_i - 1;$$

these lead to the μ states

$$\begin{aligned} \mathbf{s}_{ij} &= D^{-j}\boldsymbol{\varepsilon}_i A^{-1} G Q \\ &= D^{-j}\boldsymbol{\varepsilon}_i \Gamma B Q \\ &= D^{-j}\gamma_i \mathbf{b}_i Q \qquad 1 \leq i \leq k, 0 \leq j \leq \mu_i - 1. \end{aligned}$$

Any linear combination over F of these states $\Sigma_i \Sigma_j \psi_{ij} \mathbf{s}_{ij}$

can be written in terms of the polynomials in D^{-1}

$$\psi_i = \sum_{j=0}^{\mu_i - 1} \psi_{ij} D^{-j}$$

as

$$\sum_i \sum_j \psi_{ij} s_{ij} = \sum_{i=1}^{k} \psi_i \gamma_i b_i Q.$$

We see that such a combination can be equal to 0 only if the $\psi_i\gamma_i b_i$ and hence (since B has an anticausal inverse) the $\psi_i\gamma_i$ are anticausal sequences, i.e., polynomials in D^{-1}, but $\psi_i\alpha_i/\beta_i$ is a polynomial in D^{-1} only if β_i divides ψ_i (since gcd $(\alpha_i, \beta_i) = 1$), which is impossible since the degree of ψ_i is no greater than $\mu_i - 1$ and hence less than the degree of β_i (as polynomials in D^{-1}). A slight extension of the argument would show that all states are linearly dependent on these and hence that the state space has dimension exactly μ.

To exhibit a realization, it is necessary only to construct a linear circuit whose physical states correspond to the abstract states above, and then to arrange that each unit input set the circuit into the appropriate state, as well as give the output at time 0 directly. Such a realization can be constructed from k feedback shift registers of lengths μ_i, with feedback connections given by the denominator terms β_i, as was illustrated in Fig. 2. As the register contents at time 1^- range through all possibilities, a μ-dimensional space, the 'output' of these registers ranges through all linear combinations of sequences of the form α_i/β_i, where α_i is a polynomial in D^{-1} of degree less than or equal to $\mu_i - 1$, and the 'output' is any linear combination of the contents of these registers. Now it is not hard to see that an encoder G with invariant factors α_i/β_i with respect to $F[D^{-1}]$ can be realized by a circuit of the type illustrated in Fig. 6, where C_1 is a purely combinational circuit that generates μ_i vectors to be added (linearly over F) to the register contents, and C_2 is a purely combinational circuit that forms a linear (over F) combination of the register contents and the current inputs to give the current outputs. The only delay elements are therefore in the feedback shift registers, and their number is $\Sigma\mu_i = \mu$.

Q.E.D.

Canonical Systematic Encoders

Block codes are generated by nonsingular $k \times n$ matrices with elements in some field. Any such matrix has some $k \times k$ submatrix that is nonsingular and therefore has an inverse; premultiplication by this inverse yields an equivalent (same row space) generator matrix in systematic form, that is, with the identity matrix as a submatrix. Thus every block encoder is equivalent to a systematic encoder; consequently systematic encoders have universally been taken as the canonical class of block encoders ([9], chs. 2 and 3).

As we have seen, the elements of the generator matrix G of a convolutional code are in the ring $F_{rz}(D)$ of realizable functions. A realizable $k \times k$ matrix has a realizable inverse if and only if its determinant is a nonzero realizable function with delay zero (from our general results on inverses, or Cramer's Rule). Now every convolutional encoder is equivalent to a basic encoder, and any basic encoder must have some $k \times k$ submatrix with a determinant that is a nonzero polynomial not divisible by D, else all $k \times k$ subdeterminants would be divisible by D and the encoder would not be basic. Premultiplication by the realizable inverse of such a submatrix yields an equivalent realizable encoder in systematic form. Hence Costello's result, which follows.

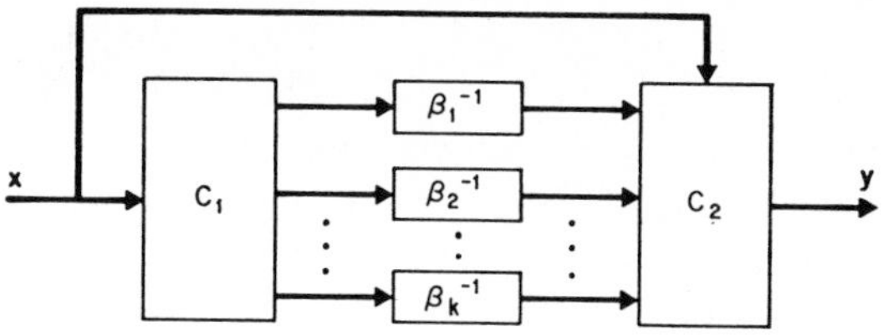

Fig. 6. Canonical realization of (n, k) encoder.

Theorem 9: Every convolutional encoder is equivalent to a systematic encoder.

There is some freedom in the choice of the k columns of the equivalent systematic encoder that are to be the k unit column vectors ε_i; once these columns are specified, assuming the ε_i are in their natural order, then the systematic encoder generating a given code is unique. Such an encoder G has a trivial delay-free feedback-free inverse, namely, the matrix G^{-1} with unit row vectors in the rows corresponding to the unit columns of G, and zeroes elsewhere, which simply strips off the k transmitted sequences that are identical to the information sequences. As in the block-code case, a dual systematic $n \times (n - k)$ encoder H^T can be formed by putting unit row vectors in the rows not corresponding to unit columns of G, and filling in the remaining $k \times (n - k)$ matrix with the negative of the remaining $k \times (n - k)$ matrix of G. (see [9], section 3.2.) (From this fact alternate proofs of the results of Appendix I can be constructed.)

To be canonical, an encoder must be realizable with μ memory elements, where μ is the minimum number of memory elements in any equivalent encoder, hence the number in an equivalent minimal encoder. Theorem 10 shows that systematic encoders have this property as well.

Theorem 10: Every systematic encoder G is canonical; that is, can be realized with the same number μ of memory elements as an equivalent minimal encoder.

Proof: From Lemma 5, dim $\Sigma_G = \mu$, since G has a polynomial inverse. From Theorem 8, G can therefore be realized with μ memory elements.

Q.E.D.

We should note that the canonical realization of even a feedback-free systematic encoder may not be the obvious realization; for example, it is well known that the most efficient realization of a conventional systematic rate-$(n-1)/n$ code, $n > 2$, with maximum generator degree ν, is Massey's [14] type-II encoder in which a single length-ν register forms all parity bits, as in Fig. 7.

Any practical decoder has some decoding constraint length λ, namely, the number of time units of delay between the time t any given information symbol is sent and the time $t + \lambda$ that a decision on that symbol must be generated by the decoder. If all previous symbols are known and the channel is memoryless, then the decoding decision may be based solely on the received word in the interval from t to $t + \lambda$, with the effects of previous symbols subtracted out (by linearity); by constancy, the decoding problem can thus be reduced to finding the initial symbols in a sequence that "starts" at time 0 on the basis of received data in times 0 through λ. Clearly only the structure of the codewords out to time λ, or modulo $D^{\lambda+1}$, enter into such a decision. We say that two encoders with the same codewords modulo $D^{\lambda+1}$ are equivalent over a decoding constraint length. Bussgang [15] observed that any $(n, 1)$ encoder is equivalent to a systematic feedback-free encoder over a decoding constraint length. Extension of this result is an easy corollary to Theorem 9.

Corollary 3: Every encoder is equivalent to a systematic feedback-free encoder over a decoding constraint length.

Proof: By Theorem 9 every encoder is equivalent to a systematic encoder, a fortiori equivalent modulo $D^{\lambda+1}$. Any realizable sequence is congruent modulo $D^{\lambda+1}$ to a polynomial of degree λ, namely, the first $\lambda + 1$ terms in such a sequence. Hence such a systematic (in fact any) encoder is equivalent to a polynomial encoder modulo $D^{\lambda+1}$. Q.E.D.

(Note that when the decoder estimates inputs at time t on the basis of received data in times t through $t + \lambda$, we can restrict our attention to the set S of codewords modulo $D^{\lambda+1}$ generated by inputs that start at time 0 or later. It may happen that two encoders equivalent over λ by our definition may not have the same sets S, due to delay in the encoder; for example $[1, 1 + D]$ appears in S when $G = [1, 1 + D]$ but not when $G = [D, D + D^2]$. This point is discussed by Costello [8] under the rubric "causal dominance;" it does not arise whenever G has a zero-delay inverse.)

Two further caveats must be entered: 1) if λ is large, the equivalent systematic encoder may be much more complex than the original encoder; 2) two encoders equivalent over a decoding constraint length behave identically only until a decoding error is made, when their error propagation characteristics may be very different.

Comparison of Minimal and Systematic Encoders

We have seen that either minimal or systematic encoders may be taken as a canonical class of convolutional encoders. Here we discuss the relative merits of each from both theoretical and practical viewpoints.

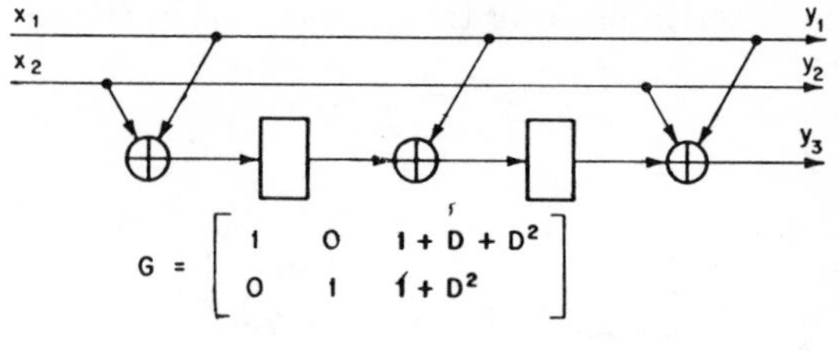

Fig. 7. Canonical realization of rate-2/3 systematic encoder.

Some codes can be generated by encoders that are both minimal and systematic, or at least feedback-free and systematic. Bucher and Heller [16] have shown that such codes are inferior to general codes of the same constraint length, in the sense that the error probability on memoryless channels with maximum-likelihood decoding is greater over a random ensemble of such codes than over the general ensemble. Bucher [17] has also shown that still further degradation in performance occurs when sequential decoding is used to decode such codes, whereas it is known that sequential decoding of general codes is asymptotically as good as maximum-likelihood decoding. Hence one does not want to confine one's attention to feedback-free systematic codes.

Both systematic and minimal encoders have delay-free feedback-free inverses. However, while the existence of former inverse is obvious, that of the latter can be demonstrated only with the aid of some algebra, and has not in the past been generally appreciated. One suspects that the main reason that nonsystematic encoders have heretofore not been used is ignorant fear of error propagation. Such fears are largely groundless, for a feedback-free inverse guarantees no catastrophic error propagation, while we have seen in Appendix I that one can find feedback-free syndrome formers as well, which in most situations can be used to ensure against ordinary error propagation.

Systematic encoders seem to be reassuring to some people by virtue of preserving the original information sequences in the codewords. The thought appears to be that at least if the decoder doesn't work, the information bits will still be there. One situation in which this consideration has real force is a broadcast situation in which information and parity sequences are sent over separate channels, and only some of the receivers have decoders, while the rest depend on the information sequence alone. If, on the other hand, all transmitted sequences are combined into one serial stream, then in order to pick out the information bits, the receiver at least has to establish phasing, and if it can do this, it is hard to see why it can not also make the feedback-free inverse linear transformation G^{-1} to recover noisy estimates of the signal. It is true that when the channel error probability is p, the error probability in these noisy estimates will be at its minimum of p when the encoder is systematic. However, when the decoder *is* working, we have seen that only minimal encoders uniformly associate short output error sequences with short error events; in general short error events may result in many output errors with systematic encoders.

Example: Minimal encoder = $[1 + D + D^2, 1 + D^2]$. Pseudo-inverse with delay D: $[1, 1]^T$. Output error probability when decoder is not working: $2p$(for p low). Most likely error event (only codeword of weight 5): $[1 + D + D^2, 1 + D^2]$. Output errors per most likely error event: 1.

Equivalent systematic encoder = $[1, 1 + D^2/1 + D + D^2]$. Inverse: $[1, 0]^T$. Output error probability when decoder is not working: p. Most likely error event: same as above. Output errors per most likely error event: 3.

It appears that if we expect the decoder to be working, we should select a minimal encoder, while if we expect it not to be, we should select a systematic encoder. This is not as frivolous as it sounds; in a sequential decoder, for example, actual (undetected) error events can be made extremely rare, the decoder failures instead occurring at times when the decoder has to give up decoding a certain segment because of computational exhaustion. During these times the decoder must put out something, and the best it can do is generally to put out the noisy estimates obtained directly from the received data, the errors in which will be minimized if the encoder is systematic. A systematic encoder (with feedback) might therefore be a good choice for a sequential decoder, depending on the resynchronization method and the performance criterion. On the other hand, a maximum-likelihood decoder (Viterbi algorithm) is subject only to ordinary error events and as a consequence should be used with a minimal encoder.

As a final practical consideration, the feedback in the general systematic encoder can lead to catastrophes if there is any chance of noise causing a transient error in the encoding circuit.

From a theoretical point of view, minimal encoders are particularly helpful in analyzing the set of finite codewords, as we saw in the main text. The fact that they are a basis for the $F[D]$-module of all such codewords means that we can operate entirely in $F[D]$, which is convenient, although throughout this paper we have seen the utility of considering larger rings. The outstanding theoretical virtue of systematic encoders is that under some convention as to which columns shall contain the identity matrix, there is a unique systematic encoder generating any code. Thus systematic encoders are most suited to the classification and enumeration of codes. Our taste is indicated by the relative placement of minimal and systematic codes in this paper, but clearly there are virtues in each class.

V. Acknowledgment

The work of Prof. J. L. Massey and his colleagues at Notre Dame, particularly the result of Olson [7], was the initial stimulus for the investigation reported here, and should be considered the pioneering work in this field. The principal results were at first obtained by tedious constructive arguments; subsequently Prof. R. E. Kalman was good enough to send along some of his work, which pointed out the usefulness of the invariant factor theorem in the guise of the Smith–McMillan canonical form, and which consequently was of great value in simplifying and clarifying the development. The close attention of Dr. A. Kohlenberg and Prof. J. Massey to the final draft was also helpful.

References

[1] J. K. Omura, "On the Viterbi decoding algorithm," *IEEE Trans. Information Theory,* vol. IT-15, pp. 177–179, January 1969.

[2] A. J. Viterbi, "Error bounds for convolutional codes and an asymptotically optimum decoding algorithm," *IEEE Trans. Information Theory,* vol. IT-13, pp. 260–269, April 1967.

[3] J. L. Massey and M. K. Sain, "Codes, automata, and continuous systems: Explicit interconnections," *IEEE Trans. Automatic Control,* vol. AC-12, pp. 644–650, December 1967.

[4] ——, "Inverses of linear sequential circuits," *IEEE Trans. Computers,* vol. C-17, pp. 330–337, April 1968.

[5] M. K. Sain and J. L. Massey, "Invertibility of linear time-invariant dynamical systems," *IEEE Trans. Automatic Control,* vol. AC-14, pp. 141–149, April 1969.

[6] D. D. Sullivan, "Control of error propagation in convolutional codes," University of Notre Dame, Notre Dame, Ind., Tech. Rept. EE-667, November 1966.

[7] R. R. Olson, "Note on feedforward inverses for linear sequential circuits," Dept. of Elec. Engrg., University of Notre Dame, Notre Dame, Ind., Tech. Rept. EE-684, April 1, 1968; also *IEEE Trans. Computers* (to be published).

[8] D. J. Costello, "Construction of convolutional codes for sequential decoding," Dept. of Elec. Engrg., University of Notre Dame, Notre Dame, Ind., Tech. Rept. EE-692, August 1969.

[9] W. W. Peterson, *Error-Correcting Codes.* Cambridge, Mass.: M.I.T. Press, 1961.

[10] C. W. Curtis and I. Reiner, *Representation Theory of Finite Groups and Associative Algebras.* New York: Interscience, 1962, pp. 94–96.

[11] R. E. Kalman, P. L. Falb, and M. A. Arbib, *Topics in Mathematical System Theory.* New York: McGraw-Hill, 1969, ch. 10.

[12] R. E. Kalman, "Irreducible representations and the degree of a rational matrix," *J. SIAM Control,* vol. 13, pp. 520–544, 1965.

[13] B. McMillan, "Introduction to formal realizability theory," *Bell Sys. Tech. J.,* vol. 31, pp. 217–279, 541–600, 1952.

[14] J. L. Massey, *Threshold Decoding.* Cambridge, Mass.: M.I.T. Press, 1963, pp. 23–24.

[15] J. J. Bussgang, "Some properties of binary convolutional code generators," *IEEE Trans. Information Theory,* vol. IT-11, pp. 90–100, January 1965.

[16] E. A. Bucher and J. A. Heller, "Error probability bounds for systematic convolutional codes," *IEEE Trans. Information Theory,* vol. IT-16, pp. 219–224, March 1970.

[17] E. A. Bucher, "Error mechanisms for convolutional codes," Ph.D. dissertation, Dept. of Elec. Engrg., Massachusetts Institute of Technology, Cambridge, September 1968.

Correction to "Convolutional Codes I: Algebraic Structure"

I am grateful to Mr. S. P. Singh of the University of Notre Dame, Notre Dame, Ind., for a counterexample that revealed an error in the above paper.[1] When Q is a ring of fractions of a principal ideal domain R that is not of the type $S^{-1}R$, the extension of the invariant factor

Manuscript received November 1, 1970.

[1] G. D. Forney, Jr., *IEEE Trans. Inform. Theory,* vol. IT-16, Nov. 1970, pp. 720–738.

theorem is true as stated except that the invariant factors of a Q matrix G with respect to R are not necessarily in Q. The only such ring of fractions considered was the ring of realizable functions, with R the ring of polynomials in $D^{-1}, F[D^{-1}]$. The most important error occurred in the last line of the proof of Lemma 5; the case of equality in Lemma 5 must be amended to read "... with equality if and only if G has both a causal and an anticausal inverse." Consequently the canonical encoders do not include all basic encoders, but only those with anticausal inverses. The earlier statement that a basic encoder with maximum degree μ among its $k \times k$ subdeterminants could be realized with μ memory elements must be augmented by the phrase, "if it has an anticausal inverse." The following basic encoder is a simple counterexample:

$$G = \begin{bmatrix} 1 & 1 & 0 \\ D & D+1 & 1 \end{bmatrix}.$$

To go beyond basic encoders to minimal encoders to obtain a canonical class is therefore a necessity, not just a matter of taste. Finally, the proof of Theorem 10 must use the fact that the identity matrix is not only a polynomial inverse, but also an anticausal inverse for a systematic matrix; and Lemma 2 must be qualified "when $Q = S^{-1}R$."

G. David Forney, Jr.
Codex Corp.
15 Riverdale Ave.
Newton, Mass. 02195

Part V: Distance Bounds, Perfect Codes, and Weight Structure

The pioneering paper of Hamming [1] included an upper bound on the minimum distance of any block code as a function of the length and the number of codewords. This bound was obtained simply by observing that if one places more spheres in a space than the volume of the space divided by the volume of one sphere, then some pair of spheres must overlap. For this reason, Hamming's bound has also become known as the "volume" bound, or the "sphere-packing" bound. The latter term is somewhat misleading, as all bounds on the minimum distances of binary codes are bounds on the number of disjoint spheres which can be packed into binary n-space.

Soon after Hamming, the first general lower bound on the number of codewords in a code with fixed length and distance was presented by Gilbert [2]. If one fixes the length and the number of codewords, then the Gilbert bound assures the existence of codes with some distance d_G, and the Hamming bound assures that no code can have distance greater than d_H, where d_H is essentially twice d_G.

The first step toward closing the gap between the Hamming and Gilbert bounds was due to Plotkin [3], who exploited the fact that the minimum distance between pairs of codewords cannot exceed the average distance between pairs of codewords. The Plotkin bound was subsequently improved by Elias, whose unpublished work combined the volume technique of Hamming with the averaging technique of Plotkin in an ingenious way. The Elias bound was rediscovered and published by Bassalygo [4]. Chapter 13 of Berlekamp [5] includes a discussion of all of these bounds in sufficient generality to include the Lee metric as well as the Hamming metric.

The Gilbert and Elias bounds remain the tightest known asymptotic lower and upper bounds on minimum distance, although there have been a number of improvements for noninfinite block lengths. Except for refinements by Varshamov [6], all of the known improvements of the Gilbert bound have been obtained by constructive techniques, which are reviewed in another chapter of this volume. Most notable among the upper bounds are those of Johnson [7]-[10].

There has been considerable effort devoted to finding all codes which have distances that meet either Plotkin's average bound or Hamming's volume bound with equality. The former problem, which amounts to finding maximally equidistant binary codes with K codewords of length $K-1$ and distance $K/2$, is equivalent to the problem of finding $K \times K$ Hadamard matrices. Hadamard matrices are known to exist only if $K = 1$, 2, or a multiple of 4. Many infinite families of Hadamard matrices are known, although there remain an infinite number of $K \geqslant 188$ for which the existence of a $K \times K$ Hadamard matrix is currently unknown. This problem had attracted the interests of combinatorial mathematicians long before Shannon and Hamming, and several papers on this problem continue to appear each year in journals such as the *Journal of Combinatorial Theory.*

On the other hand, the problem of finding all "perfect" codes, which meet Hamming's volume bound, has no pre-Shannon history. Following Hamming, however, this problem became quite popular. In addition to the single-error-correcting codes and the binary repetition codes, the only known perfect Hamming-metric codes are Golay's [11] (23, 12, 7) binary code and his (12, 6, 5) ternary code. One immediate consequence of the Elias bound is that any long perfect codes must have very high rates, but this result still left the problem undecided for a large number of possible values of the parameters. Attempts to prove the nonexistence of perfect codes led to certain Diophantine equations, which attracted considerable interest among number theorists.

The first significant step toward the proof of the nonexistence of additional perfect codes was due to Lloyd [12], whose results were extended in unpublished work by Gleason, which was later reported and extended by Assmus-Mattson [13]. In the late 1960's, a number of papers appeared which solved many special cases of the problem. The following strong result was finally proved by van Lint [14] and Tietavainen [15]: *there are no perfect Hamming-metric codes over prime-power alphabets except the binary repetition codes, the Golay codes, and single-error-correcting codes.* It remains unknown whether there are more perfect codes in the Lee metric or in the Hamming metric over alphabets whose size is not a prime power. Nor can it even be said that all perfect Hamming-metric codes over fields are known. There may be exotic nonlinear single-error-correcting codes which are even more complicated than those discovered by Vasiliev [16] and Schönheim [17].

Of course, the weight and distance structure of any code must satisfy many constraints in addition to the known general bounds on its minimum distance. The first work to explore these constraints was due to MacWilliams [18], who showed that the weight structure of any linear code may be determined from the weight structure of its orthogonal complement (i.e., the dual code) via formulas which she presented. The MacWilliams formulas have been used to determine the weight distributions of many specific classes of codes during the past decade, but it has only recently been discovered that the MacWilliams formulas are a special case of a version of Plancheral's theorem which holds for nonlinear codes as well. This result, first discovered by Zierler [19], was later obtained independently by MacWilliams, Sloane, and Goethals [20]. For electrical engineers, Zierler's original version has the great advantage of being closely analogous to familiar results about Fourier transforms. The derivation is as follows.

Let time t and frequency ω be n-dimensional binary vectors, with dot product in $GF(2)$. We now consider functions from $(GF(2))^n$ to the real numbers.

$$f : (GF(2))^n \to R.$$

If f is such a function from $(GF(2))^n$ to R, we define its Fourier transform $\hat{f}$ as

$$\hat{f}(\omega) = \sum_t (-1)^{\omega \cdot t} f(t). \qquad (1)$$

There are many functions whose Fourier transforms can be computed explicitly. One important example is the function

$$v(t) = z^{|t|} \qquad (2)$$

where z may be any real number and $|t|$ is the (real) Hamming weight of t. According to (1),

$$\hat{v}(\omega) = \sum_t (-1)^{\omega \cdot t} z^{|t|}.$$

Let $x \otimes y$ denote the componentwise product of x and y. Then, for fixed ω, we may write t as the sum of two vectors $t = r + s$ such that $r \otimes \omega = r$, $s \otimes \omega = 0$. Then $t \otimes \omega = r$, and we have

$$\hat{v}(\omega) = \sum_{\substack{r,\\ r\otimes\omega = r}} \sum_{\substack{s,\\ s\otimes\omega=0}} (-1)^{\omega \cdot r} z^{|r|+|s|}$$

$$= \left(\sum_{\substack{r,\\ r\otimes\omega = r}} (-z)^{|r|} \right) \left(\sum_{\substack{s,\\ s\otimes\omega=0}} z^{|s|} \right)$$

$$\hat{v}(\omega) = (1 - z)^{|\omega|} (1 + z)^{n - |\omega|}. \qquad (3)$$

The reason that it is reasonable to call (1) a "Fourier transform" is that if one considers the "convolution" of two functions f and g defined by

$$f * g(t) = \sum_s f(s)\, g(t - s), \qquad (4)$$

then

$$\widehat{(f * g)} = \hat{f}\hat{g}. \qquad (5)$$

In other words, the transform of the convolution is the product of the transforms. The proof of (5) is very straightforward:

$$\sum_t (-1)^{\omega \cdot t} \sum_s f(s)\, g(t - s)$$

$$= \sum_s (-1)^{\omega \cdot s} f(s) \sum_t (-1)^{\omega \cdot (t-s)} g(t - s) = \hat{f}(\omega)\, \hat{g}(\omega).$$

To obtain Plancheral's theorem, we sum (5) over all frequencies to obtain

$$\sum_\omega \hat{f}(\omega)\, \hat{g}(\omega) = \sum_s \sum_t f(t)\, g(s - t) \left(\sum_\omega (-1)^{\omega \cdot s} \right).$$

But

$$\sum_\omega (-1)^{\omega \cdot s} = \begin{cases} 2^n, & \text{if } s = 0 \\ 0, & \text{if } s \neq 0; \end{cases}$$

so

$$\sum_\omega \hat{f}(\omega)\, \hat{g}(\omega) = 2^n \sum_t f(t)\, g(t). \qquad (6)$$

Equation (6) is recognized as Plancheral's theorem on Fourier transforms. The special case when $f = g$ is called Parseval's theorem. When time and frequency are real variables, Parseval's theorem may be interpreted as a statement that the total energy of a signal is the same whether measured in the time domain or the frequency domain. Equation (6) shows that Plancheral's theorem remains valid even when time and frequency range over $(GF(2))^n$ rather than over the real numbers.

To obtain the MacWilliams identities for linear codes, we consider this function:

$$a(t) = \begin{cases} 1, & \text{if } t \text{ is a codeword} \\ 0, & \text{if } t \text{ is not a codeword.} \end{cases} \qquad (7)$$

The Fourier transform is given by

$$\hat{a}(\omega) = \begin{cases} 2^k, & \text{if } \omega \text{ is orthogonal to the code} \\ 0, & \text{if } \omega \text{ is not orthogonal to the code.} \end{cases} \qquad (8)$$

Applying (6) to (7) and (2) yields the following form of the MacWilliams identities:

$$2^n \sum_t a(t)\, z^{|t|} = \sum_\omega \hat{a}(\omega) (1 - z)^{|\omega|} (1 + z)^{n - |\omega|}. \qquad (9)$$

Of course, (9) remains valid when the code is nonlinear, but the Fourier transform $\hat{a}(\omega)$ no longer has the simple interpretation in terms of the orthogonal code as given by (8). For nonlinear codes, knowledge of the code's weight distribution is not even sufficient to determine its minimum distance. It is necessary also to investigate the differences between all pairs of codewords. Hence, the appropriate function to study is not $a(t)$, but the convolution of $a(t)$ with itself, namely,

$$A(t) = a * a(t) = \sum_s a(s)\, a(t - s).$$

Applying Plancheral's theorem to $A(t)$ and $z^{|t|}$ yields

$$2^n \sum_t A(t)\, z^{|t|} = \sum_\omega \hat{A}(\omega) (1 - z)^{|\omega|} (1 + z)^{n - |\omega|} \qquad (10)$$

where, according to (5),

$$\hat{A}(\omega) = (\hat{a}(\omega))^2. \qquad (11)$$

From (11), we may deduce that $\hat{A}$ is nonnegative for all ω.

Equating coefficients of like powers of z in (10) yields the MacWilliams identities for nonlinear codes. When the constraint that A must always be nonnegative is used in conjunction with (10), one obtains a set of linear inequalities on the distance spectrum of an arbitrary block code. Since these linear inequalities admit solutions only if the minimum distance of the code is sufficiently small, the nonlinear MacWilliams identities provide a new method for finding upper bounds on the minimum distance of nonlinear codes. Linear programming may be used to find the largest minimum distance for which the nonlinear MacWilliams identities have a solution, and this approach has been studied by Delsarte [21] and McEliece-Welch [22]. They have shown that the nonlinear MacWilliams identities imply both the Hamming bound and the Plotkin bound. McEliece and Welch have computed the solutions to the appropriate linear programs for small n and k, and they have found numerous cases where the results yield improvements over the best previously known bounds. In other cases, however, Johnson's bounds [8]-[10] remain tighter than the bounds obtained via the nonlinear MacWilliams identities. The asymptotic behavior of the bounds obtained this way is not yet known.

The first significant sequel to MacWilliam's paper was due to Pless [23], who showed that if one could obtain a sufficiently high lower bound on the minimum distance of the dual code as well as sufficient restrictions on the weights which might occur in the original code, then the MacWilliams identities have a unique solution. This result motivated additional study of weight restrictions. Solomon and McEliece [24] showed that, under certain conditions, all weights must be divisible by four. McEliece [25], [26] and Delsarte [27] subsequently generalized this result over $GF(2)$ from 2 to 2^j, and then from $GF(2)$ to $GF(p)$, and finally to $GF(q)$. Other restrictions on weight distributions have been found by other methods.

The fact that certain classes of nonbinary codes, including the Reed-Solomon codes, are maximum distance separable imposes sufficiently many restrictions that the entire weight enumerators of such codes can be derived explicitly [28]. This was first done independently by Assmus-Mattson-Turyn, Forney, and Kasami-Lin-Peterson.

The only infinite classes of binary codes whose weight distributions were known in 1967 were the relatively trivial biorthogonal Reed-Muller codes and their duals, the extended Hamming codes. Then, after deducing some strong restrictions on the weights of the second-order Reed-Muller code, Kasami [29] succeeded in obtaining the weight enumerator for many extended BCH codes which happen to be subcodes of the second-order Reed-Muller code. This historically important paper inspired a large amount of subsequent work, which has generalized and simplified the results so much that the original paper now strikes the modern reader as a laborious exercise in unnecessary calculation. Kasami's results were first extended to a few more classes of BCH codes by Berlekamp [30]. The completion of this problem required the Carlitz-Uchiyama [31] bound, a rather deep mathematical result which shows that the duals of high-rate BCH codes have very high minimum distances, much better than the BCH codes with the same low rate. Using the Carlitz-Uchiyama bound, Berlekamp [32], [33] later succeeded in obtaining the weight enumerators for *all* BCH codes which are subcodes of the second-order Reed-Muller code or supercodes of its dual. The formulas also hold for certain non-BCH codes as well, and the class of codes which have the weight enumerator given by Berlekamp's formulas was further generalized by Kasami [34].

Another interesting application of the Carlitz-Uchiyama bound was presented by Sidelnikov [35], who obtained a tight asymptotic expression for the number of primitive binary BCH codewords of any weight substantially greater than the minimum distance. The exact minimum distance of many primitive binary BCH codes remains unknown. It is frequently equal to the Bose distance (i.e., the maximum designed distance), but Kasami and Tokura [36] showed that the Solomon-McEliece theorem implies that infinitely many BCH codes have actual minimum distances which exceed the Bose distance. Furthermore, the McEliece [24] theorem implies that the difference between the actual distance and the Bose distance may be arbitrarily large. However, Berlekamp [37] showed that in any sequence of primitive binary BCH codes of fixed rate R, the ratio of the actual distance to the Bose distance approaches 1 because both are asymptotic to $2n \ln R^{-1} / \log n$. Other results on the weight distributions of various cyclic binary codes, particularly BCH codes, appear in the references cited in the survey paper by Sloane [38], which is among the reprints in this volume.

The proofs of many of the known results on the weight structure of BCH codes depends on the relationship between BCH codes and RM codes. Every extended primitive binary BCH code is a subcode of some RM code and a supercode of another RM code, and it is often possible to obtain certain information about the weight structure of the BCH code from known results about the RM code. Each of the codewords of the rth-order RM code of length 2^m corresponds to a polynomial of degree $\leqslant r$ in at most m binary variables, and this fact makes the RM codes considerably easier to study than BCH codes. The weight enumerator of the rth-order RM code of length 2^m is not yet known, but the problem has attracted a great deal of study which has led to solutions of many special cases. The weight enumerator of the second-order RM codes and some more general RM weight restrictions were first found by Berlekamp-Sloane [39], [40]. Further results on RM weight distributions have been found by van Tilborg [41], Sarwate [42], and Kasami and his colleagues [43], [44]. The most impressive result to date is the complete characterization of all codewords of weight $5 \times 2^{m-r-1}$ (this is $2\frac{1}{2}$ times the minimum distance of the rth-order code of length 2^m) by Kasami *et al.* [44]. Under the big affine group, there are no fewer than 27 substantially different types of codewords!

There has also been a considerable amount of recent research on the weight distribution of other classes of cyclic codes. Using rather deep theorems from number theory, McEliece and Rumsey [45] and Baumert and McEliece [46] obtained a recursion for the weight enumerator of all irreducible binary cyclic codes with fixed index of imprimitivity $N = (2^k - 1)/n$ where k is the code's dimension and n is its length.

For sequences of codes in which the index of imprimitivity

grows too rapidly, such as the quadratic residue codes, the McEliece formulas become intractable. Although little is known about the weight distributions of long quadratic residue codes, most of the shorter ones have been found by a variety of clever techniques. Assmus-Mattson, Karlin, and Mykkeltviet, Lam, and McEliece [47] are among those who have made significant contributions to this problem. The survey by Sloane [38] provides complete references and a more detailed discussion of this work.

The fact that many quadratic residue codes are self-dual has led to studies of the weight enumerators of self-dual codes. These investigations have been further motivated by the fact that the incidence matrix of any projective plane of order 10 can be extended into the generator matrix of a self-dual (112, 56) linear binary code. It is not known whether or not any projective planes of order 10 exist, and this is a very important unsolved problem in the area of combinatorial mathematics known as design theory. Other QR codes and symmetry codes have been used to obtain new five-designs. More details are given in the survey by Sloane [38].

The most complete and comprehensive recent survey of distance bounds and weight distributions is the paper by Sloane [38]. Since that paper also contains a number of interesting results due to its author, we include it in this volume.

It is clear that the subject of distance bounds and weight distributions remains in a very active state. Five years ago, I attempted to summarize the logical implications among the various major results known at that time; the result was a complicated figure with 33 boxes [48]. A more up-to-date version of this figure would be even more confusing. At present, the subject of weight enumeration seems to be splitting apart into a number of nearly disjoint subspecialties each of which is making rapid progress in what appears to be a different direction from any of the others.

References

[1] R. W. Hamming, "Error detecting and error correcting codes," *Bell Syst. Tech. J.*, vol. 29, pp. 147-160, 1950; *Math. Rev.*, vol. 12, p. 35.

[2] E. N. Gilbert, "A comparison of signalling alphabets," *Bell Syst. Tech. J.*, vol. 31, pp. 504-522, 1952.

[3] M. Plotkin, "Binary codes with specified minimum distances," *IRE Trans. Inform. Theory*, vol. IT-6, pp. 445-450, Sept. 1960 (also published as *Univ. Penn. Res. Div. Rep.*, pp. 51-20, Jan. 1951); *Math. Rev.*, vol. 22, p. 13361.

[4] L. A. Bassalygo, "New upper bounds for error-correcting codes," *Probl. Peredach. Inform.*, vol. 1, no. 4, pp. 41-45, 1965.

[5] E. R. Berlekamp, *Alegbraic Coding Theory*. New York: McGraw-Hill, 1968.

[6] R. R. Varshamov, "Estimate of the number of signals in error correcting codes," *Dokl. Akad. Nauk SSSR*, vol. 117, pp. 739-741, 1957; *Math. Rev.*, vol. 20, p. 1596.

[7] S. M. Johnson, "A new upper bound for error-correcting codes," *IRE Trans. Inform. Theory*, vol. IT-8, pp. 203-207, Apr. 1962; *Math. Rev.*, vol. 25, p. 1067.

[8] ——, "Improved asymptotic bounds for error-correcting codes," *IEEE Trans. Inform. Theory*, vol. IT-9, pp. 198-205, July 1963; *Math. Rev.*, vol. 29, p. 1089.

[9] ——, "On upper bounds for unrestricted binary error-correcting codes," *IEEE Trans. Inform. Theory*, vol. IT-17, pp. 466-468, July 1971.

[10] ——, "Upper bounds for constant weight error correcting codes," *Discrete Math.*, vol. 3, pp. 109-124, 1972.

[11] M. J. E. Golay, "Notes on digital coding," *Proc. IRE* (Corresp.) vol. 37, p. 657, June 1949.

[12] S. P. Lloyd, "Binary block coding," *Bell Syst. Tech. J.*, vol. 36, pp. 517-535, 1957; *Math. Rev.*, vol. 19, p. 465.

[13] E. F. Assmus, H. F. Mattson, and R. Turyn, "Cyclic codes," Air Force Cambridge Res. Lab., Final Rep., ch. II, sec. I, "Lloyd-Gleason on perfect codes," 1966.

[14] A. Tietavainen, "On the nonexistence of perfect codes over finite fields," *SIAM J. Appl. Math.*, vol. 24, no. 1, pp. 88-96, 1973.

[15] J. H. van Lint, "A survey of perfect codes," *Rocky Mountain J. Math.*, to be published.

[16] Ju. L. Vasil'ev, "On nongroup close-packed codes," *Probl. Cybern.*, vol. 8, pp. 337-339, 1962; *Math. Rev.*, vol. 29, p. 5661.

[17] J. Schönheim, "On linear and nonlinear single-error-correcting q-nary perfect codes," *Inform. Contr.*, vol. 12, pp. 23-26, 1968; *Math. Rev.*, vol. 38, p. 5517.

[18] F. J. MacWilliams, "A theorem on the distribution of weights in a systematic code," *Bell Syst. Tech. J.*, vol. 42, pp. 79-94, 1963; *Math. Rev.*, vol. 26, p. 7462.

[19] N. Zierler, unpublished correspondence, 1969.

[20] F. J. MacWilliams, N. J. A. Sloane, and J. M. Goethals, "The MacWilliams identities for nonlinear codes," *Bell Syst. Tech. J.*, vol. 51, no. 4, pp. 803-819, 1972.

[21] P. Delsarte, "Linear programming associated with coding theory," *Philips Res. Rep.*, vol. 27, pp. 272-289, 1972.

[22] R. J. McEliece and L. R. Welch, "Distance bounds from the nonlinear MacWilliams identities via linear programming," 1974, to be published.

[23] V. Pless, "Power moment identities on weight distributions in error correcting codes," *Inform. Contr.*, vol. 6, pp. 147-152, 1963.

[24] G. Solomon and R. J. McEliece, "Weights of cyclic codes," *Combinatorial Theory*, vol. 1, pp. 459-475, 1966.

[25] R. J. McEliece, "Linear recurring sequences over finite fields," Ph.D. dissertation, Dep. Math., California Inst. Technol., Pasadena, 1967.

[26] ——, "Weight congruences for p-ary cyclic codes," *Discrete Math.*, vol. 3, pp. 177-192, 1972.

[27] P. Delsarte and R. J. McEliece, "Zeroes of functions in finite Abelian group algebras," 1974, to be published.

[28] E. R. Berlekamp, *Algebraic Coding Theory*. New York: McGraw-Hill, 1968, sec. 16.5.

[29] T. Kasami, "Weight distribution of BCH codes," in R. C. Bose and T. A. Dowling Ed., *Proc. Conf. Combinatorial Math. and its Applications.* Chapel Hill, N.C.: Univ. North Carolina Press, 1968, ch. 20; *Math. Rev.*, vol. 40, p. 5325.

[30] E. R. Berlekamp, *Algebraic Coding Theory*. New York: McGraw-Hill, 1968, ch. 16.

[31] L. Carlitz and S. Uchiyama, "Bounds for exponential sums," *Duke Math. J.*, vol. 24, pp. 37-41, 1957; *Math. Rev.*, vol. 18, p. 563.

[32] E. R. Berlekamp, "Weight enumeration theorems," in *Proc. 6th Annu. Allerton Conf. Circuit and Syst. Theory*, Univ. Illinois, Urbana, 1968, pp. 161-170.

[33] ——, "The weight enumerators for certain subcodes of the second order binary Reed-Muller codes," *Inform. Contr.*, vol. 17, pp. 485-500, 1970.

[34] T. Kasami, "The weight enumerators for several classes of subcodes of the 2nd order binary Reed-Muller codes," *Inform. Contr.*, vol. 18, pp. 369-394, 1971.

[35] Sidelnikov, "On weight spectrum of binary BCH codes," *Probl. Peredach. Inform.*, vol. 7, pp. 14-22, 1971.

[36] T. Kasami and N. Tokura, "On the weight structure of Reed-Muller codes," *IEEE Trans. Inform. Theory*, vol. IT-16, pp. 752-759, Nov. 1970.

[37] E. R. Berlekamp, "Long primitive binary BCH codes have distance $d \sim 2n \ln R^{-1}/\log n$," *IEEE Trans. Inform. Theory*, vol. IT-18, pp. 415-426, May 1972.

[38] N. J. A. Sloane, "A survey of constructive coding theory, and a table of binary codes of highest known rate," *Discrete Math.*, vol. 3, pp. 265-294, 1972.

[39] E. R. Berlekamp and N. J. A. Sloane, "Restrictions on weight distribution of Reed-Muller codes," *Inform. Contr.*, vol. 14, pp. 442-456, 1969; *Math. Rev.*, vol. 39, p. 5215.

[40] N. J. A. Sloane and E. R. Berlekamp, "Weight enumerator for second-order Reed-Muller codes," *IEEE Trans. Inform. Theory*, vol. IT-16, pp. 745-751, Nov. 1970; *Math. Rev.*, vol. 42, p. 9071.

[41] H. C. A. van Tilborg, "On weights in codes," Dep. Math., Technological Univ. Eindhoven, The Netherlands, T. H. Rep. 71-WSK-03, pp. 1-76, 1971.

[42] D. Sarwate, "Weight enumeration of Reed-Muller codes and cosets," Ph.D. dissertation, Dep. Elec. Eng., Princeton Univ., Princeton, N.J., 1973.

[43] M. Sugino, Y. Ienaga, N. Tokura, and T. Kasami, "Weight distribution of (128, 64) Reed-Muller code," *IEEE Trans. Inform. Theory* (Corresp.), vol. IT-17, pp. 627–628, Sept. 1971.

[44] T. Kasami, N. Tokura, and S. Azumi, "On the weight enumeration of weights less than 2.5 d_{min} of Reed-Muller codes," *IEEE Trans. Inform. Theory*, to be published.

[45] R. J. McEliece and H. Rumsey, Jr., "Euler products, cyclotomy, and coding," *J. Number Theory*, 1972.

[46] L. D. Baumert and R. J. McEliece, "Weights of Irreducible cyclic codes," *Inform. Contr.*, vol. 20, pp. 158–175, 1972.

[47] J. Mykkeltveit, C. Lam, and R. McEliece, "On the weight enumerators of quadratic residue codes," JPL Tech. Rep. 32-1526, vol. XII.

[48] E. R. Berlekamp, "Weight enumeration theorems," in *Proc. 6th Annu. Allerton Conf. Circuit and Syst. Theory*, Univ. Illinois, Urbana, 1968, pp. 161–170.

Binary Codes with Specified Minimum Distance*

MORRIS PLOTKIN†

Summary—Two n-digit sequences, called "points," of binary digits are said to be at distance d if exactly d corresponding digits are unlike in the two sequences. The construction of sets of points, called codes, in which some specified minimum distance is maintained between pairs of points is of interest in the design of self-checking systems for computing with or transmitting binary digits, the minimum distance being the minimum number of digital errors required to produce an undetected error in the system output. Previous work in the field had established general upper bounds for the number of n-digit points in codes of minimum distance d with certain properties. This paper gives new results in the field in the form of theorems which permit systematic construction of codes for given n, d; for some n, d, the codes contain the greatest possible numbers of points.

By the use of redundancy, it is possible to encode messages for transmission in such a way that errors in transmission may be corrected, provided they are not too dense. For the special case of transmission by means of binary digits, with fixed-length words, this paper investigates the relationships among word length, number of words in the code and number of errors in a word that can be corrected. The best codes known, with respect to these relationships but not to mechanizability, are given in Table I.

In this paper, n-digit binary numbers are regarded as points in an n-dimensional space. The word "point" denotes a binary number, or more accurately, a sequence of binary digits, since the ordinary arithmetical properties of binary numbers are not utilized.[1] Two n-digit points are said to be "at distance d" if they differ in exactly d corresponding digits. For example, the points

1011101000

and

0111001001

are at distance 4, the first, second, fifth, and last digits being different for the two points. A set of n-digit points is called a "code of minimum distance d" if each point is at distance at least d from every other point of the set. The 6-digit points

000000 010101

111000 101101

100110 110011

011110 001011

TABLE I

$d = 8$	.	.	.	.	.	.	.	2	2	2	2	4	4	8	16	32
$A(n, d)$							2	2	2	2	4	4	8	16	32	
$d = 6$	.	.	.	.	.	2	2	2	4	6*	12*	24*				
					2	2	2	4	6*	12*	24*					
$d = 4$	.	.	2	2	4	8	16									2^{11}
		2	2	4	8	16									2^{11}	
$d = 2$	.	2	4	8	16	32	64	2^7	2^8	2^9	2^{10}	2^{11}	2^{12}	2^{13}	2^{14}	2^{15}
	2	4	8	16	32	64	2^7	2^8	2^9	2^{10}	2^{11}	2^{12}	2^{13}	2^{14}	2^{15}	2^{16}
				$n = 4$				$n = 8$				$n = 12$				$n = 16$

* These values differ from the corresponding values of $B(n, d)$.

form a code of minimum distance 3, as may be verified by comparing them pairwise. It is convenient to regard a set consisting of a single point as a code of minimum distance d for every positive integer d.

Clearly, for every ordered pair (n, d) of positive integers there is some maximal number $A(n, d)$[2] of n-digit points which might be selected to give a code of minimum distance d. The code exhibited above demonstrates that $A(6, 3) \geq 8$. It will be seen later than there does not exist a set of nine 6-digit points at a distance 3 or greater pairwise. The $A(n, d)$ notation describes this situation by the equation

$$A(6, 3) = 8.$$

Both the present paper and one by Hamming[1] are concerned primarily with properties of the function $A(n, d)$. [Hamming's paper would not be very different if he had used $A(n, d)$ instead of his $B(n, d)$.] Interest is attached to this function by reason of its connection with coding schemes for correction of errors in systems employing binary symbols for handling information. Consider a system for computing or transmitting n-digit binary numbers, and having the property that noise or system malfunction will affect at most x of the n-digits in any output number. There can be selected $A(n, 2x + 1)$ but no more n-digit numbers which form a code of minimum distance $2x + 1$. If the system can be designed or its operation programmed in such a manner that correct—*i.e.*, error free—operation will give rise to outputs consisting exclusively of numbers in the code, then the correct outputs will always be deducible from the actual output. There will always be exactly one code number at distance x or less from an output number. For example,

* Received by the PGIT, November 20, 1959. The work leading to this paper was sponsored by the Burroughs Corp.
† Auerbach Electronics Corp., Philadelphia, Pa.
[1] R. W. Hamming, "Error detecting and error correcting codes," *Bell Sys. Tech. J.*, vol. 29, pp. 147–160; April, 1950.

[2] This definition for $A(n, d)$ is not the same as Hamming's definition for his $B(n, d)$, in that a somewhat less restrictive class of codes is used here. $B(n, d) \leq A(n, d)$ for all n, d. $B(n, d)$ is always a power of 2; $A(n, d)$ need not be. The departure is for convenience only and does not constitute a significant innovation.

Reprinted from *IRE Trans. Inform. Theory*, vol. IT-6, pp. 445–450, Sept. 1960.

if $x = 1$ and $n = 6$, there could be used the code exhibited above to demonstrate that $A(6, 3) \geq 8$, since $d = 3 = 2x + 1$. An output of, for example, 101001 in such a system could be "corrected" to 101101, that being the code number at distance 1 or less from the actual output.

Following is a summary of Hamming's results, which are utilized in the present paper:[3]

$$A(n, 1) = 2^n,$$

$$A(n, 2) = 2^{n-1},$$

$$A(n + 1, 2k) = A(n, 2k - 1),$$

$$A(n, 2k - 1) \leq \frac{2^n}{1 + C(n, 1) + \cdots + C(n, k - 1)}$$

where

$$C(n, h) = \frac{n!}{h!\,(n - h)!}.$$

Except for the unimportant difference between $A(n, d)$ and $B(n, d)$, all definitions and results to this point are due to Hamming.

Definition: By the sum $a * b$ of two n-digit points a, b is meant that n-digit point whose jth digit is zero / unity according as the jth digits of a, b are the same / different.

For example,

if a is 1011101000

and b is 0111001001

then $a * b$ is 1100100001.

For any a, $a * a$ is the origin or null-point $00 \cdots 0$, denoted throughout by o.

Definition: $|a| = m$ means that exactly m of the digits of the point a are 1. In this notation, the distance between two n-digit points a, b is $|a * b|$.

It is clear that addition as defined above is associative: that $(a * b) * c = a * (b * c)$. If K is a code of n-digit points $a, b, c, \cdots$ of minimum distance d, then so is the code denoted by $K * x$ consisting of the points $a * x$, $b * x$, $c * x$, $\cdots$ where x is any n-digit point. For pairwise distances are preserved, since

$$|(a * x) * (b * x)| = |(a * b) * (x * x)|$$
$$= |(a * b) * 0| = |a * b|.$$

Theorem 1: If $2d > n$, then $A(n, d) \leq 2m \leq 2d/(2d - n)$, m an integer.

Proof: Let K be any code consisting of A n-digit points of minimum distance d. Let h be any point in K. Consider the code $K * h$, as defined by the notation of the preceding paragraph. Since $h * h = o$, o will be a member of $K * h$. By the minimum distance property it follows that the other $A - 1$ members of $K * h$ each have at least d digits equal to 1, so that the sum of $|k * h|$ over all k in K is at least $(A - 1)d$. This is true for each of the A possible choices of h. Letting h also run through all possible values we find that the total number N of 1's in the A^2 possible sums of two points $k * h$ for h, k both in K, must satisfy

$$A(A - 1)d \leq N = \sum |h * k|,$$

the sum over all ordered pairs (h, k).

Next, we obtain another inequality on N by considering corresponding digits of each point in K. Suppose x points in K have for their first digit 1 and the other $A - x$ have for their first digit 0. In the A^2 sums $k * h$, exactly $2x(A - x)$ will have for their first digit 1. If $y, z, \cdots$ are defined in similar manner for the second digits, third digits, $\cdots$ of the points in K, the same number $N = \sum |h * k|$ is seen to be expressible as

$$N = 2x(A - x) + 2y(A - y) + 2z(A - z) + \cdots.$$

Case 1): $A = 2m$. Each of the terms

$$2x(A - x),\ 2y(A - y),\ \text{etc.},$$

is at most $A^2/2$. Since there are n such terms,

$$N \leq nA^2/2.$$

Combining this inequality with $A(A - 1)d \leq N$,

$$2(A - 1)\, d \leq An$$

and

$$(2\,d - n)A \leq 2\,d.$$

Since $2d > n$ by hypothesis,

$$A = 2m \leq \frac{2\,d}{2\,d - n}.$$

Case 2): $A = 2m - 1$. Each of the terms

$$2x(A - x),\ 2y(A - y),\ \text{etc.},$$

is at most $(A^2 - 1)/2$. Continuing as in Case 1), it may be seen that

$$A = 2m - 1 < 2m \leq \frac{2\,d}{2\,d - n}.$$

Corollary: $A(n, n) = 2$. By the above theorem, $A(n, n) \leq 2$, and the pair $00\cdots0$, $11\cdots1$ constitute an example showing $A(n, n) \geq 2$.

Corollary: $A(4m - 1, 2m) \leq 4m$ and $A(4m - 2, 2m) \leq 2m$.

Theorem 2: $A(n, d) \leq 2A(n - 1, d)$.

Proof: Let K be a code of $A(n, d)$ n-digit points of minimum distance d. Separate the points of K into two sets according to their first digit. At least one of the two sets will contain one-half or more of the points. Deletion of the first digit in each point of that set leaves a code of minimum distance d containing at least

$$\frac{A(n, d)}{2}$$

$(n - 1)$-digit points. This proves the theorem.

[3] Hamming's proofs that the relations hold for $B(n, d)$ are valid without change for $A(n, d)$.

Corollary: Since $A(4m-1, 2m) \leq 4m$, we have

$$A(4m, 2m) \leq 8m.$$

Also, if $A(4m, 2m) = 8m$, then $A(4m-1, 2m) = 4m$ and

$$A(4m-2, 2m) = 2m.$$

Theorem 3: If $4m-1$ is a prime, then $A(4m, 2m) = 8m$.

Proof: Since we have shown $A(4m, 2m) \leq 8m$, it is sufficient to construct a code of $8m$ $4m$≡digit points of minimum distance $2m$.[4] One such construction is included in the Appendix.

Theorem 4: $A(2n, 2d) \geq A(n, 2d)\, A(n, d)$.

Proof: For this proof only the symbol $\frown$ denoting concatenation of two sets of symbols is introduced. Its meaning is illustrated by the example:

If a is 1011

and b is 00111,

then $\overset{\frown}{ab}$ is 101100111.

Clearly, $|\overset{\frown}{ab}| = |a| + |b|$ for any a, b.

Let L be a code of minimum distance d containing $A(n, d)$ n-digit points, and M a code of minimum distance $2d$ containing $A(n, 2d)$ n-digit points. From these will be constructed a code K of minimum distance $2d$, containing $A(n, d)A(n, 2d)$ points. This will prove the theorem.

Define K as the set of all points $u = (\overset{\frown}{aa}) * (\overset{\frown}{bo})$ for a in L and b in M, o being the n-digit null point. The points u will of course be $2n$-digit points. There are $A(n, d)$ $A(n, 2d)$ distinct pairs a, b. If it can be shown that two distinct pairs a_1, b_1 and a_2, b_2 give rise to points u_1, u_2, in K at a distance at least $2d$, the theorem is proved.

$$\begin{aligned} u_1 * u_2 &= (\overset{\frown}{a_1\, a_1} * \overset{\frown}{b_1\, o}) * (\overset{\frown}{a_2\, a_2} * \overset{\frown}{b_2\, o}) \\ &= (\overset{\frown}{a_1\, a_1} * \overset{\frown}{a_2\, a_2}) * (\overset{\frown}{b_1\, o} * \overset{\frown}{b_2\, o}) \\ &= \{\overset{\frown}{(a_1 * a_2)\ (a_1 * a_2)}\} * \{\overset{\frown}{(b_1 * b_2)\ o}\}. \end{aligned}$$

For a_1, b_1 and a_2, b_2 distinct pairs, three cases may occur.

1) $a_1 = a_2$, $b_1 \neq b_2$. In this case

$$|\, u_1 * u_2 \,| = |\, \overset{\frown}{o\ o} * \{\overset{\frown}{(b_1 * b_2)\ o}\} \,|$$

$$= |\, \overset{\frown}{(b_1 * b_2)\ o} \,| = |\, b_1 * b_2 \,|.$$

But, by hypothesis, b_1, b_2 are members of code M of minimum distance $2d$, so that in this case $|u_1 * u_2| \geq 2d$.

2) $a_1 \neq a_2$, $b_1 = b_2$. In this case

$$\begin{aligned} |\, u_1 * u_2 \,| &= |\, \{\overset{\frown}{(a_1 * a_2)\ (a_1 * a_2)}\} * \{\overset{\frown}{o\ o}\} \,| \\ &= |\, \overset{\frown}{(a_1 * a_2)\ (a_1 * a_2)} \,| \\ &= 2\, |\, a_1 * a_2 \,| \end{aligned}$$

and since $|a_1 * a_2| \geq d$ by hypothesis, again $|u_1 * u_2| \geq 2d$.

3) $a_1 \neq a_2$, $b_1 \neq b_2$. In this case we write

$$|\, u_1 * u_2 \,| = |\, \overset{\frown}{\{(a_1 * a_2) * (b_1 * b_2)\}\ \{(a_1 * a_2) * o\}} \,|$$

$$= |\, (a_1 * a_2) * (b_1 * b_2) \,| + |\, a_1 * a_2 \,|.$$

For any x, y, $|x * y| \geq |x| - |y|$, or

$$|\, x * y \,| + |\, y \,| \geq x.$$

Therefore, $|u_1 * u_2| \geq |b_1 * b_2| \geq 2d$.

Theorem 5: If $A(4m, 2m) = 8m$ holds for $m = x$, then it also holds for $m = 2x$.

Proof: $A(8m, 4m) \geq A(4m, 4m)A(4m, 2m) = 2A(4m, 2m)$. Also, $A(8m, 4m) \leq 16m$ by the corollary to Theorem 2. Therefore, $A(4m, 2m) = 8m$ implies $A(8m, 4m) = 16m$, which was to be proved.

Theorems 3 and 5 together prove that $A(4m, 2m) = 8m$ holds for a number of values of m. In $m \leq 20$, the values which are not reached are 7, 9, 13, 14, and 19. I know of no m for which I can show $A(4m, 2m) \neq 8m$. This suggests the conjecture $A(4m, 2m)$, $= 8m$ for all m.

From Theorems 1 to 5, in conjunction with Hamming's results, there may be deduced for a number of n, d the exact value of $A(n, d)$. With few exceptions, these n, d lie in the region $2d > n$. This is the region in which $A(n, d) \leq 4d$ and one would expect it for that reason to be the least interesting region from the point of view of practical applications. The known values of $A(n, d)$ for $d \leq 8$, $n \leq 16$ are shown in Table I. Corresponding values of $B(n, d)$ are given in Table II. The bottom two rows, $d = 1$ and $d = 2$, are given by Hamming; and values for n, d = 3, 3; 4, 4; 7, 3; 8, 4; 15, 3; 16, 4 are special cases of Golay's formula.[5] No single method can be

TABLE II

$d = 8$								2	2	2	2	4	4	8	16	32
$B(n, d)$							2	2	2	2	4	4	8	16	32	
$d = 6$						2	2	2	4	4	8	16				
					2	2	2	4	4	8	16					
$d = 4$				2	2	4	8	16	16	32	64	2^7	2^8	2^9	2^{10}	2^{11}
			2	2	4	8	16	16	32	64	2^7	2^8	2^9	2^{10}	2^{11}	2^{11}
$d = 2$		2	4	8	16	32	64	2^7	2^8	2^9	2^{10}	2^{11}	2^{12}	2^{13}	2^{14}	2^{15}
	2	4	8	16	32	64	2^7	2^8	2^9	2^{10}	2^{11}	2^{12}	2^{13}	2^{14}	2^{15}	2^{16}
				$n = 4$				$n = 8$				$n = 12$				$n = 16$

[4] Since the original writing of this report, the author has learned that such codes are a special case of a more general class that may be constructed by methods given by R. E. A. C. Paley, "On Orthogonal Matrices," *J. Math. and Phys.*, vol. 12, pp. 311–320; 1933. By virtue of Paley's work, Theorem 3 may be stated not only for $4m-1$ prime, but for $4m-1$ of the form $2^k(p^h+1)$, p an odd prime and h, k integers.

[5] M. J. E. Golay, "Notes on digital coding," Proc. IRE, vol. 37, p. 657; June, 1949.

prescribed for finding the values given for different n, d. To illustrate the procedures it will be shown that $A(13, 8) = 4$.

Because $8 - 1 = 7$ is a prime, $A(8, 4) = 16$ by $A(4m, 2m) = 8m$. This in turn implies that $A(8m, 4m) = 16m$, or $A(16, 8) = 32$.

$$A(13, 8) \geq \frac{A(14, 8)}{2} \geq \frac{A(15, 8)}{4} \geq \frac{A(16, 8)}{8} = 4$$

by Theorem 2. By Theorem 1,

$$A(13, 8) \leq 2m \leq \frac{16}{16 - 13}, \quad \text{or} \quad A(13, 8) \leqq 4.$$

Combining the two inequalities, $A(13, 8) = 4$.

For $n > 2d$, although they do not provide exact values of $A(n, d)$, Theorems 1 to 5 may be useful in obtaining bounds. Again, the method chosen will be different for different n, d. For purposes of illustration the case $n = 26$, $d = 6$ is discussed.

$$A(26, 6) = A(25, 5) \leq \frac{2^{25}}{1 + 25 + (1/2)(25)(24)} = \frac{128}{163} \cdot 2^{17},$$

$$A(26, 6) \geqq A(13, 6)A(13, 3) \geqq A(12, 6)A(14, 4)$$

$$\geq 24A(7, 4)A(7, 2),$$

and

$$A(26, 6) \geq (24)(8)(64) = 3 \cdot 2^{12}.$$

This tells us that

$$3 \cdot 2^{12} \leq A(26, 6) \leq \frac{128}{163} \cdot 2^{17}.$$

Further, it tells us how to construct a code of 3.2^{12} points for $n = 26$, $d = 6$, because all theorems of this paper bounding $A(n, d)$ from below are constructive in nature. In order to construct such a code the inequalities leading to $A(26, 6) \geq 3.2^{12}$ are retraced.

First let K_1 be the code of 8 points for $n = 4$, $d = 2$, consisting of all 4-digit points which have an even number of 1's among their digits. From K_1 and the two point code 1111,0000, there may be constructed by the method of Theorem 4 a code K_2 of (8) (2) = 16 points with $n = 8$, $d = 4$. If the sixteen points of K_2 are separated into two sets according to whether the last digit is 0 or 1, at least one of the sets will have eight or more points and deletion of the last digit will give a code K_3 of at least eight members with $n = 7$, $d = 4$. We have now got as far as $A(7, 4) = 8$ in retracing the inequalities. Next we take K_4 as the code of 64 points consisting of all 7-digit points with an even number of 1's among their digits. K_4 exemplifies $A(7, 2) = 64$. From K_3 and K_4 there may be constructed by the method of Theorem 4 a code K_5 of $A(7, 2)A(7, 4) = 2^9$ points, with $n = 14$ and $d = 4$. By merely suppressing the last digit of K_5 we get a code K_6 with the same number 2^9 of points, $n = 13$, $d = 3$.

Since $4 \cdot 3 - 1 = 11$ is a prime, the method of Theorem 3 permits construction of a code K_7 exemplifying $A(12, 6) = 24$. By the possibly wasteful process of adding an 0 at the end of each point of K_7 there may be constructed K_8, a code of 24 points with $n = 13$, $d = 6$. Finally from K_6 and K_8 there may be obtained, again by the method of Theorem 4, our desired code for $n = 26$, $d = 6$, with at least $24 \cdot 2^9 = 3 \cdot 2^{12}$ points.

Appendix

Proof That $A(4m, 2m) = 8m$ IF $4m - 1$ is a Prime

In this proof of congruences are modulo $4m - 1$ if not otherwise noted. The first and greater part of the proof consists of constructing a set $a_1, a_2, \cdots, a_{4m-1}$ of $(4m - 1)$-digit points satisfying

$$|a_j| = 2m \quad \text{and} \quad |a_j * a_k| = 2m, \qquad k \neq j$$

and

$$j, k = 1, 2, \cdots, 4m - 1.$$

After that the rest is simple.

It is a well-known theorem in elementary number theory that every odd prime p has a primitive root r: an integer such that each of $r, r^2, r^3, \cdots, r^{p-1}$ is congruent to a different one of $1, 2, 3, \cdots, p - 1$. Let r be a primitive root of the prime $4m - 1$.

An integer $x \not\equiv 0$ modulo p is called a quadratic residue of a prime p if there exists another integer y satisfying $y^2 \equiv x$ modulo p. If there exists no y satisfying $y^2 \equiv x$ modulo p then x is called a quadratic nonresidue of p. It is known from elementary number theory that exactly half of the integers $1, 2, \cdots, p - 1$ are quadratic residues and half are quadratic nonresidues and that -1 is a quadratic nonresidue of all prines of the form $4m - 1$.

The numbers $r^2, r^4, \cdots, r^{4m-2}$ are all quadratic residue of $4m - 1$, for clearly $y = r^k$ satisfies $y^2 \equiv r^{2k}$ for $k = 1, 2, \cdots, 2m - 1$. Therefore, $r, r^3, \cdots, r^{4m-3}$ must be quadratic nonresidues of $4m - 1$. The numbers $-r^2, -r^4, \cdots, -r^{4m-2}$ are also quadratic nonresidues, for if there were a w satisfying $w^2 \equiv -r^{2k}$ there would be a y—namely, the y satisfying $w = yr^k$—which satisfies $y^2 \equiv -1$, and this is known to be impossible modulo $4m - 1$. The numbers $r, r^3, \cdots, r^{4m-3}$ are, therefore, each congruent to a different one of $-r^2, -r^4, \cdots, -r^{4m-2}$, each set containing one member congruent to each of the nonresidues among $1, 2, 3, \cdots, 4m - 2$.

I shall construct the $a_1, a_2, \cdots, a_{4m-1}$ in terms of their binary digits. To that end, I first define a binary digit z_i for all integral i by:

$z_i = 1$ if i is a quadratic residue of $4m - 1$; *i.e.*, if i is congruent to one of $r^2, r^4, r^6, \cdots, r^{4m-2}$.
$z_i = 0$ if i is a quadratic nonresidue of $4m - 1$; *i.e.*, if i is congruent to one of $r, r^3, \cdots, r^{4m-3}$.
$z_i = 1$ if $i \equiv 0$.

The z_i so defined have the property, as is easily verified, that $z_i = z_{ir^2}$ for every i. It follows that $z_i = z_{ir^2} = z_{ir^4} = z_{ir^6} = \cdots$ etc., for every i. Also, since $r^2, r^4, r^6, \cdots, r^{4m-2}$ are congruent to $-r, -r^3, \cdots, -r^{4m-3}$ in some order the above equations may be expressed

$$z_i = z_{-ir} = z_{-ir^3} = z_{-ir^5} = \cdots \text{ etc., or}$$

$$z_{-i} = z_{ir} = z_{ir^3} = z_{ir^5} = \cdots \text{ etc., for every } i.$$

These equations may all be combined into

$$z_{ir^k} = \begin{cases} z_i, & k \text{ even} \\ z_{-i}, & k \text{ odd} \end{cases}$$

for any i and k.

The a_i are now defined. Let a_1 be the $(4m - 1)$-digit point whose ith digit is z_i, $i = 1, 2, \cdots, 4m - 1$. For $j = 2, 3, \cdots, 4m - 1$, a_j is obtained by cyclic permutation of the digits of a_1: let z_{j+i-1} be the ith digit of a_j. Consider the digits of a_1. The last one is z_{4m-1}, which is 1 because $4m - 1 \equiv 0$. Of the others, half of the subscripts are residues and half are nonresidues; *i.e.*, half the digits are 1's and half 0's. Therefore, $2m$ of the digits are 1's, and $2m - 1$ 0's; $|a_1| = 2m$. But since a_j is obtained by permuting the digits of a_1, we have

$$|a_j| = 2m, \qquad j = 1, 2, \cdots, 4m - 1.$$

This is one of the two conditions we shall require on the a_j, the other being

$$|a_j * a_k| = 2m \quad \text{if} \quad j \neq k.$$

We now verify that the second condition is also met.

Let

$$s_{j,k} = |a_j * a_k|, \qquad i, j = 1, 2, \cdots, 4m - 1.$$

I wish to show $s_{jk} = 2m$ for all $j \neq k$. By the cyclic construction of the a_j, it is clear that $|a_j * a_k| = |a_1 * a_{k-j+1}|$ if $k > j$. It is, therefore, sufficient to prove $s_{1,k} = 2m$, $k = 2, 3, \cdots, 4m - 1$; or $s_{1,u+1} = 2m$, $u = 1, 2, \cdots, 4m - 2$.

$s_{1,u+1} = |a_1 * a_{u+1}|$ is investigated directly by comparison of corresponding digits of a_1 and a_{u+1}. These are, in order, the pairs z_1, z_{u+1}; z_2, z_{u+2}; $\cdots$; z_{4m-1}, z_u. $s_{1,u+1}$ is the number of pairs z_j, z_{u+j} for which $z_j \neq z_{u+j}$. It is convenient to rearrange the pairs as follows (*I* utilize $z_{4m-1} = z_0$, $z_{4m-1-u} = z_{-u}$, etc.):

$$z_0, z_u; z_u, z_{2u}; \cdots; z_{-u}, z_0.$$

This rearrangement will always be possible because the first elements of the pairs are the same for both sets, $0, u, 2u, \cdots$, running through the values $1, 2, 3, \cdots$ for $u = 1, 2, \cdots, 4m - 2$.

If u is a quadratic residue of $4m - 1$, then $u \equiv r^{2k}$ and we had seen that $z_{ir^{2k}} = z_i$. We may express the pairs z_0, z_u; z_u, z_{2u}; z_{2u}, z_{3u}; $\cdots$; z_{-u}, z_0 as $z_0, z_{r^{2k}}$; $z_{r^{2k}}, z_{2r^{2k}}$; $z_{2r^{2k}}, z_{3r^{2k}}$; $\cdots$; $z_{-r^{2k}}, z_0$; or finally as z_0, z_1; z_1, z_2; z_2, z_3; $\cdots$, z_{-1}, z_0. To summarize, $s_{1,u+1}$ is the number of pairs of adjacent elements z_j, z_{j+1} in a complete cycle $z_0, z_1, z_2, \cdots, z_{-1}, z_0$ for which $z_j \neq z_{j+1}$, if u is a quadratic residue of $4m - 1$.

If u is a quadratic nonresidue of $4m - 1$, then $u \equiv r^{2k-1}$ and we had seen that $z_{ir^{2k-1}} = z_{-i}$. In this case the pairs z_0, z_u; z_u, z_{2u}; z_{2u}, z_{3u}; $\cdots$; z_{-u}, z_0 may be expressed by $z_0, z_{r^{2k-1}}$; $z_{r^{2k-1}}, z_{2r^{2k-1}}$; $z_{2r^{2k-1}}, z_{3r^{2k-1}}$; $\cdots$; $z_{-r^{2k-1}}, z_0$; and finally by z_0, z_{-1}; z_{-1}, z_{-2}; z_{-2}, z_{-3}; $\cdots$; z_1, z_0. This time it is seen that $s_{1,u+1}$ is the number of pairs of adjacent elements z_j, z_{j-1} in a complete cycle $z_0, z_{-1}, z_{-2}, \cdots, z_1, z_0$ for which $z_j \neq z_{j-1}$, if u is a quadratic nonresidue of $4m - 1$.

But the two cycles, for u a residue and for u a nonresidue, give the same value of $s_{1,u+1}$, for one cycle is the other written backwards. It remains to find $s_{1,u+1} = s$, the distance $|a_j * a_k|$ for any distinct j, k. Consider the first digit of a_j, $j = 1, 2, \cdots, 4m - 1$. It will be z_j, which has been seen to take on the value 1 for $2m$ of the j and 0 for $2m - 1$ of the j. Exactly $2m(2m - 1)$ of the

$$\frac{(4m - 1)(4m - 2)}{2}$$

different $a_j * a_k$, $j \neq k$, will have 1 for the first digit. This is true also for the second, third, etc., digits by reason of the cyclic construction of the a_j. The total number of 1's in all the different $a_j * a_k$ is therefore $\sum_{j<k} |a_j * a_k| = (4m - 1)2m(2m - 1)$. But this sum is also

$$\frac{(4m - 1)(4m - 2)}{2}$$

because s was the distance between each pair and there are

$$\frac{(4m - 1)(4m - 2)}{2}$$

pairs a_j, a_k, $j \neq k$. Combining the two,

$$s = 2m = |a_j * a_k| \quad \text{for} \quad j \neq k;$$

$$j, k = k, 2, \cdots, 4m - 1.^6$$

Now that there have been constructed the a_j with the two desired properties $|a_j| = 2m$ and $|a_j * a_k| = 2m$ if $j \neq k$, it is easy to construct a code to demonstrate $A(4m, 2m) \geq 8m$.

[6] This implies that there are $2m$ alternations—1 followed by 0 or 0 followed by 1—in the sequence $z_0 z_1 z_2 \cdots z_{-1} z_0$. Since $z_1 = z_0 = 1$ and $z_{-1} = 0$, it follows that for primes of form $4m - 1$ the quadratic residues among $1, 2, \cdots, 4m - 2$ occur in exactly m blocks, as do the nonresidues.

For $j = 1, 2, \cdots, 4m - 1$, let b_j be the $4m$-digit point obtained by adding an 0 at the end of a_j. Because $|a_j| = 2m$ and $|a_j * a_k| = 2m$ for $j = k$ it is clear that $|b_j| = 2m$ and $|b_j * b_k| = 2m$ for $j = k$. I denote by e the $4m$-digit point whose digits are all 1; by o, as before, the $4m$-digit point whose digits are all 0.

I claim that the points e, o, b_j, $e * b_j$ form a code of $8m$ $4m$-digit points of minimum distance $2m$, demonstrating that $A(4m, 2m) \geq 8m$. Clearly there are $8m$ points in the code, each of $4m$ digits. Only the minimum distance requirement need be established. Since $|b_j| = 2m$ implies $|e * b_j| = 2m$, the zeros of b_j being the 1's of $e * b_j$ and conversely, it is clear that e, o are each at distance $2m$ from each of the remaining points. Also, $|b_j * b_k| = 2m$ for $j \neq k$ implies $|(e * b_j) * (e * b_k)| = 2m$ for $j \neq k$, the two distances being equal. It remains only to check $|b_j * (e * b_k)|$. But this is equal to $|e * (b_j * b_k)|$, and is equal to $2m$ or $4m$ because $|b_j * b_k| = 2m$ or 0 depending on whether $j \neq k$ or $j = k$.

This completes the proof that the code as constructed exemplifies $A(4m, 2m) \geq 8m$. The construction of such a code for given m is quite simple in practice, compared to the proof above that the code constructed fulfills the requirements. The case $m = 3$ is illustrated.

For $m = 3$, $4m - 1 = 11$. It is readily determined that 2 is a primitive root of 11, the numbers $2, 2^2, 2^3, \cdots, 2^{10}$ being congruent modulo 11 to 2, 4, 8, 5, 10, 9, 7, 3, 6, 1, respectively. The second, fourth, etc., of these are the residues: 4, 5, 9, 3, 1; the others 2, 8, 10, 7, 6, the non-residues. The definition of z_i requires that $z_i = 1$ for $i = 1, 3, 4, 5, 9$, and also for $i = 11$; $z_i = 0$ for $i = 2, 6, 7, 8, 10$. This gives us the a_j:

a_1: 1 0 1 1 1 0 0 0 1 0 1

a_2: 0 1 1 1 0 0 0 1 0 1 1

a_3: 1 1 1 0 0 0 1 0 1 1 0

. . . .

(etc., by cyclic permutation)

. . . .

a_{10}: 0 1 1 0 1 1 1 0 0 0 1

a_{11}: 1 1 0 1 1 1 0 0 0 1 0.

The desired code of $8m = 24$ points of $4m = 12$ digits each, of minimum distance $2m = 6$ is the following:

0:	000000 000000	e:	111111 111111
b_1:	101110 001010	$e * b_1$:	010001 110101
b_2:	011100 010110	$e * b_2$:	100011 101001
b_3:	111000 101100	$e * b_3$:	000111 010011
. . . .		. . . .	
. . . .		. . . .	
b_{10}:	011011 100010	$e * b_{10}$:	100100 011101
b_{11}:	110111 000100	$e * b_{11}$:	001000 111011.

НОВЫЕ ВЕРХНИЕ ГРАНИЦЫ ДЛЯ КОДОВ, ИСПРАВЛЯЮЩИХ ОШИБКИ

Л. А. Бассалыго

В настоящей работе выводятся некоторые новые верхние оценки для кодов, исправляющих ошибки.

Рассмотрим канал, который передает сигналы * длины n из q различных элементарных символов. Эти сигналы называются кодовыми словами. Расстояние Хэмминга между двумя словами определяется как число мест, в которых эти слова отличаются друг от друга. Код исправляет все комбинации из t или меньшего числа ошибок тогда, и только тогда, когда минимальное расстояние между кодовыми словами равно по меньшей мере $d = 2t + 1$.

1. Пусть $A(n, d)$ — максимально возможное число кодовых слов длины n кода, исправляющего t или менее ошибок, а $A_r(n, d)$ — максимально возможное число таких кодовых слов, лежащих на сфере радиуса r. Чтобы получить новую верхнюю границу для $A(n, d)$ нам понадобится следующая

Лемма. *Если* $r^2 > \left(1 - \frac{1}{q}\right)(2r - d)n$, *то*

$$A_r(n, d) \leqslant \frac{\left(1 - \frac{1}{q}\right)dn}{r^2 - \left(1 - \frac{1}{q}\right)(2r - d)n}.$$

Доказательство. В доказательстве леммы для простоты вместо $A_r(n, d)$ будем писать A. Составим матрицу размерности $A \times n$ из A кодовых слов, лежащих на сфере радиуса r. Эта матрица обладает двумя свойствами:

а) в каждой строке матрицы имеется ровно r отличных от 0 символов;

б) разность двух любых строк матрицы содержит не менее d отличных от 0 символов.

Перенумеруем все q символов, причем символу 0 припишем 0-й номер. Пусть k_{ij} — число символов с номером j в i-м столбце матрицы. Очевидно, что

$$\sum_{i=1}^{n} \sum_{j=1}^{q-1} k_{ij} = Ar, \qquad \sum_{i=1}^{n} k_{i0} = A(n - r). \tag{1}$$

Рассмотрим всевозможные разности между различными строками с учетом порядка. Всего таких разностей будет $A(A - 1)$. Сумма по всем разностям чисел, отличных от нуля символов этих разностей, равна $\sum_{i=1}^{n} \sum_{j=0}^{q-1} k_{ij}(A - k_{ij})$.

Из свойства б) следует, что

$$\sum_{i=1}^{n} \sum_{j=0}^{q-1} k_{ij}(A - k_{ij}) \geqslant A(A - 1)d. \tag{2}$$

Но

$$\sum_{i=1}^{n} \sum_{j=0}^{q-1} k_{ij}(A - k_{ij}) = A^2 n - \sum_{i=1}^{n} \sum_{j=1}^{q-1} k_{ij}^2 - \sum_{i=1}^{n} k_{i0}^2. \tag{3}$$

Из (1), (2) и (3) следует

$$A^2 n - A(A - 1)d \geqslant \frac{A^2 r^2}{n(q - 1)} + \frac{A^2(n - r)^2}{n^2}. \tag{4}$$

Отсюда

$$A\left[r^2 - \left(1 - \frac{1}{q}\right)(2r - d)n\right] \leqslant \left(1 - \frac{1}{q}\right)dn$$

и, следовательно, если $r^2 > \left(1 - \frac{1}{q}\right)(2r - d)n$, то

$$A \leqslant \frac{\left(1 - \frac{1}{q}\right)dn}{r^2 - \left(1 - \frac{1}{q}\right)(2r - d)n},$$

что и требовалось доказать.

Из леммы верхняя оценка для $A(n, d)$ получается просто. Легко видеть, что

$$A(n, d) C_n^r (q - 1)^r \leqslant q^n A_r(n, d). \tag{5}$$

Полагая $r = \left[\left(1 - \frac{1}{q}\right)n - \left(1 - \frac{1}{q}\right)n\sqrt{1 - \frac{2q}{q-1}\frac{t}{n}}\right]$, из леммы получаем, что $A_r(n, d) \leqslant d$.

Следовательно,

$$A(n, d) \leqslant \frac{q^n}{C_n^{\left[\left(1-\frac{1}{q}\right)\left(1-\sqrt{1-\frac{2q}{q-1}\frac{t}{n}}\right)n\right]} (q-1)^{\left[\left(1-\frac{1}{q}\right)\left(1-\sqrt{1-\frac{2q}{q-1}\frac{t}{n}}\right)n\right]}}. \tag{6}$$

Асимптотически эта верхняя граница имеет вид

$$\frac{k}{n} \leqslant 1 - H\left[\left(1 - \frac{1}{q}\right)\left(1 - \sqrt{1 - \frac{2q}{q - 1}\frac{t}{n}}\right)\right] \lg_q 2 -$$

$$- \left(1 - \frac{1}{q}\right)\left(1 - \sqrt{1 - \frac{2q}{q - 1}\frac{t}{n}}\right) \lg_q(q - 1). \tag{7}$$

Оценка (7) лучше верхних границ, известных автору [1]. В случае $q = 2$ оценка (7) имеет вид

$$\frac{k}{n} \leqslant 1 - H\left(\frac{1 - \sqrt{1 - 4\frac{t}{n}}}{2}\right) \tag{8}$$

Эта оценка аннонсирована К. Шенноном, Р. Галлагером и П. Элайсом без доказательства в [1]. Случай $q = 2$ изображен на рисунке *

2. Используем оценку (6) для получения верхней границы для кодов, исправляющих пачки ошибок. Пачкой длины m называется вектор, ненулевые компоненты которого имеются только среди m последовательно расположенных компонент, первая и последняя из которых отлична от нуля. Код исправляет t или менее пачек ошибок, длина которых не превышает m, если разность любых двух кодовых слов нельзя представить в виде суммы менее, чем $d = 2t + 1$ пачек ошибок длины не более m.

Пусть $A(n, d, m)$ — максимально возможное число кодовых слов длины n кода, исправляющего t или менее пачек ошибок, длина которых не превышает m. Так как

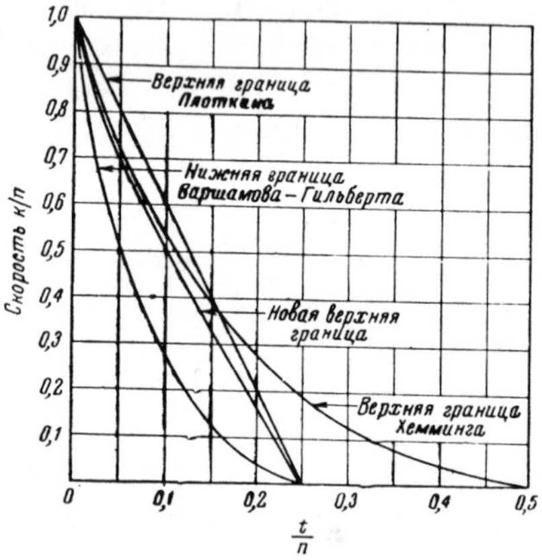

$A(n, d, m) \leqslant A(n + 1, d, m)$, то найдется такое n_1, что n_1 / m — целое, $n \leqslant n_1 \leqslant n + m - 1$ и $A(n, d, m) \leqslant A(n_1, d, m)$.

Получим верхнюю оценку $A(n, d, m)$. Для этого установим взаимно-однозначное соответствие с сохранением операции сложения между векторным пространством над полем $GF(q)$, образованным последовательностями длины m и полем $GF(q^m)$. При помощи этого соответствия установим взаимно-однозначное отображение последовательностей длины n_1 с q символами в последовательности длины n_1 / m с q^m символами следующим образом: разобьем все последовательности длины n_1 на n_1 / m последовательных кусков длины m, при этом каждому куску длины m будет соответствовать символ из $GF(q^m)$. Ясно, что если код длины n_1 с q символами исправляет t или менее пачек ошибок, длина которых не более m, то соответствующий ему при данном взаимно-однозначном отображении код длины n_1 / m с q^m символами заведомо исправляет t или менее ошибок. Обратное предположение, конечно, не верно. Следовательно, с учетом того, из какого поля берутся символы, имеем

$$A(n_1, d, m) \leqslant A\left(\frac{n_1}{m}, d\right).$$

Используя выведенную выше оценку (6), получим

$$A(n, d, m) \leqslant A(n_1, d, m) \leqslant A\left(\frac{n_1}{m}, d\right) \leqslant$$

$$\leqslant \frac{q^{n_1} d}{C_{\frac{n_1}{m}}^{\left[\left(1-\frac{1}{q^m}\right)\left(1-\sqrt{1-\frac{2q^m}{q^m-1}\frac{mt}{n_1}}\right)\frac{n_1}{m}\right]} (q^m - 1)^{\left[\left(1-\frac{1}{q^m}\right)\left(1-\sqrt{1-\frac{2q^m}{q^m-1}\frac{mt}{n_1}}\right)\frac{n_1}{m}\right]}}, \tag{9}$$

где $n \leqslant n_1 \leqslant n + m - 1$.

Асимптотически эта верхняя оценка имеет вид

$$\frac{k}{n} \leqslant 1 - H\left[\left(1 - \frac{1}{q^m}\right)\left(1 - \sqrt{1 - \frac{2q^m}{q^m - 1}\frac{mt}{n}}\right)\right]\frac{\lg_q 2}{m} -$$

$$- \left(1 - \frac{1}{q^m}\right)\left(1 - \sqrt{1 - \frac{2q^m}{q^m - 1}\frac{mt}{n}}\right)\frac{\lg_q(q^m - 1)}{m}. \tag{10}$$

Оценка (10) лучше известной автору [2].

3. Рассмотрим двоичный код ($q = 2$), исправляющий t или менее несимметрических ошибок *. Расстояние между двумя кодовыми словами введем следующим образом:

$$\bar{\rho}(x, y) = |x - y| + \left|\,|x| - |y|\,\right|,$$

где $|x|$ — число единиц в слове x.

Известен следующий результат [3]. Для того, чтобы двоичный код исправлял

* Условно сигналы, с помощью которых передаются сообщения, рассматриваются как элементы векторного пространства над полем Галуа $GF(q)$.

* На рисунке не изображена граница С. Джонсона. В просчитанных им точках она лишь немногим хуже новой верхней границы, но не имеет хорошего аналитического выражения.

* Искажений вида $(1 \to 0)$.

Reprinted with permission from *Probl. Peredach. Inform.*, vol. 1, no. 4, pp. 41-45, 1965.

или менее несимметрических ошибок, необходимо и достаточно, чтобы для любой различной пары исследуемых сигналов x и y выполнялось условие

$$\bar{\rho}(x, y) \geqslant 2t + 1 \qquad (2t + 1 = d).$$

Пусть $\bar{A}(n, d)$ — максимально возможное количество кодовых слов кода длины n исправляющего t или менее несимметрических ошибок. Так как расстояние $\bar{\rho}(x, y)$ превращается на сфере $|x| = |y| = r$ в расстояние Хэмминга, то

$$\frac{\bar{A}(n, d)}{d} \leqslant A(n, d).$$

Из (6) имеем

$$\bar{A}(n, d) \leqslant A(n, d)\,d \leqslant \frac{2^n d^2}{C_n^{\left[\frac{n}{2}\left(1-\sqrt{1-4\frac{t}{n}}\right)\right]}}. \qquad (11)$$

Асимптотически эта верхняя оценка лучше известной автору [4].

ЛИТЕРАТУРА

1. Jonson S. M. Improved asymptotic bounds for error-corresting codes. IEEE Trans. Inform. Theory, 1963, 9, *3*, 198—205.
2. Варшамов Р. Р., Мегрелышвили Р. П. Оценка числа сигналов одного класса корректирующих кодов. Автоматика и телемеханика, 1964, **25**, *7*, 1101—1103.
3. Варшамов Р. Р. Некоторые особенности линейных кодов, корректирующих несимметрические ошибки. Докл. АН СССР, 1964, **157**, *3*, 546—548.
4. Варшамов Р. Р. Оценка числа сигналов в кодах с коррекцией несимметрических ошибок. Автоматика и телемеханика, 1964, **25**, *11*, 1628, 1629.

Поступила в редакцию
23 июня 1965 г.

Binary Block Coding

By S. P. LLOYD

(Manuscript received March 16, 1956)

From the work of Shannon one knows that it is possible to signal over an error-making binary channel with arbitrarily small probability of error in the delivered information. The effects of errors produced in the channel are to be eliminated, according to Shannon, by using an error correcting code. Shannon's proof that such codes exist does not provide a practical scheme for constructing them, however, and the explicit construction and study of such codes is of considerable interest.

Particularly simple codes in concept are the ones called here close packed strictly e-error-correcting *(the terminology is explained later). It is shown that for such a code to exist, not only must a condition due to Hamming be satisfied, but also another condition. The main result may be put as follows: a close-packed strictly e-error-correcting code on n, n > e, places cannot exist unless e of the coefficient vanish in* $(1 + x)^e(1 - x)^{n-1-e}$ *when this is expanded as a polynomial in x.*

I. INTRODUCTION

In this paper we investigate a certain problem in combinatorial analysis which arises in the theory of error correcting coding. A development of coding theory is to be found in the papers of Hamming[1] and Shannon[2]; this section is intended primarily as a presentation of the terminology used in subsequent sections.

We take (0, 1) as the range of binary variables. By an *n-word* we mean a sequence of n symbols, each of which is 0 or 1. We call the individual symbols of an n-word the *letters* of the n-word. We denote by B_n the set consisting of all the 2^n possible distinct n-words. The set B_n may be mapped onto the vertices of an n-dimensional cube, in the usual way, by regarding an n-word as an n-dimensional Cartesian coordinate expression. The *distance* $d(u, v)$ between n-words u and v is defined to be the number of places in which the letters of u and v differ; on the n-cube, this is seen to be the smallest number of edges in paths along edges between the vertices corresponding to u and v. The *weight* of an n-word u is the number of 1's in the sequence u; it is the distance between u and the n-word $00\cdots0$, all of whose letters are 0.*

A *binary block code of size K on n places* is a class of K nonempty disjoint subsets of B_n where in each of the K sets a single n-word is chosen as the *code word* of the set.† Each such set is the *detection region* of the code word it contains, and we shall say that any n-word which falls in a detection region *belongs to* the code word of the detection region. The set consisting of those n-words which do not lie in any detection region we call *limbo.*‡ A *close packed* code is one for which limbo is an empty set; i.e., a code in which the detection regions constitute a partition (disjoint covering) of B_n.

A *sphere* of radius r centered at n-word u is the set $[v{:}d(u, v) \leqq r]$ of n-words v which differ from u in r or fewer places. A binary block code is *e-error-correcting* if each detection region includes the sphere of radius e centered at the code word of the detection region. We say that a binary block code is *strictly e-error-correcting* if each detection region is exactly the sphere of radius e centered at the code word of the detection region.

This paper is devoted to the consideration of close packed strictly e-error-correcting binary block codes. We shall refer to such a code as an *e-code*, for brevity. Hamming[1] observes that a necessary condition for the existence of an e-code on n places is that

$$1 + n + \frac{1}{2}n(n-1) + \cdots + \binom{n}{e} \tag{1}$$

be a divisor of 2^n. In this paper we derive an additional necessary condition. Our condition includes as a special case a condition of Golay[4] for the existence of e-codes of group type, and applies to all e-codes, whether or not they are equivalent to group codes.§

II. DISTRIBUTION OF CODE WORDS

Suppose an e-code on n places is given. Let us inquire as to the distribution of weights of code words. We denote by ν_s the number of code words of weight s, $0 \leqq s \leqq n$, and by

$$G(x) = \sum_{s=0}^{n} \nu_s x^s \tag{2}$$

the generating function for these numbers, with x a complex but otherwise free variable. We show in this section that $G(x)$ satisfies a certain inhomogeneous linear differential equation of order e.

If there exists an e-code on n places then this differential equation will have $G(x)$ as a *polynomial* solution; the necessary condition for the existence of an e-code on n places given in Section 4 is essentially a restatement of this fact.* First, however, we must derive the differential equation and obtain its solutions.

If w_α is a code word of the given e-code $(1 \leqq \alpha \leqq K)$, define the set of *j-neighbors of* w_α as the set of n-words which lie at distance exactly j from w_α; designate this set by $S_j(w_\alpha)$. ($S_0(w_\alpha)$ is the set whose only element is w_α itself.) Our derivation is based on the observation that, in an e-code on n places,

$$\bigcup_{\alpha=1}^{K} \bigcup_{j=0}^{e} S_j(w_\alpha) = B_n \text{ †} \tag{3}$$

is a partition of B_n. For, the detection regions:

$$\bigcup_{j=0}^{e} S_j(w_\alpha), \qquad 1 \leqq \alpha \leqq K$$

are disjoint, and in each such sum representing a detection region the summands are disjoint (the distance function being single valued). Furthermore, each n-word of B_n lies in some detection region (close packed property) and hence appears in one of the sets $S_j(w_\alpha)$ for some α and for some j satisfying $0 \leqq j \leqq e$.

The set

$$\bigcup_{\alpha=1}^{K} S_j(w_\alpha)$$

consists of the n-words which are j-neighbors of some (not specified) code word; let us refer to these n-words simply as *j-neighbors.* Denote by $\nu_{j,s}$ the number of j-neighbors which are of weight s (with $\nu_{0,s} = \nu_s$, as above). Applying (3) to the n-words of weight s, we see that

$$\nu_s + \nu_{1,s} + \cdots + \nu_{e,s} = \binom{n}{s}, \qquad 0 \leqq s \leqq n \tag{4}$$

is the total number of n-words of weight s. If we multiply (4) by x^s and sum on s, we have

$$G(x) + G_1(x) + \cdots + G_e(x) = (1 + x)^n \tag{5}$$

where

$$G_j(x) = \sum_{s=0}^{n} \nu_{j,s} x^s \tag{6}$$

* If B_n is regarded as a subset of the real linear vector space consisting of all sequences $\alpha = (\alpha_1, \alpha_2, \cdots, \alpha_n)$ of n real numbers, then the "weight" of an n-word is simply the ℓ_1 norm (defined as $\| \alpha \|_1 = \sum_1^n | \alpha_n |$), and our "distance" is the metric derived from this norm.

† The term "block code", due to P. Elias, serves to distinguish the codes of fixed length considered here from the codes of unbounded delay introduced by Elias, Reference 3.

‡ In a communications system[2] using such a code, the transmitter sends only code words. If, due to errors in handling binary symbols, the receiver delivers itself of an n-word other than a code word then: (a) if the n-word lies in a detection region, one assumes that the code word of the detection region was intended; (b) if the n-word lies in limbo, one makes a note to the effect that errors have occurred in handling the word but that one is not attempting to guess what they were.

§ The terms "group alphabet" (Slepian[5]), "systematic code" (Hamming[1]), "symbol code" (Golay[4]), "check symbol code" (Elias[3]), "parity check code", are roughly synonymous. More precisely, a group code is a parity check code in which all of the parity check forms are homogeneous ("even"), so that $00\cdots0$ is one of the code words; see Reference 5.

* The author is not yet able to demonstrate the converse. That is, suppose one obtains a polynomial solution $G(x)$ of (11), below, satisfying appropriate boundary conditions, and from it some coefficients ν_s, $0 \leqq s \leqq n$. It does not follow from the methods of this article that there is actually some e-code on n places for which these ν_s represent the number of code words of weight s.

† $\cup$ = set union.

is the generating function (with respect to s) for the numbers $\nu_{j,s}$.

We now express $G_j(x)$, $0 \leqq j \leqq e$, in terms of $G(x)$. Suppose code word w is of weight s; that is, w consists of s ones and $n - s$ zeros in some order. A j-neighbor of w is obtained by choosing j places out of n and changing the letters of w in these places, 0's to 1's and 1's to 0's. If, in this procedure, q of the 1's of w are changed to 0's, so that $j - q$ of the 0's are changed to 1's, then the resulting j-neighbor of w is of weight $s - q + (j - q)$. Now, there are $\binom{s}{q}$ ways of choosing q places among the s where the letters of w are 1, and there are independently $\binom{n-s}{j-q}$ ways of choosing $j - q$ places among the $n - s$ where the letters of w are 0. Thus, of the $\binom{n}{j}$ different j-neighbors of w, the number $\binom{s}{q}\binom{n-s}{j-q}$ are of weight $s + j - 2q$. We may regard each of these as contributing $1 \cdot x^{s+j-2q}$ to the generating function $G_j(x)$ of (6) (provided $0 \leqq j \leqq e$, so that there is no overlap); hence, summing over all j-neighbors of a code word and then over all code words,

$$G_j(x) = \sum_{s=0}^{n} \nu_s \sum_{q=0}^{\infty} \binom{s}{q}\binom{n-s}{j-q} x^{s+j-2q} \qquad 0 \leqq j \leqq e\,^* \tag{7}$$

From the easily verified polynomial identity

$$(x + y)^s(1 + xy)^{n-s} = \sum_{j=0}^{\infty} y^j \sum_{q=0}^{\infty} \binom{s}{q}\binom{n-s}{j-q} x^{s+j-2q}$$

(n, s integers, $0 \leqq s \leqq n$) it follows that

$$\sum_{q=0}^{\infty} \binom{s}{q}\binom{n-s}{j-q} x^{s+j-2q} = \frac{1}{2\pi i}\int_C \frac{(x+y)^s(1+xy)^{n-s}}{y^{j+1}}\,dy$$

where contour C is, say, a small circle around the origin, taken positively. Thus

$$\begin{aligned} G_j(x) &= \sum_{s=0}^{n} \frac{\nu_s}{2\pi i}\int_C \frac{(x+y)^s(1+xy)^{n-s}}{y^{j+1}}\,dy \\ &= \frac{1}{2\pi i}\int_C \frac{(1+xy)^n}{y^{j+1}}\, G\left(\frac{x+y}{1+xy}\right) dy \\ &\equiv L_jG(x) \end{aligned} \tag{8}$$

where the operator L_j is thus defined. Change of integration variable gives

$$\begin{aligned} L_jG(x) &= \frac{(1-x^2)^{n+1}}{2\pi i}\int_{C_x} \frac{G(z)\,dz}{(1-xz)^{n-j+1}(z-x)^{j+1}} \\ &= \frac{(1-x^2)^{n+1}}{j!}\frac{\partial^j}{\partial z^j}\frac{G(z)}{(1-xz)^{n-j+1}}\bigg|_{z=x} \\ &= \sum_{p=0}^{j}\binom{n-p}{j-p}\frac{x^{j-p}(1-x^2)^p}{p!}\frac{d^pG(x)}{dx^p} \end{aligned} \tag{9}$$

(with C_x a small circle enclosing x but not x^{-1}, $x^2 \neq 1$). Thus L_j may be regarded as a linear differential operator of order j, ($L_0 \equiv 1$).

Using this result, (5) may be given the form

$$\begin{aligned} (1+x)^n &= [L_0 + L_1 + \cdots + L_e]G(x) \\ &= \frac{1}{2\pi i}\int_C \frac{y^{-e-1}-1}{1-y}(1+xy)^n G\left(\frac{x+y}{1+xy}\right) dy \\ &\equiv MG(x) \end{aligned} \tag{10}$$

this last expression as a definition of operator M. Written as a differential equation, (10) is

$$\sum_{p=0}^{e} \frac{(1-x^2)^p}{p!} \sum_{r=0}^{e-p} \binom{n-p}{r} x^r \frac{d^pG(x)}{dx^p} = (1+x)^n \tag{11}$$

It is straightforward that the only singularities of this equation are regular singularities[6] at $x = \pm 1, \infty$.

III. THE DIFFERENTIAL EQUATION

In this section we discuss (11) without reference to the fact that $G(x)$ is supposed to be a generating function. That is to say, with n and e fixed but arbitrary non-negative integers, we denote by

$$G(x) = \sum_{s=0}^{\infty} \nu_s x^s \tag{12}$$

any solution of (11) regular in the unit circle.

It proves convenient to introduce certain functions $f_{n,\xi}(x)$ defined by

$$\begin{aligned} f_{n,\xi}(x) &\equiv (1+x)^\xi(1-x)^{n-\xi} \\ &= \sum_{s=0}^{\infty} \varphi_s(n,\xi)x^s \end{aligned} \tag{13}$$

where the coefficients $\varphi_s(n, \xi)$ are given by

$$\varphi_s(n,\xi) = \sum_{r=0}^{s} (-1)^r \binom{n-\xi}{r}\binom{\xi}{s-r} \tag{14}$$

Here, ξ is to be regarded as a free complex variable. By $(1+x)^\xi(1-x)^{n-\xi}$ we mean $\exp(\xi \log(1+x) + (n-\xi)\log(1-x))$, each logarithm vanishing at $x = 0$. As a function of x this function is single valued in, say, the x-plane cut on $(-\infty, -1]$ and $[1, \infty)$, and the series (13) converges to it in: $|x| < 1$.

Binomial coefficients are defined by

$$\begin{aligned} \binom{\zeta}{s} &\equiv \frac{\Gamma(\zeta+1)}{s!\,\Gamma(\zeta+1-s)} \\ &= \frac{\zeta(\zeta-1)\cdots(\zeta-s+1)}{s!} \qquad s > 0 \end{aligned}$$

when ζ is not an integer, and $\varphi_s(n, \xi)$ is seen to be a polynomial in ξ of degree s:

$$\varphi_s(n,\xi) = \frac{2^s}{s!}\xi^s + \cdots + (-1)^s\binom{n}{s} \tag{15}$$

The recurrence relation

$$\varphi_0(n,\xi) + \varphi_1(n,\xi) + \cdots + \varphi_s(n,\xi) = \varphi_s(n-1,\xi) \tag{16}$$

obtained by expanding the various factors [ ] in the identity

$$[(1+x)^\xi(1-x)^{n-\xi}][(1-x)^{-1}] = [(1+x)^\xi(1-x)^{n-1-\xi}]$$

is an important one. We note also for reference that

$$\begin{aligned} \varphi_0(n,\xi) &= 1 \\ \varphi_s(n, n-\xi) &= (-1)^s\varphi_s(n,\xi) \\ \varphi_s(n,n) &= \binom{n}{s} \\ \varphi_s(n-1,n) &= 1 + \binom{n}{1} + \cdots + \binom{n}{s} \end{aligned} \tag{17}$$

valid for all n, ξ and non-negative integers s. We see, by the way, that $\varphi_e(n-1, n)$ is simply the Hamming expression (1).

The function $f_{n,\xi}(x)$ has the property that

$$f_{n,\xi}\left(\frac{x+y}{1+xy}\right) = \frac{f_{n,\xi}(x)f_{n,\xi}(y)}{(1+xy)^n} \tag{18}$$

at least if, say, given x, $|y|$ is small enough. From this and (8) for the operator L_j it is apparent that

$$L_jf_{n,\xi}(x) = \varphi_j(n,\xi)f_{n,\xi}(x) \tag{19}$$

Similarly, using (19) and (16), or directly from (10) for the operator M,

$$\begin{aligned} Mf_{n,\xi}(x) &= [L_0 + L_1 + \cdots + L_e]f_{n,\xi}(x) \\ &= \varphi_e(n-1,\xi)f_{n,\xi}(x) \end{aligned} \tag{20}$$

If ξ_β is one of the roots of the polynomial $\varphi_e(n - 1, \xi)$ then (20) becomes

$$M(1+x)^{\xi_\beta}(1-x)^{n-\xi_\beta} = 0$$

If we assume for the moment for simplicity that $\varphi_e(n - 1, \xi)$ has e *distinct* roots ξ_β, $1 \leqq \beta \leqq e$, then (11) has as complementary function

$$\sum_{\beta=1}^{e} A_\beta(1+x)^{\xi_\beta}(1-x)^{n-\xi_\beta}$$

where the A_β are e arbitrary constants.

Fortunately, the function $(1 + x)^n = f_{n,n}(x)$ is also a member of the family (13); hence

$$M(1+x)^n = \varphi_e(n-1,n)(1+x)^n$$

* The limits $(0, \infty)$ on the q summation are merely for convenience; the binomial coefficients vanish outside the proper range, under the usual convention.

and the function

$$\frac{(1+x)^n}{\varphi_e(n-1, n)}$$

is a particular integral of (11). [We see from (17) that $\varphi_e(n-1, n)$ does not vanish in cases of interest.] Finally, when the roots of $\varphi_e(n-1, \xi)$ are distinct, the general solution of (11) must be of the form

$$G(x) = \frac{(1+x)^n}{\varphi_e(n-1, n)} + \sum_{\beta=1}^{e} A_\beta (1+x)^{\xi_\beta}(1-x)^{n-\xi_\beta} \tag{21}$$

If $\varphi_e(n-1, \xi)$ has multiple roots then the general solution will contain additional terms

$$(\text{const.})\ (1+x)^{\xi_\beta}(1-x)^{n-\xi_\beta}\left[\log\frac{1+x}{1-x}\right]^\mu \tag{22}$$

i.e., the μth derivative of $f_{n,\xi}(x)$ with respect to ξ, with μ any positive integer less than the multiplicity of root ξ_β.

Before applying these results to e-codes in detail, let us derive a certain modification of (21). First, we see from (17) that if n is a positive integer, then n is not one of the roots of $\varphi_e(n-1, \xi)$. If the roots ξ_β of $\varphi_e(n-1, \xi)$ are distinct and if A_β, $1 \leqq \beta \leqq e$, are any e numbers then a polynomial $\theta(\xi)$ of formal degree e is uniquely determined by the $e+1$ conditions:

$$\begin{aligned} \theta(\xi_\beta) &= (\xi_\beta - n)\varphi_e'(n-1, \xi_\beta)A_\beta, \qquad 1 \leqq \beta \leqq e,^* \\ \theta(n) &= 1 \end{aligned} \tag{23}$$

using, e.g., the Lagrange interpolation formula. It is obvious that $G(x)$, (21), may be expressed in terms of this polynomial as

$$G(x) = \frac{1}{2\pi i}\int_\Gamma \frac{(1+x)^\xi(1-x)^{n-\xi}\theta(\xi)}{(\xi-n)\varphi_e(n-1, \xi)}\,d\xi \tag{24}$$

where Γ is any simple closed contour surrounding the roots: $n, \xi_1, \xi_2, \cdots, \xi_e$ of the denominator of the integrand; (the numerator is an entire function of ξ provided $x^2 \neq 1$).

Analysis a little more detailed shows that even if $\varphi_e(n-1, \xi)$ has multiple roots the general solution of (11) can be represented in the form (24), again with $\theta(\xi)$ any polynomial of formal degree e such that $\theta(n) = 1$. The e constants of integration appear as the $e+1$ parameters of $\theta(\xi)$ restricted by $\theta(n) = 1$.†

Expansion of the integrand in (24) according to (13) yields the form

$$\nu_s = \frac{1}{2\pi i}\int_\Gamma \frac{\varphi_s(n, \xi)\theta(\xi)}{(\xi-n)\varphi_e(n-1, \xi)}\,d\xi \qquad s = 0, 1, 2, \cdots \tag{25}$$

for the coefficients of $G(x)$, (12).

If we denote by

$$L_jG(x) \equiv G_j(x) = \sum_{s=0}^{\infty} \nu_{j,s}x^s \tag{26}$$

the result of applying the operator L_j to any solution (24) of (11), then it is straightforward that

$$G_j(x) = \frac{1}{2\pi i}\int_\Gamma \frac{(1+x)^\xi(1-x)^{n-\xi}\varphi_j(n, \xi)\theta(\xi)}{(\xi-n)\varphi_e(n-1, \xi)}\,d\xi, \tag{27}$$

and that

$$\nu_{j,s} = \frac{1}{2\pi i}\int_\Gamma \frac{\varphi_s(n, \xi)\varphi_j(n, \xi)\theta(\xi)}{(\xi-n)\varphi_e(n-1, \xi)}\,d\xi, \qquad s = 0, 1, \cdots \tag{28}$$

(An interesting reciprocity $\nu_{j,s} = \nu_{s,j}$ is apparent from (28). In an e-code one has (number of j-neighbors of weight s) = (number of s-neighbors of weight j) only for $0 \leqq s, j \leqq e$, since $L_jG(x)$ is the generating function for j-neighbors only if $0 \leqq j \leqq e$.)

IV. BOUNDARY CONDITIONS

The coefficients ν_s, (25), of any solution of (11) satisfy the relation:

$$\nu_0 + \nu_1 + \cdots + \nu_e = \frac{1}{2\pi i}\int_\Gamma \frac{\theta(\xi)}{\xi - n}\,d\xi = 1 \tag{29}$$

by virtue of (16) and the normalizing condition $\theta(n) = 1$.

With γ an integer such that $0 \leqq \gamma \leqq e$, denote by

$$G^{(\gamma)}(x) = \sum_{s=0}^{\infty} \nu_s^{(\gamma)}x^s \tag{30}$$

a solution of (11) which satisfies the e boundary conditions

$$\begin{aligned} \nu_0^{(\gamma)} &= \nu_1^{(\gamma)} = \cdots = \nu_{\gamma-1}^{(\gamma)} = 0 \\ \nu_{\gamma+1}^{(\gamma)} &= \nu_{\gamma+2}^{(\gamma)} = \cdots = \nu_e^{(\gamma)} = 0 \end{aligned} \tag{31}$$

We must have $\nu_\gamma^{(\gamma)} = 1$ in such a solution, from (29). Thus the conditions (31) are equivalent to specifying the values of $G^{(\gamma)}(x)$ and its first $e-1$ derivatives at the ordinary point $x = 0$ of (11), so that such a solution $G^{(\gamma)}(x)$ exists and is uniquely determined.[6]

Given an e-code on n places, each n-word of B_n lies at distance e or less from exactly one code word; namely, the code word to which it belongs. In particular, the n-word $00\cdots0$ must lie at distance e or less from a single code word. That is to say, there is exactly one code word in the sphere of radius e centered at $00\cdots0$. If this code word is of weight γ, then the generating function for the given e-code can be none other than the solution $G^{(\gamma)}(x)$ of (11) defined in the preceding paragraph.

If there exists an e-code on n places in which the code word of least weight is of weight γ, then there can be derived from it an e-code on n places in which the code word of least weight is of weight γ', where γ' is any integer satisfying $0 \leqq \gamma' \leqq e$. The transformation is that of choosing certain places among n and then changing the letters of each n-word of B_n in these places, 0's to 1's and 1's to 0's. (Such a transformation corresponds to one of the operations of the orthogonal group which leaves invariant the n-cube representing B_n.) Metric properties in B_n are invariant under such a transformation, clearly, and an e-code is transformed into an e-code. Thus if there exists any e-code on n places then (11) must have $e+1$ distinct polynomial solutions $G^{(\gamma)}(x)$, satisfying boundary conditions (31) for each case $\gamma = 0, 1, \cdots, e$.

In (25) for the coefficients ν_s, move contour Γ out to a circle sufficiently large that the expansion

$$\frac{1}{(\xi-n)\varphi_e(n-1, \xi)} = \frac{e!}{2^e\xi^{e+1}} + \frac{(\text{const.})}{\xi^{e+2}} + \cdots$$

converges on Γ. Suppose that the polynomial $\theta(\xi)$, of formal degree e, is of actual degree f: $\theta(\xi) = c\xi^f + 0(\xi^{f-1})$, $c \neq 0$, where $0 \leqq f \leqq e$. Then the numerator of the integrand in (25) is of the form: $(2^sc\xi^{s+f}/s!) + 0(\xi^{s+f-1})$, and it is clear that

$$\nu_s = 0 \qquad 0 \leqq s \leqq e - f - 1$$

$$\nu_{e-f} = \frac{e!c}{2^f(e-f)!} \neq 0$$

Hence, if $\theta^{(\gamma)}(\xi)$ denotes the polynomial which gives $G^{(\gamma)}(x)$ in the representation (24), then $\theta^{(\gamma)}(\xi)$ must be of actual degree $e - \gamma$.

A particularly simple case is the one $\gamma = e$; the polynomial $\theta^{(e)}(\xi)$ must be of degree zero, and is determined by the normalization as $\theta^{(e)}(\xi) \equiv 1$. Thus

$$G^{(e)}(x) = \frac{1}{2\pi i}\int_\Gamma \frac{(1+x)^\xi(1-x)^{n-\xi}}{(\xi-n)\varphi_e(n-1, \xi)}\,d\xi \tag{32}$$

From this we have immediately the following

Theorem: If there exists an e-code on n places then the equation $\varphi_e(n-1, \xi) = 0$ in ξ has e distinct integer roots.

Proof: If there exists an e-code on n places, then there exists an e-code on n places in which the code word of least weight is of weight e. The solution (32) of (11) must be the generating function for this e-code; hence (32) must reduce to a polynomial of formal degree n. If $\varphi_e(n-1, \xi)$ had multiple roots then noncancelling logarithmic terms (22) would appear in the $G^{(e)}(x)$ of (32). Thus $\varphi_e(n-1, \xi)$ must have e distinct roots ξ_β, $1 \leqq \beta \leqq e$. Each solution $(1+x)^{\xi_\beta}(1-x)^{n-\xi_\beta}$ of the homogeneous equation appears in $G^{(e)}(x)$ with nonvanishing coefficient:

$$A_\beta = \frac{1}{(\xi_\beta - n)\varphi_e'(n-1, \xi_\beta)}$$

* The prime denotes differentiation with respect to ξ.

† If $G(x)$ of (24) is to satisfy (11) it is sufficient that $\theta(\xi)$ be any function regular within (and on) Γ and that $\theta(n) = 1$, as may be easily verified. Since $G(x)$ depends on $\theta(\xi)$ only by way of the values of $\theta(\xi)$ at the zeros of the denominator in (24), the condition that $\theta(\xi)$ be a polynomial of formal degree e serves merely to determine $\theta(\xi)$ uniquely for a given solution $G(x)$.

Since $G^{(e)}(x)$ must be a polynomial in x, it must be expressible as a polynomial in $1 + x$; hence each root ξ_β must be an integer.* (It is not necessary to require further that $0 \leqq \xi_\beta \leqq n$, since it follows easily from (14) that any real root of $\varphi_e(n - 1, \xi)$ satisfies $0 \leqq \xi_\beta \leqq n - 1$ provided n and e are integers such that $0 \leqq e \leqq n$.)

As a corollary we have that if e is odd then n must be odd. This follows from the theorem and the fact that $\frac{1}{2}(n - 1)$ is a root of $\varphi_e(n - 1, \xi)$ when e is odd, from (17).

We consider next the case $\gamma = 0$. If $00\cdots0$ is a code word, then its e-neighbors are the n-words of weight e. Furthermore, the n-words of weight less than e belong to the code word $00\cdots0$, and can be e-neighbors neither of $00\cdots0$ nor of any other code word. Hence it must be true that

$$G_e^{(0)}(x) = \binom{n}{e} x^e + 0(x^{e+1}) \tag{33}$$

With $G_e^{(0)}(x)$ represented in the form (27), divide the factor $\varphi_e(n, \xi)\theta^{(0)}(\xi)$ in the numerator by the denominator; the result will be

$$\varphi_e(n, \xi)\theta^{(0)}(\xi) = [(\xi - n)\varphi_e(n - 1, \xi)]q(\xi) + r(\xi) \tag{34}$$

with quotient $q(\xi)$ a polynomial of degree $e - 1$ and remainder $r(\xi)$ a polynomial of formal degree e. The term involving $q(\xi)$ obviously contributes nothing to $G_e^{(0)}(x)$ in (27), so that from (33) and arguments similar to those giving $G^{(e)}(x)$, above, $r(\xi)$ must be the constant

$$r(\xi) = \binom{n}{e} = \varphi_e(n, n)$$

From (34) we then obtain the values of $\theta^{(0)}(\xi)$ at the poles of the integrand in (24), and thus

$$G^{(0)}(x) = \frac{(1 + x)^n}{\varphi_e(n - 1, n)} + \sum_{\beta=1}^{e} \frac{\varphi_e(n, n)(1 + x)^{\xi_\beta}(1 - x)^{n-\xi_\beta}}{\varphi_e(n, \xi_\beta)(\xi_\beta - n)\varphi_e'(n - 1, \xi_\beta)} \tag{35}$$

Before obtaining $G^{(\gamma)}(x)$ explicitly for intermediate values of γ, we must first discuss a certain set of recursion relations holding between the coefficients ν_s of any solution of (11). These relations are

$$\sum_{s=e-\rho+1}^{e+\rho} (-1)^{e+s} k_{\rho,s}\nu_s = 0, \quad \rho = 1, 2, \cdots, \tag{36}$$

where we define $\nu_s = 0$ for $s < 0$ and where the coefficients $k_{\rho,s}$ are

$$k_{\rho,s} = \sum_{\sigma=0}^{\rho} \binom{s}{\sigma}\binom{n - s}{\rho - \sigma}\binom{\rho - 1}{e - s + \sigma} \tag{37}$$

(The derivation of (36) is given in Appendix A.) Equations (36), written out, are of the form

$$k_{1,e}\nu_e - k_{1,e+1}\nu_{e+1} = 0$$

$$k_{2,e-1}\nu_{e-1} - k_{2,e}\nu_e + k_{2,e+1}\nu_{e+1} - k_{2,e+2}\nu_{e+2} = 0$$

$$\vdots$$

from which we see that (36) may be used to determine

$$\nu_{e+1}, \nu_{e+2}, \cdots$$

recursively in terms of

$$\nu_e, \nu_{e-1}, \cdots, \nu_0$$

We see also that if

$$\nu_e = \nu_{e-1} = \cdots = \nu_{\gamma+1} = 0$$

(with γ such that $0 \leqq \gamma \leqq e - 1$) then

$$\nu_{e+1} = \nu_{e+2} = \cdots = \nu_{2e-\gamma} = 0.$$

This has the following interpretation in terms of e-codes. It is well known (and obvious) that two different code words in an e-code must be separated by distance at least $2e + 1$. Hence if the code word of least weight in an e-code is of weight γ then all other code words are of weight not less than $2e + 1 - \gamma$. In the generating function for such a code it must be the case that not only

$$G^{(\gamma)}(x) = x^\gamma + 0(x^{e+1})$$

but in fact

$$G^{(\gamma)}(x) = x^\gamma + 0(x^{2e+1-\gamma}) \tag{38}$$

Equations (36) insure that this condition is satisfied automatically.*

As a particular case of (38), we have

$$G^{(0)}(x) = 1 + 0(x^{2e+1}).$$

We see that if we apply the operator L_γ to $G^{(0)}(x)$ there will result

$$L_\gamma G^{(0)}(x) = \varphi_\gamma(n, n)x^\gamma + 0(x^{2e+1-\gamma}) \tag{39}$$

using the differential operator form for L_γ, (9). On the other hand, the function

$$\frac{L_\gamma G^{(0)}(x)}{\varphi_\gamma(n, n)} = \frac{1}{2\pi i}\int_\Gamma \frac{(1 + x)^\xi(1 - x)^{n-\xi}}{(\xi - n)\varphi_e(n - 1, \xi)}\left[\frac{\theta^{(0)}(\xi)\varphi_\gamma(n, \xi)}{\varphi_\gamma(n,n)}\right] d\xi \tag{40}$$

is a solution of differential equation (11), in view of the discussion following (24). From (39) we see that this function can be none other than $G^{(\gamma)}(x)$. Finally, applying L_γ to $G^{(0)}(x)$ in the form (35), we have explicitly

$$G^{(\gamma)}(x) = \frac{(1 + x)^n}{\varphi_e(n - 1, n)} + \frac{\varphi_e(n, n)}{\varphi_\gamma(n, n)} \sum_{\beta=1}^{e} \frac{\varphi_\gamma(n, \xi_\beta)(1 + x)^{\xi_\beta}(1 - x)^{n-\xi_\beta}}{\varphi_e(n, \xi_\beta)(\xi_\beta - n)\varphi_e'(n - 1, \xi_\beta)} \quad 0 \leq \gamma \leq e. \tag{41}$$

V. EXAMPLES

The known cases where the condition of Hamming is satisfied are the following:

Case I: $e = 0, n \geqq 1$

The Hamming expression (1) reduces to unity. In fact,

$$\varphi_0(n - 1, \xi) \equiv 1,$$

and the condition that all roots be integers is vacuous. The generating function for code words is (uniquely):

$$G^{(0)}(x) = \frac{(1 + x)^n}{\varphi_0(n - 1, n)} = (1 + x)^n$$

Each n-word of B_n is a detection region and thus a code word. There is no error correction.

Case II: $e \geqq 1, n = e$

The Hamming expression becomes the sum of all the terms in the binomial expansion of $(1 + 1)^n$. The "codes" in this class consist of a single code word surrounded by its detection region consisting of the sphere B_n of radius n. No signalling is possible, of course, but our methods still apply.

From the representation

$$\begin{aligned} \varphi_e(n, \xi) &= \frac{1}{2\pi i}\int_C \frac{(1 + x)^\xi(1 - x)^{n-\xi}}{x^{e+1}} dx \\ &= \frac{1}{2\pi i}\int_C \frac{(1 + 2v)^\xi}{v^{e+1}(1 + v)^{n-e+1}} dv \end{aligned} \tag{42}$$

(valid for all n, ξ) we have immediately

$$\varphi_n(n - 1, \xi) = 2^n\binom{\xi}{n} = \frac{2^n}{n!}\xi(\xi - 1)\cdots(\xi - n + 1)$$

and the roots are $0, 1, \cdots, n - 1$. The generating function $G^{(e)}(x)$ of (32) becomes**

$$\begin{aligned} G^{(n)}(x) &= \frac{n!}{2^n}\cdot\frac{1}{2\pi i}\int_\Gamma \frac{(1 + x)^\xi(1 - x)^{n-\xi}}{(\xi)_{n+1}} d\xi \\ &= \frac{n!}{2^n}\sum_{\xi=0}^{n} \frac{(-1)^{n-\xi}}{\xi!(n - \xi)!}(1 + x)^\xi(1 - x)^{n-\xi} \\ &= \frac{1}{2^n}[(1 + x) - (1 - x)]^n = x^n \end{aligned}$$

* The condition of Golay for the existence of group codes, obtained by different means, is essentially that $\varphi_e(n - 1, \xi)$ have at least one root an integer. Cf.: (4) of Reference 4, in view of (16), above.

* It is also necessary for the existence of an e-code that (36) determine $\nu_{e+1}, \nu_{e+2}, \cdots$ as non-negative integers when $\nu_e, \nu_{e-1}, \cdots, \nu_0$ are those of (31). This condition is discussed a little further in Appendix A.

** $(\xi)_s \equiv s!\binom{\xi}{s}$ denotes the descending factorial.

as one might expect. The explicit form for $\varphi_n(n, \xi)$ is somewhat complicated, but for ξ an integer it follows immediately from definition (13) that

$$\varphi_n(n, \xi) = (-1)^{n-\xi} \qquad \xi = 0, 1, \cdots, n$$

From (35), then,

$$G^{(0)}(x) = \frac{1}{2^n}\sum_{\xi=0}^{n}\binom{n}{\xi}(1+x)^{\xi}(1-x)^{n-\xi}$$
$$= \frac{1}{2^n}[(1+x)+(1-x)]^n = 1$$

which, again, is not surprising. The details for other values of γ seem to be more tedious, although one expects (41) to yield $G^{(\gamma)}(x) = x^{\gamma}$.

Case III: $e \geqq 1, n = 2e + 1$

The Hamming expression in this case:

$$1 + \binom{2e+1}{1} + \cdots + \binom{2e+1}{e} = 2^{2e}$$

consists of the first half of the terms in the binomial expansion of $(1 + 1)^{2e+1}$. The code words in a code of this class are any two n-words separated by distance n (i.e., two vertices at opposite corners of the n-cube). The group codes in this class are the "majority rule" codes.† From (42) we have (using the substitution $y = 4v + 4v^2$)

$$\varphi_s(n, \xi) = \frac{2^n}{2\pi i}\int_C \frac{(1+y)^{\frac{1}{2}(\xi-1)}\,dy}{[(1+y)^{\frac{1}{2}} - 1]^{s+1}[(1+y)^{\frac{1}{2}} + 1]^{n-s+1}}, \tag{43}$$

and, without difficulty,

$$\varphi_e(2e, \xi) = 2^{2e}\binom{\frac{1}{2}(\xi-1)}{e} = \frac{2^e}{e!}(\xi-1)(\xi-3)\cdots(\xi-2e+1)$$

The roots are 1, 3, $\cdots$, $2e - 1$, and from (32):

$$G^{(e)}(x) = \frac{e!}{2^e}\cdot\frac{1}{2\pi i}\int_\Gamma \frac{(1+x)^{\xi}(1-x)^{2e+1-\xi}\,d\xi}{(\xi-2e-1)(\xi-1)(\xi-3)\cdots(\xi-2e+1)}$$
$$= 2^{-2e}(1+x)[(1+x)^2 - (1-x)^2]^e = x^e + x^{e+1}$$

In the case $\gamma = 0$ we need the result

$$\varphi_e(2e+1, \xi) = 2^{2e+1}\left[\binom{\frac{1}{2}\xi}{e+1} - \binom{\frac{1}{2}(\xi-1)}{e+1}\right]$$

from (43). It is then tedious but straightforward to obtain from (35)

$$G^{(0)}(x) = \frac{1}{2^{2e}}\sum_{r=0}^{e}\binom{2e+1}{2r+1}(1+x)^{2r+1}(1-x)^{2e-2r}$$
$$= 2^{-2e-1}\{[(1-x)+(1+x)]^{2e+1} - [(1-x)-(1+x)]^{2e+1}\}$$
$$= 1 + x^{2e+1}$$

One expects to get

$$G^{(\gamma)}(x) = x^{\gamma} + x^{2e+1-\gamma}$$

from (41), but verification appears to be complicated.

Case IV: $e = 1, n = 2^t - 1\ (t = 3, 4, \cdots)$

The single error correcting codes of Hamming[1] are included here. (The examples for $t = 1$, resp. $t = 2$, appear under Case II, resp. Case III, above.) Since n is always odd the condition that $\varphi_1(n - 1, \xi) = 2\xi - n + 1$ have an integer root is automatically satisfied. For $\gamma = 1$ the generating function is

$$G^{(1)}(x) = \frac{1}{2\pi i}\int_\Gamma \frac{(1-x)^{\xi}(1-x)^{n-\xi}}{(\xi-n)(2\xi-n+1)}\,d\xi$$
$$= \frac{(1+x)^n - (1+x)^{\frac{1}{2}(n-1)}(1-x)^{\frac{1}{2}(n+1)}}{1+n}$$

from which we have

$$\nu_s^{(1)} = \frac{1}{1+n}\left\{\binom{n}{s} - (-1)^{\frac{1}{2}s}\binom{\frac{1}{2}(n-1)}{\frac{1}{2}s}\right\} \qquad s \text{ even}$$

$$\nu_s^{(1)} = \frac{1}{1+n}\left\{\binom{n}{s} - (-1)^{\frac{1}{2}(s+1)}\binom{\frac{1}{2}(n-1)}{\frac{1}{2}(s-1)}\right\} \qquad s \text{ odd}$$

For $\gamma = 0$, Eq. (35) works out as

$$G^{(0)}(x) = \frac{(1+x)^n + n(1+x)^{\frac{1}{2}(n-1)}(1-x)^{\frac{1}{2}(n+1)}}{1+n}$$

so that

$$\nu_s^{(0)} = \frac{1}{1+n}\left\{\binom{n}{s} + n(-1)^{\frac{1}{2}s}\binom{\frac{1}{2}(n-1)}{\frac{1}{2}s}\right\} \qquad s \text{ even}$$

$$\nu_s^{(0)} = \frac{1}{1+n}\left\{\binom{n}{s} + n(-1)^{\frac{1}{2}(s+1)}\binom{\frac{1}{2}(n-1)}{\frac{1}{2}(s-1)}\right\} \qquad s \text{ odd}$$

Case V: $e = 2, n = 90$

The double error correcting codes for $n = 2, 5$ are covered by Cases II, III, respectively. The discovery that

$$1 + 90 + \tfrac{1}{2}(90)(89) = 2^{12}$$

is due to Golay.[7] We have

$$2\varphi_2(n-1, \xi) = (2\xi - n + 1)^2 - (n-1)$$

with roots

$$\tfrac{1}{2}[n - 1 \pm (n-1)^{\frac{1}{2}}]$$

Since these roots are not integers when $n = 90$, there can be no 2-code for $n = 90$.* H. S. Shapiro has shown (in unpublished work) that the Hamming condition for $e = 2$ is satisfied only in the cases $n = 2, 5, 90$, so that the only nontrivial 2-codes are those equivalent to the majority rule code on 5 places.

Case VI: $e = 3, n = 23$

Golay[7] finds:

$$1 + 23 + \tfrac{1}{2}(23)(22) + (23)(22)(21)/6 = 2^{11}$$

and gives explicitly a 3-code on 23 places of group type. We have

$$6\varphi_3(n-1, \xi) = (2\xi - n + 1)[(2\xi - n + 1)^2 - (3n - 5)]$$

and when $n = 23$ we verify that the roots are the integers 7, 11, 15. Computations by the author show that for $n < 10^{10}$ the Hamming condition for $e = 3$ holds only when $n = 3, 7, 23$.

For $e = 4$ we have

$$24\varphi_4(n-1, \xi) = [(2\xi - n + 1)^2 - (3n - 7)]^2 - (6n^2 - 30n + 40)$$

For $n = 4, 9$ this reduces to the forms given under Cases II, III. Preliminary calculations by the author shows that any other solutions of the Hamming condition for $e = 4$ must be such that $n > 10^{10}$, so that the question of the existence of 4-codes (other than the majority rule code) is somewhat academic.

Computations of Mrs. G. Rowe of the Mathematical Research Department show that Cases I–VI cover all cases of the Hamming condition being satisfied in the range

$$0 \leqq e \leqq n, \qquad 1 \leqq n \leqq 150$$

Appendix A

From (13) we have

$$\left(\frac{1-x}{1+x}\right)^{n-\xi} = \frac{1}{(1+x)^n}\sum_{s=0}^{\infty}\varphi_s(n, \xi)x^s \tag{A1}$$

Applying the operator $D = -(1+x)^2 d/dx$ to both sides of (A1) ρ times, there results

$$2^{\rho}(n-\xi)_{\rho}\left(\frac{1-x}{1+x}\right)^{n-\xi-\rho} = \sum_{s=0}^{\infty}\varphi_s(n, \xi)D^{\rho}[x^s(1+x)^{-n}] \tag{A2}$$

† The two code words in such a code are 00 $\cdots$ 0 and 11 $\cdots$ 1. An n-word belongs to 00 $\cdots$ 0 if it contains more 0's than 1's, and to 11 $\cdots$ 1 if it contains more 1's than 0's.

* This settles a question raised by Golay, who shows that there is no code of group type in this case, but not that there is no code at all.

The substitution $v = (1 + x)^{-1}$ reduces D to d/dv, so that

$$\begin{aligned} D^\rho x^s(1 + x)^{-n} &= \frac{d^\rho}{dv^\rho}(1 - v)^s v^{n-s} \\ &= \sum_{\sigma=0}^{\rho} \binom{\rho}{\sigma} (-1)^\sigma (s)_\sigma (1 - v)^{s-\sigma}(n - s)_{\rho-\sigma} v^{n-s-\rho+\sigma} \\ &= \rho! \sum_{\sigma=0}^{\rho} (-1)^\sigma \binom{s}{\sigma}\binom{n - s}{\rho - \sigma} x^{s-\sigma}(1 + x)^{\rho-n} \end{aligned}$$

using Leibnitz's rule. We substitute this into Eq. (A2), multiply both sides of the result by

$$(1 + x)^{n-\rho}(1 - x)^{\rho-\tau}/\rho!$$

(with τ arbitrary), and then equate coefficients of x^t on both sides; there obtains

$$2^\rho \binom{n - \xi}{\rho} \varphi_t(n - \tau, \xi) = \sum_{s=0}^{t+\rho} (-1)^{t+s} \kappa_{\rho,s}(n, \tau; t)\varphi_s(n, \xi) \quad \text{(A3)}$$

valid for all n, ξ, τ and all non-negative integers ρ, t, where

$$\kappa_{\rho,s}(n, \tau; t) = \sum_{\sigma=0}^{\rho} \binom{s}{\sigma}\binom{n - s}{\rho - \sigma}\binom{\rho - \tau}{t - s + \sigma} \quad \text{(A4)}$$

The coefficients $\kappa_{\rho,s}(n, \tau; t)$ vanish unless $s \leqq t + \rho$; if n and $\rho - \tau$ are non-negative integers then the coefficients $\kappa_{\rho,s}(n, \tau; t)$ are positive integers provided $t - \rho + \tau \leqq s$ and vanish otherwise. In particular, (setting $\tau = 1$, $t = e$),

$$2^\rho \binom{n - \xi}{\rho} \varphi_e(n - 1, \xi) = \sum_{s=e-\rho+1}^{e+\rho} (-1)^{e+s} k_{\rho,s}\varphi_s(n, \xi), \rho = 1, 2, \cdots, \quad \text{(A5)}$$

where we define $\varphi_s(n, \xi) \equiv 0$ for $s < 0$; the $k_{\rho,s} = \kappa_{\rho,s}(n, 1; e)$ are those of (37) of the text. If we multiply ν_s of (25) by $(-1)^{e+s}k_{\rho,s}$ and sum on s there results (36), since the left hand member of (A5) is a polynomial multiple of the denominator of the integrand in (25).

If the code word of least weight in an e-code is of weight γ, then the first nontrivial one of the (37) is the one for $\rho = e + 1 - \gamma$, and it gives (since $\nu_\gamma^{(\gamma)} = 1$)

$$\begin{aligned} \nu_{2e+1-\gamma}^{(\gamma)} &= \frac{k_{e+1-\gamma,\gamma}}{k_{e+1-\gamma,2e+1-\gamma}} \\ &= \frac{(n - \gamma)(n - \gamma - 1) \cdots (n - e)}{(2e + 1 - \gamma)(2e - \gamma) \cdots (e + 1)} \end{aligned}$$

A necessary condition for the existence of an e-code on n places is that this expression be a non-negative integer in each case $\gamma = 0, 1, \cdots, e$. It is not clear, however, that this condition is independent of the one set forth in the theorem of Section IV.

Appendix B

We give here a relation due to K. M. Case[8] which shows that the statement of our main result as it appears in the Abstract heading this article agrees with the theorem proved in Section IV.

In the defining relation

$$(1 + x)^r(1 - x)^{n-r} = \sum_{s=0}^{\infty} x^s\varphi_s(n, r) \quad \text{(B1)}$$

for the coefficients $\varphi_s(n, r)$ assume that n and r are integers, multiply both sides by $(-1)^r \binom{n}{r} y^r$, and sum on r, $0 \leqq r \leqq n$. The result is

$$[(1 - x) - y(1 + x)]^n = \sum_{r=0}^{n}\sum_{s=0}^{n} (-1)^r \binom{n}{r} y^r x^s \varphi_s(n, r) \quad \text{(B2)}$$

Rearrange the left hand member and re-expand it, to get

$$\begin{aligned} [(1 - x) - y(1 + x)]^n &= [(1 - y) - x(1 + y)]^n \\ &= \sum_{s=0}^{n} (-1)^s \binom{n}{s} x^s(1 + y)^s(1 - y)^{n-s} \quad \text{(B3)} \\ &= \sum_{s=0}^{n}\sum_{r=0}^{n} (-1)^s \binom{n}{s} x^s y^r \varphi_r(n, s) \end{aligned}$$

Comparing coefficients of $x^s y^r$ in (B2) and (B3), we have, finally,

$$(-1)^r \binom{n}{r} \varphi_s(n, r) = (-1)^s \binom{n}{s} \varphi_r(n, s) \quad (n, r, s \text{ integers}), \quad \text{(B4)}$$

or, changing notation slightly,

$$\varphi_\xi(n - 1, e) = (-1)^{\xi-e} \frac{\binom{n - 1}{\xi}}{\binom{n - 1}{e}} \varphi_e(n - 1, \xi) \quad \text{(B5)}$$

(with n, e, ξ integers and $0 \leqq e, \xi \leqq n - 1$). Thus if $\varphi_e(n - 1, \xi)$ vanishes for e different integers ξ then so must $\varphi_\xi(n - 1, e)$, at least when $e < n$. But $\varphi_\xi(n - 1, e)$ is the coefficient of x^ξ in $(1 + x)^e(1 - x)^{n-1-e}$ when this is written out as a polynomial in x, by definition.

REFERENCES

1. R. W. Hamming, B.S.T.J., **29**, p. 147, 1950.
2. C. E. Shannon, B.S.T.J., **27**, p. 379, 1948.
3. P. Elias, Trans. I.R.E., **PGIT-4**, p. 29, 1954.
4. M. J. E. Golay, Trans. I.R.E., **PGIT-4**, p. 23, 1954.
5. D. Slepian, B.S.T.J., **35**, p. 203, 1956.
6. E. L. Ince, *Ordinary Differential Equations* (Dover), Ch. V, XV.
7. M. J. E. Golay, Proc. I.R.E., **37**, p. 637, 1949.
8. K. M. Case, Phys. Rev., **97**, p. 810, 1955.

ON THE NONEXISTENCE OF PERFECT CODES OVER FINITE FIELDS*

AIMO TIETÄVÄINEN†

Abstract. It is proved that there are no unknown perfect (Hamming-)error-correcting codes over finite fields.

1. Introduction. Let V be the n-dimensional vector space over the finite field $GF(q)$. For any $\mathbf{a} \in V$ we define the weight of $\mathbf{a}$ as the number of nonzero components of $\mathbf{a}$. By the (Hamming) distance $d(\mathbf{a}, \mathbf{b})$ of the elements $\mathbf{a}$ and $\mathbf{b}$ of V we mean the weight of $\mathbf{a} - \mathbf{b}$. If e is a positive integer, we define the sphere $B(\mathbf{a}, e)$ by

$$B(\mathbf{a}, e) = \{\mathbf{x} \in V | d(\mathbf{x}, \mathbf{a}) \leqq e\}.$$

A subset (say C) of V is called a code, and a subspace of V is called a linear code. The dimension n of V is the block length of C; the elements of C are code words. C is called a perfect e-(Hamming-)error-correcting code if

(i) $\bigcup_{\mathbf{a} \in C} B(\mathbf{a}, e) = V$ and

(ii) $d_{\min} = \min \{d(\mathbf{x}, \mathbf{y}) | \mathbf{x} \in C, \mathbf{y} \in C, \mathbf{x} \neq \mathbf{y}\} \geqq 2e + 1$.

The following perfect codes are known (see [4], [6], [8], [9], [15], [18], [19] and [23]):

(i) perfect single-error-correcting codes (e.g., Hamming codes);

(ii) trivial perfect codes in cases $n = e$, and $q = 2$, $n = 2e + 1$;

(iii) Golay codes in cases $e = 2$, $q = 3$, $n = 11$, and $e = 3$, $q = 2$, $n = 23$.

The nonexistence of other perfect codes has been an open problem. Cohen [5] proved that for $q \leqq 5$ there are no unknown linear perfect 2-error-correcting codes, and Alter [1], [2] extended this result to $q = 7, 8$, and 9. Van Lint [12], [13], [14] solved the problem in the general (i.e., linear and nonlinear) case for all values of q in case $e \leqq 7$. Many papers deal with the binary case $q = 2$ (see, e.g., [20], [10], [11], and references in [10]) and recently (see [22]) the nonexistence of unknown perfect codes was proved in that case for all values of e. We now solve the problem for all values of q and e by proving the following theorem.

THEOREM. *There are no unknown perfect codes over finite fields.*

The crucial lemmas in the proof of this theorem are the Elias bound for the minimum distance of code words, a necessary condition (Lemma 2 of this paper) which was found by van Lint [13]; and a refined arithmetic-mean–geometric-mean inequality. It would be desirable to know whether the generalization of our theorem to the case of nonfield alphabets is true (cf. [14]) and whether similar results can be obtained for other metrics (cf., for example, [7] and [17]).

* Received by the editors April 28, 1971, and in final revised form April 18, 1972.

† Department of Mathematics, University of Turku, Turku, Finland, and Department of Mathematics, Tampere Technical University, Tampere, Finland.

Reprinted with permission from *SIAM J. Appl. Math.*, vol. 24, pp. 88–96, Jan. 1973.

2. Preliminaries. Extending a result of Hamming [8], van Lint [12] proved the following lemma.

LEMMA 1. *If a perfect e-error-correcting code of block length n over GF(q) exists, then there is an integer k such that*

$$\sum_{i=0}^{e} \binom{n}{i} (q-1)^i = q^{n-k}, \tag{1}$$

and q^k is the cardinality of this code.

Lloyd [16] proved a theorem which gave a necessary condition for the existence of a binary perfect e-error-correcting code. This theorem was later generalized by F. J. MacWilliams and A. M. Gleason (see [3] and [15, pp. 103–112]). Using this generalization, van Lint [13], [15, Lemma 5.5.1] proved the next lemma.

LEMMA 2. *If a perfect e-error-correcting code of block length n over GF(q) exists, then there are distinct positive integers $x_1, x_2, \cdots, x_e$ such that*

$$x_1 + x_2 + \cdots + x_e = \frac{e(n-e)(q-1)}{q} + \frac{e(e+1)}{2} \tag{2}$$

and

$$x_1 x_2 \cdots x_e = e! q^{n-k-e}. \tag{3}$$

The Elias bound for the minimum distance may be given as the following two lemmas [4, Lemmas 13.61 and 13.62].

LEMMA 3. *Given an integer t and a code of block length n and cardinality q^k, there exists a critical sphere of radius t which includes K code words, where*

$$K \geqq q^{k-n} \sum_{i=0}^{t} \binom{n}{i} (q-1)^i.$$

By suitable translation of the code, this critical sphere may be centered at $(0, 0, \cdots, 0)$.

LEMMA 4. *If each of K code words has weight $\leqq (q-1)xn/q$, where $0 \leqq x \leqq 1$, then the distance between some pair of these K code words must be no greater than $x(2-x)(q-1)n/q(1-K^{-1})$.*

The special case $q = 2$ of our theorem, stated in the next lemma, was proved in [22].

LEMMA 5. *There are no unknown perfect codes over GF(2)*

Furthermore, van Lint ([11]–[14], [15, pp. 95, 96 and 116–118], see also [21]) proved the following lemmas.

LEMMA 6. *If a perfect e-error-correcting code of block length n over GF(q) exists $(e < n)$, then*

$$q \leqq (n-1)/e.$$

LEMMA 7. *If $e \geqq 4$ and $q = p^v$ with $p > e$, then there is no nontrivial perfect e-error-correcting code over GF(q).*

LEMMA 8. *If there exists an unknown perfect e-error-correcting code of block length n over GF(q), then*

(i) $e \geqq 8$ *and*

(ii) $q > 100$ *or* $n > 1000$ *or* $e > 1000$.

We also need the following refinement of the arithmetic-mean–geometric-mean inequality.

LEMMA 9. *Let* $y_1, y_2, \cdots, y_s$ *and* p *be positive integers such that* $y_{i+1}/y_i \geqq p$, *for every* i. *Then*

$$y_1 y_2 \cdots y_s \leqq R^{s-1}((y_1 + y_2 + \cdots + y_s)/s)^s, \tag{4}$$

where

$$R = 4p/(p+1)^2. \tag{5}$$

Proof (by induction). The assertion (4) is trivial for $s = 1$. Suppose now that $h \geqq 1$, $y_1 < y_2 < \cdots < y_h$,

$$y_1 y_2 \cdots y_h \leqq R^{h-1}((y_1 + y_2 + \cdots + y_h)/h)^h$$

and $y_{h+1}/y_h \geqq p$. Let $(y_1 + y_2 + \cdots + y_h)/h = Y$ and $y_{h+1} = zY$, whence $z \geqq p$. Then

$$y_1 y_2 \cdots y_{h+1} \leqq R^{h-1} z Y^{h+1}. \tag{6}$$

Let

$$f(x) = xY^{h+1}((hY + xY)/(h+1))^{-h-1} = x(h+1)^{h+1}(h+x)^{-h-1}.$$

Then f decreases on $[1, \infty)$, and hence

$$f(z) \leqq f(p) = p(1 + (p-1)/(h+1))^{-h-1} \leqq 4p(p+1)^{-2} = R.$$

Consequently,

$$zY^{h+1} \leqq R((hY + y_{h+1})/(h+1))^{h+1} = R((y_1 + y_2 + \cdots + y_{h+1})/(h+1))^{h+1}.$$

Combining this with (6), we get the assertion (4) in case $s = h + 1$.

3. Proof of Theorem in case $n \geqq \frac{1}{2}e^2 + e$. Assume the contrary: There exists an unknown perfect code with parameters e, n, and q, where $q = p^v$, p a prime, and

$$n \geqq \tfrac{1}{2}e^2 + e. \tag{7}$$

By Lemmas 5, 7 and 8, we may restrict ourselves to

$$q \geqq 3, \quad e \geqq p, \quad e \geqq 8. \tag{8}$$

For a positive integer m, define $A(m) = p^{-u}m$, where p^u is the highest power of p dividing m. Let the x_j, $1 \leqq j \leqq e$, be the numbers mentioned in Lemma 2. Denote $x_j \sim x_h$ if $A(x_j) = A(x_h)$. This relation $\sim$ defines a partition of the set $\{x_1, x_2, \cdots, x_e\}$ into disjoint subsets $X_1, \cdots, X_r$. It was proved in [22] that

$$e - r > (5e \log 2)/(4 \log e) - 1 \quad \text{for } p = 2. \tag{9}$$

We now show that generally

$$e - r \geqq [e/p] \log p / \log e. \tag{10}$$

Since $e \geqq p$ and $e - r$ is an integer, this implies

$$e - r \geqq 1. \tag{11}$$

It follows from (3) that

$$A(x_1 x_2 \cdots x_e) = A(e!). \tag{12}$$

For a real number a, let $Q(a)$ be the product of the positive integers not exceeding a and not divisible by p, and, furthermore, let $[a]$ be the largest integer not exceeding a. Then

$$\begin{aligned} A(e!) &\leqq Q(e)\cdot[e/p]! \\ &\leqq Q(e)(e/p)^{[e/p]} \\ &= Q(e)e^{[e/p](1-(\log p)/(\log e))}. \end{aligned} \tag{13}$$

On the other hand, $A(x_1 x_2 \cdots x_e)$ is greater than or equal to the product of those r least positive integers which are not divisible by p. Hence

$$A(x_1 x_2 \cdots x_e) \geqq Q(e)e^{r-e+[e/p]}. \tag{14}$$

Collecting the results (12), (13) and (14), we get the assertion (10).

Let X_i be any one of the sets $X_1, \cdots; X_r$, let $s(i)$ be the cardinality of X_i, and let

$$R_i = \left(\prod_{x \in X_i} x\right) \Big/ \left(\sum_{x \in X_i} \frac{x}{s(i)}\right)^{s(i)}.$$

Now we may apply Lemma 9 which gives the result

$$R_i \leqq R^{s(i)-1},$$

where R is defined by (5). It follows from this that

$$R_1 \cdots R_r \leqq \prod_{i=1}^{r} R^{s(i)-1} = R^{e-r}$$

or

$$x_1 x_2 \cdots x_e \leqq R^{e-r} \prod_{i=1}^{r} \left(\sum_{x \in X_i} \frac{x}{s(i)}\right)^{s(i)},$$

which implies, by the arithmetic-mean–geometric-mean inequality, that

$$x_1 x_2 \cdots x_e \leqq R^{e-r}((x_1 + x_2 + \cdots + x_e)/e)^e.$$

Using Lemma 2 and recalling (1), we get therefore

$$q^{-e}e! \sum_{i=0}^{e} \binom{n}{i}(q-1)^i \leqq R^{e-r}\left(\frac{(n-e)(q-1)}{q} + \frac{e+1}{2}\right)^e,$$

and, consequently,

$$R^{e-r} > (n-b)^{-e}e!\binom{n}{e} = \prod_{j=0}^{e-1}\left(1 + \frac{b-j}{n-b}\right), \tag{15}$$

where

$$b = e - \frac{q(e+1)}{2(q-1)}. \tag{16}$$

Let $c = [b] + 1$. Then

$$\prod_{j=0}^{e-1}\left(1+\frac{b-j}{n-b}\right) = \prod_{j=0}^{c-1}\left(1+\frac{b-j}{n-b}\right)\prod_{j=c}^{e-1}\left(1+\frac{b-j}{n-b}\right)$$

$$> \left(1+\sum_{j=0}^{c-1}\frac{b-j}{n-b}\right)\left(1+\sum_{j=c}^{e-1}\frac{b-j}{n-b}\right)$$

$$= 1 - \frac{e(e-2b-1)}{2(n-b)} - \frac{c(2b-c+1)(e-c)(c+e-2b-1)}{4(n-b)^2}$$

$$\geqq 1 - \frac{e(e-2b-1)}{2(n-b)} - \frac{(2b+1)^2(2e-2b-1)^2}{16(n-b)^2}.$$

Using (15) and (16) and recalling that $n \geqq \frac{1}{2}e^2 + e$, we thus obtain

$$R^{e-r} > 1 - \frac{e^2+e}{2(q-1)n-(q-2)e+q} - \frac{e^2(q-2)(e+1)^2 q}{16(2(q-1)n-(q-2)e+q)^2}$$

(17)

$$> 1 - \frac{1}{q-1} - \frac{(q-2)q}{16(q-1)^2} > \frac{15}{16} - \frac{1}{q-1}.$$

If $p \geqq 5$ then, by (5), (11) and (17), $5/9 \geqq R^{e-r} > 11/16$, a contradiction.

Suppose now that $p = 3$. Then $q = 3$, for in case $q \geqq 9$ (17) implies $3/4 > 13/16$, a contradiction. For $q = 3$, (17) takes the form

$$(3/4)^{[e/3]\log 3/\log e} > 29/64$$

or

$$\frac{[e/3]\log 3}{\log e} < \frac{\log(64/29)}{\log(4/3)}.$$

Hence

(18) $$e \leqq 26,$$

and it follows, by Lemma 8, that

(19) $$n > 1000.$$

The inequalities (18) and (19) imply in case $q = 3$ that

$$1 - \frac{e^2+e}{2(q-1)n-(q-2)e+q} - \frac{e^2(q-2)(e+1)^2 q}{16(2(q-1)n-(q-2)e+q)^2} > \frac{3}{4}.$$

Substituting this and the equation $R = 3/4$ in (17) and recalling that $e - r \geqq 1$, we get an impossibility.

Suppose finally that $p = 2$, whence $q \geqq 4$. Because $e \geqq 8$, we know, by (9), that $e - r \geqq 3$, and, using a similar argument as in case $p = 3$, we see that $q = 4$. Thus, by (9), we may write the inequality (17) in the form

$$(8/9)^{(5e\log 2)/(4\log e)-1} > 11/18$$

or

$$\frac{e \log 2}{\log e} < \frac{4}{5} \frac{\log (18/11)}{\log (9/8)} + 1 < 5.$$

Hence $e < 32$ and, by Lemma 8, $n > 1000$. Consequently we get, by (17), the impossibility

$$\left(\frac{8}{9}\right)^3 > 1 - \frac{e^2 + e}{6n - 2e + 4} - \frac{e^2(e + 1)^2}{2(6n - 2e + 4)^2} > 1 - \frac{1}{5} - \frac{1}{50}.$$

4. Proof of Theorem in case $n < \frac{1}{2}e^2 + e$. Suppose that, contrary to our assertion, there exists an unknown perfect code C with parameters e, n and q such that

$$n < \tfrac{1}{2}e^2 + e, \quad q \geqq 3, \quad e \geqq 8. \tag{20}$$

Then we know, by the definition of e-error-correcting codes, that

$$n \geqq d_{\min} \geqq 2e + 1. \tag{21}$$

Put $t = e + 1$ in Lemma 3. Then, by (1),

$$\begin{aligned} K &\geqq q^{k-n}\left(\sum_{i=0}^{e} \binom{n}{i}(q - 1)^i + \binom{n}{e + 1}(q - 1)^{e+1}\right) \\ &= 1 + \binom{n}{e + 1}(q - 1)^{e+1}\left(\sum_{i=0}^{e} \binom{n}{i}(q - 1)^i\right)^{-1} \\ &= 1 + \binom{n}{e + 1}(q - 1)^{e+1}\binom{n}{e}^{-1}(q - 1)^{-e}\left(1 + \frac{e}{(n - e + 1)(q - 1)} + \cdots\right)^{-1} \\ &> 1 + \frac{(n - e)((n - e + 1)(q - 1) - e)}{(e + 1)(n - e + 1)}, \end{aligned} \tag{22}$$

and hence, by (21),

$$K > 1 + \frac{(q - 2)(n - e)}{e + 1}.$$

Consequently,

$$(1 - K^{-1})^{-1} = 1 + (K - 1)^{-1} < 1 + \frac{e + 1}{(q - 2)(n - e)}. \tag{23}$$

Choosing $x = (q - 1)^{-1}n^{-1}(e + 1)q$ in Lemma 4, we get therefore

$$d_{\min} < \frac{(e + 1)(2(q - 1)n - (e + 1)q)}{(q - 1)n}\left(1 + \frac{e + 1}{(q - 2)(n - e)}\right). \tag{24}$$

Using the same method as above, but choosing $t = e + 2$, $q = 3$ and $x = 3(e + 2)/2n$, we get

$$d_{\min} < \frac{(e + 2)(4n - 3e - 6)}{2n}\left(1 + \frac{(e + 2)^2}{(2n - e)(2n - 3e + 2)}\right) \quad \text{for } q = 3. \tag{25}$$

Consider first the case $q \geqq 5$. We shall show that the inequalities (24) and (21) imply

$$F(n) = n^2 - (\tfrac{1}{2}e^2 + 3e)n + e^3 + 2e^2 > 0. \tag{26}$$

Since the zeros of F are $2e$ and $\frac{1}{2}e^2 + e$ and since we know, by (21), than $n > 2e$, n must be greater than $\frac{1}{2}e^2 + e$. This contradicts (20).

If $q \geqq 7$, then

$$1 + \frac{e + 1}{(q - 2)(n - e)} \leqq \frac{5n - 4e + 1}{5(n - e)}. \tag{27}$$

Furthermore, for all q,

$$\frac{2(q - 1)n - (e + 1)q}{(q - 1)n} < \frac{2n - e - 1}{n}. \tag{28}$$

Combining the inequalities (24), (21), (27) and (28), we get

$$(e + 1)(2n - e - 1)(5n - 4e + 1) > 5(n - e)n(2e + 1)$$

or

$$5n^2 - (3e^2 + 11e + 3)n + 4e^3 + 7e^2 + 2e - 1 > 0.$$

Since

$$(\tfrac{1}{2}e^2 - 4e + 3)n + e^3 + 3e^2 - 2e + 1 > 0,$$

this implies (26).

If $q = 5$, the inequalities (24) and (21) imply

$$12n^2 - (7e^2 + 26e + 7)n + 10e^3 + 15e^2 - 5 > 0. \tag{29}$$

If $e \leqq 40$, then, by (20), $n < 1000$, and we have the case considered by Lemma 8. Therefore $e > 40$ and hence

$$(e^2 - 10e + 7)n + 2e^3 + 9e^2 + 5 > 0. \tag{30}$$

Now (26) follows from (29) and (30).

Suppose now that $q = 4$. Then, by Lemma 6, $n > 4e$; and it follows that we may replace the assertion (26) by

$$F_1(n) = 2n^2 - (e^2 + 10e)n + 4e^3 + 8e^2 > 0, \tag{31}$$

because the zeros of F_1 are $4e$ and $\frac{1}{2}e^2 + e$. For the proof of (31) we use (22) and get

$$(1 - K^{-1})^{-1} < 1 + \frac{(e + 1)(n - e + 1)}{(n - e)(3n - 4e + 3)} \leqq \frac{3n - 3e + 5}{3n - 4e + 3}.$$

Therefore,

$$2e + 1 < \frac{(e + 1)(6n - 4e - 4)(3n - 3e + 5)}{3n(3n - 4e + 3)}$$

or

$$9n^2 - (6e^2 + 18e - 9)n + 12e^3 + 4e^2 - 28e - 20 > 0. \tag{32}$$

Since we may suppose, as in case $q = 5$, that $e > 40$, we have

$$(3e^2/2 - 27e - 9)n + 6e^3 + 32e^2 + 28e + 20 > 0. \quad (33)$$

The inequalities (32) and (33) imply the assertion (31).

Suppose finally that $q = 3$. Combining the inequalities (25) and (21), we find

$$\begin{aligned} 12n^3 - (6e^2 + 48e + 12)n^2 + (14e^3 + 63e^2 + 42e - 8)n \\ - 6e^4 - 27e^3 - 42e^2 - 36e - 24 > 0. \end{aligned} \quad (34)$$

Since we may suppose, as in case $q = 5$, that $e > 40$, we have

$$(6e + 12)n^2 + (e^3 - 21e^2 - 42e + 8)n + 15e^3 + 42e^2 + 36e + 24 > 0. \quad (35)$$

Combining the inequalities (34) and (35), we obtain

$$F_2(n) = 12n^3 - (6e^2 + 42e)n^2 + (15e^3 + 42e^2)n - (6e^4 + 12e^3) > 0.$$

Because the zeros of F_2 are $\frac{1}{2}e$, $2e$ and $\frac{1}{2}e^2 + e$, and because, by (21), $n > 2e$, n must be greater than $\frac{1}{2}e^2 + e$. This contradicts (20).

Acknowledgment. The author would like to thank Professor J. H. van Lint and the referee for many very valuable suggestions and remarks.

REFERENCES

[1] R. ALTER, *On the nonexistence of close-packed double Hamming-error-correcting codes on q = 7 symbols*, J. Comput. System Sci., 2 (1968), pp. 169–176.

[2] ———, *On the nonexistence of perfect double Hamming-error-correcting codes on q = 8 and q = 9 symbols*, Information and Control, 13 (1968), pp. 619–627.

[3] E. F. ASSMUS JR., H. F. MATTSON JR. AND R. TURYN, *Cyclic codes*, Rep. AFCRL-66-348, Applied Research Laboratory, Sylvania Electronic Systems, 1966.

[4] E. R. BERLEKAMP, *Algebraic Coding Theory*, McGraw-Hill, New York, 1968.

[5] E. L. COHEN, *A note on perfect double error-correcting codes on q symbols*, Information and Control, 7 (1964), pp. 381–384.

[6] M. J. E. GOLAY, *Notes on digital coding*, Proc. IRE, 37 (1949), p. 657.

[7] S. W. GOLOMB AND L. R. WELCH, *Algebraic coding and the Lee metric*, Error Correcting Codes, H. B. Mann, ed., John Wiley, New York, 1968, pp. 175–194.

[8] R. W. HAMMING, *Error detecting and error correcting codes*, Bell System Tech. J., 29 (1950), pp. 147–160.

[9] B. LINDSTRÖM, *On group and nongroup perfect codes in q symbols*, Math. Scand., 25 (1969), pp. 149–158.

[10] J. H. VAN LINT, *On the nonexistence of certain perfect codes*, Computers in Number Theory, Proc. Science Research Council Atlas Symposium (Oxford, 1969), 1971.

[11] ———, *1967–1969 report of the discrete mathematics group*, Rep. 69-WSK-04, Technological University, Eindhoven, 1969.

[12] ———, *On the nonexistence of perfect 2- and 3-Hamming-error-correcting codes over GF(q)*, Information and Control, 16 (1970), pp. 396–401.

[13] ———, *Nonexistence theorems for perfect error-correcting codes*, Computers in Algebra and Number Theory, SIAM-AMS, vol. 3, to appear.

[14] ———, *On the nonexistence of perfect 5-, 6- and 7-Hamming-error-correcting codes over GF(q)*, Rep. 70-WSK-06, Technological University, Eindhoven, 1970.

[15] ———, *Coding Theory*, Lecture Notes in Mathematics 201, Springer-Verlag, Berlin, 1971.

[16] S. P. LLOYD, *Binary block coding*, Bell System Tech. J., 36 (1957), pp. 517–535.

[17] F. J. MACWILLIAMS, *Error-correcting codes—a historical survey*, Error-Correcting Codes, H. B. Mann, ed., John Wiley, New York, 1968, pp. 3–13.

[18] V. PLESS, *On the uniqueness of the Golay codes*, J. Combinatorial Theory, 5 (1968), pp. 215–228.
[19] J. SCHÖNHEIM, *On linear and nonlinear single-error-correcting q-nary perfect codes*, Information and Control, 12 (1968), pp. 23–26.
[20] H. S. SHAPIRO AND D. L. SLOTNICK, *On the mathematical theory of error-correcting codes*, IBM J. Res. Develop., 3 (1959), pp. 25–34.
[21] A. TIETÄVÄINEN, *On the nonexistence of perfect 4-Hamming-error-correcting codes*, Ann. Acad. Sci. Fenn., Ser. A I, 485 (1970), pp. 3–6.
[22] A. TIETÄVÄINEN AND A. PERKO, *There are no unknown perfect binary codes*, Ann. Univ. Turku., Ser. A I, 148 (1971), pp. 3–10.
[23] JU. L. VASILEV, *On non-group close-packed codes*, Problemy Kibernet., 8 (1962), pp. 337–339 (in Russian); Probleme der Kybernetik, 8 (1965), pp. 375–378. (In German.)

A Theorem on the Distribution of Weights in a Systematic Code†

By JESSIE MACWILLIAMS

(Manuscript received September 4, 1962)

A systematic code of word length n is a subspace of the vector space of all possible rows of n symbols chosen from a finite field. The weight of a vector is the number of its nonzero coordinates; clearly any given code contains a certain finite number of vectors of each weight from zero to n. This set of integers is called the spectrum of the code, and very little is known about it, although it appears to be important both mathematically and as a practical means of evaluating the error-detecting properties of the code.

In this paper it is shown that the spectrum of a systematic code determines uniquely the spectrum of its dual code (the orthogonal vector space). In fact the two sets of integers are related by a system of linear equations. Consequently there is a set of conditions which must be satisfied by the weights which actually occur in a systematic code. If there is enough other information about the code, it is possible to use this result to calculate its spectrum.

In most systems of error correction by binary or multiple level codes the minimum distance between two code words is an important parameter. (The distance between two code words is the number of coordinate places in which they differ.) Much attention has been given to devising codes which have an assigned minimum distance.

The weight of a code word is its distance from the origin. The distance between two code words is the weight of the vector obtained by subtracting one from the other, coordinate by coordinate. If the code words form a vector space, this vector is itself a member of the code. Such codes are called systematic codes. The set of integers specifying the weight of each code word is then exactly the same collection of numbers as the set of integers specifying the distance between each pair of code words. Thus it is customary to talk about weight properties rather than distance properties of systematic codes.

In many cases, practically all that is known explicitly about the distribution of weights in a code is that the weight has a certain minimum value. Recent studies have shown that it would be useful (e.g., in the study of real life channels) to have more information. We would like to be able to answer questions of the following sort:

i. Given a method (implemented or theoretical) for constructing a systematic code, how many elements of each weight will be obtained? (It is a safe assumption that nobody will want to write out the code vectors and count them.)

ii. Given a set, $u_1, u_2, \cdots, u_s$, of positive integers, is it possible to construct a systematic code with elements of these weights only?

In theory there exists a method of answering these questions.[1,2,3] Unfortunately this method is quite difficult to apply. The purpose of this paper is to give a different method which is in some ways more useful.

We show that the spectrum of a systematic code determines uniquely the spectrum of the dual code (the orthogonal vector space). In fact, the two sets of integers are related by a system of linear equations. Our main theorem shows how to obtain this system of equations.

In Section I we give definitions and statements of the main theorem and of some corollaries. Section II contains proofs of these theorems. Section III describes how the results of Section I may be applied.

I. DEFINITIONS, NOTATION AND A STATEMENT OF THE MAIN THEOREM

Let F be a finite field of q elements; q is a prime power. Let F^n denote the direct sum of n copies of F. F^n is the set of all possible row vectors of length n, in which each coordinate is an element of F. Addition of two vectors is defined coordinate by coordinate, under the rules prevailing in F.

F^n is a vector space of dimension n over F. Choose a basis consisting of the n vectors

$$\epsilon_1 = (1, 0, 0, \cdots, 0)$$
$$\epsilon_2 = (0, 1, 0, \cdots, 0)$$
$$\cdots\cdots\cdots$$
$$\epsilon_n = (0, 0, 0, \cdots, 1).$$

An element u of F^n is then expressed uniquely as

$$u = \sum_{i=1}^{n} u_i \epsilon_i, \qquad u_i \in F.$$

We write $u = (u_1, u_2, \cdots, u_n)$.

The weight of u is defined to be the number of u_i which actually appear in this sum — i.e., the number of nonzero coordinates in the vector u.

An alphabet is any subspace of F^n; a vector belonging to the alphabet $\mathfrak{A}$ is called a letter of $\mathfrak{A}$. It may happen that every letter of $\mathfrak{A}$ has zero as the jth coordinate — this case is not excluded.

The scalar product of two vectors,

$$u = \sum_{i=1}^{n} u_i\epsilon_i, \quad v = \sum_{i=1}^{n} v_i\epsilon_i, \qquad u_i, v_i \in F,$$

is $u*v = \sum_{i=1}^{n} u_iv_i$, where the multiplication and addition are carried out in F. If F is the field of two elements 0, 1, for example, the scalar product of (1, 1, 0) with itself is $1\cdot 1 + 1\cdot 1 = 0$.

Two vectors u, v are orthogonal if their scalar product is zero. In the example above, (1, 1, 0) is orthogonal to itself.

The orthogonal complement of an alphabet $\mathfrak{A}$ is the set of all vectors of F^n which are orthogonal to every vector of $\mathfrak{A}$. It is clear that these vectors also form an alphabet, say $\mathfrak{B}$, which is called the dual alphabet of $\mathfrak{A}$. If k is the dimension of $\mathfrak{A}$, the dimension of $\mathfrak{B}$ is $m = n - k$.

The main result of this paper is as follows (the proof is given in Section II):

Let $\mathfrak{A}$ be an alphabet of dimension k, and $\mathfrak{B}$ the dual alphabet of dimension m. Let A_i, B_i denote the number of letters of weight i in $\mathfrak{A}$, $\mathfrak{B}$ respectively. Of course, $A_0 = B_0 = 1$. Set $\gamma = q - 1$. Let z be an indeterminate.

Theorem 1: The quantities defined above are related by the equation

$$\sum_{i=0}^{n} A_i(1 + \gamma z)^{n-i}(1 - z)^i = q^k \sum_{i=0}^{n} B_i z^i.$$

Remarks:

i. The formula above is symmetric in the sense that, setting $(1 - z)/(1 + \gamma z) = \hat{z}$, we obtain by straightforward algebra

$$\sum_{i=0}^{n} B_i(1 + \gamma\hat{z})^{n-i}(1 - \hat{z})^i = q^m \sum_{i=0}^{n} A_i\hat{z}^i.$$

ii. Theorem 1 is a statement about equivalent classes;[3,4] it is still true if $\mathfrak{A}$, $\mathfrak{B}$ are replaced by equivalent alphabets.

An alphabet $\mathfrak{A}$ is said to be decomposable,[3,4] with respect to the basis $\epsilon_1, \epsilon_2, \cdots, \epsilon_n$ of F^n, if it is the direct sum of two alphabets $\mathfrak{A}_1$, $\mathfrak{A}_2$, where $\mathfrak{A}_1$ contains n columns of zeros and $\mathfrak{A}_2$ occupies these columns only. For example, the alphabet

$$\begin{matrix} 0 & 0 & 0 & 0 \\ 1 & 1 & 0 & 0 \\ 0 & 0 & 1 & 1 \\ 1 & 1 & 1 & 1 \end{matrix}$$

is decomposable, with

$$\mathfrak{A}_1 = \begin{matrix} 0 & 0 & 0 & 0 \\ 1 & 1 & 0 & 0 \end{matrix} \qquad \mathfrak{A}_2 = \begin{matrix} 0 & 0 & 0 & 0 \\ 0 & 0 & 1 & 1. \end{matrix}$$

In general, $\mathfrak{A}_1$ is a k_1-dimensional alphabet in F^{n_1}, and $\mathfrak{A}_2$ a k_2-dimensional alphabet in F^{n_2}, with $n_1 + n_2 = n$, $k_1 + k_2 = k$, $k_1 \leqq n_1$, $k_2 \leqq n_2$.

The dual alphabet of a decomposable alphabet is also decomposable; in fact $\mathfrak{B} = \mathfrak{B}_1 + \mathfrak{B}_2$, where $\mathfrak{B}_i$ is the dual alphabet of $\mathfrak{A}_i$ in F^{n_i}, $i = 1, 2$. (The example above is self-dual.)

Corollary 1.1: If $\mathfrak{A}$ is decomposable, say $\mathfrak{A} = \mathfrak{A}_1 + \mathfrak{A}_2$, the equation

$$\sum_{i=0}^{n} A_i(1 + \gamma z)^{n-i}(1 - z)^i = q^k \sum B_i z^i$$

† This paper formed part of a thesis presented to the Department of Mathematics, Harvard University, in partial fulfillment of the requirements for the degree of Doctor of Philosophy.

is reducible in the obvious sense; the factors are the equations pertaining to $\mathcal{A}_i$, $\mathcal{B}_i$ in F^{n_i}, $i = 1, 2$.

For the example above we have

$$[(1 + z)^4 + (1 - z)^2]^2 = 2^2(1 + z^2)^2.$$

Corollary 1.2: A necessary condition for the existence of an alphabet containing letters of weights w_i, $i = 1, 2, \cdots, s$, and no other, is that there exists a set of integers α_i, $i = 1, 2, \cdots, s$, such that the expression

$$(1 + \gamma z)^n + \gamma \sum_{j=1}^{s} \alpha_i(1 + \gamma z)^{n-w_i}(1 - z)^{w_i},$$

when expanded in powers of z, takes the form

$$q^k + \gamma q^k \sum_{i=1}^{n} \beta_i z^i,$$

where the β_i are positive integers.

Unfortunately, this condition is not sufficient. For example,

$$(1 + z)^8 + 7(1 + z)^6(1 - z)^2 + 7(1 + z)^2(1 - z)^6 + (1 - z)^8 = 2^4(1 + 7z^2 + 7z^6 + z^8),$$

but it is not possible to construct a binary alphabet containing 7 letters of weight 2 and no letters of weight 4.

If $A_1 = A_2 = \cdots = A_{2j} = 0$, every vector of weight $\leqq j$ in F^n appears as a coset leader for $\mathcal{A}$, and conversely. Another way of saying this is that, for all pairs of distinct letters a, a', of $\mathcal{A}$ and any $i \leqq j$, the set of vectors at distance i from a is disjoint from the set of vectors at distance i from a'. In this case we can enumerate these vectors by weights as follows:

Let $f_{s,i}$ denote the number of vectors of weight s in F^n which are at distance i from some letter of $\mathcal{A}$. Write

$$(1 + \gamma z)^{n-i}(1 - z)^i = \sum_{i=0}^{n} \Psi(i, j)z^j$$

Corollary 1.3: If $\mathcal{A}$ contains no letter of weight $< 2j + 1$, then

$$\sum_{s=1}^{n} f_{s,j}x^s = \sum_{i=0}^{n} B_i\Psi(i, j)(1 + \gamma x)^{n-i}(1 - x)^i.$$

The proofs of corollaries 1.1 to 1.3 are given in Section II.

II. PROOFS

If $\mathcal{A}$ is an alphabet of F^n and $\mathcal{B}$ the orthogonal complement of $\mathcal{A}$, the weights of the letters of $\mathcal{B}$ are, of course, uniquely determined by the letters of A. However, a much stronger statement can be made: the set of integers specifying the number of letters of each weight in $\mathcal{B}$ is related by a system of linear equations to the set of integers similarly defined for $\mathcal{A}$. This section will consist of proofs of this statement and of some of its consequences.

Two proofs are given. The first is short and easy; the second is longer and more sophisticated. However, it incidentally produces a more general result and gives some insight into what is going on.

We make the following conventions for notation: $\mathcal{A}$ shall be a k-dimensional alphabet in F^n; $\mathcal{B}$ shall be the orthogonal complement of $\mathcal{A}$ of dimension $m = n - k$; γ shall denote the quantity $q - 1$. A_i, B_i denote the number of letters of weight i in $\mathcal{A}$, $\mathcal{B}$ respectively. The binomial coefficient $\binom{r}{s}$ is understood to be zero if $s > r$.

Let $\epsilon_1, \epsilon_2, \cdots, \epsilon_n$ be the usual basis of F^n. Let $s = (s_1, s_2, \cdots, s_\nu)$ be a set of ν different indices, $1 \leqq s_i \leqq n$, and let $t = (t_1, t_2, \cdots, t_{n-\nu})$ be the complementary set of indices. Denote by F_s^ν, $F_t^{n-\nu}$ the spaces generated by $\epsilon_{s_1}, \cdots, \epsilon_{s_\nu}$ and $\epsilon_{t_1}, \cdots, \epsilon_{t_{n-\nu}}$. Clearly, F_s^ν, $F_t^{n-\nu}$ are orthogonal complements in F^n. Let $|H|$ stand for the number of vectors in a space H.

Lemma 2.0:

$$|\mathcal{A} \cap F_t^{n-\nu}| = q^{k-\nu} |\mathcal{B} \cap F_s^\nu|.$$

Proof: The orthogonal complement of $\mathcal{A} \cap F_t^{n-\nu}$ is the smallest space containing $\mathcal{B}$ and F_s^ν. This is the lattice theoretic union of $\mathcal{B}$ and F_s^ν, which we write $\mathcal{B} \cup F_s^\nu$. Then

$$|\mathcal{B} \cup F_s^\nu| \cdot |\mathcal{A} \cap F_t^{n-\nu}| = q^n = q^{m+k}.$$

The number of vectors in $\mathcal{B} \dot{\cup} F_s^\nu$ is $q^m q^\nu / |\mathcal{B} \cap F_s^\nu|$. Hence

$$q^{m+\nu} |\mathcal{A} \cap F_t^{n-\nu}| = q^{m+k} |\mathcal{B} \cap F_s^\nu|$$

or

$$|\mathcal{A} \cap F_t^{n-\nu}| = q^{k-\nu} |\mathcal{B} \cap F_s^\nu|.$$

Denote by $\{(\epsilon_{s_1}, \cdots, \epsilon_{s_\nu}), a\}$ a pair consisting of ν basis vectors of F^n and a vector a of $\mathcal{A}$ which is orthogonal to each of $\epsilon_{s_1}, \cdots, \epsilon_{s_\nu}$.

Lemma 2.1:

i. For a fixed set of indices $s_1, \cdots, s_\nu$ the number of pairs

$$\{(\epsilon_{s_1}, \cdots, \epsilon_{s_\nu}), a\} \quad is \quad |\mathcal{A} \cap F_t^{n-\nu}|.$$

ii. The total number of such pairs for all choices of ν distinct basis vectors is $\sum_{i=0}^{n} A_i \binom{n-i}{\nu}$.

Proof:

i. $F_t^{n-\nu}$ consists of exactly those vectors of F^n which are orthogonal to $\epsilon_{s_1}, \cdots, \epsilon_{s_\nu}$; hence $\mathcal{A} \cap F_t^{n-\nu}$ consists of exactly those vectors of $\mathcal{A}$ which are orthogonal to $\epsilon_{s_1}, \cdots, \epsilon_{s_\nu}$.

ii. If $a \in \mathcal{A}$ is of weight i, then a is orthogonal to $n - i$ of the vectors $\epsilon_1, \cdots, \epsilon_n$. A set of ν vectors may be chosen from these $n - i$ in $\binom{n-i}{\nu}$ ways. Hence the total number of pairs

$$\{(\epsilon_{s_1}, \cdots, \epsilon_{s_\nu}), \quad a\} \text{ is } \sum_{i=0}^{n} A_i \binom{n-i}{\nu}.$$

Let $\sum_s$ indicate summation over all possible choices of ν indices $s_1, \cdots, s_\nu$; similarly, $\sum_t$ denotes summation over all the complementary sets $t_1, \cdots, t_{n-\nu}$. Lemma 2.1 is equivalent to

$$\sum_t |\mathcal{A} \cap F_t^{n-\nu}| = \sum_{i=0}^{n} A_i \binom{n-i}{\nu}.$$

Replace $\mathcal{A}$ by $\mathcal{B}$, ν by $n - \nu$ and s by t. The same argument then gives

$$\sum_s |\mathcal{B} \cap F_s^\nu| = \sum_{i=0}^{n} B_i \binom{n-i}{n-\nu}.$$

Lemma 2.2

$$\sum_{i=0}^{n} A_i \binom{n-i}{\nu} = q^{k-\nu} \sum_{i=0}^{n} B_i \binom{n-i}{n-\nu}.$$

Proof: For a fixed set s (which determines, of course, a fixed set t) we have by 2.0

$$|\mathcal{A} \cap F_t^{n-\nu}| = q^{k-\nu} |\mathcal{B} \cap F_s^\nu|.$$

Thus

$$\sum_t |\mathcal{A} \cap F_t^{n-\nu}| = q^{k-\nu} \sum_s |\mathcal{B} \cap F_s^\nu|,$$

which, by 2.1 is the same thing as

$$\sum_{i=0}^{n} A_i \binom{n-i}{\nu} = q^{k-\nu} \sum_{i=0}^{n} B_i \binom{n-i}{n-\nu}.$$

The equation of 2.2 holds for $\nu = 0, 1, \cdots, n - 1$. This is, in fact, one form of the promised set of linear equations between the quantities A_i, B_i.

We now give the second proof.

Let $\mathcal{G}$ be a finite Abelian group. A character χ of $\mathcal{G}$ is a homomorphism of $\mathcal{G}$ into the multiplicative group of complex numbers of absolute value 1. The characters of $\mathcal{G}$ form a group $\mathcal{G}^*$ which is isomorphic to $\mathcal{G}$, there being in general no canonical isomorphism.†

If $\mathcal{A}$ is a subgroup of $\mathcal{G}$, the characters such that $\chi(a) = 1$ for all a of $\mathcal{A}$ form a subgroup $\mathcal{B}^*$ of $\mathcal{G}^*$. $\mathcal{B}^*$ is precisely the character group of $\mathcal{G}$ *mod* $\mathcal{A}$.

Suppose now that $\mathcal{G}$ is the additive group of a finite field. $\mathcal{G}^*$ is just a multiplicative copy of $\mathcal{G}$, and the characters can be labeled by the ele-

† For prime fields, the proof can be given without mentioning the word character. The presentation here is an uneasy compromise with conscience — we wish to indicate possible extensions to nonprime fields without doing too much work.

ments of $\mathcal{G}$ in a symmetric way; that is, if $r, s, \cdots$ are elements of $\mathcal{G}$ we have

$$\chi_r(s) = \chi_s(r) = \chi(r, s).$$

If $\mathcal{G}$ is the additive group of a prime field of order q, we take r, s etc. to be the integers *mod* q, and set $\chi(r, s) = \zeta^{rs}$ where ζ is a primitive qth root of unity.

We have from the general theory of characters

$$\chi(r, 0) = \chi(0, s) = 1,$$

$$\sum_{r=1}^{\gamma} \chi(r, s) = -1 \qquad \text{if } s \neq 0.$$

Let $\epsilon_1, \epsilon_2, \cdots, \epsilon_n$ be the fixed basis of F^n, and $u = (u_1, u_2, \cdots, u_n)$ the coordinates of a vector of F^n with respect to this basis. The character group of F^n is, of course, a multiplicative copy of F^n. We label the characters by elements of F^n as follows:

$$\psi_u(v) = \prod_{i=1}^{n} \chi(u_i, v_i) = \psi_v(u) = \psi(u, v).$$

Let $\mathcal{A}$ be a subspace of F^n. The characters such that $\psi_b(a) = 1$ for all a of $\mathcal{A}$ form a subgroup $\mathcal{B}^*$ of the character group. $\mathcal{B}^*$ is exactly the character group of F^n *mod* $\mathcal{A}$. The elements b which label these characters form a subspace $\mathcal{B}$ of F^n, isomorphic to F^n *mod* $\mathcal{A}$. In our notation, the equation $\psi(a, b) = 1$ holds for all a of $\mathcal{A}$ and all b of $\mathcal{B}$, and given either $\mathcal{A}$ or $\mathcal{B}$, the other is uniquely† determined by this condition.

Lemma 2.3: Let $\mathcal{A}$, $\mathcal{B}$ *be related as above. Then*

i. $$\sum_{a \in \mathcal{A}} \psi(v, a) = q^k \text{ if } v \in \mathcal{B}.$$

ii. $$\sum_{a \in \mathcal{A}} \psi(v, a) = 0 \text{ if } v \notin \mathcal{B}.$$

Proof: Part *i* is obvious, since by definition $\psi(v, a) = 1$ if $v \in B$. For *ii* we observe that for $a \in \mathcal{A}$, $\psi(v, a) = \psi_a(v)$ is a character of F^n *mod* $\mathcal{B}$. If $\bar{v}$ denotes a coset of F^n *mod* $\mathcal{B}$, $\sum_{a \in \mathcal{A}} \psi_a(\bar{v}) = 0$ for $\bar{v} \neq \mathcal{B}$. Now $\psi_a(v) = \psi_a(\bar{v})$ for any v in $\bar{v}$; hence

$$\sum_{a \in \mathcal{A}} \psi(v, a) = \sum_{a \in \mathcal{A}} \psi_a(\bar{v}) = 0 \quad \text{if} \quad v \notin \mathcal{B}.$$

Lemma 2.4: If F is a prime field, then $\mathcal{A}$, $\mathcal{B}$ *are related as in* 2.3 *if and only if they are orthogonal complements.*

Proof: $\psi(a, b) = \zeta^{\sum a_i b_i}$ where ζ is a primitive qth root of unity. Hence $\psi(a, b) = 1$ implies that a is orthogonal to b. Since $\mathcal{A}$ is isomorphic to F^n *mod* $\mathcal{B}$ the dimensions of $\mathcal{A}$, $\mathcal{B}$ add up to n. Thus $\mathcal{B}$ is the orthogonal complement of $\mathcal{A}$.

Let $f(i, s)$ denote a function of the integers i, s with values in a ring R. The values of $f(i, s)$ may be added and multiplied, and these operations obey the two distributive laws. $f(i, v_i)$ denotes the same function of i and the ith coordinate of v.

Lemma 2.5:

$$\sum_{v \in F^n} \prod_{i=1}^{n} f(i, v_i) = \prod_{i=1}^{n} \sum_{r=0}^{\gamma} f(i, r).$$

Proof: If $n = 1$ the statement is

$$\sum_{s=0}^{\gamma} f(1, s) = \sum_{r=0}^{\gamma} f(1, r),$$

which is obvious. Assume the truth of the lemma for F^{n-1}.

Let $F_r^{\,n}$, $0 \leqq r \leqq \gamma$, denote the set of vectors of F^n which have last coordinate r. Clearly the $F_r^{\,n}$ are a partition of F^n. Then

$$\sum_{v \in F^n} \prod_{i=1}^{n} f(i, v_i) = \sum_{r=0}^{\gamma} \sum_{v \in F_r^n} \left[\prod_{i=1}^{n-1} f(i, v_i) f(n, r)\right]$$

$$= \sum_{r=0}^{\gamma} f(n, r) \sum_{v \in F^{n-1}} \prod_{i=1}^{n-1} f(i, v_i)$$

$$= \sum_{r=0}^{\gamma} f(n, r) \prod_{i=1}^{n-1} \sum_{r=0}^{\gamma} f(i, r) \quad \text{(by induction)}$$

$$= \prod_{i=1}^{n} \sum_{r=0}^{\gamma} f(i, r).$$

Let $z^{(r)}$ be a set of (commuting) indeterminates, $r = 0, 1, \cdots, \gamma$. To each vector $v = (v_1, v_2, \cdots, v_n)$ of F^n associate a monomial $\prod_{i=1}^{n} z^{(v_i)}$. The monomial associated with v describes how many times each field element appears as a component of v. Let R be the ring of polynomials in $z^{(0)}, z^{(1)}, \cdots, z^{(\gamma)}$ over the complex numbers. Let $u = (u_1, u_2, \cdots, u_n)$ be a fixed vector of F^n.

Lemma 2.6:

$$\sum_{v \in F^n} \psi(u, v) z^{(v_1)} z^{(v_2)} \cdots z^{(v_n)} = \prod_{j=1}^{n} \sum_{r=0}^{\gamma} \chi(u_j, r) z^{(r)}.$$

Proof: Set $f(j, v_j) = \chi(u_j, v_m) z^{(v_j)}$, which is in R. Then

$$\psi(u, v) z^{(v_1)} z^{(v_2)} \cdots z^{(v_n)} = \prod_{j=1}^{n} f(j, v_j).$$

By 2.5,

$$\sum_{v \in F^n} \psi(u, v) z^{(v_1)} z^{(v_2)} \cdots z^{(v_n)} = \prod_{j=1}^{n} \sum_{r=0}^{\gamma} f(j, r)$$

$$= \prod_{j=1}^{n} \sum_{r=0}^{\gamma} \chi(u_j, r) z^{(r)}.$$

Lemma 2.7: Let $\mathcal{A}$, $\mathcal{B}$ *be orthogonal complements in* F^n, *as usual. Then*

$$\sum_{u \in \mathcal{A}} \prod_{j=1}^{n} \sum_{r=0}^{\gamma} \chi(u_j, r) z^{(r)} = q^k \sum_{v \in \mathcal{B}} z^{(v_1)} z^{(v_2)} \cdots z^{(v_n)}$$

Proof: We evaluate the quantity

$$F(u, v) = \sum_{u \in \mathcal{A}} \sum_{v \in F^n} \chi(u, v) z^{(v_1)} z^{(v_2)} \cdots z^{(v_n)}$$

in two ways, which give the two sides of the equation.

By 2.6

$$F(u, v) = \sum_{u \in \mathcal{A}} \prod_{j=1}^{n} \sum_{r=0}^{\gamma} \chi(u_j, r) z^{(r)}.$$

Also

$$F(u, v) = \sum_{v \in F^n} z^{(v_1)} z^{(v_2)} \cdots z^{(v_n)} \sum_{u \in \mathcal{A}} \psi(u, v).$$

By 2.4 and 2.3

$$\sum_{u \in \mathcal{A}} \psi(u, v) = \begin{cases} q^k \text{ if } v \in \mathcal{B} \\ 0 \text{ if } v \notin \mathcal{B}. \end{cases}$$

Hence

$$F(u, v) = q^k \sum_{v \in \mathcal{B}} z^{(v_1)} z^{(v_2)} \cdots z^{(v_n)}.$$

Theorem 2.8: Let $\mathcal{A}$ *be a k-dimensional alphabet of* F^n, *and* $\mathcal{B}$ *the orthogonal complement of dimension* $m = n - k$. *Let* A_i, B_i *denote the number of letters of weight i in* $\mathcal{A}$, $\mathcal{B}$. *Then*

$$\sum_{i=0}^{n} A_i (1 + \gamma z)^{n-i} (1 - z)^i = q^k \sum_{i=0}^{n} B_i z^i.$$

Proof: In 2.7 set $z^{(r)} = \begin{cases} z & \text{if } r \neq 0 \\ 1 & \text{if } r = 0. \end{cases}$

If $u_j = 0$, $\chi(u_j, r)$ is 1 and $\sum_{r=0}^{\gamma} \chi(u_j, r) z^{(r)}$ becomes $(1 + \gamma z)$.

If $u_j \neq 0$ $\sum_{r=1}^{\gamma} \chi(u_j, r)$ is -1, and $\sum_{r=0}^{\gamma} \chi(u_j, r) z^{(r)}$ becomes $(1 - z)$.

Let $|u|$ denote the number of nonzero u_j.

Then $\prod_{j=1}^{n} \sum_{r=0}^{\gamma} \chi(u_j, r) z^{(r)}$ goes into $(1 + \gamma z)^{n-|u|}(1 - z)^{|u|}$;

$|u|$ is of course the weight of $u = (u_1, u_2, \cdots, u_n)$, so that the left-hand side of 2.7 becomes

$$\sum_{i=0}^{n} A_i (1 + \gamma z)^{n-i} (1 - z)^i.$$

The right-hand side of 2.7 is clearly

$$q^k \sum_{i=0}^{n} B_i z^i,$$

† That is, if F is a prime field. Otherwise we must fix the basis of F over its prime field before we claim uniqueness.

which proves the theorem.

Innumerable sets of linear equations between the quantities A_i, B_i may be obtained from theorem 2.8. The following two are sometimes useful.

Lemma 2.9: *For* $\nu = 0, 1, \cdots, n$,

$$i. \quad \sum_{i=0}^{n-\nu} A_i \binom{n-i}{\nu} = q^{k-\nu} \sum_{i=0}^{n} B_i \binom{n-i}{n-\nu}.$$

(*These are the equations of* 2.2.)

$$ii. \quad \sum_{i=\nu}^{n} A_i \binom{i}{\nu} = q^{k-\nu} \sum_{i=0}^{\nu} (-1)^i B_i \gamma^{\nu-i} \binom{n-i}{n-\nu}.$$

i. is obtained by setting $(1 + \gamma z)/(1 - z) = 1 + y$,

ii. by setting $(1 - z)/(1 + \gamma z) = 1 + y$. The algebraic details are easy to verify.

This process is reversible, i.e., (*i*) or (*ii*) imply 2.8. Before exploring the consequences of theorem 2.8, we give a different specialization of 2.7.

Theorem 2.10: Let $B_s^{(1)}$ *be the number of letters of* $\mathfrak{B}$ *which contain s coordinates equal to* 1. *Let* A_{0s} *be the number of letters u in* $\mathfrak{A}$ *of weight s for which* $\sum_{i=1}^{n} u_i = 0$. *Let* A_{1s} *be the number of letters u in* $\mathfrak{A}$ *of weight s for which* $\sum_{i=1}^{n} u_i \neq 0$. (*Clearly* $A_s = A_{0s} + A_{1s}$). *Then*

$$\sum_{s=0}^{n} B_s^{(1)} z^s = (A_{0s} - A_{1s}/\gamma)(z-1)^s (z+\gamma)^{n-s}.$$

Proof: In 2.7 set

$$z^{(r)} = \begin{cases} z & \text{if } r = 1 \\ 1 & \text{if } r \neq 1. \end{cases}$$

Then

$$q^k \sum_{\nu \in \mathfrak{B}} z^{(\nu_1)} z^{(\nu_2)} \cdots z^{(\nu_n)} \text{ becomes } q^k \sum_{s=0}^{n} B_s^{(1)} z^s,$$

$$\sum_{r=0}^{\gamma} \chi(u_i, r) z^{(r)} \text{ becomes } \chi(u_i, 1)z + \sum_{r=0}^{\gamma} \chi(u_i, r) - \chi(u_i, 1)$$

$$= \chi(u_i, 1)(z-1) + \sum_{r=0}^{\gamma} \chi(u_i, r)$$

$$= \begin{cases} \chi(u_i, 1)(z-1) & \text{if } u_i \neq 0 \\ z + \gamma & \text{if } u_i = 0 \end{cases}$$

$$\prod_{i=1}^{n} \sum_{r=0}^{\gamma} \chi(u_i, r) z^{(r)} \quad \text{becomes} \quad (z+\gamma)^{n-|u|}(z-1)^{|u|} \prod_{u_i \neq 0} \chi(u_i, 1).$$

Now if u is a letter of A, so are also the letters $2u, \cdots, \gamma u$, and these have the same weight as u. We sum first over these letters

$$\sum_{s=1}^{\gamma} \prod \chi(su_i, 1) = \sum_{s=1}^{\gamma} \chi(s\Sigma u_i, 1)$$

$$= \begin{cases} -1 & \text{if } \Sigma u_i \neq 0 \\ \gamma & \text{if } \Sigma u_i = 0 \end{cases}$$

The sum of

$$(z+\gamma)^{n-|u|}(z-1)^{|u|} \prod_{u_i \neq 0} \chi(u_i, 1)$$

over all letters in $\mathfrak{A}$ of the same weight as u is thus

$$(A_{0|u|} - (A_{1|u|}/\gamma))(z+\gamma)^{n-|u|}(z-1)^{|u|}.$$

Hence the left-hand side of 2.7 becomes

$$\sum_{s=0}^{n} (A_{0s} - (A_{1s}/\gamma))(z+\gamma)^{n-s}(z-1)^s.$$

We return now to the consequences of theorem 2.8. As remarked in the proof of 2.10, if $\mathfrak{A}$ contains a letter u, it contains also the letters $2u, \cdots, \gamma u$; that is, the number of letters of weight i in $\mathfrak{A}$ is divisible by γ for $i > 0$. We have then

Lemma 2.11: A necessary condition for the existence of an alphabet containing letters of weights w_i, $i = 1, 2, \cdots, s$, *and no other, is the existence of a set of integers* α_i, $i = 1, 2, \cdots, s$, *such that the expression*

$$(1+\gamma z)^n + \gamma \sum_{j=1}^{s} \alpha_i (1+\gamma z)^{n-w_i}(1-z)^{w_i},$$

when expanded in powers of z, takes the form

$$q^k + \gamma q^k \sum_{i=1}^{n} \beta_i z^i,$$

where the β_i *are positive integers.*

It has been pointed out before that this condition is not sufficient.

Suppose now that $\mathfrak{A} = \mathfrak{A}_1 + \mathfrak{A}_2$ is a decomposable alphabet. $\mathfrak{A}_j$ is a k_j-dimensional alphabet in F^{n_j}, $j = 1, 2$, with orthogonal alphabet B_j. $k_1 + k_2 = k$, and $n_1 + n_2 = n$. Let $A_i^{(1)}$, $A_i^{(2)}$ and $B_i^{(1)}$, $B_i^{(2)}$ be the number of letters of weight i in $\mathfrak{A}_1$, $\mathfrak{A}_2$, $\mathfrak{B}_1$, $\mathfrak{B}_2$.

Lemma 2.12:

$$\sum_{i=0}^{n} A_i (1+\gamma z)^{n-i}(1-z)^i$$

$$= \left[\sum_{i=0}^{n_1} A_i^{(1)}(1+\gamma z)^{n_1-i}(1-z)^i\right]\left[\sum_{i=0}^{n_2} A_i^{(2)}(1+\gamma z)^{n_2-i}(1-z)^i\right]$$

$$= \sum_{i=0}^{n} \left[q^{k_1} \sum_{i=0}^{n_1} B_i^{(1)} z^i\right]\left[q^{k_2} \sum_{i=0}^{n_2} B_i^{(2)} z^i\right].$$

Proof: The number of letters of weight s in $\mathfrak{B}_1 + \mathfrak{B}_2$ is

$$\sum_{\sigma+\rho=s} B_\sigma^{(1)} B_\rho^{(2)},$$

which is the coefficient of z^s in $\sum_{i=0}^{n_1} B_i^{(1)} z^i \sum_{i=0}^{n_2} B_i^{(2)} z^i$. Similarly,

$$\sum_{\sigma+\rho=s} A_\sigma^{(1)} A_\rho^{(2)}$$

is the coefficient of $(1+\gamma z)^{n-s}(1-z)^s$ in

$$\left[\sum_{i=0}^{n_1} A_i^{(1)}(1+\gamma z)^{n_1-i}(1-z)^i\right]\left[\sum_{i=0}^{n_2} A_i^{(2)}(1+\gamma z)^{n_2-i}(1-z)^i\right].$$

We define the coset leader of a coset of $\mathfrak{A}$ in F^n to be an element of least weight in the coset. The weight of a coset is defined to be the weight of its coset leader.

If $A_i = 0$ for $i = 1, 2, \cdots, 2e$ every vector of weight $\leqq e$ in F^n appears as a coset leader for $\mathfrak{A}$ and conversely. Another way of saying this is: for all pairs of distinct letters a, a' of $\mathfrak{A}$, the set of vectors at distance $i \leqq e$ from a is disjoint from the set of vectors at distance i from a'.

Let $c_1^{(i)}, c_2^{(i)}, \cdots, c_\nu^{(i)}$ be the cosets of A of weight i; we assume that $\nu = \gamma^i \binom{n}{i}$, i.e., that all vectors of weight i appear as coset leaders. Let $f_{s,i}$ be the number of vectors of weight s contained in the set-theoretic union $\bigcup_{j=1}^{\nu} c_j^{(i)}$. The polynomial $\sum_{s=0}^{n} f_{s,i} x^s$ is called the enumerator (by weight) of this set of vectors. We propose to show that theorem 2.8 gives a convenient expression for this enumerator. We need the following preliminary lemma.

Lemma 2.13: Let u be a fixed vector of weight i. Let d_{st} *be the number of vectors of weight s which are at distance t from u. Then*

$$\sum_{s=0}^{n} \sum_{t=0}^{n} d_{st} x^s y^t = (1+\gamma xy)^{n-i}[x + y + (\gamma - 1)xy]^i.$$

Proof: Suppose first that $u = (u_1, u_2, \cdots, u_i)$ is a vector of weight i in F^i. We show that under these circumstances

$$\sum_{s=0}^{i} \sum_{t=0}^{i} d_{st} = [x + y + (\gamma - 1)xy]^i.$$

This is obvious for $i = 1$; we suppose it true for $i - 1$. Let $v = (v_1, v_2, \cdots, v_{i-1})$ be a vector of weight s distant t from $(u_1, u_2, \cdots, u_{i-1})$ in F^{i-1}. From v we obtain:

i. One vector $(v_1, v_2, \cdots, v_{i-1}, 0)$, weight s, distant $t + 1$ from u.

ii. One vector $(v_1, v_2, \cdots, v_{i-1}, u_i)$ weight $s + 1$ distant t from u.

iii. $\gamma - 1$ vectors $(v_1, v_2, \cdots, v_{i-1}, v_i)$ $v_i \neq 0$, $v_i \neq u_i$ which have weight $s + 1$, and are distant $t + 1$ from u.

Hence the enumerator for i is obtained by multiplying that for $i - 1$ by $[x + y + (\gamma - 1)xy]$, and the lemma is proved for $n = i$.

We now apply induction to $n - i$. Let u be a vector of weight i in F^n, and u' a vector of weight i in F^{n-1} obtained from u by omitting one zero coordinate. Let v' be a vector of F^{n-1} which has weight s and is distant t from u'. From v' we obtain in F^n

i. One vector of weight s, distant t from u, by adding a zero coordinate to v'.

ii. γ vectors of weight $s + 1$, distant $t + 1$ from u, by adding a nonzero coordinate to v'.

This corresponds to multiplication by $(1 + \gamma xy)$. Hence the lemma is proved.

Lemma 2.14: Suppose that $A_i = 0$, $i = 1, 2, \cdots, 2e$, *and take* $t \leq e$. *Then the enumerator,* $\sum_{s=0}^{n} f_{s,t}x^s$, *of vectors in cosets of weight* t *of* F^n *mod* $\mathcal{A}$ *is the coefficient of* y^t *in*

$$\sum_{i=0}^{n} A_i(1 + \gamma xy)^{n-i}(x + y + (\gamma - 1)xy)^i.$$

Proof: The cosets of weight t in F^n *mod* $\mathcal{A}$ are disjoint, and contain all vectors of F^n which are at distance t from some letter of $\mathcal{A}$.

Set $(1 + \gamma z)^{n-i}(1 - z)^i = \sum_{r=0}^{n} \Psi(i, n, r)z^r$. Let $\mathcal{B}$, B_i have their usual meaning. Assume the conditions of 2.14.

Lemma 2.15:†

$$q^m \sum_{s=0}^{n} f_{s,t}x^s = \sum_{i=0}^{n} B_i \Psi(i, n, t)(1 + \gamma x)^{n-i}(1 - x)^i$$

Proof: Set

$$z = \frac{x + y + (\gamma - 1)xy}{1 + \gamma xy},$$

then

$$1 + \gamma z = \frac{(1 + \gamma x)(1 + \gamma y)}{(1 + \gamma xy)}, \qquad 1 - z = \frac{(1 - x)(1 - y)}{1 + \gamma xy}.$$

Make this substitution in the equation

$$\sum_{i=0}^{n} B_i(1 + \gamma z)^{n-i}(1 - z) = q^m \sum_{i=0}^{n} A_i z^i$$

we obtain

$$\sum_{i=0}^{n} B_i(1 + \gamma x)^{n-i}(1 - x)^i(1 + \gamma y)^{n-i}(1 - y)^i$$

$$= q^m \sum_{i=0}^{n} A_i(1 + \gamma xy)^{n-i}(x + y + (\gamma - 1)xy)^i.$$

Equating coefficients of y^t gives us

$$\sum_{i=0}^{n} B_i\Psi(i, n, t)(1 + \gamma x)^{n-i}(1 - x)^i = q^m \sum_{s=0}^{n} f_{s,t}x^s$$

III. APPLICATIONS

The easiest application of theorem 1 is to a generalized Hamming alphabet, that is, a close-packed 1-error correcting alphabet over a field of q elements. Such an alphabet exists for $n = (q^m - 1)/\gamma$, all $m > 1$.[3]

The dual alphabet is of dimension m, and contains $(q^m - 1)$ letters of weight q^{m-1}. The spectrum of a generalized Hamming alphabet is thus given by the expansion of

$$(1 + \gamma z)^n + (q^m - 1)(1 + \gamma z)^{n-u}(1 - z)^u$$

where $n = (q^m - 1)/\gamma$, $u = q^{m-1}$.

† A similar formula ($q = 2$) is found by Lloyd[5] for close-packed codes which are not assumed to be group codes.

TABLE I — DISTRIBUTION OF WEIGHTS IN THE TWO GOLAY CODES

The first table is for the 3-error-correcting (23, 12) alphabet over Z_2, the second for the 2-error-correcting (11, 6) alphabet over Z_3. In both cases, i stands for weight, B_i for the number of letters of weight i in the dual alphabet, A_i for the number of letters of this weight in the Golay alphabet.

i	B_i	A_i
0	1	1
7	0	23 × 11
8	23 × 22	23 × 22
11	0	23 × 56
12	23 × 56	23 × 56
15	0	23 × 22
16	23 × 11	23 × 11
23	0	1

i	B_i	A_i
0	1	1
5	0	2 × 66
6	2 × 66	2 × 66
8	0	2 × 165
9	2 × 55	2 × 55
11	0	2 × 12

Theorem 1 may, in fact, be used to calculate the number of letters of each weight in any close-packed code. The results for the two Golay[6] codes are given in Table I.

Anything which is known about the structure of an alphabet or its dual may be used with theorem 1 to limit the number of possible weight distributions. Such items of information are a very diversified character, and no general method has been developed. However, the results obtained by hand calculation indicated that it is probably worthwhile to make a systematic computer study of the known classes of alphabets.

ACKNOWLEDGMENTS

The author would like to thank her friends and associates for their consistently helpful suggestions; and is especially indebted to H. E. Elliott for the elegant proof of lemma 2.5 and to J. B. Kruskal for theorem 2.10. The criticisms and suggestions of Professor A. M. Gleason of Harvard University produced new and better proofs of practically every other theorem in this paper.

REFERENCES

1. Slepian, D., A Class of Binary Signaling Alphabets, B.S.T.J., **35**, January, 1956, pp. 203–234.
2. Bose, R. C., and Kuebler, R. R., Jr., A Geometry of Binary Sequences Associated with Group Alphabets in Information Theory, Ann. Math. Stat. **31**, 1960, p. 113.
3. MacWilliams, F. J., Error-correcting Codes for Multiple-Level Transmission, B.S.T.J., **40**, January, 1961, pp. 281–308.
4. Slepian, D., Some Further Theory of Group Codes, B.S.T.J., **39**, September, 1960, pp. 1219–1252.
5. Lloyd, S. P., Binary Block Coding, B.S.T.J., **36**, March, 1957, pp. 517–535.
6. Golay, N. J. E., Notes on Digital Coding, Proc. I.R.E., **37**, 1949, p. 657.
7. Peterson, W. W., *Error Correcting Codes*, M.I.T. Press and John Wiley and Sons, 1961.

Power Moment Identities on Weight Distributions in Error Correcting Codes

VERA PLESS

Air Force Cambridge Research Laboratories, Bedford, Massachusetts

A series of identities relating the weight distribution in any code space to the weight distribution in the orthogonal code space has been given by Mrs. J. MacWilliams (1962). Here we derive a series of power moment identities from the MacWilliams identities. An earlier result of Assmus and Mattson is shown to be equivalent to the third power moment identity.

A unique solution to the power moment identities is given under certain conditions. Applications are given.

I. INTRODUCTION

A. SUMMARY

A series of identitites relating the weight distribution in any code space to the weight distribution in the orthogonal code space has been given by Mrs. J. MacWilliams (1962). Here we derive a series of power moment identities from the MacWilliams identities. An earlier result of Assmus and Mattson (1961) is shown to be equivalent to one of these identities. For $q = 2$, their result was also proven by Zierler (1962).

We study states of partial information on weight distributions such that the power moment identities determine uniquely the remaining values. Applications are made in determining the weight distributions of certain cyclic codes.

B. DEFINITIONS

These are the definitions and notations relevant primarily to Section II.

Let V be a vector space of dimension n over $GF(q)$, the Galois field of q elements, q a prime power. Let A be a subspace of V, of dimension k. Two vectors $v = (v_0, v_1, \cdots, v_{n-1})$ and $w = (w_0, w_1, \cdots w_{n-1})$ in V are said to be orthogonal if $\sum_{i=0}^{n-1} v_i w_i$ is 0 in $GF(q)$. Let B be the set of all vectors in V orthogonal to each vector in A. B is called the orthogonal of A.

If $v = (v_0, \cdots, v_{n-1})$ is a vector in V, the weight of v is defined to be the number of nonzero v_i. A_i will denote the number of vectors of weight i in A and B_i will denote the number of vectors in weight i in B.

The following are the facts relevant to Section III.

We recall that a code space is called cyclic if it is invariant under the coordinate permutation corresponding to the mapping $i \to i + 1$ (modulo n) of the coordinate indices. For n prime, a cyclic code is called a quadratic residue code if, for $0 < b < n$, the coordinate permutation corresponding to the mapping $i \to bi$ (modulo n) leaves the code invariant if and only if b is a quadratic residue of n. For the rest of this paragraph and in section III we assume that $q = 2$ and n is a prime, which we will refer to by p, such that $p \equiv -1$ modulo 8. These codes were originally defined in a different context by Gleason (1961) and then in this fashion by Prange. Both Gleason (1961) and Prange from their different points of view demonstrated the following facts about quadratic residue codes A of dimension $(p - 1)/2$.

(a) There are exactly two quadratic residue codes of dimension $(p - 1)/2$. These are equivalent and hence have the same weight distributions.

(b) $A_i = 0$ unless i is a multiple of 4.

(c) If B is the orthogonal of A, B is a quadratic residue code of dimension $(p + 1)/2$, and B is obtained from A by adjoining the all one vector. This implies

(d) $B_i = 0$ unless $i = 0$ or 3 modulo 4.

(e) Prange showed that if the minimum weight in B is m, m must be odd, and $m + 1$ is then the minimum weight in A. From this it follows that the maximum weight in A is $n - m$.

II. THE POWER MOMENTS

Let $\gamma = q - 1$. MacWilliams (1962) binomial moments are

$$\sum_{j=0}^{n} \binom{j}{\nu} A_j = q^{k-\nu} \sum_{j=0}^{n} (-1)^j B_j \gamma^{\nu-j} \binom{n-j}{n-\nu}, \qquad B_1$$

$$\sum_{j=0}^{n} \binom{n-j}{\nu} A_j = q^{k-\nu} \sum_{j=0}^{n} \binom{n-j}{n-\nu} B_j \qquad B_2$$

where $\binom{a}{b} = 0$ if $b > a$; ν can take any integral values from 0 to n so that (B_1) (or (B_2)) is a set of $n + 1$ equations.

The power moment identities which can be derived from these are the following:

$$\sum_{j=0}^{n} (j^r) A_j = \sum_{j=0}^{n} (-1)^j B_j \left(\sum_{\nu=0}^{r} \nu!\, S(r,\nu) q^{k-\nu} \gamma^{\nu-j} \binom{n-j}{n-\nu} \right), \qquad P_1$$

$$\sum_{j=0}^{n} (n-j)^r A_j = \sum_{j=0}^{n} B_j \left(\sum_{\nu=0}^{r} \nu!\, S(r,\nu) q^{k-\nu} \binom{n-j}{n-\nu} \right). \qquad P_2$$

Here $S(r, \nu)$ is a Stirling number of the second kind,

$$S(r,\nu) = \left[\frac{\Delta^\nu x^r}{\nu!} \right]_{x=0} = \frac{1}{\nu!} \sum_{i=1}^{\nu} (-1)^{\nu-i} \binom{\nu}{i} i^r.$$

We get an infinite system of equations for either P_1 or P_2 as r can take any positive values or zero. P_1 and P_2 are valid for $r > n$ provided we understand that $\binom{a}{b} = 0$ if $b < 0$. If $r < n$, then $\sum_{j=0}^{n}$ on the right sides of P_1 and P_2 can be replaced by $\sum_{j=0}^{r}$.

We will give the derivation of P_2 from B_2. P_1 follows from B_1 similarly. From Jordan (1950) we know that

$$(n-j)^r = \sum_{\nu=0}^{r} \nu! \binom{n-j}{\nu} S(r,\nu)$$

so that

$$\sum_{j=0}^{n} (n-j)^r A_j = \sum_{j=0}^{n} \left(\sum_{\nu=0}^{r} \nu! \binom{n-j}{\nu} S(r,\nu) \right) A_j$$

$$= \sum_{\nu=0}^{r} \nu!\, S(r,\nu) \left(\sum_{j=0}^{n} \binom{n-j}{\nu} A_j \right)$$

$$= \sum_{\nu=0}^{r} \nu!\, S(r,\nu) \left(q^{k-\nu} \sum_{j=0}^{n} \binom{n-j}{n-\nu} B_j \right) \qquad \{\text{by } B_2\}$$

$$= \sum_{j=0}^{n} B_j \left(\sum_{\nu=0}^{r} \nu!\, S(r,\nu) q^{k-\nu} \binom{n-j}{n-\nu} \right).$$

It can also be shown that the binomial identities can be derived from the power moment identities.

Since we are interested in solutions for codes, from now on we assume $A_0 = B_0 = 1$.

The following is a list of the first few equations in P_1 for $q = 2$.

$$r = 0 \qquad \sum A_j = 2^k, \qquad (1)$$

$$r = 1 \qquad \sum jA_j = 2^{k-1}n - 2^{k-1}B_1, \qquad (2)$$

$$r = 2 \qquad \sum j^2 A_j = 2^{k-2}n(n+1) - 2!\,2^{k-2}nB_1 + 2!\,2^{k-2}B_2, \qquad (3)$$

$$r = 3 \qquad \sum j^3 A_j = 2^{k-3}(n^3 + 3n^2) - 2^{k-3}(3n^2 + 3n - 2)B_1 + 3!\,2^{k-3}nB_2 - 3!\,2^{k-3}B_3, \qquad (4)$$

$$r = 4 \qquad \sum j^4 A_j = 2^{k-4}(n^4 + 6n^3 + 3n^2 - 2n) - 2^{k-2}(n^3 + 3n^2 - 9n + 7)B_1 + 2^{k-2}(3n^2 + 3n - 4)B_2 - 4!\,2^{k-4}nB_3 + 4!\,2^{k-4}B_4. \qquad (5)$$

The Assmus-Mattson (1961) result on the sum of the squares of the weights in a code can be stated as follows. Consider the matrix M whose rows are all the elements in A over a general $GF(q)$, q a prime. Assmus and Mattson assume no column of M is 0 (this is equivalent to $B_1 = 0$). Let $c_1, \cdots, c_s$ be a largest set of columns of M such that no two are multiples of each other. Let j_i be the number of columns of M which are nonzero multiples of c_i. Then

$$B_2 = \gamma \sum_{i=1}^{s} \binom{j_i}{2}, \qquad n = \sum_{i=1}^{s} j_i.$$

Reprinted with permission from *Inform. Contr.*, vol. 6, pp. 147–152, June 1963.

TABLE 1
WEIGHT DISTRIBUTIONS FOR SOME QUADRATIC RESIDUE CODES

	7	23	31	47	71
A_0	1	1	1	1	1
A_4	7				
A_8		22 × 23	15 × 31		
A_{12}		56 × 23	280 × 31	276 × 47	35 × 71
A_{16}		11 × 23	589 × 31	7590 × 47	2,345 × 71
A_{20}			168 × 31	49588 × 47	186,186 × 71
A_{24}			5 × 31	81720 × 47	4,340,910 × 71
A_{28}				35420 × 47	37,861,505 × 71
A_{32}				3795 × 47	129,893,225 × 71
A_{36}				92 × 47	181,404,764 × 71
A_{40}					103,914,580 × 71
A_{44}					24,093,685 × 71
A_{48}					2,170,455 × 71
A_{52}					71,610 × 71
A_{56}					670 × 71
A_{60}					7 × 71

The equation in P_1 for $r = 2$ is

$$\begin{aligned}\sum i^2 A_i &= q^{k-1}\gamma n + q^{k-2}\gamma^2 n(n-1) + B_2(q^{k-2}2)\\ &= q^{k-1}\gamma n + q^{k-2}\gamma^2 n(n-1) + \gamma q^{k-2}2\sum_{i=1}^{s}\binom{j_i}{2}\\ &= q^{k-1}\gamma n + q^{k-2}\gamma^2 n(n-1) + \gamma q^{k-2}\left(\sum_{i=1}^{s} j_i^2 - \sum_{i=1}^{s} j_i\right)\\ &= q^{k-1}\gamma n - q^{k-2}\gamma^2 n - \gamma q^{k-2}\sum_{i=1}^{s} j_i + q^{k-2}\gamma^2 n^2 + q^{k-2}\gamma\sum_{i=1}^{s} j_i^2\\ &= q^{k-2}\gamma n(q-\gamma-1) + q^{k-2}\gamma^2 n^2 + q^{k-2}\gamma\sum_{i=1}^{s} j_i^2\\ &= \gamma q^{k-2}\left(n^2\gamma + \sum_{i=1}^{s} j_i^2\right)\end{aligned}$$

which is the Assmus-Mattson identity.

III. APPLICATIONS

THEOREM. *A unique solution to P_1 (hence also to B_1, B_2, P_2) exists under the following conditions. Only s A_i's are unknown and B_1, B_2, $\cdots$, B_{s-1} are known.*

PROOF: The first s equations in P_1 are s equations in s unknowns, whose coefficient matrix is van der Monde. Hence there is a unique solution for the unknown A_i's in the first s equations and each additional equation brings in exactly one additional B_i which can be uniquely solved for.

As an application of the preceding theorem consider the quadratic residue codes, for $q = 2$, of dimension $(p - 1)/2$, $p \equiv -1$ modulo 8. Let A denote such a code and B denote its orthogonal. Mattson-Solomon (1961) give the following bound on the minimum odd weight m in B, $m^2 - m + 1 \geqq p$. By remark (e) we know that the minimum weight in A is $m + 1$. For $p = 7, 23, 31$, or 47 this makes enough B_i's zero so that the conditions of the theorem are fulfilled and the unique solution is precisely the known weight distribution of these codes. Since Muir and Metzler (1930) give simple expressions for van der Monde determinants and also for determinants which are close to van der Monde, the computation of the weights is not arduous.

As an example consider the $p = 23$ case. $m = 7$ so that the only nonzero A_i possible are A_8, A_{12}, and A_{16}. The first three equations in P_1 are

$$\begin{aligned}A_8 + A_{12} + A_{16} &= 23 \times 89,\\ 8A_8 + 12A_{12} + 16A_{16} &= 2^{10} \times 23,\\ 8^2A_8 + 12^2A_{12} + 16^2A_{16} &= 2^9 \times 23 \times 24.\end{aligned}$$

Hence $A_8 = 22 \times 23$, $A_{12} = 56 \times 23$, $A_{16} = 11 \times 23$.

When $p = 71$, the general known properties of quadratic residue codes together with the power moment identities do not yield a unique weight distribution for A, but do eliminate all except four possibilities. The correct weight distribution was then determined by examining sets of vectors in the code. All computations were done by hand. This new result is listed along with the cases $p = 7, 23, 31, 47$. These cases had been known previously, the calculation for $p = 47$ being due to Gleason (1961).

ACKNOWLEDGMENTS

I wish to thank John Pierce for aid in solving P_1 and Eugene Prange for helpful suggestions and illuminating discussions.

RECEIVED: September 28, 1962; revised January 16, 1963

REFERENCES

ASSMUS, E. F., AND MATTSON, H. F. (1961), Error correcting codes: An axiomatic approach. ARM No. 269, Applied Research Laboratory, Sylvania Electronic Systems, Waltham, Mass.
GLEASON, A. M. (1961–1962), private communications.
JORDAN, C. (1950), "Calculus of Finite Differences," 2nd ed. Chelsea, New York.
MACWILLIAMS, F. J. (1962), A theorem on the distribution of weights in a systematic code. *Bell System Tech., J.* **42**, 79–94.
MATTSON, H. F., AND SOLOMON, G. (1961), A new treatment of Bose-Chaudhuri codes. *J. Soc. Ind. Appl. Math.* **9**, 654–669.
MUIR, T., AND METZLER, W. (1930), "Theory of Determinants." Privately published, Albany, N. Y.
PRANGE, E. (1961–1962), private communications.
ZEILER, N. (1962), A note on the mean square weight for group codes. *Inform. and Control* **5**, 87–89.

Weight Distributions of Bose-Chaudhuri-Hocquenghem Codes

TADAO KASAMI[1], *Osaka University*

1. INTRODUCTION

The weight distribution problem of a code is to find the number of code vectors of each weight in the code. The weight distribution is one of the important properties of the structure of a code and gives the complete information on the probability of an undetected error when the code is used for error detection only.

To the author's knowledge, explicit formulas of weight distribution have been known only for the Hamming codes [13] and the Reed-Solomon codes [1, 3, 8]. The weight distributions of all cyclic codes of length 31 were computed by Prange [16] and a number of weight distributions for BCH codes and their dual codes of length 63 to 1023 found by digital computation have been tabulated by Peterson [14].

In this paper, several methods useful for finding the weight distributions of binary Bose-Chaudhuri-Hocquenghem codes [2] (BCH codes) of length $2^m - 1$ are presented. Explicit weight distribution formulas for several classes of BCH codes and some other cyclic codes are derived.

Let C be a binary linear code of length n and let k denote the number of information digits. Let a_j denote the number of vectors of weight j in C and b_j denote the number of vectors of weight j in the dual code of C. A series of identities by which each a_j can be calculated from the b_j's has been given by MacWilliams [9]. Thus it is enough to consider the case where $k \leq n - k$. The following power moment identities have been derived from MacWilliams identities by Pless [15].

$$(1) \qquad \sum_{j=0}^{n} j^l a_j = \sum_{j=0}^{n} (-1)^j b_j \left(\sum_{\nu=0}^{l} \nu! G_l^\nu 2^{k-\nu} \binom{n-j}{n-\nu} \right),$$

where G_l^ν is a Stirling number of the second kind [5].

Simple formulas for even j,

$$(2) \qquad j a_j = (n + 1 - j) a_{n+1-j}$$

$$(3) \qquad j b_j = (n + 1 - j) b_{n+1-j}$$

have been proved to hold for the BCH codes of length $2^m - 1$ by Prange and Peterson [14] as a simple consequence of the fact that it is possible to extend these codes by adding one more check digit in such a way that the extended code is invariant under a doubly transitive group of permutations on the components of a code vector.

For odd d, by a d-BCH code is meant a binary BCH code of length $2^m - 1$ which has $\beta, \beta^2, \ldots, \beta^{d-1}$ but not β^d as roots of its generator polynomial, where β is a primitive element of $GF(2^m)$. For even d, let a d-BCH code be a code consisting of the code vectors of even weight in a $(d-1)$-BCH code. A t-error-correcting BCH code [2] is a $(2t+1)$-BCH code.

A cyclic code of length $2^m - 1$ can be derived from the νth order Reed-Muller code of length 2^m [12] by deleting the first component of each code vector and permuting the remaining components suitably. The resulting code will be called the νth order modified Reed-Muller code. This code has been proved to be a subcode of a $(2^{m-\nu} - 1)$-BCH code [7, 8]. It is shown in section 3 that the possible values of weights of code vectors of the second order Reed-Muller code are very sparse. By using this fact as well as the power moment identities and the invariant property, explicit weight distribution formulas are obtained for the following subcodes of the second order modified Reed-Muller code: the dual code of every double-error-correcting BCH code, the dual code of triple-error-correcting BCH code for any odd $m \geq 5$ and several even m's, $(2^{m-1} - 2^{m/2-1})$-BCH codes for even $m \geq 4$, $(2^{m-1} - 2^{m/2})$-BCH codes for even $m \geq 4$, $(2^{m-1} - 2^{(m-1)/2})$-BCH codes for odd $m \geq 3$, $(2^{m-1} - 2^{(m+1)/2})$-BCH codes for odd $m \geq 5$, $(2^{m-1} - 2^{(m+3)/2})$-BCH codes for odd $m \geq 11$ and some other classes of cyclic codes.

2. INVARIANT PROPERTIES

Let $n = 2^m - 1$. The extended code of C is the code with an overall parity check added to C as the first digit. The first component in a code vector is numbered 0, and for $i > 1$ the ith component is numbered α^{i-2}, where α is a primitive element of $GF(2^m)$. Let v be a vector of the extended code. For a ($\neq 0$) and b in $GF(2^m)$, permute the component of v in position X to position $aX + b$. Then, the resulting vector will be denoted by $\pi_{ab} v$. If the extended code of C is invariant under the doubly transitive group of permutations $\Pi = \{\pi_{ab} \mid a \neq 0, b \in GF(2^m)\}$, then C is a cyclic code by definition. Peterson [14] proved that the extended codes of $(2t+1)$-BCH codes are invariant under permutation group Π.

Let i be a positive integer less than 2^m. Then i can be expressed in binary form:

$$i = \sum_{j=0}^{m-1} \delta_j 2^j .$$

Let $I(i)$ denote the set of all nonzero integers i' such that

$$i' = \sum_{j=0}^{m-1} \delta'_j 2^j ,$$

where $0 \leq \delta'_j \leq \delta_j$ for $0 \leq j < m$.

Theorem 1 [7, 8]. *Let C be a cyclic code of length $2^m - 1$ generated by a polynomial $g(X)$. The extended code of C is invariant under the permutation group Π if and only if* (1) $g(1) \neq 0$ *and* (2), *for every root α^i of $g(X)$, $g(\alpha^{i'}) = 0$ for i' in $I(i)$.*

Let C_0 be a cyclic code of length $2^m - 1$ generated by $g(X) = (X^{2^m-1} - 1)/(h_0(X) \ldots h_p(X))$, where $h_0(X) = X - 1$, $h_i(X)$ is an irreducible polynomial of degree m_i and $h_i(\alpha^{j_i}) = 0$ $(0 \leq i \leq p)$. Suppose that $g(X)$ satisfies the condition of Theorem 1. Let $v(X)$ be the polynomial representation [1] of a code vector of C_0. If $g(\alpha^j) = 0$, then $v(\alpha^j) = 0$. Obviously, $v(\alpha^{j_i}) \in GF(2^{m_i})$ $(0 \leq i \leq p)$. Conversely, for any set of β_i in $GF(2^{m_i})$ $(0 \leq i \leq p)$, there exists a unique code vector[2] $v(\beta_0, \ldots, \beta_p : X)$ in C_0 such that $v(\beta_0, \ldots, \beta_p : \alpha^{j_i}) = \beta_i$ $(0 \leq i \leq p)$ (Mattson and Solomon [10]). Let $\bar{v}(\beta_0, \ldots, \beta_p)$ denote the vector with an overall parity added to code vector $v(\beta_0, \ldots, \beta_p : X)$ as the first component. $\bar{v}(\beta_0, \ldots, \beta_p)$ is a vector of the extended code C_{ex} of C_0. Let $X_1, \ldots, X_w$ be the location numbers of nonzero components of $\bar{v}(\beta_0, \ldots, \beta_p)$. By definition, w is an even integer and

$$\sum_{f=1}^{w} X_f^{j_i} = v(\beta_0, \ldots, \beta_p : \alpha^{j_i}) = \beta_i , \qquad (1 \leq i \leq p) .$$

[1] This work was supported in part by the Joint Services Electronics Program (U. S. Army, U. S. Navy, and U. S. Air Force) under Contract No. DA 28 043 AMC 00073(E), and in part by the National Science Foundation under Grant NSF GK-690, and in part by the Air Force Cambridge Research Laboratories under Contract AF 19(628)4379.

[2] Vector $v(X)$ means the vector represented by polynomial $v(X)$.

Reprinted with permission from *Proc. Conf. Combinatorial Mathematics and Its Applications,* R. C. Bose and T. A. Dowling, Eds. Chapel Hill, N.C.: University of North Carolina Press, 1968.

If $g(\alpha^l) = 0$, then

(4) $$\sum_{f=1}^{w} X_f^l = 0 .$$

Otherwise, $h_q(\alpha^l) = 0$ for some q and, consequently, $l \equiv j_q 2^\nu$ (mod $2^{m_q} - 1$) for some $0 \le \nu < m_q$. Hence,

(5) $$\sum_{f=1}^{w} X_f^l = \beta_q^{2^\nu} .$$

For any a ($\neq 0$) and b in $GF(2^m)$, there exists $\bar{v}(\beta_0', \dots, \beta_p')$ in C_{ex} such that

$$\bar{v}(\beta_0', \dots, \beta_p') = \pi_{ab}\bar{v}(\beta_0, \dots, \beta_p) .$$

By definition,

(6) $$\beta_i' = \sum_{f=1}^{w} (aX_f + b)^{j_i} , \qquad (1 \le i \le p) .$$

If

$$j = 2^{\sigma_1} + 2^{\sigma_2} + \dots + 2^{\sigma_t} \qquad (0 \le \sigma_1 < \sigma_2 < \dots < \sigma_t < m) ,$$

(7) $$(aX_f + b)^j = (a^{2^{\sigma_1}} X_f^{2^{\sigma_1}} + b^{2^{\sigma_1}}) \dots (a^{2^{\sigma_t}} X_f^{2^{\sigma_t}} + b^{2^{\sigma_t}}) = \sum_{l \in I(j)} a^l X_f^l b^{j-l} .$$

It follows from (4) through (7) that

(8) $$\beta_i' = \sum_{q=1}^{p} \sum_{\nu \in E_{iq}} a^{j_q 2^\nu} \beta_q^{2^\nu} b^{j_i - j_q 2^\nu} , \qquad (1 \le i \le p)$$

where E_{iq} is the set of integer ν's such that the remainder of $j_q 2^\nu/(2^m - 1)$ is in $I(j_i)$ and that $0 \le \nu < m_q$.

Lemma 2. *Assume that, for given β_i and β_i' in $GF(2^{m_i})$ $(1 \le i \le p)$, there are a ($\neq 0$) and b in $GF(2^m)$ which satisfy* (8). *Then, if the weight of $v(0, \beta_1, \dots, \beta_p\colon X)$ is w, the weight of $v(0, \beta_1', \dots, \beta_p'\colon X)$ is either w or $n + 1 - w$.*

Proof. It follows from the assumption that there exists β_0' in $GF(2)$ such that $\bar{v}(\beta_0', \dots, \beta_p') = \pi_{ab}\bar{v}(0, \beta_1, \dots, \beta_p)$. Obviously, the weights of $\bar{v}(0, \beta_1, \dots, \beta_p)$ and $\bar{v}(\beta_0', \dots, \beta_p')$ are equal to w. If $\beta_0' = 0$, the weight of $v(0, \beta_1', \dots, \beta_p'\colon X)$ is w. If $\beta_0' = 1$, the weight of $v(1, \beta_1', \dots, \beta_p'\colon X)$ is $w - 1$ by definition. C_0 contains the all-one vector $e(X) = 1 + X + \dots + X^{2^m - 2}$. Since $e(\alpha^j) = \sum_{f=0}^{2^m-2} \alpha^{jf} = 0$ $(0 < j < 2^m - 1)$, $v(0, \beta_1', \dots, \beta_p'\colon X) = v(1, \beta_1', \dots, \beta_p'\colon X) + e(X)$. Therefore, the weight of $v(0, \beta_1', \dots, \beta_p'\colon X)$ is $n + 1 - w$.

Since C_0 contains the all-one vector $\boldsymbol{e} = (1, \dots, 1)$,

$$a_{n-j} = a_j , \qquad \text{for any } j.$$

Consequently, it is enough to consider the code C consisting of all the code vectors of even weight in C_0. In C, $\beta_0 = 0$. Since the symmetry property (2) holds for C_0 by Prange's Theorem [14, 16], it also holds for C. Hence, it is sufficient to find $a_j + a_{n+1-j}$ for even j $(0 < j \le (n+1)/2)$. Thus the following power moments are convenient.

$$I_l = \sum_{j \neq 0} (j - [(n+1)/2])^l a_j ,$$

where $[x]$ denotes the integer part of x.

If n is odd and $b_1 = b_2 = 0$, then

(9) $$I_2 = 2^{k-2}(n+1) - 2^{-2}(n+1)^2 ,$$

(10) $$I_4 = 2^{k-4}[3(n+1)^2 - 2(n+1)] - 2^{-4}(n+1)^4 + 3 \cdot 2^{k-1}(b_3 + b_4) .$$

If n is odd and $b_i = 0$ $(1 \le i \le 4)$,

(11) $$I_6 = 15 \cdot 2^{k-6}[(n+1)^3 - 2(n+1)^2] + 2^{k-2}(n+1) - 2^{-6}(n+1)^6 + 6!\,2^{k-6}(b_5 + b_6) .$$

The proof of (9), (10) and (11) is given in Appendix 1.

3. MODIFIED REED-MULLER CODES

Let V_j denote a j-dimensional vector space over $GF(2)$ and x_i $(1 \le i \le m)$ be a variable over $GF(2)$. For $1 \le \nu \le m$, let P_ν be the set of polynomials over $GF(2)$ of variables $x_1, \dots, x_m$ of degree ν or less. For $0 \le j < 2^m - 1$, let

$$\alpha^j = \sum_{i=0}^{m-1} v_{ji}\alpha^i , \qquad v_{ji} \in GF(2) .$$

For $f(x_1, \dots, x_m) \in P_\nu$, let $v(f)$ denote a vector in V_{2^m} of which the first component is $f(0, \dots, 0)$ and the jth component $(j > 1)$ is $f(v_{j-2\,0}, v_{j-2\,1}, \dots, v_{j-2\,m-1})$. Then the νth order Reed-Muller code of length 2^m is the set of vectors $\{v(f) \mid f \in P_\nu\}$.[3] Delete the first component of each vector of the νth order Reed-Muller code of length 2^m. Then the resulting set of vectors in $V_{2^m - 1}$ will be called the νth order modified Reed-Muller code. Let

$$y_j = u_{j0} + \sum_{i=1}^{m} u_{ji}x_i \in P_1 \qquad (1 \le j \le m) .$$

If vectors $(u_{j1}, u_{j2}, \dots, u_{jm})$ $(1 \le j \le l)$ are linearly independent, $y_1, \dots, y_l$ will be said to be independent. For $f(x_1, \dots, x_m) \in P_\nu$, there is f' in P_ν such that $f(y_1, \dots, y_m) = f'(x_1, \dots, x_m)$. Therefore, if $v(f)$ is a code vector of the νth order Reed-Muller code, then $v(f')$ is also a code vector. It follows from this fact that a modified Reed-Muller code is cyclic [7, 8]. Let $w(j)$ denote the number of ones in the binary expression of j.

Theorem 3 [7, 8]. *Let $g(X)$ be the generator polynomial of the νth order modified Reed-Muller code of length $2^m - 1$. Then α^j is a root of $g(X)$ if and only if $0 < w(j) < m - \nu$.*

This theorem implies that the νth order modified Reed-Muller code is a subcode of a $(2^{m-\nu} - 1)$-BCH code.

For the polynomial $f(x_1, \dots, x_m)$, let $|f|_m$ denote the number of m-tuple's $(v_1, \dots, v_m)$ such that

$$f(v_1, \dots, v_m) = 1 .$$

By definition, $|f|_m$ is the weight of vector $v(f)$. If the y_j's $(1 \le j \le m)$ in P_1 are independent and $f'(x_1, \dots, x_m) = f(y_1, \dots, y_m)$, then

(12) $$|f'|_m = |f|_m .$$

This follows from the fact that $y_j = u_{j0} + \sum_{i=1}^{m} u_{ji}x_i$ $(1 \le j \le m)$ defines a one-to-one mapping from V_m onto itself.

Lemma 4. *Assume that* (1) $m \ge 2$, (2) $f(x_1, \dots, x_m) \in P_2$, (3) f *does not depend on x_i $(i < i_0 \le m)$ but on x_{i_0}. Then there exist independent $y_j^{(i)}$'s $(1 \le i \le t;\ 1 \le j \le l_i)$ in P_1 such that*

(1) $$y_1^{(1)} = x_{i_0}$$

(2) $$f(x_1, \dots, x_m) = u_0 + \sum_{i=1}^{t} \Big(\sum_{j=1}^{l_i - 1} y_j^{(i)} y_{j+1}^{(i)} + u_i y_{l_i}^{(i)} \Big) ,$$

where $u_i \in GF(2)$ $(0 \le i \le t)$.

Proof. If $m = 2$, it is easy to check that this lemma holds. Suppose that this lemma holds for $2 \le m < m'$. Consider the case of $m = m'$. Let

(13) $$f(x_1, \dots, x_m) = F_0(x_2, \dots, x_m) + x_1F_1(x_2, \dots, x_m) ,$$

where $F_0 \in P_2$ and $F_1 \in P_1$. If $F_1 = 1$,

$$f(x_1, \dots, x_m) = x_1 + F_0(x_2, \dots, x_m) .$$

Then apply the induction hypothesis to $F_0(x_2, \dots, x_m)$. Suppose

[3] The order of the digit positions is different from the original one [12, 13].

that F_1 is not a constant. Since x_1 and $F_1(x_2, \ldots, x_m)$ are independent, there exist independent $y_1, \ldots, y_m$ such that

(1) $y_1 = x_1$,

(2) $y_2 = F_1(x_2, \ldots, x_m)$,

(3) $y_2, \ldots, y_m$ are polynomials of $x_2, \ldots, x_m$ of the first degree.

Then it follows from (13) that

$$f(x_1, \ldots, x_m) = y_1 y_2 + f'(y_2, \ldots, y_m),$$

where $f'(y_2, \ldots, y_m) = F_0(x_2, \ldots, x_m)$. Now apply the induction hypothesis to $f'(y_2, \ldots, y_m)$.

Let $G_0(l)$ and $G_1(l)$ be defined by the following:

$$G_0(l) = |x_1x_2 + x_2x_3 + \ldots + x_{l-1}x_l|_l, \quad l \geq 2$$

$$G_0(1) = 0,$$

$$G_1(l) = |x_1x_2 + x_2x_3 + \ldots + x_{l-1}x_l + x_l|_l, \quad l \geq 1.$$

Note that

(14) $$|f(x_1, \ldots, x_m)|_m = |f(x_1, \ldots, x_{m-1}, 0)|_{m-1} + |f(x_1, \ldots, x_{m-1}, 1)|_{m-1}$$

(15) $$|1 + f(x_1, \ldots, x_m)|_m = 2^m - |f(x_1, \ldots, x_m)|_m.$$

It is easy to check that, for $l \geq 2$,

(16) $$\begin{aligned} G_0(l) &= |x_1x_2 + \ldots + x_{l-2}x_{l-1}|_{l-1} \\ &\quad + |x_1x_2 + \ldots + x_{l-2}x_{l-1} + x_{l-1}|_{l-1} \\ &= G_0(l-1) + G_1(l-1), \end{aligned}$$

(17) $$\begin{aligned} G_1(l) &= |x_1x_2 + \ldots + x_{l-2}x_{l-1}|_{l-1} \\ &\quad + |x_1x_2 + \ldots + x_{l-2}x_{l-1} + x_{l-1} + 1|_{l-1} \\ &= G_0(l-1) + 2^{l-1} - G_1(l-1). \end{aligned}$$

By (16) and (17), for $l \geq 3$

$$\begin{aligned} G_0(l) &= G_0(l-2) + G_1(l-2) \\ &\quad + G_0(l-2) + 2^{l-2} - G_1(l-2) \\ &= 2G_0(l-2) + 2^{l-2}, \end{aligned}$$

$$\begin{aligned} G_1(l) &= G_0(l-2) + G_1(l-2) + 2^{l-1} \\ &\quad - (G_0(l-2) + 2^{l-2} - G_1(l-2)) \\ &= 2G_1(l-2) + 2^{l-2}. \end{aligned}$$

Hence,

(18) $$2^{l-1} - G_i(l) = 2(2^{l-3} - G_i(l-2)), \quad i = 0, 1.$$

On the other hand, it is easy to check that

$$G_i(2) = 1, \quad i = 0, 1,$$

$$G_0(1) = 0,$$

$$G_1(1) = 1.$$

Therefore, it follows from (18) that, for even $l \geq 2$,

(19) $$G_0(l) = G_1(l) = 2^{l-1} - 2^{l/2-1},$$

and that, for odd $l \geq 1$,

(20) $$G_0(l) = 2^{l-1} - 2^{(l-1)/2},$$

(21) $$G_1(l) = 2^{l-1}.$$

Lemma 5. *Suppose that $m \geq 2$ and $f(x_1, \ldots, x_m) \in P_2$. Then $|f|_m$ is of the form:*

$$2^{m-1} + \varepsilon 2^l,$$

where $m/2 - 1 \leq l \leq m - 1$ and ε is either 0, 1 or -1.

Proof. If $m = 2$, this lemma is obvious. Assume that this lemma holds for $2 \leq m < m'$ and consider the case of $m = m'$. By Lemma 4, there exist independent $y_1, \ldots, y_m$ in P_1 such that, for some h $(1 \leq h \leq m)$,

$$f(x_1, \ldots, x_m) = f_0(y_1, \ldots, y_h) + f_1(y_{h+1}, \ldots, y_m),$$

$$f_0(y_1, \ldots, y_h) = y_1y_2 + y_2y_3 + \ldots + y_{h-1}y_h + uy_h,$$

where $u \in GF(2)$ and, if $h = m$, f_1 is a constant. If $h = m$, this lemma follows from (15), (19), (20) and (21). Otherwise, it follows from the induction hypothesis that

(22) $$|f_0|_h = 2^{h-1} + \varepsilon_0 2^{l_0},$$

(23) $$|f_1|_{m-h} = 2^{m-h-1} + \varepsilon_1 2^{l_1},$$

where ε_i $(i = 0, 1)$ is either 0, 1 or -1 and

(24) $$h/2 - 1 \leq l_0 \leq h - 1$$

(25) $$(m-h)/2 - 1 \leq l_1 \leq (m-h) - 1.$$

It is easy to check that

$$|f|_m = |f_0|_h(2^{m-h} - |f_1|_{m-h}) + |f_1|_{m-h}(2^h - |f_0|_m).$$

By (22) and (23),

$$\begin{aligned} |f|_m &= (2^{h-1} + \varepsilon_0 2^{l_0})(2^{m-h-1} - \varepsilon_1 2^{l_1}) \\ &\quad + (2^{h-1} - \varepsilon_0 2^{l_0})(2^{m-h-1} + \varepsilon_1 2^{l_1}) \\ &= 2^{m-h-1} - \varepsilon_0\varepsilon_1 2^{l_0+l_1+1}. \end{aligned}$$

By (24) and (25),

$$m/2 - 1 \leq l_0 + l_1 + 1 \leq m - 1.$$

Since $\varepsilon_0\varepsilon_1$ is either 0, 1 or -1, the lemma holds.

The following theorem follows from the definition of Reed-Muller codes and Lemma 5.

Theorem 6. *The weight of a code vector of the second order Reed-Muller code of length 2^m is of the form*

$$2^{m-1} + \varepsilon 2^l,$$

where $m/2 - 1 \leq l \leq m - 1$ and ε is either 0, 1 or -1.

4. SUBCODES OF THE SECOND-ORDER MODIFIED REED-MULLER CODES

In what follows, the weight distributions of subcodes of the second order modified Reed-Muller code will be considered.

Lemma 7. *A cyclic code C with overall parity check is a subcode of the second-order modified Reed-Muller code of length $2^m - 1$, if and only if the generator $g(X)$ is of the form*

$$g(X) = (X^{2^m-1} - 1)/(h_1(X) \ldots h_p(X)),$$

where $h_1(X), \ldots, h_p(X)$ are different irreducible polynomials and there are integers μ_i $(1 \leq i \leq p)$ such that

$$0 \leq \mu_1 < \mu_2 < \ldots < \mu_p \leq m/2,$$

$$h_i(\alpha^{-2^{\mu_i}-1}) = 0.$$

Proof. It follows from Theorem 3 that if $h_i(\alpha^j) = 0$ $(0 < j < 2^m - 1)$, then $m - 2 \leq w(j) < m$. Hence, $j = 2^m - 1 - j'$, where $1 \leq w(j') \leq 2$. If $w(j') = 1$, let $\mu_i = 0$.

If $\mu_1 = 0$, the extended code of C is invariant under permutation group Π by Theorem 1 and, consequently, the symmetry properties (2) and (3) hold for C. If $\mu_i = m/2$, the degree of $h_i(X)$ is $m/2$. Otherwise, the degree of $h_i(X)$ is m. (Refer to [6].) Hence, if $\mu_p = m/2$, then $k = (2p-1)m/2$, and otherwise $k = pm$.

Theorem 8. *Suppose that code C satisfies the condition of Lemma 7 and that $p \geq 2$. If $\mu_1 = 0$, let $\mu = \mu_2$. Otherwise, let $\mu = \mu_1$. Then the weight of a nonzero code vector of C is of the form*

$$2^{m-1} + \varepsilon 2^l ,$$

where $m/2 - 1 \leq l \leq m - 1 - \mu$ and ε is either 0, 1 or -1.

Proof. Let C_0 be the cyclic code of length $2^m - 1$ generated by $g(X)/(X-1)$. Then C is the set of the code vectors of even weight in C_0. It is easy to check that, for $1 \leq \mu_i \leq m/2$, $2^{m-1} - 2^{m-1-\mu_i} - 1$ is the smallest among the positive exponents of the roots of $h_i(X)$. Hence, the minimum distance of C_0 is at least $2^{m-1} - 2^{m-1-\mu} - 1$ by the BCH bound [2]. Since C_0 contains the all-one vector $\boldsymbol{e}$, there is no code vector of weight j with $2^{m-1} + 2^{m-1-\mu} < j < 2^m - 1$. Thus, this theorem follows from Theorem 6 and Lemma 7.

For $0 \leq i \leq [(m-1)/2]$, let

$$\bar{a}_i = a_{2^{m-1}-2^{[(m-1)/2]+i}} + a_{2^{m-1}+2^{[(m-1)/2]+i}} . \tag{26}$$

From Theorem 8, it follows that, for even l,

$$I_l = \sum_{i=0}^{[m/2]-\mu} 2^{l[(m-1)/2]+li}\bar{a}_i . \tag{27}$$

Lemma 9. *Let $m \geq 3$. If $\mu_1 = 0$, $\mu_i = [m/2] - p + i$ $(2 \leq i \leq p)$ and $[m/2] - [m/3] + 2 \geq p$, then the code C is a $(2^{m-1} - 2^{m-[m/2]+p-3})$-BCH code.*

Proof. For $0 < j < 2^m - 1$, let $j_{\min}$ denote the smallest exponent of the roots of the minimum polynomial of α^j. If $g(\alpha^j) = 0$ and $w(j) \leq m - 3$, then $j_{\min} \leq 2^m - 1 - (2^{m-1} + 2^{m-[m/3]-1} + 2^{[m/3]-1})$. Hence,

$$j_{\min} < 2^{m-1} - 2^{m-[m/3]-1} - 1 \leq 2^{m-1} - 2^{m-[m/2]+p-3} - 1 .$$

If $g(\alpha^j) = 0$ and $w(j) = m - 2$, then

$$j_{\min} \leq 2^m - 1 - (2^{m-1} + 2^{m-1-[m/2]+p-1}) < 2^{m-1} - 2^{m-[m/2]+p-3} - 1 .$$

On the other hand, if $g(\alpha^j) \neq 0$,

$$j_{\min} \geq 2^m - 1 - (2^{m-1} + 2^{m-[m/2]+p-3}) = 2^{m-1} - 2^{m-[m/2]+p-3} - 1 .$$

Thus, this lemma follows from the definition of $(2^{m-1} - 2^{m-[m/2]+p-3})$-BCH codes.

It is easy to check that $b_1 + b_2 = 0$, if and only if the exponent of $g(X)$ is equal to $2^m - 1$. Hereafter, this condition will be assumed. Let $(l_1, \ldots, l_f)$ denote the greatest common divisor of $l_1, \ldots$ and l_f.

Lemma 10. *Let $\mu_1 = 0$. Then $b_3 + b_4 \neq 0$, if and only if $(m, \mu_2, \ldots, \mu_p) > 1$.*

Proof. From (3)

$$4b_4 = (2^m - 4)b_{2^m-4} .$$

Since the dual code of C contains the all-one vector $(1, \ldots, 1)$,

$$b_{2^m-1-3} = b_3 .$$

Hence, $b_3 + b_4 \neq 0$ if and only if $b_3 \neq 0$. Assume that α^{j_1}, α^{j_2} and α^{j_3} are the location numbers of non-zero components of a code vector of weight 3 in C. Then,

$$\alpha^{j_1} + \alpha^{j_2} = \alpha^{j_3} , \tag{28}$$

$$\alpha^{j_1(2^{\mu_i}+1)} + \alpha^{j_2(2^{\mu_i}+1)} = \alpha^{j_3(2^{\mu_i}+1)} , \qquad (1 < i \leq p) . \tag{29}$$

From (28),

$$\begin{aligned} \alpha^{j_3(2^{\mu_i}+1)} &= (\alpha^{j_1} + \alpha^{j_2})^{2^{\mu_i}+1} \\ &= \alpha^{j_1(2^{\mu_i}+1)} + \alpha^{j_1 2^{\mu_i}}\alpha^{j_2} + \alpha^{j_1}\alpha^{j_2 2^{\mu_i}} + \alpha^{j_2(2^{\mu_i}+1)} . \end{aligned} \tag{30}$$

By subtracting (29) from (30),

$$\alpha^{j_1 2^{\mu_i}}\alpha^{j_2} + \alpha^{j_1}\alpha^{j_2 2^{\mu_i}} = 0 ,$$

$$\alpha^{(j_1-j_2)(2^{\mu_i}-1)} = 1 .$$

Thus, for $1 < i \leq p$,

$$(j_1 - j_2)(2^{\mu_i} - 1) \equiv 0 \pmod{2^m - 1} .$$

Since $j_1 - j_2 \not\equiv 0 \pmod{2^m - 1}$, the "only if part" of the lemma follows. The converse can be proved similarly.

5. WEIGHT DISTRIBUTION FORMULAS

Several cases will be considered in detail.

(a) $p = 2$ and $\mu_1 = 0$.

If $\mu_2 = m/2$, then $k = 3m/2$, and otherwise, $k = 2m$. For examples, $(2^{m-1} - 2^{m-[m/2]-1})$-BCH codes and the duals of double-error-correcting BCH codes belong to this case. Since the order of the permutation group Π is $2^m(2^m-1)$ and the number of code vectors is 2^{2m} or $2^{3m/2}$, Lemma 2 is very useful. By using Lemma 2 and power moment identities (9) and (10), the weight distribution formula is derived for any μ_2.

Theorem 11. *Let $p = 2$ and $\mu_1 = 0$.*

(1) *If $(m, \mu_2) = (m, 2\mu_2) = c$, then*

$$a_{2^{m-1}\pm 2^{(m+c)/2-1}} = (2^{m-c-1} \mp 2^{(m-c)/2-1})(2^m - 1) ,$$

$$a_{2^{m-1}} = (2^m - 2^{m-c} + 1)(2^m - 1) ,$$

$$a_j = 0 , \qquad \textit{for other nonzero } j.$$

(2) *If $2(m, \mu_2) = (m, 2\mu_2) = c$ and $c \neq m$, then*

$$a_{2^{m-1}\pm 2^{(m+c)/2-1}} = 2^{(m-c)/2-1}(2^{(m-c)/2} \mp 1)(2^m - 1)/(2^{c/2} + 1) ,$$

$$a_{2^{m-1}\pm 2^{m/2-1}} = 2^{(m+c)/2-1}(2^{m/2} \mp 1)(2^m - 1)/(2^{c/2} + 1) ,$$

$$a_{2^{m-1}} = ((2^{c/2} - 1)2^{m-c} + 1)(2^m - 1) ,$$

$$a_j = 0 , \qquad \textit{for other nonzero } j.$$

(3) *If $2(m, \mu_2) = (m, 2\mu_2) = m$, then*

$$a_{2^{m-1}\pm 2^{m/2-1}} = (2^{m-1} \mp 2^{m/2-1})(2^{m/2} - 1) ,$$

$$a_{2^{m-1}} = 2^m - 1 ,$$

$$a_j = 0 , \qquad \textit{for other nonzero } j.$$

The proof is given in [6].

The following theorem is due to Pless [15]:

Theorem 12. *If only u a_j's are unknown, and $b_1, b_2, \ldots, b_{u-1}$ are known, then a unique solution to (1) exists.*

(b) m = odd and $k = 2m$.

Theorem 13. *Let C be any binary linear code for which $b_1 = b_2 = 0$, $n = 2^m - 1$ and $k = 2m$, where m is an odd integer.*

(i) *Let j_0 denote the smallest j such that*

$$a_j + a_{2^m-j} \neq 0\,, \qquad 0 < j < 2^{m-1}\,.$$

Then,

$$j_0 \leq 2^{m-1} - 2^{(m-1)/2}\,.$$

If j_0 is identical with the upperbound $2^{m-1} - 2^{(m-1)/2}$, the weight distribution is the same as the weight distribution of the dual code of a double-error-correcting BCH code:

$$a_{2^{m-1}\pm 2^{(m-1)/2}} = (2^{m-2} \mp 2^{(m-3)/2})(2^m - 1)\,,$$

$$a_{2^{m-1}} = (2^{m-1} + 1)(2^m - 1)\,,$$

$$a_j = 0\,, \qquad \textit{for other nonzero } j.$$

(ii) *If C is a subcode of the second order modified Reed-Muller code for which $b_3 = b_4 = 0$, C has the weight distribution mentioned above.*

Proof. By (9) and (10),

$$I_2 = 2^{2m-2}(2^m - 1)\,,$$

$$I_4 = 2^{3m-3}(2^m - 1) + 3\cdot 2^{2m-1}(b_3 + b_4)\,.$$

Thus,

$$(31) \qquad I_4 - 2^{m-1}I_2 = \sum_{j_0 \leq j < 2^{m-1}} (2^{m-1} - j)^2[(2^{m-1} - j)^2 - 2^{m-1}](a_j + a_{2^m-j}) = 3\cdot 2^{2m-1}(b_3 + b_4)\,.$$

Hence,

$$(32) \qquad (2^{m-1} - j_0)^2 \geq 2^{m-1}\,.$$

If $j_0 = 2^{m-1} - 2^{(m-1)/2}$, it follows from (31) that for $j_0 < j < 2^{m-1}$,

$$a_j + a_{2^m-j} = 0$$

and that

$$b_3 + b_4 = 0\,.$$

Since only $a_{2^{m-1}\pm 2^{(m-1)/2}}$ and $a_{2^{m-1}}$ are unknown, part (i) follows from Theorem 12. The weight distribution of the dual code of a double-error-correcting BCH code is given by letting $\mu_2 = 1$ in Theorem 11 (1).

Consider part (ii). By Theorem 8, for $2^{m-1} - 2^{(m-1)/2} < j < 2^{m-1}$,

$$a_j + a_{2^m-j} = 0\,.$$

Since $b_3 + b_4 = 0$, for $j \neq 0$, $2^{m-1} \pm 2^{(m-1)/2}$ and 2^{m-1},

$$a_j + a_{2^m-j} = 0\,.$$

Thus part (ii) follows from Theorem 12.

The results in cases (a) and (b) can be applied to the cross-correlation problem of two maximum length sequences [4, 6].

(c) $(2^{m-1} - 2^{m/2})$-BCH codes for even $m \geq 4$.

Let $p = 3$, $\mu_1 = 0$, $\mu_2 = m/2 - 1$ and $\mu_3 = m/2$. Then $k = 5m/2$. Lemma 9 shows that this code is a $(2^{m-1} - 2^{m/2})$-BCH code. Therefore,

$$(33) \qquad \bar{a}_i = 0 \qquad (i \geq 2)\,.$$

Since $m/2 - 1$ is relatively prime to $m/2$, it follows from Lemma 10 that

$$b_3 + b_4 = 0\,.$$

By (9) and (10),

$$I_2 = 2^{2m-2}(2^{3m/2} - 1)\,,$$

$$I_4 = 2^{7m/2-4}(2^{m/2} - 1)(3\cdot 2^{m/2} + 2)\,.$$

By solving (27) for $l = 2$ and 4,

$$\bar{a}_0 = 2^m(2^{m/2} - 1)(2^m + 2^{m/2+1} + 4)/3\,,$$

$$\bar{a}_1 = 2^{m+2}(2^{m/2+1} - 1)(2^m - 1)/3\,.$$

By using (2), the following theorem is obtained.

Theorem 14. *For even $m \geq 4$, a $(2^{m-1} - 2^{m/2})$-BCH code has the following weight distribution:*

$$a_{2^{m-1}\pm 2^{m/2}} = 2^{m/2-2}(2^{m/2-1} \mp 1)(2^{m/2+1} - 1)(2^m - 1)/3$$

$$a_{2^{m-1}\pm 2^{m/2-1}} = 2^{m/2-1}(2^{m/2} \mp 1)(2^{m/2} - 1)(2^m + 2^{m/2+1} + 4)/3\,,$$

$$a_{2^{m-1}} = (2^{m/2} - 1)(2^{2m-1} + 2^{3m/2-2} - 2^{m-2} + 2^{m/2} + 1)\,,$$

$$a_j = 0\,, \qquad \textit{for other nonzero } j.$$

(d) $p = 3$ and $m =$ odd ≥ 5.

In this case, $k = 3m$. By (9), (10) and (11),

$$(34) \qquad I_2 = 2^{2m-2}(2^{2m} - 1)\,,$$

$$(35) \qquad I_4 = 3\cdot 2^{4m-4}(2^m - 1)\,,$$

$$(36) \qquad I_6 = 2^{4m-5}(7\cdot 2^m - 8)(2^m - 1) + 6!\,2^{3m-6}(b_5 + b_6)\,.$$

(d1) Assume that $b_i = 0$ $(1 \leq i \leq 6)$. The dual code of a triple-error-correcting BCH code is this case.

By (27),

$$(37) \qquad 2^{m-1}\bar{a}_0 + 2^{m-1}\sum_{i\geq 1} 2^{2i}\bar{a}_i = 2^{2m-2}(2^{2m} - 1)\,,$$

$$(38) \qquad 2^{2(m-1)}\bar{a}_0 + 2^{2(m-1)}\sum_{i\geq 1} 2^{4i}\bar{a}_i = (3\cdot 2^{4m-4})(2^m - 1)\,,$$

$$(39) \qquad 2^{3(m-1)}\bar{a}_0 + 2^{3(m-1)}\sum_{i\geq 1} 2^{6i}\bar{a}_i = 2^{4m-5}(7\cdot 2^m - 8)(2^m - 1)\,.$$

By subtracting 2^{m-1} times (37) from (38),

$$(40) \qquad 2^{2(m-1)}\sum_{i\geq 1} 2^{2i}(2^{2i} - 1)\bar{a}_i = 2^{3m-3}(2^m - 1)(2^{m-1} - 1)\,.$$

By subtracting 2^{m-1} times (38) from (39),

$$(41) \qquad 2^{3(m-1)}\sum_{i\geq 1} 2^{4i}(2^{2i} - 1)\bar{a}_i = 2^{4m-2}(2^m - 1)(2^{m-1} - 1)\,.$$

By subtracting 2^{m+1} times (40) from (41),

$$\sum_{i\geq 1} (2^{4i} - 2^{2i+2})(2^{2i} - 1)\bar{a}_i = 0\,.$$

Since $(2^{4i} - 2^{2i+2})(2^{2i} - 1) > 0$ for $i > 1$,

$$(42) \qquad \bar{a}_i = 0\,, \qquad (i > 1)\,.$$

From (37) and (38),

$$\bar{a}_0 = 2^{m-1}(2^m - 1)(5\cdot 2^{m-1} + 4)/3\,,$$

$$\bar{a}_1 = a^{m-3}(2^m - 1)(2^{m-1} - 1)/3\,,$$

Since Equation (2) holds for the dual code of a triple-error-correcting BCH code, it follows from Equation (42) and Theorem 12 that Equation (2) holds for other cases. Hence the weight distribution can be calculated easily.

(d2) Now consider the case where $a_j + a_{2^m-j} = 0$ for $0 < j < 2^{m-1} - 2^{(m+1)/2}$ and $b_i = 0$ for $1 \leq i \leq 4$. For example, let $\mu_1 = 0$, $\mu_2 = (m-3)/2$ and $\mu_3 = (m-1)/2$. Lemma 9 shows that this code is a $(2^{m-1} - 2^{(m+1)/2})$-BCH code. Since $(m-1)/2$ is relatively prime to m, it follows from Lemma 10 that $b_i = 0$ for $1 \leq i \leq 4$.

Theorem 12 and Equation (42) imply that the weight distribution for case **(d2)** is the same as the one for case **(d1)**. Consequently, for $1 \leq i \leq 6$,

$$b_i = 0\,.$$

Thus the following theorem holds.

Theorem 15. *Let m be an odd integer greater than 4.*

(i) *A $(2^{m-1}-2^{(m+1)/2})$-BCH code and the dual code of a triple-error-correcting BCH code have the following weight distribution:*

$$a_{2^{m-1}\pm 2^{(m+1)/2}} = 2^{(m-5)/2}(2^{(m-3)/2} \mp 1)(2^m-1)(2^{m-1}-1)/3\,,$$

$$a_{2^{m-1}\pm 2^{(m-1)/2}} = 2^{(m-3)/2}(2^{(m-1)/2} \mp 1)(2^m-1)(5\cdot 2^{m-1}+4)/3\,,$$

$$a_{2^{m-1}} = (2^m-1)(9\cdot 2^{2m-4}+3\cdot 2^{m-3}+1)\,,$$

$$a_j = 0\,, \qquad \textit{for other nonzero } j.$$

(ii) *These weight distribution formulas hold also for every subcode with $k = 3m$ of the second order modified Reed-Muller code that satisfies one of the following conditions:*

(1) $b_i = 0$ *for* $1 \le i \le 6$.

(2) $a_j + a_{2^m-j} = 0$ *for* $0 < j < 2^{m-1} - 2^{(m+1)/2}$ *and* $b_i = 0$ *for* $1 \le i \le 4$.

(e) $(2^{m-1}-2^{(m+3)/2})$-BCH codes for odd $m \ge 11$.

Let $p = 4$, $\mu_1 = 0$, $\mu_2 = (m-5)/2$, $\mu_3 = (m-3)/2$ and $\mu_4 = (m-1)/2$. Then $k = 4m$. From Theorem 8, $\bar{a}_i = 0$ $(i > 3)$. Lemma 9 shows that, for $m \ge 11$, this code is a $(2^{m-1}-2^{(m+3)/2})$-BCH code. The dual code is a subcode of the dual code of a $(2^{m-1}-2^{(m+1)/2})$-BCH code, which has minimum weight 7 by Theorem 15. Consequently,

$$b_i = 0 \qquad (1 \le i \le 6)\,.$$

By solving (27) for $l = 2$, 4 and 6 and using symmetry property (2), the following theorem is obtained.

Theorem 16. (i) *For odd $m \ge 7$, let $p = 4$, $\mu_1 = 0$, $\mu_2 = (m-5)/2$, $\mu_3 = (m-3)/2$ and $\mu_4 = (m-1)/2$. Then,*

$$a_{2^{m-1}\pm 2^{(m-1)/2}} = (2^{m-1} \mp 2^{(m-1)/2}) \times (151\cdot 2^{2m-3} + 25\cdot 2^m + 2^5)(2^m-1)/45\,,$$

$$a_{2^{m-1}\pm 2^{(m+1)/2}} = (2^{m-2} \mp 2^{(m-1)/2}) \times (23\cdot 2^{m-5}+1)(2^{m-1}-1)(2^m-1)/9\,,$$

$$a_{2^{m-1}\pm 2^{(m+3)/2}} = (2^{m-6} \mp 2^{(m-7)/2}) \times (2^{m-3}-1)(2^{m-1}-1)(2^m-1)/45\,,$$

$$a_{2^{m-1}} = 2^{4m} - 1 - \sum_{j\ne 0,2^{m-1}} a_j\,,$$

$$a_j = 0\,, \qquad \textit{for other nonzero } j.$$

(ii) *For $m \ge 11$, the code in (i) is a $(2^{m-1}-2^{(m+3)/2})$-BCH code.*

(f) The dual codes of triple-error-correcting BCH codes for even $m \ge 6$.

Let $\mu_1 = 0$, $\mu_2 = 1$, and $\mu_3 = 2$. Then $k = 3m$. It is easy to check that this code is the dual code of a tripple-error-correcting BCH code. Hence, $b_i = 0$ $(1 \le i \le 6)$. From (27), (34), (35) and (36),

(43) $$2^{m-2}\bar{a}_0 + 2^m\bar{a}_1 + 2^{m-2}\sum_{i\ge 2} 2^{2i}\bar{a}_i = 2^{2m-2}(2^{2m}-1)\,,$$

(44) $$2^{2m-4}\bar{a}_0 + 2^{2m}\bar{a}_1 + 2^{2m-4}\sum_{i\ge 2} 2^{4i}\bar{a}_i = 3\cdot 2^{4m-4}(2^m-1)\,,$$

(45) $$2^{3m-6}\bar{a}_0 + 2^{3m}\bar{a}_1 + 2^{3m-6}\sum_{i\ge 2} 2^{6i}\bar{a}_i = 2^{4m-5}(7\cdot 2^m-8)(2^m-1)\,.$$

By eliminating $\bar{a}_0$ and $\bar{a}_1$,

$$\bar{a}_2 + 28\bar{a}_3 + \sum_{i\ge 4} c_i\bar{a}_i = 2^{m-4}(2^{m-2}-1)(2^m-1)/15\,,$$

where $c_i > 28$.

On the other hand, it can be shown that, for $j \ge 3$, $\bar{a}_j$ is divisible by $2^{m-4}(2^m-1)$. The proof is given in Appendix 2. Hence, it is easy to check that, for $6 \le m \le 10$, $\bar{a}_i = 0$ $(i \ge 3)$. Consequently, the weight distributions of the dual codes of triple-error-correcting BCH codes for $m = 6$, 8 and 10 can easily be found by solving Equations (43), (44) and (45) and using symmetry property (2).

APPENDIX 1

The Proof of (9), (10) and (11).

Assume that code C has no code vector with odd weight. Add the all-one vector $(1, \ldots, 1)$ to the basis of C, and let C' denote the resulting code with $k+1$ information digits. Add an overall parity check to C' and denote the resulting code of length $n+1$ by C''. Let a'_j (or a''_j) denote the number of code vectors of weight j of code C' (or C'') and let b'' denote the number of code vectors of weight j of the dual of code C''. Then, for even j

(A1) $$a'_j = a'_{n-j} = a_j\,, \qquad a''_j = a'_j + a'_{j-1} = a_j + a_{n+1-j}\,, \qquad j \ne 0\,.$$

It is easy to check that for odd j

(A2) $$b''_j = 0\,,$$

and that for even $j \ne 0$

(A3) $$b''_j = b_j + b_{j-1}\,.$$

By identity (1),

(A4) $$\sum_{j=0}^{n+1} j^l a''_j = \sum_{h=0}^{n+1} (-1)^h b''_h \left[\sum_{\nu=0}^{l} \nu!\, G_l^\nu 2^{k+1-\nu}\binom{n+1-h}{n+1-\nu}\right],$$

where G_l^ν is a Stirling number of the second kind and $\binom{n+1-h}{n+1-\nu} = 0$ for $h > \nu$. By (A1) through (A4),

(A5) $$\begin{aligned} I_p &= 2^{-1}\sum_{j=0}^{n+1}(j-(n+1)/2)^p a''_j - (n+1)^p 2^{-p} \\ &= 2^{-1}\sum_{l=0}^{p}(-1)^{p-l}\binom{p}{l}(n+1)^{p-l}2^{-p+l}\sum_{h=0}^{p}(-1)^h b''_h \sum_{\nu=0}^{i}\nu! \\ &\qquad \times G_l^\nu 2^{k+1-\nu}\binom{n+1-h}{n+1-\nu} - (n+1)^p 2^{-p} \\ &= 2^{k-p}\sum_{h=0}^{p}(-1)^h b''_h J_{ph} - (n+1)^p 2^{-p}\,, \\ &= 2^{k-p}\sum_{h=0}^{p/2}(b_{2h}+b_{2h-1})J_{p2h} - (n+1)^p 2^{-p}\,, \qquad \text{for even } p, \end{aligned}$$

where

$$J_{ph} = \sum_{l=0}^{p}(-1)^{p-l}\binom{p}{l}(n+1)^{p-l}2^l\sum_{\nu=0}^{l}\nu!\, G_l^\nu 2^{-\nu}\binom{n+1-h}{n+1-\nu}\,.$$

By using the formula

$$(n+1)n\ldots(n-\nu+2) = \sum_{f=1}^{\nu} S_\nu^f (n+1)^f\,,$$

where S_ν^f is a Stirling number of the first kind [5], we have

$$J_{p0} = \sum_{l=0}^{p}(-1)^{p-l}\binom{p}{l}(n+1)^{p-l}2^l\sum_{\nu=0}^{l} G_l^\nu 2^{-\nu}\sum_{f=0}^{\nu} S_\nu^f (n+1)^f\,.$$

By noting that $S_\nu^f = 0$ for $f > \nu$ and $G_l^\nu = 0$ for $\nu > l$,

(A6) $$\begin{aligned} J_{p0} &= \sum_{q=0}^{p}(n+1)^{p-q}\sum_{l=0}^{p}(-1)^l\binom{p}{l}2^l\sum_{\nu=0} 2^{-\nu}G_l^\nu S_\nu^{l-q} \\ &= \sum_{q=0}(n+1)^{p-q}\sum_{i=0}^{p}2^i\sum_{l=0}^{q}(-1)^l\binom{p}{l}G_l^{l-i}S_{l-i}^{l-q}\,. \end{aligned}$$

Since $G_l^l = 1$, we have

$$J_{pp} = p! \,. \tag{A7}$$

By (A5), (A6), (A7) and a straightforward but tedious calculation, equations (9), (10) and (11) can be derived. Code C has been assumed to have no code vector with odd weight. However, it follows from identity (1) that the form of I_p depends on only n and k. Therefore, (9), (10) and (11) hold for the general case.

APPENDIX 2

The notations in section 2 will be used. Let $p = 3$, $j_1 = 2^m - 1 - 3$, $j_2 = 2^m - 1 - 5$ and $j_3 = 2^m - 1 - 1$. It is easy to check that (1) E_{iq} $(i \neq q,\ i = 1, 2)$ is empty, (2) $E_{ii} = \{1\}$ $(1 \leq i \leq 3)$ and (3) $E_{31} = \{0, m-1\}$ and $E_{32} = \{0, m-2\}$. From (8) it follows that

$$\beta_1' = a^{-3}\beta_1$$

$$\beta_2' = a^{-5}\beta_2$$

$$\beta_3' = a^{-3}\beta_1 b^2 + a^{-3\cdot 2^{m-1}}\beta_1^{2^{m-1}} b^{2^{m-1}} + a^{-5}\beta_2 b^4 + a^{-5\cdot 2^{m-2}}\beta_2^{2^{m-2}} b^{2^{m-2}} + a^{-1}\beta_3 \,.$$

Let $Z = b^{2^{m-2}}$ and

$$f(Z) = a^{-5}\beta_2 Z^{2^4} + a^{-3}\beta_1 Z^{2^2} + a^{-3\cdot 2^{m-1}}\beta_1^{2^{m-1}} Z^2 + a^{-5\cdot 2^{m-2}}\beta_2^{2^{m-2}} Z \,.$$

If β_1 or β_2 is not equal to zero, the zeros of $f(Z)$ form a σ-dimensional subspace of $GF(2^m)$, where $1 \leq \sigma \leq 4$. Thus, for fixed a, β_1, β_2, and β_3, the number of elements of $\{\beta_3' | \beta_3' = f(Z) + a^{-1}\beta_3,\ Z = b^{2^{m-2}},\ b \in GF(2^m)\}$ is divisible by 2^{m-4}. Now, suppose that $\boldsymbol{v}(0, \beta_1, \beta_2, \beta_3 \colon X)$ has weight j with $2^{m/2+2} \leq |j - 2^{m-1}| < 2^{m-1}$. If β_1 or β_2 is equal to zero, $\boldsymbol{v}(0, \beta_1, \beta_2, \beta_3 \colon X)$ is a code vector of a code considered in case (**a**) and the weight of the nonzero vector $\boldsymbol{v}(0, \beta_1, \beta_2, \beta_3 \colon X)$ is greater than $2^{m-1} - 2^{m/2+1} - 1$ and smaller than $2^{m-1} + 2^{m/2+1} + 1$. Hence, β_1 and β_2 can not be zero.

Let l be a positive integer such that

$$3l \equiv 5l \equiv 0 \pmod{2^m - 1} \,.$$

Then,

$$2l \equiv 0 \pmod{2^m - 1} \,.$$

Hence, l must be a multiple of $2^m - 1$. Thus the number of pairs $\{(\beta_1', \beta_2') | \beta_1' = a^{-3}\beta_1,\ \beta_2' = a^{-5}\beta_2,\ a \neq 0,\ a \in GF(2^m)\}$ is equal to $2^m - 1$. Consequently, it follows from Lemma 2 that $a_j + a_{2^m - j}$ is divisible by $2^{m-4}(2^m - 1)$.

Acknowledgement

The author is grateful to Professor W. W. Peterson for many valuable suggestions and to Professors R. T. Chien and J. S. Lin for their helpful discussions.

References

1. Assmus, E. F., Mattson, H. F. and Turyn, R. "Cyclic Codes," Scientific Report, AFCRL-65-322, 1965, Air Force Cambridge Research Labs., Bedford, Mass.
2. Bose, R. C. and Ray-Chaudhuri, D. K. "On a Class of Error-Correcting Binary Group Codes," *Inf. and Control*, **3** (1960), 68–79.
3. Forney, G. D., Jr. *Concatenated Codes*, MIT Press, Cambridge, Mass., 1966.
4. Gold, R. and Kopitzka, E. "Study of Correlation Properties of Binary Sequences," Report of Magnavox Research Lab., Vol. 1–15, 1965.
5. Jordan, C. *Calculus of Finite Differences*, 2nd ed., Chelsea, New York, 1950.
6. Kasami, T. "Weight Distribution Formula for some Classes of Cyclic Codes," Report of Coordinated Science Lab., University of Illinois, 1966.
7. Kasami, T. and Lin, S. "Some Codes which Are Invariant under a Doubly-Transitive Permutation Group and their Connection with Balanced Incomplete Block Designs," Scientific Report AFCRL-66-142, Air Force Cambridge Research Labs., Bedford, Mass., 1966.
8. Kasami, T., Lin, S. and Peterson, W. W. "Some Results on Cyclic Codes which Are Invariant under the Affine Group," Scientific Report, AFCRL-66-622, Air Force Cambridge Research Labs., Bedford, Mass., 1966. Partly published in *Inf. and Control*, **11** (1967), 475–496, in *J. Inst. Elec. Commun. Engrs., Japan*, **50** (1967), 1617–1672, and in *IEEE Trans.*, **IT-14** (1968), 189–199.
9. MacWilliams, F. J. "A Theorem on the Distribution of Weights in a Systematic Code," *Bell System Tech. J.*, **42** (1962), 79–94.
10. Mattson, H. F. and Solomon, G. "A New Treatment of Bose-Chaudhuri Codes," *J. Soc. Indust. Appl. Math.*, **9** (1961), 654–669.
11. Mattson, H. F. and Assmus, E. F. "Research Program to Extend the Theory of Weight Distribution and Related Problems for Cyclic Error-Correcting Codes," Applied Research Lab., Sylvania Electronic Lab., Waltham, Mass., 1964.
12. Muller, D. E. "Application of Boolean Algebra to Switching Circuit Design and to Error Detection," *IRE Trans.*, **EC-3** (1954), 6–12.
13. Peterson, W. W. *Error Correcting Codes*, John Wiley and Sons, New York, 1961.
14. Peterson, W. W. "On the Weight Structure and Symmetry of BCH Codes," *J. Inst. Elec. Commun. Engrs., Japan*, **50** (1967), 1183–1190.
15. Pless, V. "Power Moment Identities on Weight Distributions in Error Correcting Codes," *Inf. and Control*, **6** (1963), 147–152.
16. Prange, E. Unpublished paper.

BOUNDS FOR EXPONENTIAL SUMS

By L. Carlitz and S. Uchiyama

1. Let $F(x) = a_0x^r + \cdots + a_r$ be a polynomial with integral coefficients and put

$$S = \sum_{u=1}^{p} e^{2\pi i F(u)/p}$$

where p is a prime. Some years ago Mordell [4; 67] noted the conjecture

$$|S| \le (r-1)p^{\frac{1}{2}} \tag{1}$$

and remarked that it was presumably a consequence of the Riemann hypothesis proved by Weil [8] for the zeta-function of algebraic function fields. Indeed Hasse [3; 53] had indicated the connection in 1935.

In this note we wish to point out, in the first place, how (1) can be proved from Weil's result in a comparatively simple way. Since it is no more difficult, we consider the slightly more general situation in which the coefficients of $F(x)$ are numbers of the finite field $k = GF(q)$, $q = p^n$. For $\alpha \in k$, define

$$e(\alpha) = e^{2\pi i t(\alpha)/p}, \qquad t(\alpha) = \alpha + \alpha^p + \cdots + \alpha^{p^{n-1}} \tag{2}$$

and let

$$S = \sum_{\alpha \in k} e(F(\alpha)), \tag{3}$$

the summation extending over all numbers of k.

Let $K_0 = k[x]$ denote the domain of polynomials in the indeterminate x with coefficients in k. Let $P = P(x)$ denote an irreducible polynomial in K_0 of degree m. If $A = A(x)$ is an arbitrary polynomial in K_0, define $\rho(A, P)$ as the unique number of k such that

$$\rho(A, P) \equiv A + A^q + \cdots + A^{q^{m-1}} \pmod{P}. \tag{4}$$

Thus the congruence $U^q - U \equiv A \pmod{P}$ is solvable with $U \in K_0$ if and only if $\rho(A, P) = 0$. Next define

$$\lambda(A, P) = e\{\rho(A, P)\}, \tag{5}$$

with $e(\alpha)$ defined in (2). The definitions (4) and (5) are extended as follows. If M is an arbitrary polynomial in K_0, $M = P_1 \cdots P_h$, put

$$\rho(A, M) = \sum \rho(A, P_i), \qquad \lambda(A, M) = \prod \lambda(A, P_i);$$

the P_i are irreducible polynomials, not necessarily distinct.

Now consider the equation

$$y^p - y - F = 0, \tag{6}$$

where F is an arbitrary member of the field $K = k(x)$. Assume $F \ne C^p - C + b$, $C \in K$, $b \in k$; then the left member of (6) is absolutely irreducible, that is, irreducible over $\Gamma(x)$, where Γ is the algebraic closure of k. Let $\Omega = K(y)$; then Ω is a cyclic extension of K of degree p. We next construct the zeta function for the field Ω in the case $F \in K_0$. There is no loss in generality in assuming that $p \nmid \deg F$. We find that (compare [3; 51])

$$\begin{aligned} Z_\Omega(s) &= \sum_A |NA|^{-s} \\ &= (1 - q^{1-s})^{-1} \prod_{u=1}^{p-1} L(s, \lambda^u), \end{aligned} \tag{7}$$

where A runs through the ideals of Ω,

$$L(s, \lambda^u) = \sum_M \frac{\lambda^u(F, M)}{|M|^s} \qquad (|M| = q^{\deg M}), \tag{8}$$

and M runs through the (primary) polynomials of K_0. (We remark that (7) differs slightly from the zeta-function defined by Hasse; the precise connection is described by [5; 30] but will not be required here.)

In the next place we show that

$$\sum_{\deg M = m} \lambda^u(F, M) = 0 \qquad (m \ge \deg A;\ u = 1, \cdots, p-1). \tag{9}$$

To prove (9) we make use of the following result [1, Theorem 11.4]; see also [2]. Let P' denote the derivative of the irreducible polynomial $P = P(x)$. Then if $\deg P = m$ and

$$FP' \equiv c_1x^{m-1} + c_2x^{m-2} + \cdots + c_m \pmod{P},$$

it follows that $c_1 = \rho(F, P)$. This is easily extended to $\rho(F, M)$; indeed if $\deg M = m$ and

$$FM' \equiv c_1x^{m-1} + \cdots + c_m \pmod{M}, \tag{10}$$

then $c_1 = \rho(F, M)$. Now it is easily seen that for $F = x^r$, $c_1 = \rho(x^r, M)$ is the sum of the r-th powers of the roots of $M(x)$. Expressing this power sum in terms of the coefficients of $M = x^m + b_1x^{m-1} + \cdots + b_m$, it is clear that $c_1 = -rb_r$ plus a polynomial in $b_1, \cdots, b_{r-1}$. Hence for $r = \deg F$, $p \nmid r$, we have

$$\rho(F, M) = -rb_r + \phi(b_1, \cdots, b_{r-1}),$$

where ϕ denotes a polynomial in $b_1, \cdots, b_{r-1}$. In view of (5), (9) follows at once. Then (8) becomes

$$L(s, \lambda^u) = \sum_{i=0}^{r-1} f_i q^{-is}. \tag{11}$$

By the Riemann hypothesis the roots of $Z_\Omega(s)$ have absolute value $q^{\frac{1}{2}}$; hence by (7) the same is true of the roots of $L(s, \lambda^u)$. Consequently (8) and (11) yield

$$\Big|\sum_{\alpha \in k} \lambda(F, x - \alpha)\Big| \le (r-1)q^{\frac{1}{2}},$$

which is the same as

$$\Big|\sum_{\alpha \in k} e(F(\alpha))\Big| \le (r-1)q^{\frac{1}{2}}. \tag{12}$$

We have proved (12) under the assumption that F is a polynomial of degree r such that

$$F \ne C^p - C + b \qquad (b \in k,\ C \in K_0).$$

In particular (12) holds for $r \le p - 1$.

2. We next discuss briefly the Kloosterman sum

$$T = \sum_{\alpha \in k;\, \alpha \ne 0} e\left(\alpha + \frac{c}{\alpha}\right) \qquad (c \ne 0). \tag{13}$$

In place of (6) we now consider

$$y^p - y = x + \frac{c}{x} \qquad (c \ne 0),$$

and again put $\Omega = K(y)$. Note that $(x) = \mathfrak{P}^p$, where $\mathfrak{P}$ is a prime ideal of Ω of the first degree. Corresponding to (9) we now have

$$\sum_{\deg M = m} \lambda\left(x + \frac{c}{x}, M\right) = 0 \qquad (m > 2), \tag{14}$$

where we put $\lambda(x + cx^{-1}, M) = 0$ for M divisible by x. Using (10) with $M = x^m + b_1x^{m-1} + \cdots + b_m$, $b_m \ne 0$, we get

$$\rho(x, M) = -b_1, \qquad \rho\left(\frac{1}{x}, M\right) = -\frac{b_{m-1}}{b_m},$$

so that

$$\rho\left(x + \frac{c}{x} M\right) = -b_1 + \frac{cb_{m-1}}{b_m}.$$

Thus

$$\begin{aligned} \sum_{\deg M = m} \lambda\left(x + \frac{c}{x}, M\right) &= \sum_{\substack{\deg M = m \\ x \nmid M}} e\left(-b_1 + \frac{cb_{m-1}}{b_m}\right) \\ &= 0 \qquad (m > 2), \end{aligned}$$

which proves (14). Then exactly as in the proof of (12) we find that the sum (13) satisfies

$$|T| \le 2q^{\frac{1}{2}}. \tag{15}$$

This result was proved in a less elementary way by Weil [9].

The method used above evidently applies to more general exponential sums.

3. As an application of the results in §1, we may mention the following. Let us again denote by k the finite field $GF(q)$, $q = p^n$, and write $K_0 = k[x]$. If $f(x)$ is a primary polynomial in K_0 of degree r, $1 \le r < p$, we denote by $V(f)$ the number of distinct values assumed by $f(x)$, $x \in k$.

It is known that

$$\begin{aligned} \sum_{\deg f = r} V(f) &= \sum_{j=1}^{r} (-1)^{j-1} \binom{q}{j} q^{r-j} \\ &= c_r q^r + O(q^{r-1}), \end{aligned} \tag{16}$$

where the summation on the left-hand side is extended to all primary polynomials in K_0 of degree r without constant term, and where

Received June 13, 1956.

Reprinted with permission from *Duke Math. J.*, vol. 24, pp. 37-41, Dec. 1957.

$$c_r = \sum_{j=1}^{r} (-1)^{j-1}/j!.$$

In other words, we have

$$V(f) = c_r q + O(1)$$

on the average. Moreover, it can be shown that

$$\sum_{\deg f = r} V^2(f) = c_r^2 q^{r+1} + O(q^r), \tag{17}$$

and hence

$$\sum_{\deg f = r} (V(f) - c_r q)^2 = O(q^r)$$

by (16) and (17). We have thus

$$V(f) = c_r q + O(q^{\frac{1}{2}})$$

for *almost all* primary polynomials f in K_0 of degree r.

These results were noted in [6] and [7], where, in the latter, the inequality (17) was obtained under a certain hypothesis; but, by virtue of (12), we are now able to prove it without any hypothesis (cf. [7]).

References

1. L. Carlitz, *On certain functions connected with polynomials in a Galois field*, this Journal, vol. 1(1935), pp. 137–168.
2. L. Carlitz, *On certain equations in relative-cyclic fields*, this Journal, vol. 2(1936), pp. 650–659.
3. H. Hasse, *Theorie der relativ-zyklischen algebraischen Funktionenkörper, insbesondere bei endlichem Konstantenkörper*, Journal für die reine und angewandte Mathematik, vol. 172(1935), pp. 37–54.
4. L. J. Mordell, *Thoughts on number theory*, Journal of the London Mathematical Society, vol. 21(1946), pp. 58–74.
5. F. K. Schmidt, *Analytische Zahlentheorie in Körpern der Charakteristik p*, Mathematische Zeitschrift, vol. 33(1931), pp. 1–32.
6. S. Uchiyama, *Note on the mean value of V(f)*, Proceedings of the Japan Academy, vol. 31(1955), pp. 199–201.
7. S. Uchiyama, *Note on the mean value of V(f). III*, Proceedings of the Japan Academy, vol. 32(1956), pp. 97–98.
8. A. Weil, *On the Riemann hypothesis in function fields*, Proceedings of the National Academy of Sciences, vol. 27(1941), pp. 345–347.
9. A. Weil, *On some exponential sums*, Proceedings of the National Academy of Sciences, vol. 34(1948), pp. 204–207.

Duke University
Hokkaido University

A SURVEY OF CONSTRUCTIVE CODING THEORY, AND A TABLE OF BINARY CODES OF HIGHEST KNOWN RATE

N.J.A. SLOANE

Bell Telephone Laboratories, Inc., Murray Hill, N.J., U.S.A.

Received 6 November 1971

Abstract. Although more than twenty years have passed since the appearance of Shannon's papers, a still unsolved problem of coding theory is to construct block codes which attain a low probability of error at rates close to capacity. However, for moderate block lengths many good codes are known, the best-known being the BCH codes discovered in 1959. This paper is a survey of results in coding theory obtained since the appearance of Berlekamp's "Algebraic coding theory" (1968), concentrating on those which lead to the construction of new codes. The paper concludes with a table giving the smallest redundancy of any binary code, linear or nonlinear, that is presently known (to the author), for all lengths up to 512 and all minimum distances up to 30.

§1. Introduction

This paper is a survey of recent developments in the design of block codes for the correction of random errors.

In 1948, Shannon [98] showed that there exist codes which attain a low probability of error at rates close to capacity. Gilbert [31] in 1952 obtained a lower bound on d/n for the best codes of a given rate. Since then a great deal of effort has been made to construct arbitrarily long codes which meet or even come close to the Gilbert bound, but so far without success (except for codes with rates approaching 0 or 1).

For moderate block lengths, however, many good codes have been discovered. The best-known are the Reed–Muller (RM) codes ([79], [93], [B1, Ch. 15]), Bose–Chaudhuri–Hocquenghem (BCH) codes ([19], [45], [B1, Ch. 7 and Ch. 10]), and quadratic residue (QR) codes ([B1, Ch. 15], [111, §4.4]). A systematic description of these and other codes discovered prior to 1968 will be found in [B1].

In this paper we describe some of the developments in coding theory that have taken place since the appearance of [B1]. We concentrate on those papers which construct new codes, or give new properties of codes previously known. Important topics not considered here are the weight enumeration of codes [7, 8, 10, 14, 15, 32, 41, 53–57, 73, 81, 86, 94, 100, 106, 114], the classification of cosets of a code [9, 16, 94, 103], decoding techniques and burst error correction [20, 85, 95, 109, 110], synchronization recovery [105], and source coding and rate distortion theory [6, 47]. See also the recent survey of Goethals [35], which is of broader scope than the present work, and the book by Van Lint [111]. We have had little access to Russian work, and refer the reader to the surveys by Kautz and Levitt [58] and Dobrushin [30].

The paper is arranged as follows. §2 deals with cyclic and related codes, including BCH, irreducible, perfect, abelian group, Goppa, Srivastava, and circulant codes. §3 deals with nonlinear codes and codes formed by combining other codes. §4 gives a table containing the best binary codes known to the author. An extensive bibliography concludes the paper. A shortened preliminary version of this paper has appeared in [102].

§2. Cyclic codes

Most of the codes considered to date have been linear and cyclic, for the excellent reasons that such codes are simpler to implement and to analyze.

An (n, k) *linear code* $\mathcal{C}$ over the field $GF(q)$ consists of q^k vectors (called *codewords*) of length n over $GF(q)$ such that (a) the sum, taken componentwise in $GF(q)$, of any two codewords is again a codeword, and (b) the componentwise product of any codeword and any element of $GF(q)$ is also a codeword. The *redundancy* of the code is $r = n-k$ and the *rate* is $R = k/n$. The *minimum distance* is denoted by d.

The *dual* code $\mathcal{C}^\perp$ of $\mathcal{C}$ consists of all vectors u of length n over $GF(q)$ such that $u \cdot v = 0$ for all $v \in \mathcal{C}$, the scalar product being evaluated in $GF(q)$. Thus $\mathcal{C}^\perp$ is an $(n, n-k)$ linear code. If $\mathcal{C} = \mathcal{C}^\perp$, $\mathcal{C}$ is *self-dual*.

An application of self-dual codes to a famous unsolved problem of geometry is given in [74].

A code is *shortened* by omitting all codewords except those having prescribed values for certain components, and then deleting those components [B1, p. 336].

A *cyclic code* is a linear code with the property that a cyclic shift of any codeword is also a codeword. BCH, QR, and shortened RM codes are all cyclic.

2.1. *Are long cyclic codes bad?* Gilbert [31] showed that there exist arbitrarily long linear codes with a fixed rate $R = k/n$ for which d/n is bounded away from zero. In fact Koshelev [59] and Kozlov [60] have shown that most linear codes meet the Gilbert bound.

On the other hand for BCH codes of fixed rate R, $d/n \to 0$ as $n \to \infty$ ([67], [B1, Ch. 12]). In fact Berlekamp [12] has recently shown that for BCH codes

$$d \sim \frac{2n \ln R^{-1}}{\log n}$$

as $n \to \infty$. But it is not known whether long cyclic codes are also bad.

Berman [18] has shown that cyclic codes of fixed rate and with block lengths n which are divisible by a fixed set of primes (and only by these primes) have bounded minimum distance.

Kasami [52] has shown that good linear codes cannot be too symmetric, by showing that any code with given d/n which is invariant under the affine group must have rate $R \to 0$ as $n \to \infty$. (This includes BCH codes.)

More recently McEliece [77] showed that it is not the symmetry alone that makes a code bad, by showing that there exist arbitrarily long block codes (not necessarily linear) which are invariant under large permutation groups and which meet the Gilbert bound. Also Weldon [23], [118] has shown that there exist very, but not arbitrarily, long circulant and quasi-cyclic codes which meet the Gilbert bound. (A quasi-cyclic code is a linear code with the property that a cyclic shift of any codeword by a certain prescribed number of places is also a codeword. Circulant codes are defined in §2.9.) Weldon's proof would apply to arbitrarily long codes if the conjecture were proved that there are an infinite number of primes for which 2 is a primitive root.

Kasami [52] and Chen [22] have shown that there exist arbitrarily long shortened cyclic codes which meet the Gilbert bound.

Thompson (see [75]) has shown that self-dual linear binary codes in which all weights are divisible by 4 meet the Gilbert bound.

To give a rough summary of these results, a good family of codes can be linear, or have many symmetries, but not both.

2.2. BCH *codes.* We recall the definition of a BCH code. Let q be a prime power, let m be the order of q modulo n, and let α be a primitive n^{th} root of unity in $GF(q^m)$. Then the BCH *code* of length n, designed distance $d = d_{BCH}$, and symbols from $GF(q)$, has the parity check matrix

$$H = \begin{pmatrix} 1 & \alpha & \alpha^2 & \dots & \alpha^{n-1} \\ 1 & \alpha^2 & \alpha^4 & \dots & \alpha^{2(n-1)} \\ & & & & \vdots \\ 1 & \alpha^{d-1} & \alpha^{2(d-1)} & \dots & \alpha^{(n-1)(d-1)} \end{pmatrix}$$

If $n = q^m - 1$ the code is called *primitive*. By the BCH bound any such code has actual minimum distance

$$d_{\min} \geq d_{BCH} .$$

It was conjectured [55], [B1, p. 295] that for primitive BCH codes $d_{\min} = d_{BCH}$, but in 1969 Kasami and Tokura [K2] showed that for $m > 6$, $m \neq 8, 12$, there are binary primitive BCH codes of length $n = 2^m - 1$ for which

$$d_{\min} > d_{BCH} .$$

On the other hand Berlekamp [8] showed that if the extended binary BCH code of length $n = 2^m$ has $d_{BCH} = 2^{m-1} - 2^i$ for some $i \geq \frac{1}{2}m - 1$,

Reprinted with permission from *Discrete Math.*, vol. 3, pp. 265–294, Sept. 1972.

then $d_{min} = d_{BCH}$. The results of [8] have been further generalized by Kasami [53]. But the precise determination of conditions on n and d_{BCH} for $d_{min} = d_{BCH}$ to hold remains an unsolved problem.

Leont'ev [66] showed that a BCH code of length $n = 2^m - 1$ and designed distance $2t + 1$ is not quasi-perfect for $2 < t < \sqrt{n}/\ln n$ and $m \geq 7$.

V.M. Sidel'nikov [100] showed that for $t < \sqrt{n}/10$, the number of words of weight w in the binary BCH code of designed distance $2t + 1$ is $(n+1)^{-t}\binom{n}{w}(1+\epsilon)$, where $|\epsilon| < Cn^{-0.1}$, for most values of w.

Anderson ([1], [111, p. 127]) obtained the following bound for the dual of a BCH code, using deep number-theoretic results of Weil, Carlitz and Uchiyama. The minimum distance of the dual of the binary BCH code of length $n = 2^m - 1$ and designed distance $d_{BCH} = 2t + 1$ is at least

$$2^{m-1} - 1 - (t-1)^{\frac{1}{2}m}$$

2.3. *Extensions of* BCH *codes.* Wolf [117] showed that two columns may be added to the parity check matrix of a BCH code to give the new parity check matrix

$$H' = \begin{pmatrix} 1\ 0 & \\ 0\ 0 & \\ \dots & H \\ 0\ 1 & \end{pmatrix},$$

while preserving the minimum distance of the code. In some cases the redundancy is also unchanged, in which case we have a new code with

$$n' = n + 2, \quad k' = k + 2, \quad r' = r, \quad d' = d.$$

This happens for example when the original code is a Reed–Solomon code over GF(q), with $n = q - 1$ and $r = d - 1$. Then the parameters of the new code are $n' = q + 1$, and $r' = d' - 1 = d - 1$.

It is easy to show that for any code $d \leq r + 1$. Codes with $d = r + 1$ are called *maximum distance separable* or MDS codes (see [101], [B1, p. 309], [111, p. 72]). Such codes have also been called *optimal*. Reed–Solomon codes are MDS and so are the new family of doubly extended Reed–Solomon codes.

Assmus and Mattson [3] have shown that MDS codes whose block length n is a prime number are very common, by showing that every cyclic code of prime length n over GF(p^i) is MDS, for all i, for all except a finite number of primes p.

Wolf [118] has obtained a further extension of BCH codes, by replacing α in H' by the $m \times m$ matrix A over GF(q), where

$$A = \begin{pmatrix} 0 & 1 & 0 & & 0 \\ 0 & 0 & 1 & \dots\dots & 0 \\ 0 & 0 & 0 & & 0 \\ \vdots & \vdots & \vdots & & \vdots \\ 0 & 0 & 0 & \dots\dots & 1 \\ a_0 & a_1 & a_2 & & a_{m-1} \end{pmatrix}$$

and where $M(x) = x^m - a_{m-1}x^{m-1} - \dots - a_0$ is the minimal polynomial of α over GF(q). Call the new parity check matrix H''. Then if H generates a primitive BCH code over GF(q) of length $n = q^m - 1$, designed distance d_{BCH} and redundancy $r = m(d_{BCH} - 1)$, so that this is a maximally redundant BCH code, then H'' is the parity check matrix for a code over GF(q) with $n' = m(q^m + 1)$, $r = m(d_{BCH} - 1)$ and $d \geq d_{BCH}$. The rate of the code has thus been considerably increased.

For example, if the original code is over GF(5) with $n = 5^2 - 1 = 24$, $d_{BCH} = 5$, $k = 16$, $R = 0.67$, the new code has $n = 52$, $k = 44$, $d \geq 5$ and $R = 0.87$.

Other extensions of BCH codes are mentioned in §3.3.

2.4. *The minimum distance of cyclic codes.* The BCH bound for a cyclic code says that if the generator polynomial $g(x)$ has $d_{BCH} - 1$ consecutive roots then the minimum distance is $\geq d_{BCH}$.

Goethals [33] and Kasami [51] have given improvements on the BCH bound for codes of composite length. Hartmann [38–43] has given many further generalizations of the BCH bound, including extensions of Kasami's results. We will just state two of Hartmann's theorems. Here β denotes a primitive n^{th} root of unity. The first is a bound on the minimum odd weight.

Theorem. *Let $k|n$. If for some $\bar{d} \leq n/k$, $g(\beta^{ki}) = 0$ for all $i = 1, 2, \dots, \bar{d}$, then the minimum odd weight is at least $\bar{d}$.*

Example. By the BCH bound the (33, 13) BCH code has $d_{BCH} = 5$, and also $d_{even} \geq 10$. But by the theorem $d_{odd} \geq 11$, and so the minimum weight is ≥ 10.

The second theorem is an example of Hartmann's generalizations of the BCH bound to the case where $g(x)$ has several sets of consecutive roots.

Theorem. *If $g(\beta^{m_0+i+\delta(j-1)}) = 0$ for $i = 0, 1, 2, \dots, d-1$ and $j = 1, 2, \dots, r$ with $(\delta, n) = 1$, so that $g(x)$ has r sets of $d-1$ roots each, then $d_{min} \geq d + r$.*

Kasami and Tokura [K2] have shown that for any even $m \geq 6$ there exist binary cyclic codes of length $2^m - 1$ having more codewords than the corresponding BCH codes. The first such example is a (63, 28) $d = 15$ cyclic code, compared with the (63, 24) $d = 15$ BCH code.

Chen [C3] used an IBM 360/50 to calculate the minimum distance of all binary cyclic codes of lengths ≤ 65. He found three codes of length 63 having more codewords than the corresponding BCH codes. These are the (63, 28) code just mentioned, a (63, 46) $d = 7$ code given previously by Peterson [86], and a (63, 21) $d = 18$ code. The BCH codes closest to the last two are (63, 45) $d = 7$ and (63, 18) $d = 21$ codes.

2.5. *Irreducible cyclic codes.* A cyclic code over GF(q) is called irreducible if its check polynomial $h(x)$ is irreducible over GF(q) [111, p. 45]. The simplest examples of irreducible codes are maximal length shift register ($2^m - 1$, m) codes (also known as shortened first order RM codes).

Baumert, McEliece, and Rumsey [5], [78], generalizing earlier work of Delsarte and Goethals [29], have given a method for finding the weight enumerators of all irreducible cyclic codes. For example in the binary case, let N be a fixed odd number and let k be the smallest positive number such that $2^k \equiv 1 \pmod N$. Then there are binary irreducible cyclic ($n = (2^{km} - 1)/N$, km) codes $\mathcal{C}_m$, for $m = 2, 3, \dots$. Each $\mathcal{C}_m$ consists of the zero vector plus N cycles of n codewords each. Let $w_0, w_1, \dots, w_{N-1}$ be the weights of these cycles. Then the generating function $w_0 + w_1 y + \dots + w_{N-1} y^{N-1}$ is given by

$$2^{m-1}(E(y))^m \pmod{y^N - 1}$$

where $E(y)$ is independent of m. For example when $N = 7$ we obtain (9, 6), (73, 9), (585, 12), (4681, 15), ... codes with minimum distances respectively equal to 2, 28, 280, 2320, ... (The complete weight distributions are given in [78].)

The weight distributions of several other families of cyclic codes have been given by Oganesyan and Yagdzhan [81], [92]. We just mention one of these, which consists of codes with check polynomials of the form $h(x) = \Pi_{i=0}^{f} p_i(x)$, where $p_0(x)$ is irreducible of degree k_0 and period e_0, $m_0 = (2^{k_0} - 1)/e_0$ is prime, 2 is a primitive root of m_0, $p_i(x)$ is a primitive polynomial of degree k_i and period e_i, and the numbers e_i are relatively prime.

2.6. *Perfect codes.* An e-error-correcting code over GF(q) is called *perfect* if every vector is at a distance of at most e from the nearest codeword.

Examples of perfect codes are various trivial codes containing 1, 2, or q^n codewords; the Hamming $d = 3$ codes over any field; and the two Golay codes, the (11, 6) $d = 5$ code over GF(3) and the (23, 12) $d = 7$ code over GF(2).

A long-standing conjecture that no other perfect codes exist over finite fields has recently been proved by Tietäväinen, using earlier work of Lloyd and Van Lint [107], [108], [112].

The (12, 6) and (24, 12) extended Golay codes have many important combinatorial properties. Their symmetry groups are the Mathieu groups M_{12} and M_{24}; their low weight vectors form the Steiner systems $S(5, 6, 12)$ and $S(5, 8, 24)$; and the lattices Λ_{12} and Λ_{24} (the Leech lattice) can be constructed from them [62], [65]. A series of uniqueness theorems have been proved: Pless [87] showed the uniqueness of the Golay codes, Stanton [104] the uniqueness of the Mathieu groups, Witt [116] the uniqueness of the associated Steiner systems, and Conway [24], [25] the uniqueness of the Leech lattice.

[illegible] showed that the Nordstrom–Robinson code (§3.2) is contained in the (24, 12) code. Berlekamp [11] has studied the symmetry groups of the principal subcodes of the (24, 12) code.

Since the Nordstrom–Robinson code is the first member of Preparata's family of nonlinear double-error correcting codes (§3.2), it is natural to ask if the others can be extended to give codes analogous to the Golay code. However, Preparata [91] has shown that this is impossible, in one way at least.

Turyn [2] showed that the (24, 12) Golay code may be obtained as the set of vectors of the form $(a+x, b+x, a+b+x)$ $a, b \in \mathcal{C}_1, x \in \mathcal{C}_2$, where $\mathcal{C}_1$ and $\mathcal{C}_2$ are two different first order RM codes. The same construction was used in [S4] to obtain an infinite family of linear codes with $d/n = \frac{1}{3}$. The first three codes of the family are the (24, 12) Golay code, and (48, 15) $d = 16$ and (96, 18) $d = 32$ codes. As the length increases the rate approaches zero. A generalization of the (12, 6) Golay code is described in the next section.

Parker and Nikolai [82] described an unsuccessful search for simple transitive groups analogous to the Mathieu groups.

2.7. *Abelian group codes.* Let $\mathcal{C}$ be a binary cyclic (n, k) code. If codewords are represented by polynomials, $c_0 c_1 \ldots c_{n-1} \leftrightarrow c_0 + c_1 x + \ldots + c_{n-1} x^{n-1}$, then it is well known that the codewords in $\mathcal{C}$ form an ideal in the ring of polynomials modulo $x^n - 1$ [P1, Ch. 8].

MacWilliams [71], [72], Berman [17], [18] and others [21], [27] have studied the following generalization of cyclic codes. Let $G = \{g_1, \ldots, g_n\}$ be a multiplicative abelian group, and let R denote the set of all formal sums

$$c_1 g_1 + c_2 g_2 + \ldots + c_n g_n,\ c_i = 0 \text{ or } 1$$

with the obvious addition and multiplication. R is a vector space of dimension n over GF(2). An ideal $\mathcal{A}$ is a linear subspace of R such that if $A \in \mathcal{A}, g \in G$ then $gA \in \mathcal{A}$. Then $\mathcal{A}$ is a natural generalization of a cyclic code, and is called an abelian group code.

Many properties of cyclic codes carry over to abelian group codes, such as the existence of a generator codeword whose multiples generate the code.

Berman [18] has shown that for fixed rate and for block lengths n which are divisible by a fixed set of primes (and only by these primes), as $n \to \infty$ abelian group codes have higher minimum distance than cyclic codes.

2.8. *Goppa and Srivastava codes.* Goppa [G4] has recently described a new family of linear noncyclic codes, some of which meet the Gilbert bound.

Let integers m, t be given satisfying $3 \leq m < mt < 2^m$. Let

$$Z = \{z \in GF(2^{mt}) | \text{ degree of minimal polynomial of } z \text{ is } mt\},$$

and let α be a primitive element of $GF(2^m)$. Then for any $z \in Z$ the binary Goppa code $\mathcal{C}(m, t, z)$ is the $(n = 2^m, k \geq 2^m - mt)$ code with the $mt \times 2^m$ parity check matrix

$$H = \left[\frac{1}{z-0}, \frac{1}{z-1}, \frac{1}{z-\alpha}, \ldots, \frac{1}{z-\alpha^{2^m-2}}\right].$$

Goppa has shown that the minimum distance of $\mathcal{C}(m, t, z)$ is (i) at least $2t+1$ for all $z \in Z$, (ii) equal to that given by the Gilbert bound for some $z \in Z$. Unfortunately it is not known how to choose $z \in Z$ so as to make this happen.

Srivastava codes [B1, §15.1] resemble Goppa codes. Helgert [H2], [H3] has recently found a number of good Srivastava codes.

2.9. *Circulant codes.* In 1964 Leech [61] showed that the generator matrix for the (23, 12) Golay code can be written as

$$\left[\begin{array}{ccc|c} 1 & & & C \\ & \ddots & & \\ & & 1 & 1 \ldots 1 \end{array}\right]$$

where C is a circulant matrix, that is, each row is a cyclic shift of the previous row by one place. In this case the first row of C can be taken to be 1 1 0 1 1 1 0 0 0 1 0, having a 1 at position 0 and at the quadratic residues of 11.

Then in 1965 in an important paper ([K1], see also [46], [49], [50]) Karlin found a large number of binary codes generated by circulants, many having a higher rate than the best codes previously known. Examples are (27, 14), (30, 16), (34, 12), and (53, 14) codes, having minimum distances respectively equal to 7, 7, 11 and 17.

This approach also simplified the calculation of the minimum distance, and Karlin was able to determine the minimum distance of a number of binary quadratic residue codes, e.g., the (79, 40) $d = 15$ and (89, 45) $d = 17$ QR codes. Karlin also asserts that the QR codes of lengths 103 and 107 both have minimum distance equal to 19.

Pless [88], [89] has constructed self-dual $(2q+2, q+1)$ codes over GF(3) for every odd prime power $q \equiv -1 \pmod 3$. These are circulant codes, with a generator matrix of the form

$$\left[\begin{array}{cccc|cccc} 1 & & & & 1 & & & \\ & 1 & & & 1 & & C & \\ & & 1 & & 1 & & & \\ & & & 1 & 0 & 1 & 1 & 1 \end{array}\right]$$

where C is a circulant matrix. The first five are (12, 6) (the ternary Golay code), (24, 12), (36, 18), (48, 24) and (60, 30) codes, with minimum distances 6, 9, 12, 15 and 18 respectively. These five codes have rate $\frac{1}{2}$ and $d = \frac{1}{4}n + 3$. But unfortunately later codes in the family have smaller distances. Nevertheless circulant codes are a very promising area for research.

§3. Nonlinear codes and codes formed by combining other codes

Since the number of codewords in a nonlinear code need not be a power of the alphabet size, it is convenient to have a new notation:

An (n, M, d) code $\mathcal{C}$ is a set of M codewords of length n, with symbols from GF(q) and minimum distance d. The dimension of this code is $k = \log_q M$, the redundancy is $r = n - \log_q M$, and the rate is $R = k/n$. Now k and r need not be integers.

A *coset* of $\mathcal{C}$ is an arbitrary translation $a + \mathcal{C}$ of the codewords of $\mathcal{C}$ (where a is any vector of length n). If $\mathcal{C}$ is linear, then two cosets of $\mathcal{C}$ are either equal or disjoint, but this need not be true if $\mathcal{C}$ is nonlinear.

An (n, M, d) code is said to be *optimal* if it has the largest possible number of codewords for the given values of n and d. This is using optimal in a very naive sense, of course, since it omits any consideration of encoding and decoding. But it can be argued that once good codes have been found, the techniques for their implementation will be developed later, as has happened with BCH codes [B1, Ch. 7].

It seems reasonable to expect that optimal codes will often be nonlinear, and that even near-optimal linear codes will have a complicated structure. As a well-known verse by J.L. Massey [70] says, "... good codes just might be messy."

Nonlinear codes have been successfully used to construct dense sphere packings in Euclidean space [63–65].

3.1. *Codes derived from Hadamard and conference matrices.* An $n \times n$

Hadamard matrix $\mathcal{H}_n$ is a matrix of +1's and −1's such that $\mathcal{H}_n \mathcal{H}_n^t = nI$ (where I is a unit matrix). Replacing +1's by 0's and −1's by 1's converts $\mathcal{H}_n$ into a *binary Hadamard* matrix H_n.

It was shown by Plotkin [P1, p. 79], [P3], [B1, p. 316] that the $(n, 2n, \frac{1}{2}n)$ code consisting of the rows of H_n and its complement is optimal. When n is a power of 2 this is a (linear) first order Reed–Muller code, while in the other cases this is a nonlinear Hadamard code.

Many other nonlinear codes can be obtained by manipulating Hadamard matrices. Levenshtein ([L2], see also [58, p. 206], [83]) showed that optimal codes for all d and all $n \leq 2d$ can be obtained in this way (provided the requisite Hadamard matrices exist) by showing that such codes meet the Plotkin bound [P3]. Patel [84] has determined the optimal linear codes in the same region.

Recently [S2] good nonlinear codes with n slightly greater than $2d$ have been obtained from conference matrices. An $n \times n$ *conference matrix* T_n is a matrix with 0's on the diagonal and ±1's elsewhere, satisfying $T_n T_n' = (n-1)I$. Whenever a symmetric T_n exists, an $(n-1, 2n, \frac{1}{2}(n-2))$ binary nonlinear code can be constructed. The first few codes obtained are the (9, 20, 4) optimal code of Julin [J1], a (13, 28, 6) code which is inferior to Nadler's code [80], the (17, 36, 8) code given in [S1], and (25, 52, 12), (29, 60, 14), (37, 76, 18), (41, 84, 20) codes.

3.2. *Preparata and Kerdock codes.* Nonlinear double-error-correcting codes were constructed by Nadler ([80], a (12, 32, 5) code), Green ([37], (13, 64, 5)), and Nordström and Robinson ([N1], $(15, 2^8, 5)$). Van Lint [114] has given a simple construction of the Nadler code.

Preparata [P4] gave a high-rate generalization of the Nordström–Robinson code. For every even $m \geq 4$ he constructed a nonlinear $(2^m - 1, 2^{2^m - 2m}, 5)$ code. These codes are optimal, contain twice as many codewords as double-error-correcting BCH codes, and have straightforward encoding and decoding algorithms.

Semakov and Zinov'ev [96], [97] and Goethals and Snover [36] have independently obtained the weight distribution of the Preparata codes.

Kerdock [K5] has given a corresponding low-rate generalization of the Nordström–Robinson code. He showed that for every even $m \geq 4$ it is possible to take the union of the $(2^m, 2^{m+1}, 2^{m-1})$ first order RM code and $2^{m-1} - 1$ of its cosets to obtain a

$$(2^m, 2^{2m}, 2^{m-1} - 2^{\frac{1}{2}(m-2)})$$

nonlinear code. For $m = 4$ this is the extended Nordström–Robinson code; for $m = 6$ this is a $(64, 2^{12}, 28)$ code containing four times as many codewords as the best extended cyclic code of that length and distance.

The Preparata and Kerdock codes are "duals" in the sense that their weight distributions satisfy the MacWilliams identity [B1, p. 401]. The reason for this is not yet understood.

The next four sections describe constructions for combining two, three, or four codes to obtain new codes.

3.3. *Constructions X and X4.* Construction X combines three codes to form a fourth. Suppose we are given an (n_1, M_1, d_1) code $\mathcal{C}_1$ and an $(n_1, M_2 = bM_1, d_2)$ code $\mathcal{C}_2$, with the property that $\mathcal{C}_2$ is the union of b disjoint cosets of $\mathcal{C}_1$,

$$\mathcal{C}_2 = (x_1 + \mathcal{C}_1) \cup (x_2 + \mathcal{C}_1) \cup \ldots \cup (x_b + \mathcal{C}_1),$$

for some set of vectors $S = \{x_1, x_2, \ldots, x_b\}$. Let

$$\mathcal{C}_3 = \{y_1, y_2, \ldots, y_b\}$$

be any (n_3, b, Δ) code.

Let π be an arbitrary permutation of $\{1, 2, \ldots, b\}$; so that $x_i \to y_{\pi(i)}$ defines a one-one mapping from S onto $\mathcal{C}_3$.

The new code is then defined to be

$$\mathcal{C}_4 = (x_1 + \mathcal{C}_1, y_{\pi(1)}) \cup (x_2 + \mathcal{C}_1, y_{\pi(2)}) \cup \ldots$$
$$\ldots \cup (x_b + \mathcal{C}_1, y_{\pi(b)}).$$

Simply stated, $\mathcal{C}_2$ is divided into cosets of $\mathcal{C}_1$, and a different codeword of $\mathcal{C}_3$ is attached to each coset. See fig. 1.

Then $\mathcal{C}_4$ is an $(n_1 + n_3, M_2, d = \min\{d_1, d_2 + \Delta\})$ code. Similarly construction $X4$ combines four codes to form a fifth. See [S4] for details and further examples.

Example 1. Take $\mathcal{C}_1$ to be a Preparata code, $\mathcal{C}_2$ a Hamming code, $\mathcal{C}_3$ an even weight code. Then, after showing that the Hamming code is a union of cosets of the Preparata code, one obtains $(2^m + m - 1, 2^{2^m - m - 1}, 5)$ codes for $m \geq 4$. Using construction $X4$ one can do even better, and extend the Preparata code by the addition of $\sqrt{n+1}$ information symbols at the cost of adding one check symbol [S4].

Example 2. Using BCH codes one obtains new codes having at least as many codewords as those given by the Andryanov–Saskovets construction [B1, p. 333]. In some cases e-error-correcting BCH codes may be extended by the addition of about $n^{1/e}$ information symbols at the cost of adding one check symbol [S4].

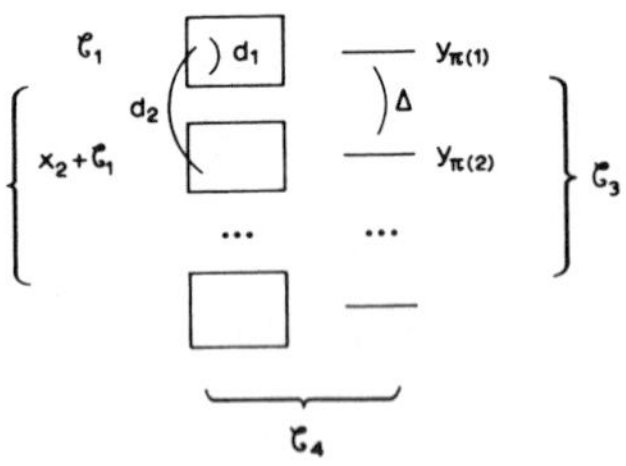

Fig. 1.

3.4. *Constructions Y1, Y2, Y3.* The following constructions were suggested by Goethals [34].

Construction Y1. Let $\mathcal{C}_1$ be an $(n, 2^{k_1}, d_1)$ linear code and let $\mathcal{C}_2$ be its $(n, 2^{r_1}, d_2)$ dual code, with coordinates chosen so that there is a minimum weight codeword 1 ... 10 ... 0 in $\mathcal{C}_2$. Let S be the subgroup of $\mathcal{C}_1$ in which the first $d_2 - 1$ coordinates are zero. Then the d_2^{th} coordinates of S are also zero. If the initial d_2 zeros are deleted from S we are left with an

$$(n - d_2, 2^{k_1 - d_2 + 1}, d_1)$$

linear code.

Construction Y2. Let T be the union of S and all of the $d_2 - 1$ cosets of S in $\mathcal{C}_1$ with coset leaders 110^{n-2}, 1010^{n-3}, ..., $10^{d_2-2}10^{n-d_2}$. By deleting the first d_2 coordinates of T we obtain an

$$(n - d_2, d_2 2^{k_1 - d_2 + 1}, d_1 - 2)$$

nonlinear code.

Construction Y3. Taking all the cosets with coset leaders of weight 2 we obtain an

$$(n - d_2, \left(1 + \binom{d_2}{2}\right) 2^{k_1 - d_2 + 1}, d_1 - 4)$$

nonlinear code.

Many examples of codes obtained by these constructions are given in [S4].

3.5. *Construction Z.* This combines two codes to form a third [P3], [44], [S1], [68].

Let $\mathcal{C}_1 = (n_1, M_1, d_1)$ and $\mathcal{C}_2 = (n_2, M_2, d_2)$ be arbitrary codes over GF(q). Let σ denote the zero vector of length $|n_1 - n_2|$. Then the new code $\mathcal{C}_3$ is defined as follows.

(i). If $n_1 \le n_2$, $\mathcal{C}_3 = \{(x, (x, \sigma) + y) | x \in \mathcal{C}_1, y \in \mathcal{C}_2\}$, and is an $(n_1 + n_2, M_1 M_2, d = \min(2d_1, d_2))$ code. (In the definition of $\mathcal{C}_3$, the comma denotes concatenation and + denotes vector addition in GF(q).)

(ii). If $n_1 > n_2$, $\mathcal{C}_3 = \{(x, x + (y, \sigma)) | x \in \mathcal{C}_1, y \in \mathcal{C}_2\}$, and is a $(2n_1, M_1 M_2, d = \min(2d_1, d_2))$ code.

$\mathcal{C}_3$ is linear if $\mathcal{C}_1$ and $\mathcal{C}_2$ are. A number of applications are given in [S1], including the construction of an infinite family of nonlinear single-error-correcting codes which contain more codewords than shortened Hamming codes. Other examples, of which those in table 1 are typical, will be seen in table 2.

3.6. *Assmus and Mattson's rate $\frac{1}{3}$ cyclic codes.* Let p be a prime of the form $8N + 5$ for which 2 is a primitive root (e.g., $p = 5, 13, 29, 37, \ldots$). Assmus and Mattson [4] showed how to concatenate three different versions of the $(p, p-1)$ even weight code to obtain a linear binary cyclic $(3p, p-1)$ code, denoted by $3E$, with minimum distance at least $2\sqrt{3p}$. Let $3E^*$ be the cyclic $(3p, p)$ code consisting of the codewords of $3E$ together with their complements. The first few examples of $3E^*$ are $(15, 5)$ $d = 7$, $(39, 13)$ $d = 12$, $(87, 29)$ $d = 24$, and $(111, 37)$ $d = 24$ codes.

3.7. *Other constructions.* Other techniques for constructing codes have been given in [26], [28], [76], [99]. However the codes obtained appear to contain fewer, or at best as many, codewords as known cyclic codes.

Table 1

$\mathcal{C}_1$			$\mathcal{C}_2$			$\mathcal{C}_3$		
n_1	r_1	d_1	n_2	r_2	d_2	n	r	d
23	11	7	11	4.830	4	34	15.830	7
19	12	8	19	18	19	38	30	16
14	13	14	264	101	27	278	114	27

§4. A table of the best codes presently known

4.1. For a given value of the length n and minimum distance d, let M be the maximum number of codewords of any binary (n, M, d) code, linear or nonlinear, that is presently known (to the author). Then table 2 gives the redundancy $r = n - \log_2 M$ of this code as a function of n and d, for all $n \le 512$ and $d \le 30$. These codes are of considerable theoretical interest in themselves, provide a basis for judging new codes, and as a lower bound to the densest possible codes complement Johnson's table of upper bounds [48]. Previous tables of codes are to be found in [B1], [C1], [C3], [G3], [L1], [P1], [P3], [W1], and [86].

4.2. *Types of codes.* The codes are classified as follows.

B = Bose–Chaudhuri–Hocquenghem code [B1, Ch. 7].
C = Cyclic linear code [P1, Ch. 8].
D = Goppa code ([G4] and §2.8 above).
G = Group or linear code [P1, Ch. 3].
H = Hadamard code [L2].
J = Code from conference matrix [S2].
K = Circulant code [K1].
N = Nonlinear code.
P = Nordström–Robinson–Preparata code [N1], [P4].
Q = Quadratic residue code ([B1, §15.2], [111, §4.4]).
R = Reed–Muller code [B1, §15.3].
S = Srivastava code [B1, §15.1], [H2].
XA, XC, XP = Codes from construction X applied to BCH codes (the generalized Andryanov–Saskovets construction), to cyclic codes, and to Preparata codes respectively ([S4] and §3.3 above).
X4 = Codes from construction $X4$ ([S4] and §3.3 above).
Y1, Y2, Y3 = Codes from constructions $Y1$, $Y2$, $Y3$ ([S4] and §3.4 above).
Z = Codes from construction Z ([S1] and §3.5 above).

Types B, C, D, G, K, Q, R, S, XA, XC, Y1 are linear, H, J, N, P, XP, Y2, Y3 are nonlinear, and X4, Z may be linear or nonlinear.

In table 2, codes for which the reference [S3] is given are new. With one exception these are all examples of constructions mentioned in the text. The exception is an $(85, 18)$ $d = 25$ which was obtained using construction 39 of Hatcher [44].

4.3. Since an $(n, M, d$ odd$)$ code is equivalent to an $(n+1, M, d+1)$ code, only codes for odd d need be given. An (n, M, d) code may be punctured to give $(n-i, M, d-i)$ codes for $0 < i < d$, or shortened to give $(n-i, M2^{-i}, d)$ codes for $0 < i < \log_2 M$. In table 2 all such modified codes carry the name of the original code. An (n, M, d) of redundancy r may be thought of as an $(n+i, M, d)$ code of redundancy $r + i$, for $i \ge 1$, in which case the name is left blank.

4.4. Some nonlinear codes in table 2 have redundancy r which is not a whole number. In such cases the number of codewords is quickly found as follows: If $r = R + a$, where R is a whole number and $0 < a < 1$, the number of codewords is $M = i2^{n-R-5}$ where i is given by

a	.046	.093	.142	.193	.245	.300	.356	.415	.541	.608	.678	.752	.830	.913
i	31	30	29	28	27	26	25	24	22	21	20	19	18	17

4.5. In many cases the reference is to a place where the complete weight distribution of the code may be found, rather than to the original determination of the minimum distance.

4.6. Although many of these codes may be optimal, in the sense of having the smallest possible redundancy, very few of them are known to be optimal. (Compare [48].) The reader is invited to try and improve on them. Those of type Z and distances 25–29 are especially weak. The author is eager to hear of any improvements.

4.7. In the first section of table 2, for minimum distance $d = 3$, when $n = 3, 4, 5$ and $3 \cdot 2^{m-2} \le n < 2^m$, $m \ge 3$, the codes shown are (shortened) Hamming codes [H1]. When $2^m \le n < 3 \cdot 2^{m-1}$, $m \ge 3$, the codes shown are nongroup codes, discovered by Golay [G2] and Julin [J1] for $n = 8, 9, 10, 11$ and by Sloane and Whitehead [S1] for $n \ge 16$.

Acknowledgments

I should like to thank H.L. Berger for his help in collecting data for table 2, E.R. Berlekamp for information about recent Russian work, F.J. MacWilliams for helpful comments on the manuscript, and H.J. Helgert for supplying a number of good codes.

TABLE 2
BINARY CODES OF LENGTH N,
MINIMUM DISTANCE D, AND
SMALLEST KNOWN REDUNDANCY R.

NOTE: AN ASTERISK * DENOTES A CODE WHICH HAS BEEN ADDED TO THE TABLE SINCE THE ORIGINAL VERSION OF THIS PAPER WAS PUBLISHED. NEW REFERENCES WILL BE FOUND AT THE END OF THE PAPER.

DISTANCE D = 3

N	R
4- 7	3
8	3.678
9	3.752
10- 11	3.830
12- 15	4
16- 17	4.678
18- 19	4.752
20- 23	4.830
24- 31	5
32- 35	5.678
36- 39	5.752
40- 47	5.830
48- 63	6
64- 71	6.678
72- 79	6.752
80- 95	6.830
96-127	7
128-143	7.678
144-159	7.752
160-191	7.830
192-255	8
256-287	8.678
288-319	8.752
320-383	8.830
384-511	9
512	9.678

DISTANCE D = 5

N	R	TYPE	REF
7- 8	6	G	[L1]
9- 11	6.415	H	[P3]
12- 15	7	P	[N1]
16- 19	8	XP	[S4]
20	8.678	X4	[S4]
21- 23	9	G	[W1]
24- 32	10	B	[G3]
33- 63	11	P	[P4]
64- 70	12	X4	[S4]
71- 73	13	S	[H2]
74-128	14	B	[S4]
129-255	15	P	[P4]
256-271	16	X4	[S4]
272-277	17	S	[H2]
278-512	18	B	[S4]

DISTANCE D = 7

N	R	TYPE	REF
10- 11	9	G	[L1]
12- 15	10	R	[P1]
16	10.830	J	[S1]
17- 23	11	Q	[G1]
24	12		
25- 27	13	K	[K1]
28- 30	14	K	[K1]
31- 32	15	D	[G4]
33	15.752	Z	[S3]
34- 35	15.830	Z	[S3]
36- 63	16	N	[G5] *
64	17		
65- 67	18	XC	[S4]
68- 70	19	XA	[S3]
71- 83	20	S	[H2]
84-128	21	D	[G4]
129-255	22	N	[G5] *
256-257	23- 24		
258-264	25	XA	[S3]
265-311	26	S	[H2]
312-512	27	D	[G4]

DISTANCE D = 9

N	R	TYPE	REF
13- 14	12	G	[L1]
15	13		
16	13.415	H	[L2]
17- 19	13.678	H	[L2]
20	14.678		
21	15.415	H	[L2]
22	16	G	[B2] *
23- 25	16.678	Y2	[S4]
26	17.678		
27- 29	18	B	[P1]
30- 35	18.415	Y2	[S4]
36	19.415		
37- 41	20	Q	[B1]
42- 45	21	Q	[P2]
46	22		
47- 49	22.193	Y2	[S4]
50- 52	23	S	[H2]
53- 73	24	B	[K3]
74- 75	25- 26		
76- 90	27	S	[H2]
91-128	28	B	[S4]
129-135	29	XA	[S3]
136	30		
137-159	31	Y1	[H4] *
160-256	32	B	[S4]
257-264	33	XA	[S3]
265	34		
266-311	35	S	[H3]
312-512	36	B	[S4]

DISTANCE D = 11

N	R	TYPE	REF
16- 17	15	G	[L1]
18	16		
19	16.415	H	[L2]
20	17	B	[G3]
21- 23	17.415	H	[L2]
24	18.300	J	[S2]
25- 26	19	G	[H3]
27- 31	20	B	[P1]
32	21		
33- 35	22	Y1	[S4]
36- 47	23	Q	[P2]
48- 50	24- 26		
51- 63	27	B	[P1]
64- 67	28	XA	[S3]
68- 70	29- 31		
71- 74	32	XC	[S4]
75- 79	33	G	[H4] *
80- 95	34	Y1	[H4] *
96-128	35	D	[G4]
129-135	36	XA	[S3]
136-137	37- 38		
138-191	39	Y1	[H4] *
192-256	40	D	[G4]
257-264	41	XA	[S3]
265-266	42- 43		
267-311	44	S	[H3]
312-512	45	D	[G4]

DISTANCE D = 13

N	R	TYPE	REF
19- 20	18	G	[C2]
21- 22	19- 20		
23	20.415	H	[L2]
24	21	G	[C1]
25- 27	21.193	H	[L2]
28	22.093	J	[S2]
29	23	R	[P1]
30	24		
31	24.830	H	[L2]
32	25.752	J	[S2]
33- 37	26	XA	[S3]
38	27		
39- 43	28	C	[B1]
44- 45	29- 30		
46- 55	31	Y2	[S4]
56	32		
57- 63	33	B	[P1]
64- 70	34	XA	[S3]
71- 72	35- 36		
73- 77	37	Q	[K1]
78- 79	38- 39		
80- 85	40	Q	[K1]
86- 96	41	S	[H3]
97-128	42	B	[S4]
129-135	43	XA	[S3]
136-138	44- 46		
139-191	47	Y1	[H4] *
192-256	48	B	[S4]
257-264	49	XA	[S3]
265-267	50- 52		
268-327	53	Y1	[H4] *
328-512	54	B	[S4]

DISTANCE D = 15

N	R	TYPE	REF
22- 23	21	G	[C2]
24- 25	22- 23		
26	23.415	H	[L2]
27	24	G	[L2]
28	24.678	H	[L2]
29- 31	25	R	[P1]
32	26		
33	26.830	H	[L2]
34	27.752	J	[S2]
35	28	Z	[S3]
36- 37	29	Z	[S3]
38- 41	30	XC	[S4]
42	31		
43- 47	32	G	[S4]
48- 50	33	C	[B1]
51- 55	34	C	[B1]
56- 63	35	C	[K2]
64- 66	36	XC	[S4]
67- 68	37- 38		
69- 71	38.830	Y2	[S4]
72- 79	39	Q	[K1]
80- 81	40- 41		
82- 87	42	Q	[K1]
88- 91	43- 46		
92- 99	47	Q	[K1]
100	48		
101-128	49	D	[G4]
129-135	50	XA	[S3]
136-139	51- 54		
140-191	55	Y1	[H4] *
192-256	56	D	[G4]
257-264	57	XA	[S3]
265-268	58- 61		
269-327	62	Y1	[H4] *
328-512	63	D	[G4]

DISTANCE D = 17

N	R	TYPE	REF
25- 26	24	G	[C2]
27- 28	25- 26		
29	26.415	H	[L2]
30	27.415		
31	28	G	[L2]
32	28.415	H	[L2]
33- 35	28.830	H	[L2]
36	29.752	J	[S2]
37	30.678	H	[L2]
38	31.608	J	[S2]
39	32.541	H	[L2]
40	33.541		
41	34	N	[H3]
42	35	G	[H3]
43- 46	36	G	[H3]
47- 49	37	G	[K4]
50	38		
51- 53	39	K	[K1]
54- 55	39.142	Y3	[S4]
56	40	Y1	[S4]
57- 62	41	C	[C3]
63- 66	42	XC	[S4]
67- 71	43	Y1	[S4]
72- 89	44	Q	[K1]
90- 93	45- 48		
94-101	49	Q	[K1]
102	50		
103-105	51	K	[K1]
106-107	52- 53		
108-125	54	B	[P1]
126	55		
127-128	56	B	[S4]
129-135	57	XA	[S3]
136-140	58- 62		
141-191	63	Y1	[H4] *
192-256	64	B	[S4]
257-264	65	XA	[S3]
265-269	66- 70		
270-331	71	Y1	[H4] *
332-512	72	B	[S4]

Table 2

DISTANCE D = 19

N	R	TYPE	REF	
28- 29	27	G	[C2]	
30- 32	28- 30			
33	30.415	H	[L2]	
34	31	C	[C3]	
35	31.678	H	[L2]	
36	32.415	H	[L2]	
37- 39	32.678	H	[L2]	
40	33.608	J	[S2]	
41	34.541	H	[L2]	
42	35.541			
43	36.415	H	[L2]	
44	37	G	[H3]	
45- 48	38	G	[H3]	
49- 51	39	G	[K4]	
52- 53	40- 41			
54- 59	42	N	[G5]	*
60- 61	43	B	[P1]	
62- 63	44	C	[C3]	
64	45			
65- 66	46	XC	[S4]	
67- 68	47	XC	[S4]	
69- 70	48	XC	[S4]	
71- 74	49	XC	[S4]	
75- 83	50	Y1	[S4]	
84-103	51	Q	[K1]	
104	52			
105-107	53	K	[K1]	
108-109	54- 55			
110-127	56	B	[P1]	
128-131	57- 60			
132-139	61	XC	[S4]	
140-144	62- 66			
145-191	67	Y1	[H4]	*
192-255	68	B	[P1]	
256-260	69	XA	[S3]	
261-268	70- 77			
269-270	78	Z	[S3]	
271	79			
272-339	80	Y1	[H4]	*
340-512	81	D	[G4]	

DISTANCE D = 21

N	R	TYPE	REF	
31- 32	30	G	[C2]	
33- 35	31- 33			
36	33.415	H	[L2]	
37	34.415			
38	35	G	[L2]	
39	35.678	H	[L2]	
40	36.193	H	[L2]	
41- 43	36.541	H	[L2]	
44	37.541			
45	38.415	H	[L2]	
46	39.356	J	[S2]	
47- 48	40	C	[C3]	
49- 51	41- 43			
52- 57	43.415	Y2	[S4]	
58- 61	44	N	[G5]	*
62- 63	45	B	[P1]	
64- 70	46- 52			
71- 77	53	XC	[S4]	
78	54			
79- 85	55	K	[K1]	
86- 91	56- 61			
92-105	62	Y1	[H4]	*
106-127	63	B	[P1]	
128-135	64	XA	[S3]	
136-144	65- 73			
145-146	74	Z	[S3]	
147-191	75	Y1	[H4]	*
192-255	76	B	[P1]	
256-264	77	XA	[S3]	
265-273	78- 86			
274-275	87	Z	[S3]	
276	88			
277-339	89	Y1	[H4]	*
340-512	90	B	[S4]	

DISTANCE D = 23

N	R	TYPE	REF	
34- 35	33	G	[C2]	
36- 38	34- 36			
39	36.415	H	[L2]	
40	37.415			
41	38	G	[L2]	
42	39			
43	39.415	H	[L2]	
44	40	C	[C3]	
45- 47	40.415	H	[L2]	
48	41.356	J	[S2]	
49- 50	42	C	[C3]	
51- 53	43- 45			
54- 63	46	N	[G5]	*
64	47			
65- 66	48	XA	[S3]	
67- 74	49- 56			
75- 87	57	K	[K1]	
88- 97	58- 67			
98-103	68	G	[H4]	*
104-111	69	Y1	[H4]	*
112-127	70	B	[P1]	
128-135	71	XA	[S3]	
136-145	72- 81			
146-150	82	G	[H4]	*
151-207	83	Y1	[H4]	*
208-255	84	B	[P1]	
256-264	85	XA	[S3]	
265-274	86- 95			
275-276	96	Z	[S3]	
277	97			
278-339	98	Y1	[H4]	*
340-512	99	D	[G4]	

DISTANCE D = 25

N	R	TYPE	REF	
37- 38	36	G	[C2]	
39- 42	37- 40			
43	40.415	H	[L2]	
44	41.415			
45	42	G	[L2]	
46	42.678	H	[L2]	
47	43.415	H	[L2]	
48	44	G	[L2]	
49- 51	44.300	H	[L2]	
52	45.245	J	[S2]	
53	46.193	H	[L2]	
54	47.193			
55	48.093	H	[L2]	
56- 61	49	N	[K5]	
62- 63	50- 51			
64- 66	52	K	[K1]	
67	53			
68- 70	54	XA	[S3]	
71	55			
72- 73	56	XC	[S4]	
74- 79	57- 62			
80- 86	63	B	[H4]	*
87- 95	64- 72			
96-101	73	G	[H4]	*
102-109	74	Y1	[H4]	*
110-125	75	B	[P1]	
126-127	76- 77			
128-135	78	XA	[S3]	
136-146	79- 89			
147-156	90	G	[H4]	*
157-207	91	Y1	[H4]	*
208-255	92	B	[P1]	
256-264	93	XA	[S3]	
265-275	94-104			
276-277	105	Z	[S3]	
278	106			
279-339	107	Y1	[H4]	*
340-512	108	B	[S4]	

DISTANCE D = 27

N	R	TYPE	REF	
40- 41	39	G	[C2]	
42- 45	40- 43			
46	43.415	H	[L2]	
47	44.415			
48	45	C	[C3]	
49	46			
50	46.678	H	[L2]	
51	47.193	H	[L2]	
52	47.830	H	[L2]	
53- 55	48.193	H	[L2]	
56	49.193			
57	50.093	H	[L2]	
58- 63	51	N	[K5]	
64- 69	52- 57			
70- 74	58	XC	[S4]	
75- 80	59- 64			
81- 88	65	B	[H4]	*
89- 97	66- 74			
98-103	75	G	[H4]	*
104-111	76	Y1	[H4]	*
112-127	77	B	[P1]	
128-131	78- 81			
132-139	82	XC	[S4]	
140-150	83- 93			
151-160	94	K	[K4]	
161-164	95- 98			
165-207	99	Y1	[H4]	*
208-255	100	B	[P1]	
256-264	101	XA	[S3]	
265-276	102-113			
277-278	114	Z	[S3]	
279	115			
280-347	116	Y1	[H4]	*
348-512	117	D	[G4]	

DISTANCE D = 29

N	R	TYPE	REF	
43- 44	42	G	[C2]	
45- 48	43- 46			
49	46.415	H	[L2]	
50	47.415			
51	48.415			
52	49	G	[L2]	
53	49.678	H	[L2]	
54	50.415	H	[L2]	
55	51.193	H	[L2]	
56	51.678	H	[L2]	
57- 59	52.093	H	[L2]	
60	53.046	J	[S2]	
61	54	R	[P1]	
62	55			
63	55.913	H	[L2]	
64	56.913			
65- 67	57	XA	[S3]	
68- 74	58- 64			
75- 78	65	XC	[S4]	
79- 82	66	C	[H4]	*
83	67			
84- 85	68	B	[H4]	*
86- 91	69- 74			
92- 93	75	G	[S4]	
94- 97	76- 79			
98	79.415	Z	[S3]	
99-101	80	G	[H4]	*
102-111	81	Y1	[H4]	*
112-125	82	B	[K2]	
126-127	83- 84			
128-135	85	XA	[S3]	
136-148	86- 98			
149-150	99	Z	[S3]	
151-155	100-104			
156-158	105	Z	[S3]	
159-170	106	G	[H4]	*
171-207	107	Y1	[H4]	*
208-255	108	B	[P1]	
256-264	109	XA	[S3]	
265-277	110-122			
278-279	123	Z	[S3]	
280	124			
281-347	125	Y1	[H4]	*
348-512	126	B	[S4]	

References

Abbreviations: BSTJ = Bell System Technical Journal, IC = Information and Control, JCT = Journal of Combinatorial Theory, PGIT = IEEE Transactions on Information Theory.

1. *References for the table of codes*

[B1] E.R. Berlekamp, Algebraic coding theory (McGraw-Hill, New York, 1968) (see especially pp. 360, 432–433).

[C1] L. Calabi and E. Myrvaagnes, On the minimal weight of binary group codes, PGIT 10 (1964) 385–387.

[C2] J.T. Cordaro and T.J. Wagner, Optimum $(n, 2)$ codes for small values of channel error probability, PGIT 13 (1967) 349–350.

[C3] C.L. Chen, Computer results on the minimum distance of some binary cyclic codes, PGIT 16 (1970) 359–360.

[F1] A.B. Fontaine and W.W. Peterson, Group code equivalence and optimum codes, PGIT 5 (1959) (Special Suppl.) 60–70.

[G1] M.J.E. Golay, Notes on digital coding, Proc. IRE, 37 (1949) 657.

[G2] M.J.E. Golay, Binary coding, PGIT 4 (1954) 23–28.

[G3] H.D. Goldman, M. Kliman and H. Smola, The weight structure of some Bose–Chaudhuri codes, PGIT 14 (1968) 167–169.

[G4] V.D. Goppa, A new class of linear error-correcting codes, Prob. Peredači Inform. 6 (1970) 24–30 (in Russian).

[H1] R.W. Hamming, Error detecting and error correcting codes, BSTJ 29 (1950) 147–160.

[H2] H.J. Helgert, Srivastava codes, PGIT 18 (1972) 292–297.

[H3] H.J. Helgert, personal communication.

[J1] D. Julin, Two improved block codes, PGIT 11 (1965) 459.

[K1] M. Karlin, New binary coding results by circulants, PGIT 15 (1969) 81–92.

[K2] T. Kasami and N. Tokura, Some remarks on BCH bounds and minimum weights of binary primitive BCH codes, PGIT 15 (1969) 408–413.

[K3] T. Kasami, S. Lin and W.W. Peterson, Polynomial codes, PGIT 14 (1968) 807–814.

[K5] A.M. Kerdock, A class of low-rate nonlinear codes, IC 20 (1972) 182–187.

[K4] M. Karlin, personal communication.

[L1] A.E. Laemmel, Efficiency of noise reducing codes, in: W. Jackson, ed., Communication theory (Butterworth, London, 1953) 111–118.

[L2] V.I. Levenshtein, The application of Hadamard matrices to a problem in coding, Problems of Cybernetics 5 (1964) 166–184.

[L3] V. Lum and R.T. Chien, On the minimum distance of Bose–Chaudhuri–Hocquenghem codes, SIAM J. Appl. Math. 16 (1968) 1325–1337.

[N1] A.W. Nordström and J.P. Robinson, An optimum nonlinear code, IC 11 (1967) 613–616.

[P1] W.W. Peterson, Error-correcting codes (M.I.T. Press, Cambridge, Mass., 1961) (see especially pp. 71, 166–167).

[P2] V.S. Pless, Power moment identities on weight distributions in error correcting codes, IC 6 (1963) 147–152.

[P3] M. Plotkin, Binary codes with specified minimum distances, PGIT 6 (1960) 445–450.

[P4] F.P. Preparata, A class of optimum nonlinear double-error correcting codes, IC 13 (1968) 378–400.

[S1] N.J.A. Sloane and D.S. Whitehead, A new family of single-error correcting codes, PGIT 16 (1970) 717–719.

[S2] N.J.A. Sloane and J.J. Seidel, A new family of nonlinear codes obtained from conference matrices, Ann. New York Acad. Sc. 175 (1970) 363–365.

[S3] A new code.

[S4] N.J.A. Sloane, S.M. Reddy and C.L. Chen, New binary codes, PGIT 18 (1972) 503–510.

[T1] N. Tokura, K. Taniguchi and T. Kasami, A search procedure for finding optimum group codes for the binary symmetric channel, PGIT 13 (1967) 587–594.

[W1] T.J. Wagner, A search technique for quasi-perfect codes, IC 9 (1966) 94–99.

2. *Further references cited in text*

[1] D.R. Anderson, A new class of cyclic codes, SIAM J. Appl. Math. 16 (1968) 181–197.

[2] E.F. Assmus Jr., H.F. Mattson Jr. and R.J. Turyn, Research to develop the algebraic theory of codes (Sci. Rept. AFCRL-67-0365, Air Force Cambridge Res. Lab., Bedford, Mass., 1967).

[3] E.F. Assmus Jr. and H.F. Mattson Jr., New 5-designs, JCT 6 (1969) 122–151.

[4] E.F. Assmus Jr. and H.F. Mattson Jr., Some $(3p, p)$ codes, in: Information processing 68 (North-Holland, Amsterdam, 1969) 205–209.

[5] L.D. Baumert and R.J. McEliece, Weights of irreducible cyclic codes, to appear.

[6] T. Berger, Rate distortion theory (Prentice-Hall, Englewood Cliffs, N.J., 1971).

[7] E.R. Berlekamp, Weight enumeration theorems, in: Proc. 6th Allerton Conf. on Circuit and Systems Theory, Urbana (Univ. of Illinois Press, Chicago, Ill. 1968) 161–170.

[8] E.R. Berlekamp, The weight enumerators for certain subcodes of the second order binary Reed–Muller codes, IC 17 (1970) 485–500.

[9] E.R. Berlekamp, Some mathematical properties of a scheme for reducing the bandwidth of motion pictures by Hadamard smearing, BSTJ 49 (1970) 969–986.

[10] E.R. Berlekamp, A survey of coding theory for algebraists and combinatorialists (Intern. Centre for Mech. Sci., Udine, Italy, 1970).

[11] E.R. Berlekamp, Coding theory and the Mathieu groups, IC 18 (1971) 40–64.

[12] E.R. Berlekamp, Long primitive binary BCH codes have distance $d \sim 2n \ln R^{-1}/\log n$ ..., PGIT 18 (1972) 415–426.

[14] E.R. Berlekamp and N.J.A. Sloane, Weight enumerator for second order Reed–Muller codes, PGIT 16 (1970) 745–751.

[15] E.R. Berlekamp and L.R. Welch, Weight distributions of the cosets of the (32, 6) Reed–Muller code, PGIT 18 (1972) 203–207.

[16] E.R. Berlekamp, F.J. MacWilliams and N.J.A. Sloane, Gleason's theorem on self dual codes, 18 (1972) 409–414.

[17] S.D. Berman, On the theory of group codes, Cybernetics 3 (1967) 25–31.

[18] S.D. Berman, Semisimple cyclic and abelian codes II, Cybernetics 3 (1967) 17–23.

[19] R.C. Bose and D.K. Ray-Chaudhuri, On a class of error correcting binary group codes, IC 3 (1960) 68–79, 279–290.

[20] H.O. Burton, A survey of error correcting tecniques for data on telephone facilities, in: Proc. Intern. Commun. Conf., San Francisco, Calif., 1970.

[21] P. Camion, Abelian codes, Math. Res. Center, Univ. of Wisconsin, Rept. 1059 (1970).

[22] C.L. Chen, The existence of arbitrarily long pseudo-cyclic codes that meet the Gilbert bound, in: Proc. 5th Ann. Princeton Conf. Inform. Sci. (1971) 242.

[23] C.L. Chen, W.W. Peterson and E.J. Weldon Jr., Some results on quasicyclic codes, IC 15 (1969) 407–423.

[24] J.H. Conway, A group of order 8,315,553,613,086,720,000, Bull. London Math. Soc. 1 (1969) 79–88.

[25] J.H. Conway, A characterization of Leech's lattice, Invent. Math. 7 (1969) 137–142.

[26] G. Dagnino, On a new class of binary group codes, Calcolo 5 (1968) 277–294.

[27] P. Delsarte, Automorphisms of abelian codes, Philips Res. Rept. 25 (1970) 389–402.

[28] P. Delsarte, Majority logic decodable codes derived from finite inversive planes, IC 18 (1971) 319–325.

[29] P. Delsarte and J.M. Goethals, Irreducible binary cyclic codes of even dimension, Univ. North Carolina at Chapel Hill, Inst. Statist., Mimeo Ser. No. 600.27, 1970.

[30] R.L. Dobrushin, Survey of Soviet research in information theory, to appear.

[31] E.N. Gilbert, A comparison of signaling alphabets, BSTJ 31 (1952) 504–522.

[32] A.M. Gleason, Weight polynomials of self-dual codes and the MacWilliams identities, in: Proc. Intern. Congr. Mathematicians, Nice (1970) 140–144.

[33] J.M. Goethals, Factorization of cyclic codes, PGIT 13 (1967) 242–246.

[34] J.M. Goethals, On the Golay perfect binary code, JCT 11 (1971) 178–186.

[35] J.M. Goethals, Some combinatorial aspects of coding theory, in: Proc. Combinat. Symp., Fort Collins, 1971, to appear.

[36] J.M. Goethals and S.L. Snover, Nearly perfect binary codes, Discrete Math. 3 (1972) 65–88 (this issue).

[37] M.V. Green, Two heuristic techniques for block-code construction (Abstract), PGIT 12 (1966) 273.

[38] C.R.P. Hartmann, On the minimum distance structure of cyclic codes and decoding beyond the BCH bound, Ph. D. Thesis, Univ. of Illinois, 1970; also Coord. Sci. Lab. Rept. R-458, Univ. of Illinois, 1970.

[39] C.R.P. Hartmann, A note on the minimum distance structure of cyclic codes, PGIT 18 (1972) 439–440.

[40] C.R.P. Hartmann, A generalization of the BCH bound, submitted to IC.

[41] C.R.P. Hartmann, On the weight structure of cyclic codes of composite length, in: Proc. Fourth Hawaii Inter. Conf. System Sci., (1971) 117–119.

[42] C.R.P. Hartmann and K.K. Tzeng, A bound for cyclic codes of composite length, PGIT 18 (1972) 307.

[43] C.R.P. Hartmann, K.K. Tzeng and R.T. Chien, Some results on the minimum distance structure of cyclic codes, PGIT 18 (1972) 402–409.

[44] T. Hatcher, On minimal distance, shortest length, and greatest number of elements for binary group codes (Parke Mathematical Labs., Carlisle, Mass., Tech. Memo. 6, 1964).

[45] A. Hocquenghem, Codes correcteurs d'erreurs, Chiffres, 2 (1959) 147–156.

[46] C.W. Hoffner II and S.M. Reddy, Circulant bases for cyclic codes, PGIT 16 (1970) 511–512.

[47] F. Jelinek, Free encoding of memoryless time-discrete sources with a fidelity criterion, PGIT 15 (1969) 584–590.

[48] S.M. Johnson, On upper bounds for unrestricted binary error-correcting codes, PGIT 17 (1971) 466–478.

[49] M. Karlin, Decoding of circulant codes, PGIT 16 (1970) 797–802.

[50] M. Karlin, Weight/moment relationships in $(Q + E)$ circulants, unpublished.

[51] T. Kasami, Some lower bounds on the minimum weight of cyclic codes of composite length, PGIT 14 (1968) 814–818.

[52] T. Kasami, An upper bound on k/n for affine-invariant codes with fixed d/n, PGIT 15 (1969) 174–176.

[53] T. Kasami, The weight enumerators for several classes of subcodes of the second order binary Reed–Muller codes, IC 18 (1971) 369–394.

[54] T. Kasami, Some results on the weight structure of Reed–Muller codes, to appear.

[55] T. Kasami, S. Lin and W.W. Peterson, Some results on weight distributions of BCH codes, PGIT 12 (1966) 274.

[56] T. Kasami and N. Tokura, On the weight structure of Reed–Muller codes, PGIT 16 (1970) 752–759.

[57] T. Kasami, N. Tokura and S. Azumi, On the weight distribution of Reed–Muller codes, Inst. Electron. Comm. Eng., Japan, PGIT Rept. (1971) (in Japanese).

[58] W.H. Kautz and K.N. Levitt, A survey of progress in coding theory in the Soviet Union, PGIT 15 (1969) 197–244.

[59] V.N. Koshelev, Some properties of random group codes of large length, Probl. Peredačy Inform. 1 (1965) 45–48.

[60] M.V. Kozlov, The correcting capacities of linear codes, Soviet Physics – Doklady 14 (1969) 413–415.

[61] J. Leech. Some sphere packings in higher space, Can. J. Math. 16 (1964) 657–682.

[62] J. Leech, Notes on sphere packings, Can. J. Math. 19 (1967) 251–267.

[63] J. Leech and N.J.A. Sloane, New sphere packings in dimensions 9–15, Bull. Amer. Math. Soc. 76 (1970) 1006–1010.

[64] J. Leech and N.J.A. Sloane, New sphere packings in more than thirty-two dimensions, in: Proc. second Chapel Hill Conference on Comb. Math., Chapel Hill, N.C. (1970) 345–355.

[65] J. Leech and N.J.A. Sloane, Sphere packing and error-correcting codes, Can. J. Math 23 (1971) 718–745.

[66] V.K. Leont'ev, A hypothesis on Bose–Chaudhuri codes, Probl. Peredačy Inform. 4 (1968) 66–70.

[67] S. Lin and E.J. Weldon Jr., Long BCH codes are bad, IC 11 (1967) 445–451.

[68] C.L. Liu, B.G. Ong and G.R. Ruth. A construction scheme for linear and nonlinear codes, in: Proc. 5th Ann. Princeton Conf. Inform. Sci. (1971) 245–247.

[69] R.W. Lucky, J. Salz and E.J. Weldon Jr., Principles of data communication (McGraw Hill, New York, 1968).

[70] F.J. MacWilliams, Error-correcting codes – An historical survey, in: H.B. Mann, ed., Error correcting codes (Wiley, New York, 1968).

[71] F.J. MacWilliams, Codes and ideals in group algebras, in: R.C. Bose and T.A. Dowling, eds., Combinatorial mathematics and its applications (Univ. North Carolina Press, Chapel Hill, 1969) Ch. 18.

[72] F.J. MacWilliams, Binary codes which are ideals in the group algebra of an abelian group, BSTJ 49 (1970) 987–1011.

[73] F.J. MacWilliams, C.L. Mallows and N.J.A. Sloane, Generalizations of Gleason's theorem on weight enumerators of self-dual codes, PGIT 18 (1972), to appear.

[74] F.J. MacWilliams, N.J.A. Sloane and J.G. Thompson, On the existence of a projective plane of order 10, JCT, to appear.

[75] F.J. MacWilliams, N.J.A. Sloane and J.G. Thompson, Good self-dual codes exist, Discrete Math. 3 (1972) 153–162 (this issue).

[76] A.S. Marchukov, Summation of the product of codes, Probl. Peredačy Inform. 4 (1968) 8–15.

[77] R.J. McEliece, On the symmetry of good nonlinear codes, PGIT 16 (1970) 609–611.

[78] R.J. McEliece and H. Rumsey, Jr., Euler products, cyclotomy, and coding, in: Space programs summary (Jet Propulsion Lab., Calif. Inst. Technol.) Vol. 37-65-III (1970) 22–27; and J. Number Theory 4 (1972) 302–311.

[79] D.E. Muller, Application of boolean algebra to switching circuit design and error detection, IRE Trans. Electronic Computers, EC3 (1954) 6–12.

[80] M. Nadler, A 32-point $n = 12, d = 5$ code, PGIT 8 (1962) 58.

[81] S.Sh. Oganesyan and V.G. Yagdzhyan, Weight spectra for some classes of cyclic error-correcting codes, Probl. Peredači Inform. 6 (1970) 31–37 (in Russian).

[82] E.T. Parker and P.J. Nikolai, A search for analogues of the Mathieu groups, Math. Comp. 12 (1958) 38–43.

[83] A.M. Patel, Maximal codes with specified minimum distance, IBM Tech. Rept. TR 44.0085 (1969).

[84] A.M. Patel, Maximal group codes with specified minimum distance, IBM J. Res. Devel. 14 (1970) 434–443.

[85] W.K. Pehlert Jr., Analysis of a burst-trapping error correction procedure, BSTJ 49 (1970) 493–519.

[86] W.W. Peterson, On the weight structure and symmetry of BCH codes, Air Force Cambridge Res. Lab., Bedford, Mass., Rept. AFCRL-65-515, 1965.

[87] V.S. Pless, On the uniqueness of the Golay codes, JCT 5 (1968) 215–228.

[88] V.S. Pless, On a new family of symmetry codes and related new five-designs, Bull. Am. Math. Soc. 75 (1969) 1339–1342.

[89] V.S. Pless, Symmetry codes over GF(3) and new five-designs, JCT 12 (1972) 119–142.

[90] V.S. Pless, A classification of self-orthogonal codes over GF(2), Discrete Math. 3 (1972) 209–246 (this issue).

[91] F.P. Preparata, A new look at the Golay (23, 12) code, PGIT 16 (1970) 510–511.

[92] Proc. Second Intern. Symp. Inform. theory, Tsahkadsor, Armenia, September 1971.

[93] I.S. Reed, A class of multiple-error-correcting codes and the decoding scheme, PGIT 4 (1954) 38–49.

[94] D.V. Sarwate and E.R. Berlekamp, On the weight enumeration of Reed–Muller codes and their cosets, to appear.

[95] J.E. Savage, The complexity of decoders, II, Computational work and decoding time, PGIT 17 (1971) 77–85.

[96] N.V. Semakov and V.A. Zinov'ev, Balanced codes and tactical configurations, Probl. Peredači Inform. 5 (1969) 28–36. (in Russian).

[97] N.V. Semakov, V.A. Zinov'ev and G.V. Zaicev, Uniformly packed codes, Probl. Peredači Inform. 7 (1971) 38–50 (in Russian).

[98] C.E. Shannon, A mathematical theory of communication, BSTJ 27 (1948) 379–423, 623–656.

[99] S.G.S. Shiva, Certain group codes, Proc. IEEE 55 (1967) 2162–2163.

[100] V.M. Sidel'nikov, Weight spectra of binary Bose–Chaudhuri–Hocquenghem codes, Probl. Peredači Inform. 7 (1971) 14–22 (in Russian).

[101] R. Singleton, Maximum distance Q-nary codes, PGIT 10 (1964) 116–118.

[102] N.J.A. Sloane, A survey of recent results in constructive coding theory, in: National Telemetering Conf. NTC'71 Record (IEEE, New York, 1971) 218–227.

[103] N.J.A. Sloane and R.J. Dick, On the enumeration of cosets of first order Reed–Muller codes, IEEE Intern. Conf. Communications (Montreal, 1971) 7 (1971) 36–2 to 36–6.

[104] R. Stanton, The Mathieu groups, Can. J. Math. 3 (1951) 164–174.

[105] J.J. Stiffler, Theory of synchronous communications (Prentice-Hall, Englewood Cliffs, N.J., 1971).

[106] M. Sugino, Y. Ienaga, N. Tokura and T. Kasami, Weight distribution of (128, 64) Reed–Muller code, PGIT 17 (1971) 627–628.

[107] A. Tietäväinen, On the nonexistence of perfect codes over finite fields, SIAM J, to appear.

[108] A. Tietäväinen and A. Perko, There are no unknown perfect binary codes, Ann. Univ. Turku, Ser. AI 148 (1971) 3–10.

[109] S.Y. Tong, Burst-trapping techniques for a compound channel, PGIT 15 (1969) 710–715.

[110] S.Y. Tong, Performance of burst-trapping codes, BSTJ 49 (1970) 477–491.

[111] J.H. van Lint, Coding theory, Lecture Notes in Math. 201 (Springer, Berlin, 1971).

[112] J.H. van Lint, A survey of recent work on perfect codes, Rocky Mountain J. Math., to appear.

[113] J.H. van Lint, A new description of the Nadler code, PGIT to appear.

[114] H.C.A. van Tilborg, Weights in the third-order Reed–Muller codes, Jet Propulsion Lab., Calif. Inst. Technol., Tech. Rept. 32–1526, IV, 1971.

[115] E.J. Weldon, Jr., Long quasi-cyclic codes are good (abstract) PGIT 16 (1970) 130.

[116] E. Witt, Über Steinersche Systeme, Abh. Math. Sem. Univ. Hamburg 12 (1938) 265–275.

[117] J.K. Wolf, Adding two information symbols to certain nonbinary BCH codes and some applications, BSTJ 48 (1969) 2405–2424.

[118] J.K. Wolf, Nonbinary random error-correcting codes, PGIT 16 (1970) 236–237.

Author Index

Editor's Biography

Elwyn R. Berlekamp (S'62–M'65–F'72) was born in Dover, Ohio, on September 6, 1940. A winner of the December 1961 William Lowell Putnam Intercollegiate Mathematics Competition, he received the B.S., M.S., and Ph.D. degrees in electrical engineering from the Massachusetts Institute of Technology, Cambridge, in 1962, 1962, and 1964, respectively.

From September 1964 to February 1967 he was an Assistant Professor of Electrical Engineering at the University of California, Berkeley. From 1967 to 1971 he was a member of the Mathematics Research Center at Bell Laboratories, Murray Hill, N.J. Since 1971 he has been Professor of Mathematics and Electrical Engineering-Computer Science at the University of California, Berkeley.

Prof. Berlekamp has been a member of the Editorial Boards of *Information and Control,* the *American Mathematical Monthly*, and *Utilitas Mathematicae* (a new Canadian journal of applied mathematics, computer science, and statistics). His book, *Algebraic Coding Theory*, received the IEEE Group on Information Theory's best research paper award for information theory publications which appeared in 1967 and 1968. He received the 1971 "Outstanding Young Electrical Engineer" award from Eta Kappa Nu. In 1973 he was the President of the IEEE Information Theory Group.